Lecture Notes in Civil Engineering

Volume 181

Lecture Notes in Civil Engineering (LNCE) publishes the latest developments in Civil Engineering - quickly, informally and in top quality. Though original research reported in proceedings and post-proceedings represents the core of LNCE, edited volumes of exceptionally high quality and interest may also be considered for publication. Volumes published in LNCE embrace all aspects and subfields of, as well as new challenges in, Civil Engineering. Topics in the series include:

- Construction and Structural Mechanics
- Building Materials
- Concrete, Steel and Timber Structures
- Geotechnical Engineering
- Earthquake Engineering
- Coastal Engineering
- Ocean and Offshore Engineering; Ships and Floating Structures
- Hydraulics, Hydrology and Water Resources Engineering
- Environmental Engineering and Sustainability
- Structural Health and Monitoring
- Surveying and Geographical Information Systems
- Indoor Environments
- Transportation and Traffic
- Risk Analysis
- Safety and Security

To submit a proposal or request further information, please contact the appropriate Springer Editor:

- Pierpaolo Riva at pierpaolo.riva@springer.com (Europe and Americas);
- Swati Meherishi at swati.meherishi@springer.com (Asia - except China, and Australia, New Zealand);
- Wayne Hu at wayne.hu@springer.com (China).

All books in the series now indexed by Scopus and EI Compendex database!

More information about this series at http://www.springer.com/series/15087

Volodymyr Onyshchenko ·
Gulchohra Mammadova ·
Svitlana Sivitska · Akif Gasimov
Editors

Proceedings of the 3rd International Conference on Building Innovations

ICBI 2020

Springer

Editors
Volodymyr Onyshchenko
Yuri Kondratyuk Poltava Polytechnic National University
Poltava, Ukraine

Gulchohra Mammadova
Azerbaijan University of Architecture and Construction
Baku, Azerbaijan

Svitlana Sivitska
Yuri Kondratyuk Poltava Polytechnic National University
Poltava, Ukraine

Akif Gasimov
Azerbaijan University of Architecture and Construction
Baku, Azerbaijan

ISSN 2366-2557 ISSN 2366-2565 (electronic)
Lecture Notes in Civil Engineering
ISBN 978-3-030-85042-5 ISBN 978-3-030-85043-2 (eBook)
https://doi.org/10.1007/978-3-030-85043-2

This Springer imprint is published by the registered company Springer Nature Switzerland AG
The registered company address is: Gewerbestrasse 11, 6330 Cham, Switzerland

Contents

List of Abbreviations

AAE	auditory accessibility elements
Azeri	MİDA – State Agency for Housing Construction
AZN	Azerbaijanian Manat
CAS	Conditional Alternative Sets
CFS	cold-formed structures
CSIS	Center for Strategic and International Studies
CSPED	Computer System for Processing Enterprise Data
FE	finite element
GCD	greatest common divider
GDP	gross domestic product
IDP	internally displaced person
IT	information technology
LED	light-emitting diode
LLC	Limited Liability Company
PC	light thin-walled reinforced concrete elements
PNS	Positional Numeral System
PO	light steel thin-walled plates
POF, PCF	free edges of unloaded sides
POH, PCH	hinged support of unloaded sides
POR, PCR	rigid clamping of unloaded sides
SHW	solid household waste
SRC	system of residual classes
TEA	tactile elements of accessibility
UK	United Kingdom
USCCS	ultralight steel and concrete composite structures
VAE	visual accessibility elements
VF	ventilated facades

Constructions and Building Materials. Structural Mechanics

Stability of Thin-Walled Steel Elements of Composite Steel and Concrete Structures

Yurii Avramenko, Alina Zyhun, Samira Akbarova, and Oleg Yurin

Abstract This paper investigates the behavior of new type of steel and concrete composite structures. With a view of a more effective use of CFS in modern construction (increasing the carrying capacity while reducing the thermal conductivity) the authors propose to fill the open cavities of thin-walled steel structures with light insulating concrete to create ultralight steel and concrete composite elements. The paper presents the results of a theoretical analysis of the local stability of thin-walled steel elements supported by lightweight concrete. The study of the influence of structural parameters on the local stability of steel elements of reinforced concrete structures. The comparison of experimental results with the results of calculations is performed in accordance with the theory of elasticity and the proposed method.

Keywords Local stability · Cold-formed structures · Composite steel and concrete structures · Plates · Finite element

1 Introduction

Over the past couple of decades, constructions using steel thin-walled cold-formed profiles (cold-formed structures—CFS) [2, 4], also known as LiteSteel structures, are becoming more common in buildings.

Among the most common types of cross-sections of cold-formed profiles, should be allocated the C-sections used primarily as load-bearing structures of ceilings and walls of low-rise buildings, as well as Z-shaped (Z-profiles), the scope of which are purlins and wall coverings (see Fig. 1).

In general, the use of CFS in the load-bearing structures has a number of features caused primarily by their thickness and specific forms of section. Consequently, a

Y. Avramenko (✉) · A. Zyhun · O. Yurin
National University «Yuri Kondratyuk Poltava Polytechnic», Pershotravnevyj Ave. 24., Poltava 36011, Ukraine

S. Akbarova
Azerbaijan University of Architecture and Construction, Ayna Sultanova 11, Baku, Azerbaijan

V. Onyshchenko et al. (eds.), *Proceedings of the 3rd International Conference on Building Innovations*, Lecture Notes in Civil Engineering 181,
https://doi.org/10.1007/978-3-030-85043-2_1

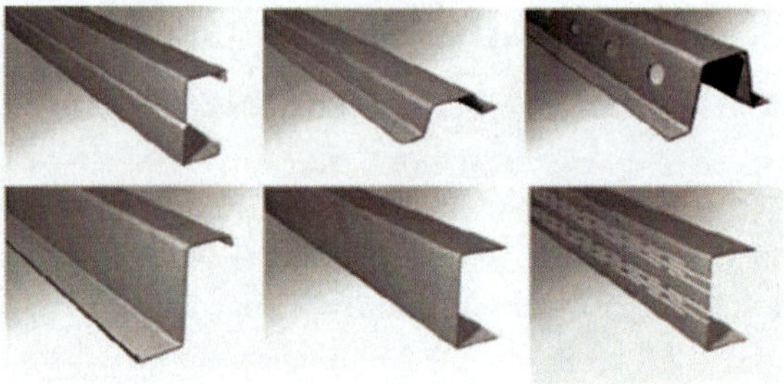

Fig. 1 Example of structures failures as local buckling of CFS members

Fig. 2 Example of structures failures as local buckling of CFS members

significant feature of this type of elements is their possibility of local buckling of flanges and walls under bending due to axial compression; the bending torsion owing to the behavior of flexural and compressed members with eccentricities; significant thermal conductivity of solid profiles, which leads to forming of heat-conducting inclusions etc.

For the most structures under operating conditions, local loss of stability is unacceptable and can be the cause of structures failures, as shown in Fig. 2.

Structural failure is the value of critical stresses caused by local bending of steel thin-walled cold-formed elements (plates) of lightweight composite steel and concrete structures.

In this respect, one of the ways to increase the carrying capacity of this type of structural elements can be to fill them with an ultralight weight concrete. In this case, the complex structure formed in this manner allows using the advantages of each of the components in the most efficient way by combining a carrier and heat-insulating functions.

Thus, the main objectives of these investigations are theoretical and experimental validation of CFS using efficiency combined with ultralight weight concrete (ultra-light steel and concrete composite structures) in order to increase the local stability of thin-walled steel profiles under complex types of deformation.

The authors suggest to fill the open cavities of thin-walled steel structures with light insulating concrete to create ultralight steel and concrete composite elements for a more effective using of CFS in modern construction (increasing the carrying capacity while reducing the thermal conductivity).

2 Methods of Study

The investigation approach used in this study, develops a numerical model to represent a behavior of ultralight steel and concrete composite structures (USCCS) under loading. The intended outcome of this modeling effort and experimental investigations was to establish a fundamental understanding of the local stability of CFS that is supported by concrete filling. The numerical modeling was used to confirm the existing, experimental data which are limited, on the behavior and ultimate capacity of light steel–concrete composite structures.

Experimental studies were conducted to verify the proposed approach, in which quantitative and qualitative data. It was the data about obtaining of the influence of steel elements thickness and concrete composition of reinforced concrete structures on local stability and stress–strain state of compressed steel and reinforced concrete structures under static load and their bearing capacity and the nature of the destruction. For example, there are concrete cubes 100 × 100 × 100 mm, which are supported on both sides by steel plates, 1 mm thick.

Separate plates (compartments) of steel structural elements are calculated and simplifications are introduced, which consist in the improvement of conditions of plate edges fixing and contour loads are introduced into the schematization. Freely supported plates are considered in most available studies and freely supported contour is provided in practical calculations. Actual fastening conditions can occupy any intermediate position between free support and fixed support, but the difference in the magnitude of the critical load can be quite significant.

The commercial finite element (FE) computer package [1, 8] was used to generate the numerical models. The accuracy and validity of the FE simulation and analysis were investigated through comparison of the experimental data and the results of theoretical analysis.

3 Materials Properties

The authors suggest to use the ultralight weigh concrete such a polystyrene concrete to fill the open cavities of thin-walled steel structures. Expanded polystyrene concrete is an especially lightweight (ultra-light) concrete with a cementing agent and a porous aggregate, such as expanded polystyrene aggregate, which can be used as an insulating and structural material.

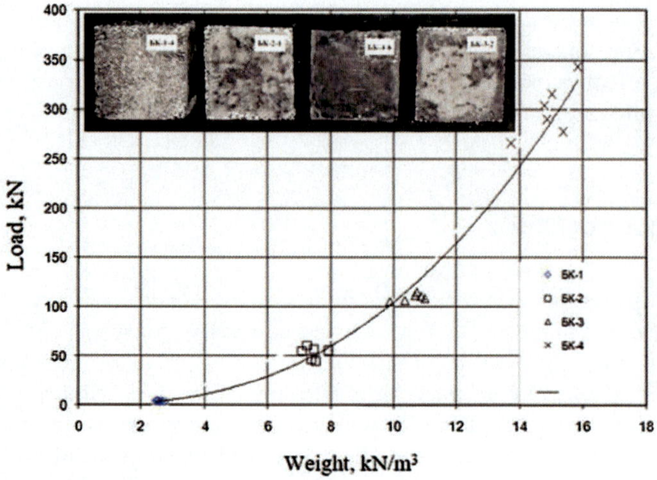

Fig. 3 The initial stress state of the plate on an elastic basis

In this regard, polystyrene concrete is mostly used for building envelopes of framed structures with a different number of floors, in facade systems, as heat and sound insulation [5–7, 9, 10, 12–16] between floors and roofing, etc.

It is obvious that the strength and deformation properties of polystyrene concrete largely depend on its composition, specifically depending in particular the water-cement ratio and the percentage of porous aggregate and cement. Therefore, the specific weight and strength of materials are functionally related [11].

Figures 3 shows the received experimetal relationship between specific weight and strength of standard concrete cubes with polystyrene concrete with different compositions.

Thus, the analysis of experimental results allows concluding that optimal specific weight of light polystyrene concrete for ultralight steel concrete composite structures is 9 kN/m^3 with a strength under axial load 3500 kN/m^2.

4 Review of Analytical Model

One of the main tasks of the USCCS research is to solve the issue of local stability of their steel parts. Unlike traditional cold-formed structures, they rely on an elastic base, which is ultralight concrete. These elements should be considered as thin plates that come into contact with an elastic base, the modulus of elasticity of which is much smaller. This is a theoretical calculation model. At the same time, the stiffness of both components of the USCCS usually differs insignificantly.

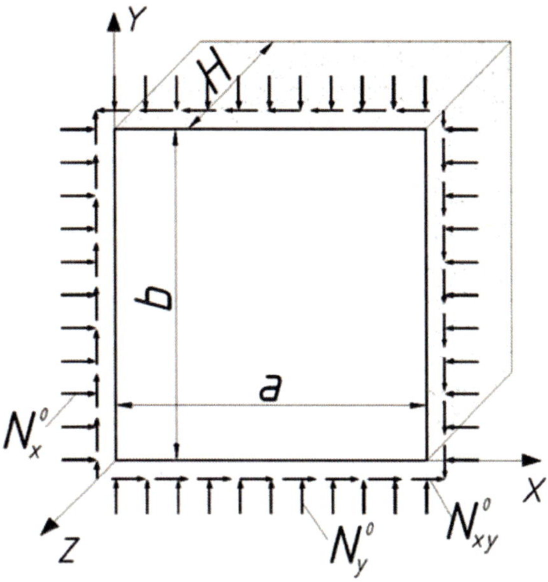

Fig. 4 The initial stress state of the plate, supported by an elastic base

Thus, the solution of the problem of local stability of steel elements of USCCS is reduced to the joint integration of the equations of the theory of shells and three-dimensional theory of elasticity, considering the boundary conditions on the contact surface of media (steel and concrete) and the conditions of these edges.

From this point of view, the rectangular steel plate is an integral part of the thin-walled rod and works in this direction in compression. The practical importance of the study of this necessitates the clarification of known engineering techniques [3], based on the analysis of methods for solving the basic equation of neutral equilibrium.

$$Z_0 = N_x^0 \omega_{xx} + 2N_{xy}^0 \omega_{xy} + N_y^0 \omega_{yy}. \tag{1}$$

Based on the assumption of a linear distribution of displacements in the thickness of the aggregate (ultralight polystyrene concrete), the equation of stability of a thin-walled steel plate on an elastic basis was considered by determining the specified relative to the deflection of the plate ω (Fig. 4).

$$K_1 K_2 L_1(\omega) - L_2(K_1 \omega_{xx} + K_2 \omega_{yy}) + \nabla_0^4 Q_z - \nabla_0^4 (N_x^0 \omega_{xx} + 2N_{xy}^0 \omega_{xy} + N_y^0 \omega_{yy}) = 0 \tag{2}$$

where L_1, L_2 are differential operators of shell theory, considering to account of the anisotropy of plates.

The most common cases of bearing capacity of the plate along the contour, compressed in one direction, with the corresponding boundary conditions of the

Table 1 Comparative analysis of theoretical and experimental bearing capacity of prototypes

Samples series	Load-bearing capacity, kN			
	Lightweight steel thin-walled structural elements			
	Experimental value	Theory of elasticity	Estimated value	Finite element
POF	2,40	3,04	3,34	3,02
POH	13,30	17,57	17,54	12,03
POR	48,00	42,96	37,52	36,40
	Lightweight composite steel and concrete structural elements			
PCF	27,00	–	28,23	27
PCH	34,00	–	52,34	45
PCR	92,00	–	74,87	68

unloaded sides were considered. It gives the division of such structures into lamellar elements.

From systems of homogeneous algebraic equations, considering the deformation of a plate with a rigidly bound elastic base, expressions were obtained to determine the critical load that is shown in Table 1, which was confirmed by experimental studies.

Hinged support:

$$N_{cr} = \min_{(k)}\left\{\frac{a^2b}{k^2\pi^2}\left[F\left(\frac{k\pi}{a}, \frac{\pi}{b}\right) + q_{zk1}\right]\right\} \quad (3)$$

Fixed support:

$$N_{cr} = \frac{a^2b}{3\pi^2}\min_{(k)}\left\{\frac{1}{k^2}\left[F_1\left(\frac{k\pi}{a}, \frac{2\pi}{b}\right) + 2q_{zk0} + 2q_{zk2}\right]\right\} \quad (4)$$

Free support:

$$N_{cr} = \min_{(k)}\left\{\frac{a^2b}{k^2\pi^2}\left[F\left(\frac{k\pi}{a}\right) + q_{zk0}\right]\right\} \quad (5)$$

The critical load (N_{cr}) is the minimization of the right part of the expression with respect to $k = 1, 2, 3\ldots$, and the load function has the form:

$$\begin{aligned} F\left(\tfrac{k\pi}{a}, \tfrac{n\pi}{b}\right) &= \left[K_1\left(\tfrac{k\pi}{a}\right)^2 + K_2\left(\tfrac{n\pi}{b}\right)^2\right]\frac{\Delta_{kn}^{(1)}}{\Delta_{kn}^{(2)}} + K_2\varepsilon_2\frac{\Delta_{kn}}{\Delta_{kn}^{(2)}}; \\ F_1\left(\tfrac{k\pi}{a}, \tfrac{2\pi}{b}\right) &= \left[3K_1\left(\tfrac{k\pi}{a}\right)^2 + K_2\left(\tfrac{2\pi}{b}\right)^2\right]\frac{\Delta_k^{(3)}}{\Delta_k^{(4)}} + 3K_1\varepsilon_2\frac{\Delta_k^{(5)}}{\Delta_k^{(4)}}, \end{aligned} \quad (6)$$

where K_1, K_2 are the stiffness parameters that characterize the transverse displacements.

The value of the components is according to the following formulas, taking into account the compatible deformation of the plate with an elastic base:

$$\begin{aligned}
&\Delta_{kn} = \alpha_k^4 + 2(2\mu_1 + \nu_1)\alpha_k^2\beta_n^2 + \mu_{21}\beta_n^4;\ \ \Delta_{kn}^{(1)} \\
&= \alpha_k^4 + (1 - \nu_1\nu_2 - 2\nu_2\mu_2)\alpha_k^2\beta_n^2 + \mu_1\beta_n^4; \\
&\Delta_{kn}^{(2)} = \Delta_{kn}^{(1)} + (\varepsilon_2 + \varepsilon_1\mu_2)\alpha_k^2 + (\varepsilon_1 + \varepsilon_2\mu_1)\beta_n^2 + \varepsilon_1\varepsilon_2; \\
&\Delta_{kn}^{(3)} = 3D_1D_3\left(\tfrac{k\pi}{a}\right)^4 + [(3 - v_1v_2)D_2 - 2v_2D_3]D_1\left(\tfrac{2k\pi^2}{ab}\right) + D_2D_3\left(\tfrac{2\pi}{b}\right)^4; \\
&\Delta_{kn}^{(4)} = \Delta_{kn}^{(3)} + 3(D_1K_2 + D_3K_1)\left(\tfrac{k\pi}{a}\right)^2 + (D_3K_2 + 3D_2K_1)\left(\tfrac{2\pi}{a}\right)^2 + 3K_1K_2; \\
&\Delta_{kn}^{(5)} = 3D_1\left(\tfrac{k\pi}{a}\right)^4 + 2(v_2D_1 + D_3)\left(\tfrac{2k\pi^2}{ab}\right) + D_2\left(\tfrac{2\pi}{b}\right)^4,
\end{aligned} \tag{7}$$

where, the isotropy of the plate is taken into account:

$$\mu_1 = \frac{G(1 - \nu^2)}{E}, \tag{8}$$

where D_I is the cylindrical stiffness of the plate; ν and E are the Poisson's ratio and the modulus of elasticity of the plate material; α and β are the parameters of the plate formation wave; q_{zkn} - component that takes into account the reaction of the elastic filler.

$$q_z = 2G_c\sqrt{\alpha^2 + \beta^2}\frac{f_1 + \chi g_1}{f_2 + \chi g_2}, \tag{9}$$

where G_c is the shear modulus of the elastic base; ν_c and E_c - Poisson's ratio and modulus of elasticity of the base material.

$$\begin{aligned}
&f = (1 - 2\nu c)sh^2\eta * +\gamma_+\gamma_-H^2 + 2(1 - \nu_c)(1 - 2\nu c)\left(1 - \gamma_+/\gamma_-\right); \\
&f_1 = (1 - \nu c)(sh2\eta * -2\gamma_-H); \\
&f_2 = (3 - 4\nu c)ch^2\eta * +\gamma_+\gamma_-H^2 + (1 - 2\nu c)\left[3 - 4\nu c - 2(1 - \nu c)\gamma_+/\gamma_-\right]; \\
&g_1 = sh^2\eta * -\gamma_+\gamma_-H^2 + 2(1 - \nu c)(1 - 2\nu c)\left(\gamma_+/\gamma_- + \gamma_-/\gamma_+ - 2\right); \\
&g_2 = (1 - \nu c)(sh2\eta * +2\gamma_-H); \\
&\chi = -2Gc\frac{2\alpha\beta d_{xy} - \alpha^2 d_x - \beta^2 d_y + 2Gc\gamma_+^3 th\,\eta *}{B\gamma_+^2 d_{\alpha\beta} - Gc\left(2\alpha\beta d_{xy} + \alpha^2 d_y + \beta^2 d_x\right)\gamma_+ th\,\eta *}.
\end{aligned} \tag{10}$$

Thus, the problem is reduced to the iterative refinement of expressions (3)–(5), that is the reduction of bearing capacity depending on the nature of the waveform k, n.

Using the obtained dependences to determine the influence of the properties of the reinforcing base, the dependences of the interaction N (ν, E) were considered. The modulus of elasticity E_s and the Poisson's ratio ν_s varied from 1000 MPa to 100,000 MPa and 0.1… 0.3, respectively, which is covered a significant number of physical and mechanical properties of concrete.

A decrease in critical forces is initially observed in the case of a three-layer structural element for a filler with a very low modulus of elasticity E_s at low H. This is due to the large shear deformations of the aggregate, which reduce the bending stiffness of the wall so much that it becomes less than the stiffness of the bearing layers (the stiffness of the walls at $H = 0$). In fact, with very small E_c modules, the load-bearing layers operate almost independently of each other when the stability is lost.

Meanwhile, it is physically clear that critical efforts must continue to grow. The results of the solutions show that this increase at low modulus of elasticity of the aggregate is very small, because the parts of the aggregate, far enough from the carrier layers, do not have a supporting effect on the latter from each other.

Under different conditions of fixing the unloaded edges of a rectangular plate with a thickness of 1 mm, the obtained dependences are curvilinear. It consists of increasing modulus of elasticity increases the load-bearing capacity of the structural element, and increasing the value of ν_s, in turn, does not significantly reduce the value of N. ν. It (ν) can be significant for structures whose layers are orthotropic and have very different Poisson's ratios.

Based on the condition that the greater strength of concrete significantly increases the stability not only of thin-walled elements of reinforced concrete structures, but also perceives considerable effort, which is not always justified. Because under such conditions the weight of the structure increases significantly, Poisson's ratio ν_s was taken as 0,15, modulus of elasticity $E_s = 5000$ MPa, as the physical and mechanical characteristics that best correspond to lightweight concretes. It is used in subsequent experimental studies (Fig. 5).

A detailed analysis revealed that at relatively large values of H (0.1… 0.3 and more), which exceed the geometric dimensions of the plate, remote parts of the aggregate do not have a supportive effect. It indicates irrational use of elastic base material at certain limits conditions and sizes, regardless of the selected material of reinforcing base.

The main task of experimental research was to obtain quantitative and qualitative data on the influence of steel elements thickness and concrete composition of reinforced concrete structures on local stability and stress–strain state of compressed steel and reinforced concrete structures under static load, also with their bearing capacity and fracture nature.

Modeling of light steel thin-walled plates (PO) and light thin-walled reinforced concrete elements (PC) in software complexes by the finite element method allowed to take into account the features of loss of stability of thin-walled steel plate and to investigate their work in more detail.

Different boundary conditions were applied to the corresponding faces of thin-walled steel plates to bring the models as close as possible with the help of a software package. It consists of free edges (POF, PCF), hinged support (POH, PCH), and rigid clamping (POR, PCR) of unloaded sides.

An additional layer of two-dimensional orthotropic material with a thickness of 0.1 mm was introduced to model the reaction properties of the elastic base in conjunction with the steel component of the structure. This approach allowed obtaining and

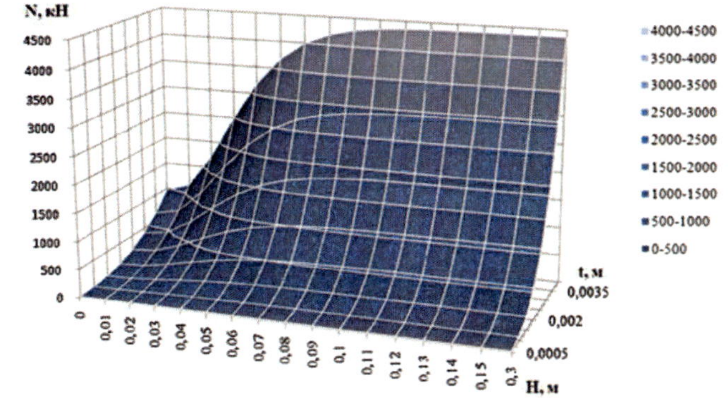

(a) Hinged support

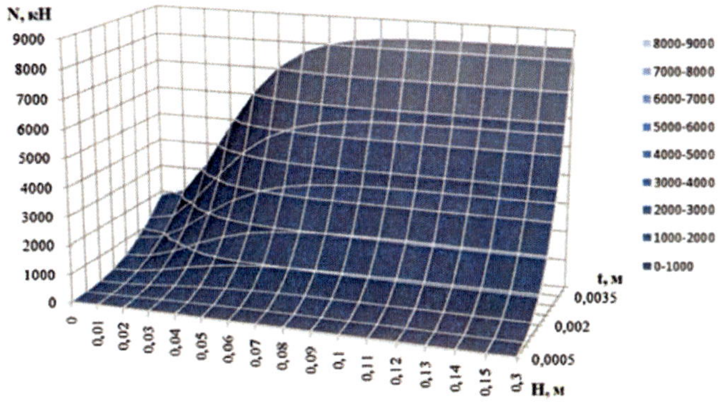

(b) Fixed support

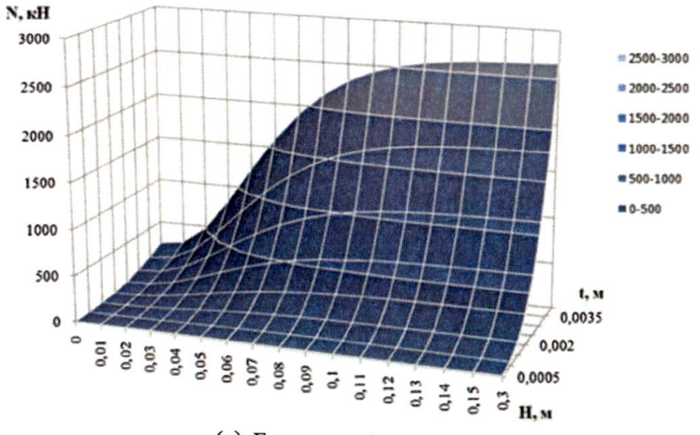

(c) Free support

Fig. 5 Interaction surfaces N (νs, Es) under different boundary conditions of longitudinal unloaded sides.

varying the amount of "gluing" of concrete to the plate, and as a result, it is the actual work of the structure.

As a result of finite element calculations using a computer "Destruction" of the sample models is as follows: the sample with a hinged support of the steel plate along the contour is formed by one half-wave in the transverse and longitudinal directions at a critical load of 45 kN. In the model, which is a reinforced concrete structural element with rigid clamping of the unloaded sides of the steel plate - two half-waves in the longitudinal direction and one in the transverse at a value of 68 kN. And one half-wave in the longitudinal direction at 26.7 kN with free support.

5 Conclusions

Therefore, as expected by theoretical calculations, despite the relatively small modulus of elasticity of the reinforcing base (lightweight concrete), the bearing capacity of the samples increases significantly, which indicates the feasibility of lightweight concrete component in lightweight steel structures to prevent local loss of stability.

The difference between the critical experimental and theoretical load for the samples of the series of POF and PCF was not more than 8.9%; for samples of the POR and PCR series about 14%, because in the theoretical calculation insignificant stiffening ribs were not taken into account for modeling the clamping of the longitudinal sides. The significant difference in the critical load for the samples of the POH and PCH series that is due to experimental studies, because it is quite difficult to obtain an ideal hinge. It is obvious that for samples with a concrete component this deviation is smaller.

References

1. ANSYS (2011). User's manual revision 14.0. Canonsburg, PA: ANSYS, Inc.
2. Bock M, Real E (2015) Effective width equations accounting for element interaction for cold-formed stainless steel square and rectangular hollow sections. Structures 2(2015):81–90
3. EN 1994–1–1 (2004). Eurocode 4: design of composite concrete and steel structures – part 1–1: general rules and rules for buildings. Brussels: European Committee for Standardization (CEN)
4. EN 1993–1–3 (2006). Eurocode 3: design of steel structures. Part 1–3: General rules. Supplementary rules for cold-formed members and sheeting. Brussels: European Committee for Standardization (CEN)
5. Ewadh HM, Basri NA (2012) Effectiveness of polystyrene beads as aggregate replacement material to recycle solid waste: a study on workability and absorption results of concrete. Int J Sci Eng Res 3(8):1–4
6. Liew JYR, Koh CG, Sohel KMA (2009) Impact tests on steel-concrete-steel sandwich beams with lightweight concrete core. Eng Struct 31(9):2045–2059
7. Park SG, Chisholm DH (1999) Polystyrene aggregate concrete. Study Report. SR 85, Building Research Association of New Zealand, Judgeford

8. Shanmugam NE, Kumar G (2002) Thevendran V (2002) Finite element modelling of double skin composite slabs. Finite Elements Anal Des 38(7):579–599
9. Tamut T, Prabhu R, Venkataramana K, Yaragal SC (2014) Partial replacement of coarse aggregates by expanded polystyrene bead in concrete. Int J Res Eng Technol 03(02):238–241
10. Leshchenko MV, Semko V (2015) Thermal characteristics of the external walling made of cold-formed steel studs and polystyrene concrete. Mag Civil Eng 60(8):44–55
11. Hasenko A, Semko O, Drobotia O, Sirobaba V (2020) Experimental and numerical studies of nodes of light steel-reinforced concrete structures. In: Paper presented at the Proceedings of the 2020 Session of the 13th Fib International PhD Symposium in Civil Engineering, pp 73–178
12. Onischenko VA, Soloviev VV, Chernenko LA, Malyshev VV, Bondus SN (2014) Acidic-basic interactions in tungstate melts based on tungsten electroplating out of them. Materialwissenschaft Und Werkstofftechnik 45(11):1030–1038. https://doi.org/10.1002/mawe.201400222
13. Kochkarev D, Azizov T, Galinska T (2018) Bending deflection reinforced concrete elements determination. Paper presented at the MATEC Web of Conferences, p 230. https://doi.org/10.1051/matecconf/201823002012
14. Pichugin SF (2018) Reliability estimation of industrial building structures. Mag Civil Eng 83(7):24–37. https://doi.org/10.18720/MCE.83.3
15. Zotsenko M, Vynnykov Y, Lartseva I, Sivitska S (2018) Ground base deformation by circular plate peculiarities. In: Paper presented at the MATEC Web of Conferences, p 230. https://doi.org/10.1051/matecconf/201823002040
16. Cherniha R, Pliukhin O (2013) New conditional symmetries and exact solutions of reaction-diffusion-convection equations with exponential nonlinearities. J Math Anal Appl 403(1):23–37. https://doi.org/10.1016/j.jmaa.2013.02.010

Some Properties of Fiber-Reinforced Road Concrete Using Iron Ore Dressing Wastes

Rasul Akhmednabiev, Lyudmila Bondar, Oksana Demchenko, and Vladimir Shulgin

Abstract The results of studies of fine-grained road fiber-reinforced concrete using iron ore dressing wastes as fillers at the Poltava mining and processing plant are presented. Ore dressing wastes are crushed rocks of the quartzite group, which are not worse in strength than granites. The waste has a rocky torn surface, which contributes to good adhesion to the cement stone. The uniformity of the of polypropylene fibers distribution in the volume of the concrete mixture was assessed by the change in the degree of separation. The dependences of homogeneity of a concrete mixture with different amounts and volumetric concentration and length of polypropylene fibers on the duration of mixing have been obtained. The graphs were built and the optimal duration of mixing the concrete mixture was determined.

The influence of polypropylene fibers on the strength of concrete of different grades has been investigated. The research results indicate that polypropylene fibers do not have an adhesive bond with cement stone and, therefore, cannot significantly increase the strength of concrete in comparison with steel fibers. However, polypropylene fibers can have a positive effect on the crack resistance and dynamic strength of road concretes.

Keywords Iron ore dressing wastes · Polypropylene fibers · Degree of separation · Volumetric concentration of fibers · Homogeneity of concrete mix

1 Introduction

With the development of scientific and technological progress, mankind extracts more and more minerals from the earth. The careers are getting deeper and the environmental impacts are becoming more tangible. As you know, almost all the fossils that mankind extracts from the earth have associated rocks or waste is generated after their use. One way or another, a large amount of waste accumulates on the ground. Some wastes are mineral raw materials suitable for further use in activities

R. Akhmednabiev (✉) · L. Bondar · O. Demchenko · V. Shulgin
National University "Yuri Kondratyuk Poltava Polytechnic",
Pershotravnevyj Ave. 24., Poltava 36011, Ukraine

V. Onyshchenko et al. (eds.), *Proceedings of the 3rd International Conference on Building Innovations*, Lecture Notes in Civil Engineering 181,
https://doi.org/10.1007/978-3-030-85043-2_2

of mankind, for example, ash and slag from thermal power plants, waste of ferrous and non-ferrous metals. In some developed countries, such waste is used almost 100%, and in Ukraine—only about 15%.

On the territory of Ukraine there are six largest mining and processing plants, which form the lion's share in the Ukrainian production of concentrate, pellets and agglomerate.

At Poltava Mining and Processing Plant, dry magnetic separation wastes are sold in the form of crushed stone, and 335 million m^3 of wet magnetic separation waste is stored in a sludge storage facility. Annually, Poltava MPP produces about 11 million m^3 of wet magnetic separation waste, which is currently not used.

Review of Literary Sources. A feature of the large-tonnage waste of mining and processing plants is that technogenic raw materials have crystalline structures and do not contain organic impurities [1]. They are crushed material in which its texture and structure are disturbed, as well as, in many respects, the morphology of individuals and aggregates of minerals. Due to the extraction of ore minerals from the grinding products, the mineral and chemical composition of the dressing wastes differs significantly from the composition of the original ore. The storage of waste in storage facilities is accompanied by gravitational differentiation of the crushed material. As a result, the mineral and chemical composition of stale waste in specific areas of the storage facilities differ markedly [2].

As you know, along with by-products of iron ore mining and finely ground waste of wet magnetic enrichment at mining and processing plants of Ukraine, dry magnetic enrichment units of crushed mass operate in the technological line for crushing ferruginous quartzites, resulting in crushed stone. The main physical and technical properties of crushed stone from dry magnetic enrichment waste of ferruginous quartzite are practically not inferior to granite and therefore it is recommended for use as a coarse aggregate in the preparation of concrete for monolithic structures and as a ballast material in construction and road works [3].

Waste of wet magnetic separation by grain size composition is a finely ground material containing 70 … 75% of grains with a particle size of 0.085 … 0.16 mm and 7.7 … 17.3% of grains with a particle size of 0.16.,.. 3, 0 mm, specific surface area 30 … 40 m^2/l [4].

The experience of crushed stone use and sand of ore dressing wastes as concrete aggregates shows that they allow, in many cases, to increase the strength and density of concrete, its frost resistance, impact and abrasion resistance, in comparison with traditional aggregates [3, 5].

Several authors investigated the possibility of using other industrial waste in the technology of building materials production [6, 7].

Concretes with the use of aggregates from iron ore dressing wastes have an increased density due to the presence of iron residues in their compositions. Such concretes are more suitable for road construction. However, road concretes must be able to maintain their operational characteristics under conditions of constant exposure to dynamic loads and corrosive environments inherent in roads.

Since the beginning of the last century, a method has been known to increase the dynamic strength of concrete by introducing fibers into their composition. This method is called fiber reinforcement. At present, experience has been accumulated in the use of various types of fibers, both steel and polymer, and even of plant origin, for fiber reinforcement of concrete.

Materials Used in the Work. Portland cement PC 500 of Ivano-Frankivsk plant was used as a binder. As a fine aggregate—waste of wet magnetic separation of the Poltava Mining and Processing Plant with a modulus of 1.01; as a large aggregate, crushed stone of 5–10 mm fraction from dry magnetic separation waste of the same plant. As a fiber reinforcement—polypropylene fibers with a diameter of 0.1 mm of various lengths. A superplasticizer based on modified polycarboxylates "Fluid Premia 196" was used in the work.

Research Methods: The properties of mineral aggregates were investigated by standard methods. The activity of the cement was determined according to the DSTU method.

It is known that the greatest difficulty in the technology of dispersed concrete reinforcement is the provision of a uniform distribution of fibers in the volume of the material. Some researchers have developed special equipment for feeding fiber into a mixer, as well as mixers for mixing dispersion-reinforced mixtures. In this work, the uniformity of fiber distribution in the volume of the mixture was estimated by the change in the degree of separation, which is determined by the equation.

$$S = \left(\frac{1}{V}\right)\left(\frac{1}{\rho_{m.a}}\right)\sum_{n}\sum_{m}[\rho_i - \rho_{i.a}] \cdot \Delta V_i \qquad (1)$$

where $\rho_{m.a}$—the average density of the concrete mix, kg/m^3; ρ_i—density of component $\boldsymbol{i}$, kg/m^3; $\rho_{i.a}$—density of component i in the entire volume of the mixture, kg/m^3; V—the total geometric volume of the mixture, m^3; $\boldsymbol{m}$—the number of samples; $\boldsymbol{n}$—the number of mixes. In this case, the fibers were considered a component of the mixture.

The study of the effect of fiber length and their number on the concrete properties was carried out using mathematical design of the experiment. In this case, the fiber length varied from 10 to 30 mm, and its volume content—from 1 to 2%. the amount of superplasticizer varied within 1 ... 2% of the cement mass. Experimental conditions are shown in Table 1.

All the studies were carried out on the same composition of the concrete mixture: cement—500 kg; crushed stone 1120 kg; sand (iron ore dressing waste) 640 kg; water—200 *l*. Compressive strength of concrete 54.3 MPa.

Research Results. It is known, the degree of separation is a value equal to the specific average deviation of the density of the mixture components from their average density and characterizing the uneven distribution of the components in the entire volume of

Table 1 Experimental conditions

Factors		Levels of variation			Interval of variation
Natural view	Coded view	−1	0	1	
Fiber length, mm	X_1	10	20	30	10
Volume content, %	X_2	1	1,5	2,0	0,5
Additive content, %	X_3	0,8	1,2	1,6	0,4

the mixture. The change in the degree of separation or mixing process over time is expressed by a decaying exponent.

$$S = a + (S_{max} - a)e^{-kt} \quad (2)$$

where ***a, k***—parameters depending on the nature of the components, on the design of the mixing equipment, the mixing mode and to be determined by the least squares methods*;* ***t*** is the mixing time of the mixture, s; S_{max}—some value of the degree of separation corresponding to the initial stage of the mixing process. When the components occupy a separate volume, and the deviation of the density of these components at a given place from their average density in the entire volume is maximum.

$$S_{max} = 2[1 - 1/V(\sum_m \rho_i V_i^2/\sum_m \rho_i V_i)] \quad (3)$$

where V—the total volume of the mixture components, m^3;

ρ_i—density of component ***i,*** kg/m^3; V_i—the volume of component ***i,*** in the mixture, m^3; ***m***—the number of components in the mixture.

The experimental values of the degree of separation were determined from the equation

$$S = \left(\frac{1}{V}\right)\left(\frac{1}{\rho_{a.m}}\right)\sum_n \sum_m [(\rho_i - \rho_{i.a})] \cdot \Delta V_i \quad (4)$$

where $\rho_{a.m}$—is the average density of the concrete mix, kg/m^3;

ρ_i—density of the i-th component, kg/m^3;

$\rho_{i.a}$—is the average density of the ***i***-th component in the entire volume of the mixture, kg/m^3;

V—the total geometric volume of the mixture, m^3;

n—the number of mixes;

m—the number of samples taken from the mixture.

Fiber was loaded into a laboratory compulsory-type concrete mixer along with the components of the concrete mixture. The effect of fiber length on the uniformity of distribution was determined at a volume content of 2%. The effect of the volumetric content on the uniformity of distribution was determined with a fiber length of 20 mm.

At certain time intervals according to the experiment, the mixture was unloaded from the mixer into a laboratory flat bucket with an area 1 m^2. Samples were taken from the mixture in a checkerboard-nesting manner. The samples were weighed and the volume was measured, after which the cement was washed out of the samples with water. The remaining mass was dried and separated into individual components. Thus, the data necessary to assess the uniformity of fibers distribution in the mixture were determined.

Taking the degree of separation as a random variable, to determine the position of the found values of S, the method of mathematical statistics was used and the numerical characteristics were found.

- mathematical expectation

$$M = \sum\nolimits_{i=1}^{m} \frac{S_i}{m}, \tag{5}$$

- standard deviation

$$\sigma = \sqrt{D} = \sqrt{\sum\nolimits_{i=1}^{m} (S_1 - M)^2/(m-1)}, \tag{6}$$

- the coefficient of variation

$$\varphi = \left(\frac{\sigma}{M}\right) \cdot 100 \tag{7}$$

The calculation results are shown in Table 2.

Based on the results of mathematical processing of the obtained data, the graphs the separation degree dependence on the length of the fiber (Fig. 1) and the volumetric content of the fiber (Fig. 2) were constructed.

Analysis of the graphs shows that fiber length has a significant effect on the uniformity of fiber distribution in the volume of the mixture. With increasing fiber length, the degree of separation increases.

Obviously, with increasing fiber length, the number of contact points between them increases, which leads to the formation of lumps.

The graphs in Fig. 2 demonstrate the nature of the change in the degree of separation depending on the volumetric fiber content. As can be seen, with an increase in the volumetric content of fiber, the degree of separation increases.

Analysis of the graphs in Figs. 1 and 2 shows that the change in the degree of separation over time is expressed by a decaying exponential and an increase in the mixing time does not improve the homogeneity of the concrete mixture above a certain value.

Three periods of mixture formation can be distinguished on the graphs: the first period is characterized by a sharp decrease in the degree of separation, the second—by the stabilization of the mixing process in time, the third period is characterized by a slight increase in the degree of separation, i.e. a decrease in the homogeneity of

Table 2 Results of mathematical processing of data for experimental determination of the degree of separation

Characteristics of random variables	Mixing duration, min					
	1	2	3	4	5	6
Fiber volume content 1% with a length of 10 mm						
M	9.4	4.2	3.2	2.6	2.67	2.71
σ	1.34	0.47	0.32	0.23	0.28	0.30
φ	14.0	11.2	10.0	8.09	10.5	11.07
Fiber volume content 1% with a length of 20 mm						
M	11.9	7.6	5.0	4.1	4.18	4.2
σ	1.98	0.98	0.61	0.43	0.46	0.465
φ	16.8	12.9	12.2	10.4	10.7	10.8
Fiber volume content 1% with a length of 30 mm						
M	13.4	10.2	7.9	5.7	6.9	7.0
σ	2.52	1.43	0.96	0.63	0.84	0.85
φ	18.8	14.1	12.4	11.09	12.1	12.29
Fiber volume content 1.5% with a length of 20 mm						
M	11.8	6.3	5.4	4.65	4.68	4.71
σ	1.77	0.94	0.69	0.52	0.59	0.63
φ	15.4	14.7	12.7	11.1	12.6	13.35
Fiber volume content 2% with a length of 20 mm						
M	13.6	8.3	7.5	6.55	6.61	6.63
σ	2.34	1.3	0.97	0.81	0.95	0.96
φ	17.3	15.8	13.06	12.36	14.05	14.47

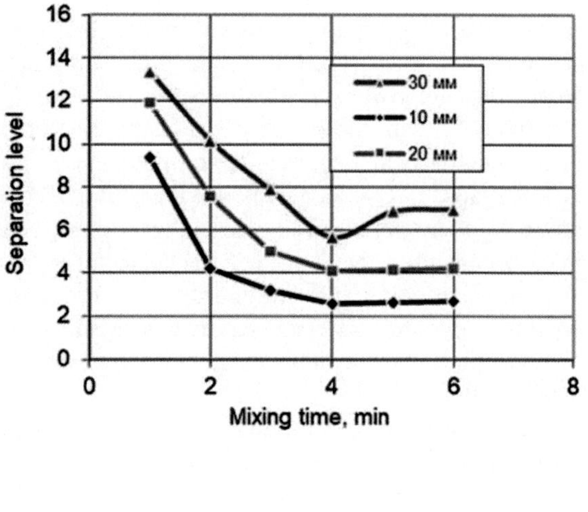

Fig. 1 Change in the degree of separation of concrete mixture depending on the mixing

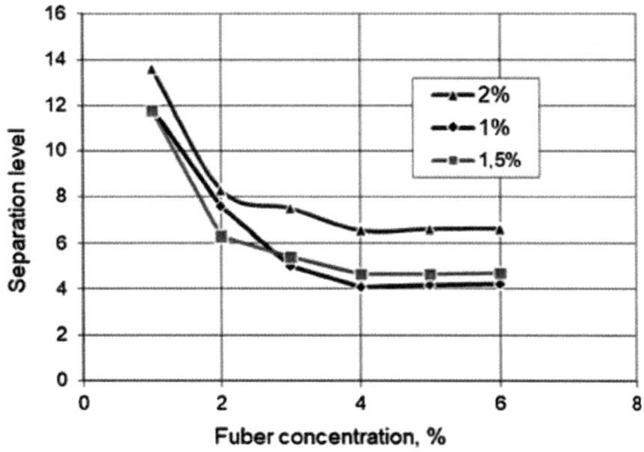

Fig. 2 Variation of separation degree depending on fiber concentration and mixing time

the mixture. Obviously, with prolonged mixing, the process of formation of lumps from fibers is enhanced. Therefore, the mixing time of the components of the mixture should be limited to the second period.

The paper studies the effect of the amount and length of fiber on the strength of heavy concrete. In the study, the method of mathematical planning of the experiment was applied. The conditions for planning the experiment are given in Table 1. The studies were carried out on cube specimens with an edge size of 100 mm and prisms with dimensions of 10 × 10 × 40 cm. The specimens were made in metal molds and stored in laboratory conditions for 28 days. The components of the concrete mixture were loaded into a forced action concrete mixer of the following sequence: crushed stone, sand, fiber, cement, water with an additive. The mixing of the concrete mixture lasted 5 min.

After solidification, cubic specimens were tested for compressive strength, and prisms for bending.

The test results were processed by the STATISTICA-10 program and are shown in Fig. 3.

The graph shows that with an increase in the fiber length within the experiment, the concrete strength first increases and then decreases. decreases. Obviously, polypropylene fibers, although they do not have adhesion strength with a cement matrix, contribute to an increase in strength due to friction forces at the fiber-cement stone interface. It should be noted that the reduction in strength is only 5–7% compared to unreinforced concrete.

The concentration of fiber in the bulk of concrete, within the experiment, has an even smaller effect on the strength of concrete. Although it should be noted that with an increase in the volume concentration of fiber, the strength of concrete, within the experiment, decreases. So, for example, with an increase in the volumetric fiber concentration from a minimum 0.8% to a maximum 2.0%, the strength of concrete decreases by 5.0 MPa from unreinforced concrete, which is almost 10%.

From the point of view of the theory of composite materials, polypropylene fibers cannot increase the strength of concrete, since the strength of polypropylene is an

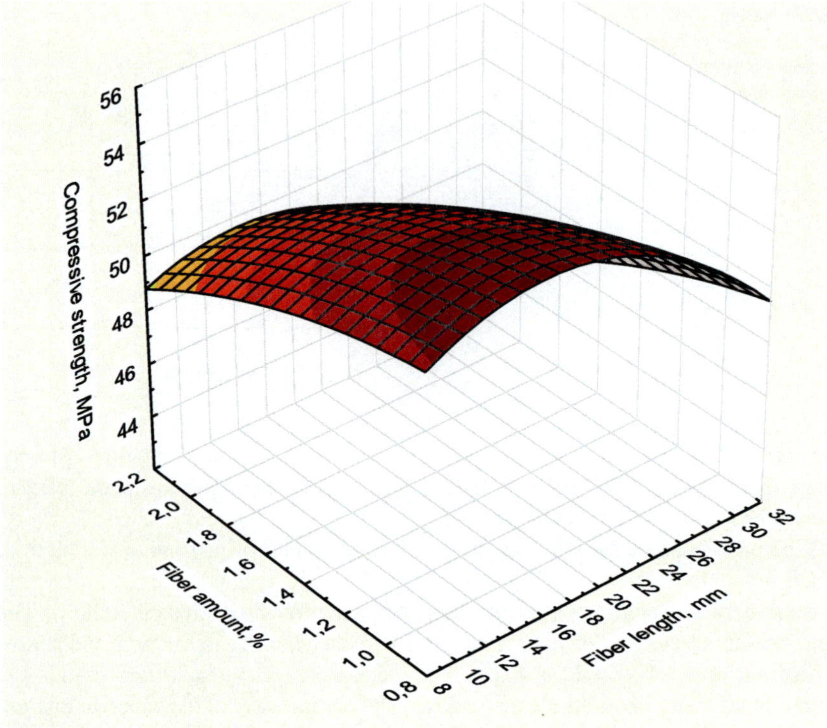

Fig. 3 Surface of influence of length and fiber content on concrete strength

order of magnitude lower than the strength of the concretes under study [8]. Insignificant fluctuations in the strength of concrete with the introduction of polypropylene fibers can be explained by the small work of the forces of friction of the surface of the fibers against the surrounding cement stone. As you know, polypropylene is not wetted with water, i.e. it is a hydrophobic material, in addition, it is absolutely inert material in relation to the environment of cement hardening. Proceeding from this, there can be no question that adhesion forces similar to steel reinforcement can arise between propylene fibers and cement stone. All interactions between these materials are based solely on frictional forces. It is obvious that the use of polypropylene fibers for dispersed concrete reinforcement aims at increasing the crack resistance and resistance of concrete to dynamic loads.

It is known that under the action of repetitive loads in the body of brittle materials, which are especially road concretes, microcracks appear, which, with prolonged exposure to dynamics, combine into a macrocrack, which leads to the destruction of structures [9–15]. The theory of composite materials also gives us a mechanism for the occurrence and development of microcracks in materials, in particular in brittle materials. The more brittle the material, the smaller the radius at the crack mouth and the higher the stress concentration. If the crack encounters an obstacle that increases the radius of curvature at the crack mouth, then the stress concentration decreases sharply and additional work of external forces will be required to achieve the same stress concentration. In this aspect, the use of polypropylene fibers for concrete reinforcement makes sense. But it should be noted that the fibers should not lead to a significant decrease in the strength of concrete.

However, such studies are carried out in this work and their results will be published in future publications.

2 Conclusion

Based on the results of the carried out studies, the following conclusions can be drawn:

Waste from iron ore dressing is a worthy replacement for traditional aggregates of heavy concrete; it was found that the degree of separation of concrete mixtures filled with polypropylene fibers in the initial period drops sharply and then begins to increase; the mixing time of concrete mixtures with polypropylene fibers should not exceed 4–5 min in order to avoid disturbing the homogeneity of the mixture; it was found that polypropylene fibers contribute to a decrease in strength by 5–10% within the experiment.

References

1. Chuikova IL (2011) Increasing the efficiency of building composites using technogenic raw materials by regulating the processes of structure formation. Dissertation for competition Doctor of Technical Sciences/Belgorod
2. Vilkul YG (2013) Processing and complex use of mineral raw materials from technogenic deposits. In: Azaryan AA, Kolosov VA (eds) Girnichiy Visnik, vip, p 96
3. Bondarenko SV (2012) Use of by-products from mining and processing plants. In: Sevastyanova OE, Bondarenko AS Starodubov (eds) Readings
4. DSTU B V. 2.7–33–2001 Quartz-iron sand and fine fraction for construction work on waste mining and processing plants of Ukraine
5. Bolshakov VI Waste from mining and processing plants and their use in construction. In: Bondarenko SV (eds) Dnepropetrovsk , p 96
6. Bondar V, Shulgin V, Demchenko O, Bondar L (2017) Experimental study of properties of heavy concrete with bottom ash from power stations. In: MATEC Web of Conferences, vol 116, p 02007. https://doi.org/10.1051/matecconf/20171160200, Transbud-2017 (Scopus) (Web of Science)
7. Demchenko OV, Shulgin VV, Petrash RV (2018) Experimental study on light concrete properties using bottom ash of thermal power stations. International Journal of Engineering Technology 7(3.2):1–5. (Scopus), https://doi.org/10.14419/ijet.v7i2.26.14366, Article ID: 14366, https://www.sciencepubco.com/index.php/ijet/article/view/14366
8. Croc F, Brautman L (1970) Modern composite materials. Moscow Mir, p 406
9. Storozhenko L, Butsky V, Taranovsky O (1998) Stability of compressed steel concrete composite tubular columns with centrifuged cores. J Constr Steel Res 46(1–3):484. https://doi.org/10.1016/S0143-974X(98)80098-9

10. Piskunov VG, Gorik AV, Cherednikov VN (2000) Modeling of transverse shears of piecewise homogeneous composite bars using an iterative process with account of tangential loads 2. Resolving equations and results. Mech Comp Mater 36(6):445–452. https://doi.org/10.1023/A:1006798314569
11. Piskunov VG, Goryk AV, Cherednikov VN (2000) Modeling of transverse shears of piecewise homogeneous composite bars using an iterative process with account of tangential loads. 1. Construction of a model. Mech Comp Mater 36(4):287–296. https://doi.org/10.1007/BF02262807
12. Onyshchenko VO, Filonych OM, Bunyakina NV, Senenko NB (2020). A new agent for removing concrete residues from the surfaces of polypropylene molds in the manufacture of paving slabs and its advantages. https://doi.org/10.1007/978-3-030-42939-3_67
13. Loburets AT, Naumovets AG, Senenko NB, Vedula YS (1997) Surface diffusion and phase transitions in strontium overlayers on W(112). Zeitschrift Fur Physikalische Chemie 202(1–2):75–85. https://doi.org/10.1524/zpch.1997.202.part_1_2.075
14. Onischenko VA, Soloviev VV, Chernenko LA, Malyshev VV, Bondus SN (2014) Acidic-basic interactions in tungstate melts based on tungsten electroplating out of them. Materialwissenschaft Und Werkstofftechnik 45(11):1030–1038. https://doi.org/10.1002/mawe.201400222
15. S. V. Piddubnyi, H. O. Tatarchenko, V. M. Sokolenko, (2020) Express method for the evaluation of the frost resistance of silicate building materials. Materials Science 56(2):240–246

Calculation of Composite Bending Elements

Taliat Azizov, Dmitry Kochkarev, Tatiana Galinska, and Oleksiy Melnyk

Abstract The article describes the method of calculation of stone beams strengthened with side reinforced concrete plates. The method comprises an imaginative dissection of a composite beam into stone and reinforced concrete parts and further inspection of the conditions of strain compatibility in the place where both parts of the combined structure are connected. The paper proves that it is not difficult to determine the conditions of strain compatibility. It is shown that the degree of a stone beam and reinforced concrete plates joint action depends on the quantity and diameter of bonds, their location, as well as on the strength and deformability characteristics of the stone beam and reinforced concrete plates. The article suggests that local deformations in bonds are better determined using empirical formulas. Using a stone beam strengthened with reinforced concrete plates as an example, the authors proved the efficiency of such strengthening and analyzed the factors of joint influence. The research suggests that in case of high bond stiffness, the composite structure can be calculated as monolithic, keeping in mind the usage of two different materials.

Keywords Composite beam · Bonds · Strain compatibility equations · Stone elements · Reinforced concrete plate

T. Azizov · O. Melnyk
Pavlo Tychyna Uman State Pedagogical University, Sadova St., 2, Uman 20300, Ukraine

D. Kochkarev
National University of Water and Environment Engineering, Soborna St., 11, Rivne 33028, Ukraine

T. Galinska (✉)
National University «Yuri Kondratyuk Poltava Polytechnic», 24, Pershotravnevij Ave, Poltava, Ukraine
e-mail: galinska@i.ua

V. Onyshchenko et al. (eds.), *Proceedings of the 3rd International Conference on Building Innovations*, Lecture Notes in Civil Engineering 181,
https://doi.org/10.1007/978-3-030-85043-2_3

1 Introduction

Problem statement and research analysis. Scientific research [1–7] suggests different ways to arrange clips for strengthening stone structures are proposed. Those are single or double-sided clips that are attached to the wall, usually through clamps or ruffs. All these structures are designed to strengthen stone structures against the effect of compressive forces. It is known that the calculation of stone structures is almost the same as the calculation of reinforced concrete structures. The calculation of bent structures is analyzed by [7, 8]. Research by [9] presents the calculation method which takes into account the nonlinear properties of concrete, Nevertheless all those studies do not consider cases when the element section contains materials with different characteristics, although it affects their stress-strain state.

Thus, research does not consider bent stone structures strengthened with side reinforced concrete plates, as well as strengthening structures made of light concrete blocks with reinforced concrete plates, including light concrete bent structures.

The possibility of using one-sided and two-sided reinforced concrete plates to strengthen bent stone structures, including structures made of light concrete blocks, is hindered by the lack of methods for calculating such structures. Both [10, 11] suggest the method and algorithm for calculating the above-described combined structures. However, those papers do not cover the issues of rigidity of anchors connecting the stone and reinforced concrete parts of the composite construction.

2 Method for Calculating

In this regard, the purpose of this article is to develop a method for calculating bent stone structures strengthened with side reinforced concrete plates.

Results and discussion. Consider a bendable element consisting of three layers (in a vertical plane) connected by bonds at individual points. Moreover, the side plates have the same thickness (Fig. 1). Due to the same thickness of the side plates, the action of the composite design will be subject to a flat bend. Such a structure can be calculated as one that is only affected by a bending load without torsion and oblique bending.

For the calculation, as in [11], we mentally divide the composite structure into two separate layers (two beams), layers 1 and 2 (Fig. 1). The first layer is the stone part of the composite beam. The second layer is a plate that has twice the thickness of each of the side plates.

Unknown vertical Si and horizontal Ti forces, where i is the number of the connection, will act in both the first and second cut-off layers (beams). These unknowns can be determined by composing the equations of strain (displacements) compatibility for layer 1 and layer 2 of the composite structure.

The components of the displacements will include:

Vertical displacements at the *i* point consist of the following components:

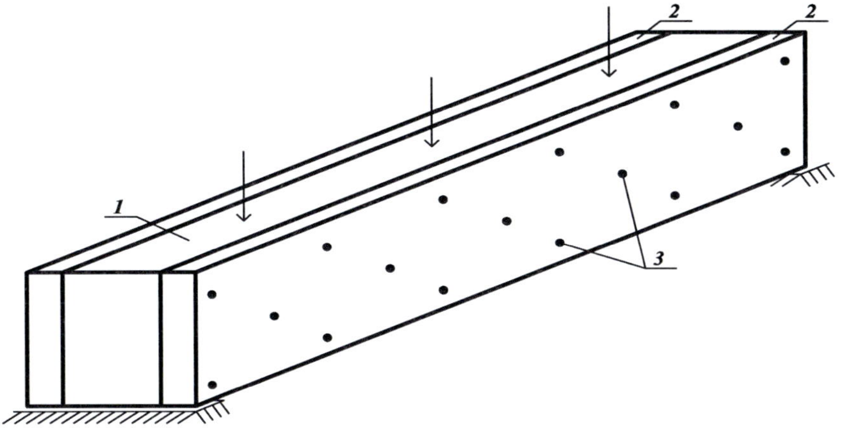

Fig. 1 Diagram of a two-layer bendable element. 1—stone beam; 2—concrete plates; 3—bonds

Fig. 2 Diagram of forces in bonds

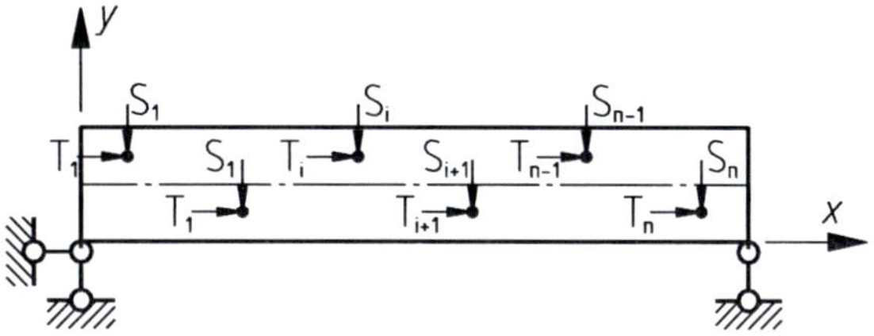

1. from bending by external load;
2. from bending by vertical forces S_i (Fig. 2);
3. from bending moments created by horizontal forces T_i. In this case, if the force is above the neutral axis of the beam, the moment is positive. If it is lower, the moment is negative (Fig. 2);
4. from local deformation at the point of application of the vertical force S_i (bond deformation and concrete crumpling under the bond).
   Horizontal movements at the *i* point are made up of such components as (if the point is above or below the neutral axis, then the displacements can be either positive or negative):
5. from bending by external load;
6. from bending by vertical forces S_i;
7. from bending moments created by horizontal forces T_i in the XOY plane (Fig. 2);
8. from compression (stretching) by T_i forces;
9. from local deformation at the point of application of the horizontal force T_i.

All the components of displacements, except for displacements according to clauses 4 and 9, are determined by the known formulas for the resistance of materials [12].

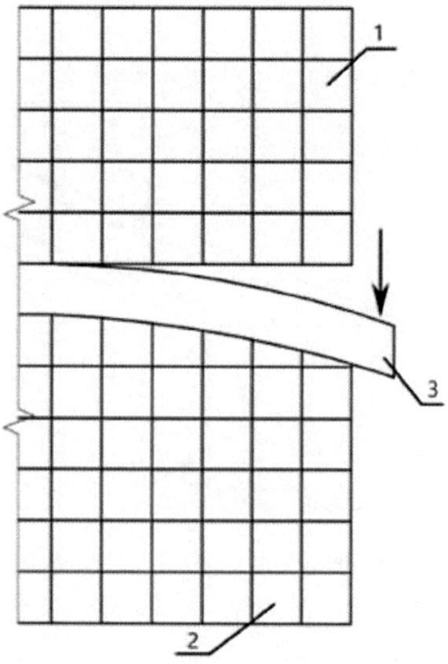

Fig. 3 Diagram of deformation of a reinforcing bar under the action of a transverse load 1—finite elements of concrete (stone); 2—finite elements of a reinforcing bar

Let us focus separately on displacements from local deformation. At first glance, the most accurate solution to this problem seems to be by modeling with three-dimensional finite elements using the well-known software *Ansys*, *Abacus*, *Lira*, etc. However, a detailed analysis reveals almost insurmountable obstacles. The most important of them is the correct modeling of the connection between the concrete and the anchor when it is acting on a load perpendicular to its axis. The fact is that when the anchor is working on a transverse load, one part of the concrete under the metal rod is crushed, and the opposite part "moves away" from the concrete almost without any resistance (Fig. 3).

In addition, it is known that the concrete in the contact zone with the reinforcement has mechanical characteristics that differ from the characteristics of the main part of the reinforced concrete element. According to M. M. Kholmyansky [13], microcracks are formed in the contact zone of reinforcing bars and concrete, which are difficult to account for by calculation. The deformability and strength of the contact layer may differ significantly from similar characteristics of the main concrete.

Modeling using one-way bonds (from the side of reinforcing bars separation from concrete) is also not acceptable due to the above-mentioned properties of the contact layer. The above facts are easily verified by experimental research.

In this regard, the displacements of the anchor from the transverse load should be determined experimentally. The recommendations [14] provide a formula obtained from processing experimental data, from which it is possible to determine the transverse displacement of a reinforcing bar loaded with a load perpendicular to its axis:

$$a_{loc} = 1000\frac{Q^2}{d_s^3 E_c^2} + \frac{Q}{d_s E_c} \tag{1}$$

where d_s and E_c are, respectively, the diameter of a reinforcing bar and the modulus of concrete deformations; Q is the force applied to the reinforcing bar in the direction perpendicular to its axis.

The empirical formula (1) integrally takes into account all the facts of the complex stress-strain state of the reinforcement-concrete system.

Equating the displacements v_i and y_i for the beam of the first layer (the stone part) to the same displacements for the beam of the second layer, we get the equation of strain compatibility at the *i* point of the structure. By composing such equations for all ***n*** points of the composite beam, it is not difficult to obtain a system of ***2*n*** equations with ***2*n*** unknowns $T_1 \ldots T_n$, $S_1 \ldots S_n$. The calculation algorithm based on a system of equations for the compatibility of deformations allows for a detailed analysis of the gain efficiency when varying different factors. At the same time, the software (the authors wrote it in *Pascal*) is quite simple, but the effectiveness of its use is high in terms of selecting the thickness of the reinforced concrete plate required by the designer, the number and diameter of anchors connecting the cage to the strengthened stone bending element.

After determining the unknowns, each of the two beams (stone and reinforced concrete plate) is calculated as a statically definable structure, which is affected by external forces and unknown forces $T_1 \ldots T_n$, $S_1 \ldots S_n$ in bonds determined from the solution of the system of compatibility equations.

The degree of compatibility of stone beams and side reinforced concrete plates depends on the number of anchors, connecting the parts of the composite beams, position of anchors, their diameter, and the mechanical characteristics of the stone part and reinforced concrete plates. Let us conditionally call this degree of compatibility "the gain coefficient". This coefficient ***k*** is equivalent to the ratio of the maximum bending moment in a non-strengthened stone beam from the action of an external load to the maximum bending moment in the stone part of the strengthened beam. In other words, if the calculation based on joint action shows that the maximum bending moment M_{st} acts in the stone part of the combined beam, and the maximum bending moment in the beam from the external load is ***M***, then the gain coefficient ***k*** will be determined in the following way:

$$k = M/M_{st} > 1 \tag{2}$$

It should be noted that when composing a system of equations for the strain compatibility, the malleability from local deformation of the first layer (the stone part) and the second layer (reinforced concrete plates) will be different.

The degree of influence of the bond stiffness on the operation of the composite structure will be shown by an example. Let's assume that there is a stone beam consisting of D500 aerated concrete blocks. Beam width $b_1 = 200$ mm, its height is 300 mm, the span is 3000 mm. Aerated concrete blocks are strengthened with side reinforced concrete plates arranged symmetrically on both sides. We will vary the thickness of reinforced concrete plates b2/2 (meaning that the total thickness of one plate is equal to b_2). Let the bonds be arranged in two horizontal rows, located at a distance of 50 mm from the top and bottom of the axis of the composite beam. The

Table 1 Results of the composite beam calculation

Variant	b_2	n	Gain coefficient k when d_s equals			
			3	6.5	10	14
1	10	2	1,5	1,61	1,64	1,65
2	10	3	1,8	1,93	1,96	1,97
3	10	5	1,79	1,87	1,88	1,88
4	20	2	1,75	2,04	2,11	2,15
5	20	3	2,4	2,9	3,01	3,06
6	20	5	2,44	2,73	2,78	2,8
7	40	2	2,0	2,59	2,79	2,89
8	40	3	3,24	4,93	5,47	5,74
9	40	5	3,45	4,45	4,68	4,77

bond beams are distributed symmetrically along the length of the beam, their number $\boldsymbol{n}$ is variable. The diameter of bonds $\boldsymbol{d_s}$ also varies. Let a uniformly distributed load $\boldsymbol{q} = 10$ kN/m act on the stone part of the beam. Table 1 shows the calculation results of such a composite beam.

As can be seen from the table, the reinforcement efficiency is affected by the diameter of the anchors $\boldsymbol{d_s}$, the thickness of the concrete plate b2, and the quantity of anchors $\boldsymbol{n}$. A significant increase in the anchor diameter has little effect on increasing the gain coefficient. For example, with an anchor diameter of 14 and 10 mm, the gain coefficients differ slightly. The difference in the gain coefficients for different anchor diameters is greater the thicker the side reinforced concrete plate.

We can also conclude that for small spans, the most effective is a small number of anchors connecting a stone beam with a reinforced concrete plate, which differs from the requirements for the location of anchors when installing reinforced concrete clips [5]. These tables also show that the reinforced concrete plate can significantly strengthen the stone element (in the example given it is five or more times), which indicates a fairly high efficiency of using such structures.

It should be noted that the results shown in Table 1 were obtained without taking into account the crack formation in the composite beam and the nonlinear properties of materials. Research on the nonlinear properties of the material of a clip and a stone beam, as well as on their cracking can be performed iteratively, changing the equivalent stiffness of the stone part and the reinforced concrete plate at each iteration.

If the connections connecting the stone part and reinforced concrete plates are malleable, then the zero point of the stress plot in the stone part and in the reinforced concrete plates will be located at different distances in height (Fig. 4). In this case, it is possible that the zero stress point in the stone element is higher than the zero stress point in concrete and vice versa. The ratio depends on the degree of malleability of anchors, deformability of the stone and reinforced concrete parts, and their reinforcement. All these factors are easily determined by iterative calculation, when at each

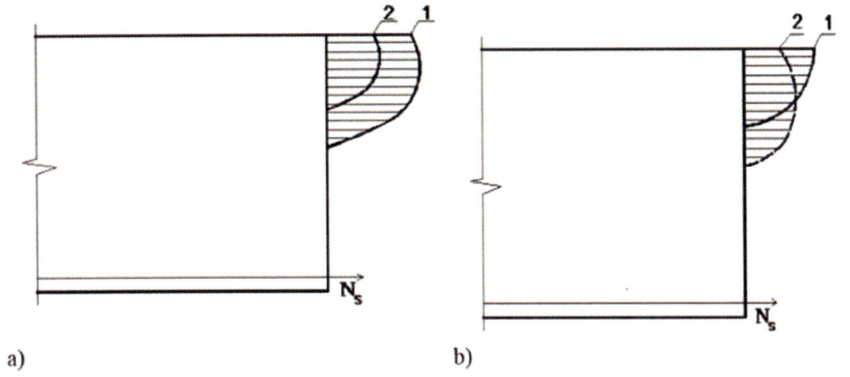

Fig. 4 Possible forms of stress plot in the compressed zone of the composite element. a)—The zero point of stress in the stone element is higher than the zero point of stress in concrete; b)—Vice versa. 1-stresses in the reinforced concrete plate; 2-stresses in the stone part

iteration, after determining the forces in the links, the stone and reinforced concrete beams are considered separately.

As a result of the flexibility of bonds, in addition to the various locations of the zero points of the stress plot, deflections of the stone part and reinforced concrete plates are different as well. There is a shift of the stone part relative to the reinforced concrete one. This factor is also easily determined using a computer-based calculation software developed by a method based on the compilation of a system of equations for the compatibility of deformations of two layers and the determination of unknown forces in the bonds. This theoretical fact of displacement of the stone part relative to the reinforced concrete plate has been empirically tested in [15–20].

If the stiffness of the anchors is high and local deformation can be ignored, then the composite beam can be considered as a monolithic consisting of two different materials in cross section. According to expression (1), the flexibility of the bonds depends on the diameter of the bond and the modulus of deformations of the concrete and stone part. The condition when the pliability of the bonds can be ignored is easily obtained by varying the stiffness of the bonds according to the software developed on the basis of the suggested methodology. The authors claim that the calculation of the composite structure, which can be considered monolithic, should be carried out using the method of calculated resistances of reinforced concrete [5], which is the subject of further research.

3 Conclusions

The paper suggests a method for determining the forces in the bonds connecting two parts of a composite structure, based on the compilation of conditions for the compatibility of deformations at the location of the bonds. The degree of joint action of stone beams and reinforced concrete plates is dependent on the number and diameter of bonds, their location and the strength and deformation characteristics of stone beams and reinforced concrete plates. With a high stiffness of the bonds, the composite structure can be calculated as a monolithic one, but taking into account the presence of two different materials in the cross section.

Further research is needed to develop a method for calculating the composite structure in the case when the malleability of anchors can be ignored, using the method of calculated resistances of reinforced concrete.

References

1. Azizov T, Melnik O, Myza O (2019) Strength and deformation of combined beams with side reinforced plates. https://doi.org/10.4028/www.scientific.net/MSF.968.234
2. Azizov T, Melnyk O, Orlova O, Kalenchuk-Porkhanova A, Vakal L (2017) Calculation of reinforced concrete ceilings with normal cracks accounting the chebyshev approximation. In: Paper Presented at the MATEC Web of Conferences, p 116. https://doi.org/10.1051/matecconf/201711602002
3. Azizov T, Jurkowska N, Kochkarev D (2019). Basis of calculation on torsion for reinforced concrete structures with normal cracks. In: Paper presented at the Proceedings of the Fib Symposium 2019: Concrete - Innovations in Materials, Design and Structures, pp 1718–1725
4. Kochkarev D, Azizov T, Galinska T (2020) Design of effective statically indeterminate reinforced concrete beams. https://doi.org/10.1007/978-3-030-42939-3_10
5. Kochkarev D, Azizov T, Azizova A, Galinska T (2021) Designing of standard cross sections of composite bending reinforced concrete elements by the method of design resistance of reinforced concrete. https://doi.org/10.1007/978-3-030-57340-9_25
6. Azizov T, Kochkarev D, Galinska T (2020) Reinforced concrete rod elements stiffness considering concrete nonlinear properties. Lect Notes Civil Eng 47:1–6. https://doi.org/10.1007/978-3-030-27011-7_1
7. Kochkarev D, Azizov T, Galinska T (2018) Bending deflection reinforced concrete elements determination. In: Paper Presented at the MATEC Web of Conferences, vol 230, p 02012. https://doi.org/10.1051/matecconf/201823002012
8. Kholmyanskij MM (1981) Kontakt armatury‘ s betonom. Strojizdat, p 184
9. Kamiani ta armokamiani konstruktsii (2011) Osnovni polozhennia: DBN V.2.6–162:2010. Minrehionbud Ukrainy, p 97 s
10. Piskunov VG, Gorik AV, Cherednikov VN (2000) Modeling of transverse shears of piecewise homogeneous composite bars using an iterative process with account of tangential loads 2. Resolving equations and results. Mech Comp Mater. 36(6):445–452. https://doi.org/10.1023/A:1006798314569
11. Piskunov VG, Goryk AV, Cherednikov VN (2000). Modeling of transverse shears of piecewise homogeneous composite bars using an iterative process with account of tangential loads. 1. Construction of a model. Mech Comp Mater 36(4):287–296. https://doi.org/10.1007/BF02262807
12. Pavlikov A, Kochkarov D, Harkava O (2019) Calculation of reinforced concrete members strength by new concept. In: CONCRETE. Innovations in Materials, Design and Structures: Proceedings of the fib Symposium 2019, Kraków, Poland, 27–29 May 2019, pp 820–827. http://reposit.pntu.edu.ua/handle/PoltNTU/6064
13. Pavlikov A, Kochkarev D, Harkava O (2020) Analysis of eccentrically loaded members of circular cross section by nonlinear deformation model. https://doi.org/10.1007/978-3-030-42939-3_19

14. Storozhenko L, Yermolenko D, Gasii G (2018). Investigation of the deformation state of a composite cable space frame structures with a photogrammetric method. Int J Eng Technol 7(3.2):442–446. https://doi.org/10.14419/ijet.v7i3.2.14568
15. Cherniha R, Pliukhin O (2013) New conditional symmetries and exact solutions of reaction-diffusion-convection equations with exponential nonlinearities. J Math Anal Appl 403(1):23–37. https://doi.org/10.1016/j.jmaa.2013.02.010
16. Kozak O, Zhuravskyi O, Delyavskyy MI (2019). Effect of the pre-stressed reinforcement curvature on the bearing capacity of inclined sections of monolithic beams. In: Paper presented at the AIP Conference Proceedings, p 2077. https://doi.org/10.1063/1.5091888
17. Panchenko OV, Ivanyts'kyi YL, Kun' PS, Zhuravs'kyi OD (2018) Determination of the durability of ferroconcrete bridge beams reinforced by composite strips. Mater Sci 53(5):660-665. https://doi.org/10.1007/s11003-018-0121-3
18. Semko O, Fenko O, Hasenko A, Harkava O, Kyrychenko V (2018) Influence of external and internal cooling at solidification on strength of brittle duralumin in compression. In: Paper Presented at the MATEC Web of Conferences, p 230. https://doi.org/10.1051/matecconf/201823002029
19. Hasenko A, Semko O, Drobotia O, Sirobaba V (2020) Experimental and numerical studies of nodes of light steel-reinforced concrete structures. In: Paper presented at the Proceedings of the 2020 Session of the 13th Fib International PhD Symposium in Civil Engineering, pp 173–178
20. Semko O, Hasenko A, Kyrychenko V, Sirobaba V (2020) The rational parameters of the civil building steel frame with struts. https://doi.org/10.1007/978-3-030-42939-3_25

Thermo-Technical Issues of Ventilated Facades in Azerbaijan

Samira Akbarova and Yurii Avramenko

Abstract When utilization the technology of ventilated facades for any buildings it is often necessary to solve complex technical and technological issues for designers and operators. By implement architectural solutions and ideas it is essential to calculate major thermo-technical characteristics of the facades, the location of each bracket, structural elements, complex joints in order to avoid or minimize thermal bridges. Ventilated facades usually have very complicated geometric shapes and configurations. This article reflects an expert assessment of the thermo- technical issues of using ventilated facades in harsh wind operating conditions in terms of reducing the heat bridges. The effect of wind on the ventilated facade systems for climate indicators of the city of Baku is considered. Measurements of the wind speed along the facade height and air velocity in the cavity had been carried out with the multi-functional measuring instruments Testo. An increase in wind speed leads to the growth of the cavity air velocity that is achieved 1–1.5 m/s and as result, it causes declining in the thermal resistance of the external wall. The resulting approximation equation accurately reflects actual air velocity in the air gap. The main consequences of the thermal-technical issues are revealed. Findings can be used for calculated- experimental control of the facade thermal characteristics in order to fulfill a multi-disciplinary energy audit of the building.

Keywords Ventilated façade · Thermo-technical issues · Wind speed · Ventilated air gap · Air permeability · Operating conditions

---

S. Akbarova (✉)
Azerbaijan University of Architecture and Construction, Ayna Sultanova 11, Baku, Azerbaijan
e-mail: samira.akbarova@azmiu.edu.az

Y. Avramenko (✉)
National University «Yuri Kondratyuk Poltava Polytechnic», Pershotravnevyj Ave. 24., Poltava 36011, Ukraine

V. Onyshchenko et al. (eds.), *Proceedings of the 3rd International Conference on Building Innovations*, Lecture Notes in Civil Engineering 181,
https://doi.org/10.1007/978-3-030-85043-2_4

1 Introduction

Although the facade cladding technology with a ventilated air gap is very popular today and it has been sufficiently studied by specialists and scientists around the world, still there are some issues in point of their application to specific projects in regions with a humid windy climate [1]. When using the technology of ventilated facades, it is often necessary to solve complex technical and technological issues for designers, architects and operators [2]. Oftentimes by implementing architectural solutions and ideas, it is essential to calculate major thermo-technical characteristics of the facade, in particular, the location of each fasten and bracket, structural elements, fixing substructures, complex joints in order to avoid or minimize any kind of thermal bridges [3, 4]. Ventilated facades (VF) usually have a sophisticated geometric shape and configuration [5]. Sometimes a complicated configuration thought out by an architect requires accurate calculations and even the necessity to develop a separate project for facades with a ventilated air gap [6, 7]. For example, mounting of window spaces window sills, the stained-glass windows, a combination of any transparent elements with an opaque wall, the presence of oval or sharp external corners of the buildings should be considered separately and they should be designed precisely. In addition that, sometimes there may be deviations along the wall height up to 4 cm per floor that is why in order to avoid the deviations along the facade height no more than the normalized 3–4 mm per floor it is needed to design its own project for facades with the ventilated air gap [8].

This article reflects an expert assessment of the thermo-technical issues of using ventilated facades in harsh operating conditions in order to minimize thermal bridges. The effect of wind on VF for climate indicators of Baku is considered. The main characteristic consequences of the thermo-technical issues are given.

2 Methods

2.1 *Facade Design Solutions*

The building facade system with a ventilated air gap has a number of significant differences from the known wall structures without a ventilated air gap and the thermal insulation layer. Figure 1a shows a general view of the wall facade system with the ventilated air gap. The thermal insulation layer is located in the air gap of the facade system, the system has a metal substructure and a front panel that determines the architectural appearance of the building. Figure 1b represents the percentage of various types of front panels for VF utilized in Azerbaijan.

Ventilated facades are called ventilated because there is a ventilation gap between the front panel and the thermal insulation layer if during installation the ventilation gap is overlapped moisture of the air can destroy the insulation [9]. The minimum gap thickness is between 30–70 mm and the maximum is regulated by fire safety.

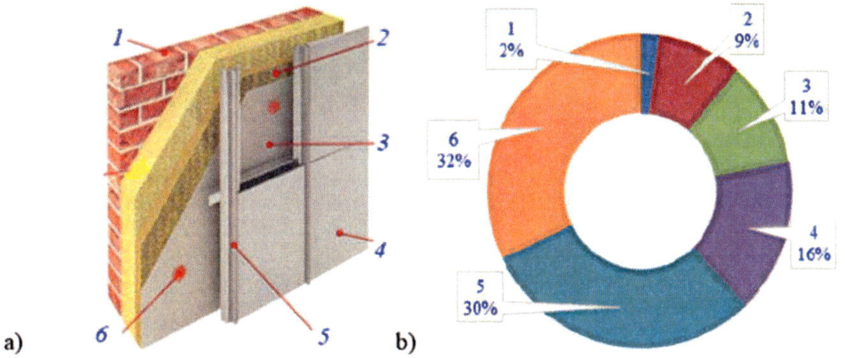

Fig. 1 a) Constructive scheme of a vertical cross-section of VF: 1- main masonry; 2- thermal insulation; 3- ventilated air gap; 4- front panel; 5- fixing substructure; 6- moisture protected layer; b) Different types of front panels, used in Baku, in %: 1-polyurethane, 2- composite panels, 3- porcelain stoneware, 4- fiber-cement panels, 5- metal cassettes, 6- natural stone

The main mistake in the implementation of VF is the installation of some devices on the building envelope, such as air conditioners, which are strictly prohibited by the manufacturers. If necessary, they can be mounted directly on the supporting main wall, having previously cut out the corresponding space in the front panels and thermal insulation layer [10, 11]. The next issue is when during the restoration of the historical buildings it is unwanted to apply VF. For the buildings, which were built in the early eighteenth century, the wall structures with VF are not used in Baku, because the installation technology involves the utilizing of a metal frame for the installation of insulation and cladding panels since the weight load on the exterior walls can be significant. The use of ventilated facades for old buildings of the nineteenth century with brickwork is excluded, that kind of buildings and houses there are in Baku, Ganja, Nakhchivan, Shusha cities of Azerbaijan.

2.2 *Climate Conditions*

In Baku, ventilated facades have been widely used for about 25–30 years. There is an opinion that ventilated facades can be mounted year-round under any climatic conditions, but not in Baku, not without reason Baku is called the city of winds. The location of the city on the Absheron Peninsula with an amphitheater to the Caspian Sea determines the constant presence of northerly strong winds all year round. Here, 100–150 days per year the wind speed exceeds 15 m/s, and the average annual wind speed is 6–9 m/s, and on some days hurricane winds are also possible at speeds up to 40 m/s. In addition that, the difference in air temperature and humidity can be significant in different parts of the city. Increased humidity during all year dictates a more careful choice of fasteners made of expensive galvanized steel with a polymer coating, telescopic brackets with movable insert parts, reliable thermal suppressors [3] and so forth. Due to wind effects for the decoration of high-rise buildings above 7 floors with a height of more than 20 m, lightweight cladding front panels are preferred. For the lower floors, less than seven, there are used expensive natural stones, various composite materials. Ceramic granite is also very popular in Baku for VF (Fig. 1b).

3 Results

It is widely known that wind influence is one of the dominant driving force for building facade thermal performance. The air movement in the air gap of the façade occurs under the influence of wind pressure and buoyancy. In general, air velocity should be calculated using the formula below [12]:

$$V_C = \sqrt{\frac{k \cdot V_{10}^2 + 0.08 \cdot L \cdot (t_C - t_{ext})}{\sum_i \varepsilon_i}} \tag{1}$$

V_{10} - average speed of wind, m/s, on the terrain according to the meteorological data on the height $H = 10$ m;
L - height difference between the air cavity inlet and outlet, m;
$\sum_i \xi_i$ – sum of the local aerodynamic drag coefficients;
$\xi = 5.5$ - for input where airflow is bent;
$\xi = 4.5$- for output where airflow is sharply bent;
$(t_c - t_{ext})$ - difference between the mean air temperature in the cavity and the mean external temperature that is taken between (1.5–2.5), ^{0}C;
k - coefficient of the wind speed change, based on building's height, H, m [13–15]:

$$k \approx 0.123 \cdot \sqrt{H} \tag{2}$$

The speed of wind can be calculated based to any building's height, H, more than 10 m using the formula [13]:

$$V_H = V_{10} \cdot \left(\frac{H}{10}\right)^{\beta} \tag{3}$$

β - local terrain exponent, coefficient, that depends on the area type;
$\beta = 0.4$, $\beta = 0.28$, $\beta = 0.16$ for urban area with the buildings height H > 25 m, H = 10-25 m, H < 10 m respectively [12].

Table 1 Engineering calculation results based on formulas (1–3)

No	L, m	V_H, m/s			V_C, m/s		
1	7.2	7	8	9	1.39	1.54	1.68
2	14.4	8.1	9.26	10.41	1.55	1.72	1.88
3	21.6	9.53	10.89	12.24	1.76	1.94	2.12
4	28.8	10.69	12.21	13.74	1.91	2.12	2.31
5	36.0	11.68	13.35	15.02	2.05	2.26	2.47
6	43.2	12.57	14.36	16.16	2.16	2.39	2.61
7	46.8	12.98	14.83	16.69	2.21	2.45	2.67

Table 2 Experimental data of average wind speed and average air velocity in air gap

No	L, m	V_f, m/s			V, m/s		
1	7.2	7	8	9	1	1.13	1.2
2	14.4	7.8	8.8	10	1.11	1.18	1.26
3	21.6	8.8	10	11.3	1.18	1.26	1.34
4	28.8	9.6	11	12.3	1,2	1.32	1.4
5	36.0	10.2	11.7	13.2	1.3	1.36	1.45
6	43.2	10.8	12.4	13.8	1.3	1.4	1.48
7	46.8	11.1	12.7	14.2	1.33	1.43	1.51

Results of calculated air velocity (V_C) and wind speed (V_H) by the formulas 1–3 are given in Table 1.

Measurements of air velocity in the air gap for seven different points through the open joints of the cladding panels of VF and wind speed along the building height had been carried out by testo. The next empirical equation allows to approximate a full-scale experimental values of the air velocity in the air gap (V) depending on wind speed (V_f) according to the façade height (Table 2):

$$V = 0.4 \cdot V_f^{0.7} \tag{4}$$

Analysis of experimental and calculated results shows that the higher wind speed causes greater air velocity in the air gap along the facade height (Fig. 2).

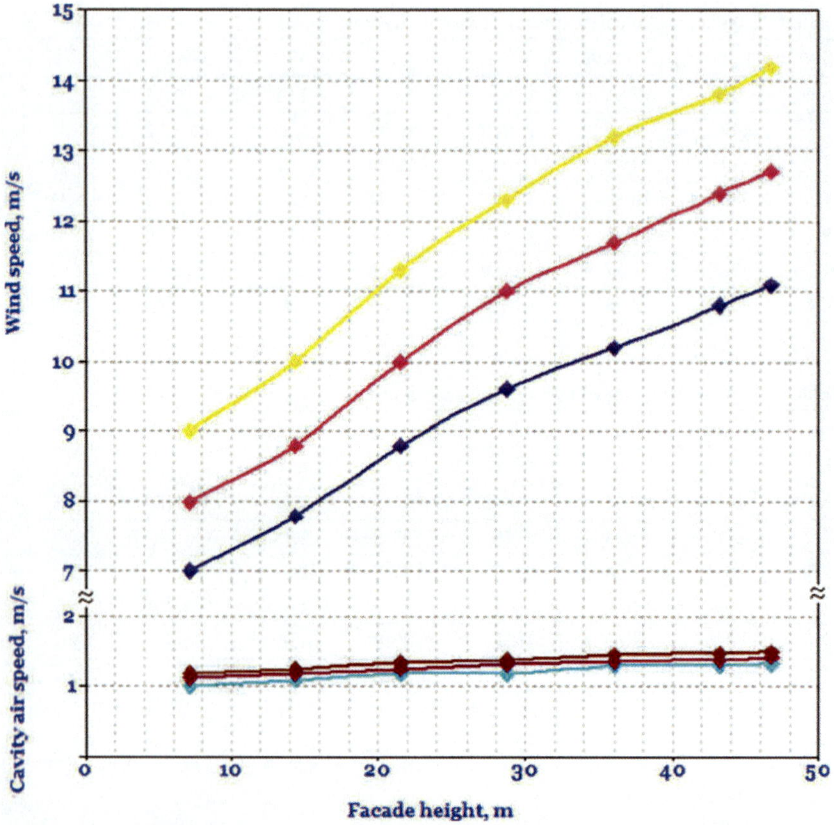

Fig. 2 Dependence between wind speed and air velocity in the cavity along facade height according to experimental data Table 2: (dark blue) and (light blue) for $V_{10} = 7$ m/s, (red) and (dark red) for $V_{10} = 8$ m/s, (yellow) and (dark red) for $V_{10} = 9$ m/s

4 Discussion

It must be admitted that during recent 10 years, almost all restored buildings in Azerbaijan were arranged with ventilated facades but without any thermal insulation layer, which should be attached to the outside of the main masonry. But today, for multi-story multi-functional buildings, mineral wool with vapor barrier and hydro-wind protection layer are used for wall insulation, an example of this can be the buildings of Panorama Park Baku class A residential complex.

In Baku, the external walls with air gaps are used everywhere especially for the reconstruction of the housing stock for buildings of 5–9 floors. VF is used for high-rise buildings and skyscrapers up to 100 m high, which determines the features of their thermo-physical properties. Today, some experts believe that ventilated facades have no obvious thermo- technical issues. But the many years of experience in their design and operation gives full reason to local experts to assert the existence of tasks that must be solved throughout the entire life cycle of the building.

The purpose of this article was to study the thermal- technical characteristics of a naturally ventilated facade system under extreme wind conditions. Analysis of the results of the full-scale experimental study on the thermal performance of the air cavity of ventilated façade allows us to summarize:

1-an increase of wind speed along facade height leads to the growth of the cavity air velocity that is achieved 1–1.5 m/s (Table 2) and as result, it causes declining in the thermal resistance of the external wall and enlargement of the heat bridges. The resulting approximation Eq. (4) more accurately reflects actual air velocity in the cavity, allows to calculate air velocity in the air gap depending on wind speed and may be used for express calculation methods;

2-the main thermo-technical engineering issues arising from the operation of ventilated facade systems with the air gap can be a regular accumulation of moisture in the thickness of the structure and in the air gap and thermo-technical heterogeneity of the facade and wall structure as a whole. The main reasons for that are poor ventilation of the air gap, significant resistance to vapor permeation of the windproof membrane, increased vapor permeability of the main masonry of the wall and the presence of the thermally conductive inclusions due to incorrect design. As the consequence, there are reduced durability of the used materials, deterioration of the building appearance, reduced thermal resistance of the structure and low value of the coefficient of heat engineering uniformity.

5 Conclusions

The main thermo-technical issue related to VF is the fact that there are unaccounted thermal conductive elements and joints in its structure as the formation of low-temperature zones is possible at the points of inclusion of thermal conductive elements. These issues can be solved through the use of modern materials for anchoring fasteners, as well as optimizing the thermal and humidity regime for each individual case of the wall structure [16]. The article summarizes that the issues encountered during the operation stage of such systems lead to a decrease in the thermo-physical properties of the facade and require further research.

It should be highlighted that the market VF in Azerbaijan has stabilized somewhat in the recent 10 years. Nowadays here there are new type front panels with inbuilt LED which give unlimited illumination possibilities. And besides, since the assortment has expanded and there is a variety of front panels, for example, they can match the structure of wood, be with a mirror surface like polished metal, etc. it increases bold architectural solutions. It is important that local specialists will continue to improve the methodology of a complete and accurate thermal engineering calculation of VF based on national construction norms and standards in combination with on-site experiments and theoretical studies.

References

1. Gagarin V, Kozlov V, Tsykanovsky E (2015) Thermal protection of facades with ventilated air gap. J ABOK 8:70–81
2. Gagarin V, Kozlov V, Lushin K (2015) Calculation of the velocity of air in the air gap facade systems, where natural ventilation. Int J Appl Eng Res 10(23):43438–43441. http://www.ripublication.com
3. Bienvenido-Huertas D, Rodríguez-Álvaro R, Moyano J, Marín D, Rico F (2019) A comparative study of the methods to assess the thermal transmittance in opaque Wall in the Mediterranean winter. Inf Constr 71(554):e288. https://doi.org/10.3989/ic.62542
4. Mammadov N, Akbarova S, Rzayeva N (2018) Building energy auditing is a tool to improve their energy efficiency. In: Ingenieurtag. Brandenburq Technical University, Germany, pp 210–219. https://opus4.kobv.de/opus4btu/frontdoor/deliver/index/docId/4724/file/4_Ingenieurtag_2018.pdf
5. Halawa E, Ghaffarianhoseini A, Ghaffarianhoseini A, Trombley J, Hassan N, Baig M (2017) A review on energy conscious designs of building façades in hot and humid climates: Lessons for (and from) Kuala Lumpur and Darwin. Renew Sustain Energy Rev 82:2147–2161. https://doi.org/10.1016/j.rser.2017.08.061
6. Petrichenko M, Musorina T, Statsenko E, Ostrovaia A, Tarasov1 V (2016) Functionality of ventilated facades: protection of insulation. In: International Scientific Conference Week of Science in SPbPU. https://doi.org/10.1051/matecconf/20165301036
7. Gagarin V, Guvernjuk S, Kozlov V, Ledenev P, Tsykanovsky E (2010) Results of researches of properties of hinged facade systems with the ventilated air layer in the frame of the grant of the russian fundamental researches fund «Aerothermophysics of nontight bodies in low speed air streams». Constr Sci J 3:261–278
8. Prada A, Baratieri M, Gasparella A (2013) Analysis of the impact of ventilated cavities on the performance of opaque components. In: 1st IBPSA- Italy International Conference
9. Gagarin VG, Kozlov VV (2010) Teplofizicheskie-raschety-pri-proektirovanii-navesnyh-teploizolyacionnyh-fasadnyh-sistem-s-vozdushnym-zazorom. https://docplayer.ru/43635078-Teplofizicheskie-raschety-pri-proektirovanii-navesnyh-teploizolyacionnyh-fasadnyh-sistem-s-vozdushnym-zazorom.html, Accessed 11 Jan 2021
10. Mammadov N, Akbarova S (2018) Multi-disciplinary energy auditing of educational buildings in azerbaijan: case study at a university campus. In: IFAC, International Federation of Automatic Control. International Conference, vol 51, no 30, pp 311–315. https://www.sciencedirect.com/journal/ifac-papersonline, Accessed 16 Jan 2021, https://doi.org/10.1016/j.ifacol.2018.11.308
11. Tabunschikov Y (2014) Road map of green construction in russia: problems and growth perspectives. J Sustain Build Technol (3). www.abok.ru/for_spec/articles.5786
12. SNiP 2.01.07–85* (1985) (Russian acronym, Construction Norms and Regulations). Loads and exposures. In Russian
13. Gagarin V, Guvernjuk S, Kozlov V, Ledenev P, Tsykanovsky E (2010) Results of researches of properties of hinged facade systems with the ventilated air layer in the frame of the grant of the russian fundamental researches fund «Aerothermophysics of nontight bodies in low speed air streams». Constr Sci 3:261–278
14. Semko O, Yurin O, Skliarenko S, Avramenko Y (2017) Thermophysical aspects of cold roof spaces. In: MATEC Web of Conferences. Transbud – 2017: Structure, Materials and Infrastructure, vol 116–2017, p 02030. https://doi.org/10.1051/matecconf/201711602030
15. Leshchenko MV, Semko V (2015) Thermal characteristics of the external walling made of cold-formed steel studs and polystyrene concrete. Mag Civil Eng 60(8):44–55. https://doi.org/10.5862/MCE.60.6
16. Katunska J, Bullová I, Špaková M (2017) Analysis of air flow in the ventilated insulating air layer of the external wall. J Civ Eng. https://doi.org/10.1515/sspjce-2016-0015

Calculation of Wooden Elements Based on the Nonlinear Deformation Model

Natig Baghirzade

Abstract Currently, construction norms recommend developing calculation methods based on real nonlinear deformation diagrams of materials in general and structural elements in particular. Recent studies based on nonlinear models have shown that the calculation of compression elements, especially based on nonlinear deformation models, is accompanied not only by quantitative solutions, but also by qualitatively new results. In this regard, the development of a method for calculating compressed wooden elements based on models of nonlinear deformations is an important task of practical importance. The article develops an effective numerical method for determining the strained state and the load-bearing capacity of wooden elements operating on off-center compression in general, using a deformation diagram approximated by a quadratic parabola during wood compression. Applying the hypothesis of flat sections when constructing a method, the level of deformation on the edge and the relative height of the compression zone of the cross section linearly determined by comparing the curved axis of the shaft with a sine wave under eccentric compression. The method of numerical solution of the established system of solvent equations is explained. The proposed solution methodology also allows you to build a load-bending graph for a compressed element..

Keywords Nonlinear strain model · Eccentricity · Tension · Deformation · Load-bending graph

1 Introduction

The calculation of compressive elements based on linear models leads to different levels of errors in determining the strained state and load-bearing capacity, depending on the eccentricity of the compressive force and the flexibility of the compressed element. Therefore, there is a need to apply a nonlinear deformation model in the development of methods for calculating compressed building elements, especially

N. Baghirzade (✉)
Ministry of Emergency Situations of the Republic of Azerbaijan, State Agency for Control of the Construction Safety, General Administrative Office for State Expertise, Baku, Azerbaijan

V. Onyshchenko et al. (eds.), *Proceedings of the 3rd International Conference on Building Innovations*, Lecture Notes in Civil Engineering 181,
https://doi.org/10.1007/978-3-030-85043-2_5

wooden ones, since only on the basis of such models can we obtain results that are realistic and consistent with experimental studies. In flexible compressive elements, the load-bearing capacity is often determined by the stability condition. The paper develops an effective numerical method for constructing the load - bending relationship, which is indicated by the use of a diagram approximated by the law of a quadratic parabola when stretching-compressing the wood.

2 Main Part

2.1 Purpose of the Article

The aim of the article is to develop an effective numerical method for determining the parameters and load-bearing capacity that characterize the strained state of wooden elements working on compression, based on a model of nonlinear deformation. The article presents a system of nonlinear solvent equations for any value of the eccentricity of the compressive force and the flexibility of the element, as well as a method for numerical solution of the system using any diagram approximated by the law of the quadratic parabola when stretching or compression of the wood.

2.2 Research Methodology

The solution of the problem for the level of arbitrary load is assumed by the correctness of the hypothesis about the distribution of the longitudinal strain over the cross section. Then, under eccentric compression, the curved axis of the shaft is approximated by a sinusoid, and the deflection of the most stressed part of the shaft is expressed by these two parameters. The system of nonlinear equations for solving the problem is obtained from the equilibrium equations for the considered area. Based on the level of deformation of the extruded surface, the relative height of the compression zone of the corresponding cross-section, the deflection parameter and the load parameter are determined. By varying the strain in equal steps in the strain region, the parameters characterizing the strained state and the coordinates of the "load-bend" graph are determined, on the basis of which the bearing capacity of the compressed element can be defined.

2.3 Results

The wood tension–compression strain diagram can be approximated based on the following diagram, which includes both the compression work and the tensile work.

[1–3]:

$$\sigma = E \cdot \varepsilon - B \cdot \varepsilon^2 \tag{1}$$

In general, the accuracy of the flat cross-section hypothesis for an off-center compressed wooden element is assumed to be [2–5], then we can write that to change the relative strain along a rectangular cross-section,

$$\varepsilon_z = \frac{\varepsilon_f}{\xi} \cdot \left(\overline{z} + \xi - \frac{1}{2} \right); \quad \xi = \frac{x}{h}; \quad \overline{z} = \frac{z}{h} \tag{2}$$

(1) According to the diagram, the neutral axis does not pass through the center of gravity of the cross-section even in pure bending, because the resistance of the wood material to tension and compression is different, so the distribution of strain over the cross-section depends on (2). Now, considering Eq. (2) in (1), we can write down the distribution of normal stresses in the cross-section.:

$$\sigma_z = E \cdot \frac{\varepsilon_f}{\xi} \cdot \left(\overline{z} + \xi - \frac{1}{2} \right) - B \cdot \left(\frac{\varepsilon_f}{\xi} \right)^2 \cdot \left(\overline{z} + \xi - \frac{1}{2} \right)^2 \tag{3}$$

Regardless of whether the neutral axis is inside or outside the cross-section, the following expressions can be written for the main vector of normal stresses in the cross-section [1, 2] and the head moment of this vector relative to the center of gravity of the cross-section. based on Eq. (3):

$$N = b \cdot h \cdot \int_{-\frac{1}{2}}^{\frac{1}{2}} \sigma_z d\overline{z} = b \cdot h \cdot \int_{-\frac{1}{2}}^{\frac{1}{2}} \left[E \cdot \frac{\varepsilon_f}{\xi} \cdot \left(\overline{z} + \xi - \frac{1}{2} \right) - B \cdot \left(\frac{\varepsilon_f}{\xi} \right)^2 \cdot \left(\overline{z} + \xi - \frac{1}{2} \right)^2 \right] d\overline{z}$$

$$= b \cdot h \cdot \xi \cdot \left\{ \frac{E \cdot \varepsilon_f}{2} \cdot \left[1 - \left(\frac{\xi - 1}{\xi} \right)^2 \right] - \frac{B \cdot \varepsilon_f^2}{3} \cdot \left[1 - \left(\frac{\xi - 1}{\xi} \right)^3 \right] \right\}$$

$$M = b \cdot h^2 \cdot \int_{-\frac{1}{2}}^{\frac{1}{2}} \sigma_z \cdot \overline{z}\, d\overline{z} = b \cdot h^2 \cdot \int_{-\frac{1}{2}}^{\frac{1}{2}} \overline{z} \cdot \left[E \cdot \frac{\varepsilon_f}{\xi} \cdot \left(\overline{z} + \xi - \frac{1}{2} \right) - B \cdot \left(\frac{\varepsilon_f}{\xi} \right)^2 \cdot \left(\overline{z} + \xi - \frac{1}{2} \right)^2 \right] d\overline{z}$$

$$= b \cdot h^2 \cdot \left\{ \xi \cdot \frac{E \cdot \varepsilon_f}{4} \cdot \left[1 - \left(\frac{\xi - 1}{\xi} \right)^2 \right] - \frac{B \cdot \varepsilon_f^2}{6} \cdot \xi \cdot \left[1 - \left(\frac{\xi - 1}{\xi} \right)^3 \right] \right.$$

$$\left. + \xi^2 \cdot \frac{E \cdot \varepsilon_f}{6} \cdot \left[\left(\frac{\xi - 1}{\xi} \right)^2 \cdot \frac{\xi + 2}{\xi} - 1 \right] - \xi^2 \cdot \frac{B \cdot \varepsilon_f^2}{12} \cdot \left[\left(\frac{\xi - 1}{\xi} \right)^3 \cdot \frac{\xi + 3}{\xi} - 1 \right] \right\}$$

In the above expressions, we go to the strain level of the extruded fiber. For this reason, it is assumed that when $\varepsilon = \varepsilon_{f,\max}$ is, the tangent modulus is zero, and the stress value corresponding to this deformation is equal to the compressive strength of the wood R. From these conditions it follows that according to the accepted diagram (1) $\varepsilon_{f,\max} = \frac{E}{2B}$ and $R = \frac{E^2}{4B}$. Then take into account that the above equations are based on this is $\varepsilon_f = \beta \cdot \frac{E}{2B}$, including for the strain level $\beta = \frac{\varepsilon_f}{\varepsilon_{f,\max}} = \varepsilon_f \cdot \frac{2B}{E}$. Then we can write that

$$N = b \cdot h \cdot R \cdot N_*(\beta,\ \xi)\ ;$$
$$N_*(\beta,\ \xi) = \xi \cdot \beta \cdot \left[1 - \left(\frac{\xi - 1}{\xi}\right)^2\right] - \xi \cdot \frac{\beta^2}{3} \cdot \left[1 - \left(\frac{\xi - 1}{\xi}\right)^3\right] \tag{4}$$

$$M = b \cdot h^2 \cdot R \cdot M_*(\beta,\ \xi)\ ;$$
$$M_*(\beta,\ \xi) = \frac{\xi \cdot \beta}{2} \cdot \left[1 - \left(\frac{\xi - 1}{\xi}\right)^2\right] - \frac{\xi \cdot \beta^2}{6} \cdot \left[1 - \left(\frac{\xi - 1}{\xi}\right)^3\right]$$
$$+ \frac{\xi^2 \cdot \beta}{3} \cdot \left[\left(\frac{\xi - 1}{\xi}\right)^2 \cdot \frac{\xi + 2}{\xi} - 1\right] - \frac{\xi^2 \cdot \beta^2}{12} \cdot \left[\left(\frac{\xi - 1}{\xi}\right)^3 \cdot \frac{\xi + 3}{\xi} - 1\right] \tag{5}$$

They satisfy the dimensionless parameters and conditions $0 \le \beta \le 1$ and $\xi > 0$ used in these equations, which are then used to solve practical problems. Now suppose that the ends of the compressing element are fastened by a hinge and that this element is compressed by an eccentric system. We express the equation of the deformable axis of the compressed element in the form of a sinusoid [2], that is, we assume that $y(x) = f \cdot \sin \frac{\pi \cdot x}{l}$, it has the value of curvature $\chi = \left|y''\left(\frac{l}{2}\right)\right| = f \cdot \frac{\pi^2}{l^2}$ in the most stressed middle section. On the other hand, the value of the curvature in the mean cross-section is based on the flat cross-section hypothesis is $\chi = \frac{\varepsilon_f}{\xi \cdot h} = \frac{\beta}{\xi \cdot h} \cdot \frac{E}{2B}$. Therefore, the deflection of the middle section can be expressed by a parameter that determines the level of deformation in the extruded fiber of this section and the position of the neutral axis. Therefore, the deflection of the middle section can be expressed by a parameter that determines the level of deformation in the extruded fiber of this section and the position of the neutral axis.:

$$f \cdot \frac{\pi^2}{l^2} = \frac{\beta}{\xi \cdot h} \cdot \frac{E}{2B} \text{ or } f_0 = \frac{f}{h} = \frac{\beta}{\xi} \cdot \frac{l^2}{\pi^2 \cdot h^2} \cdot \frac{E}{2B} = \rho_0 \tag{6}$$

Where,

$$\rho_0 = \frac{l^2}{\pi^2 \cdot h^2} \cdot \frac{E}{2B} \tag{7}$$

Now we write the equilibrium equations of the middle part of the compressed shaft:

$$\begin{cases} b \cdot h \cdot R \cdot N_*(\beta,\ \xi) = P \\ b \cdot h^2 \cdot R \cdot M_*(\beta,\ \xi) = P \cdot (e + f) \end{cases} \tag{8}$$

Here you can write the equilibrium equations based on the two parameters as follows, taking into account Eq. (6) and $\sigma_0 = \frac{P}{b \cdot h \cdot R}$ including the notation (9):

$$N_*(\beta,\ \xi) = \sigma_0 \text{ and } M_*(\beta,\ \xi) = \sigma_0 \cdot \left(e_0 + \rho_0 \cdot \frac{\beta}{\xi}\right) \tag{9}$$

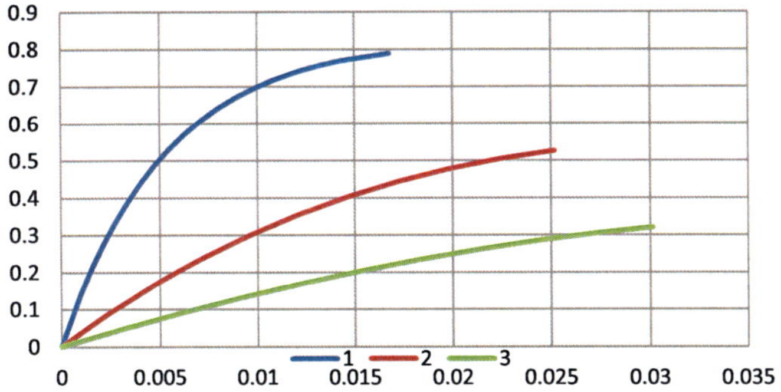

Fig. 1 Graph "load-bending" of a compressed shaft, the ends of which are fastened pivotally; 1 $- e_0 = 0,05$; $2 - e_0 = 0,2$; $3 - e_0 = 0,5$

From these two equations, you can write the following equation, which eliminates the compressive force parameter σ_0 and creates a relationship between the parameters β and ξ [1]

$$\Omega(\beta,\ \xi) = M_*(\beta,\ \xi) - N_*(\beta,\ \xi) \cdot \left(e_0 + \rho_0 \cdot \frac{\beta}{\xi}\right) = 0 \quad (10)$$

The peculiarity of this equation is that for each accepted value of the strain level from this equation, the value of the parameter that determines the position of the neutral axis can be set to one value. The deflection parameter can then be calculated based on Eq. (6) and finally the force parameter based on Eq. (9). Thus, solutions can be obtained for each value of the strain level $(\beta,\ \xi,\ f_0,\ \sigma_0)$ by applying the semi-inverse method. By changing the strain parameter in the range of its variation, a relationship can be established between the deflection parameter and the load parameter, on the basis of which the bearing capacity of the compressed shaft can be determined. To implement the algorithm, the corresponding software module was implemented and numerical experiments with its application were carried out. For example, in the following figure, where $E = 27 \cdot 10^3\ MPa$, $B = 3645 \cdot 10^3\ MPa$, $l = 4,5\ m$, $h = 0,6\ m$ at different values of the eccentricity $\sigma_0 \div f_0$, that is," load-bending " graph is built in dimensionless parameters (Fig. 1).

2.4 Scientific Novelty

An effective numerical method has been developed for determining the tensile deformation state and bearing capacity for an arbitrary assessment of the eccentricity of the compressive force and elasticity of an element when approximating the deformation diagram under tension–compression of wood by a quadratic parabola.

2.5 *Practical Importance*

The developed method of numerical calculation makes it possible to construct semi-empirical dependences for approximate engineering calculations based on numerical experiments, as well as to determine the strained state and the load-bearing capacity of compressed wooden elements based on a real nonlinear deformation diagram. the material and its results can be used to improve regulatory documents.

3 Conclusions

Based on the research, it can be said that the influence of the eccentricity of the compressive force on the bearing capacity is large and should be determined using this nonlinear deformation model.

References

1. Hajiev MA, Aliev RD, Guliyev FM (2019) Calculation of bending reinforced concrete elements for strength according to normal sections. Baku, p 121
2. Sanzharovsky RS, Veselov AA (2002) Theory of calculation of building structures for stability and modern standards. Moscow-Saint Petersburg, ACU, p 126
3. Pyatikrestovsky KP (2011) Power resistance of spatial wooden structures under short-term and long-term loads: dis. by Doctor of Technical Sciences, p 320
4. Pyatikrestovsky KP, Khunagov HS (2013) Nonlinear deformations of statically indeterminate wooden structures. Izvestiya vuzov. Construction, no 11–12, p 21–30
5. Varenik KA (2016) Calculation of centrally compressed wooden elements taking into account creep: dis. by Candidate of Technical Sciences. St. Petersburg, p 167

Consideration of the Actual Performance in Reliability of Channel Frames

Kseniia Chichulina, Viktor Chichulin, Serhiy Shkirenko, and Tetiana Kushnirova

Abstract The article deals with reliability calculation of a rigid node of a steel frame of a building. Some parameters of the actual operation of the flange connection of the crossbar to the column are taken into account. It is proposed to use conditionally rigid frame nodes. The reliability of a node is determined from the conditions of interrelated operation of individual elements. Calculated coefficients for designing this type of node are proposed. The paper presents flanged nodes with high-strength bolts. Such connections increase the manufacturability of structures. The use of these compounds requires the detailed study of their actual operation. During calculations, it is necessary to determine the relationships between the load-bearing capacity of the node elements. The measure of the statistical relationship between these values is calculated using the correlation coefficient. For steel frames, it is proposed to use conditionally rigid pliable nodes that provide a redistribution of forces between those loaded on crossbars and columns.

Keywords Steel frame · Rigid nodes · Connection · Reliability · Calculation · Correlation coefficient

1 Introduction

Modern construction requires exclusion from the practice of designing components for assembly welding due to their significant complexity. Therefore, for both domestic and European standards, flanged units with high-strength bolts have recently been recommended for use. Such connections make it possible to reduce the installation time of structures. Introducing new connections requires a more complete understanding of their actual operation. The need to reduce the weight of structures requires a more complete definition of the relationships between the load-bearing capacity of the node elements. The correlation coefficient can be used to determine the degree of statistical dependence between these values.

K. Chichulina (✉) · V. Chichulin · S. Shkirenko · T. Kushnirova
National University "Yuri Kondratyuk Poltava Polytechnic", Poltava, Ukraine
e-mail: Chichulinak@ukr.net

V. Onyshchenko et al. (eds.), *Proceedings of the 3rd International Conference on Building Innovations*, Lecture Notes in Civil Engineering 181,
https://doi.org/10.1007/978-3-030-85043-2_6

Current structural codes for steel and stainless steel structures such as Eurocode 3 [1] are based on the traditional two-step member-based design approach, in which internal actions are first obtained from a structural analysis, usually elastic, and the strength of each member and connection is subsequently checked using the structural design standard. However, the most recent versions of these standards already incorporate preliminary versions of the direct, or one-step, system-based design alternative, which is based on the design-by-analysis concept and allows evaluating the strength of structures directly from numerical simulations, although the standards in their current form do not provide reliability requirements for structural systems. Therefore, it is necessary to build a rigorous structural reliability framework to investigate acceptable target reliability indices for structural systems and to provide adequate system safety factors and system resistance factors. The paper [2] presents an extension of the DDM to the analysis of stainless steel structures, in which system reliability calibrations are presented for six stainless steel portal frames under gravity loads covering the three most common stainless steel families and different failure modes using advanced numerical simulations. From the derived reliability calibrations, suitable system safety factors and system resistance factors are proposed for the direct design of stainless steel frames in the European, US and Australian design frameworks under gravity loads. The paper [3] presents an algorithm to determine the reliability index of moment-resisting steel frames with bolted end-plate connections, by considering the effect of the uncertainties of capacity and seismic demand. To reflect the probabilistic response of structures, the horizontal and vertical distance of the bolt from edge of end-plate, the pretension force and the ultimate strength of the bolts, yield strength and the elastic modulus of plates for beam and column, which are the main parameters of construction errors in bolted end-plate connection, considered as random variables for the moment-rotation behavior of connections. All the components of the connections have been regarded as random parameters. Then, the probabilistic moment-rotation curve of connections was determined by means of the Latin Hypercube sampling method. Afterward, by conducting the nonlinear dynamic analysis in the 2-dimensional 5, 10, and 15 stories frames and considering the idealized random behavior of the connections as well as the random effect of different records of the earthquake, the results obtained with the aid of Monte Carlo sampling technique. Design-by-analysis methods for steel structures receive considerable attention from professional engineers, researchers and standard-writing groups. Designing by analysis, termed as the Direct Design Method (DDM), is premised on the use of geometric nonlinear inelastic finite element analysis to determine the ultimate strength of steel structural frames and subsequently incorporating the system resistance factor (ϕ_s) to account for the effects of uncertainties in geometric parameters, stiffness and strength. The paper [4] outlines the DDM in the context of cold-formed compact Hollow Steel Sections (HSS), including the reliability analysis framework at system level underpinning the Method. The system resistance factors for a series of representative 3D frames with hollow locally stable cross-sections are derived. The paper [5] presents accurate and efficient numerical procedure for evaluating the system reliability of steel frames with semi-rigid connections. The ultimate strength and behaviour of the frame were predicted using a refined plastic hinge model due to its computational efficiency, whilst the nonlinear behaviour of semi-rigid connections was captured using a three-parameter power model. The statistical properties for the three-parameter power model were obtained based on available experimental results. The sensitivity of reliability to the model error was also studied. Monte Carlo simulation was used to estimate the probability of failure and the reliability index of a system. Two example frames subjected to combined gravity and wind loads were examined and their system reliability indices for both strength and serviceability limit states were evaluated based on the randomness in loadings, material and geometric properties and semi-rigid connections. The results indicate that the frame reliability is strongly affected by semi-rigid connections. The design of steel frames by advanced analysis (second-order inelastic analysis with imperfections) of overall system behaviour is permitted in the American steel specification and the Australian Standard. In both specifications, the strength of a structural frame can be determined by a rigorous system nonlinear analysis in lieu of checking member resistances to the specific provisions of the Specification, provided that the limit states covered by the Specification equations are detected by inelastic analysis, and comparable or higher level of structural reliability is achieved by the inelastic analysis rather than by member-based design (DDM). The paper [6] describes the framework for developing reliability-based system resistance factors suitable for use with DDM. In [7] a numerical study to evaluate the reliability of 3D steel buildings with moment-resisting frames (MRFs) at the perimeter and gravity frames (GFs) at the interior under the action of seismic loading, is carried out. To this aim, the median values (MV) of the maximum drifts (MD) and fragility curves are obtained. The reliability, expressed as the mean annual rate of exceeding (υ) particular values of MD, is calculated by using the probabilistic seismic hazard analysis format. Three models, representing steel buildings of low-, mid- and high-rise are considered. The interior connections are assumed to be, first perfectly pinned (PP) as considered in practical design, and then as semi-rigid (SR). The Beam-Line Theory together with The Model of Richard are used to represent the SR connections. Results indicate that the MV demands and the seismic fragility can be significantly reduced when the stiffness and dissipation of energy at interior connections are considered; reductions higher than 40% and 60%, respectively, are observed in many of the cases. The structural reliability has also been significantly improved; reductions close to 20% in the expected maximum drift are observed in many cases. It is concluded that the effects of the stiffness and energy dissipation of the connections of GFs on the seismic response (in terms of MV), on the fragility curves, and on the mean annual rate of exceedance (υ), are significant so they should not be ignored. Even for shear connections which contribution in practical design is totally ignored, the effect is considerable. The high strength bolted end-plate connection is main type of the connections used widely in industrial construction [8]. There are two kinds of end-plate connections in steel portal frames: flush and extend end-plate connections. The main initial imperfection of bolted end-plate connection lies in which the thickness of end-plate and column flange can't meet the code provisions based on the field investigation. Considering the effect of initial imperfection, the actual behavior of end-plate connections in steel portal frames is seldom fully rigid. The true behavior of the connections is usually semi-rigid. Neglecting the real behavior of connections in the analysis may lead to unrealistic predictions of the response and reliability of steel portal frames. The paper [8] considers the effects of semi-rigid behavior of the connections in the finite element analysis and reliability analysis of steel portal frames. Assuming that the loads and the resistance of members are random variables, then the Monte Carlo simulation technique is used to estimate the failure probability of steel portal frame system. The results confirm that the thickness of end-plate has significant effect on safety of steel portal frame. In general, one of the founders of the school of reliability of steel structures is S. Pichugin [9, 10]. His works present methodological and theoretical approaches to calculating the reliability of steel frame nodes.

2 Main Body

The main rigid connection of the crossbar to the column for frame-link frames of single-story and multi-storey buildings is the connection by means of a flange with high-strength bolts of class 6.6–10.9 (see Fig. 1). In such a unit, the bolts work for stretching, the flange works for bending, and the transverse force from the beam is taken by a table welded to the column, or the bolts of the flange itself, which are more often calculated to work in reserve for cutting. Bolts are designed for the maximum tensile force and are accepted (more often) of the same size, so there is a certain reserve of load-bearing capacity in the joint, which must be taken into account when considering the actual operation of the node.

Node calculations take into account the various failure states that the connection may be in. The calculation of such a node presented in the European design standards [1] is close to considering the operation of node elements in the form of logical connection of elements.

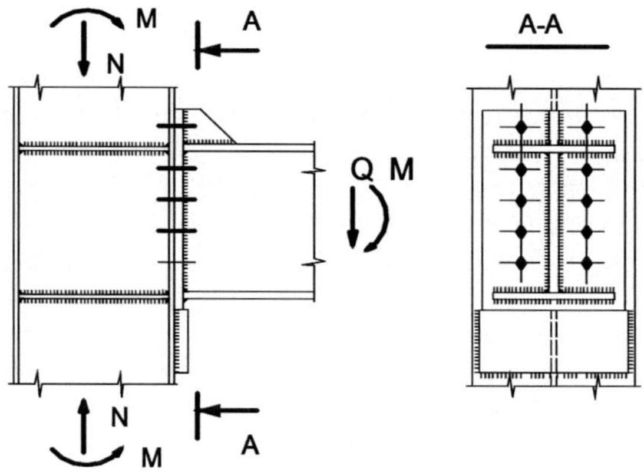

Fig. 1 Typical rigid beam-to-column coupling connection (rigid node)

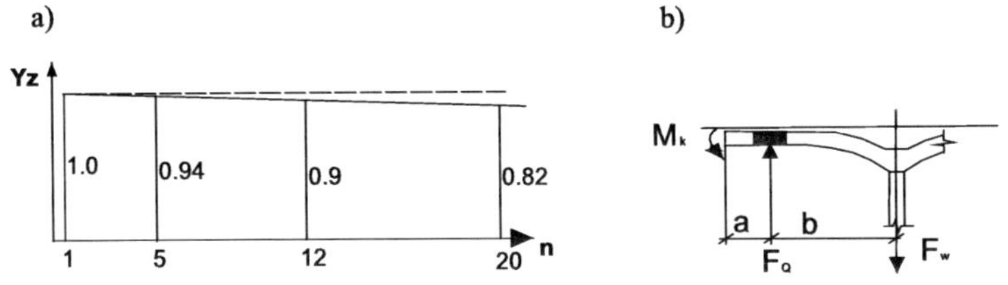

Fig. 2 For evaluating the reliability of nodes: **a** dependence graph of the coefficient of reduction of the calculated resistance γ_z on the number of statically independent elements of the node *n*; **b** diagram of the operation of the beam flange

When considering the operation of the elements of the node, it can be noted that plastic deformations of the unit provide for deformations of bolts and deformations from bending the flange in the plane of action of bolts along with the main forces in the wall of the beam F_w, from the action of the moment in the beam and the tensile force in the bolt F_a, it is necessary to take into account the action of the moment in the cantilever part of the flange M_k. The reaction of this moment adds tensile force to the bolt due to shoulder "a" (see Fig. 2b). The difference in the tensile force in the bolt is determined by the formula (1):

$$\Delta F_a = R_y W_k (a + b)/ab, \quad (1)$$

where *Ry* is the calculated resistance of the flange steel;

W_k is the moment of resistance of the console at the bolt operation site.

Unlike steel structures, in reinforced concrete structures, the rigidity of the crossbar and column is regulated by reinforcement. Up to the calculated moment (*M*), the node operates as a normal rigid connection, with an increase in the load and reaching the maximum moment, then the node operates as a conditionally articulated one. It is proposed to use conditionally rigid pliable units for steel frames, which provide redistribution of forces (bending moment) between loaded crossbars and columns.

If you limit the value of the moment transmitted to the column from the beam by adjusting the stiffness of the node, you can get significant steel savings when designing columns of single-story and multi-story buildings. The pliability of the node can be ensured by elastic plastic operation of the bolts in the flange connection, or by using calculated springs that are not compressed on the bolts of the flange connection, or by installing elastic gaskets in a closed steel housing. All these damping devices must be pre-calculated for compression, taking into account the stiffness limit of the node. Damping devices should act as a force leveler in the rows of bolts in the flange connection.

When calculating for reliability, the condition for trouble-free operation of the frame node can be represented as:

$$\tilde{x}_i = \tilde{\sigma}_{mi} - \tilde{\sigma}_{qi} \geq 0, \quad (2)$$

where $\tilde{\sigma}_{mi}$ is the random parameter of the strength of the i - th element of the node; $\tilde{\sigma}_{qi}$ is the random load parameter of the i-th node element.

For nodes where block diagrams can be represented as a serial connection of independent elements, the uptime conditions have the following form:

$$Y(x_1, x_2 \ldots x_n) = \prod_{i=1}^{n} X_i. \tag{3}$$

Let's write an expression for the load-bearing capacity reserve of two elements of the same node:

$$\tilde{Y}_1 = \tilde{X} - a_1\tilde{q}, \tag{4}$$

$$\tilde{Y}_2 = \tilde{Z} - a_2\tilde{q}, \tag{5}$$

where $\tilde{X},\ \tilde{Z}$ is the random characteristic of the element's strength (the yield strength of steel for the main parts, the yield strength of metal welds and bolts), for generality, we consider them different for different parts;

$\tilde{q}$ is the load on the structure as a whole, for example, snow load on the surface, which creates stress in the elements of nodes $a_1\tilde{q},\ a_2\tilde{q}$ (the operation of the structure is considered linear).

We describe the correlation moment for functions (4) and (5) based on calculations.

$$\begin{aligned}
K_{Y_1Y_2} &= M\left[(Y_1 - \overline{Y}_1)(Y_2 - \overline{Y}_2)\right] = M\left[(X - a_1 q - \overline{X} + a_1\overline{q})\right]\left[(Z - a_2 q - \overline{Z} + a_2\overline{q})\right] \\
&= M\left\{\left[(X - \overline{X}) - a_1(q - \overline{q})\right]\left[(Y - \overline{Y}) - a_2(q - \overline{q})\right]\right\} \\
&= M\left[(X - \overline{X})(Z - \overline{Z}) - a_1(q - \overline{q})(Z - \overline{Z}) - a_2(q - \overline{q})(X - \overline{X}) + a_1 a_2 (q - \overline{q})^2\right] \\
&= M\left[(X - \overline{X})(Z - \overline{Z})\right] - M\left[a_1(q - \overline{q})(Z - \overline{Z})\right] - M\left[a_2(q - \overline{q})(X - \overline{X})\right] + M\left[a_1 a_2 (q - \overline{q})^2\right] \\
&= K_{XZ} - a_1 K_{qZ} - a_2 K_{qX} + a_1 a_2 \hat{q}^2.
\end{aligned} \tag{6}$$

Standards for linear functions (4) and (5) are written as:

$$\hat{Y}_1 = \sqrt{\hat{X}^2 + a_1^2\hat{q}^2}; \qquad \hat{Y}_2 = \sqrt{\hat{Z}^2 + a_2^2\hat{q}^2}, \tag{7}$$

Taking into account the obtained expressions we write down the formula of the correlation coefficient for this case in general form:

$$r_{Y_1Y_2} = \frac{K_{Y_1Y_2}}{\hat{Y}_1\hat{Y}_2} = \frac{K_{XZ} - a_1 K_{qZ} - a_2 K_{qZ} + a_1 a_2 \hat{q}^2}{\sqrt{\hat{X}^2 + a_1^2\hat{q}^2}\sqrt{\hat{Z}^2 + a_2^2\hat{q}^2}}. \tag{8}$$

Let's generalize the formula for the correlation coefficient. For these possible cases, due to the independence and strength of the node elements, the corresponding correlation moments (the second and third terms of the formula numerator (8)) are converted to zero. Next, we will consider all possible cases.

Case 1. Both parts are made of the same material, so $X = Z$ and the correlation moment K_{XZ} remaining in the numerator is converted to variance, and formula (8) takes the following form:

$$r_{Y_1Y_2} = \frac{K_{Y_1Y_2}}{\hat{Y}_1\hat{Y}_2} = \frac{\hat{X}^2 + a_1a_2\hat{q}^2}{\sqrt{\hat{X}^2 + a_1^2\hat{q}^2}\sqrt{\hat{Z}^2 + a_2^2\hat{q}^2}}. \tag{9}$$

Case 2. Parts are made of different materials, as a result of which the random variables X and Z become independent, the correlation moment K_{XZ} turns to zero, and formula (8) is written as:

$$r_{Y_1Y_2} = \frac{K_{Y_1Y_2}}{\hat{Y}_1\hat{Y}_2} = \frac{a_1a_2\hat{q}^2}{\sqrt{\hat{X}^2 + a_1^2\hat{q}^2}\sqrt{\hat{Z}^2 + a_2^2\hat{q}^2}}. \tag{10}$$

Let's link the standards of strength and stress from the total load with the ratios:

$$a_1\hat{q} = p_1\hat{X},\ a_2\hat{q} = p_2\hat{Z}. \tag{11}$$

At the same time, in practical calculations, the strength standard significantly exceeds the stress standard from an external load, so usually $p_1 < 1$, $p_2 < 1$. Taking into account the relations that were introduced, we transform the formula (9):

$$\begin{aligned} r_{Y_1Y_2} &= \frac{K_{Y_1Y_2}}{\hat{Y}_1\hat{Y}_2} = \frac{\hat{X}^2 + a_1a_2\hat{q}^2}{\sqrt{\hat{X}^2 + a_1^2\hat{q}^2}\sqrt{\hat{Z}^2 + a_2^2\hat{q}^2}} = \frac{\hat{X}^2 + p_1p_2\hat{q}^2}{\sqrt{\hat{X}^2 + p_1^2\hat{q}^2}\sqrt{\hat{Z}^2 + p_2^2\hat{q}^2}} \\ &= \frac{\hat{X}^2(1 + p_1p_2)}{\overset{\leftrightarrow}{X}\sqrt{1 + p_1^2}\hat{X}\sqrt{1 + p_2^2}} = \frac{1 + p_1p_2}{\sqrt{1 + p_1^2}\sqrt{1 + p_2^2}} = \frac{1 + p_1p_2}{\sqrt{1 + p_1^2 + p_2^2 + p_1^2p_2^2}}. \end{aligned} \tag{12}$$

Similarly, we transform the formula (10):

$$\begin{aligned} r_{Y_1Y_2} &= \frac{K_{Y_1Y_2}}{\hat{Y}_1\hat{Y}_2} = \frac{a_1a_2\hat{q}^2}{\sqrt{\hat{X}^2 + a_1^2\hat{q}^2}\sqrt{\hat{Z}^2 + a_2^2\overset{\leftrightarrow}{q}^2}} = \frac{p_1p_2\hat{X}\hat{Z}}{\sqrt{\hat{X}^2 + p_1^2\hat{X}^2}\sqrt{\hat{Z}^2 + a_2^2\hat{Z}^2}} \\ &= \frac{p_1p_2\hat{X}\hat{Z}}{\hat{X}\sqrt{1 + p_1^2}\hat{Z}\sqrt{1 + p_2^2}} = \frac{p_1p_2}{\sqrt{1 + p_1^2}\sqrt{1 + p_2^2}} = \frac{p_1p_2}{\sqrt{1 + p_1^2 + p_2^2 + p_1^2p_2^2}}. \end{aligned} \tag{13}$$

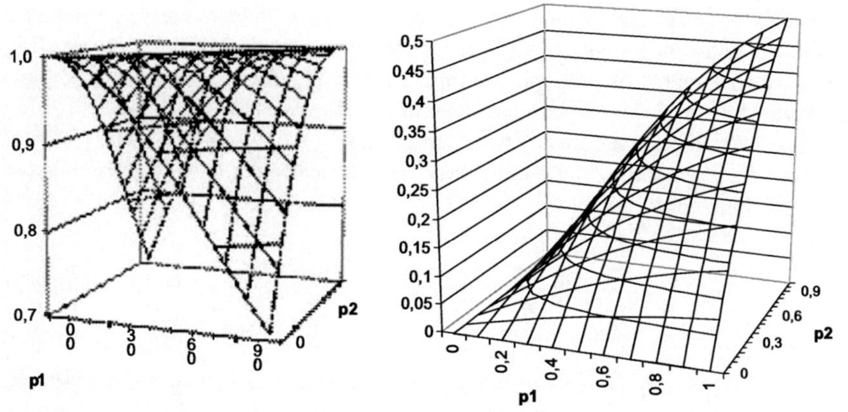

Fig. 3 Numerical values of the correlation coefficient for different values of p_1 and p_2

The resulting formulas have the same denominator and differ by one in the numerator. Numerical values of the correlation coefficient for different values of p_1 and p_2 are shown in the graphs in Fig. 3.

3 Conclusion

If relatively small correlation coefficients $r_{ij} < 0.5$ are obtained, element failures can be considered independent. If random characteristics are independent for node elements, the probability of failure of the node as a whole will be approximately calculated as the sum of all the probabilities of failure of elements designed according to the norms, it will be slightly higher than the probability of failure of one element. To take into account this factor of randomness of failure, it is proposed to use the reduction factor of the calculated resistance γ_z, which depends on the number of independent elements of the node n, represented by the graph (see Fig. 1a) for engineering practice, it is proposed to use the coefficient $\gamma_z = 0{,}9$.

References

1. Design of Steel Structures. Eurocode 3 (2016) Design of Steel Structures. Part 1–1: General Rules and Rules for Buildings, 2nd edn. Wiley, USA
2. Arrayago I, Rasmussen KJR (2021) System-based reliability analysis of stainless steel frames under gravity loads. Eng Struct 231. https://doi.org/10.1016/j.engstruct.2020.111775
3. Radmehr M, Homami P (2020) The seismic reliability analysis of moment resisting frames with bolted end-plate connection. J Constr Steel Res 171:106134. https://doi.org/10.1016/j.jcsr.2020.106134

4. Zhang WL, Rasmussen K (2018) System reliability-based Direct Design Method for space frames with cold–formed steel hollow sections. Eng Struct 166:79–92. https://doi.org/10.1016/j.engstruct.2018.03.062
5. Thai H, Uy B, Kang W, Hicks S (2016) System reliability evaluation of steel frames with semi-rigid connections. J Constr Steel Res 121:29–39. https://doi.org/10.1016/j.jcsr.2016.01.009
6. Zhang H, Shayan S, Rasmussen KJR, Ellingwood BR (2016) System-based design of planar steel frames I: Reliability framework. J Constr Steel Res 123:135–143. https://doi.org/10.1016/j.jcsr.2016.05.004
7. Llanes-Tizoc MD, Reyes-Salazar A, Ruiz SE, Valenzuela-eltrán F, Bojorquez E, Chávez R (2020) Reliability analysis of steel buildings considering the flexibility of the connections of the GFs. Structures 27:2170–2181. https://doi.org/10.1016/j.istruc.2020.08.014
8. Duan HJ, Zhao JC, Song ZS. (2011) Effects of initial imperfection of bolted end-plate connections in the reliability of steel portal frames. Procedia Eng 14:2164–2171. https://doi.org/10.1016/j.proeng.2011.07.272
9. Sergiy P (2019) Scientific school «reliability of building structures»: new results and prespectives. Acad J Ser Ind Mach Build Civil Eng 2(53):5–12. https://doi.org/10.26906/znp.2019.53.1880
10. Pichugin SF (2018) Reliability estimation of industrial building structures. Mag Civil Eng 83(7):24–37. https://doi.org/10.18720/MCE.83.3

Discontinuous Solutions of Concrete Elements Strength Problems Using the Principle of Virtual Velocitiess

Oksana Dovzhenko, Volodymyr Pohribnyi, Volodymyr Kyrychenko, and Olena Malovana

Abstract To determine the strength of concrete and reinforced concrete elements, discontinuous solutions are used in the theory of plasticity. Concrete is considered as a rigid-plastic body. Plastic strain is considered to be localized in thin layers on the failure surface in the compressed zone. Neighboring areas are taken absolutely rigid. The principle of virtual velocities is used. Velocity jumps on the failure surface are considered, which are expressed in terms of variable parameters: the angle between the velocity vector and the direction of the failure plane, the geometric parameters of the failure surface and the ratio of velocities in orthogonal directions. The basic dependences for establishing the shear strength of concrete and reinforced concrete elements obtained on the basis of the mathematical apparatus of the theory of plasticity are given. The proposed dependences on the concrete elements loaded in the direction of the principal stresses were checked. The obtained results are compared with the values of stresses under the conditions of concrete strength under plane stress state and plane strain, which are considered as plasticity conditions. An example of solving the strength problem for short reinforced concrete cantilevers is given. The prospects of application of the variational method using the principle of virtual velocities for the top estimation of strength of concrete elements on the general theoretical basis are specified.

Keywords Plasticity theory · Functional · Velocity jumps · Shear · Plane stress state · Plane strain

1 Introduction

Based on the mechanics of a deformable solid, a number of boundary strength problems have been successfully solved [1–4]. To determine the ultimate state of elements made of plastic materials, the method of characteristic lines is widely used, which is also used for bodies made of quasi-brittle materials. Among them problems of

O. Dovzhenko (✉) · V. Pohribnyi · V. Kyrychenko · O. Malovana
National University «Yuri Kondratyuk Poltava Polytechnic», Poltava, Ukraine

V. Onyshchenko et al. (eds.), *Proceedings of the 3rd International Conference on Building Innovations*, Lecture Notes in Civil Engineering 181,
https://doi.org/10.1007/978-3-030-85043-2_7

indentation in the base of the stamp and wedge, unilateral pressure on the wedge, bending and stretching of the strip weakened by cutouts and holes, bending of the beam, dragging of the strip, body impact on a rigid obstacle at plate stress state and plate strain are known [5, 6]. For massive and plane elements made of concrete, the plastic properties of which are limited, the value of the ultimate load is set by local compression of the half-space under the action of rectangular and curvilinear stamps, compression of truncated wedge, array with spherical and cylindrical cavity, concrete pipe or ring [7].

It should be noted that along with the use of continuous fields of stresses and velocities, in many cases it is convenient and efficient to use approximate discontinuous solutions [8, 9]. In the limit state, jumps of discontinuity in stresses and velocities are quite common. The extreme principles of the minimum properties of the actual velocities and the maximum properties of the actual stresses are applied in the case of discontinuous fields [5, 10]. Their main uses for rigid-plastic body are justified in [11–13].

The application of discontinuous fields and the principle of virtual velocities leads to fairly simple solutions. The value of the load corresponds to the minimum power of plastic strain. At the same time, it is necessary to consider that for materials which are not compressed, jumps of discontinuity (abrupt change) of velocities take place only in the tangential direction. For concrete as a material with a significant difference in tensile and compressive strengths, velocity jumps are possible both in the tangent and in the normal component of the velocity vector. This is due to the changed volume (dilatancy) in the ultimate state. The effect of jumps in the normal direction to the failure plane for concrete is taken into account in studies [4, 14].

The localization of plastic strain on the failure surface, as a prerequisite for the implementation of the displacement of separate parts of the elements, indicates the prospects of using the theory of plasticity to solve boundary value problems of concrete and reinforced concrete under the shear action.

The need to improve the mathematical apparatus and provide it in a user-friendly form is an urgent task for further application of the principle of virtual velocities in solving strength problems of concrete and reinforced concrete elements.

2 Functional of the Principle of Virtual Velocities

When solving the strength problems of steel elements, in addition to the consideration in the failure zone of the plasticity areas, the use of discontinuous and discrete solutions is known [5, 8, 9, 15]. The plastic properties of concrete are significantly reduced compared to steel. Therefore, discontinuous solutions are most consistent with the behavior of concrete in the ultimate state.

Concrete is considered as a rigid-plastic body. Plastic strain is localized only on the failure surface; other areas are considered rigid, which is possible if the stresses in them do not reach the yield strength [5].

The functional of the principle of virtual velocities [16] in the components of the stress and strain tensor at rigid-plasticity has the form

$$J = \int_{V} (TH' + \sigma\xi' + \rho(\omega_i - g_i)\upsilon_i')dV - \int_{S_F} f_i^* \upsilon' dS, \tag{1}$$

where T and H'—intensity of shear stresses and shear strain velocities, σ and ξ'—hydrostatic (average) pressure and volumetric strain, υ'—kinematically possible velocities, V—body volume, ρ, ω_i and g_i—mass density of deformable material, components of the vector of motion acceleration and mass force, S_F—the surface on which the forces are set f_i^*.

Only the values of strain parameters, velocities of rigid blocks displacement and velocity jumps on rupture surfaces vary. The values of T and σ are determined by the history of strain over time [16].

In the case when plastic strains are localized only in a thin layer on the failure surface, we have

$$J = \int_{S_l} (TH' + \sigma\xi')\Delta n dS - \int_{S_F} f_i \upsilon_i' dS, \tag{2}$$

where Δn—the thickness of the plasticity layer, S_l—surface area of failure.

The surface of the velocity jumps of discontinuity is considered as the limit position of the thin layer that surrounds it, and when $\Delta n \to 0$ it turns into this surface. The displacement velocity is continuous, but changes sharply and linearly in thickness.

For concrete as a compressible material, there is a changed volume (dilatancy), jumps of discontinuity occur in both tangential and normal directions to the surface S_l. Derivatives in the direction of the surface S_l from the components of the displacement velocity will turn to infinity. In this case, the functional of the principle of virtual velocities has a finite value and provides an opportunity to obtain solutions to the problems of the theory of plasticity.

Extreme principles for a rigidly plastic body together with extreme properties of ultimate loads make it possible to determine the ultimate load of elements by applying the upper and lower estimates [5, 16].

Functional J is investigated at steady state using a variational equation $\delta J = 0$, which is equivalent to solving a boundary value problem.

Let's calculate the ultimate value of the subintegral expression under the condition $\Delta n \to 0$. To do this, let's place on S_l an orthogonal coordinate system t m n and direct n along the normal to S_l, and t along the velocity jumps of discontinuity.

By the extreme theorem, which states that among all kinematically possible velocity fields υ_i', the real field will be the one for which expression (1) takes a minimum value, so we have:

$$\int\limits_V (TH + \sigma\xi)dV - \int\limits_{S_F} f_i v_i dS \le \int\limits_V (TH' + \sigma\xi')dV - \int\limits_{S_F} f_i v_i' dS \quad (3)$$

and corresponds to the upper estimate of the power of plastic strain.

In the absence of volumetric plastic zones and surface where the velocities are given, inequality (3) takes the form

$$f_i \le \left(\int\limits_{S_l} (TH' + \sigma\xi')\Delta n dS \right) / v_i'. \quad (4)$$

As a condition of concrete strength the Balandin-Geniyev equation is taken, which has a simple notation and is confirmed experimentally

$$T^2 + m\sigma - T_{sh}^2 = 0, \quad (5)$$

where $m = f_c - f_t$, $T_{sh}^2 = f_c f_{ct}/3$, where f_c—strength of concrete to axial compression, f_{ct}—strength of concrete to axial tension.

The mathematical apparatus of the theory of ideal plasticity is used, the strength condition is considered as a condition of plasticity.

The intensity of the shear strain velocities *H'* is equal to

$$H' = \sqrt{\begin{array}{c} 2\left[\left(\frac{\partial v_t'}{\partial t} - \frac{\partial v_m'}{\partial m}\right)^2 + \left(\frac{\partial v_m'}{\partial m} - \frac{\partial v_n'}{\partial n}\right)^2 + \left(\frac{\partial v_n'}{\partial n} - \frac{\partial v_t'}{\partial t}\right)^2\right]/3 \\ + \left(\frac{\partial v_t'}{\partial m} - \frac{\partial v_m'}{\partial t}\right)^2 + \left(\frac{\partial v_m'}{\partial n} - \frac{\partial v_n'}{\partial m}\right)^2 + \left(\frac{\partial v_n'}{\partial t} - \frac{\partial v_t'}{\partial n}\right)^2 \end{array}}, \quad (6)$$

where derivatives are limited, except $\partial v_n'/\partial n = \lim(\Delta v_n'/\Delta n) \to \infty$, $\partial v_t'/\partial n = \lim(\Delta v_t'/\Delta n) \to \infty$, $\Delta v_m' = 0$, the direction t is chosen along the velocity jumps of discontinuity and therefore its component is continuous.

Therefore

$$\lim_{\Delta n \to 0} \left(TH'\Delta n\right) = T\sqrt{4\Delta v_n'^2/3 + \Delta v_t'^2}, \quad (7)$$

$$\begin{aligned} \lim_{\Delta n \to 0} (\sigma\xi'\Delta n) &= \lim_{\Delta n \to 0} [\sigma(\partial v_n'/\partial n + \partial v_m'/\partial m + \partial v_t'/\partial t)\Delta n] \\ &= \lim_{\Delta n \to 0} \left\{\sigma[\Delta v_n' + (\partial v_m'/\partial m)\Delta n + (\partial v_t'/\partial t)\Delta n]\right\} = \sigma\Delta v_n', \end{aligned} \quad (8)$$

here $\Delta v_n'$ and $\Delta v_t'$—velocity jumps, respectively, in the normal and tangential to the jumps of discontinuity surface directions.

Then the equation $\delta J = 0$ has the form

$$\delta\left[\int_{S_t}\left(T\sqrt{4\Delta\upsilon_n'^2/3+\Delta\upsilon_t'^2}+\sigma\Delta\upsilon_t'^2\right)-\int_{S_F} f_i\upsilon_i' dS\right]=0. \tag{9}$$

Considering:
- associated flow law

$$\xi=\lambda\frac{\partial\Xi}{\partial\sigma}, \tag{10}$$

where Ξ—plastic potential;
– the relationship between the intensity of shear stresses and the shear strain velocity

$$T=0.5H/\lambda, \tag{11}$$

here $\lambda=\xi/m$,
and writing the strength condition (5) in the form of $\sigma=(T_{sh}-T)/m$, the following is obtained

$$\delta\left\{\int_{S_l}\left\{\left[\frac{d^2}{m}+\frac{m}{4}\left(\frac{\Delta\upsilon_t'}{\Delta\upsilon_n'}\right)^2\right]\Delta\upsilon_n'\right\}dS-\int_{S_F} f_i\Delta\upsilon_i' dS\right\}=0, \tag{12}$$

where $d^2=(f_c^2-f_c f_{ct}+f_{ct}^2)/3$—characteristics of concrete strength, γ—the angle between the surface of the velocity jumps of discontinuity and the direction of force f_i.

At a plane stress state, providing $\Delta n\to 0$ the velocities derivatives are limited except $\frac{\partial\upsilon_n'}{\partial n}=-\frac{\partial\upsilon_m'}{\partial m}=\lim(\Delta\upsilon_n'/\Delta n)\to\infty$, $\frac{\partial\upsilon_t}{\partial n}=\lim(\Delta\upsilon_t'/\Delta n)\to\infty$ the equation $\delta J=0$ is written as

$$\delta\int_{S_t}\left[d\sqrt{4\Delta\upsilon_n'^2+\Delta\upsilon_t'^2}-m\Delta\upsilon_n'\right]dS-\int_{S_F} f_i\upsilon_i' dS=0. \tag{13}$$

3 The Problem of Strength of Concrete Elements Loaded in the Direction of the Principal Stresses. Checking the Mathematical Apparatus

To verify the proposed dependences at plane strain, the obtained results of the ultimate stress values on the failure surface are compared, determined on the basis of Eq. (9), with the stress on the condition of concrete strength (5), which in coordinates $|\tau_n|-\sigma_n$

Fig. 1 Condition of concrete strength at plane strain

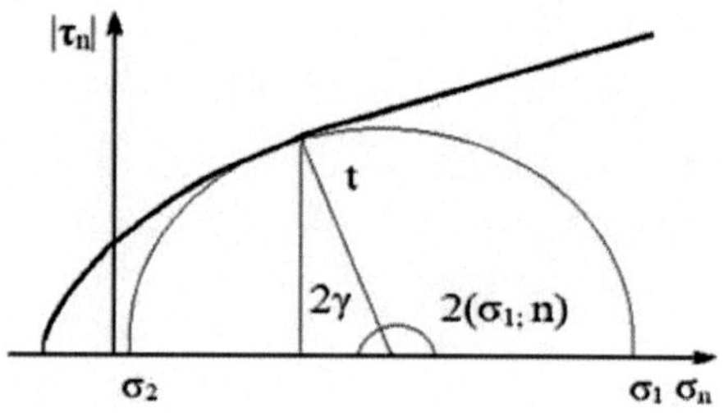

is written by Eq. (14) and is presented in Fig. 1.

$$|\tau_n| = \varphi(\sigma_n) = \sqrt{m(\sigma_n + \frac{1}{4}m + \frac{1}{12}n^2/m)}. \tag{14}$$

Let's consider a concrete element of unlimited length, loaded with surface force $f_1 = \sigma_1$ in the direction of greater principal stresses σ_1, and on the sides loaded with the surface force $f_2 = \sigma_2$ (Fig. 2).

In the orthogonal coordinate system at velocities υ_1 and υ_2 in the direction of axes 1 and 2, respectively, through them we express the jumps of velocities $\Delta\upsilon_n = \upsilon_2 \cos\gamma - \upsilon_1 \sin\gamma$, $\Delta\upsilon_t' = \upsilon_1' \cos\gamma' + \upsilon_2' \sin\gamma'$, where γ—the angle between the direction of the principal stresses σ_1 and the plane of rupture $S_l = A/\sin\gamma$, here A—the cross-sectional area of the element.

Then Eq. (12) is written as

$$\delta\left\{\left[\frac{d^2}{m} + \frac{m}{4}\left(\frac{\upsilon_1' \cos\gamma' + \upsilon_2' \sin\gamma}{\upsilon_2' \cos\gamma' - \upsilon_1' \sin\gamma'}\right)^2\right]\frac{\upsilon_2' \cos\gamma' - \upsilon_1' \sin\gamma'}{\sin\gamma'} - f_1\upsilon_1' \pm f_2\upsilon_2'\right\} = 0, \tag{15}$$

Fig. 2 Kinematic scheme of concrete plate failure

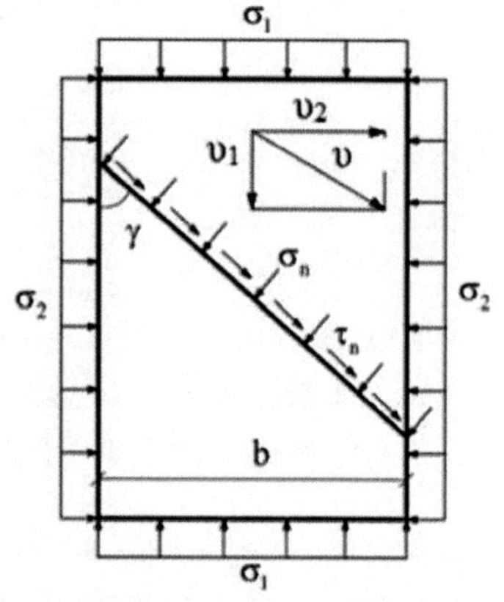

where the sign "+" before f_2 is used if the velocity vector is opposite to the direction of force (during compression), the sign "−" is if the directions coincide.

After maintaining the parameters $k = \upsilon_2/\upsilon_1$ and $\tan\gamma$ expression (15) has the form

$$\sigma_1 = \frac{1}{\tan\gamma}\left\{\left[\frac{d^2}{m} + \frac{m}{4}\left(\frac{1 + k\tan\gamma}{k - \tan\gamma}\right)^2\right](k - \tan\gamma) \pm k\sigma_2\right\}. \quad (16)$$

The value of the shear and normal stresses on the failure surface are determined as

$$\tau_n = \frac{m(1 + k\tan\gamma)}{2(k - \tan\gamma)}, \quad (17)$$

$$\sigma_n = \frac{m}{4}\left[\left(\frac{1 + k\tan\gamma}{k - \tan\gamma}\right)^2 - \frac{1}{3}\left(\frac{n}{m}\right)^2 - 1\right]. \quad (18)$$

Using the angle ψ between the rigid disk velocity vector and the velocity rupture surface, the velocity jumps $\Delta\upsilon_n = \upsilon\sin\psi$, $\Delta\upsilon_t = \upsilon\cos\psi$ and the rupture surface area $S_l = A/\sin(\pi/4 - \psi/2)$ are recorded. Considering the relationship between the angles $\psi = \pi/2 - 2\gamma$ we obtain

$$A\left\{\left[\frac{d^2}{m} + \frac{m}{4}\left(\frac{\cos\psi'}{\sin\psi'}\right)^2\right]\frac{\upsilon\sin\psi'}{\sin(\pi/4 - \psi'/2)} - f_1\upsilon_1' \pm f_2\upsilon_2'\right\} = 0. \quad (19)$$

Taking into account that $\upsilon_1 = \upsilon\cos(\gamma + \psi) = \upsilon\cos(\pi/4 + \psi/2)$ and $\upsilon_2 = \upsilon\sin(\pi/4 + \psi/2)$ taking into account the geometry of the failure surface we have

$$f_1 = \left[\frac{d^2}{m} + \frac{m}{4}\left(\frac{\cos\psi'}{\sin\psi'}\right)^2\right]\frac{\sin\psi'}{\cos^2(\pi/4 + \psi'/2)} \pm f_2\tan^2(\pi/4 + \psi'/2) = 0. \quad (20)$$

After conversion the following is received

$$f_1 = \frac{2\left[\frac{d^2}{m} + \frac{m}{4}\left(\frac{\cos\psi'}{\sin\psi'}\right)^2\right]\sin\psi' \pm f_2(1 + \sin\psi')}{1 - \sin\psi'} = 0. \quad (21)$$

When used as a parameter of variation $\tan\psi'$ we have the following

$$\sigma_1 = \frac{2\left(\frac{d^2}{m} + \frac{m}{4\tan^2\psi'}\right)\tan\psi' + \sigma_2\left(\sqrt{1 + \tan^2\psi'} + \tan\psi'\right)}{\sqrt{1 + \tan^2\psi'} - \tan\psi'}. \quad (22)$$

Table 1 The values of the angles γ and ψ and the parameters of the stress state at plane strain by $f_{ct}/f_c = 0.1$

γ,°	Ψ,°	τ_n/f_c	σ_n/f_c	σ_1/f_c	σ_2/f_c	σ_3/f_c
25.37	39.25	0.551	0	1.161	−0.261	0.9
27.91	34.17	0.663	0.151	1.402	−0.2	1.051
31.64	26.72	0.894	0.551	2.002	0	1.451
33.85	22.3	1.097	1	2.636	0.264	1.9
36.48	17.24	1.45	2	3.968	0.931	2.9
37.72	14.55	1.733	3	5.241	1.66	3.9
38.58	12.84	1.976	4	6.475	2.423	4.9
39.66	10.67	2.388	6	8.881	4.02	6.9
→ 45	→ 0	→ ∞	→ ∞	→ ∞	→ ∞	→ ∞

The values of shear and normal stresses on the failure surface are determined from the equations

$$\tan \tau_n = \frac{m}{2 \tan \psi}, \tag{23}$$

$$\sigma_n = \frac{m}{4}\left[\frac{1}{\tan^2 \psi} - \frac{1}{3}\left(\frac{n}{m}\right)^2 - 1\right]. \tag{24}$$

The results of the calculation by formulas (16)–(18), (22)–(24) are given in Table 1.

Obtained in Table 1 the values of the stress parameters are similar to those determined by the strength condition (5), (14).

Let's consider the strength of a concrete slab (Fig. 1) under the conditions of plane stress state.

Expressing the jumps of velocities through υ_1 and υ_2 $\Delta\upsilon_n = \upsilon_2 \cos\gamma - \upsilon_1 \sin\gamma$, $\Delta\upsilon_t' = \upsilon_1' \cos\gamma' + \upsilon_2' \sin\gamma'$, taking into account the geometric parameters of the failure surface S_l, Eq. (13) is solved with respect to f_1

$$\sigma_1 = \frac{d\sqrt{4(k' - \tan\gamma')^2 + (1 + k' \tan\gamma')^2} - m(k' - \tan\gamma') \pm k'\sigma_2}{\tan\gamma'}, \tag{25}$$

where the sign "+" before σ_2 means compression, the sign "−" means lateral tension.

According to the obtained parameters k and γ values of shear stresses are set on the failure surface

$$\tau_n = \frac{d(1 + k\tan\gamma)}{\sqrt{4(k - \tan\gamma)^2 + (1 + k\tan\gamma)^2}}, \tag{26}$$

and normal stresses

$$\sigma_n = m - \frac{4d(k - \tan\gamma)}{\sqrt{4(k - \tan\gamma)^2 + (1 + k\tan\gamma)^2}}. \tag{27}$$

When recording the velocity jumps through an angle ψ $\Delta\upsilon_n = \upsilon \sin\psi$, $\Delta\upsilon_t = \upsilon\cos\psi$, Eq. (13) takes the form

$$f_1 = \frac{2\left[d\sqrt{4\sin^2\psi' + \cos^2\psi'} - m\sin\psi'\right] \pm (1 + \sin\psi') f_2}{1 - \sin\psi'}. \tag{28}$$

After transformation (28) we have

$$\sigma_1 = \frac{2\left[d\sqrt{3\sin^2\psi' + 1} - m\sin\psi'\right] \pm (1 + \sin\psi')\sigma_2}{1 - \sin\psi'}. \tag{29}$$

When used as a parameter of variation $\tan\psi'$ we have

$$\sigma_1 = \frac{2\left[d\sqrt{4\tan^2\psi' + 1} - m\tan\psi'\right] \pm \left(\sqrt{1 + \tan^2\psi'} + \tan\psi'\right)\sigma_2}{\sqrt{1 + \tan^2\psi'} - \tan\psi'}. \tag{30}$$

The values of ψ set the values of shear and normal stresses on the failure surface.

$$\tau_n = \frac{d}{\sqrt{1 + 4\tan^2\psi}}, \tag{31}$$

$$\sigma_n = m - 4d\frac{\tan\psi}{\sqrt{1 + 4\tan^2\psi}}. \tag{32}$$

The calculation results are given in Table 2.

The values of stresses in Table 2 correspond to the given condition of strength at a plane stress state in coordinates $\sigma_1 - \sigma_2$ (5) and $\tau_n - \sigma_n$ (33), which is shown in Fig. 3.

$$|\tau_n| = \varphi(\sigma_n) = \sqrt{d^2 - \frac{1}{4}(\sigma_n - m)^2}. \tag{33}$$

Table 2 The values of the angles γ and ψ and the stress parameters at the plane stress state for $f_{ct}/f_c = 0.1$

Characteristic points of the strength condition	γ,°	Ψ,°	τ_n/f_c	σ_n/f_c	σ_1/f_c	σ_2/f_c
Uniaxial compression	37.27	15.47	0.482	0.367	1	0
Maximum shear stresses	45	0	0.551	0.9	1.451	0.349
Maximum compression stresses	90	−90	0	2.002	2.002	1.451

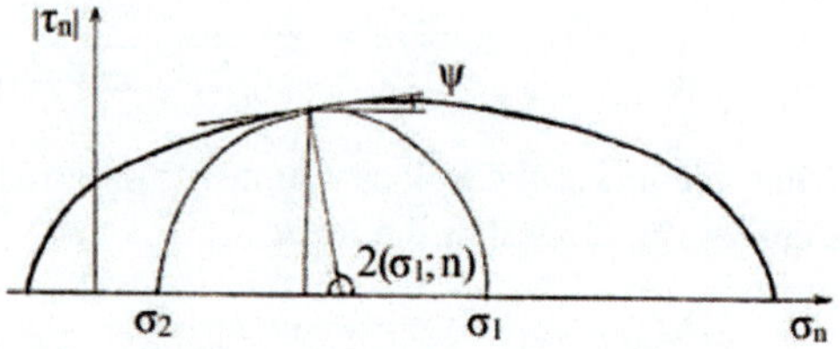

Fig. 3 Ultimate circumferential circles of the Mora $|\tau_n| = \varphi\ (\sigma_n)$ at a plane stress state

The intervals of stress states by the shear implementation in concrete are given in [17].

For the taking into account the specifics of the stress–strain state of the elements when solving strength problems, additional components are introduced into the basic dependences (16), (22), (25) and (29). In the absence of tensile areas in the failure zone, it is convenient to apply the basic dependences (22) and (29).

As an example, let's consider the strength problem of a reinforced concrete short cantilever.

Kinematic failure schemes of short cantilevers are given in Fig. 4.

Failure cases of the entire section by shear and within an inclined compressed strip are considered.

The influence of reinforcement is taken into account by lateral compression of the corresponding level of reinforcement intensity. The value of stresses in the tensile zone of the cantilever is taken to be equal to the tensile strength of concrete.

The calculated dependence by the shear on the entire section, which is realized at $a/h \leq 0.15$ and $c/h \leq 0.3$, has the form

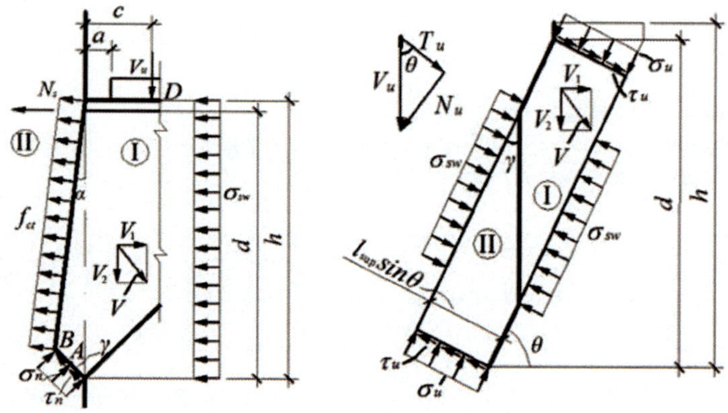

a b

Fig. 4 For calculating the strength of reinforced concrete cantilevers: **a** in failure case along the entire cross section; **b** shear within the compressed inclined strip; I and II—rigid disks

$$\frac{V_u}{\delta h} = d\sqrt{4(k' - \tan\gamma')^2 + (1 + k'\tan\gamma')^2} - m(k' - \tan\gamma')\frac{\tan\alpha + f_{ct}\tan\gamma'(k' + \tan\alpha)}{\tan\alpha + \tan\gamma'} + k'(\sigma_s + \sigma_{sw}), \tag{34}$$

where δ—cantilever thickness; $\sigma_s = \rho_s f_y$, here $\rho_s = \frac{A_s}{\delta h}$; $\sigma_{sw} = \rho_{sw} f_y$, here $\rho_{sw} = \frac{A_{sw}}{\delta s_w}$.

In addition to (34), equilibrium equations are additionally used

$$\sum M_A = 0; \quad \sum M_B = 0; \quad \sum M_D = 0. \tag{35}$$

At failure under the shear within an inclined strip we have

$$\frac{V_u}{\delta l_{\text{sup}}} = \frac{2\left[d\sqrt{1 + 3\sin^2\psi'} - m\sin\psi'\right] + (1 - \sin\psi')\sigma_{sw}}{(1 - \sin\psi')(1 + \cot\theta)}, \tag{36}$$

where l_{sup}—the width of the load transfer platform, θ—the angle of the strip to the vertical.

The strength problems of concrete wedges as models of compressed zone over dangerous inclined crack in beams and near support sections of three-hinged frames, compressed elements at local application of loading in the conditions of plane and volume stress state, concrete and reinforced concrete keyed joints are solved [18–20], which found experimental confirmation [21–26].

4 Conclusions

1. To calculate the strength of concrete and reinforced concrete elements, fairly simple dependences are proposed which are established from discontinuous solutions of the variational method in the theory of plasticity using the principle of virtual velocities.
2. The results of solving the strength problem are obtained by basic dependences for concrete elements loaded from the ends and sides in the direction of the principal stresses, similar to the given conditions of concrete strength, stress values at plane stress state and plane strain.
3. The specificity of the stress–strain state of the elements for different cases is taken into account by introducing into the basic dependencies of the components that reflect the influence of determining factors (reinforcement, compression, load application schemes, shapes, etc.).
4. The prospects of application of discontinuous solutions for strength estimation of concrete and reinforced concrete elements under the shear are proved.

References

1. Nielsen MP, Hoang LC (2011) Limit Analysis and Concrete Plasticity. CRC Press, Taylor & Francis Group
2. Sorensen JH, Hoang LC, Olesen JF, Fischer G (2017) Test and analysis of a new ductile shear connection design for RC shear walls. Struct Concr 18:189–204
3. Yu MH (2002) Advances in strength theories for materials under complex stress state in the 20th century. Mech Rev 55(3):169–218
4. Mitrofanov V (2006) Teoriya Ideal'noy Plastichnosti kak Elementarnaya Mekhanika Psevdoplasticheskogo Predel'nogo Sostoyaniya Betona: Osnov y, Ogranicheniya, Prakticheskiye Aspekty. Sovershenstvovaniye. City utilities 72:6–26 (In Russian)
5. Kachanov L (2013) M: Osnovy Teorii Plastichnosti. Ripol Klassik, Moscow (In Russian)
6. Ivanova SV, Pavlova TN (2010) Uprugo-Plasticheskoye Sostoyaniye Tonkikh i Tolstykh Plit, Oslablennykh Otverstiyami: Uchebnoye Posobiye: CHGPU, Cheboksary. (In Russian)
7. Geniyev GA, Kissyuk VN, Tyupin GA (1976) Teoriya Plastichnosti Betona i Zhelezobetona. Strojizdat. Strojizdat, Moscow (In Russian)
8. Karpov VV, Sal'nikov AYu (2009) Variatsionnyye Metody i Variatsionnyye Printsipy Mekhaniki pri Raschete Stroitel'nykh Konstruktsiy. SPb. (In Russian)
9. Morar' GA (1990) Metod Razryvnykh Resheniy v Mekhanike Deformiruyemykh Tel. Shtiinia, Kishinev. (In Russian)
10. Bukhan'ko AA, Kocherov YeP, Khromov AI (2012) Deformatsionno-Energeticheskiy Podkhod k Otsenke Prochnosti Elementov Konstruktsiy. Trudy Mezhdunarodnoy Konferentsii Zhivuchest' i Konstruktsionnoye Materialovedeniye T.1, pp 46–55. IMASHRAN, Moskva. (In Russian).
11. Hill R (1950) The Mathematical Theory of Plasticity. Oxford
12. Prager W, Hodge P (1951) Theory of Perfectly Plastic Solids. New York, London
13. Markov AA (1947) O Variatsionnykh Printsipakh v Teorii Plastichnosti. Prikladnaya Matematika i Mekhanika vyp. 3 (T. XI):s. 339–350. (In Russian)
14. Dovzhenko O, Pohribnyi V, Yurko I (2018) Concrete and reinforced concrete strength under action of shear, crushing and punching shear. In: IOP Conferencer Series: Materials Science and Engineering, vol 463, p 022026
15. Rabotnov YuN (1988) Mekhanika Deformiruyemogo Tverdogo Tela. Nauka, Moscow. (In Russian)
16. Kolmogorov VL (1986) Mekhanika Obrabotki Metallov Davleniyem. Metallurhyia, Moscow (In Russian)
17. Dovzhenko O, Pogrebnyi V, Yurko I (2018) Shear failure form realization in concrete. News the national academy of sciences of the republic of Kazakhstan. Ser Geol Tech Sci 2(428):55–62
18. Pohribny, V, Dovzhenko O, Maliovana O (2018) The ideal plasticity theory usage peculiarities to concrete and reinforced concrete. Int J Eng Technol 7(3.2):19–26
19. Dovzhenko O, Pohribnyi V, Pents V, Mariukha D (2018) Bearing capacity calculation of reinforced concrete corbels under the shear action. MATEC Web Conf 230:02005
20. Pohribnyi V, Dovzhenko O, Kuznietsova I, Usenko D (2018) The improved technique for calculating the concrete elements strength under local compression. MATEC Web Conf 230:02025
21. Han Q, Zhou Y-L, Zhong Z-L, Du X-L (2017) Seismic capacity evaluation of exterior shear keys of highway bridges. J Bridge Eng 22(2):04016119
22. Kim HS, Chin WJ, Cho JR, Kim YJ, Yoon H (2015) An experimental study on the behavior of shear keys according to the curing time of UHPC. Engineering 7:212–218
23. Jiang H, Chen L, Ma ZJ, Feng W (2005) Shear behavior of dry joints with castellated keys in precast concrete segmental bridges. J Bridg Eng 20(2):04014062
24. Dovzhenko O, Pohribnyi V, Karabash L (2018) Experimental study on the multikeyed joints of concrete and reinforced concrete elements. Int J Eng Technol 7(3.2):354–359

25. Piskunov VG, Gorik AV, Cherednikov VN (2000) Modeling of transverse shears of piecewise homogeneous composite bars using an iterative process with account of tangential loads. 2 resolving equations and results. Mech Comp Mater 36(6):445–452
26. Piskunov VG, Goryk AV, Cherednikov VN (2000) Modeling of transverse shears of piecewise homogeneous composite bars using an iterative process with account of tangential loads. 1 construction of a model. Mech Comp Mater 36(4):287–296

Peculiarities of Tansformations in Systems of Coordination of Nitrate Precursors of REE and Alkali Metals During Formation of Polyfunctional Layered Oxide Materials

Oleksandr Dryuchko, Yanping Yuan, Liangliang Sun, and Katerina Kytaihora

Abstract The information on alkaline coordination nitrates of rare-earth elements of the cerium subgroup - precursors of promising modern multifunctional materials - on the conditions of their formation and existence, nature of chemical bonding, composition, structure, shape of Ln coordination polyhedra, type of ligand coordination, existence of isotype series is generalized. on stoichiometry of structure, structure, the found out characteristic properties. The obtained data (as primary information) are the basis for detection, identification, control of the phase state of processing objects in the preparatory stages, selection of compatibility criteria of components in the formation of single-layer and layered nanostructured oxide composite systems of lanthanides and transition elements with catalytic activity and photocatalyst., a coating capable of self-cleaning with hydrophilic properties; development of various combined methods of their activation and establishment of technological and functional dependencies; controlled modification of the properties of the obtained target products. To increase the photocatalytic activity of coating samples based on highly dispersed TiO_2 anatase modification, a methodology for chemical modification of oxidation centers in their surface layer with heat treatment in contact with thermolysis products of alkaline coordination nitrates of lanthanides is proposed. The effective test photocatalytic destruction of vapors of organic substrates on the example of ethanol is revealed.

Keywords Alkaline coordination nitrates of lanthanides · Formation conditions · Crystal structure of compounds · Characteristic properties · Transformations with physical activation

O. Dryuchko (✉)
National University «Yuri Kondratyuk Poltava Polytechnic», Poltava 36011, Ukraine

Y. Yuan · L. Sun · K. Kytaihora
Southwest Jiaotong University, Chengdu, China

V. Onyshchenko et al. (eds.), *Proceedings of the 3rd International Conference on Building Innovations*, Lecture Notes in Civil Engineering 181,
https://doi.org/10.1007/978-3-030-85043-2_8

1 Introduction

Currently, the search for methods and complex technologies to create new and improve existing regulations for the production of perfect multifunctional oxide materials of transition and rare earth elements with the structure of defective perovskite, garnet with reproducible properties by low-temperature methods of "soft chemistry" and precursors. They have a complex structure and in scientific and technological terms are difficult objects that are intensively studied [1–9], including with the participation of authors [see, 10]. Therefore, modern materials science, which is based on them, requires regulatory solutions simple in configuration, low-speed, energy efficient, characterized by scale, with the ability to reproduce products with a given homogeneity, stability, a set of predefined characteristics.

There are many methods for the synthesis of these oxide multicomponent materials [2–12], based on different physical and chemical principles. The main ones are:

- high-temperature method of solid-phase chemical reactions;
- condensation liquid-phase methods for obtaining nanosized oxide materials based on:

- various variants of mixing of initial components (chemical precipitation (coprecipitation); sol – gel; hydrothermal; complexonate homogenization; solvent replacement; synthesis under the action of microwave radiation);
- rapid thermal decomposition of precursors in solution (spray drying; rapid expansion of supercritical fluid solutions; cryochemical);
- self-igniting (glycine-nitrate; Pechini method; cellulose (fabric, paper) technology; pyrolysis of polymer-salt films).

The choice of using a particular method depends on the chemical nature of the obtained compounds, the size and morphological characteristics of the particles of the synthesis products, the conditions and method of formation of the latter; the material and condition of the surface, the shape of the samples on which the coating is applied; capabilities of available technical equipment, etc. These methods are used both independently of each other and in combination.

Recently, titanium dioxide has attracted special attention due to new unique prospects for its use in the form of nanostructured materials and nanocomposites with controlled morphological, physicochemical and optical properties. TiO_2, which has high chemical and thermal stability, as well as impurity levels in the electronic structure of the material, created by a given type of doping, is unique for building on its basis new effective functional materials used in photocatalysis and photovoltaics, sensory, catalysis, for liquid chromatography and other areas.

The essence of the FC properties of TiO_2 is that in the volume of a semiconductor particle under the influence of electromagnetic radiation, electron - hole pairs are generated, which, upon reaching the surface of a TiO_2 particle, enter redox reactions with adsorbed molecules. For titanium dioxide, the process is as follows:

$$(Ox_1)_{ads} + (Red_1)_{ads} \frac{TiO_2}{h\upsilon > E_g} \rightarrow Ox_2 + Red_2$$

In this case, part of the electrons and holes can undergo recombination in the bulk or on the surface of TiO_2. For the effective occurrence of photocatalytic processes, it is necessary that redox reactions involving an electron–hole pair be more effective than recombination processes.

The advantages of the photocatalytic method of purification are well known: 1) the ability to oxidize almost any organic matter and also a number of inorganic, such as CO, H_2S, HCN, NH_3, NO_x, etc.; 2) the method works at room temperature and atmospheric pressure; 3) it is possible to oxidize even small concentrations of pollutants, cleaning of which by other methods is economically unprofitable; 4) for the implementation of the method of photocatalytic purification does not require additional reagents, because the oxidant is oxygen.

However, for the photocatalytic method of air purification, a number of disadvantages are known, such as: 1) a relatively low purification rate; 2) the need to use ultraviolet light sources in the case where the photocatalyst is titanium dioxide; 3) low adsorption capacity of most simple photocatalysts; 4) the possibility of formation of intermediate products during the oxidation of high concentrations of pollutants. Therefore, the development of new photocatalytic systems that would overcome these shortcomings is the subject of an urgent series of modern studies.

When creating and heat treatment of materials based on titanium dioxide, it is necessary to take into account the possibility of surface and bulk defects of the TiO_2 crystal lattice (due to the existence of phase diversity in the range of $O/Ti < 2$ ratios on the phase diagram of the Ti - O system [13]) and detection their significant effect on the photochemical properties and photoinduced hydrophilic ability of the synthesis products [14].

Available information on the state and possible directions of improvement of technologies for creating such materials, current requirements for their stability and reproducibility of properties, expansion of their use [1–14], manifestation of high activity of low-crystallized particles of structural components formed by solvent thermolysis [15], new information on reactivity and transformation of layered perovskite-like oxides [16], stabilization of photocatalytic and sensory-active crystalline modification of anatase due to NO_3^- ions [17], doping of Ln_2O_3 [18, 19] in obtaining TiO_2 from solutions [20] initiated the continuation of our study. Today, ways to control the technical parameters of the target products through the choice of composition, synthesis conditions and method of processing.

One of the most promising classes of complex oxide materials of rare earth elements and titanium are nanostructured layered perovskite-like compounds and solid solutions based on them. Depending on the composition and structure, they have a wide range of physicochemical properties. Presented in this paper perovskite-like layered titanates belong to the homologous series $(Me, Ln)_{n+1}$ Ti_nO_{3n+1}, where Ln - La – Nd, Me - Li – Cs, n is the number of nanolayers of perovskite (Ruddlesden-Popper phases; with a thickness of approximately one layer 0.5 nm). Accordingly, $MeLnTiO_4$ in its structure contains one nanolayer of perovskite, $Me_2Ln_2Ti_3O_{10}$ -

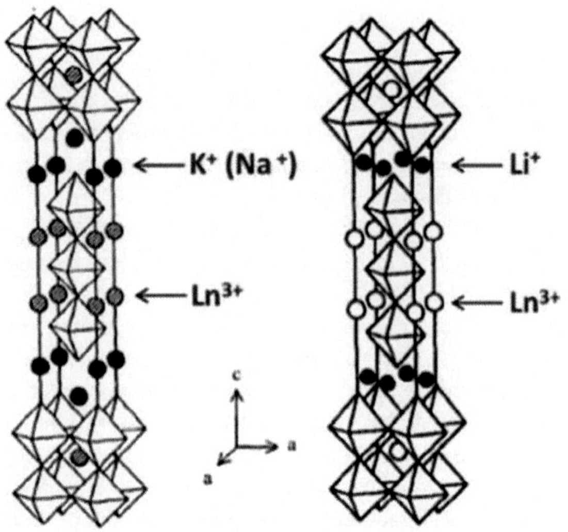

Fig. 1 Extended elementary cells of Ruddlesden-Popper phases

three. Perovskite-like nanolayered titanates were obtained and studied by the authors of [21–24]. Figure 1 shows an extended unit cell for $Me_2Ln_2Ti_3O_{10}$ (Ln - La, Nd; Me - Li, K) [23] and $Na_2Ln_2Ti_3O_{10}$ [24].

So oxides of $K_2Nd_2Ti_3O_{10}$ (as an example) crystallize in a tetragonal structure. The spatial group for these compounds is defined as I4/mmm. The thickness of layered oxides of this type is characterized by three titanium-oxygen octahedra $[Nd_2Ti_3O_{10}]$, alternating with each other and separated by alkali metal cations, in this case potassium cations, between the layers. The lattice parameter c (≈30 Å) indicates the displacement of adjacent perovskite layers by ½. The neodymium cation is located in the center of the perovskite lattice and is characterized by a 12-coordinated oxygen environment. The alkali metal cation is located in the interlayer space and is usually 9-coordinated [25].

The layered structure, consisting of lamellar particles, is stored by the product when kept in humid air and confirmed by scanning electron microscopy [26]. The identified stability of the systems recommends them as promising photocatalysts in conditions close to their use and is important for other innovative areas of their applications.

Analysis of recent publications shows that titanium dioxide is mainly used in thin film form, which most effectively implements its properties necessary for photocatalysis, solar energy, sensors, self-cleaning coatings and more.

And the practical implementation of modern technologies of the already proposed variant of the composite photocatalyst [27], the structure of the granules of which is represented by three layers: adsorbent, silicon dioxide and photocatalyst - titanium dioxide anatase modification, will simultaneously solve problems: 1) effective adsorption of both polar and nonpolar (for example, pollutants); 2) exclusion of the influence of the electrically conductive properties of the sorbent on the recombination of photogenerated electron–hole pairs; 3) to ensure complete absorption of incident light by the particles of the photocatalyst, not the adsorbent; 4) detection of photocatalytic activity under visible light.

Today, thanks to the technological methods of reactions of "soft chemistry", it is possible to create substances with various structural features, to obtain metastable compounds by a sequence of low-temperature topochemical syntheses. Such reactions with a change in the structure and morphology of the particles take place at low temperatures while maintaining the basic structural features in perovskite-like layered oxide compounds. Depending on the nature and stoichiometry of their cations, they can exhibit a variety of physical and chemical properties: superconductivity, colossal magnetoresistance, ferroelectricity, catalytic and photocatalytic activity, the ability to ion exchange in solutions and melts, and the ability to hydrate.. Therefore, the study of the peculiarities of transformations of intermediate precursors - alkaline coordination nitrates of REE, their reactivity during the synthesis of layered perovskite-like oxide phases directly affects the possible areas of further application of the latter.

For layered perovskite-like compounds, such researchers include, in particular, ion exchange [25], intercalation and deintercalation [28], various substitution and condensation processes [29], fission processes [30], and mutual transformations of one structure into another [31] (for example, the transition from Ruddlesden-Popper phases to Dion-Jacobson phases; transition within one type of phase with increasing or decreasing number of layers).

The most common reactions of "soft" chemistry are ion exchange reactions, during which the weakly bound cations of the interlayer space are replaced, while the perovskite layers are quite stable mainly due to metal–oxygen covalent bonds and play the role of a framework in the layered structure.. This allows the substitution reactions of some interlayer cations to others, without affecting the basic structure of the layered oxide. Such reactions can be used to obtain a wide range of new perovskite-like structures.

For effective management of properties of the received products deep understanding of physical and chemical processes, the phenomena occurring during their formation is necessary. And their complex research with application of modern physical and chemical methods allows to improve our knowledge of characteristic features of fast-moving processes, stages of evolution of structure and microstructure of technological objects.

The Purpose and Tasks of the Study. To study the cooperative processes of interaction between structural components in the systems of nitrate precursors of REE representatives of the cerium subgroup and elements of IA subgroup of the periodic table (Li, Na, K) in conditions similar to the regulations for creating multifunctional oxide materials for various purposes using photocatalytically active TiO_2 to form reliable ideas and obtaining objective knowledge about the features of transformations and the total behavior of the constituent elements in the preparatory stages of processing of technological objects with thermal activation (25–1000 °C), necessary for improving and developing methodologies and regulations of modern manufacturing technologies.

To achieve this goal, the following tasks were gradually solved in the work:

1) study of the mechanisms of transformations in the systems of coordination nitrate precursors of REE and alkali metals during the formation of multifunctional photocatalytically active layered oxide materials;
2) development of methodology and production of samples of photocatalysts based on TiO_2 anatase modification and composite with the structure of three-layer titanate $K_2Ln_2Ti_3O_{10}$ with their two-stage application and molding on structured metal carriers;
3) study of the kinetic patterns of reactions of test photocatalytic oxidation of vapors of organic substances (for example, ethanol) in a static reactor;
4) establishment of technological and functional dependencies; controlled modification of the properties of the obtained target products.

Experiment Methodology. To evaluate the possibility of controlling the processes of multistage formation of complex oxide compositions with multifunctional properties and substantiation of phase formation mechanisms as a model using a set of physicochemical methods, water-salt nitrate systems $MeNO_3$ - Ln $(NO_3)_3$ - H_2O, (Me - Li, Na, K; Ln - La-Sm) at 25–100 °C. The choice of the composition of the objects of study, temperature sections are determined by a number of factors.

First, among the elements of the rare earth series, the highest complexing ability is found by representatives of the cerium subgroup; among them the largest changes in the composition, structure, and properties of their compounds are the elements of its middle, Pr and Nd. The selected components of the systems specify the technical characteristics of the target product or are modifiers of its properties. And the presence of a large number for the use of potential electronic analogues (representatives of natural series of rare earth, alkaline elements) causes significant variability and breadth of the range of modification of their characteristics. Temperature cross sections are due to the areas of existence of crystal hydrate forms of the source components.

Secondly, according to the research of the authors [32], three layered potassium titanates $K_2Ln_2Ti_3O_{10}$ (Ln - La, Nd) of the above components, obtained by ceramic technology, in suspension form in aqueous-alcoholic solutions under the action of UV radiation show the highest among known species phases photo catalytic activity for the decomposition of H_2O and the release of hydrogen. This fact is associated with the morphology of their particles and the ability to reversibly intercalate water molecules in the interlayer space, which can lead to both an increase in the effective specific surface area of the photocatalyst and contribute to the spatial distribution of redox centers.

To determine the nature of chemical interaction and phase equilibria in water-salt systems of the studied nitrates (precursors of multicomponent oxide polyfunctional materials) in full concentration ratios in the temperature range of solutions used the method of additives described in [33, 34] and based on the study of solutions one of the properties of the most "sensitive" to the detection of phase transformations in systems, which is both a parameter of their state, and also simple experimental methods currently available. The method allows to find the limits of self-development to which the isolated system of the set structure goes in concrete conditions in an equilibrium state.

The equilibrium of the phases was reached within 2–3 days. Hydrated and anhydrous nitrates of these "c.f.a". elements were used as starting salts.

Chemical analysis of liquid and solid phases, "residues" was performed on the content of Ln^{3+} and nitrogen. The Ln^{3+} content was determined trilonometrically in the presence of xylenol orange as an indicator (acetate buffer solution, pH = 5–6) [35]; nitrogen - by distillation [36]; Me^{+} ions - calculated by the difference, based on the total nitrate content and partly on the dry residue.

The obtained data for individual ions were converted into salt content and plotted on the solubility diagrams according to the principle of conformity. Graphical representation of the composition of solid phases formed in the system was performed according to Screenmakers [33, 36], confirmation of their individuality and characterization - chemical, crystal-optical, X-ray phase, X-ray structural, IR spectroscopic, thermographic, and other methods.

Crystal-optical determinations of the compounds were performed by the immersion method using a MIN-8 microscope. Phase analysis was performed on a DRON-3 M diffractometer (Cu Kα radiation, Ni filter) by the "powder" method. Diffractograms were deciphered from the PDF file JCPDS. Determination of symmetry, parameters of unit cells and measurement of the intensity of diffraction reflections from single crystals was performed on an automatic X-ray single crystal diffractometer CAD - 4F "Enraf - Nonius" (Mo Kα - radiation, graphite monochromator; $\omega/2\theta$ - method). All calculations for the determination and refinement of atomic structures were performed using complexes of crystallographic programs SHELX, XTL – SM, AREN. The IR absorption spectra of the synthesized compounds in the region of 400–4000 cm^{-1} were recorded on a UR – 20 spectrophotometer using a standard vaseline oil suspension technique. Thermograviometric analysis was performed on a Q – 1500 D derivatograph at temperatures from 293 to 1273 K in air with a heating rate of 10°/min and the developed device for DTA.

At the stage of research of photocatalytic oxidation of vapors of organic substrates for formation of samples of composite photocatalysts was used Titanium Oxide Micro Powder (TiO_2, Anatase, 1500 nm, 99,9%) US Research Nanomaterials, Inc.

To study the effect of the formed samples-photocatalysts on the kinetics of formation of gaseous intermediates, the study of photocatalytic oxidation of organic vapors (on the example of ethanol) by static method in a developed sealed chamber-container equipped with replaceable hinged sample holder small and volley doses of injections of substrates, a fan - a mixer of the internal gas environment, an additional internal heater, a sensor for measuring the concentration of CO_2, hinged investigated "passive" plates-adsorbers, lighting system.

A portable multifunctional electronic gas analyzer AZ 7755 (AZ Instrument Corp., Taiwan) was used to measure the CO_2 concentration in the study medium, which allows simultaneous measurement of temperature, relative humidity and has the ability to connect to an external interface.

The following were used as illuminators: a low-pressure fluorescent lamp with a power of 8 W and a bactericidal lamp of the same power with a wavelength of 254 nm.

2 Results of the Research and Their Discussion

2.1 Mechanisms of Transformations in Systems of Coordination Nitrate Precursors of REE And Alkali Metals Accompanying Formation of Polyfunctional Photocatalytically Active Layered Oxide Materials

Generalized and important for practical use information on alkaline coordination nitrates of rare earth elements of the cerium subgroup - precursors of promising modern multifunctional materials - on the conditions of their formation and existence, nature of chemical bond, composition, structure, type of ligand coordination, existence of isotype series on stoichiometry, composition the structures of the revealed properties are systematized by authors [10] and in the most obvious form are resulted in Tables 1, 2, 3 and 4. The choice of this form of data presentation is the most informative and useful in the development of innovative projects allows to predict the causal fundamental patterns of behavior of structural components in similar production processes, to choose the right modes, stages, methods of forming and obtaining target products with reproducible structurally sensitive characteristics.

The revealed regularities in the nature of the behavior of structural components in rubidium, cesium nitrate systems La - Sm, in objects based on REE of the yttrium subgroup (Y, Gd - Lu) indicate the possibility of only limited or special application of the latter predecessors in the study area. There are a number of objective and economic reasons for this. These are the features of the electronic structure of their atoms, lower manifestation of chemical activity and complex-forming ability of these Ln^{3+} in comparison with the elements of the cerium subgroup, weaker effect of the considered influencing factors on the studied processes. To clarify the general patterns and build a holistic objective picture of the behavior of such technological precursors, the authors studied the systems of natural series Y, La - Lu, Li - Cs. Analysis of the results of the study was published in previous works of the authors [10, 37, 38].

In ternary REE-containing nitrate precursor systems, which are integral components of more complex multicomponent systems, exchange transformations begin from the moment the components are dissolved in water. It was found that the Ln^{3+} cerium subgroup in the studied conditions are active complexing agents, form anionic coordination compounds of Me^{+} of all alkali metals, and their stability and complex of inherent properties are potent technological factors that significantly affect the

Table 1 Isothermal concentration limits of crystallization of alkaline coordination nitrates of neodymium from solutions of water-salt systems $MeNO_3 - Nd(NO_3)_3 - H_2O$ (Me – Li, Na, K)

The composition of the compounds	t, °C	Compositions of saturated solutions corresponding to transition and eutonic points, wt. %		The nature of solubility
		$MeNO_3$	$Nd(NO_3)_3$	
$Li_3[Nd_2(NO_3)_9]\cdot 3H_2O$	65	19,32 13,95	59,61 65,63	Incongruent
	100	24,03 9,68	54,68 72,51	Congruent
$Na_2[Nd(NO_3)_5]\cdot H_2O$	50	16,55 8,60	51,62 62,58	Incongruent
	65	20,44 3,03	50,40 70,17	Incongruent
	100	25,27 4,15	47,28 76,96	Incongruent
$K_2[Nd(NO_3)_5 (H_2O)_2]$	50	27,26 21,34	51,62 54,91	Incongruent
$K_3[Nd_2(NO_3)_9]\cdot H_2O$	50	21,34 11,49	54,91 63,31	Incongruent
	65	32,57 8,67	47,88 70,44	Congruent
	100	40,15 3,39	45,02 76,44	Congruent

nature of transformations in systems as intermediates. and the results of processes in general.

The obtained information allows to model the behavior of structural components at the preparatory stages of formation of modern multifunctional photocatalytically active materials according to innovative technological regulations with the use of nitrate REE-containing precursors.

The available identified trends in phase formation in the model systems studied are thermodynamically the most probable limits of transformations in technological objects in the conditions of formation and production of target products. And possible real deviations in such systems are caused by inhomogeneity of the reaction medium in composition, content of reacting components, conditions of their location, finiteness of transformations, diffusion features, heat capacity, viscosity, nature of transformations at the boundaries of formed hetero phases, using their applied principles and methods. activation and other specific factors. And the identified processes of complexation in aqueous solutions of nitrates contribute to the homogenization of systems of structural components at the molecular level in complex or combined processing.

The analysis of the obtained data indicates the flow of competing processes of replacement of water molecules by nitrate ions in the systems in the immediate

Table 2 Radiographic data of neodymium lithium, sodium, potassium coordination nitrates

$Li_3[Nd_2(NO_3)_9]·3H_2O$				$Na_2[Nd(NO_3)_5]·H_2O$		$K_2[Nd(NO_3)_5 (H_2O)_2]$				$K_3[Nd_2(NO_3)_9]·H_2O$			
d, Å	I/I_0,%	d, Å	I/I_0, %	d, Å	I/I_0, %	d, Å	I/I_0, %	d, Å	I/I_0, %	d, Å	I/I_0, %	d, Å	I/I_0, %
8,36	30	2,013	19	7,84	63	5,42	66	2,056	24	9,48	85	2,281	21
7,64	77	1,979	19	7,54	47	5,27	90	1,993	27	7,74	92	2,249	69
6,68	13	1,944	28	7,07	73	4,94	45	1,947	26	7,65	32	2,189	54
6,00	38	1,931	26	5,18	27	4,11	72	1,777	11	5,36	39	2,108	68
5,75	79	1,855	15	4,23	100	3,88	15			5,27	40	2,082	37
5,42	98	1,778	15	3,80	57	3,80	42			4,94	30	2,058	15
5,26	74	1,726	28	3,15	23	3,66	12			4,76	46	2,012	49
4,76	51	1,708	34	3,09	67	3,53	42			4,49	87	1,909	21
4,64	100			3,02	86	3,35	38			4,26	27	1,837	27
4,35	43			2,629	20	3,18	13			4,06	100	1,757	19
4,19	47			2,391	17	3,05	73			3,89	40	1,729	16
3,94	51			2,346	17	2,873	17			3,78	22	1,714	18
3,90	40			2,307	8	2,843	33			3,73	39		
3,56	26			2,234	13	2,783	14			3,36	19		
3,32	19			2,178	27	2,750	17			3,27	26		
3,22	47			1,979	13	2,724	40			3,18	55		
2,978	34					2,664	14			3,07	16		
2,772	19					2,639	19			3,04	17		
2,617	28					2,594	100			2,844	20		
2,545	43					2,463	37			2,755	12		

(continued)

Table 2 (continued)

$Li_3[Nd_2(NO_3)_9]\cdot 3H_2O$				$Na_2[Nd(NO_3)_5]\cdot H_2O$		$K_2[Nd(NO_3)_5\ (H_2O)_2]$				$K_3[Nd_2(NO_3)_9]\cdot H_2O$			
d, Å	I/I_0,%	d, Å	I/I_0, %	d, Å	I/I_0, %	d, Å	I/I_0, %	d, Å	I/I_0, %	d, Å	I/I_0, %	d, Å	I/I_0, %
2,385	21					2,392	19			2,730	18		
2,328	17					2,374	44			2,647	55		
2,305	23					2,314	15			2,592	16		
2,226	19					2,235	25			2,508	87		
2,135	49					2,188	10			2,468	78		
2,111	43					2,099	21			2,349	32		

Note: d, Å - interplanar distances; I/I_0,% - the relative intensities of the reflexes

Table 3 Coordination polyhedra of Ln atoms in alkaline rare earth nitrates

Compound	D_1	D_2	D_3	C. N.	Polyhedron type, its symmetry
$Li_3[La_2(NO_3)_9]·3H_2O$		2,66 2,65	2,66 2,65	12 12	Ic $4L_33L_2$ Ic L_33L_2
$Li_3[Nd_2(NO_3)_9]·3H_2O$		2,61 2,60	2,61 2,60	12 12	Ic $4L_33L_2$ Ic L_33L_2
$Na_2[Nd(NO_3)_5]·H_2O$		2,62 2,61 2,61	2,62 2,61 2,61	12 12 12	Ic L_33L_2 Ic L_33L_2 Ic L_33L_2
$K_2[La(NO_3)_5 (H_2O)_2]$	2,70	2,68	2,68	12	Ic L_2
$K_2[Nd(NO_3)_5 (H_2O)_2]$	2,62	2,61	2,61	12	Ic L_2
$K_3[Pr_2(NO_3)_9]$		2,63	2,63	12	Ic L_33L_2
$K_3[Nd_2(NO_3)_9]$		2,61	2,61	12	Ic L_33L_2
$(NH_4)_2[La(NO_3)_5 (H_2O)_2]·H_2O$ $(NH_4)_2[La(NO_3)_5 (H_2O)_2]·2H_2O$	2,54 2,59	2,70 2,70	2,67 2,68	12 12	Ic L_35L_2 Ic L_35L_2
$(NH_4)_2[Pr(NO_3)_5 (H_2O)_2]·2H_2O$	2,54	2,66	2,64	12	Ic L_35L_2

D_1, D_2, D_3 – average distance (Å) ‹Ln–O $_{H2O}$›, ‹Ln–O $_{NO3}$›, ‹Ln–O›; Ic – icosahedron.

vicinity of the Ln^{3+} complexing agent. The degree of completeness of substitution depends on the nature of Ln^{3+}, the presence of Me^+, the properties of electron-donor oxygen atoms and the spatial structure of ligands, the concentration of anions, the amount of solvent. A significant influence of temperature factor on these processes is revealed. There are differences in the complexing ability of the elements of cerium and yttrium subgroups, Y, as well as among REE in the middle of the first subgroup. The obtained results indicate the gradual processes of complexation. The presence of certain values of the temperature of the beginning of the release into the solid phase of complex compounds - the existence of an energy barrier and the need for some activation energy to carry out such transformations. In the formation of nitrate complexes, the requirements of symmetry are largely met, and the planar small-sized ligand NO_3^- is "convenient" for the formation of a highly symmetric environment of the Ln_3^+ ion. Lanthanides have a tendency to form three types of NO_3^- ligand coordination. This leads to the formation of both isolated complexes and their polymerization into dual-core, chains, frameworks.

All established coordination compounds were synthesized in single crystal form and characterized by a set of physicochemical methods. Table 2 shows the X-ray diffraction characteristics of the newly formed phases for the possibility of their identification and detection during processing.

The authors conducted a crystal chemical analysis of alkaline rare earth nitrate compounds, which is based on the results of our own research [38] and on literature data [39]. Particular attention is paid to the structure of coordination polyhedra Ln, which largely determines the basic properties of the corresponding compounds.

Table 4 The temperature values of the detected effects during the heat treatment of the representatives of the established groups of alkali coordination nitrates REE

Compounds; spatial group of crystals	Representatives	Temperature interval of formation, °C	The nature of solubility	Dehydration	Melting in crystallization water	Polymorphic transitions	Melting anhydrous form	The composition of the products of conversion at 980 °C
$Li_3[Ln_2(NO_3)_9]\cdot 3H_2O$ cubic.; $P2_13$	La –Sm	65–100	Congr.	65 183 216	183	–	274	$LiLnO_2$
$Na_2[Ln(NO_3)_5]\cdot H_2O$ monocl.; $P2_1/a$	La –Sm	50–100	Congr.	81 148 236	–	271	328	$NaLnO_2$
$K_2[Ln(NO_3)_5(H_2O)_2]$ rhomb.; Fdd2	La – Nd	50–100	Incongr.	95,111	95	219	314	$KLnO_2$, Ln_2O_3
$K_3[Ln_2(NO_3)_9]\cdot H_2O$ cubic.; $P4_332$	La –Sm	50	Congr.	126	–	–	347	Ln_2O_3
$K[Ln(NO_3)_4(H_2O)_2]$ prim. rhomb.; $P2_1cn$	Y, Gd - Lu	50–100	Congr.	138,172	138	–	–	Ln_2O_3

Analyzing the shape, symmetry and other properties of polyhedra (see Table 3), we get the opportunity to approach the understanding of the individual features of REE.

Data on average Ln - O distances in rare earth nitrates of alkaline cations discussed in this paper are in good agreement with the expected tendency to decrease

Ln - O distances according to lanthanide compression and to increase these distances with increasing coordination number for fixed REE ion. The distances Ln - O (H_2O), as a rule, are among the shortest contacts in polyhedra. This fact can be explained on the basis of the presence of competing interactions with bonds of the type Ln - O (NO_3^-). Coordination polyhedra are composed, as a rule, of oxygen atoms of bidentate coordinated nitrate groups, less often in combination with oxygen atoms of water molecules. In all considered compounds coordination numbers of representatives of lanthanides of cerium subgroup 12.

When summarizing the structural data of alkaline coordination nitrates, REE draws attention to the extremely limited range of coordination polyhedra for such high CN REEs. In Fig. 2 shows a general view of the identified types of icosahedrons found in the structures of the studied compounds. In cases where water is not part of the coordination sphere of the Ln atom, the polyhedra are constructed in exactly the same way (Fig. 2a). Their equivalence is that the shortened O – O ribs (common to NO_3^- – ligands) occupy the same positions. Exceptions are one of the 2 independent polyhedra Ln in two isostructural compounds $Li_3[Ln_2(NO_3)_9]{\cdot}3H_2O$ (Ln - La, Nd) (Fig. 2b). If the coordination saturation occurs with the participation of water molecules, the shape of the 12-vertices changes slightly. Topologically, it is still the same icosahedron. However, the distribution of shortened ribs in this case is different (Fig. 2c).

The analysis of the Ln-polyhedra considered by us gives grounds to note the tendency of REE in this class of compounds to the organization of a symmetric coordination environment. In some cases, despite the fact that Ln^{3+} ions are

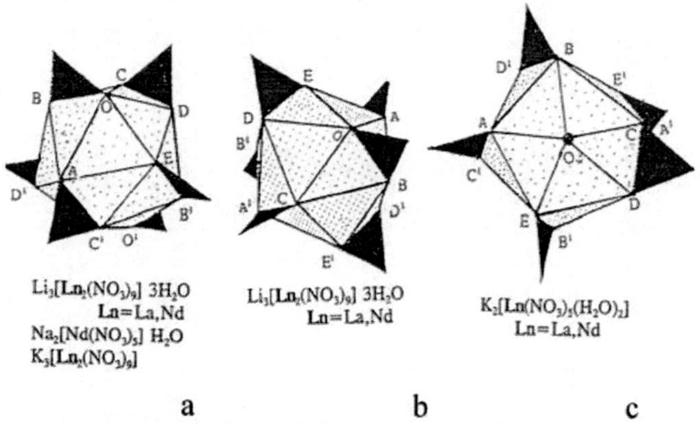

Fig. 2 Schematic representation and general view of Ln-icosahedrons found in the structures of the REEs of the cerium subgroup and lithium, sodium, and potassium

located in common positions, their coordination polyhedra have at least one noncrystallographic axis of symmetry 2 (see Table 3).

The following data from the thermographic study of lithium, sodium, potassium coordination nitrates of rare earth elements of the cerium subgroup (Fig. 3, Table 4) clarify the nature and patterns of thermal transformations of these compounds in the temperature range 25–1000 °C, establish their heat resistance, intervals, staged processes, phase formation depending on the composition, content, nature of the components, the method of packing coordination polyhedra in spatial construction, conditions and method of processing, etc. This makes it possible to predict the behavior of the applied potential predecessors in real technological systems in similar conditions. For comparison and analysis there is information about the low

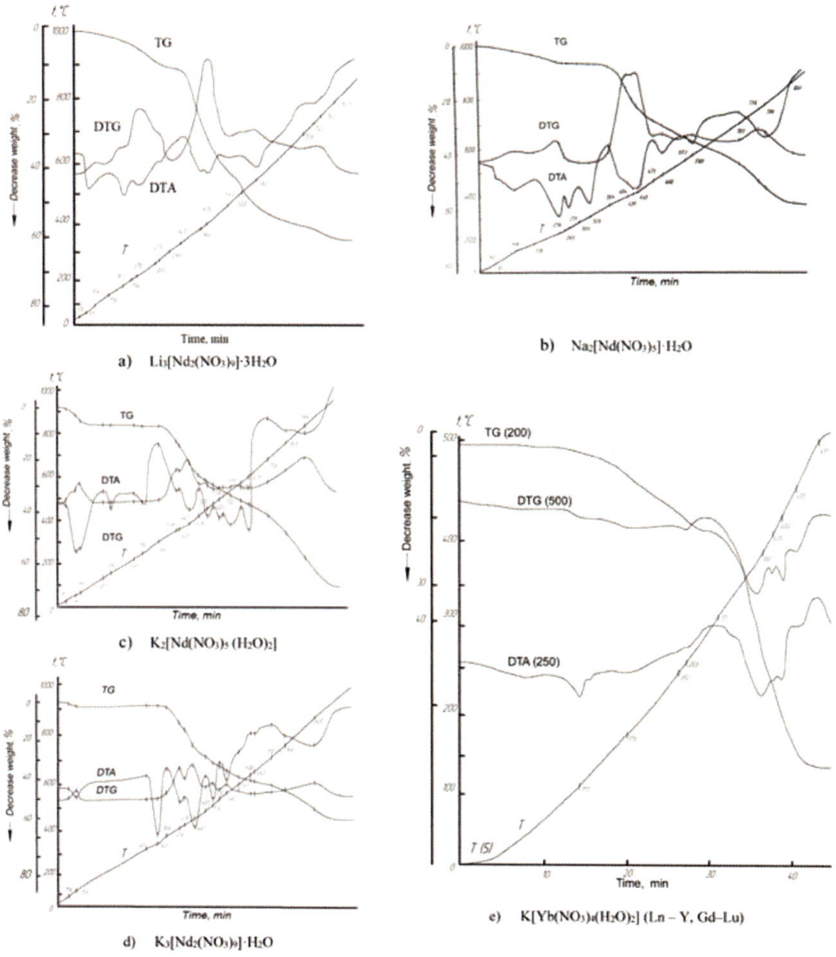

Fig. 3 Derivatograms of compounds

stability and heat resistance of potassium coordination nitrates Y, Gd - Lu composition $K[Ln(NO_3)_4(H_2O)_2]$, which leads to limitations in the use of this type of precursors in technological transformations to modify the properties target products.

It was found that the coordination numbers of the Ln^{3+} cerium subgroup - 12, found in low-temperature associated forms, remain unchanged until the formation of stable high-temperature multicomponent oxide phases $MeLnO_2$, $Me_2Ln_2Ti_3O_{10}$. This indicates the feasibility of using alkaline coordination nitrate REE-containing precursors in such technological innovative solutions.

2.2 *Influence of Formed Samples-Photocatalysts (with Structure of Three-Layer Titanate $Me_2Ln_2Ti_3O_{10}$ (Me - K; Ln - Nd) on Kinetics of Oxidation of Vapors of Organic Substances (on an Example of Ethanol)*

The logical practical application and testing of the obtained set of empirical knowledge about the combined behavior and properties of constituent components in the studied systems was the development of one of the possible ways of applying and forming composite photocatalytically active coatings on structured metal carriers (to minimize the contribution of adsorption component). submicron sizes of the anatase phase of TiO_2 and similar particles with modified properties of their surface layer and subsequent comparative test for activity in the processes of photodestruction of vapors of organic substances (for example, ethanol) in the air under the influence of UV radiation. Modification of the activity of oxidation centers of objects was carried out by two-stage heat treatment of samples with separate stages of processes: application and fixing of coatings on the basis of water-suspension systems TiO_2 and soluble nitrate precursors Ln and alkali metals taken in given ratios; formation of diffusion flows at the interfacial boundaries of the components of heterogeneous composite systems and regulation of the composition and processing conditions of nitrate precursor melts (in order to create favorable conditions for the association of titanates with a layered structure of $Me_2Ln_2Ti_3O_{10}$). Such composite systems are crystallization-condensed curing structures.

(Requirements for structured functionally active such coatings, their preparation procedures and characteristics are discussed in several review articles [40, 41]).

In the preparatory stages of heat treatment of the original water-suspension systems with refueling of the component composition of soluble nitrate precursors in a ratio corresponding to the formation of a three-layer perovskite-like titanate $Me_2Nd_2Ti_3O_{10}$ (Me_2O $Nd_2O_3{\cdot}3TiO_2$, Me – Li, Na, K; Ln – La–Nd) the solvent is removed from the supersaturated dispersion medium, and according to the studied polythermal solubility diagram [11] (transformation in the region of concentration congruence), the alkaline crystals of alkaline coordination nitrates Ln are gradually isolated. Further heating of these intermediate newly formed phases leads to their melting and subsequent decomposition with the release of oxides of nitrogen and oxygen.

In [42] it was found that the transition of submicron titanium dioxide powder from anatase to rutile occurs at 750–850 °C, and the role of primary structural elements in such dispersed objects is played by powder particles.

TiO_2 (anatase) in the studied multicomponent heterogeneous system behaves chemically indifferent with respect to the constituent structural components until the moment of nucleation in the products of melt thermolysis of nitrate coordination precursors of weakly crystallized chemically active particles of double oxides $MeLnO_2$ ($Me_2O{\cdot}Ln_2O_3$) [15]. With increasing activation energy of the system (t > 520 °C) and, accordingly, the energy of thermal motion of structural elements there is a possibility of their convergence at shorter distances, there is a strengthening of the coordination chemical bond between the constituent cations of the corresponding metals and oxygen anions with the formation of perovski or to some extent, cation-ordered three-layer oxide structures of $Me_2Ln_2Ti_3O_{10}$ with a set of inherent properties.

In the study of the photocatalytic activity of the synthesized materials was performed on the example of the test reaction of ethanol vapor oxidation in a static reactor. The process occurs with the formation in the gas phase of an intermediate product - acetaldehyde, which is eventually completely oxidized to CO_2. For the formed composite photocatalysts, an increase in the oxidation rate of the substrate was observed in comparison with pure TiO_2 (see Fig. 4). As a result, this led to a decrease in the time of removal of the substrate and the intermediate from the gas phase and a decrease in the maximum concentration of acetaldehyde in the latter.

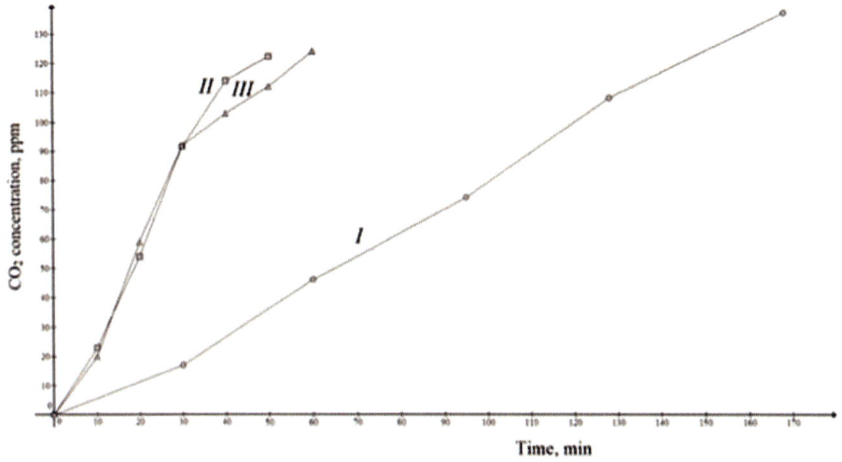

Fig. 4 Kinetics of ethanol vapor oxidation processes under ultraviolet irradiation of titanium dioxide (anatase) (I) photocatalysts and compositions (II, III) modified with the structure of layered perovskite-like oxides $K_2Nd_2Ti_3O_{10}$

The rate of oxidation of the substrate depended on the method of coating, the nature of the applied property modifiers, the sequence of stages and modes of processing, the compatibility of the coating and the base material and the nature of its preparation, the applied composition and content of impregnating systems. To be able to compare the activity of the developed photocatalysts, the portion of the mass of TiO_2 in the original refueling of water-suspension systems samples with pure titanium dioxide and samples with applied modified compositions took the same.

Cation-ordered three-layer composite materials $K_2Nd_2Ti_3O_{10}$ under appropriate conditions can act as alternatives to TiO_2 anatase modification, the relative specific activity of which is 6,2 times higher compared to this characteristic of titanium dioxide.

The obtained information allows to optimize the conditions of formation of cation-ordered layered titanates; find out the conditions and identify the temperature range of application of this class of compounds. The obtained own and literary physicochemical, thermochemical and structural data, as well as the results of their interpretation are a stage in the development of experimental and theoretical scientific databases on layered compounds and processes with their participation.

(Note. Test oxidation of substrates of vapors of organic substances (for example, ethanol) formed by photocatalysts was carried out by irradiation with a bactericidal lamp (254 nm, 8 W) by static method in the laboratory chamber V chamber = 40 dm^3; at T = 292 K, respectively: a) with the introduction of a large simulated dose of C_2H_5OH 2 ml (experiment I) in the presence of TiO_2 anatase modification applied to a structured nonwoven base and b) small doses of 0,1 ml (experiment II), 0,2 ml (experiment III) C_2H_5OH followed by evaporation, in the presence of a functionally active composition $K_2Nd_2Ti_3O_{10}$, applied to steel sheets and moistened with H_2O.

In Experiment I, the coating was formed by drying the anatase modification applied to the structured nonwoven base of an aqueous submicron dispersion of TiO_2; in experiments II, III composite coatings with the structure $K_2Nd_2Ti_3O_{10}$ are formed on sheet steel carriers by a two-stage procedure of application and fixation of precursors and subsequent slow heating of systems at a rate of 3–5 °/min. up to 550 °C and exposure for 4 h).

The studied class of layered compounds is a promising basis for creating functional materials with unique properties, which are determined by the two-dimensional nature of the interlayer space, distortion of the structure of titanium-oxygen octahedra of the perovskite layer and high mobility of alkali metal cations. Such their properties can be used in innovative fields of science, technology, energy, electronics, ecology (see Fig. 4, Table 5).

Table 5 Evaluation of the conditional activity of photocatalyst samples: I sample - based on TiO_2 (anatase) and II sample - based on the composition of the three-layer oxide $K_2Nd_2Ti_3O_{10}$ in the decomposition of ethanol (Fig. 4, for areas with proportional trends)

Cation ordered three-layer $K_2Nd_2Ti_3O_{10}$	TiO_2 (modification of anatase)
$S = 218\ cm^2$	$S = 395\ см^2$
t = 18,8 °C	t = 19,0 °C
$\tau_\Sigma = 40$ min	$\tau_\Sigma = 128$ min
$\Delta CO_2 = 114$ ppm	$\Delta CO_2 = 108$ ppm
$v_{avg.} = 2{,}85$ ppm/min	$v_{avg.} = 0{,}84$ ppm/min
$v_{avg.\ act.} = 13{,}1 \cdot 10^{-3}$ $ppm/cm^2 \cdot min$	$v_{avg.\ act.} = 2{,}1 \cdot 10^{-3}$ $ppm/cm^2 \cdot min$

3 Scientific Novelty

A comprehensive systematic study of the interaction of structural components in systems of rare earth nitrates and IA groups of elements of the periodic table - precursors of modern multicomponent oxide functional materials based on them - established the formation of a wide class of alkaline coordination nitrates of lanthanides. The identified objective patterns have a fundamental and applied value, deepen the understanding of:

- chemical and physical properties of Ln, their complexing ability,
- possibility of formation and existence in similar systems of associated new phases, their atomic-crystalline structure, stability and stability,
- the influence of the nature of lanthanides and alkali metals on the structure of complex anions and compounds in general,
- individuality of Ln complexes,
- existence of isotypic in composition and structure of groups of compounds in the natural series of lanthanides and alkali metals,
- the role of NO_3^- - groups in the stereochemistry of this class of nitrates,
- the role of water in the formation of the immediate environment of Ln^{3+} ions-complexing agents.

4 Practical Significance

The obtained system of knowledge about transformation processes in REE-containing nitrate precursor systems and crystal chemical properties of Ln coordination nitrate samples acquires special value in the formation of nanostructured layered perovskite-like compounds of lanthanides and transition elements (including titanium, described in [7, 8, 11, 32, 43–51], others), solid solutions based on them; in establishing technological and functional dependencies between the method of preparation, variability of the method of activation of technological systems, methodology of manufacturing the target product and its phase composition, lattice parameters, specific surface area, morphology of constituent particles, activity of self-cleaning compositions with photocatalytic and hydrophilic properties. and special structural elements; in the practical implementation of innovative projects of water decomposition for the purposes of hydrogen production (as an alternative fuel), decomposition of toxic organic substances in solutions and air, incomplete oxidation of carbohydrates; in obtaining other perovskite-like phases by ion exchange reactions and in other areas.

5 Conclusions

1. The results of the study show that the processes of obtaining oxide REE-containing structural and functional materials for different purposes using nitrates of elements of different electronic structure by chemical mixing of the source components in the joint separation of products from the liquid phase by sequential or co-precipitation followed by heat treatment. Data on their composition, content and behavior in each case require prior systemic empirical knowledge in full concentration ratios in a given temperature range.
2. Differences in the behavior of structural components in the systems of lanthanides of cerium and yttrium subgroups, in their nature of interaction, stages, features and patterns of flow.
3. The new knowledge is the basis for:
   - finding ways to increase the activity of Ln-forms;
   - elucidation of the nature of successive thermal transformations in nitrate REE-containing multicomponent systems of different aggregate states during their heat treatment; conditions of formation and existence, properties of intermediate phases; influencing factors; possible ways to control the receipt of the target product;
   - in case of creation of modern perfect low-cost technologies of formation of functional materials of various function with reproducible properties.

References

1. Mazurenko EA, Gerasemchuk AI, Trunova EK et al (2004) Ukr Chem J 70(7):32
2. Belous AG (2009) Ukr Chem J 75(7):3
3. Zhang Q, Schmidt N, Lan J et al (2014) Chem Com 50:5593
4. Gavrilenko OM, Pashkova OV, Bilous AG (2005) Ukr Chem J 71(8):73
5. Fortal'nova EA, Gavrilenko ON, Belous AG et al (2008) Ros Chem J LII(5):43
6. Belous AG (2008) Ukr Chem J 74(1):3
7. Yakovleva IS, Isupova LA (2012) Pat. RU 2440292 S 2, C 01 B 3/18 Method for obtaining perovskites. (RU) – publication
8. Schaak RE, Mallouk TE (2002) Chem Mater 14(4):1455

9. Rodionov IA, Siljukov OI, Zvereva IA (2012) J Gen Chem 4:548
10. Dryuchko OG, Storozhenko DO, Bunyakina NV et al (2018) Bull NTU "KhPI" Ser Chem Chem Technol Ecol 39(1315):3
11. Dryuchko O, Storozhenko D, Vigdorchik A et al (2018) Mol Cryst Liq Cryst 672(10):199
12. Zhang X, Zhang A, Zhang H, et al (2012) Pat. CN 102382490 A. Preparation and application of a hydrophilic self-cleaning coating with photocatalytic activity. (CN) – publication
13. Varma A, Mukasyan AS, Rogachev AS et al (2016) American chemical society. Chem Rev 116:14493
14. Seeber A et al (2010) Mater Res 13(1):99
15. Fujishima A, Zhang X, Tryk DA (2008) Surf Sci Rep 63(12):515
16. Kudrenko EO, Shmyt'ko YM, Strukova HK (2008) Fyzyka tverdoho tela 50(5):924
17. Rodionov IA, Zvereva IA (2016) Russ Chem Rev 85(3):248
18. Lagaly G (1986) Solid State Ionics 22:43
19. Machida M, Miyazaki K, Matsushima S et al (2003) J Mater Chem 13(6):1433
20. Silyukov O, Chislov V, Burovikhina A et al (2012) J Therm Anal Calorim 110(1):187
21. Gopalakrishnan J, Sivakumar T, Ramesha K et al (2000) Chem Phys 9:6237
22. Schaak RE, Mallouk TE (2001) J Solid State Chem 161(2):225
23. Zhu WJ, Feng HH, Hor PH (1996) Mater Res Bull 31(1):107
24. Toda M, Watanabe K, Sato J (1996) Mater Res 31(11):1427
25. Richard M, Brohan L, Tournoux M (1994) J Solid State Chem 112:345
26. Gopalakrishnan J, Bhat V (1987) Inorg Chem 26(26):4299
27. Utkina T, Chislov M, Silyukov O et al (2016) J Therm Anal Calorim 125:281
28. Kozlov DV, Selishchev DS, Kolinko PA et al (2013) Pat. RU 2478413 C1 IPC B01D 21/06 B01J 21/08 B01J 21/18 B01J 23/38 B01D 53/86 C02F 1/30. Composite photocatalyst for water and air purification. (RU) – publ. [Google Scholar]
29. Kato M, Kajita T, Hanakago R et al (2006) Phys C Supercond 445:26
30. Thangadurai V, Subbanna GN, Gopalakrishnan J (1998) J Chem Commun 7(c):1299
31. Ranmohotti KGS, Josepha E, Choi J et al (2011) Adv Mater 23(4):442
32. Sivakumar T, Lofland S, Ramanujachary K et al (2004) J Solid State Chem 177(7):2635
33. Rodionov IA, Silyukov OI, Utkina TD (2012) Russ J Gen Chem 82:1191
34. Anosov VYa, Ozerova MI, Fialkov YuYa (1976). Basics of physical and chemical analysis, p 503. Nauka Publication, Moscov
35. Goroshchenko YaG (1978). Physical and chemical analysis of homogenous and heterogenous systems, p 490. Naukova dumka Publication, Kiev
36. Busev AI, Tiptsova VG, Ivanov VM (1978) Management on analytical chemistry of rare elements, p 432. Chemistry Publication, Moscov
37. Kreshkov AP (1976) Fundamentals of analytical chemistry. Quantitative analysis, 2nd edn, p 480. Chemistry Publication, book, Moscov
38. Storozhenko DO, Dryuchko OG, Bunyakina NV, Ivanytska IO (2015) Innov Corr Mater Sci 5(2):80
39. Vigdorchik AG, Malinovskiy YuA, Dryuchko AG et al (1992) Crystallography 37(4):882
40. Eriksson B, Larrson LO, Niinisto L et al (1980) Acta Chem Scand 34(8):567
41. Meille V (2006) Appl Catal 315:1
42. Avila P, Montes M, Miró EE (2005) Chem Eng J 109:11
43. Cromer DT, Herrington K (1955) J Am Chem Soc 77(18):4708
44. Mendoza E, Padmasree KP, Montemayor SM et al (2012) J Mater Sci 47:6076
45. Onyshchenko VO, Filonych OM, Bunyakina NV, Senenko NB (2020). https://doi.org/10.1007/978-3-030-42939-3_67
46. Onischenko VA, Soloviev VV, Chernenko LA, Malyshev VV, Bondus SN (2014). Materialwiss Werkstofftech 45(11):1030. https://doi.org/10.1002/mawe.201400222
47. Loburets AT, Naumovets AG, Senenko NB, Vedula YS (1997). Zeitschrift Fur Physikalische Chemie 202(1–2):75. https://doi.org/10.1524/zpch.1997.202.part_1_2.075
48. Popova AV, Kremenetsky VG, Solov'ev VV, Chernenko LA, Kremenetskaya OV, Fofanov AD, Kuznetsov SA (2010). Russ J Electrochem 46(6):671. https://doi.org/10.1134/S1023193510060121

49. Cherniha R, Pliukhin O (2013). J Math Anal Appl 403(1):23. https://doi.org/10.1016/j.jmaa.2013.02.010
50. Cherniha R, Serov M (2003). Nonlinear systems of the burgers-type equations: Lie and Q-conditional symmetries, ansätze and solutions. J Math Anal Appl 282(1):305–328. https://doi.org/10.1016/S0022-247X(03)00155-0
51. Kochkarev D, Azizov T, Galinska T (2018). Bending deflection reinforced concrete elements determination. Paper presented at the MATEC Web of Conferences, 230. https://doi.org/10.1051/matecconf/201823002012

Carrying Out of Researches for Determining the Term of Effective Exploitation of Thermal Insulation Materials for 100 Years (Part 2)

Gennadiy Farenyuk and Olena Oleksiienko

Abstract The study aims at finding indicators of longevity of the mineral wool products while being used as an thermal insulation layer of building envelope of construction projects with operational life up to 100 years. Subject of the study is valuation of changes in air permeability, compression strength at 10% linear deformation, operational characteristics of material—thermal conductivity, specimen appearance and geometry under cyclic climate effects that simulate environmental factors which affect the state of thermal insulating materials of building envelopes while being used in construction projects. It was conducted a study of life characteristics of Rockwool mineral wool products by Batts Lights (Rockmin) and Ruth Batts (Dachrock max) as the thermal insulating layer of building envelopes with the life time up to 100 years. During the study simulation of climatic impacts on the thermal insulating products was conducted in conditions of normal use of products in building envelopes and in conditions of probable failures of building envelopes. The results of the study show that the stability criteria of physical parameters of the Rockwool mineral wool product by Batts Lights (Rockmin) and Ruth Batts (Dachrock max) do not exceed acceptable limits after massive cyclic climatic impact. This makes possible to estimate the conditional life time of the mentioned Rockwool mineral wool products being not less than 100 years.

Keywords Thermal insulation · Durability · Climate effects · Thermal conductivity · Air permeability · Compression strength

1 Introduction

Modern building envelopes are constructed from materials of different thermophysical properties and terms of durability. The widespread use of new thermal insolating materials without proper verification in the practice of domestic house building causes

G. Farenyuk · O. Oleksiienko (✉)
State Enterprise "State Research Institute of Building Structures", 37 Preobrazhenska Street 5/2, Kyiv, Ukraine
e-mail: mb-elena@ukr.net

V. Onyshchenko et al. (eds.), *Proceedings of the 3rd International Conference on Building Innovations*, Lecture Notes in Civil Engineering 181,
https://doi.org/10.1007/978-3-030-85043-2_9

thermal failures of building envelopes in the operation of buildings. The concept of the period of efficient operation of products was introduced to characterize the insulating materials and to determine the possibility of their use in the modern structural schemes of insulating cover of building in the national regulatory framework—DBN V.2.6–31 [1].

The period of efficient operation (estimated durability) of thermal insulating products is the estimated operational period during which the products retain their thermal insulating properties at the level of project performance, as evidenced by the results of laboratory tests and specified in the provisional warranty (lifetime).

The characteristics of an efficient operation period are based on the concept of material resource i.e. the parameter that characterizes the change of thermophysical properties of the material under service conditions that are simulated in the course of the tests.

The basis of the method of determination of the operational resources is the hypothesis of existence of material structure damage linear accumulation area under single or total climate effects. The accepted concept affords ground to determine the resource parameters by fixation and analysis of the nature of changes of general physical parameters that determine the thermal performance of building envelope within the field of linear accumulation of damages in both temperature and humidity effects. This is the approach that is also adopted in the standards for methods of testing roofing materials in accordance with the DSTU B EN 1296 [2] and insulating materials in accordance with the ISO 11,561 [3] and [4–11].

2 Methodology and Research

The tests are carried out for the products that perform insulation functions in constructions i.e. thermal insolating and air insolating functions. Accordingly, the change of the following thermophysical properties of materials such as thermal conductivity, air permeability are recorded during the test.

The research methodology is based on that fact that the tested material is subjected to cyclic climate effects that simulate operating conditions of the material in building envelopes after which the changes of the thermophysical properties of the materials are determined.

There is an operation imitation of climatic factors under normal use of the thermal insolating materials in building envelopes in the methodological basis when duration of effective operation of the material is estimated in accordance with the results of the thermophysical characteristics measurement depending on the constructive solution of thermal insulation of external building envelopes and in terms of possible structure damages and appropriate action of climatic factors directly to the insulation product.

As for the thermal insulating materials the period of efficient operation is measured by change of thermal conductivity in standard tests, as well as changes in an air permeability coefficient of material and change of the linear dimensions and strength characteristics.

Determination of stability of thermal insulating products to cyclic climate effects (effective operation period). Stability factors of external appearance of the test samples were determined visually and recorded for each test sample. After 180 cycles of climate effects of freezing—thawing—heating it was determined on the base of visual inspection of the test samples that the appearance of the samples of Light Batts (Rockmin) and Ruth Batts (Dachrock max) brands does not change. Change of the material color and its continuity was not fixed visually in the course of peeling and tearing of the material fibers. Visual appearance of one of the test samples is shown in Fig. 1.

Change of geometrical dimensions of the samples is within 5% for Light Batts (Rockmin) brand and 2% for Ruth Batts (Dachrock max) brand that allows evaluating positively the stability of geometric dimensions and exterior view to the action of cyclic climatic factors.

The thermal conductivity of ROCKWOOL mineral wool products of Light Batts (Rockmin) brand—number of cycles diagram is shown in Fig. 2, Ruth Batts (Dachrock max) brand in Fig. 3 respectively.

Fig. 1 Exterior view of the samples before (sample at the left of the Figure) and after (sample at the right of the Figure) 180 cycles of climate impacts

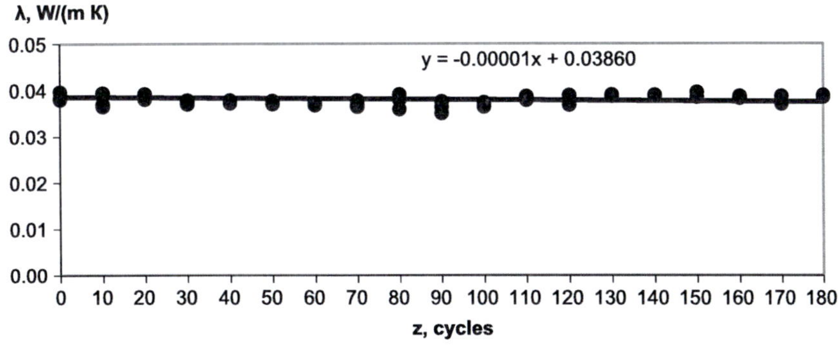

Fig. 2 The dependence of thermal conductivity of ROCKWOOL mineral wool products of Light Batts (Rockmin) brand from cyclic effects

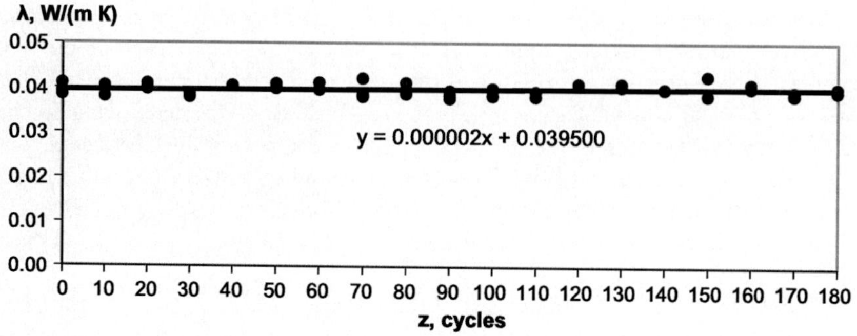

Fig. 3 The dependence of thermal conductivity of ROCKWOOL mineral wool products of Ruth Batts (Dachrock max) brand from cyclic effects

The dependence of thermal conductivity of ROCKWOOL mineral wool on the number of cycles of freezing—thawing—heating is approximated by the formulas:

- for Light Batts (Rockmin) brand:

$$\lambda(z) = 0,0386 - 0,00001 \cdot z \pm 0,0009 \quad (1)$$

- for Ruth Batts (Dachrock max) brand:

$$\lambda(z) = 0,0395 + 0,000002 \cdot z \pm 0,0004. \quad (2)$$

Resource rate determined by the "Eg. (17)" [10] is:

for Light Batts (Rockmin) brand—$r = -0{,}0018$;
for Ruth Batts (Dachrock max) brand—$r = 0{,}0004$.
Fulfillment of condition is checked by the formula "Eg. (23)" [10]:
for Light Batts (Rockmin) brand:

$$\frac{r}{\lambda_0}k_z = \frac{-0,0018}{0,0386} \cdot 5 = -0,23 \leq 0,2 \quad (3)$$

for Ruth Batts (Dachrock max) brand:

$$\frac{r}{\lambda_0}k_z = \frac{0,0004}{0,0395} \cdot 5 = 0,05 \leq 0,2 \quad (4)$$

The test results of breathability coefficient *i*, kg/m^2 h Pa, with $\Delta p = 10\ Pa$ depending on the number of cycles of freezing—thawing—heating the ROCKWOOL mineral wool of Light Batts (Rockmin) and Ruth Batts (Dachrock max) brands shown in Figs. 4 and 5 respectively.

The ROCKWOOL mineral wool breathability coefficient dependence on the number of cycles of freezing—thawing—heating is approximated in the formulas:

for Light Batts (Rockmin) brand:

$$i(z) = 0,4511 - 0,00005 \cdot z \quad (5)$$

for Ruth Batts (Dachrock max) brand:

$$i(z) = 0,0617 + 0,00002 \cdot z \quad (6)$$

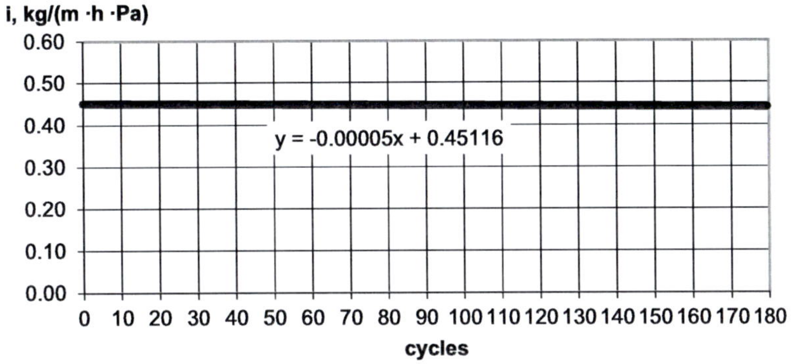

Fig. 4 Cyclic impact dependence of breathability coefficient of the ROCKWOOL mineral wool products of Light Batts (Rockmin) brand

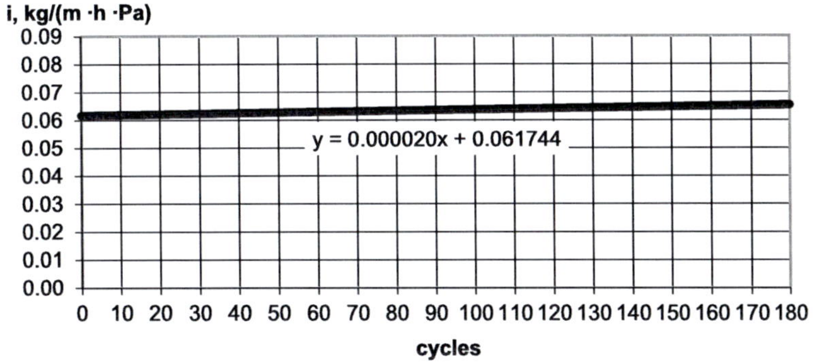

Fig. 5 Cyclic impact dependence on breathability coefficient of the ROCKWOOL mineral wool products of Ruth Batts (Dachrock max) brand

Resource rate determined by the "Eg. (17)" [10] is:

for Light Batts (Rockmin) brand—$r = -0{,}009$.
for Ruth Batts (Dachrock max) brand—$r = 0{,}0036$.
Fulfillment of condition is checked by the formula "Eg. (24)"[10]:
for Light Batts (Rockmin) brand:

$$\frac{r}{i_0}k_z = \frac{-0,009}{0,4511} \cdot 5 = -0,1 \leq 0,35 \tag{7}$$

for Ruth Batts (Dachrock max) brand:

$$\frac{r}{i_0}k_z = \frac{0,0036}{0,0617} \cdot 5 = 0,29 \leq 0,35 \tag{8}$$

The results of compressive strength test with 10% deformation depending on the number of cycles of freezing—thawing—heating of the ROCKWOOL mineral wool of Light Batts (Rockmin) and Ruth Batts (Dachrock max) brands are shown in Figs. 6 and 7 respectively.

The compressive strength with 10% deformation of the ROCKWOOL mineral wool dependence on the number of cycles of freezing—thawing—heating is approximated in the formulas:

for Light Batts (Rockmin) brand:

$$\sigma^{10}(z) = 0,0004878 - 0,0000001 \cdot z \tag{9}$$

for Ruth Batts (Dachrock max) brand:

$$\sigma^{10}(z) = 0,034 - 0,000011 \cdot z \tag{10}$$

Resource rate determined by the "Eg. (17)" [10] is:

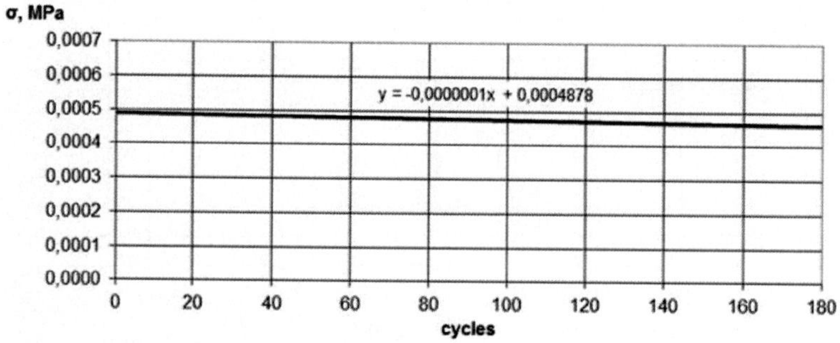

Fig. 6 Cyclic impact dependence on compressive strength with 10% deformation of the ROCKWOOL mineral wool products of Light Batts (Rockmin) brand

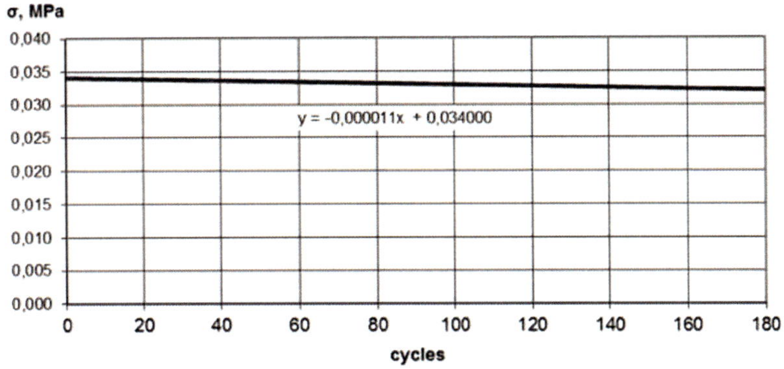

Fig. 7 Cyclic impact dependence on compressive strength with 10% deformation of the ROCK-WOOL mineral wool products of and Ruth Batts (Dachrock max) brand

for Light Batts (Rockmin) brand—$r = -0{,}0018$.
for Ruth Batts (Dachrock max) brand—$r = -0{,}00{,}198$.
Fulfillment of condition is checked by the "Eg. (25)" [10]:
for Light Batts (Rockmin) brand:

$$\left|\frac{r}{\sigma_0^{10}}k_z\right| = \left|\frac{-0,000018}{0,0004878}\cdot 5\right| = |-0,18| = 0,18 \leq 0,35 \tag{11}$$

for Ruth Batts (Dachrock max) brand:

$$\left|\frac{r}{\sigma_0^{10}}k_z\right| = \left|\frac{-0,00198}{0,034}\cdot 5\right| = |-0,29| = 0,29 \leq 0,35 \tag{12}$$

Thus, the conditions by the "Eg. (23), (24) and (25)" [10] are fulfilled.

Summary data on the results of testing the efficient operation period of the ROCK-WOOL mineral wool of Light Batts (Rockmin) and Ruth Batts (Dachrock max) brands are listed in Table 1.

Table 1 Results of testing the material efficient operation period

Brand	Density, kg/m^3	$\frac{r}{\lambda_0}k_z$	$\frac{r}{i_0}k_z$	$\frac{r}{\sigma_0^{10}}k_z$	Efficient operation period
Light Batts (Rockmin)	37	−0,23	−0,1	0,18	Not less than 100 years
Ruth Batts (Dachrock max)	145	0,05	0,29	0,29	Not less than 100 years

Stability of the Rockwool mineral wool performance figure to the effects of climatic moisture. The test results of thermal conductivity and changes of geometrical parameters of the Rockwool mineral wool samples by simulating the impact of climatic factors are listed in Appendix A.

As a result of the tests a criterion of thermal conductivity change was calculated for each sample by the "Eg. (26)" [10] the greatest value of which for each mode is presented in Table 2.

According to the results of the tests it is specified that the requirements of criterion "Eg. (26)" [10] for stability of thermal conductivity under cyclic wetting–drying effects that simulate conditions of the thermal insulating products in the course of the probable failure of building envelope in the process of upkeep of buildings with the corresponding soaking of thermal protection by liquid water and its natural drying are performed.

Stability of the Rockwool mineral wool performance figures to the effects of solar radiation and fog. As a result of the tests carried out in order determined in point [10] a criterion of thermal conductivity change was calculated for each sample by the "Eg. (26)" [10] the greatest value of which for testing mode is presented in Table 3.

Table 2 Results of testing the Rockwool mineral wool for resistance to climatic moisture

Brand	Drying temperature, °C	The greatest criterion	Specified specification, no more than	Conformity
Light Batts (Rockmin)	+20	0,006	0,1	+
	0	0,002		+
	−10	0,012		+
Ruth Batts (Dachrock max)	+20	0,012		+
	0	0,032		+
	−10	0,017		+

Table 3 Results of tests on mineral wool Rockwool resistance to climatic moisture

Brand	Average thermal conductivity in initial conditions, W/(m K)	Average thermal conductivity after climate effects, W/(m K)	The greatest value of criterion	Normative characteristic not more than	Conformity
Light Batts (Rockmin)	0,0386 ± 0,0009	0,0393 ± 0,0006	0,058	0,1	+
Ruth Batts (Dachrock max)	0,0395 ± 0,0004	0,0409 ± 0,0003	0,054		+

The received experimental data show that the requirements of criterion “Eg. (26)” [10] for stability of thermal conductivity under cyclic effects of radiation and high air humidity simulating condition of thermal insulating products during the probable failure of building envelopes in the course of operation of construction projects with related direct effect of solar radiation and humid air on the thermal insulating layer are carried out.

Thus, the stability of thermophysical characteristics of the tested products under conditions not foreseen for the thermal insolating mineral wool products in building envelope but which may possibly occur in the course of building operation and are defined as failures that can be repaired, is high enough for the established evaluation criteria.

3 Conclusion

The researches of identification of life characteristics of the Rockwool mineral wool products of Batts Lights (Rockmin) and Ruth Batts (Dachrock max) brands for the use of products as the thermal insulating layer of building envelopes with the life time up to 100 years were conducted. The researches were conducted with simulation of climatic impacts on the thermal insulating products under conditions of normal use of products in building envelopes and under conditions of probable failures of building envelopes. The evaluation of life time was carried out by the nature of changes of the thermal insolating product performance criteria—air permeability, thermal conductivity, compressive strength at 10% linear deformation and changes of appearance and geometric parameters of the samples under cyclic climatic impacts and in terms of possible damages of building envelope during operation of buildings.

According to the results it is found that the stability criteria of physical parameters of the Rockwool mineral wool product of Batts Lights (Rockmin) and Ruth Batts (Dachrock max) brands after the complex of cyclic climatic impacts are within acceptable limits, which allows to assess the conditional life time of the mentioned Rockwool mineral wool products for not less than 100 years.

References

1. DBN V.2.6–31 Construction of buildings and structures. Thermal insulating of buildings
2. DSTU B EN 1296:2011 Materials for coverings and rolls, sheets and sheets at a glance. The method of piece old for completing the tempered temperature (EN 1296: 2000, IDT)
3. ISO 11561 Ageing of thermal insulating materials – Determination of the long-term change in thermal resistance of closed-cell plastics (accelerated laboratory test methods)
4. DSTU B V.2.6–37 Construction of buildings and structures. Methods for determination of the building envelope and their components air permeabilty values in laboratory environment
5. DSTU-N B V.2.6–88 Construction of buildings and structures. Construction of the outer walls with facade thermal insulating. Guidance on technical testing, technical inspection and monitoring

6. DSTU B V.2.7–182 Building materials. Methods of period determination for effective operation and thermal conductivity of construction insulating materials at rated and standard duties
7. DSTU B V.2.7–38 (GOST 17177–94) Building materials. Building insulation materials and products. Test methods
8. DSTU B V.2.7–105 (GOST 7076–99) Construction materials and products. Method of determination of thermal conductivity and thermal resistance at steady thermal mode
9. DSTU B V.2.6–36 Construction of the outer walls with facade thermal insulating and plaster finishing. General specifications
10. Leshchenko MV, Semko V (2015) Thermal characteristics of the external walling made of cold-formed steel studs and polystyrene concrete. Mag Civil Eng 60(8):44–55
11. Farenyuk G, Oleksiienko O (2020) Development of methods for determining the term of effective exploitation of thermal insulation materials for 100 years. Actual problems of engineering mechanics. Key Eng Mater 864:80–92

Investigation of the Impact of Organizational and Technological Factors on Construction Processes in the Construction of High-Rise Monolithic Reinforced Concrete Buildings

Sahib Farzaliyev and Roman Pahomov

Abstract The article is devoted to the analysis of the main features of organizational and technological solutions to improve the efficiency of construction of high-rise monolithic reinforced concrete buildings, their timely commissioning, to enhance the organization of production, to reduce the cost of construction products.

In order to study the factors that influence organizational and technological solutions in the construction of high-rise monolithic reinforced concrete buildings; the influence force and nature of factors, affecting the feasibility of the study, were determined by using the method of mathematical and statistical analysis of Kolmogorov's theory. Usually, it is more useful to define the parameters of the model for small sample sizes by using the criterion of Kolmogorov and to give recommendations for improving these parameters.

Therefore, in order to improve the efficiency of organizational and technological solutions in the construction of monolithic reinforced concrete buildings, it is necessary to determine the factors affecting its production and prepare scientifically justified recommendations.

Keywords High-rise monolithic reinforced concrete works · Organizational and technological factors · Kolmogorov's statistics · Distribution hypothesis · Cost · Efficiency

1 Introduction

In the construction industry, along with the architectural volume-planning and constructive solutions of monolithic reinforced concrete frame buildings, features

S. Farzaliyev (✉)
Azerbaijan University of Architecture and Construction, 5 Ayna Sultanova, Baku, Azerbaijan

R. Pahomov
National University «Yuri Kondratyuk Poltava Polytechnic»,
Pershotravnevyj Avenue 24, Poltava 36011, Ukraine

V. Onyshchenko et al. (eds.), *Proceedings of the 3rd International Conference on Building Innovations*, Lecture Notes in Civil Engineering 181,
https://doi.org/10.1007/978-3-030-85043-2_10

such as high earthquake resistance and durability are increasing its development day by day. The demand for monolithic reinforced concrete buildings also necessitates solving organizational and technological issues aimed at improving its efficiency. Issues such as reducing labor costs and the cost of monolithic work also make these solutions relevant [8].

One of the indicators of improving the efficiency of construction production is the technological, possible implementation and organization of the constructive solution. The solution of these issues is one of the factors affecting the increase in labor costs and determining the economic efficiency of construction with monolithic reinforced concrete structures.

2 Main Part

2.1 Purpose of the Article

The main purpose of the article is to determine the factors that have the greatest impact on technical and economic indicators, using the method of mathematical and statistical analysis of the Kolmogorov theory to improve the efficiency of the construction of high-rise monolithic reinforced concrete buildings and to make recommendations for improving these parameters.

2.2 Research Methodology

The methodological basis of the research is a systematic approach aimed at studying and solving the works of domestic and foreign scientists dealing with the problems of improving the efficiency of construction of high-rise monolithic reinforced concrete buildings.

During the research, methodological guidelines of leading scientific institutions dealing with the problem of improving the efficiency of the construction of monolithic reinforced concrete buildings were used, as well as actual materials of a number of construction contractors performing large volumes of monolithic reinforced concrete works.

The research methodology used was the method of expert assessments, probability theory and mathematical statistics, correlation and regression analysis, graph-analytical method, methods of economic and mathematical modeling.

2.3 Results

Many scientists have contributed to improving the efficiency of organizational and technological solutions for monolithic reinforced concrete works [1, 2, 4, 5]. However, the results of these scientific analyses show that these recommendations are not enough to improve the efficiency of organizational and technological solutions for monolithic reinforced concrete works. Therefore, to improve the effectiveness of organizational and technological solutions, it is necessary to identify factors that affect this, and develop scientifically based recommendations.

Although the impact on the construction process varies, the joint consideration of organizational and technological factors has proven its value in practice. The construction processes of high-rise monolithic reinforced concrete frame buildings are complex and are influenced by a number of organizational and technological factors such as usage of the automated systems, frequent updates of the formwork systems, the attraction of new machines, and air and climate conditions. This shows that the assessment of these factors affecting the solution of organizational and technological issues of monolithic reinforced concrete works is a very difficult and complex process. Therefore, there is a need for analysis in various areas to identify these aspects that affect organizational and technological factors [3, 6].

For this purpose, a number of factors influencing organizational and technological solutions were investigated on the basis of data from ArchiCo LLC, which builds social facilities for IDPs in the 12th micro-district of Sumgayit.

As a result of the research, it was found that a group of factors in the construction industry influences organizational and technological decisions in the construction of high-rise reinforced concrete buildings (Table 1).

The goal of this study is to determine the level of influence of these factors on the effectiveness of decisions made in the organization and technology of work in the construction of monolithic reinforced concrete buildings. As usual, the methods of mathematical statistics and correlation and regression analysis are used for this purpose [9].

The first task of mathematical statistics is to show the collection and grouping of statistical data obtained from observations and special experiments, and the second is to develop the methods for analyzing statistical data in accordance with the goals of the study [9].

This includes:

1. Estimation of unknown probabilities of events, unknown distribution function, distribution parameters and random quantity dependent on one or more random quantities.
2. Testing statistical hypotheses about an unknown type of distribution law or distribution parameters whose type is unknown.

Thus, the task of mathematical statistics is to create methods for collecting and processing statistical indicators to obtain scientific and practical results.

Table 1 Group of factors affecting organizational and technological solutions

Technological complexity of the object	1. Volume of monolithic reinforced concrete works of the building
	2. Object configuration
	3. Floor of the building
	4. Characteristics of constructive solutions
	5. Volume of precast concrete structures
	6. Share of grouped works
	7. Volume of monolithic reinforced concrete structures made with the use of individual molds
Volume of monolithic reinforced concrete structures	
The area where the object is located	8. Natural construction conditions
	9. Non-natural construction conditions
	10. Removal of the object from the construction base
	11. The impact of climatic conditions on construction
	12. Demand for construction products in the area
Type of construction	13. Innovation of the project solution
	14. Level of different types of products
	15. Level of product with different options
Level of mechanization of construction	16. Level of grouping of the mold system
	17. Level of consolidation of construction machinery
	18. Level of use of construction machinery
Organization of construction and labor management	19. The level of specialization of the main processes
	20. Systematization of work on processes
	21. The level of application of the method of flow
	22. The level of application of supervisory control
Level of use of employees	23. Complectation level of labor resources
	24. The rate of use of working time
	25. The level of compliance of the employee with the production norm

The purpose of our study is to determine the nature of factors affecting organizational and technological solutions in the construction of high-rise monolithic reinforced concrete buildings and their role in improving its efficiency.

If the distribution law is unknown, but there are reasons to assume that it belongs to a certain type (we denote it by A), then the null A hypothesis is checked that the total set is distributed according to the law A. The test of the hypothesis under consideration is carried out using the criterion of agreement of a specially selected random variable. The consensus criterion refers to a hypothetical statistical model and statistical criteria designed to test experimental data that is affected by this model.

The following consent criteria are commonly used in statistical studies:

Criteria of Pearson, Kolmogorov, Smirnov [11]. It is more appropriate to use the Kolmogorov criterion to test the hypothesis of the normal distribution of the set of heads, when not large sample sizes and model parameters are known.

Our goal is to determine the degree of influence of these factors on a particular indicator, which can be a measure of the effectiveness of decisions made in the organization and technology of their work in the construction of monolithic reinforced concrete buildings. For this purpose, methods of mathematical statistics and correlation and regression analysis are often used [10–12].

Modern mathematical statistics can be defined (accepted) as the science of decision-making under conditions of uncertainty. The determination of the regularities of mass random events is based on the study of the results of statistical observation based on the methods of probability theory. The use of mathematical and statistical analysis allows us to identify the strength and nature of the dependence of technical and economic indicators on various factors, on their basis to develop measures and practical recommendations for improving these indicators.

The first task of mathematical statistics is to show the accumulation and grouping of statistical data obtained as a result of observations and specially designed experiments, and the second is to develop methods for analyzing statistical data in accordance with the objectives of the study [11]. This includes:

Estimation of unknown probabilities of events, distribution functions, distribution parameters, estimation of the dependence of a random variable, the type of which is known, on one or more random variables.

2. Testing hypotheses about the distribution law, the type of which is unknown, or about the prices of the distribution parameters, the type of which is known.

Thus, the task of mathematical statistics is to determine the methods of collecting statistical indicators and processing them to obtain scientific and practical results.

The purpose of our research work is to determine the nature of the studied factors and their impact on the organizational and technological solutions of construction processes used in the construction of high-rise monolithic reinforced concrete buildings. If the distribution law is unknown, but there is reason to believe that it belongs to a certain type (let's denote it by a), then check the null hypothesis a that the total number is distributed according to the law A. The test of the hypothesis in question is carried out using a specially selected random variable—the criterion of agreement.

The coherence criterion is a statistical criterion designed to test a hypothetical statistical model and the experimental data described by this model.

In statistical studies, the following consent criteria are usually used: $\chi 2$, the Pearson, Kolmogorov, and Smirnov criteria [9]. It is considered appropriate to use the Kolmogorov test to test the hypothesis of the normal distribution of the set of heads, when not large sample sizes and model parameters are known. It should be borne in mind that the Kolmogorov criterion does not prove the correctness of the hypothesis, like any other criterion. It is only at a perceived level that it determines the importance of agreeing with observational data.

The research paper assumes that the experimental data express a random sample. In this case, the theoretical model describes the distribution law. Speaking about the law of theoretical distribution, it should be noted that in this law it is necessary to distinguish between simple and complex hypotheses. A simple hypothesis shows a certain probability law (probability distribution) that determines the sample estimates. A complex hypothesis, however, involves not a single distribution, but any set of distributions (for example, a parametric family).

In the case of a simple hypothesis, the measured data becomes numbers, in other words, one-dimensional random variables are considered. The distribution of one-dimensional random variables can be fully described by their distribution functions. In the case of a complex hypothesis, it is checked whether the sample under study obeys a certain parametric distribution law (for example, the law of normal distribution). At the same time, the parameters of the distribution law remain undefined.

As mentioned above, the Kolmogorov's statistic, usually denoted by D, is used in the literature to determine the distribution characteristics of the set of indicators characterizing the factors under study and the determinant [7].

Let's assume that any n-dimensional sample is taken and denote the real distribution function, which observations obey via G (x), and the hypothetical distribution function—via F (x). Then the hypothesis H that the real distribution function is F (x) can be written as follows:

$$H : G(\cdot) = F(\cdot) \tag{1}$$

The hypothesis H can be tested in such a way: if H is true, then there will be some similarity in the distribution functions Fn and F, and the difference between them will be decreased with increasing n.

To express the similarity between these distribution functions, it is possible to use any distance between them. For example, F_n and F_i can be compared using the following quantity, called regular metrics.:

$$D_n = \sup_{-\infty < x < \infty} |F_n(x) - F(x)| \tag{2}$$

The following algorithm is used to test hypotheses:

To use Kolmogorov's statistics to test the simple hypothesis H: G = F in the original sample, the value of the D_n statistics is necessary and can be calculated by the formula, shown below:

$$D_n = \underset{1 \le k \le n}{mak} \left[\frac{k}{n} - F\left(x_{(k)}\right),\ F\left(x_{(k)}\right) - \frac{k-1}{n} \right] \tag{3}$$

Here $x_1, x_2, \ldots x_n$ are the elements of a series of variations constructed by initial selection. The a D_n quantity obtained is compared with the crisis values taken from the corresponding tables. If the value of D_n obtained in practice exceeds the value of the crisis taken (within the selected significance level), then the H hypothesis is refuted.

In a normal distribution, most values of a random variable are grouped symmetrically relative to mathematical expectations. This corresponds to the nature of measurements in which gross errors or omissions are less likely. Based on this assumption, a formula was defined for finding the required amount of statistical data in mathematical statistics:

$$n = \frac{t^2 \cdot \sigma^2}{\delta^2} \tag{4}$$

here; t is found from the ratio F (t) = v/2; (F (t) is a Laplace function, the table value),
ν—level of reliability;
σ—a priori studied random variable, i.e. the standard deviation known before experimental measurements;
δ—the given accuracy.

$t = 2$ and $\delta = 0.05$ are sufficient values to solve the problem. Based on the analysis of data for similar samples, it is assumed that $\sigma = 0.173$. Taking into account the accepted level of reliability the required amount of statistical data in this case will be as follows:

$$n = \left(2^2 x\ 0, 173\right)/0, 05^2 = 47, 9$$

Rounding this value assumes that n = 50.

Thus, enough information is collected to analyze the studied factor to ensure the required degree of reliability of the results to be calculated.

The interval sequence of cost allocation for monolithic reinforced concrete works is shown in Table 2 below.

When studying the indicator that determines the specific cost of monolithic reinforced concrete works, the calculation found a statistical value of Kolmogorov $D_{50} = 0.017$. At the significance level of 1% of the table indicator, the statistical value is $D_{table} = 0.224$. When $D_n < D_{table}$, we can conclude that the initially studied indicators agree with the accepted hypothesis of zero relative to the normal nature of the distribution.

Figure 1 below shows a graph of the density of the distribution function of the studied indicators of the unit cost of monolithic reinforced concrete works.

Thus, the density graph of the distribution function of the studied indicators of the unit cost of monolithic reinforced concrete works shows that the number of rows that quantify the studied indicators is described in accordance with the law with the normal distribution curve. This fact confirms that a number of studied organizational and

Table 2 The interval of value cost allocation

The sequence number of intervals	Value range	The number of intervals	Frequency of occurrence (density)
1	500	2000–2500	1
2	500	2500–3000	1
3	500	3000–3500	2
4	500	3500–4000	2
5	500	4000–4500	4
6	500	4500–5000	6
7	500	5000–5500	8
8	500	5500–6000	7
9	500	6000–6500	7
10	500	6500–7000	5
11	500	7000–7500	3
12	500	7500–8000	2
13	500	8000–8500	1
14	500	8500–9000	1
Total			**50**

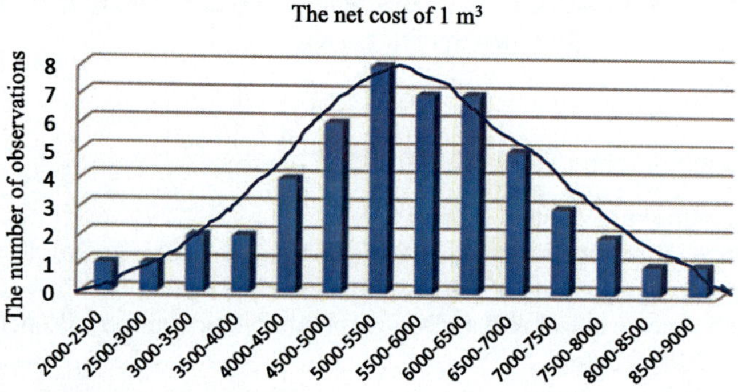

Fig. 1 Density Graph of the distribution function of the studied indicators of the unit cost of monolithic reinforced concrete works

technological factors affect the efficiency of monolithic reinforced concrete works and the unit cost.

2.4 Scientific Novelty

As a result of the research, the most significant factors affecting the organizational and technological solutions of construction processes in the construction of monolithic reinforced concrete buildings were identified and classified. The dependence of the influence of these factors on the effectiveness of organizational and technological solutions is established. A graphoanalytic model is developed to determine the zones of effective action of factors, based on the identification of patterns of influence of factors that affect the effectiveness of organizational and technological solutions in the construction of monolithic reinforced concrete buildings.

2.5 Practical Importance

Research results can be used in the development of concepts of perspective urban planning, organizational and technological solutions for the construction of high-rise monolithic reinforced concrete buildings in the creation of cheap housing market.

3 Conclusions

The analysis of factors affecting the effectiveness of organizational and technological solutions of construction processes in the construction of high-rise monolithic reinforced concrete frame buildings was carried out on the basis of data from LLC "Archiko". It was found out that to ensure the degree of reliability of factors affecting the effectiveness of organizational and technological solutions, approximately n = 50 data indicators are required.

References

1. Ataev SS (1988) Tehnologiya industrialnogo stroitelstva iz monolitnogo betona. Stroyizdat, p 255
2. Afanasev AA (1990) Vozvedenie zdaniy i sooruzheniy iz monolitnogo zhelezobetona. Stroyizdat, p 368
3. Granov GS, Safarov GM, Tagirbekov KP (2001) Ekonomiko-matematicheskoe modelirovanie v reshenii organizatsionno-upravlencheskih zadach v stroitelstve. ASV, p 325
4. Tsay TN, Shirshikov BF, Baetov BI (1987) Otsenka organizatsionno-tehnicheskogo urovnya podgotovki stroitelnogo proizvodstva. Ekonomika stroitelstva 8:59–61
5. Shchukina MN (2007) Modern high-rise construction (440 p). Monograph. SOP "ETC 205 Moscomarchitecture", p 206
6. Mordich AI, Vigdorchic PI, Belevich VN (1999) New universal frame system of 210 high-rise buildings. Conc Reinforced Conc 1:2–4

7. Farzaliyev SA (2018) İmproving the efficiency of construction of high-rise monolithic reinforced concrete buildings with the introduction of innovative technologies and technologies. Int J Eng Technol 23
8. Gmurman VE (1997) Teoriya veroyatnostey i matematicheskaya statistika: Ucheb. posobie dlya vuzov. Infra, p 302
9. Berezhnaya EV, Berezhnoy VI Matematicheskie metodyi modelirovaniya ekonomicheskih sistem. Finansyi i statistika, p 324
10. Barkalov SA, Mescheryakova OK, KolpachYov VN, Kurochka PN (2002) Osnovyi nauchnyih issledovaniy po organizatsii i upravleniyu stroitelnyim proizvodstvom. Gosudarstvennyiy arhitekturno-stroitelnyiy universitet, Voronezh, p 416
11. Drobyazko S, Bondarevska O, Klymenko D, Pletenetska S, Pylypenko O (2019) Model for forming of optimal credit portfolio of commercial bank. J Manag Inf Decis Sci 22(4):501–506
12. Iasechko S, Haliantych MK, Skomorovskyi VB, Zadorozhnyi V, Obryvkina O, Pohrebniak O (2020) Contractual relations in the information sphere. Syst Rev Pharm 11(8):301–303. https://doi.org/10.31838/srp.2020.8.46

Desalınatıon of Salt Water by Chemıcal Method

Gulnar Feyziyeva and Olena Stepova

Abstract Ways to improve the efficiency of chemical methods of water desalination process by analyzing anion was determined that the first step to further expand the scope of application of this method anionit filter regeneration NaOH instead of cyanide, which is much cheaper than the Ca (OH)2 will need to use cyanide. As it is known, when the highly acid cationites regenerate with acid, the best regeneration is Na, the good regeneration is Mg and the worst regeneration is Ca ions. The results of the raw water composition show that by using filters operating with the reverse flow principle, the first stage H-cationite filters at stoichiometric value or regenerating with an acid close to it and can achieve high exchange capacity. An effective technology for regeneration of anionite filters with lime solution has been developed and studied in AUAC. In this technology, wastewater is not produced and regeneration products are removed from the system in the form of a calcium sulfate suspension. Anionit filters waste with lime and H-cationite 3–5% solution of sulfuric acid regeneration technology developed filters. Alkali and acid regeneration technologies have been proposed stexiometrik private consumption is equal to the price.

Keywords Reverse osmosis · Salinity · Regeneration · İon exchange · Exchange capacity · Filter · Lime suspension · Desalination technology

1 Introduction

The production of desalinated water for thermal power plants (TPPs) can be carried out using either chemical, reverse osmosis or thermal methods. The question of which of the above methods to desalinate raw water for these TPPs depends on mainly the salinity of the raw water, ionic composition and the required quality of desalinated water. When the salinity of raw water is up to 0.6–0.7 g/dm^3, according

G. Feyziyeva (✉)
Azerbaijan University of Architecture and Construction, Ayna Sultanova 11, Baku, Azerbaijan

O. Stepova
National University Yuri Kondratyuk Poltava Polytechnic, Poltava, Ukraine

V. Onyshchenko et al. (eds.), *Proceedings of the 3rd International Conference on Building Innovations*, Lecture Notes in Civil Engineering 181,
https://doi.org/10.1007/978-3-030-85043-2_11

to current norms, economically, the chemical method is used. If the concentration of salts in the primary water is higher than the above value, reverse osmosis or thermal methods are used. The amount of acid and alkali used by the traditional method to purify water from chemical salt is 2–3 and 1.8–2.5 times higher than stoichiometric amounts, respectively. Although the consumption of reagents is much higher than the theoretical amount for the regeneration of ion exchangers, only some of their total exchange capacities are used. For example, although the total exchange capacity of strongly acidic KU-2–8 cationites is 1700 (g-eq)/m^3, the working exchange capacity in the H-cationization process is on average 450–500 (g-eq)/m^3, that is, this value covers 25–30% of the total exchange capacity [1, 2].

The chemical desalination technology of the water discussed in the article allows reducing the amount of acid and alkali used for regeneration of ionites up to a stoichiometric value, which means an average reduction of reagent consumption by 2–2.5 times. Also, due to the 1.5–2.5 times increase in the exchange capacity of ion exchangers, the decrease in ionite consumption in the considered productivity, as well as the number of filters used in the device, is significantly reduced. As a result, capital investment in the construction of a water treatment plant is sharply decreasing.

From the above, we can observe that it is possible to desalinate water with higher salinity rate (1500–2000 mg/dm^3) by using the technologies developed at the Azerbaijan University of Architecture and Construction (AUAC), by extending the area of use of chemical desalination. In order to further expand the usage area of chemical desalination of the water, the regeneration of the first stage anionite filters can be carried out with the most common $Ca(OH)_2$ alkali rather than NaOH alkali [1].

First stage anionite filters can be reproduced with a 0.5–1.5% lime suspension or lime solution. The disadvantages of regenerating anionite filters with lime suspension are the complexity of the regeneration process, the low utilization rate of exchange capacity of anionite, the deterioration of the filtrate quality due to contamination with calcium ions and the pollution of water basins with the treated regenerated solution of anionite filters.

The low solubility (44–46 mg-eq/dm^3) of lime in water allows it to effectively regenerated anion filters. When the anion filters are regenerated with lime solution, the water consumption for the preparation of the solution increases, a large amount of wastewater is formed and the water basins are contaminated.

Although the value of Ca(OH)2 is sharply lower than the value of NaOH, it is not used to replenish anionites for the above reasons that have arisen during the regeneration of anionite with Ca(OH)2. In order to regenerated the first stage anionite filters from Ca(OH)2 alkali, a new technology should be developed to overcome the shortcomings mentioned above.

2 Methods

An effective technology for regeneration of anionite filters with lime solution has been developed and studied in AUAC [1, 3]. In this technology, wastewater is not produced and regeneration products are removed from the system in the form of a calcium sulfate suspension.

The schematic diagram of the developed technology is shown in Fig. 1. The processed regeneration solution of (1) anionite filter operating in the sulfate ion capture mode accumulates into (2) tank and is drawn from there with (3) pump and passed through a lime suspension loaded device such as (4) saturator. Here, calcium sulfate collapses and the solution is saturated with calcium hydroxide. Then a solution of lime containing a certain amount of calcium sulfate is released from anionite at a speed of 10–15 m/h, where hydroxide ions turn into sulfate ions, that is, the regeneration of the anionite's filter takes place and the solution of calcium sulfate is obtained from the anionite. The same solution is discharged again from a lime suspension loaded device, where calcium sulfate precipitates and the solution saturates with calcium hydroxide, after that the solution sends to the anionite filter. Thus, the same solution circulates between the anionite filter and the lime suspension loaded device. In this case, sulfate ions are transferred from the anionite filter to the device loaded with lime suspension and separated in the form of calcium sulfate, OH-ions are transferred from the apparatus loaded with the lime suspension to the anionite filter and regenerate it. It is also possible to repair the circulating regeneration solution in ordinary clarifier [4].

Anionite filter (1) is washed with the water of the 1st stage H-cationite filter and the solid part and the diluted part are collected in the processed (2) regeneration tank of the anionite's filter and (5) tank, respectively. From there, it is transmitted to the primary water clarifier with the help of (6) pump.

The first stage anionite filter which is loaded with weak-based anionite can work in two modes. In the first case of the first stage anionite filter, mainly SO_4 ions are retained, and in the second case, both SO_4 and Cl ions are retained in the filter.

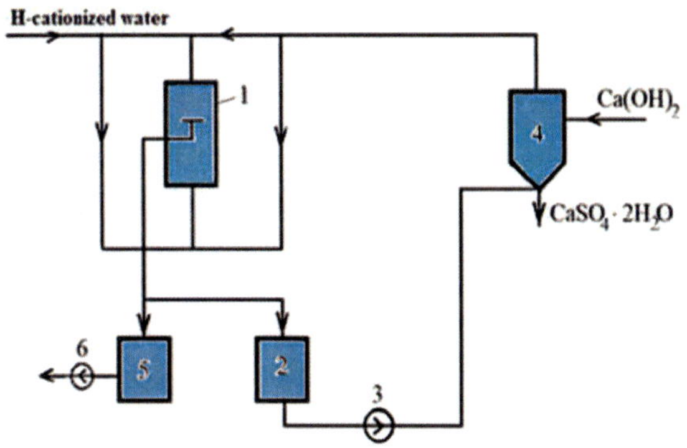

Fig. 1 Scheme of regeneration of anionite filters with lime solution

Since the SO_4 ions are primarily kept in the first stage anionite filter, Cl ions are retained in the second stage anionite filter which is loaded with high-base anionite. Since Silicate and CO_2 ions are also retained in the second stage anionite filters, it must be regenerated with NaOH alkali. As a result, NaOH alkali consumption for the regeneration of the second stage anionite's filter will be several times higher than regeneration of only silicate and CO_2 ions. In addition, due to the specific consumption of NaOH alkali for regeneration of the second stage anionite filter is several times higher than the stoichiometric amount, the processed regeneration product consists primarily of alkaline solution, which is sufficient for regenerating the first stage anionite filter for many types of water.

In case of keeping SO_4 and Cl ions together in the first stage anionite filter loaded with weak base anionite, it is necessary to release the sulfate salt solution from it and to be converted to SO_4 before regenerating it with the lime solution. As a sulfate salt solution, a neutral sulfate salt solution can be used, which is arisen by the regeneration of cationite filters with a sulfuric acid solution. Depending on the ionic types of salts in primary water and the regeneration regime of cationites, the processed regeneration solution cannot consist of a mixture of sodium and coarse salts.

The scheme of the proposed device (in "Shimal" Steam Power Plants) for the desalination of brine with a total salinity of 1702 mg/dm3 and ion content of Ca = 5,2; Mg = 11,6; Na = 9,36; Cl = 5,35; SO4 = 16,56; HCO3 = 4,25 mq-eq/dm3 is shown in Fig. 2.

The raw brine is mixed with the wash water from the regeneration solutions of the first and second stage anionite filters and passed through (1) mechanical filter, (2) first stage cationite and (3) anionite filters in top-down direction order to (4)

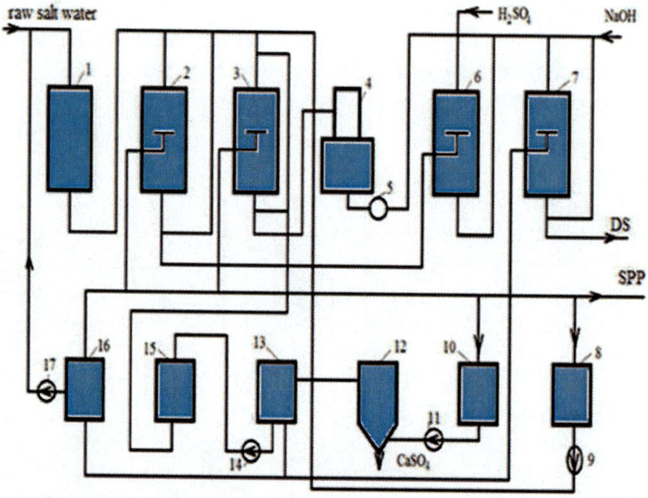

Fig. 2 The scheme of the desalination unit of saltwater by regenerating the first stage anionite filter with the lime solution

decarbonizer. Decarburized water is desalinated from by passing through (6) second stage cationite and (7) anionite filters in top-down direction with the help of (5) pump. Desalinated water (DW) is supplied to consumers. The first stage continues in accordance with the operating conditions of cationite and anionite filters until the Na- and Cl ions begin to pass to the filtrate [5].

As it is known, when the highly acid cationites regenerate with acid, the best regeneration is Na, the good regeneration is Mg and the worst regeneration is Ca ions [1]. Ca cations in the water whose ionic content is shown above make up only 20% of the total cations. Na- and Mg cations are 4 times more than Ca-cations and approximately 2 times more than Na cations.

The results of the raw water composition show that by using filters operating with the reverse flow principle, the first stage H-cationite filters" at stoichiometric value or regenerating with an acid close to it and can achieve high exchange capacity.

H-cationite filters are regenerated as shown below. Firstly, 2–3% sulfuric acid is taken from the middle drainage by releasing from above and below the second stage H-cationite filter, then it is also released from the first stage H-cationite filter in the same way. The solid portion of the processed regeneration product (SPP) is collected in (8) tank. In the next regenerations, the cationite layer, which is in the upper part of the middle drainage system of the first stage H-cationite filter, is regenerated with SPP taken from (8) tank and SPP is discarded. After that, H_2 and H_1-cationite filters are regenerated with 3–5% sulfuric acid solution as before.

Since calcium ions are kept with cationite, which is mainly located in the upper part of the middle drainage system during H-cationization and its contents are taken from (8) tank are almost completely transformed into Na and Mg form when it is regenerated with SPP as its content consists mainly of sodium and magnesium salts. In this case, the concentration of sulfuric acid solution can be increased by 4–5% or more. As a result, the metabolic capacity of cationite can be sharply increased.

To convert SO_4 and Cl ions held in the first stage anionite filter to SO_4 form completely it is necessary to be passed the salt solution which is accumulated in the tank (8) and which is the SPP of the first stage cationite filter consisting mainly of sodium and magnesium salts [6–9].

The first stage anionite filter is regenerated with lime solution in a straight reverse flow scheme. The solid part of the SPP is taken from the medium drainage system and collected into (10) tank by filtering through (12) device loaded with lime solution with the help of (11) SPP pump collected in (13) tank. Calcium sulfate collapses in the unit (12) and the solution is saturated with calcium hydroxide. Then, the lime solution containing some calcium sulfate is passed through (15) mechanical filter with the help of (14) pump and passed through it to regenerate the first stage anionite filter, so the cycle is repeated.

The second stage anionite filter is regenerated with NaOH alkali in a straight reverse flow scheme, the solid portion of the TPP is collected in (13) tank and the liquid portion in (16) tank and mixed with raw brine before mechanical filtration with the help of (17) pumps.

Thus, the area of usage of desalination by chemical method can be significantly extended using the scheme as shown in Fig. 2.

3 Conclusions

By analyzing the ways to increase the efficiency of anionization in the chemical desalination process of water, it was determined that the regeneration of the first stage anionite filter should be done by using $Ca(OH)_2$ alkali instead of NaOH alkali to further expand the area of usage of this method. Regeneration technologies have been developed with lime solution of anionite filters waste-free, and 3–5% sulfuric acid of H-cationite filters. In the proposed regeneration technologies, specific alkali and acid consumption are equal to stoichiometric values.

References

1. Feyziev H (2009) Highly efficient methods of water softening, desalination and demineralization. Takhsil CCI, Baku
2. Feyziev H, Feyziyeva G (2016) T: Desalination of sea and salt waters. Khazar, Baku
3. Kopylov A, Lavygin V, Ochkov V (2003) T: Water treatment in the energy sector. M: Publishing House MEI, Moscow
4. Feyziev H (1986) Method of regeneration of anionite filters of a chemical desalting plant. A.S. 1264966 USSR, MKI4S 02F//42//BOIS 49/00. Discoveries, Inventions. No. 39
5. Grosheva G, Jalilov M (1989) T: Industrial test of anion exchanger regeneration with lime solution. Energetic, No. 9
6. Feyziyeva G, Jalilov M, Jalilova A, Azimova M (2020) A new Technology to Prevent the Formation of Limescale in Hot Water Supply Systems. Trans Tech Publications Ltd., Switzerland
7. Onischenko VA, Soloviev VV, Chernenko LA, Malyshev VV, Bondus SN (2014) Acidic-basic interactions in tungstate melts based on tungsten electroplating out of them. Materialwiss Werkstofftech 45(11):1030–1038. https://doi.org/10.1002/mawe.201400222
8. Loburets AT, Naumovets AG, Senenko NB, Vedula YS (1997) Surface diffusion and phase transitions in strontium overlayers on W(112). Zeitschrift Fur Physikalische Chemie 202(1–2):75–85. https://doi.org/10.1524/zpch.1997.202.part_1_2.075
9. Onyshchenko VO, Filonych OM, Bunyakina NV, Senenko NB (2020) A new agent for removing concrete residues from the surfaces of polypropylene molds in the manufacture of paving slabs and its advantages https://doi.org/10.1007/978-3-030-42939-3_67

Research of Methods of Calculation of Thermal Characteristics of Enclosing Designs in Summer Conditions of Operation

Olena Filonenko, Gennadiy Farenyuk, Olexandr Semko, and Nataliia Mahas

Abstract The methods of calculation of heat resistance of enclosing constructions are analyzed in the work. The dynamic thermal performance of a building envelope describes the thermal behavior of a component when it is subjected to varying boundary conditions. The result of the calculation according to the national method is represented by values that do not allow to include them in the assessment of the overall energy efficiency of buildings. To verify the results of a theoretical study of the dynamics of the temperature regime of the enclosing structure, experimental measurements were carried out.

Keywords Heat transfer resistance · Heat loss · Thermal insulation

1 Introduction

The rooms in which people are located must provide a certain level of comfort. The state of comfort is a subjective feeling that occurs in people under the complex influence of such basic factors: acoustic environment, color, temperature, humidity, air mobility, and the like.

The main factors that form the microclimate of the premises are: temperature, speed of movement and air humidity, as well as radiation temperature, that is, the average temperature of the surfaces of enclosing structures and objects.

Outside air temperature fluctuations (the effect of solar radiation, daily change in the outside air temperature) put additional requirements on the temperature regime of the enclosing structures in the summer.

The performance of the enclosing structures largely depends on the magnitude of temperature fluctuations on their inner surface. With a significant amplitude of fluctuations in the temperature of the enclosing structure in the summer, there may be

O. Filonenko (✉) · O. Semko · N. Mahas
National University "Yuri Kondratyuk Poltava Polytechnic", Poltava, Ukraine

G. Farenyuk
State Enterprise "States Cientific-Research Institute of Building Constructions", Kiev, Ukraine

V. Onyshchenko et al. (eds.), *Proceedings of the 3rd International Conference on Building Innovations*, Lecture Notes in Civil Engineering 181,
https://doi.org/10.1007/978-3-030-85043-2_12

a periodic increase in the temperature of the internal air, followed by the return of heat into the house and overheating of its premises. To assess the comfort indicators of the microclimate in the room, the heat resistance of the inner surface of the enclosing structures is calculated.

The purpose of the calculation is to provide the enclosing structures with the necessary heat-shielding qualities that guarantee the maintenance of an almost constant temperature on the inner surface of the structure with periodic changes in the parameters of the external environment.

The influence of local influences on thermal sensations in non-uniform environmental conditions is considered in the works [1, 2].

The assessment of the climatic parameters of the indoor environment of the room is carried out by the reaction of a person. Models based on a multi-node model are being developed, taking into account the countercurrent heat exchange of blood, a detailed view factor has been added to calculate the radiation heat transfer of a human body to the environment, including the load of solar radiation on people. The model can calculate human thermal reactions in transient and inhomogeneous environments [3]. The study of the thermal parameters of multilayer structures is presented in the works [4–10]. When designing modern structures, the use of extreme principles is promising [11–15].

For a detailed study of the internal parameters of the premises, it is necessary to clarify the methods for studying the thermal characteristics of the enclosing structures.

The purpose of the article is to compare various methods for calculating the dynamic thermal characteristics of the building envelope and experimental measurements.

Research methods are based on engineering methods for determining thermal characteristics.

2 Theoretical Study of the Dynamic Thermal Characteristics of the Enclosing Structures

The dynamic thermal performance of the building envelope describes the thermal behavior of a component when it is subjected to variable boundary conditions.

Dynamic thermal performance is used to calculate the indoor temperature of a room, heating or cooling equipment, and the effects of intermittent heating or cooling.

According to UNE-EN ISO 13,786: 2011 [8] dynamic thermal characteristics of any component:

The periodic thermal conductivity L_{mn} is a complex number that shows the relationship between periodic heat flux and periodic temperature in sinusoidal conditions. The value is determined by the formula:

$$\widehat{\Phi}_m = L_{mm}\widehat{\theta}_m - L_{mn}\widehat{\theta}_n \tag{1}$$

where $\widehat{\theta}_{m,n}$ and $\widehat{\Phi}_m$ are complex amplitudes of temperature and heat flow.

The heat capacity is the modulus of the ratio of periodic thermal conductivity and angular frequency. The value is determined by the formula

$$C_m = \frac{1}{\omega}|L_{mm} - L_{mn}| \tag{2}$$

where $\omega = 2\pi/T$ and T is the period of variation in seconds.

In Ukrainian standards, the influence of fluctuations in the outside air temperature on the temperature regime of the enclosing structures in the summer is taken into account by the following indicators:

- the amplitude of fluctuations in the outside air temperature $A_{t зроз}$, °C is the maximum temperature deviation on the inner surface of the opaque enclosing structure from the average daily value when exposed to solar radiation in summer operating conditions;
- the value of attenuation of the amplitude of fluctuations in the temperature of the outside air ν;
- the coefficient of heat assimilation of the material of individual layers s, W/(m^2 K) - this is a physical parameter that reflects the ability of the material to perceive heat when the temperature fluctuates on its surface. It is determined by the ratio of the amplitude of heat flux fluctuations (W) to the amplitude of temperature fluctuations (K) on a unit surface area of the material (m^2) over a period of 24 h.

Method of calculation according to UNE-EN ISO 13,786: 2017 [16]:

The method given in this International Standard is based on heat conduction in building components composed of several plane, parallel, homogeneous layers, under regular sinusoidal boundary conditions and one dimensional heat flow.

That means that at any location in the component, the temperature variations can be modeled by

$$\theta_n(x,t) = \overline{\theta}(x) + \frac{\widehat{\theta}_{+n}(x)e^{j\omega t} + \widehat{\theta}_{-n}(x)e^{-j\omega t}}{2} \tag{3}$$

and the variations of the density of heat flow rate are

$$q_n(x,t) = \overline{q}(x) + \frac{\widehat{q}_{+n}(x)e^{j\omega t} + \widehat{q}_{-n}(x)e^{-j\omega t}}{2} \tag{4}$$

With $\widehat{\theta}_{\pm}(x) = \left|\overline{\theta}(x)\right|e^{\pm j\psi}$ and $\widehat{q}_{\pm}(x) = \left|\widehat{q}(x)\right|e^{\pm j\varphi}$.

Temperature and density of heat flow rate variations are those around the mean values *0* and *cj* of these variables, which are linked by

$$\overline{q} = U\left(\overline{\theta}_i - \overline{\theta}_e\right) \tag{5}$$

where U is the thermal transmittance of the component.

3 Methodological Aspects of Experimental Research

To check the results of a theoretical study of the dynamics of the temperature regime of the enclosing structure, experimental measurements were carried out. The essence of the methods for the experimental determination of the heat resistance of enclosing structures is to find the dynamics of changes in the temperature regime of the inner surface of the enclosing structure according to the results of thermal measurements in the summer period.

In natural conditions, tests were carried out in the summer period of the year, with a clear sky, when thermal conditions close to the calculated ones are established.

The purpose of experimental research:

when testing opaque enclosing structures, determine the amplitude of fluctuations in the temperature of the inner surface;

to verify theoretical studies.

Experimental studies were performed according to the method of DSTU V.2.6–100: 2010.

The study measured:

- temperatures of internal and external surfaces of a protection;
- indoor and outdoor air temperatures;
- humidity of internal and external air;
- heat flow through the wall.

Registration and display of controlled parameters (heat fluxes, temperatures and humidity) with reference to time and date was performed by the heat flux density and temperature meter.

The study was conducted from 07/27/2015 to 08/21/2015, as the hottest period of 2015. The object of research is an external solid brick wall with a thickness of 510 mm with an internal lime-sand plaster with a thickness of 20 mm.

The tests were carried out on the building envelope in use. Tests of the wall in the intermediate floor room with the orientation of the external enclosing structure to the south-west and the absence of shading by the surrounding buildings and trees.

The results of measuring the temperature of the indoor and outdoor air, the temperature on the inner surface of the structure and the heat flow during the calculated day are shown in Fig. 1. The values were recorded every 10 min.

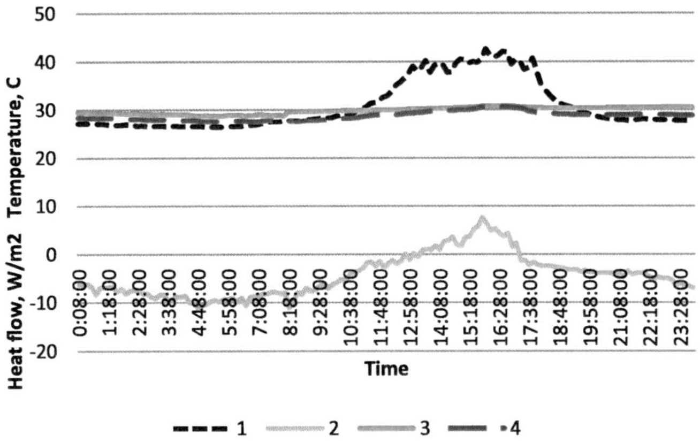

Fig. 1 The graph of the temperature distribution of the internal (4) and external (1) air, the temperature of the internal surface of the enclosure (3) and the heat flow (2) for the calculated day

4 Comparison of Theoretical Calculation Results and Experimental Data

To compare the theoretical and experimental studies, the dynamic parameters were calculated during the day with the initial data according to the measured values (Tables 1 and 2).

A comparison of the measured and calculated parameters is shown in the graph in Fig. 2. The result of statistical processing of the values measured in experimental conditions allows to accept them for the subsequent comparison with the calculated data (Fig. 3).

Table 1 Initial data for calculating dynamic parameters

Layer name	Thermal conductivity [W/m.K]	Gross density [kg/m^3]	Spec. heat capacity C [J/kg.K]	Heat absorption coefficient s, [W/(m^2•K)]	Layer thickness d [m]	R [m^{2}K/W]
Rsi						0,13
Plaster	0,810	1600,0	840	9,76	0,0200	0,025
Brickwork	0,810	1800,0	880	10,12	0,5100	0,630
Rse						0,04
		U-value:	1,2131		W/m^{2}K	
		Total thickness:	0,530		m	

Table 2 The result of the theoretical calculation of dynamic parameters

№	Name quantities	Value	Units
1	External thermal admittance:	7,418	W/(m^{2}K)
2	Time shift external side:	2,19	h
3	Internal thermal admittance:	4,563	W/(m^{2}K)
4	Time shift internal side:	1,31	h
5	Periodic thermal transmittance:	0,082	W/(m^{2}K)
6	Time shift periodic thermal transmittance:	−16,48	h
7	External areal heat capacity:	101,809	kJ/(m^{2}K)
8	Internal areal heat capacity	62,811	kJ/(m^{2}K)
9	Decrement factor f:	0,068	

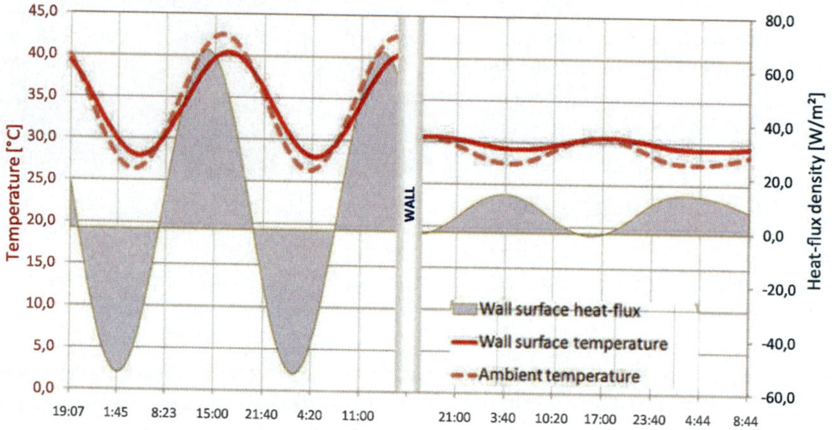

Fig. 2 The result of the theoretical calculation of the dynamic parameters of the wall

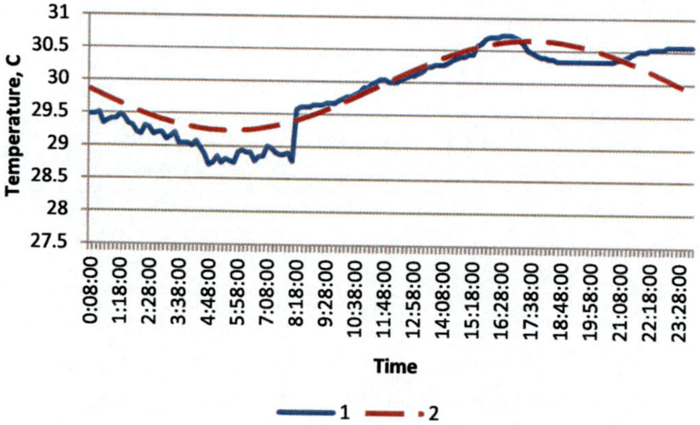

Fig. 3 The result of the calculation (2) and measurement (1) of the temperature on the inner surface of the wall during the day

The method for determining the coefficient of heat assimilation of the surface of the fence Y, set forth in the Ukrainian building codes, based on the theory of Professor Vlasov.

The coefficient of heat absorption of the material of the fence S characterizes its ability to perceive heat when the temperature on the surface fluctuates. The value of this coefficient depending on the thermophysical properties of the material (λ, c, ρ) and the cyclic frequency of temperature fluctuations $w = 2\pi / Z$:

$$S_{24} = \sqrt{\frac{2 \cdot \pi \cdot \lambda \cdot c \cdot \rho}{Z}}. \tag{6}$$

where λ is the coefficient of thermal conductivity, W/(m · K); s - heat capacity, J/(kg · K); ρ - material density, kg/m^3.

It is obvious that the coefficient S in a given design time period T depends only on the properties of the material, therefore it can be considered a physical characteristic of the material of the fence.

Usually, in the thermal engineering calculations of fences, in addition to the floor, the oscillation period $T = 24$ h is taken, then $S_{24} = 0,51\sqrt{\lambda \times c \times \rho}$.

Calculation of $Y_{вн}$ begins with determining the conditional thickness of the first layer, starting the numbering of layers from the inner surface of the fence.

a) when the inner layer of the enclosing structure has thermal inertia $D \geq 1$, then

$$Y_{вн} = s_1.$$

b) if the thermal inertia of the first layer of the enclosing structure D < 1, and the first and second layers of the structure $D_1 + D_2 \geq 1$, then the heat absorption coefficient of the inner surface is calculated by the formula

$$Y_{вн} = \frac{R_1 s_1^2 + s_2}{1 + R_1 s_2}; \tag{7}$$

where R_1, s_1, s_2 - thermal resistance and coefficient of heat absorption, respectively, of the first and second layers;

To compare the two methods, a calculation was performed with the same initial data.

The calculation results according to Ukrainian regulatory documents are presented in Table 3.

Table 3 Calculation of thermal inertia of each layer of the structure

Layer number	Designation	Value	Calculation
1	D_1	1,861	$D_1 = R_1 s_1 = 0{,}025 \cdot 9{,}76 = 0{,}244$
2	D_2	0,85	$D_2 = R_2 s_2 = 0{,}63 \cdot 10{,}12 = 6{,}376$

The thermal inertia of the first layer of the enclosing structure D < 1, and the first and second layers of the structure D1 + D2 ≥ 1, so the heat transfer coefficient of the inner surface is calculated by formula (7) and is: $Y_{вн} = 9,98 \mathrm{W/m^2 \cdot K}$

Coefficient of heat absorption B_j by the inner surface of the j-th opaque enclosing structure of the room is 4,65 W/m^{2}K.

5 Conclusions

The calculation results according to the UNE-EN ISO 13,786 and DSTU-N B V.2.6–190:2013 methods give values that characterize various physical phenomena in the structure. Therefore, they cannot be compared. But the UNE-EN ISO 13,786 methodology more accurately allows you to study the dynamic thermal characteristics, and use the obtained parameters in calculating the energy efficiency of the building as a whole. At the same time, the Ukrainian calculation method allows only checking compliance with the requirements of DBN V.2.6–31 [17], in terms of heat resistance in summer operating conditions of enclosing structures. In addition, the given requirements require revision in connection with changes in climatic parameters and increased requirements for thermal protection of buildings.

References

1. Schellen L, Loomans MGLC, de Wit MH, Olesen, BW, van Marken Lichtenbelt WD (2012) The influence of local effects on thermal sensation under non-uniform environmental conditions — gender differences in thermophysiology, thermal comfort and productivity during convective and radiant cooling. Phys Behav 107(2):252–261. https://doi.org/10.1016/j.physbeh.2012.07.008
2. Barna E, Bánhidi L (2012) Combined effect of two local discomfort parameters studied with a thermal manikin and human subjects. Energ Build 51:234–241. https://www.sciencedirect.com/science/article/abs/pii/S0378778812002708
3. Veselý M, Zeiler W (2014) Personalized conditioning and its impact on thermal comfort and energy performance – a review. Renew Sustain Energy Rev 34:401–408. https://doi.org/10.1016/j.rser.2014.03.024
4. Filonenko O (2018) Definition of the parameters of thermal insulation in the zone of building foundation according to the ground freezing depth. Energ Effi 11(3):603–626. https://doi.org/10.1007/s12053-017-9600-x
5. Farenyuk G, Oleksiienko O (2020) Development of methods for determining the term of effective exploitation of thermal insulation materials for 100 years. Actual problems of engineering mechanics. Key Engineering Materials 864:80–92
6. Filonenko O, Maliushytskyi O, Mahas N (2018) Design features of thermomodernization of combined roofs. Int J Eng Technol 7(3.2):111–115. https://doi.org/10.14419/ijet.v7i3.2.14385
7. Filonenko O, Yurin O, Kodak O (2018) Thermal modernization of the panel buildings external walls. Int J Eng Technol 7(3.2):116–122. https://doi.org/10.14419/ijet.v7i3.2.14386
8. Leshchenko MV, Semko V (2015) Thermal characteristics of the external walling made of cold-formed steel studs and polystyrene concrete. Mag Civil Eng 60(8):44–55. https://doi.org/10.5862/MCE.60.6

9. Yurin O, Galinska T (2017) Study of heat shielding qualities of brick wall angle with additional insulation located on the outside fences. In: MATEC Web of Conferences, vol 116 https://doi.org/10.1051/matecconf/201711602039
10. Yurin O, Azizova A, Galinska T (2018) Study of heat shielding qualities of a brick wall corner with additional insulation on the brick. In: MATEC Web of Conferences, vol 230 https://doi.org/10.1051/matecconf/201823002039
11. Pohribnyi V, Dovzhenko O, Maliovana O (2018) The ideal plasticity theory usage peculiarities to concrete and reinforced concrete. Int J Eng Technol 7(3.2):19–26
12. Dovzhenko O, PohribnyiV, Klymenko Y, Oreškovič M, Maliovana O (2019) Strength calculation of risers near the supports of reinforced concrete three-hinged frames based on the concrete plasticity theory. In: IOP Conference Series: Materials Science and Engineering, vol 708, pp 012046
13. Hasenko A, Semko O, Drobotia O, Sirobaba V (2020) Experimental and numerical studies of nodes of light steel-reinforced concrete structures. In: Proceedings of the 2020 Session of the 13th Fib International PhD Symposium in Civil Engineering, pp 173–178
14. Semko O, Hasenko A, Kyrychenko V, Sirobaba V (2020) The rational parameters of the civil building steel frame with struts. In: Onyshchenko V, Mammadova G, Sivitska S, Gasimov A (eds) Proceedings of the 2nd International Conference on Building Innovations. ICBI 2019. Lecture Notes in Civil Engineering, vol 73, pp 235–243 Springer, Cham. https://doi.org/10.1007/978-3-030-42939-3_25
15. Pichugin SF (2018) Reliability estimation of industrial building structures. Mag Civil Eng 83(7):24–37. https://doi.org/10.18720/MCE.83.3
16. UNE-EN ISO 13786:2017 Thermal performance of building components – Dynamic thermal characteristics – Calculation methods.
17. DBN B.2.6–31:2016. Thermal insulation of buildings. – К.: Мінрегіон України, 2016. – 70

Flexural Strength of Steel-Reinforced Concrete Composite Structural Span Elements

Tatiana Galinska, Dmytro Ovsii, and Alexandra Ovsii

Abstract The analytical model of calculation of flexural strength of steel-reinforced concrete composite truss (structural) running elements is proposed in the work. This model makes it possible to calculate the strength of the calculated sections of truss run elements taking into account their stress–strain state at the time of maximum bearing capacity. Comparison of experimental test data of steel-reinforced concrete truss beams and elements, which were performed by scientists of the world, with theoretical calculations of the proposed model confirmed the possibility of its use in the practice of their design. The following analytical dependencies can be used to solve two practical problems: checking the flexural strength and designing the optimal cross sections of structural trusses of steel-reinforced concrete running elements.

Keywords Steel-reinforced concrete · Composite · Structural span elements · Flexural strength

1 Introduction

Steel-reinforced concrete structural constructions in the form of spatial truss elements with an upper reinforced concrete slab or beam are most used in the construction of bridge crossings and span structures of a coating of considerable length.

Examples of the use of steel-reinforced concrete composite truss structures in bridges are given by K. Flaga and K. Furtak in [1] and Jose J. Oliveira Pedro et al. in [2, 3]. A survey analysis of the evolution of the construction of composite truss bridges, which was carried out by a group of scientists led by L. Yongjian in [4, 5], showed the economic efficiency of their construction.

At the same time, despite the widespread use of composite truss bridges in construction practice, generalized norms and standards for their design are absent in most leading countries of the world, as noted by A. Sharma, P.K.Singh, K.K. Pathak

T. Galinska (✉) · D. Ovsii · A. Ovsii
National University «Yuri Kondratyuk Poltava Polytechnic», 24, Pershotravnevij Ave., Poltava, Ukraine
e-mail: galinska@i.ua

V. Onyshchenko et al. (eds.), *Proceedings of the 3rd International Conference on Building Innovations*, Lecture Notes in Civil Engineering 181,
https://doi.org/10.1007/978-3-030-85043-2_13

in [6]. Therefore, today there is a need to develop and improve a practical methodology for calculating the bending strength of steel–concrete truss elements, which are structural components of composite bridges and spans of considerable lengths.

The goal of research is to develop the general procedure and analytical dependencies of the calculation of the flexural strength of steel-reinforced concrete (SRC) composite structural span elements (CSSE), depending on the stress–strain state of their design section at the breaking moment.

The scientific developments of the authors of the article are associated with their preliminary studies, which are set out in the works [7, 8, 9], and are also a further development of the research of leading scientists Piskunov V. [10, 11], Pavlikov A. & Kochkarev D. [12, 13, 19], Storozhenko L. [14, 15], Semko O. [16] with their co-authors.

2 Analytical Model of Calculation of the Flexural Strength of SRC CSSE

The analytical model for the calculation of the flexural strength of steel-reinforced concrete (SRC) composite structural span elements (CSSE) is the continuation of the authors' scientific research results aimed to improve their calculation procedure. The main theoretical methodological prerequisites for the calculation of the SRC CSSE have been previously developed by the authors in the following academic papers [17, 18].

In order to work out the analytical model for the calculation of the flexural strength of steel-reinforced concrete (SRC) composite structural span elements (CSSE) we have defined the following prerequisites:

- strain distribution in section at plastic (Composite-PSD) or elastic–plastic (Composite-SC) stages is carried out jointly by linear dependencies. The steel profile in the section of SRC truss elements has rigid vertical or inclined connections with the top concrete slab;
- the criterion for the limit state at the breaking moment of the design section of the SRC CSSE is the extremum criterion for achieving deformations of the compressive zone of the concrete with the limit value ε_{cu}, at which the flexural strength of M_{Rb} elements will be maximum:
  case a: $M_{plRb}(\varepsilon_{cu}, \varepsilon_a > \varepsilon_{au}) = max-$ is the plastic stage of destruction of SRC CSSE (Composite-PSD);
  case b: $M_{Rb}(\varepsilon_{cu}, \varepsilon_a = \varepsilon_{au}) = max-$ boundary stage of destruction of SRC CSSE – the border set between plastic stage (Composite-PSD) and elastic–plastic stage of destruction (Composite-SC);
  case c: $M_{Rb}(\varepsilon_{cu}, \varepsilon_a < \varepsilon_{au}) = max-$ elastic–plastic stage of destruction of SRC CSSE (Composite-SS);
  The extremum criterion for the destruction of $M_{Rb}(\varepsilon_{cu}, \varepsilon_a = \varepsilon_{au}) = max$ was formulated similarly to the criterion $N(\varepsilon_{cu}, \varepsilon_s = \varepsilon_{su}) = max$, which was

recommended by Mitrofanov V.P. in article [20] in order to calculate optimal compression RC elements (Fig. 1).

- the effort N_c in the compression area of the SRC CSSE's cross-section is determined by mathematical relations, proposed by the scientists James K. Wight and James G. Macgregor and is now the basis for calculating the flexural strength of RC beams in Eurocode 2 [21] and SRC beams in Eurocode 4 [22];
- the analysis of cross sections of SRC truss elements revealed that most part of cross-sections can be corrected to generalized characteristic design sections, which will be reduced to their vertical axis. The steel profile in the cross section of the SRC truss elements can be in the form of an I-beam (a), a rectangle (a square) (b) or a circular pipe (c). This steel profile can be arranged in conjunction or at some distance from the top concrete cap.
- typological analysis of cross sections of SRC truss elements has showed that all their cross-sections for generalization of the calculation methodology for the flexural strength of SRC CSSE can be transformed to one deduced design section. The transformed design cross section of the SRC CSSE consists of a top concrete slab and an I- girder (steel profile) (Fig. 2);
- as a result of the generalization, we have selected four isolated cases of strain–stress state of the design section of the SRC CSSE at determining the flexural strength (Fig. 3). Differentiation of cases of the SRC CSSE's limit state depending

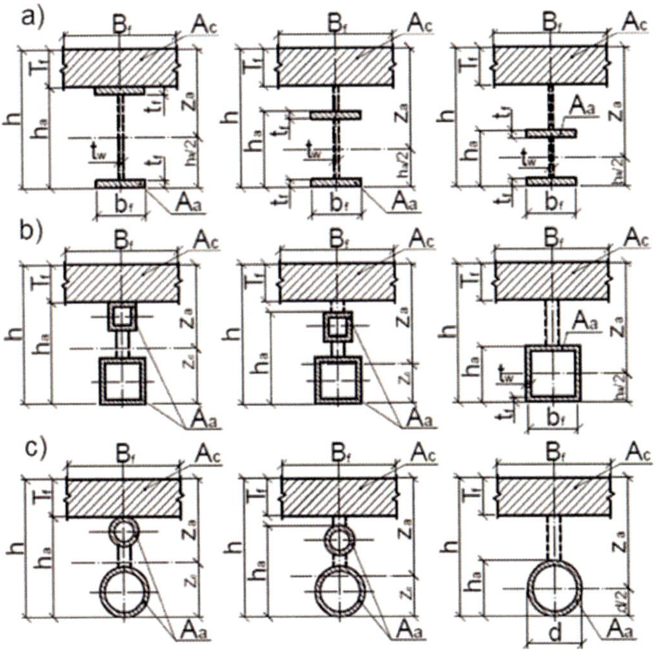

Fig. 1 Typological series of characteristic design sections of SRC truss elements

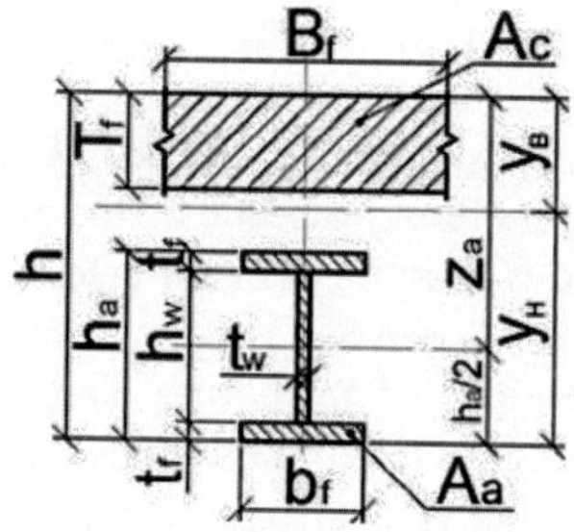

Fig. 2 The overall transformed cross section of the SRC CSSE

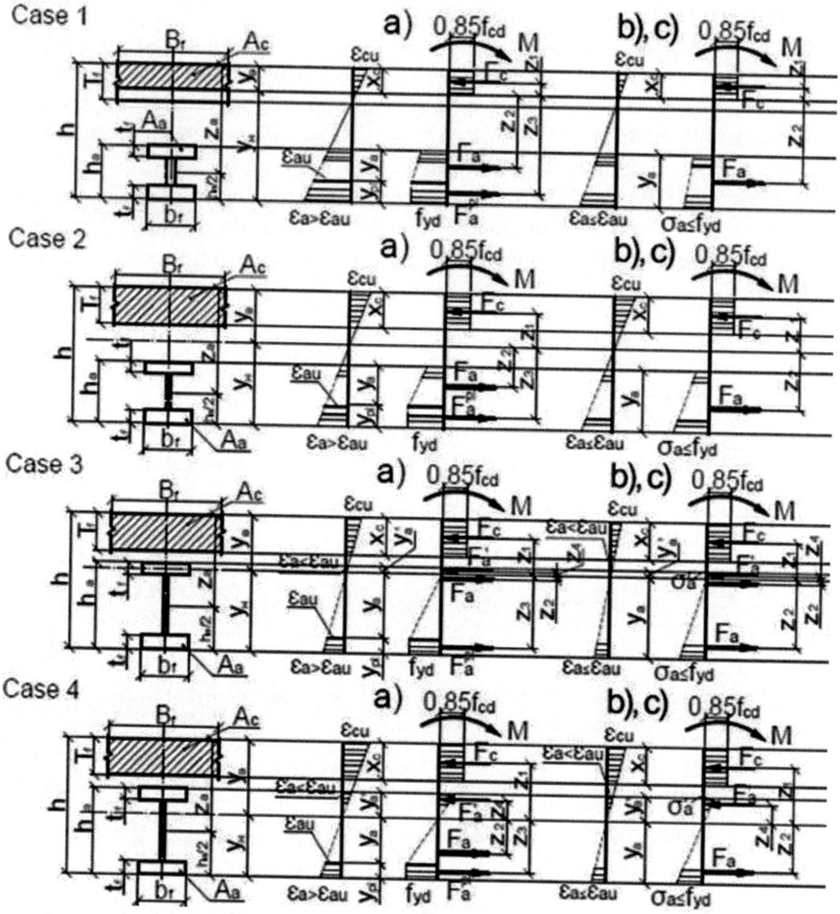

Fig. 3 Cases of the limit stress–strain state of the SRC CSSE's design transformed cross section at determining their flexural strength

on the position of the neutral axis in their section allows us to work out a stepwise algorithm of the analytical calculation model of their flexural strength and to obtain the basic calculated correspondences.

2.1 *The algorithm for SRC CSSE's Flexural Strength Calculation*

The purpose of the mathematical method of design of the SRC CSSE's flexural strength is to determine the limit value of the bending beam moment M_{Rb}, which takes up its design section, and to compare it with the external moment M from the load action: $M_{Rb} \geq M$ or $M_{plRb} \geq M$.

The succession of determining the SRC CSSE's flexural strength according to the proposed mathematical method of design (analytical method of calculation) is shown in the block diagram (Fig. 4).

At the first stage of calculation the flexural strength of the SRC CSSE at present parameters, dimensions of the design section and strength of the composites (ε_{cu}, ε_{au}, E_C, E_a, f_{cd}, f_y. A_C, A_a and h_w are calculated using Eqs. (1), (2) and (3)), we will find the following:

$$A_C = B_f \cdot T_f \tag{1}$$

$$A_a = 2 \cdot t_f \cdot b_f + h_w \cdot t_w \tag{2}$$

$$h_w = h_a - 2 \cdot t_f \tag{3}$$

- the result of the product $a_a \cdot \mu$ Eq. (4) and the value of internal forces N_{cf} Eq. (5) and $N_{pl,a}$ Eq. (6):

$$N_{cf} = 0,85 \cdot f_{cd} \cdot B_f \cdot T_f \tag{4}$$

$$N_{pl,a} = A_a \cdot f_y \tag{5}$$

$$a_a \cdot \mu = E_a \cdot A_a/(E_C \cdot A_C) \tag{6}$$

- let us check the constraint:

$$a_a \cdot \mu \geq a_a \cdot \mu_{onm},$$

where $a_a \cdot \mu_{onm}$ is the optimal value of the product, at which the maximum flexural strength will be equal to the mathematical relation: $M_{Rb}(\varepsilon_{cu}, \varepsilon_a = \varepsilon_{au}) = max$, when the deformations in the extreme fiber of the compressive zone of the concrete

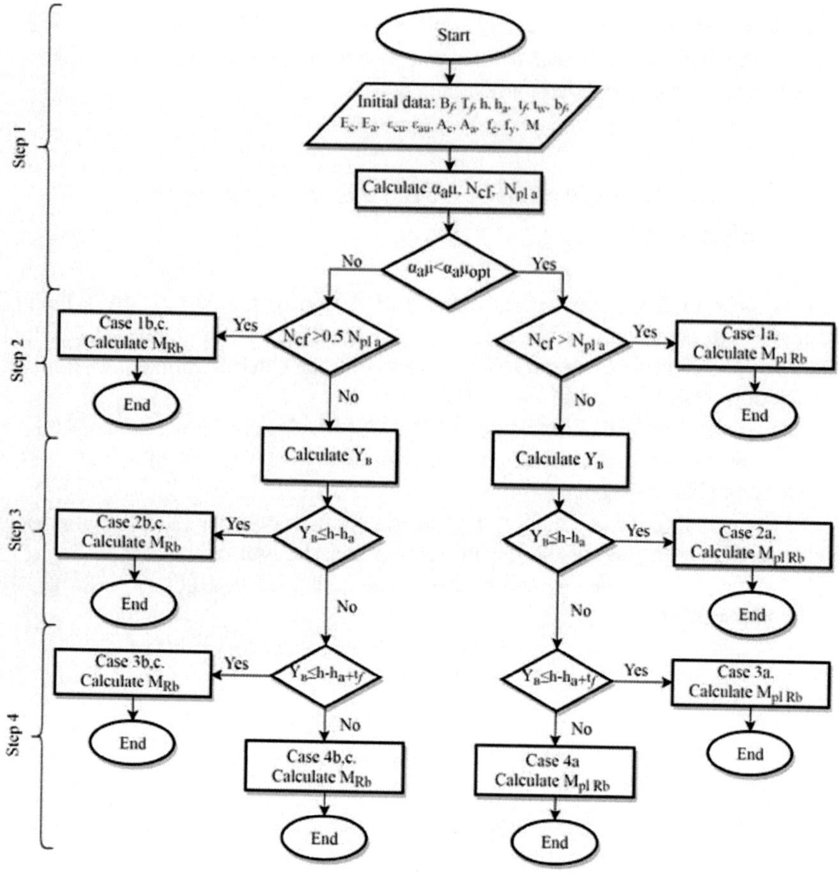

Fig. 4 The block diagram of the succession of determining the SRC CSSE's flexural strength

reach the value $\varepsilon_C = \varepsilon_{cu}$, and in the steel segment they reach the value $\varepsilon_a = \varepsilon_{au}$. The values of the product quantities $a_a \cdot \mu_{onm}$ can be determined from the data in the tables given in the authors' research works [7, 8], or using the mathematical relations given in Eqs. (7), (8), (9), (10) and (11):

$$a_a = E_a / E_C \tag{7}$$

$$\mu_{onm} = [2 \cdot \Delta_Z \cdot \Delta_\varepsilon \cdot (1 + \Delta_h) - \Delta_\varepsilon - 1] / \{a_a \cdot [2 \cdot \Delta_Z \cdot (1 + \Delta_h) - \Delta_h \cdot (1 + \Delta_\varepsilon)]\} \tag{8}$$

$$\Delta_Z = Z_a / \left(T_f + h_a/2\right) \tag{9}$$

$$\Delta_\varepsilon = \varepsilon_u / \varepsilon_{au} \tag{10}$$

$$\Delta_h = h_a / T_f \tag{11}$$

At the second stage of calculation, we determine the possibility of the arrangement of the neutral horizontal axis of the section at the height (T_f) of the concrete slab on exposure to the external moment (M), under the following condition:

$$N_{cf} > N_{pl,a}.$$

If the condition $N_{cf} > N_{pl,a}$, is satisfied, then the neutral horizontal axis of the element's section falls within the limits of the height of the top concrete cap T_f. Next, we determine the height of the compressive zone of the concrete (x_C) and the value of the bending beam moment (M_{Rd}) for the SRC CSSE's design section in strain–stress state in cases 1a or 1b, c (Fig. 3) according to the mathematical relations, stated below in Table 1.

At the third stage of calculation, under condition $N_{cf} < N_{pl,a}$, we determine in the design section of the SRC CSSE the value Y_B (height) from the upper fiber under compression to the neutral axis and compare it with the condition:

$$Y_B \leq h - h_a.$$

Table 1. Analytical dependences of the calculation of flexural strength M_{Rd} of the SRC CSSE depending on the case of strain–stress state of their design section

№ case	Analytical dependences of the calculation of flexural strength of the SRC CSSE in the determinate case of the boundary strain-stress state of its design section at the breaking moment.
Case 1a	When the conditions $\alpha_a \cdot \mu < \alpha_a \cdot \mu_{onm}$; $N_{cf} \geq N_{pl\,a}$;$\rightarrow x_c \leq T_f$ are met. The height of the compressive zone of the concrete will be: $x_c = A_a \cdot f_y / (0{,}85 \cdot f_c \cdot B_f)$. Flexural strength will be: $M_{plRb} = 0{,}5 \cdot A_a \cdot f_y \cdot (z_a - x_c / 2)$; $z_a = h - h_a / 3$.
Case 2a	When the conditions $\alpha_a \cdot \mu < \alpha_a \cdot \mu_{onm}$; $N_{cf} < N_{pl\,a}$;$\rightarrow x_c = T_f$ are met. The distance from the top edge of the compressive zone of the concrete to the neutral section line will be: $Y_B = A_a \cdot f_y / (0{,}85 \cdot f_c \cdot B_f)$. When the following conditions $Y_B \leq h - h_a$ are met. Flexural strength will be: $M_{pl\,Rb} = A_a \cdot f_y \cdot z_a + 0{,}85 \cdot f_c \cdot B_f \cdot T_f \cdot z_c$; $z_c = Y_B - T_f / 2$; $z_a = (h - h_a / 2) - Y_B$;

(continued)

Table 1. (continued)

№ case	Analytical dependences of the calculation of flexural strength of the SRC CSSE in the determinate case of the boundary strain-stress state of its design section at the breaking moment.
Case 3a	When the conditions $\alpha_a \cdot \mu < \alpha_a \cdot \mu_{onm}; N_{cf} < N_{pl\,a}; \rightarrow x_c = T_f$ are met. The distance from the top edge of the compressive zone of the concrete to the neutral section line in the first approximation will be: $Y_B = A_a \cdot f_y / \left(0{,}85 \cdot f_c \cdot B_f\right).$ When the condition $h - h_a < Y_B \le h - h_a + t_f$ is met. The distance from the top edge of the compressive zone of the concrete to the neutral section line in the second approximation will be: $Y_B = (A_a \cdot f_y - 0{,}85 \cdot f_c \cdot B_f \cdot T_f) / (2 \cdot f_y \cdot b_f) + h - h_a;$ Flexural strength will be: $M_{plRb1} = A_a \cdot f_y \cdot z_{a1} - 2 \cdot f_a \cdot b_f \cdot (Y_B + h_a - h) \cdot z_{a2};$ $z_{a1} = 0{,}5 \cdot \left(2h - h_a - T_f\right);$ $z_{a2} = 0{,}5 \cdot \left(Y_B - h_a + h - T_f\right);$ $M_{plRb2} = 0{,}85 \cdot f_c \cdot B_f \cdot T_f \cdot z_C + 2 \cdot f_a \cdot b_f \cdot \left(Y_B + h_a - h\right) \cdot z_{a3},$ $z_c = 0{,}5 \cdot \left(2h - h_a - T_f\right);$ $z_{a3} = 0{,}5 \cdot \left(h - Y_B\right).$
Case 4a	When the conditions $\alpha_a \cdot \mu < \alpha_a \cdot \mu_{onm}; N_{cf} < N_{pl\,a}; \rightarrow x_c = T_f$ are met: The distance from the top edge of the compressive zone of the concrete to the neutral section line in the first approximation will be: $Y_B = A_a \cdot f_y / \left(0{,}85 \cdot f_c \cdot B_f\right).$ When the condition $Y_B > h - h_a + t_f$ is met. The distance from the top edge of the compressive zone of the concrete to the neutral section line in the second approximation will be: $Y_B = (A_a \cdot f_y - 0{,}85 \cdot f_c \cdot B_f \cdot T_f - 2 \cdot f_y \cdot b_f \cdot t_f) / \left(2 \cdot f_y \cdot t_w\right) + h - h_a + t_f.$ Flexural strength will be: $M_{plRb1} = A_a \cdot f_y \cdot z_{a1} - 2 \cdot f_a \cdot t_f \cdot b_f \cdot z_{a4} - 2 \cdot f_a \cdot t_w \cdot \left(Y_B + h_a - h - t_f\right) \cdot z_{a5};$ $z_{a4} = h - h_a - 0{,}5 \cdot \left(T_f - t_f\right);$ $z_{a5} = 0{,}5 \cdot \left(Y_B - h_a + h - T_f - t_f\right);$ $M_{plRb2} = 0{,}85 \cdot f_c \cdot B_f \cdot T_f \cdot z_C + 2 \cdot f_a \cdot b_f \cdot t_f \cdot z_{a6} + 2 \cdot f_a \cdot t_w \cdot \left(Y_B + h_a - h - t_f\right) \cdot z_{a7};$ $z_C = 0{,}5 \cdot \left(2 \cdot h - h_a - T_f\right);$ $z_{a6} = 0{,}5 \cdot \left(h_a - t_f\right);$ $z_{a7} = 0{,}5 \cdot \left(h - Y_B - t_f\right).$

(continued)

Table 1. (continued)

№ case	Analytical dependences of the calculation of flexural strength of the SRC CSSE in the determinate case of the boundary strain-stress state of its design section at the breaking moment.
Case 1b.c	When the conditions $\alpha_a \cdot \mu > \alpha_a \cdot \mu_{onm}; N_{cf} \geq 0,5 \cdot N_{pl\,a}; \rightarrow x_c \leq T_f$ are met. The height of the compressive zone of the concrete will be: $x_c = 0,5 \cdot A_a \cdot f_y / \left(0,85 \cdot f_c \cdot B_f\right)$. Flexural strength will be: at $\varepsilon_a = \varepsilon_{au}$; $M_{Rb} = 0,5 \cdot A_a \cdot f_y \cdot \left(z_{a8} - x_c / 2\right)$; at $\varepsilon_a < \varepsilon_{au}$; $M_{Rb} = 0,5 \cdot A_a \cdot \sigma_a \cdot \left(z_{a8} - x_c / 2\right)$; $z_{a8} = h - h_a / 3$; $\varepsilon_a = \left(\varepsilon_{cu} \cdot \left(h - x_c\right) / x_c\right)$; $\sigma_a = \varepsilon_a \cdot E_a$.
Case 2b.c	When the following conditions $\alpha_a \cdot \mu > \alpha_a \cdot \mu_{onm}; N_{cf} < 0,5 \cdot N_{pl\,a}; \rightarrow x_c = T_f$ are met. The distance from the top edge of the compressive zone of the concrete to the neutral section line will be: $Y_B = 0,5 \cdot A_a \cdot f_y / \left(0,85 \cdot f_c \cdot B_f\right)$. When the condition $Y_B \leq h - h_a$ is met. Flexural strength will be: at $\varepsilon_a = \varepsilon_{au}$; $M_{Rb} = 0,5 \cdot A_a \cdot f_y \cdot z_{a9} + 0,85 \cdot f_c \cdot B_f \cdot T_f \cdot z_c$; at $\varepsilon_a < \varepsilon_{au}$; $M_{Rb} = 0,5 \cdot A_a \cdot \sigma_a \cdot z_{a9} + 0,85 \cdot f_c \cdot B_f \cdot T_f \cdot z_c$; $z_c = Y_B - T_f / 2$; $z_{a9} = \left(h - h_a / 3\right) - Y_B$; $\varepsilon_a = (\varepsilon_{cu} \cdot (h - Y_B) / Y_B)$; $\sigma_a = \varepsilon_a \cdot E_a$.
Case 3b.c	When the conditions $\alpha_a \cdot \mu < \alpha_a \cdot \mu_{onm}; N_{cf} < 0,5 \cdot N_{pl\,a}; \rightarrow x_c = T_f$ are met. The distance from the top edge of the compressive zone of the concrete to the neutral section line in the first approximation will be: $Y_B = 0,5 \cdot A_a \cdot f_y / \left(0,85 \cdot f_c \cdot B_f\right)$. When the condition $h - h_a < Y_B \leq h - h_a + t_f$ is met. The distance from the top edge of the compressive zone of the concrete to the neutral section line in the second approximation will be: $Y_B = (0,5 \cdot A_a \cdot f_y - 0,85 \cdot f_c \cdot B_f \cdot T_f) / (\sigma_{a2} \cdot b_f) + h - h_a$; $\varepsilon_{a2} = \varepsilon_{cu} \cdot t_f / Y_B$; $\sigma_{a2} = \varepsilon_{a2} \cdot E_a$. Flexural strength will be: at $\varepsilon_{a1} = \varepsilon_{au}$, $M_{Rb1} = 0,5 \cdot A_a \cdot f_y \cdot z_{a10} - \sigma_{a2} \cdot b_f \cdot \left(Y_B + h_a - h\right) \cdot z_{a11}$; $z_{a10} = h - h_a / 3 - T_f / 2$; $z_{a11} = Y_B / 3 - T_f / 2 - 2 \cdot \left(h_a - h\right) / 3$; $M_{Rb2} = 0,85 \cdot f_c \cdot B_f \cdot T_f \cdot z_C + \sigma_{a2} \cdot b_f \cdot \left(Y_B + h_a - h\right) \cdot z_{a12}$; $z_C = z_{a10}$; $z_{a12} = \left(h - Y_B + h_a\right) / 3$; at $\varepsilon_{a1} < \varepsilon_{au}$, $M_{Rb1} = 0,5 \cdot A_a \cdot \sigma_{a1} \cdot z_{a10} - \sigma_{a2} \cdot b_f \cdot \left(Y_B + h_a - h\right) \cdot z_{a11}$; $\varepsilon_{a1} = \left(\varepsilon_{cu} \cdot \left(h - Y_B\right) / Y_B\right)$; $\sigma_{a1} = \varepsilon_{a1} \cdot E_a$.

(continued)

Table 1. (continued)

№	Analytical dependences of the calculation of flexural strength of the SRC CSSE in case the determinate case of the boundary strain-stress state of its design section at the breaking moment.
Case 4b,c	When the conditions $\alpha_a \cdot \mu > \alpha_a \cdot \mu_{onm}; N_{cf} < 0,5 \cdot N_{pl\,a}; \rightarrow x_c = T_f$ are met. The distance from the top edge of the compressive zone of the concrete to the neutral section line in the first approximation will be: $Y_B = 0,5 \cdot A_a \cdot f_y / \left(0,85 \cdot f_c \cdot B_f\right).$ When the condition $Y_B > h - h_a + t_f$. is met. The distance from the top edge of the compressive zone of the concrete to the neutral section line in the second approximation will be: $Y_B = (0,5 \cdot A_a \cdot f_y - 0,85 \cdot f_c \cdot B_f \cdot T_f - \sigma_{a2} \cdot b_f \cdot t_f) / (\sigma_{a3} \cdot t_w) + h - h_a + t_f;$ $\varepsilon_{a2} = \varepsilon_{cu} \cdot (Y_B + h_a - h) / Y_B;\ \sigma_{a2} = \varepsilon_{a2} \cdot E_a$ $\varepsilon_{a3} = \varepsilon_{cu} \cdot (Y_B + h_a - h - t_f) / Y_B;\ \sigma_{a3} = \varepsilon_{a3} \cdot E_a$ Flexural strength will be: at $\varepsilon_{a1} = \varepsilon_{au}, M_{Rb1} = 0,5 \cdot A_a \cdot f_y \cdot z_{a13} - \sigma_{a2} \cdot t_f \cdot b_f \cdot z_{a14} - \sigma_{a3} \cdot t_w \cdot \left(Y_B + h_a - h - t_f\right) \cdot z_{a15};$ $z_{a13} = (2 \cdot h + Y_B) / 3 - T_f / 2;$ $z_{a14} = h - h_a + t_f / 3 - T_f / 2;$ $z_{a15} = (2 \cdot (h - h_a) + Y_B) / 3 + t_f - T_f / 2;$ $M_{Rb2} = 0,85 \cdot f_c \cdot B_f \cdot T_f \cdot z_C + \sigma_{a2} \cdot b_f \cdot t_f \cdot z_{a16} + \sigma_{a3} \cdot t_w \cdot \left(Y_B + h_a - h - t_f\right) \cdot z_{a17};$ $z_C = z_{a13};\ z_{a16} = h_a - \left(h + t_f - Y_B\right) / 3;$ $z_{a17} = 2 \cdot h_a / 3 - t_f;$ at $\varepsilon_{a1} < \varepsilon_{au}, M_{Rb1} = 0,5 \cdot A_a \cdot \sigma_{a1} \cdot z_{a13} - \sigma_{a2} \cdot t_f \cdot b_f \cdot z_{a14} - \sigma_{a3} \cdot t_w \cdot \left(Y_B + h_a - h - t_f\right) \cdot z_{a15};$ $\varepsilon_{a1} = \left(\varepsilon_{cu} \cdot \left(h - Y_B\right) / Y_B\right);\ \sigma_{a1} = \varepsilon_{a1} \cdot E_a.$

If the conditions $N_{cf} < N_{pl,a}$, and are satisfied, then the neutral horizontal axis of the element's section falls within the limits of the difference of the heights of. Next, we determine the height of the compressive zone of the concrete (Y_B) and the value of the bending beam moment (M_{Rd}) for the design section of the SRC CSSE in the strain–stress state in cases 2a or 2b,c (Fig. 3) according to the mathematical relations, stated above in Table 1.

At the fourth stage of calculation, when the value (height) $Y_B \leq h - h_a$, the value of the bending beam moment (M_{Rd}) for the design section of the SRC CSSE is determined in the strain–stress state in cases 3a or 3b,c (Fig. 4), 4a or 4b,c according to the mathematical relations, stated above in Table 1.

The value of flexural strength (M_{Rd}) of the design section of the SRC CSSE is compared with the value of the torque from external forces (M). The strength of the SRC CSSE section will be ensured with the requirement that:

$$M_{Rd} \geq M.$$

If the condition of the SRC CSSE's flexural strength is not satisfied, then it is necessary to increase the size of their design section and to take the materials of their components with larger values of strength characteristics (shear properties), and then to do the calculation again.

2.2 *Analytical Dependences of SRC CSSE's Flexural Strength Depending on the Case of Strain–Stress State of the Design Section at the Breaking Moment*

Analytical dependences of calculation flexural strength of steel-reinforced concrete (SRC) composite structural span elements (CSSE), depending on the case of strain–stress state of their design section at the breaking are shown in Table 1.

3 Comparisons between Experimental and Analytical Results of Bending Strength Calculation SRC CSSE

In order to compare the proposed analytical model of calculation of flexural strength of SRC CSSE, we have used the results of experimental studies of national and foreign scientists, such as: T. P. Kuch (specimens of beams B-1, B-2–1, B-2–2-1, B-2–3, B-3–1, B-3–2-1, B-3–3) [23]; F. S Shkoliar (specimens of beams B-1.1, B-1.2, B-2.1, B-2.2, B-3.1,) [24]; L. Luo (Lisheng Luo) (specimens of beams B-1… B-3) [25]; J. Bujnak (a specimen beam without marking) [26].

When comparing the results, we have determined arithmetic mean ($\overline{X}$), root-mean-square deviation (σ_{n-1}) and coefficient of variation (ν).

The comparisons of experimental results (M^{test}) and analytical findings of the calculation of flexural strength of SRC truss elements (M^{calc}) are given in Table 2.

Comparison of experimental and theoretical strength values of 16 specimens of reinforced concrete truss beams, which components are bond, leads to the following statistical indicators:

- for partial factors for concrete $\gamma_C = 1,0$ and for material property, also accounting for model uncertainties and dimensional variations $\gamma_M = 1,0$- $\overline{X} = 1,094$; $\sigma_{n-1} = 0,006$; $\nu = 0,52\%$;
- for partial factors for concrete $\gamma_C > 1,0$ and for material property, also accounting for model uncertainties and dimensional variations $\gamma_M > 1,0$- $\overline{X} = 1,261$; $\sigma_{n-1} = 0,014$; $\nu = 1,13\%$.

Table 2 Comparison of experimental bending moments with theoretical meanings

№	Author	Specimen	M^{test} kNm	$M^{calk}_{\gamma=1,0}$ kNm	$\frac{M^{test}}{M^{calk}_{\gamma=1,0}}$	$M^{calk}_{\gamma>1,0}$ kNm	$\frac{M^{test}}{M^{calk}_{\gamma>1,0}}$
1	*T. Kuch*	B-1	38,5	37,0	1,04	33,7	1,14
2		B-2–1	27,8	27,3	1,02	25,5	1,09
3		B-2–2-1	28,3	27,8	1,02	26,1	1,08
4		B-2–3	29,8	28,1	1,06	26,5	1,12
5		B-3–1	46,5	45,0	1,03	38,8	1,20
6		B-3–2-1	50,3	46,5	1,09	40,6	1,23
7		B-3–3	52,0	47,4	1,10	41,9	1,24
8	*F. Shkoliar*	B-1.1	33,0	31,8	1,04	26,2	1,26
9		B-1.2	32,0	31,8	1,01	26,2	1,22
10		B-2.1	21,0	17,5	1,20	15,1	1,39
11		B-2.2	20,0	17,5	1,14	15,1	1,32
12		B-3.1	30,5	25,5	1,19	21,2	1,43
13	*L. Luo*	B-1	385,6	364,6	1,06	300,6	1,28
14		B-2	405,9	364,6	1,11	300,6	1,35
15		B-3	408,7	364,6	1,12	300,6	1,36
16	*J. Bujnak*	-	357,8	281,9	1,27	243,4	1,47

4 Conclusions

The algorithms and analytical dependences of the method for calculating the flexural strength of steel-reinforced concrete (SRC) composite structural span elements (CSSE) are presented in the academic paper. Comparative analysis of the experimental findings with theoretical calculations of flexural strength of SRC CSSE has showed their adequate convergence, which allows to apply the proposed analytical dependencies in the design practice.

Further scientific research in this area will be aimed at improvement of the proposed method for calculating the SRC CSSE and its comparison with a wider range of experimental studies of the elements.

Notations

A_a	Cross-sectional area of the structural steel section
A_c	Cross-sectional area of concrete
B_f	Width of concrete slab
b_f	Width of the flange of a steel section
E_a	Modulus of elasticity of structural steel
E_c	Modulus of elasticity for concrete
f_c	Characteristic value of the cylinder compressive strength of concrete at 28 days
f_y	Nominal value of the yield strength of structural steel
h	Overall depth of cross-section
h_a	Depth of the structural steel section
N_{cf}	Design value of the compressive normal force in the concrete flange with full shear connection
$N_{pl\ a}$	Design value of the plastic resistance of the structural steel section to normal force
M	Bending moment
M_{plRd}	Design value of the plastic resistance moment of the composite section with full shear connection
M_{Rd}	Design value of the resistance moment of a composite section
T_f	Depth of the concrete slab
t_f	Thickness of a flange of the structural steel section
t_w	Thickness of the web of the structural steel section
Y_B	Distance between the neutral axis and the extreme fiber of the concrete slab in compression
z_a	Distance between the center of mass of the steel section and the middle of the compressed section of concrete slab
α_a	Ratio of the elastic modules of structural steel and concrete;
$\alpha_a \cdot \mu_{onm}$	Product of the coefficients of the ratio of elastic moduli and optimal structural steel profile (table value)
ε_{au}	Yield strain of structural steel
ε_{cu}	Yield strain of concrete
μ	Ratio of the cross-sectional areas of steel and concrete;

References

1. Flaga K, Furtak K (2015) Examples of solutions for steel-concrete composite structures in bridge engineering. Civil Environ Eng Reps 16(1):51–68. https://doi.org/10.1515/ceer-2015-0004

2. Reis AJ, Pedro JO (eds) (2011) Composite truss bridges: new trends, design and research. In: 6th European Conference on Steel and Composite Structures, Budapest, 2011. https://www.researchgate.net/publication/281280725
3. Pedro JO, Reis AJ (eds) (2013) Composite steel-concrete cable-stayed bridges: developments and future trends. In: 5th International Conference on Structural Engineering, Mechanics and Computation, Cape Town, 2013. https://www.researchgate.net/publication/281280573
4. Liu Y, Xiong Z, Luo Y, Cheng G, Liu G, Yang J (2015) Double-composite rectangular truss bridge and its joint analysis. J Traffic Transp Eng (English Edition) 2(4):249–257. https://doi.org/10.1016/j.jtte.2015.05.005
5. Zhijuan T, Yongjian L, Lei J, Weiqing Z, Yinping M (2019) A review on application of composite truss bridges composed of hollow structural section members. J Traffic Transp Eng (English Edition) 6(1):94–108. https://doi.org/10.1016/j.jtte.2018.12.001
6. Sharma A, Singh PK, Pathak KK (2017) Under slung steel truss bridge with composite RCC deck bridge. Int J Eng Res Mech Civil Eng (IJERMCE) 2(11):43–47. https://www.technoarete.org/common_abstract/pdf/IJERMCE/v4/i11/Ext_37051.pdf
7. Azizov T, Kochkarev D, Galinska T (2020) Reinforced concrete rod elements stiffness considering concrete nonlinear properties. Lecture Notes in Civil Engineering, vol 47, pp 1 –6. https://doi.org/10.1007/978-3-030-27011-7_65
8. Kochkarev D, Azizov T, Galinska T (2018) Bending deflection reinforced concrete elements determination. In: MATEC Web of Conferences, vol 230, p 02012 https://doi.org/10.1051/matecconf/201823002012
9. Kochkarev D, Galinska T, Tkachuk O (2018) Normal sections calculation of bending reinforced concrete and fiber concrete element. Int J Eng Technol 7(3.2): 176–182. https://doi.org/10.14419/ijet.v7i3.2.14399
10. Piskunov VG, Gorik AV, Cherednikov VN (2000) Modeling of transverse shears of piecewise homogeneous composite bars using an iterative process with account of tangential loads 2. resolving equations and results. Mech Compos Mater 36(6):445–452. https://doi.org/10.1023/A:1006798314569
11. Piskunov VG, Goryk AV, Cherednikov VN (2000) Modeling of transverse shears of piecewise homogeneous composite bars using an iterative process with account of tangential loads: 1. construction of a model. Mech Compos Mater 36(4):287–296. https://doi.org/10.1007/BF02262807
12. Pavlikov A, Kochkarov D, Harkava O (2019) Calculation of reinforced concrete members strength by new concept. In: CONCRETE. Innovations in Materials, Design and Structures: Proceedings of the fib Symposium 2019, pp 820–827, Kraków, Poland, 27–29 May 2019. http://reposit.pntu.edu.ua/handle/PoltNTU/6064
13. Pavlikov A, Kochkarev D, Harkava O (2020) Analysis of eccentrically loaded members of circular cross section by nonlinear deformation model. In: Onyshchenko V, Mammadova G, Sivitska S, Gasimov A (eds) Proceedings of the 2nd International Conference on Building Innovations. ICBI 2019. Lecture Notes in Civil Engineering, vol 73, pp 171–181. Springer, Cham. https://doi.org/10.1007/978-3-030-42939-3_19
14. Storozhenko L, Yermolenko D, Gasii G (2018) Investigation of the deformation state of a composite cable space frame structures with a photogrammetric method. Int J Eng Technol 7(3.2):442–446. https://doi.org/10.14419/ijet.v7i3.2.14568
15. Storozhenko L, Butsky V, Taranovsky O (1998) Stability of compressed steel concrete composite tubular columns with centrifuged cores. J Constr Steel Res 46(1–3):484. https://doi.org/10.1016/S0143-974X(98)80098-9
16. Semko O, Hasenko A, Kyrychenko V, Sirobaba V (2020) The rational parameters of the civil building steel frame with struts https://doi.org/10.1007/978-3-030-42939-3_25
17. Galinska TA, Muravl'ov VV, Ovsiy NA (2013) Methodical bases of calculation of the strength the normal section of steel-concrete beams with of concrete upper belt and the outer (remote) reinforcement on the basis of the calculated deformation model. Resursoekonomni materialy, konstruktsii, budivli ta sporudy 27:41–55. http://nbuv.gov.ua/UJRN/rmkbs_2013_27_9

18. Galinska T, Ovsii D, Ovsii M (2018) The combining technique of calculating the sections of reinforced concrete bending elements normal to its longitudinal axis, based on the deformation model. Int J Eng Technol 7(3.2):123–127. https://doi.org/10.14419/ijet.v7i3.2.14387
19. Kochkarev D, Azizov T, Galinska T (2020). Design of effective statically indeterminate reinforced concrete beams https://doi.org/10.1007/978-3-030-42939-3_10
20. Mitrofanov VP (2009) Extreme strength criterion and design of RC elements. J fib: Struct Concr 10(4):163–172. https://doi.org/10.1680/stco.2009.10.4.163
21. Comité Européen de Normalisation (CEN), (2004a) "Eurocode 2: Design of Concrete Structures-Part 1-1: General Rules and Rules for Buildings", European Standard BS EN 1992-1-1: 2004, European Committee for Standardization (CEN), Brussels, Belgium, 227 pp
22. Comité Européen de Normalisation (CEN), (2004b) "Eurocode 4: Design of Composite Steel and Concrete Strucures-Part 1-1:General Rules and Rules for Buildings", European Standard BS EN 1994-1-1: 1994. European Committee for Standardization (CEN), Brussels, Belgium, 118 pp
23. Kuch TP (2012) Stress-strain state and load-bearing capacity of reinforced concrete beam structures with exposed pipe reinforcement (Extended abstract of PhD dissertation)
24. Shkoliar FS (2015) Tensely-deformed state and bearing capacity of reinforced concrete beams with remote working reinforcement (PhD dissertation). https://pntu.edu.ua/page/spetsializovani-vcheni-radi-po-zakhistu-disertatsij.html
25. Luo L, Zhang X (2019) Flexural response of steel-concrete composite truss beams. Adv Civil Eng 1502707 https://doi.org/10.1155/2019/1502707
26. Bujnak J, Michalek P, Baran W (2018) Experimental and theoretical investigation of composite truss beams. In: MATEC Web of Conferences, vol 174, p 04001. https://doi.org/10.1051/matecconf/201817404001

Preparation Technique of Experimental Specimens of Steel and Concrete Composite Slabs

Leonid Storozhenko, Grygorii Gasii, Myron Hohol, and Olena Hasii

Abstract The composite frames are used widely as the roof system for different buildings or structures as well as covering spaces as a shell. The technique of experimental investigation of steel and concrete composite space frames (SCCS frame) is compiled because it is a novel type of structure and has not sufficiently behavior investigated yet. The current investigation is aimed at prefabrication technique description. This research is a complete study of prefabrication techniques and also the strength of materials for making prototypes of the SCCS frame, which consist of steel lattice and steel and concrete composite slab (SCC slab) as well as include top and bottom chords. The research includes testing specimens of the concrete cubes and prism and reinforcement bars. It has long been known that concrete is often and widely used for slabs and surfaces in compression or bending. However, the SCC slab has another way of reinforcement than an ordinary reinforced concrete slab.

Keywords Steel · Concrete · Slab · Stress · Strain

1 Introduction

Currently, SCC structures are the widespread structure types all over the world. The SCC structures are used in various fields of building [5, 10].

This is why the investigation of the SCC structures is an important issue. The SCC slabs are a part of the SCCS frames [26]. The SCCS frame is the novel type of roof system that was designed and was patented recently. The SCCS frames consist of steel lattice and SCC slab as well as include top and bottom chords. The SCCS

L. Storozhenko · G. Gasii (✉)
National University «Yuri Kondratyuk Poltava Polytechnic», Poltava, Ukraine
e-mail: gasiigm@gmail.com

M. Hohol
Lviv Polytechnic National University, Lviv, Ukraine

O. Hasii
Poltava University of Economics and Trade, Poltava, Ukraine

V. Onyshchenko et al. (eds.), *Proceedings of the 3rd International Conference on Building Innovations*, Lecture Notes in Civil Engineering 181,
https://doi.org/10.1007/978-3-030-85043-2_14

frame has the design as shown in [26]. The SCC slab plays the role of the top chord. It delivers bearing and shielding roles. The SCCS frames are used as the roof of civil and industrial buildings or structures or covering large areas as a shell.

The experimental investigating technique of SCCS frames was compiled because of its are a novel type of structure and has not yet sufficiently behavior investigated but the current investigation is aimed for prefabrication technique describe.

So, the purpose of this research is a complete the study of prefabrication technique and also the strength of materials for making samples. The research includes testing specimens of the reinforcement bars, concrete cubes as well as concrete prisms.

2 Method

The methods of experimental research, as well as theoretical research, taking into account previous research [2, 3, 5, 8, 9, 11, 13–36], are used in this study. To study the mechanical properties of materials, standard and generally accepted methods are used [4] etc.

3 Results and Discussions

3.1 The Size and Design of Experimental Specimens of the SCC Slabs

To making the SCC slab is used concrete because it has suitable mechanical properties for this structure type [25]. It has long been known that concrete is often and widely used for slabs and shells in compression or bending [19]. Nevertheless, the SCC slab has another way of reinforcement than an ordinary reinforced concrete slab [12]. Ordinary reinforced concrete slabs include reinforcement mesh that is made of steel bars that caused a significant cross-section height and total weight. Those are not suitable options for space surfaces.

That is why the SCC slabs are reinforced with meshes that are made from steel wire instead of steel bars. This method of reinforcement allows getting sufficiently homogeneous material.

Nevertheless, it caused a significant flexible thin-walled structure. That is why the SCC slab has ribs for providing rigidity. These ribs are reinforced with a few steel bars. To explorer, the stress-strained state of the SCC slab was manufactured experimental specimens (Fig. 1a).

Experimental specimens of the SCC slabs have a square shape with a side length of 1000 mm and slab thickness of 20 mm and rib height of 50 mm.

3.2 *Materials of Experimental Specimens of SCC Slabs*

To making the SCC slab are used steel wire meshes and a concrete mixture. There is the composition of material for concrete mixture in Table 1.

Concrete mixture was produced with concrete mixer CM-250 in the industrial department (Fig. 1b).

The steel wire meshes were made of steel wire 0.9 mm in diameter and cell size 12 mm (Fig. 2).

Steel bars 6 mm in diameter are used for reinforcement of ribs.

To prevent plain bars under tension from slipping in concrete solid, both ends of the bar are bent. In addition, reinforcement should be bent due to the design requirement.

The detailing of the bent segments is listed in design codes or acceptance specifications for construction quality.

Figure 2 illustrates bent segments that are used in the SCC slabs [6]. Figure 3 shows the dimensions and reinforcement scheme of SCC slabs.

Table 1 Quantities of materials per cubic meter of the concrete mixture

Material	Portland cement	Fine aggregates (sand) 1.1 mm	Water
Mixture proportions	1	1.75	0.4
Mass of material, kg	700	1225	280

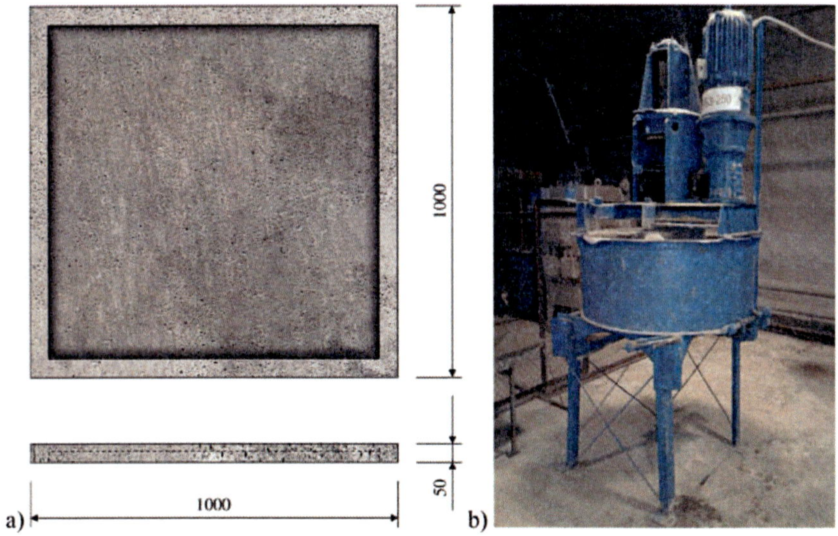

Fig. 1 The SCC slabs **a** and Concrete mixer CM-250 **b**

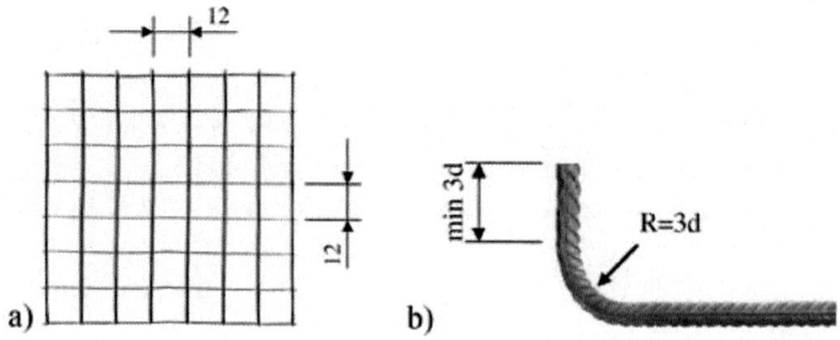

Fig. 2 Fragment of the welded wire mesh with cell size 12 mm **a** and Bent steel bar **b**

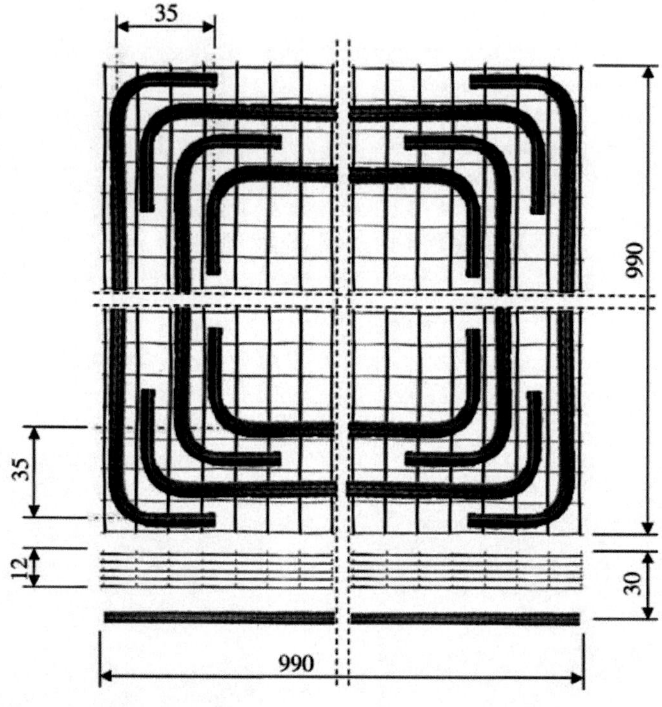

Fig. 3 Dimensions (mm) and reinforcement layout

3.3 *Mechanical Properties of Materials for Experimental Specimens of SCC Slabs*

The concrete strength depends on the quality of the concrete mixture, water-cement ratio, cement strength, etc. [1, 7]. Test cubes of 150 mm for obtaining the compressive strength of concrete were carried out.

In Ukraine as in all over the world, the cube strength is used as the main index of concrete strength.

A compression test of concrete cubes is an easy way to obtain concrete strength. On top of that, the strength is measured is quite accurate.

Moulds are used to making concrete cubes. No voids must be in the concrete mixture during the moulds are filled. A vibrating table to concrete mixture compaction in moulds was used.

The load-carrying capacity of the table is about 300 kg. All cubes moulds that were filled simultaneously were accommodated on the vibrating table. The vibration was adjusted to 2000 vibrations per minute.

The concrete cubes are taken out from moulds after 24 h and for curing, it put in water. The concrete cubes are tested after 28 days of curing. For the test are used the compression testing machine PG-125.

To fail the concrete cubes, the load is applied gradually 130 kg/cm^2 per minute.

The compressive strength of concrete is obtained according to [4]. The average value of the mechanical properties of concrete cubes is 17.86 MPa.

The concrete prisms to determine the prism strength are tested.

Concrete prisms were manufactured by the same technique as concrete cubes are used, except moulds. Before, prisms testing on side surfaces had been set strain gauges. Strain gauges set on the side surfaces of prisms with clue BF-2.

The average value of the mechanical properties of concrete prisms is 16.30 MPa. The stress–strain curves based on the testing results are mapped (Fig. 4).

Reinforcement strength was obtained by tests of specimens of 600 mm length with press PM-500.

However, the strengths of different specimens, despite that they were of the same type, were different. It was due to the inherent variability of reinforcement materials. The stress–strain curves of steel bars, which were used for reinforcing experimental specimens of SCC slabs, were obtained from tension tests, in which the loads are monotonically applied without any unloading until the bars fail in a short time.

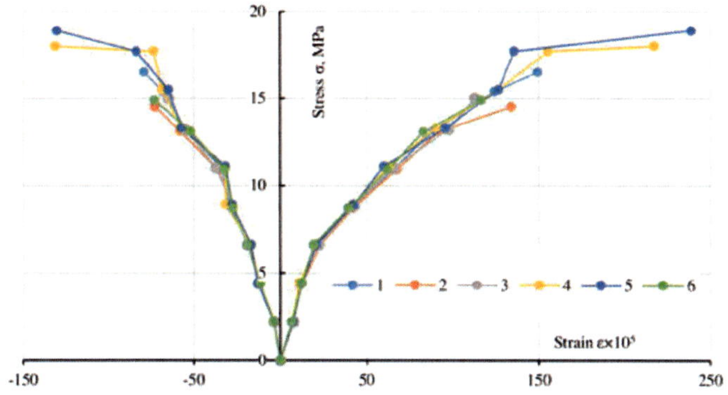

Fig. 4 The stress–strain curves for concrete prisms

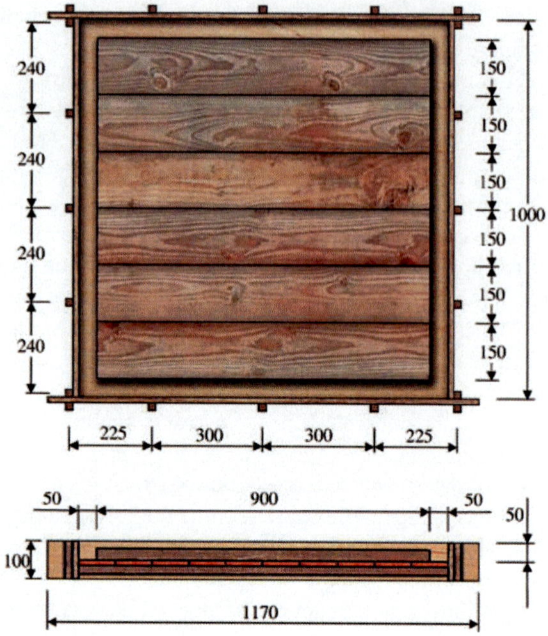

Fig. 5 Timber moulds for slab manufacture

Therefore, measured strength with a certain degree of confidence was taken as the strength of reinforcement. The degree of confidence is more than 95%. The average strength is 279.74 MPa.

3.4 Preparation of Experimental Specimens of SCC Slabs

For experimental specimens preparation was used timber moulds (Fig. 5). Preparation of experimental specimens of SCC slabs includes 4 steps. The first step was to set moulds on a vibrating table. Then were put reinforcement bars and welded wire meshes in mould and concreting.

4 Conclusions

The fabrication technique of experimental specimens of steel and concrete composite slabs is described. The mechanical properties of materials are obtained by testing. The strengths of concrete and steel bars are determined. The average value of the mechanical properties of concrete cubes is 17.86 MPa. The failure load N is from 384 to 430 kN. The average value of the mechanical properties of concrete prisms is 16.30 MPa. The failure load N is from 331 to 427 kN. The average strength of the reinforcement is 279.74 MPa. This strength is sufficient for the structures under study. thus, the proposed technology for the production of concrete mix is acceptable. It allows you to get a durable material. This also applies to fittings. The adopted reinforcement provides the necessary rigidity and strength of the slab ribs. The technology of concreting samples is quite simple to manufacture and does not require special equipment, which makes it cost-effective.

References

1. Alvarez Rabanal FP, Guerrero-Muñoz J, Alonso-Martinez M, Martinez-Martinez JE (2016) Bending and shear experimental tests and numerical analysis of composite slabs made up of lightweight concrete. J Eng. 2016:Article ID 6819190. https://doi.org/10.1155/2016/6819190
2. Charif A, Mourad SM, Khan MI (2016) Flexural behavior of beams reinforced with steel bars exceeding the nominal yield strength. Latin Am J Solids Struct 13(5):945–963
3. Dass G, Talwar M (2017) Review paper on Ferrocement in construction. Int J Adv Res Comput Sci 8(4):384–386
4. DSTU B V.2.7–224:2009 (2010). Betony. Pravyla Kontroliu Mitsnosti
5. Gasii GM (2014) Technological and design features of flat-rod elements with usage of composite reinforced concrete. Metal Min Ind 4:23–25
6. Gu X, Jin X, Zhou Y (2016) Basic Principles of Concrete Structures. Springer, Berlin
7. Hedaoo NA, Gupta LM, Ronghe GN (2012) Design of composite slabs with profiled steel decking: a comparison between experimental and analytical studies. Int J Adv Struct Eng 4(1):1–15
8. Fawzi NM, Jalawi A (2020) Quality control of production lightweight ferrocement plate using sustainable materials. Key Eng Mater 857:10–14
9. Kaish ABMA, Jamil M, Raman SN, Zain MFM, Nahar L (2018) Ferrocement composites for strengthening of concrete columns: a review. Constr Build Mater 160:326–340
10. Leon RT, Perea T, Rassati GA, Lange J (2011, February) Composite construction in steel and concrete VI. American Society of Civil Engineers
11. Liu B, Zhang Z, Zhang M, Wang X (2019) Experimental study of the mechanical performance of corrugated steel plate-concrete composite structures. Int J Steel Struct 19(3):733–746
12. Lysenko E F (1981) Armocementnye konstrukcii: ucheb. Vishcha shkola.
13. Madadi A, Eskandari-Naddaf H, Shadnia R, Zhang L (2018) Digital image correlation to characterize the flexural behavior of lightweight ferrocement slab panels. Constr Build Mater 189:967–977
14. Madadi A, Eskandari-Naddaf H, Shadnia R, Zhang L (2018) Characterization of ferrocement slab panels containing lightweight expanded clay aggregate using digital image correlation technique. Constr Build Mater 180:464–476
15. Madadi A, Tasdighi M, Eskandari-Naddaf H (2019) Structural response of ferrocement panels incorporating lightweight expanded clay and perlite aggregates: experimental, theoretical and statistical analysis. Eng Struct 188:382–393
16. Mughal UA, Saleem MA, Abbas S (2019) Comparative study of ferrocement panels reinforced with galvanized iron and polypropylene meshes. Constr Build Mater 210:40–47
17. Murali G, Amran M, Fediuk R, Vatin N, Raman SN, Maithreyi G, Sumathi A (2020) Structural behavior of fibrous-ferrocement panel subjected to flexural and impact loads. Materials 13(24):5648
18. Naderpour H, Eidgahee DR, Fakharian P, Rafiean AH, Kalantari SM (2020) A new proposed approach for moment capacity estimation of ferrocement members using group method of data handling. Eng Sci Technol Int J 23(2):382–391

19. Nervi PL (1965) Aesthetics and Technology in Building. Harvard University Press, Cambridge
20. Patil SS, Gaidhankar DG, Kulkarni MS (2020) Ferrocement panels under flexure by partial replacement of cement With marble powder. IJITEE 9(8):255–260
21. Shaaban IG, Shaheen YB, Elsayed EL, Kamal OA, Adesina PA (2018) Flexural characteristics of lightweight ferrocement beams with various types of core materials and mesh reinforcement. Constr Build Mater 171:802–816
22. Shaheen YB, Abusafa HM (2017) Structural behavior for rehabilitation ferrocement plates previously damaged by impact loads. Case Stud Constr Mater 6:72–90
23. Shaheen YB, Etman ZA, Gomaa O (2019) Structural behavior of thin ferrocement plates with and without stiffeners subjected to compression loading. Asian J Civil Eng 20(2):237–260
24. Soman M, Mohan J (2018) Rehabilitation of RC columns using ferrocement jacketing. Constr Build Mater 181:156–162
25. Storozhenko LI, Tymoshenko VM, Nyzhnyk OV, Gasii GM, Murza SO (2008) Research and Design of Composite Steel and Concrete Space Structural Designs. ASMI, Poltava
26. Storozhenko LI, Gasii GM, Gapchenko SA (2015) Space Steel and Concrete Grid-Gable Roofs. ASMI, Poltava
27. Wang W, Song J, Hou M, Liu G, Wang W (2019) Experimental study and numerical simulation of replaceable corrugated steel plate-concrete composite shear walls. Soil Dyn Earthq Eng 127:105827
28. Yang Y, Liu R, Huo X, Zhou X, Roeder CW (2018) Static experiment on mechanical behavior of innovative flat steel plate-concrete composite slabs. Int J Steel Struct 18(2):473–485
29. Yardim Y (2018) Review of research on the application of ferrocement in composite precast slabs. Periodica Polytech Civil Eng 62(4):1030–1038
30. Leshchenko MV, Semko V (2015) Thermal characteristics of the external walling made of cold-formed steel studs and polystyrene concrete. Mag Civil Eng 60(8):44–55. https://doi.org/10.5862/MCE.60.6
31. Kochkarev D, Azizov T, Galinska T (2018) Bending deflection reinforced concrete elements determination. In: MATEC Web of Conferences, vol 230. https://doi.org/10.1051/matecconf/201823002012
32. Korobko B (2016) Investigation of energy consumption in the course of plastering machine's work. Eastern-Eur J Enterp Technol 4(8–82):4–11. https://doi.org/10.15587/1729-4061.2016.73336
33. Loburets AT, Naumovets AG, Senenko NB, Vedula YS (1997) Surface diffusion and phase transitions in strontium overlayers on W(112). Zeitschrift Fur Physikalische Chemie 202(1–2):75–85. https://doi.org/10.1524/zpch.1997.202.part_1_2.075
34. Pavlenko A, Koshlak H (2015) Design of processes of thermal bloating of silicates. Metall Min Ind 7(1):118–122
35. Pichugin SF (2018) Reliability estimation of industrial building structures. Mag Civil Eng 83(7):24–37. https://doi.org/10.18720/MCE.83.3
36. Popova AV, Kremenetsky VG, Solov'ev VV, Chernenko LA, Kremenetskaya OV, Fofanov AD, Kuznetsov SA (2010) Standard rate constants of charge transfer for nb(V)/Nb(IV) redox couple in chloride-fluoride melts: experimental and calculation methods. Russ J Electrochem 46(6):671–679. https://doi.org/10.1134/S1023193510060121

Calculation of a Compressed Reinforced Concrete Element with a Circular Cross Section Using a Three-Line Concrete Compression Scheme

Ulviye Hajiyeva Mukhlis kizi

Abstract The application of a nonlinear deformation model allows the maximum use of the ultimate strength of materials. In addition, only on the basis of such models that reflect reality as much as possible, it is possible to correctly evaluate the deformation solutions of a reinforced concrete element. The article develops an effective numerical engineering method for calculating compressed reinforced concrete elements of circular cross-section, which has been studied much less than traditional rectangular cross-sections. When constructing the proposed calculation method, it is assumed that the hypothesis of flat sections becomes true for such a complex cross-section, when cracks occur in the zones of stretching of the reinforced concrete element. For the most sharpened section by approximation of curved axis of the centrifugal compressed element axis sinus is also formulated the equations of equilibrium deformations of the shaft and ultimately the solution of the problem is reduced to solving a system of nonlinear algebraic equations taking into account deformation of the compressible edge of the cross section and the relative height of compression zone of the cross section. On the basis of the article, a three-line curves numerically with any accuracy determines the parameters characterizing the state of tensile deformation by increasing the strain values in all changes of deformation with defined steps. The load capacity of the compressible element is determined based on the "load-bending" graph, constructed numerically for the compressible element.

Keywords Nonlinear deformation model · Eccentricity · Tension · Deformation · Load-bending graph · Three-line diagram

1 Introduction

When developing methods for calculating elements of construction structures, preference is given to methods based on real nonlinear schemes of materials because, only on the basis of these types of techniques, along with creating conditions for

U. H. M. kizi (✉)
Azerbaijan University of Architecture and Construction, 5 Ayna Sultanova, Baku, Azerbaijan

V. Onyshchenko et al. (eds.), *Proceedings of the 3rd International Conference on Building Innovations*, Lecture Notes in Civil Engineering 181,
https://doi.org/10.1007/978-3-030-85043-2_15

maximum use of the strength resources of materials, it is possible to more reliably and accurately determine the cases of tension deformation formed in structural elements under the influence of external loads. Modern building norms suggest the development of methods for solving the calculation of compressed reinforced concrete elements by approximating a curved diagram of deformation under short-term static loads of concrete with a three-line diagram. In the modern scientific literature, simpler cross-sectional shapes are usually studied. However, in construction practice, columns with a circular cross-section are also widely used. That is why the development of a method for determining the strain under tension and the load capacity under centrifugal compression of these elements is of great theoretical and practical importance. An effective numerical method for determining the state of strain under tension and the load-bearing capacity of reinforced concrete elements of circular cross-section compressed during concrete compression using a three-line deformation diagram is developed.

2 Main Part

2.1 *Purpose of the Article*

The main goal is to develop an effective method of engineering calculations based on a nonlinear deformation model for determining the strain state of tension and load-bearing capacity with an arbitrary assessment of the eccentricity of flexibility and compression force of round transversely compressed reinforced concrete elements used as columns in construction practice. When constructing the solution method, a three-line scheme of concrete deformation under tension was used, and a two-line scheme of reinforcement was used. The goal was the numerical construction of the load-bending diagram, which plays an important role in determining the load-bearing capacity for load-bearing elements.

2.2 *Research Methodology*

Having accepted the hypothesis of flat cross-sections for a reinforced concrete element, in the tension zone of which cracks have formed, applying a three-line diagram of concrete in compression and two-line compression and tension reinforcement, he distribution of normal stress on the cross section depending on the deformation of the compressible edge of the cross section and the relative height of compression zone is determined, and then the inclined axis of the compressible shaft with a sinusoidal semi-axis is approximated and the value of the maximum bending of the shaft is expressed by the same parameter. On the basis of the deformed scheme of the compressed shaft, the deformation equations for the most tensed section are compiled. The obtained nonlinear systems of equations are interpreted in the numerical solution algorithm taking into account the above parameters. The proposed numerical method is based on the fact that the parameters characterizing the state of the tensile strain and the load capacity of the compressible shaft are determined on the basis of the "load-bending" diagram.

2.3 Results

The complete scheme of concrete under compression is expressed by the fractional-rational Sargi diagram according to the Eurocode proposal: [1–4, 7–9]:

$$\sigma_b = R_b \cdot \frac{k \cdot \frac{\varepsilon_b}{\varepsilon_R} - \left(\frac{\varepsilon_b}{\varepsilon_R}\right)^2}{1 + (k - 2) \cdot \frac{\varepsilon_b}{\varepsilon_R}} \tag{1}$$

Eurocod also allows you to use a three-line diagram instead of this diagram, which is defined as follows [4, 5].

In this diagram $\sigma_A = 0,6 \cdot R_b, \sigma_B = R_b, \varepsilon_A = \frac{0,6 \cdot R_b}{E_b}, \varepsilon_B = 0,002 = \varepsilon_R,$ $\varepsilon_C = 0,0035 = \varepsilon_{b,\,ult}.$

Obviously, the analytical expression of the above diagram is written as follows (Fig. 1):

$$\sigma_b = \begin{cases} \frac{\sigma_A}{\varepsilon_A} \cdot \varepsilon_b\,; \quad when\ 0 \le \varepsilon_b \le \varepsilon_A \\ \sigma_A + \frac{\sigma_B - \sigma_A}{\varepsilon_B - \varepsilon_A} \cdot (\varepsilon_b - \varepsilon_A)\,; \quad when\ \varepsilon_A \le \varepsilon_b \le \varepsilon_B \\ \sigma_B\,; \quad when\ \ \varepsilon_B \le \varepsilon_b \le \varepsilon_C \end{cases} \tag{2}$$

Taking into account the correctness of the hypothesis of flat sections for any load step for an element with a circular cross-section and designating the value of

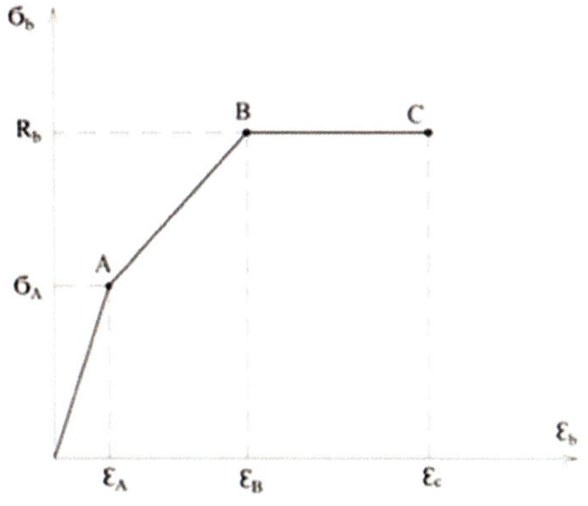

Fig. 1 Three-line diagram of concrete compression

deformation along the edge of the compressible face ε_b and the relative height of the compressible face of the section $\xi = \frac{x}{R}$, we can write down the dependence of the height of the deformation on the change along the height of the section $\overline{z} = \frac{z}{R}$, R- the is the radius of the cross-section. Taking this into account in (2), we obtain a dependence expressing the law of variation of the normal strains formed in concrete over the section, along the section height from the deformation value at the edge of the compressible face and the relative height of the section compression zone:

$$\sigma_{bz} = R_b \cdot \begin{cases} \frac{\overline{\sigma}_A}{\varepsilon_A} \cdot \frac{\varepsilon_b}{\xi} \cdot (\overline{z} + \xi - 1)\,; \; when \;\; 0 \le \varepsilon_b \le \varepsilon_A \\ \overline{\sigma}_A + \frac{\overline{\sigma}_B - \overline{\sigma}_A}{\varepsilon_B - \varepsilon_A} \cdot \left(\frac{\varepsilon_b}{\xi} \cdot (\overline{z} + \xi - 1) - \varepsilon_A\right); \; when \;\; \varepsilon_A \le \varepsilon_b \le \varepsilon_B \\ \overline{\sigma}_B\,; \;\; when \;\; \varepsilon_B \le \varepsilon_b \le \varepsilon_C \end{cases}$$

In this equation, the values of characteristic stresses expressed in portions of the compressive strength of concrete $\sigma_i = \overline{\sigma}_i \cdot R_b$ were used. Then, based on the corresponding formulas of the resistance of the materials for the normal force and torque that these stresses form in the cross section, it can be written that,

$$N_b = \int_{A_b} \sigma_{bz} \cdot dA = 2R_b \cdot R^2 \cdot \int_{q(\xi)}^{1} \overline{\sigma}_{bz}(\varepsilon_b, \xi) \cdot \sqrt{1 - \overline{z}^2} d\overline{z} = R_b \cdot R^2 \cdot N_*(\varepsilon_b, \xi) \quad (3)$$

$$M_b = \int_{A_b} \sigma_{bz} \cdot z \cdot dA = 2R_b \cdot R^3 \cdot \int_{q(\xi)}^{1} \overline{\sigma}_{bz}(\varepsilon_b, \xi) \cdot \overline{z} \cdot \sqrt{1 - \overline{z}^2} d\overline{z} = R_b \cdot R^3 \cdot M_*(\varepsilon_b, \xi) \quad (4)$$

In the above equations $q(\xi) = \begin{cases} -1\,; \;\; when \; \xi \ge 2 \\ 1 - \xi\,; \;\; when \;\; 0 < \xi < 2 \end{cases}$

Let's assume that the armature shafts are distributed according to the specified law on the cross section and express the deformation ε_b and ξ parameters of each armature on the basis of the hypothesis of flat cross sections. For an arbitrary fixture $\varepsilon_{sj} = \frac{\varepsilon_b}{\xi} \cdot \left(\frac{R_{sj}}{R} \cdot \sin\phi_{sj} + \xi - 1\right)$, then it can be written for the tension that arises in the armatue which is physical stream pitch,

$$\sigma_{sj} = \begin{cases} E_{sj} \cdot \frac{\varepsilon_b}{\xi} \cdot \left(\frac{R_{sj}}{R} \cdot \sin\phi_{sj} + \xi - 1\right); \\ when \; \left|\frac{\varepsilon_b}{\xi} \cdot \left(\frac{R_{sj}}{R} \cdot \sin\phi_{sj} + \xi - 1\right)\right| \le \varepsilon_{s,ax} \\ R_{sj} \cdot sign(\varepsilon_{sj}); \;\; when \; \left|\frac{\varepsilon_b}{\xi} \cdot \left(\frac{R_{sj}}{R} \cdot \sin\phi_{sj} + \xi - 1\right)\right| > \varepsilon_{s,ax} \end{cases} \quad (5)$$

Once the strains that occur on the armature shafts are known, we can write down the following known equations for the normal force and torque that occur in the cross section, due to the armature shafts:

$$N_s = R_s \cdot A_s \cdot \sum_{j=1}^{n} \overline{\sigma}_{sj} \cdot \overline{A}_{sj}; \quad M_s = R_s \cdot A_s \cdot R \cdot \sum_{j=1}^{n} \overline{\sigma}_{sj} \cdot \overline{A}_{sj} \cdot \overline{R}_{sj} \cdot \sin \varphi_{sj} \tag{6}$$

Approximating the curved axis of the compressed shaft with the semi-axis of the sinus [4, 6], we can

$$f_0 = \frac{f}{R} = \vartheta_0 \cdot \frac{\varepsilon_b}{\xi} \qquad \vartheta_0 = \frac{l_0^2}{\pi^2 \cdot R^2}$$

express the maximum bend with parameters ε_b and ξ:, where, l_0 is the calculated length of the compressed shaft. We write down the equllibrium equations for the most strained section, taking into account these:

$$N_*(\varepsilon_b, \xi) + \lambda_s \cdot \sum_{j=1}^{n} \overline{\sigma}_{sj} \cdot \overline{A}_{sj} = \overline{P}_0 \tag{7}$$

$$M_*(\varepsilon_b, \xi) + \lambda_s \sum_{j=1}^{n} \overline{\sigma}_{sj} \cdot \overline{A}_{sj} \cdot \overline{R}_{sj} \cdot \sin \phi_{sj} = \overline{P}_0 \cdot (\overline{e}_0 + f_0) \tag{8}$$

In these equations $\lambda_s = \dfrac{R_s \cdot A_s}{R_b \cdot R^2}, \ \overline{P}_0 = \dfrac{P}{R_b \cdot R^2}, \ \overline{e}_0 = \dfrac{e}{R}$ (9)

1–$e_0 = 0,05$; 2–$e_0 = 0,1$; 3–$e_0 = 0,5$; 4–$e_0 = 0,8$

From these equations, by calculating the compression force parameter, we obtain an equality in terms of elements that creates a relationship between the value of deformation on the compression face and the relative height of the compression zone of the cross-section:

$$\begin{aligned} & M_*(\varepsilon_b, \xi) + \lambda_s \sum_{j=1}^{n} \overline{\sigma}_{sj} \cdot \overline{A}_{sj} \cdot \overline{R}_{sj} \cdot \sin \phi_{sj} \\ & - \left[N_*(\varepsilon_b, \xi) + \lambda_s \cdot \sum_{j=1}^{n} \overline{\sigma}_{sj} \cdot \overline{A}_{sj} \right] \cdot (\overline{e}_0 + f_0) = 0 \end{aligned} \tag{10}$$

Since the interval of variation of the strain is obvious, changing it with a certain step, for each accepted strain estimate, it is possible to determine the relative height of the compressible cross-section zone corresponding to the accepted strain estimate from (10), the force parameter based on equality (9) and, finally, the strain on the reinforcement shafts based on equality (5). To perform calculations based on the above formulas, a corresponding software module was developed in the algorithmic language ABC Turbo Pascal, using this software module, numerical experiments were carried out and the letters "load-slope" for various values of eccentricity were

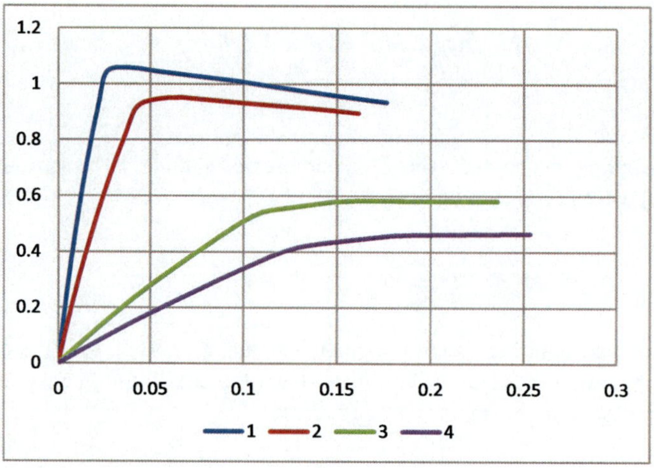

Fig. 2 Graphics "load-bending" of the centrifugal compression shaft

constructed in Fig. 2, while the element whose ends are hinged ($r_s = 18\ sm$, and $R = 20\ sm$, $R_b = 11,5\ MPa$) is reinforced with 12 Ø 12 reinforcement with a regular distribution over the cross section. The downward arm is also implemented in this diagram, while the real strain diagram is replaced by a three-line diagram, as seen in this diagram.

2.4 *Scientific Novelty*

A numerical method for determining the parameters and the "load-bending" diagram that characterize the state of deformation of a deformable multilayer reinforced concrete element of circular cross-section operating on centrifugal compression is developed, and its effectiveness is proved by numerical examples.

2.5 *Practical Importance*

Developed numerical method is of great practical importance, as it allows to determine the parameters characterizing the deformation of a tension round a broadband compressed reinforced concrete element, and a load element for an arbitrary estimate of the eccentricity of the force and flexibility of the shaft. On the basis of the developed methodology, conditions are created for the compilation of semi-automatic expressions for approximate engineering calculations.

3 Conclusions

The load-bearing capacity of reinforced concrete elements for eccentric compression is often expressed in terms of its value determined on the basis of the "load-bearing" schedule, depending on the value of flexibility, and it is important to apply this calculation method to the calculation of reinforced concrete elements.

References

1. Hajiev MA, Aliev RD, Guliyev FM (2019) Calculation of bending reinforced concrete elements for strength according to normal sections. Baku, p 121
2. Sanzharovsky RS, Veselov AA (2002) Theory of calculation of building structures for stability and modern standards. Moscow-Saint Petersburg, ACU, p 126
3. Bibi EV (2012) Guide for designers to the Eurocode 2. Design of reinforced concrete structures. Moscow MSUC Publishing House, p 292
4. Beglov AD, Sanzharovsky RS (2011) Eurostandards and nonlinear theory of reinforced concrete. ACB Publishing House, Saint Petersburg, p 309
5. Kodysh EN, Nikitin IK, Trekin NN (2011) Calculation of reinforced concrete structures made of heavy concrete for strength, crack resistance and deformation. Moscow, ACU, p 352
6. Hajiev MA (2005) Stability of compressed reinforced concrete rods reinforced with a fiberglass cage during prolonged loading. Reports of the NAS. Azerbaijan, Baku, 5:10-15
7. Kochkarev D, Azizov T, Galinska T (2018) Bending deflection reinforced concrete elements determination. In: MATEC Web of Conferences, vol 230. https://doi.org/10.1051/matecconf/201823002012
8. Kozak O, Zhuravskyi O, Delyavskyy M (2019) Effect of the pre-stressed reinforcement curvature on the bearing capacity of inclined sections of monolithic beams. In: AIP Conference Proceedings, vol 2077 https://doi.org/10.1063/1.5091888
9. Panchenko OV, Ivanyts'kyi YL, Kun' PS, Zhuravs'kyi OD (2018) Determination of the durability of ferroconcrete bridge beams reinforced by composite strips. Mater Sci 53(5):660–665. https://doi.org/10.1007/s11003-018-0121-3

Structural—Parametric Synthesis of Steel Combined Trusses

Myron Hohol, Grygorii Gasii, Volodymyr Pents, and Dmytro Sydorak

Abstract The aim of the research is structural—parametric synthesis of steel combined trusses with a span of 30 m according to the criteria of minimum mass and potential energy. It is shown that the bearing capacity of the structure is used most fully when using the criterion of rational design for strength. It is shown that the largest reserves of regulation of the stress deformation state and improvement of structural forms are hidden in the combined structures. The use of combined trusses instead of traditional ones makes it possible to achieve the greatest savings in metal, reducing the complexity of manufacturing and reducing installation time. It is confirmed that the rational, technological structure of the steel combined truss with a span of 30 m, which has a minimum cost of materials, minimum labor and time for its manufacture, there will be a truss 2 m high with six panels of the upper belt. The rational, on a steel expense, angle of inclination of racks of steel combined trusses in the range 78–82° is defined.

Keywords Combined truss · Rational truss · Stress deformation state · Weight · Upper belt panel · Rational angle

1 Introduction

Improvement and development of new progressive structural forms of light steel structures of coatings in order to increase their efficiency and ensure competitiveness compared to foreign counterparts, is becoming increasingly important. These include various combined systems. Analysis of existing design and construction experience

M. Hohol (✉) · D. Sydorak
Lviv Polytechnic National University, S. Bandera str. 12, Lviv 79013, Ukraine
e-mail: gogolmyron@i.ua

G. Gasii · V. Pents
National University «Yuri Kondratyuk Poltava Polytechnic», Pershotravnevyj Ave. 24, Poltava, Ukraine

V. Onyshchenko et al. (eds.), *Proceedings of the 3rd International Conference on Building Innovations*, Lecture Notes in Civil Engineering 181,
https://doi.org/10.1007/978-3-030-85043-2_16

shows that compared to traditional beam and frame structures, combined systems have a number of advantages [5, 11, 18].

The weight of buildings can be reduced up to five times compared to reinforced concrete, metal consumption up to three times due to the use of light steel structures, increase labor productivity by 1.5–2 times, and reduce construction time by 30–60% [13]. The reduction in the weight of the combined structures, depending on the materials used, can reach up to 83% [6].

Therefore, parametric study of combined steel trusses, which will provide a rational topology and assessment of their economic efficiency is currently an urgent problem.

Of particular importance is the task of their rational calculation and design. Rational design, which, in contrast to the optimal, does not involve the existence of any target functionality, and is expressed in the heuristic requirements for the stress deformation state (SDS) of the structure (uniformity, uniform stress, maximum rigidity, or minimum mass), which guarantees the improvement of its qualities by the most natural criterion of rational design for strength. The load-bearing capacity of the structure is used most fully.

The aim of the article is structural—parametric synthesis, topology research for design of rational (with high technical and economic indicators) steel combined trusses with a span of 30 m according to the criteria of minimum mass and potential energy, improvement of their constructive forms and comparison of efficiency in comparison with typical. The objectives of the study are: (a) determine the rational: the number of panels of the combined truss and the angle of the racks; (b) to investigate the stress deformation state of the stiffening beam with a rational topology of the combined truss.

1.1 Analysis of Basic Research and Publications

Beam-type steel structures and trusses are the most common structural elements in floors and coatings for industrial, civil and agricultural purposes. At present, the Ukrainian Research and Design Institute of Steel Structures named after V.M. Szymanowski developed a state standard for steel trusses from bent-welded profiles of rectangular cross-section [3]. This standard takes into account modern achievements in the field of steel structures, meets domestic and foreign requirements for such systems, and also provides weight restrictions, for example, the weight of the truss with a span of 30 m, for a load of 1.3 t/m should not exceed 2000 kg [3].

However, the following typical rafter trusses [3, 15] have the following disadvantages: imperfect complex topology due to a large number of elements and nodes (for example, a typical farm with a run of 30 m consists of 39 elements and has 21 nodes); incomplete use of the strength characteristics of steel in the elements along the length of the truss due to compliance with manufacturability in the manufacture; increased material and labor intensity and cost.

The problem can be solved: by developing and creating effective structural forms that meet the requirements of advanced manufacturing technologies: achieving the greatest savings in metal, reducing labor intensity and reducing installation time.

It is established that the largest reserves of SDS regulation and improvement of structural forms are hidden in combined structures [16]. Ways to increase the efficiency of combined structures are substantiated: low element, the concentration of the bulk of the system in the stiffening beam, the absence of external spacing, taking into account the deformed state of the stiffening beam based on the energy variation method of Lagrange, exclusion of power control methods, ensuring an even stress state of the stiffening beam only by rational selection of stiffness of the system elements [5].

Analysis of the development of structural forms and methods of calculation of combined steel structures shows that such structures are the most promising and they have great potential, but to date they have not found mass use [2, 5]. The use of combined structures allows to reduce steel costs to 21–27% and the cost of their manufacture, which as a result reduces their prime cost by 31% in comparison with typical designs of similar span. The main advantage of combined structures is the concentration of materials and the ability to design them low-element [1, 20].

Thus, the most promising in terms of improving the efficiency of structural forms are structures that the formation of which includes elements of beams and trusses, i.e. combined. The main load-bearing element of these structures is a stiffening beam (its weight reaches up to 70% of the total mass), from the metal content of which largely depend on the technical and economic performance of the entire system.

Currently, most attention has been paid to the study of rational combined steel trusses with a maximum span of 18–24 m [2, 21]. However, it is known that combined designs are not always rational and economical [13, 14, 17, 22–25].

The analysis of literature sources showed [5, 7] that in the special literature the problem of rational design of combined steel trusses and research of their rational topology (especially with a run of 30 m) is insufficiently covered.

Therefore, to solve this and the above problems it is necessary to investigate and develop new rational structural forms of steel combined trusses of the increased span to 30 m, which would have minimal steel costs, low labor intensity, were low-element and the most technological.

2 Presentation of the Main Research Material

We will study the replacement of traditional typical rafter steel trusses (for example, a 30 m span) combined with the same dimensions as the typical ones.

To evaluate the constructive forms it is necessary to adopt efficiency criteria. These include: volume of material; the cost of construction in the "case", or reduced value; the cost of manufacture, transport and installation. The criterion of value is of a conjunctural nature and real prices are fair at short notice. Given that the cost of steel is approximately 70% of the cost in the "case" of the structure [19], the criterion of the minimum cost should be replaced by a simpler criterion of the minimum mass of the material, which can be written as

$$M = \sum M_i = min, \text{ or } M_i = \sum_{i=1}^{m} A_i l_i \rho = min, \quad (1)$$

where M—is the mass of the structure; M_i—is the mass of the element as a component of the structure, ρ—is the density of the material.

Consider, as a standard, a typical rafter truss with bent-welded profiles of rectangular cross section with a span of 30 m and a height of 2 m with support on the upper belt (Fig. 1a) [3, 4]. It is known that the efficiency and manufacturability of trusses mainly depends on the structural shapes and dimensions [12], as well as the number of its elements [9, 10].

Let's analyze the design parameters of the following sprung trusses with a span of 30 m when working on an evenly distributed load: typical (reference)—Fig. 1a [3]; proposed combined trusses with support on the upper belt with a different number of panels of the upper belt (10 for the truss in Fig. 1b; 8 for the truss in Fig. 1c; 6 for the truss in Fig. 1d; 4 in Fig. 1e and 3 in Fig. 1f) with a lattice with descending struts and regular placement of racks, and the same angle of inclination of the racks, equal to $\beta = 67.5°$. For comparison, the overall dimensions of all trusses are the same as in the typical (Fig. 1).

The mass of each of the above trusses was determined by performing a comparative calculation on a symmetrically evenly distributed load q = 13 kN/m, with the same dimensions on the PC in the program "LIRA-CAD 2016 R5", Table 1.

From the Table 1, the mass of the trusses (Fig. 1d and g) exceeds the mass limit of 2000 kg according to [1], they are not rational for this span and load, so we will not consider them in the future. Indicators by mass of the other three truss (Fig. 1b, c, d) are less than 2000 kg and less than the mass of the reference truss (Fig. 1a).

As can be seen from Table 1, replacement of a typical (reference) truss (Fig. 1a) with a combined truss with the number of upper belt panels eight (Fig. 1c), which has a smaller number of elements (typical 39 and has 21 nodes, and combined—22 and has 16 nodes), also has a reduced mass by about 7% (Table 1). Combined truss with the number of panels of the upper belt six (Fig. 1d), compared with the combined truss (Fig. 1c), has an even smaller number of elements—16 and nodes 12, but not much more mass from the truss (Fig. 1c), by 2.5%. Thus, it is the most technological among those considered.

Also, in [5], based on the energy principles of structural mechanics, a functional dependence was obtained between the deformation energy U_b when bending a conventional beam and the deformation energy U when bending a continuous rigidity beam on intermediate elastic supports with even extremal diagram M_x, which simulates the upper belt of combined trusses, and the number of its spans n (excluding the energy of the supports):

$$U \cong \frac{U_b}{4,6n^4}., \text{ or } \lim U = \lim_{n\to\infty} \frac{U_b}{4,6n^4} = 0. \quad (2)$$

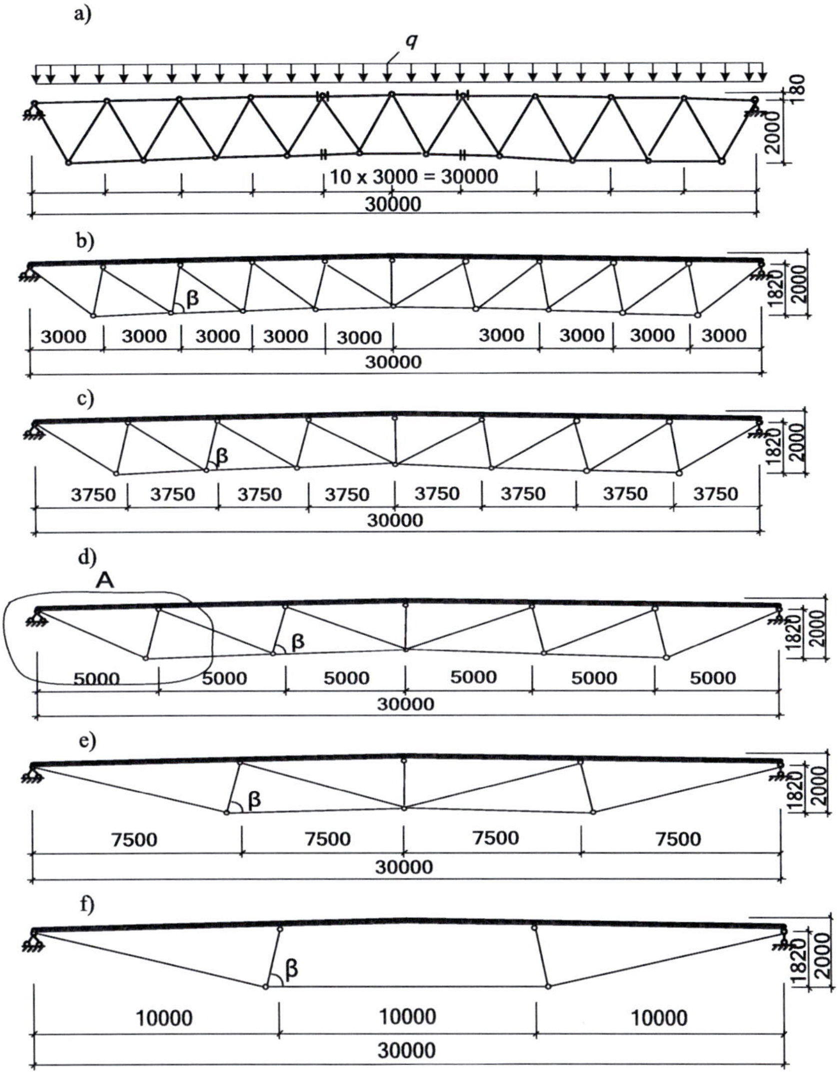

Fig. 1 Schemes of rafter steel trusses

Therefore, with the increase in the number of "n" spans of the stiffening beam (corresponding to n−1 number of its reinforcements by the struts of the sprung system) to infinity, the deformation energy of such a beam is reduced to zero, the beam turns into a rigid rod on a rigid basis in which there are no deformations of a bend (Fig. 2).

It was established [5, 8] that the mass of the beam span on two supports during its transformation into an inseparable beam on intermediate elastic supports (upper belt

Table 1 Weight and deflections of trusses

Truss	Deflection, mm	Mass, kg	Truss	Deflection, mm	Mass, kg
Figure 1a	147 mm (1/204 L)	1859,6	Figure 1d	148 mm (1/203 L)	1775,1
Figure 1b	147 mm (1/204 L)	1813,7	Figure 1e	148 mm (1/203 L)	2135.1
Figure 1c	148 mm (1/203 L)	1732,0	Figure 1f	137 mm (1/219 L)	2194.7

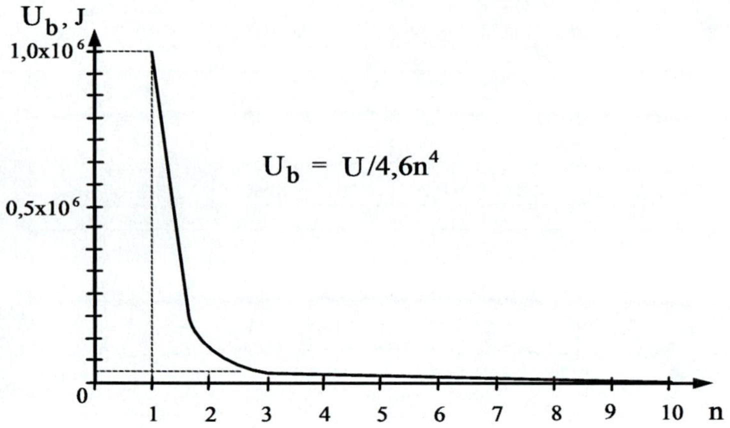

Fig. 2 Dependence of the potential energy of deformation of the beam U_b, when transformed into inseparable, on the number of its span n [5]

Fig. 3 Angles of the racks of the six-panel rational combined truss (node A for Fig. 1d)

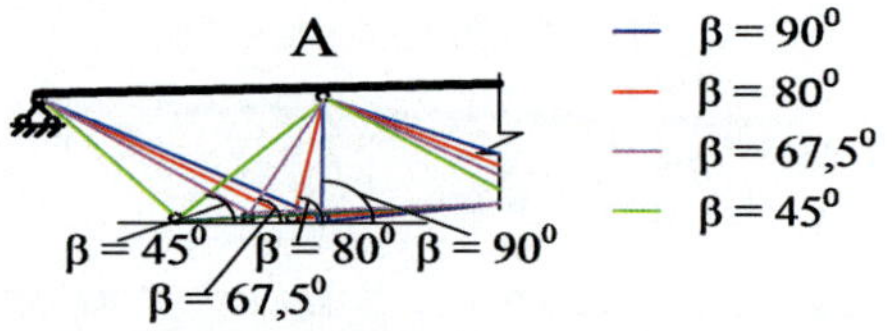

of combined trusses) decreases intensively with a small number of spans, ie small values of "n" (Fig. 2).

Based on this, we can conclude that it would be rational to combine a truss with the number of panels of the upper belt six in Fig. 1d, the upper belt of which has only five intermediate supports, and the trusses in other schemes—more, have seven and nine supports. In the case of a larger number of such supports, the mass of the beam decreases slowly, while each new support increases proportionally the mass of the reinforcement system and adds to the cost of structure and increases the complexity of manufacture and installation.

Thus, such a truss (Fig. 1d) is low-element, with the concentration of the bulk of the system in the stiffening beam—the upper belt, that is, it is a rational, technological structure that has a minimum cost of materials, minimum labor and time for its manufacture.

We analyze the influence of the angle of inclination β of the racks (Fig. 1d), on the mass of a rational truss height of 2 m, with the number of panels of the upper belt—6 pieces. To do this, determine the mass of the combined trusses at different angles of the racks in the range of 45–90°: $\beta = 45°$; $\beta = 67.5°$; $\beta = 80°$; $\beta = 90°$ (Fig. 3). The calculation was performed on a symmetrically evenly distributed load q = 13 kN/m, on a PC in the program "LIRA-CAD 2016 R5", Table 2.

As can be seen from Table 2 and the diagram in Fig. 4, the mass of the combined truss with the angle of inclination of the racks 80^0 is the smallest and the difference with the largest mass at an angle of 90° is 5%. The mass of the other two trusses, at angles of 67.5° and 45°, is greater than the minimum. Therefore, we can assume that the rational angle of the racks of the combined truss, the cost of steel, is approximately 78–82°, given the impact of a limited range of metal profiles and the accuracy of calculations.

Table 2 Weight and deflections of combined trusses

Truss	Deflection, mm	Mass, kg	Truss	Deflection, mm	Mass, kg
Figure 1d, Fig. 3 (angle 90°)	150 mm, (1/200 L)	1873,3	Figure 1d, Fig. 3 (angle 67,5°)	148 mm, (1/203 L)	1775,1
Figure 1d, Fig. 3 (angle 80°)	150 mm, (1/200 L)	1756,3	Figure 1d, Fig. 3 (angle 45°)	146 mm, (1/205 L)	1799.0

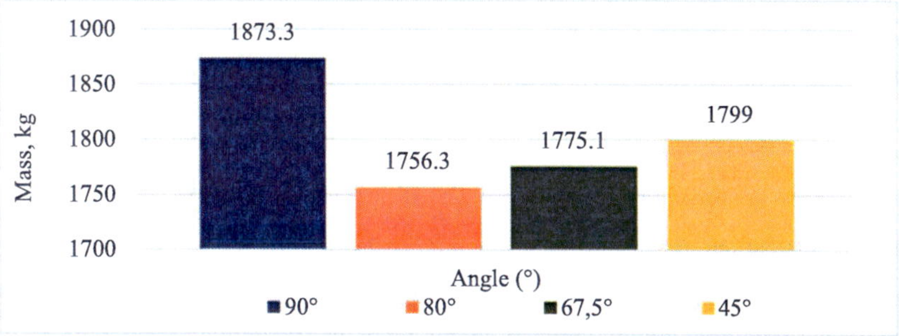

Fig. 4 Diagram of the influence of the angle of inclination of the racks β on the mass of the combined trusses

3 Conclusions and Prospects

It is shown that a rational combined steel truss with a span of 30 m has 16 elements and 12 nodes, and a typical truss of the same girder, respectively—39 elements and 21 nodes, therefore, such a combined truss should be considered low-element. Also, compared to a typical, rational combined truss has about 5% less mass, that is, it is a technological structure that has a minimum cost of materials and, accordingly, the minimum cost of labor and time for its manufacture. The combined truss with a span of 30 m and a height of 2 m will be rational, both in terms of steel costs and their energy consumption, then, when it will have no more than six panels of the upper belt, and the upper belt itself will have no more than five intermediate supports. The influence of the angle of inclination of racks of combined trusses on their mass is investigated. The rational angle of inclination of racks of steel combined trusses which makes 78–82° is defined.

References

1. BendsoeM, Sigmund O (2003) Topology optimization: theory, methods and applications. Springer Verlag, Heidelberg
2. Bilyk S, Tonkacheiev V (2018) Determining sloped-load limits inside von Mises truss with elastic support. Materiali in Tehnologije 52(2):105–109
3. DSTU B B.2.6–74: 2008 (2009) Constructions of houses and buildings. Steel truss trusses made of rectangular bent welded profiles. Ministry of Regional Development of Ukraine, Kiev
4. European Committee for Standardization (CEN) (2005) EN 1993-1-8, Eurocode 3: Design of steel structures – Part 1–8: Design of joints
5. Gogol M (2018) Stress regulation in steel combined structures: Monograph. Steel, Kyiv
6. Gaganelis G, Mark P (2019) Downsizing weight while upsizing efficiency: an experimental approach to develop optimized ultra-light UHPC hybrid beams. Struct Concrete 20:1883–1895 (2019). https://doi.org/10.1002/suco.201900215
7. Hudz S, Storozhenko L, Gasii G, Hasii O (2020) Features of operation and design of steel sloping roof purlins. Lecture notes in civil engineering, vol 73, pp 65–73 (2020)
8. Gogol M, Zygun A, Maksiuta N (2018) New effective combined steel structures. Int J Eng Technol 3.2(7):343–348 (2018). https://doi.org/10.14419/ijet.v7i3.2.14432
9. Gogol M, Kropyvnytska T, Galinska T, Hajiyev M (2020) Ways to improve the combined steel structures of coatings. In: Proceedings of the 2nd international conference on building innovations. Lecture notes in civil engineering, vol 73, pp 53–58. Springer Nature, Switzerland AG (2020). https://doi.org/10.1007/978-3-030-42939-3_6.
10. Lavrinenko L, Zotina A (2019) Effective parameters of low-element sprung trusses with the use of I-beams with corrugated walls. Build Struct Theo Pract 4:56–69. https://doi.org/10.32347/2522-4182.4.2019.56-59.
11. Lim J, Jeonga J, Kimb I, Ahna S (2015) Arch under-tension footbridges http://www.cospi.co.kr/upload/board/2015051884822_pap88lim.pdf
12. Madrazo-Aguirre F, Wadee M, Ruiz-Teran A (2015) Non-linear stability of under-deck cable-stayed bridge decks. Int J Non-Linear Mech 77:28–40. https://doi.org/10.1016/j.ijnonlinmec.2015.07.001
13. Pichugin S, Gasenko A, Dmitrenko A, Kramar A (2013) Technical and economic comparison of light steel structures of summer stage cover. Resour Saving Mater Struct Build Struct 25:576–582

14. Pichugin S (2018) Reliability estimation of industrial building structures. Mag Civil Eng 83(7):24–37. https://doi.org/10.18720/MCE.83.3
15. Recommendations for the design of steel trusses with belts from wide-flange I-beams and lattice from bent-welded profiles (1988). Tsniiproekstalkonstruktsiya them. N.P. Melnikov, Moscow
16. Ruiz-Teran A, Aparicio A (2010) Developments in under-deck and combined cable-stayed bridges. Bridge Eng 163:67–78 (2010). https://doi.org/10.1680/bren.2010.163.2.67
17. Semko O, Hasenko A, Kyrychenko V, Sirobaba V (2020) The rational parameters of the civil building steel frame with struts. Lecture Notes in Civil Engineering, vol 73, pp 235–243
18. Shimanovsky O, Gogol M (2018) A new approach to designing effective combined steel trus. In: International Scientific and Practical Conference «Technology, Engineering and Science–2018», London, pp 16–18
19. Vasilkin A, Shcherbina S (2015) The automated solution of a problem of determination of optimum height of a steel farm on criterion of a minimum of weight at a farm height variation. Integration, partnership and innovations in construction science and education. In: Collection of materials of the International conference Ministry of Science and Education of RF Moscow State University of Civil Engineering, pp 131–134
20. Yuriev A, Zinkova V (2020) Optimization of loading of metal trusses. Bull BSTU im. V.G. Shukhov 2:56–61
21. Yuriev A Variational formulations of structural synthesis problems. Collection of scientific papers Sworld 4(14), 67–70
22. Loburets AT, Naumovets AG, Senenko NB, Vedula YS (1997) Surface diffusion and phase transitions in strontium overlayers on W(112). Zeitschrift Fur Physikalische Chemie 202(1–2):75–85. https://doi.org/10.1524/zpch.1997.202.part_1_2.075
23. Zotsenko ML, Mykhailovska OV, Sivitska SP (2020) Construction features durable storage of toxic waste in boreholes.https://doi.org/10.1007/978-3-030-42939-3_33
24. Vynnykov Y, Manhura A, Zimin O, Matviienko A (2019) Use of thermal and magnetic devices for prevention of asphaltene, resin, and wax deposits on oil equipment surfaces. Min Miner Deposits 13(2):34–40. https://doi.org/10.33271/mining13.02.034
25. Pavlenko A, Koshlak H (2015) Design of processes of thermal bloating of silicates. Metall Min Ind 7(1):118–122

Matrix Memory Device

Mirfatma Javadova and Ilona Chernytska

Abstract The article considers a matrix memory element, which is a storage matrix built using diodes based on complex semiconductors. A decoder device using memory cells based on Cu_2Se diodes, which provides the necessary signal based on various combinations of input signals. MD (memory device) can find application in information processing systems when creating a storage device with the required capacity and capable of working in the interval from −60 °C to +125 °C. As a multi-functional device, the requirement for the electoral device is that it is advantageous in terms of energy saving and for the same reason, the decoders are built on the elements with the lowest energy consumption. Matrix memory element can be used built using diodes based on complex semiconductors and matrix memory device can be used in information distribution and processing networks, computer technology, automation systems and in other areas of technology. İn information processing systems when creating a memory device with the required capacity. Based on the abovementioned, the devices are modern, economical, highly technological for the development and applicable to modern automation devices and switching equipment.

Keywords Information transfer · Complex semiconductor · Reliability · Digital switches · Electricity

1 Introduction

The development of microelectronics has allowed the use of semiconductor elements (bipolar transistors and MOS structures) to build memory devices. The use of bipolar transistors and MOS structures can significantly increase the speed of MD reduce their mass, dimensions and increase reliability [1].

M. Javadova
Azerbaijan University of Architecture and Construction, Ayna Sultanova 11, Baku, Azerbaijan

I. Chernytska (✉)
National University «Yuri Kondratyuk Poltava Polytechnic», Pershotravnevyj Avenue 24, Poltava 36011, Ukraine

V. Onyshchenko et al. (eds.), *Proceedings of the 3rd International Conference on Building Innovations*, Lecture Notes in Civil Engineering 181,
https://doi.org/10.1007/978-3-030-85043-2_17

Table 1 Characteristics of MD made on various electronic technological basis

RAM type	Applicabe elements	Characteristics			
		Production time, ns	Typical information capacity, byte	Density of information placement, byte	Energy consumption during storage of information
Semi conductor	Bipolyar Transistors	50 ÷ 300	103 ÷ 105	Up to 200	Yes
	MOS Structures	250 ÷ 100	103 ÷ 106	200 ÷ 300	Yes
Magnetic	Ferrite cores	350 ÷ 1200	106 ÷ 108	10 ÷ 20	No

The Table 1 shows the data characterizing the MD made on a different elemental – technological basis. The table shows that on bipolar transistors it is advisable to build a MD with an information capacity of less than 105 bits, but with high speed. The MD on MOS structures have a capacity of $10^3 \div 10^6$ bits at moderate speed.

2 Methods

The main disadvantage of MD on mentioned structures is the energy consumption during storage of information. In this regard, at present the development of memory is carried out in two directions. MD are developed on integrated circuits with the improvement of their capabilities, as well as on new elements which do not consume energy when storing information.

A special memory device can be built using diodes based on complex semiconductors. It‘s possible to consider the memory matrix built using diodes based on complex semiconductors. Matrix memory device can be used in information distribution and processing networks, computer technology, automation systems and in other areas of technology. Advantageously, MD can find application in information processing systems when creating a memory device with the required capacity.

XM-based diodes have a memory, moreover, information can be stored in them even in the absence of power, the duration of information storage by XM diodes is checked for 1.5 years. The ability of XM diodes to withstand high temperatures and radiation resistance suggests that memory cells developed on the basis of XM diodes can compete with modern memory elements currently used in memory devices [2].

A characteristic feature of the proposed memory device is that it is built on diodes based on complex semiconductors with two stable states, Fig. 1. With N capacitance, this storage device also provides reception, storage and delivery of N number n bit numbers. A disadvantage of the known matrix memory device is the inability to use the operating currents less than the full value supplied by the vertical and horizontal buses, which leads to disruption of the elements installed before and after the selected

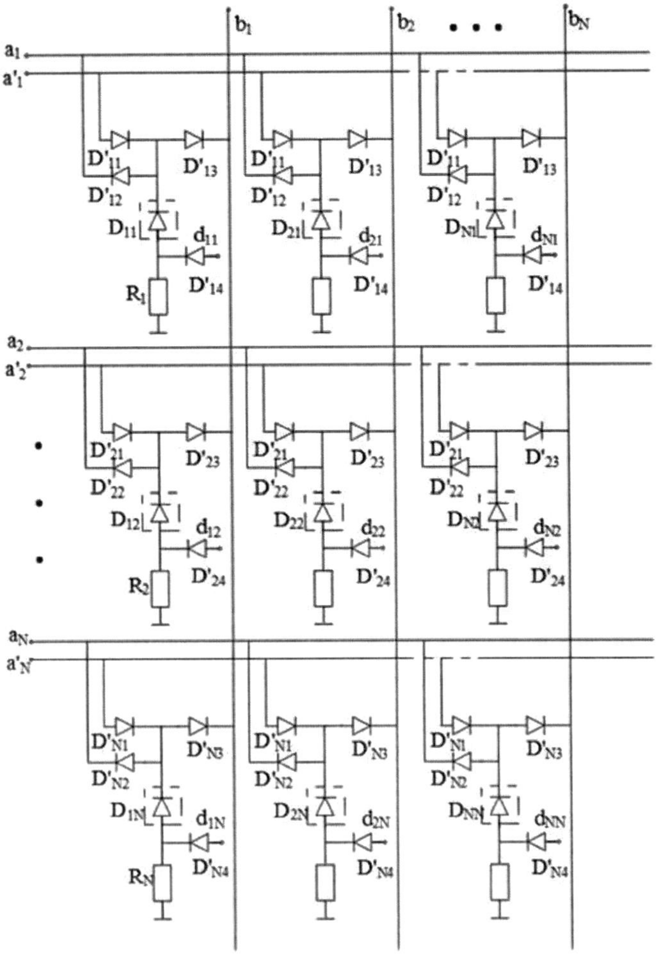

Fig. 1 Matrix memory device

elements. Therefore, in such storage devices of a matrix type, at the outputs of each storage element, additional elements are used to record information.

The matrix storage device contains b_1,b_2, …b_N vertical bus; D_{11}, D_{12}, …D_{1N}; D_{21}, D_{22}, …D_{2N}; D_{N1}, D_{N2}, …D_{NN} diodes based on complex semiconductors; D'_{11}, D'_{12}, D'_{13}, D'_{14}; D'_{21}, D'_{22}, D'_{23}, D'_{24};... D'_{N1}, D'_{N2}, D'_{N3}, D'_{N4} decoupling diodes in one column d_{11}, d_{12}, …d_{1N}; d_{21}, d_{22},... d_{2N};... d_{2N},... d_{N1}, d_{N2},.., d_{NN} output pins of diodes based on complex semiconductors for each column. Matrix memory device operates as follows. The memory device simultaneously records all the digits of the same number.

Voltage across one vertical bus and n horizontal buses (n is the number of bits of one number) is supplied simultaneously. A voltage with a value is supplied via the vertical bus, where U is the total voltage value $\frac{2}{3}U$ for switching the memory element. The voltage with the value $-\frac{1}{3}U$ is supplied via the horizontal bus a, the voltage with the value $+\frac{1}{3}U$ is supplied through the horizontal bus a1. To record n bits of a single number in the memory device, one of the vertical buses b1, b2, … bn is supplied with voltage with magnitude $\frac{2}{3}U$, while voltage for each horizontal bus is either with magnitude $+\frac{1}{3}U$ or magnitude $-\frac{1}{3}U$. When recording '1' in the memory device, voltage is applied simultaneously to the voltage of the vertical bus $\frac{2}{3}U$ via the corresponding horizontal bus $+\frac{1}{3}U$. If it is necessary to record '0' in the storage device, then a voltage with magnitude is supplied via the corresponding horizontal bus $+\frac{1}{3}U$. Thus, when recording '1', the storage element—a diode based on complex semiconductors is powered by a voltage with a value

$$\frac{2}{3}U + \frac{1}{3}U = U \tag{1}$$

The value $\frac{1}{3}U$ is so small that the diode based on complex semiconductors does not respond to this and remains in the same zero '0' state. When reading the recorded information in the storage device via the corresponding vertical bus, a voltage signal with a value $-\frac{2}{3}U$ is provided, and at the same time, horizontal signals of a voltage signal each with a value $-\frac{1}{3}U$. The reliability of the memory cells in the mode of reading, storing and overwriting information is determined by the stability of the voltage and switching current of the memory element. Researches have shown that the switching voltage and switching current of these elements is practically not depends of the ambient temperature in the range of $-60 \div +125$ °C. The distribution curves of the value of the turn-off voltage obtained at a frequency of 100 kHz and for a pulse duration of a logical "0" 0.5 μs are shown in Fig. 2. Figure 3 shows the distribution curve of switching current. As follows from Fig. 2, the most likely value of the turn-off voltage is 6.3 V, which allows the use of memory cells in conjunction with standard TTL ICs with a voltage of 5 V.

However, the value of the turn-off voltage is characterized by a certain spread both from one switching cycle to another and for a separate memory element, which leads to the need to use statistical analysis methods to establish guaranteed acceptable signal values that ensure non-destructive reading of information. Assuming that the experimentally obtained distributions of the turn-off voltage values obey the normal

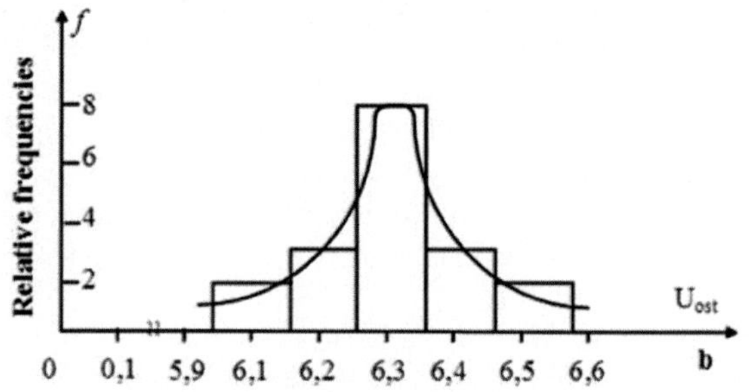

Fig. 2 Turn-off voltage distribution curve

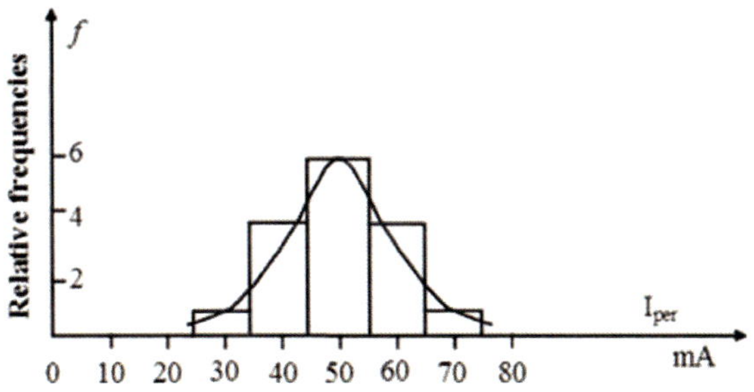

Fig. 3 Distribution curve of the switching current

law, numerical methods can be used to obtain statistical estimates of the mathematical expectation of the variance and the standard deviation of given value.

It is proved that the expansion of relations between the processes carried out in many areas of the national economy, as well as the implementation of the reception and transmission of information at numerous points, leads to the development of a two-way amount of information between various objects. At the modern stage of development of switching technology, the main means of providing two-way transmission of information between different objects is a digital switch [3].

The developed device of the decoder using memory cells based on Cu2Se diodes provides the output of a specific signal based on various combinations of input signals. Such devices are used in converters, control and selective circuits. The decoder, as a selective device, performs the function of selecting a device connected to its output. As a multifunctional device, it is beneficial from the point of view of energy saving and for the same reason, the decoders are built on the elements with the lowest energy consumption. As known, these elements, even at rest mode, do not consume electricity, while at the same time saving the information recorded in it.

3 Conclusions

A decoder device using memory cells based on Cu2Se diodes, which provides the necessary signal based on various combinations of input signals. Switching devices and memory elements called as a memory cell are capable of operating in the temperature range from −60 °C to +125 °C [4, 5]. Decoder devices developed using memory cells based on Cu2Se diodes, which provides the necessary signal based on various combinations of input signals. ("Patent" No. İ 2006 0041, 2006). Based on the above-mentioned, it can be confirmed that that the devices are economical and highly technological and suitable for the development and use of modern automation devices and switching equipment.

References

1. Abbasov G, Ibragimov M, Radjabov M (2002) On the construction of a switch using diodes based on complex semiconductors (in English). Physics number 3. ELM Publishing House
2. Abbasov H, Ibrahimov M, Rajabov M (2002) Symmetrical switching device. AzACU. Scientific works
3. Mehdiyeva S, Abbasov H, Cavadova M (2003) About the implementation of processes in determined three-stage switching systems. Az ACU. Scientific works. 1
4. Ermakov O (2004) Applied Optoelectronics. Technosphere
5. Cherniha R, Serov M (2003) Nonlinear systems of the burgers-type equations: lie and Q-conditional symmetries, ansätze and solutions. J Math Anal Appl 282(1):305–328. https://doi.org/10.1016/S0022-247X(03)00155-0

Methods of Probabilistic Assessment of Building Enclosing Structures Thermal Reliability

Alla Kariuk, Victor Pashynskyi, Mykola Pashynskyi, and Fidan Mammadova

Abstract Probabilistic method for determining statistical characteristics of heat transfer resistance and enclosing structures inner surface temperature, which takes into account random nature of enclosing structures layers thickness, thermal characteristics of materials, indoor and ambient temperatures are developed. All random variables have normal distributions, and the air temperature is presented in a form of a sequence of 12 normally distributed random variables for each month of the year. Obtained statistical characteristics of the enclosing structure inner surface temperature for each of the heating period months enable estimating the duration of thermal failure (in minutes per year) by the criteria of comfort near the enclosure and formation of condensate on its inner surface. Relative durations of thermal failures can be roughly considered as annual failure probabilities. Developed techniques allow performing a comparative assessment of different enclosing structures thermal reliability level in the given climatic conditions of operation according to the criteria of sufficient heat transfer resistance, comfort in the room and possibility of condensation on the enclosures inner surface. Performed examples of calculations indicate a sufficient level of residential buildings walls thermal reliability with facade insulation, made in accordance with current design standards of Ukraine, and impossibility of normal operation of 510 mm thick brick walls without additional insulation, which were massively erected in the second half of last century.

Keywords Enclosing structures · Thermal reliability · Materials random properties · Random temperature effects

A. Kariuk (✉)
National University «Yuri Kondratyuk Poltava Polytechnic», Pershotravnevyj Avenue 24, Poltava 36011, Ukraine

V. Pashynskyi · M. Pashynskyi
Central Ukrainian National Technical University, University Avenue 8, Kropyvnytskyi 25006, Ukraine

F. Mammadova
Azerbaijan University of Architecture and Construction, Ayna Sultanova 11, Baku, Azerbaijan

V. Onyshchenko et al. (eds.), *Proceedings of the 3rd International Conference on Building Innovations*, Lecture Notes in Civil Engineering 181,
https://doi.org/10.1007/978-3-030-85043-2_18

1 Introduction

Enclosing structures should ensure not only buildings energy efficiency, but also the comfort of being indoors. Current design standards [1] establish four criteria for ensuring thermal reliability in enclosing structures design:

- specific annual energy consumption of the building should not exceed the maximum allowable value;
- heat transfer resistance must be not less than the minimum allowable value, depending on the type of enclosing structure and the temperature zone of its operation;
- temperature difference between the indoor air temperature and the enclosing structure inner surface temperature must not exceed the permissible value, which ensures the comfort of being near the enclosures;
- minimum value of the inner surface temperature in the areas of heat-conducting inclusions must be not less than the dew point temperature, which ensures impossibility of condensation on the enclosure inner surface.

Random nature of materials thermal characteristics, structural layers thickness, ambient air temperature and room temperature necessitates the use of probabilistic methods for assessing the level of enclosing structures thermal reliability of [2, 3].

Among a significant number of studies on atmospheric air temperature, we place emphasis on the works [4, 5], that substantiates possibility of atmospheric air temperature probabilistic representation in the form of a quasi-stationary random process or a sequence of 12 random variables with normal distribution and statistical characteristics of these models for almost 500 observation points of Ukraine. Similar studies of atmospheric air temperature made it possible to develop zoning according to climatic indicators [1, 6, 7].

Calculated values of indoor air temperature are set by norms [1] depending on the premises purpose. For civil buildings they are equal to 20–22 °C. In [3], the results of processing the systematic measurements of indoor air temperature, according to which the standard of indoor air temperature varies within 0.4…0.8 °C are presented.

The urgency of energy saving problem has led to a significant amount of research in the field of thermal engineering. Research and probabilistic representation of thermal insulation building materials characteristics was carried out in [8–11] and other works of the authors. It is established that coefficients of thermal conductivity can be represented in the form of normally distributed random variables with 0.03–0.14 variation coefficients. The average and calculated values of different thermal conductivity of building materials in different operating conditions are given in the standard [12]. Thermal characteristics of enclosing structures were analyzed in articles [2, 3, 11, 13, 14] and many other works. Considerable attention has also been given to the optimal values of thermal characteristics selection [15, 16] and the improvement of the buildings walls structures in order to increase their thermal reliability [17, 18].

One of the first approaches to assessing reliability of load-bearing structures by the criterion of strength is developed by O.R. Rzhanitsyn and methodically explicated in the monograph [19]. The essence of this method is to determine probability of structural element failure-free operation, taking into account bearing capacity random values and force from the current load. In the recent decades, methods for estimating the reliability indicators of load-bearing structures have been developed [20–23], which are based on potential input in form of random processes, sequences of maximum values and other probabilistic models.

The works [2, 3, 10, 11] are dedicated to probabilistic assessment of thermal reliability of enclosing structures. The disadvantage of these studies is incomplete consideration of calculated parameters statistical variability, in particular the outside and inside air temperature, thickness of enclosing structure layers and thermal conductivity of the materials used. In the article [3] it is offered to use not probability, but duration of thermal failure condition as numerical indicator of thermal reliability. This to some extent corresponds to the norms approach [24] to the calculation of structures according to the requirements of the serviceability limit state.

In general, the literature analysis showed that development of methods for assessing thermal reliability with a comprehensive account of the random nature of enclosing structures structural and thermal characteristics, temperature in the premises and ambient air temperature remains an urgent task.

The purpose of the study is to develop a method for assessing enclosing structures thermal reliability in the winter time, taking into account random nature of the influencing factors.

2 Prerequisites and Initial Data

The given below methods for assessing enclosing structures thermal reliability are based on the following prerequisites:

- the basis is made with calculation methods and formulas of thermal engineering given in the current design norms;
- enclosing structure layers thickness of the δ_i and coefficients of materials thermal conductivity λ_i are considered to be normally distributed random variables;
- outdoor air temperature, τ_{out} is a sequence of normally distributed random variables corresponding to individual months of the year;
- room temperature τ_{in} is a normally distributed random variable, the average value of which is set according to design standards, while the standard is set according to the results of field observations in residential buildings;
- dew point temperature τ_c is a random variable, statistical characteristics of which depend on the temperature and relative humidity in the premises;
- heat transfer coefficients of the enclosing structure inner α_{in} and outer α_{out} surfaces are considered to be deterministic values and are accepted according to the design norms;

Table 1 Wall layers construction and thermal conductivity characteristics

Wall construction	Wall layers thickness δ_i, m			S_δ m	Thermal conductivity characteristics		
	Type 1	Type 2	Type 3		R_i	M_λ	S_λ
Outer protective layer	0,01	0,01	–	0,002	0,011	0,860	0,043
Mineral wool plate insulation	0,12	0,08	–	0,002	2,449	0,047	0,004
Laying of hollow ceramic bricks	0,51	0,51	0,51	0,006	0,797	0,591	0,030
Internal plaster	0,02	0,02	0,02	0,006	0,025	0,749	0,037

- the minimum allowable value of heat transfer resistance of the enclosing structure R_{min} is a deterministic value adopted in accordance with the requirements of design standards.

Methods for thermal reliability assessing are illustrated by examples of calculation of residential buildings brick walls in the climatic conditions of Poltava and Kropyvnytskyi:

Type 1—brick wall, insulated in accordance with the requirements of the norms [1] regarding the minimum required heat transfer resistance $R_{min} = 3{,}3$ $m^2\cdot K/W$.

Type 2—brick wall, insulated in accordance with the requirements of the norms [1] regarding the minimum required heat transfer resistance $R_{min} = 2{,}475$ $m^2\cdot K/W$ reduced by 25% which is allowed provided ensuring overall energy efficiency of the building;

Type 3—brick wall of a residential building without additional insulation, which was built en masse in the second half of the twentieth century;

Wall constructions and statistical characteristics of thickness and thermal conductivity of each layer are given in Table 1. Mathematical expectations of layer thickness δ_i are accepted as equal to their design value, and S_δ standards are estimated by normative tolerances on product and layer dimensions taking into account 0.95 provision. Heat transfer supports of R_i wall layers are calculated according to calculated values of materials thermal conductivity from the norms [12], and their M_λ mathematical expectations and S_λ standards—from the results of experimental and statistical studies, performed by authors.

3 Enclosing Structure Heat Transfer Resistance

Calculated value the structure heat transfer resistance is determined by a well-known formula:

$$R = \alpha_{in} + \alpha_{out} + \sum_{i=1}^{N} \frac{\delta_i}{\lambda_i}, \quad (1)$$

where N –number of enclosing structure layers,

δ_i – thickness of i-th enclosing structure layer;

λ_i –calculated thermal material conductivity of the i-th layer of the structure;

α_{in}, α_{out} – heat transfer coefficients of enclosing structure inner and outer surfaces according to the data [12];

Random nature of δ_i enclosing structure layers thickness and λ_i materials thermal conductivity coefficients determines the randomness of heat transfer resistance resulting value (1). Considering the nonlinear operation of dividing random variables δ_i and λ_i in the formula (1), statistical characteristics of the resulting random variable R are obtained by the method of random variables functions linearization described in [25]. For this purpose, partial derivative functions (1) are determined from the random parameters δ_i and λ_i:

$$\frac{\partial R}{\partial \delta_i} = \frac{1}{\lambda_i}; \qquad \frac{\partial R}{\partial \lambda_i} = -\frac{\delta_i}{\lambda_i^2}. \quad (2)$$

Then, according to the method of linearization [25], mathematical expectation and standard of heat transfer resistance are equal to the following:

$$M_R = \alpha_{in} + \alpha_{out} + \sum_{i=1}^{N} \frac{M_{\delta i}}{M_{\lambda i}}, \quad S_R = \sqrt{\sum_{i=1}^{N} \left[\left(\frac{S_{\delta i}}{M_{\lambda i}} \right)^2 + \left(\frac{S_{\lambda i} M_{\delta i}}{M_{\lambda i}^2} \right)^2 \right]}, \quad (3)$$

where N – number of enclosing structure layers;

$M_{\delta i}$ and $S_{\delta i}$ – mathematical expectation and the standard of thickness of the i-th layer;

$M_{\lambda i}$ та $S_{\lambda i}$ – mathematical expectation and the standard of the i-th layer thermal conductivity.

As a result of calculations according to formulas (1)… (3) and the data from Table 1, the following calculated values of R, mathematical expectations M_R and standards S_R of heat transfer resistance of the wall of three types are obtained:

for wall type 1 – R = 3,440 $m^2 \times K/W$, M_R = 3,612 $m^2 \times K/W$, S_R = 0,226 $m^2 \times K/W$;

for wall type 2 – R = 2,623 $m^2 \times K/W$, M_R = 2,761 $m^2 \times K/W$, S_R = 0,158 $m^2 \times K/W$;

for wall type 3 – R = 0,819 $m^2 \times K/W$, M_R = 0,873 $m^2 \times K/W$, S_R = 0,036 $m^2 \times K/W$.

Calculated values of heat transfer resistance of walls of types 1 and 2 correspond to the established norms [1] and the above-mentioned minimum permissible values for the walls of civil buildings in the first temperature zone of Ukraine. For wall type 3, the calculated value is 4 times lower than modern standards [1]. Mathematical expectations of heat transfer resistance of all walls are slightly higher than their calculated values.

4 Temperature of the Enclosing Structure Inner Surface

In the stationary mode of heat transfer, the temperature of enclosing structure inner surface can be determined from the known dependences of thermal engineering:

$$\tau_w = \frac{1}{R \cdot \alpha_{in}}[\tau_{in}(R \cdot \alpha_{in} - 1) + \tau_{out}], \quad (4)$$

where R – enclosing structure heat transfer resistance;

α_{in} – heat transfer coefficient of the enclosing structure inner surface;

τ_{in}, τ_{out} – indoor and outdoor air temperatures.

According to the accepted preconditions, all variables of formula (4) are normally distributed random variables. Therefore, the temperature of the enclosure structure inner surface is also a random variable, statistical characteristics of which can be determined by the method of linearization [25], as it was done for heat transfer resistance. Its mathematical expectation M_w and standard S_w are equal to the following:

$$M_w = \frac{1}{M_R \cdot \alpha_{in}}[M_{in}(M_R \cdot \alpha_{in} - 1) + M_{out}], \quad (5)$$

$$S_w = \frac{1}{M_R \cdot \alpha_{in}}\sqrt{S_{out}^2 + (M_{in} \cdot \alpha_{in} - 1)^2 S_{in}^2 + \frac{(1 - M_R \cdot \alpha_{in})^2}{M_R^2} \cdot S_R^2}, \quad (6)$$

where α_{in} – heat transfer coefficient of the enclosing structure inner surface;

M_R, S_R – mathematical expectation and standard of enclosure heat transfer resistance;

M_{in}, S_{in} – mathematical expectation and indoor air temperature standard;

M_{out}, S_{out} – mathematical expectation and outdoor air temperature standard.

According to formulas (5) and (6) it is possible to establish statistical characteristics of enclosing structure internal surface temperature for each of the heating period months. For this purpose it is necessary to substitute to formulas (5) and (6) mathematical expectations and standards of external air temperature in a certain month of the year, statistical characteristics of internal air temperature and enclosing structure heat transfer resistance (3). A simplified version of formula (6) without the third summand under the radical was obtained in [3], where the heat transfer resistance was considered a deterministic value.

5 Duration of the Thermal Failure State According to the Criterion of Comfort

This indicator is determined due to the probability that the difference Δ between the indoor air temperatures τ_{in} and the inner wall surface τ_w will exceed the permissible value Δ_{max} established by the norms [1]. According to the theorems on the numerical characteristics of random variables [25], the mathematical expectation and the standard of this difference are equal to the following:

$$M_\Delta = M_{in} - M_w, \quad S_\Delta = \sqrt{S_{in}^2 + S_w^2}, \tag{7}$$

where M_{in}, S_{in} – mathematical expectation and indoor air temperature standard;

M_w, S_w – mathematical expectation and temperature standard of the enclosing structure inner surface by (5), (6).

Absolute duration of the thermal failure state (in minutes per month) in the j-th month of the year is equal to the following:

$$Q_j = 43920\,[1 - F(\Delta_{max})], \tag{8}$$

where $F(\Delta_{max})$ – function of normal distribution of temperature difference Δ with numerical characteristics (8);

43,920 – number of minutes per month with an average duration of 30.5 days.

If changes in outdoor air temperature during the year are presented as a sequence of 12 normally distributed random variables, the annual duration of the Q_{year} thermal failure state is equal to the sum of the values (8) during the months relating to the heating period.

$$Q_{year} = \sum_j Q_j \tag{9}$$

and is expressed in minutes per year. For most of Ukraine, 7 months from October to April should be taken into account.

An alternative indicator is the relative duration of the thermal failure state Q_{rel}, calculated by dividing (9) by the length of the year in minutes

$$Q_{rel} = Q_{year}/525600. \tag{10}$$

Relative duration of thermal failure state (10) is a dimensionless quantity, which can be considered approximately the probability of thermal failure during the year.

Absolute (9) and relative (10) durations of the thermal failure state according to the comfort criterion are determined for three types walls described above in the climatic conditions of Poltava and Kropyvnytskyi. The following initial data are taken into account:

Table 2 The walls thermal failure state duration by the comfort criterion

City	Absolute duration of the state of thermal failure of wall types			Relative duration of the thermal failure state of wall types		
	Type 1	Type 2	Type 3	Type 1	Type 2	Type 3
Poltava	16,3	53,5	41,100	$3{,}10 \times 10^{-5}$	$1{,}02 \times 10^{-4}$	$7{,}82 \times 10^{-2}$
Kropyvnytskyi	14,4	46,1	36,000	$2{,}75 \times 10^{-5}$	$8{,}77 \times 10^{-5}$	$6{,}86 \times 10^{-2}$

- statistical characteristics of outdoor air temperature for the cold months of the year from October to April M_{out} and S_{out} according to the data [4, 5];
- mathematical expectation of indoor air temperature for residential premises M_{in} = 20 °C, equal to the normative value according to [1] and the standard S_{in} = 0,6 °C according to the data [3];
- statistical characteristics of heat transfer resistance M_R and S_R, defined and given above for walls of three types;
- the maximum allowable value of the difference between the temperature of the indoor air and enclosing structure inner surface in residential buildings in accordance with the rules [1] is equal to Δ_{max} = 4 °C.

The results of the calculations are shown in the Table 2.

The data in Table 2 indicate the close levels of thermal reliability of the walls being used in climatic conditions of Poltava and Kropyvnytskyi, located at a distance of about 200 km. Insignificant probable duration of the thermal failure state according to the comfort criterion indicates a sufficient level of thermal reliability of type 1 walls, the insulation of which fully meets the requirements of the norms [1]. The duration of thermal failures of type 2 walls is 3… 4 times longer, yet the discomfort per one hour a year can also be considered acceptable. Brick walls of type 3 without additional insulation, which were erected in the last century, can create uncomfortable conditions in the room for 25… 28 days per year. This duration of thermal failure state is clearly unacceptable, so the walls of type 3 are subject to mandatory thermal modernization by additional facade insulation.

6 Duration of the Thermal Failure State According to the Criterion of Condensate Formation

This figure is equal to probability that actual temperature of the enclosure structure inner surface τ_w will fall below the dew point τ_d. Statistical characteristics of the temperature of the enclosing structure inner surface are determined by formulas (5), (6). In [3] it was shown that the dew point temperature for living premises conditions can be considered a random variable with a mathematical expectation and a standard M_d = 10,6 °C, S_d = 1,7 °C.

Table 3 Duration of the thermal failure state of the walls by the criterion of condensate formation

City	Absolute duration of thermal failure state of wall types			Relative duration of thermal failure state of wall types		
	Type 1	Type 2	Type 3	Type 1	Type 2	Type 3
Poltava	0,20	0,38	126	$3{,}77 \times 10^{-7}$	$7{,}21 \times 10^{-7}$	$2{,}40 \times 10^{-4}$
Kropyvnytskyi	0,19	0,35	102	$3{,}54 \times 10^{-7}$	$6{,}62 \times 10^{-7}$	$1{,}95 \times 10^{-4}$

According to the theorems on the numerical characteristics of random variables [25], mathematical expectation and the standard of the temperature difference $\Delta = \tau_w - \tau_d$ in the j-th month of the year are equal to:

$$M_\Delta = M_w - M_d; \quad S_\Delta = \sqrt{S_w^2 + S_d^2}. \tag{11}$$

Absolute duration of the thermal failure state (in minutes per month) in the j-th month of the year is equal to probability that the temperature difference $\Delta = \tau_w - \tau_d$ will be less than zero:

$$Q_j = 43920\ F(0). \tag{12}$$

Annual duration of thermal failure state by the criterion of condensate formation Q_{year} (in minutes per year) is determined by formula (9), and the relative duration—by (10). The results of calculations of thermal failure absolute and relative duration according to the criterion of condensate formation of three types of walls in the climatic conditions of Poltava and Kropyvnytskyi are given in Table 3.

The data in Table 3 indicate practical impossibility of condensation on the first and second types walls inner surface. The total duration of conditions for condensate formation on the type 3walls inner surface is 2 h per year. It is possible that condensate formation can damage the finishing layers, thus level of thermal reliability of the 3 type wall should be considered insufficient.

Tables 2 and 3 show relative durations of thermal failures of the 1 and 2 type walls are close and even lower than failure probabilities of load-bearing structures recommended by current design standards of Ukraine [24]. Relative durations of thermal failures of type 3 walls exceed the values recommended by these standards.

7 Conclusions from Research Results

1. The technique is developed for defining statistical resistance characteristics of heat transfer and temperature of an internal surface of enclosing structures taking into account random character of the structures sizes and thermal characteristics of materials, as well as casual temperatures of internal and atmospheric air.

2. As indicators of thermal reliability it is expedient to use absolute and relative duration of thermal failures state on criteria of comfort and formation of condensate on enclosing structures internal surfaces. The proposed probabilistic methods for determining these indicators enables performing a comparative assessment of thermal reliability level of different enclosing structures in the given climatic conditions of operation.
3. The performed examples of calculations indicate a sufficient level of thermal reliability of the residential buildings walls with facade insulation, made in accordance with current design standards of Ukraine, and the impossibility of normal operation of 510 mm thick brick walls without additional insulation, which were built en masse in the second half of the last century
4. To more accurately determine the indicators of thermal reliability of enclosing structures, it is necessary to perform experimental and statistical studies in order to obtain and probabilistically represent the thermal conductivity of structural and insulating building materials.

References

1. DBN B.2.6-31 (2016) Thermal insulation of buildings. Ministry of Construction of Ukraine, Kyiv, p 3. (in Ukrainian)
2. Farenyuk G (2019) The determination of the thermal reliability criterion for building envelope structures. Tehnički glasnik. 13(2):129–133 (2019). https://doi.org/10.31803/tg-20181123111226
3. Pashynskyi VA, Plotnikov OA (2017) Probabilistic method for analyzing thermal reliability of units of enclosing structures. Bull Belarusian-Russian Univ 3(56):129–135
4. Kariuk A, Koshlatyi O, Mishchenko R (2018) The Statistical characteristics and calculated values for air temperature in building's cladding design. Int J Eng Technol 7(3.2):608–613. https://doi.org/10.14419/ijet.v7i3.2.14600
5. Pashynskyi V, Pashynskyi M, Pushkar N, Skrynnik I (2019) Method of administrative-territorial zoning of the design parameters of air temperature. Electron J Faculty of Civil Eng Osijek-e-GFOS 19:50–57. https://doi.org/10.13167/2019.19.5
6. European climate zones and bio-climatic design requirements (2016) Project report BEAR-iD, Nobatek, p 31
7. DSTU-N B V.1.1–27 (2010) 2010 Protection from dangerous geological processes, harmful operational influences, fire. Construction Climatology, Kyiv. (in Ukrainian)
8. Li Y, Sun Y, Zhuang Y, Duan LiMin, Xie Ke (2020) Thermal conductivity characteristics of thermal insulation materials immersed in water for cold-region tunnels. Adv Mater Sci Eng 2020:1–15. https://doi.org/10.1155/2020/9345615
9. Son H-J, Ko A-R, Lee W-K (2015) Characteristics of thermal conductivity for insulating brick made from converter slag. J Korea Soc Waste Manage 32:378–382. https://doi.org/10.9786/kswm.2015.32.4.378
10. Semko, V, Pichugin SF, Leshchenko M (2017) Probabilistic analysis of thermal performance of the wall from light-gauge thin-walled steel structures. Acad J Ser Ind Mach Build Civil Eng. 48:144–155. https://doi.org/10.26906/znp.2017.48.788
11. Leshchenko MV, Semko V (2015) Thermal characteristics of the external walling made of cold-formed steel studs and polystyrene concrete. Mag Civil Eng 60(8):44–55. https://doi.org/10.5862/MCE.60.6

12. DSTU B V.2.6-189 (2013) Methods of heat-insulating material selection for warming of buildings. K.: Ministry of Regional Development of Ukraine, p 51. (in Ukrainian)
13. Košir M, Pajek, L, Hudobivnik, B, Dovjak M, Iglič N, Bozicek D, Kunič R (2017) Non-stationary thermal performance evaluation of external façade walls under central European summer conditions. In: ISES conference proceedings, pp 1–10. https://doi.org/10.18086/swc.2017.15.03
14. Ahmed K, Carlier M, Feldmann C, Kurnitski J (2018) A New method for contrasting energy performance and near-zero energy building requirements in different climates and countries. Energies MDPI Open Access J 11(6):1–22. https://doi.org/10.3390/en11061334
15. Scartezzini, J-L, Duman Ö, Koca A, Acet R, Çetin M, Gemici Z (2015) A study on optimum insulation thickness in walls and energy savings based on degree day approach for 3 different demo-sites in Europe. In: Future buildings and districts sustainability from nano to urban scale, pp 155–160. https://doi.org/10.5075/epfl-cisbat2015-155-160
16. Kaynakli Ö, Kaynakli F (2016) Determination of optimum thermal insulation thicknesses for external walls considering the heating, cooling and annual energy requirements. Uludağ Univ J Faculty Eng 21:227–241. https://doi.org/10.17482/uujfe.27323
17. Kariuk A, Rubel V, Pashynskyi V, Dzhyrma S (2020) Improvement of residential buildings walls operation thermal mode. In: Proceedings of the 2nd international conference on building innovations. ICBI 2019. Lecture Notes in Civil Engineering, vol 73, pp 75–81. https://doi.org/10.1007/978-3-030-42939-3_9
18. Pashynskyi M, Dzhyrma S, Pashynskyi V, Nastoyashchiy V (2020) Providing the thermal reliability of window junctions during the thermal modernization of civil buildings. Electron J Faculty Civil Eng Osijek-e-GFOS 21:45–54. https://doi.org/10.13167/2020.21.4
19. Rzhanitsin AR (1978) The theory of building structures design for reliability, Stroyizdat, p 240. https://dwg.ru/lib/1942
20. Pichugin SF (2018) Reliability estimation of industrial building structures. Mag Civil Eng 83(7):24–37. https://doi.org/10.18720/MCE.83.3
21. Belentsov YA (2020) The method of calculation of building structures by reliability level. Stroitel'nye Materialy 786:54–59. https://doi.org/10.31659/0585-430X-2020-786-11-54-59.
22. Chybowski L, Zolkiewski S (2015) Basic reliability structures of complex technical systems. https://doi.org/10.1007/978-3-319-16528-8_31
23. Pownuk A (2021) Calculation of reliability of structures using random sets
24. DBN B.1.2-14-2018 (2009) System for ensuring the reliability and safety of construction sites. General principles of ensuring the reliability and structural safety of buildings and structures. Ministry of Regional Development of Ukraine, Kyiv. (in Ukrainian)
25. Wentzel ES (2006) Probability theory: textbook for universities, 10th edn. Higher. sch., p 575

Theoretical Study of the Dynamic System «Vibration Platform – Polymer Concrete» Stress–Strain State

Oleksandr Maslov, Dmytro Savielov, Yuliia Salenko, and Mirfatma Javadova

Abstract The analysis of the dynamic system "vibration platform polymer concrete" for the theoretical determination of the polymer concrete stress–strain state compacted by the movable frame of vibration platform was carried out. Polymer concrete, which is compacted by vibration load, was presented as a distributed parameter system. As a result of expression substitution that describes the law of the vibrating platform movable frame motion in the operating mode into the relationship between stress and deformation for the conditions of a uniaxial stress state, the law of stresses changes that arise in the polymer concrete compacted layer was found. The obtained law made it possible to determine the stresses changes in the base and on the surface of the polymer concrete compacted layer, which is deformed on the vibrating platform movable frame. The graphs are based on the found theoretical relationships. They clearly illustrate the changes in the vibration amplitude of the vibrating platform movable frame depending on the height of the product, the stresses that arise at the base and on the surface of the polymer concrete compacted layer, as well as the peculiarities of the stresses changing along the compacted layer height depending on the relative density of the polymer concrete layer for the selected impact vibration mode.

Keywords Vibration platform · Polymer concrete · Vibrations · Deformation

1 Introduction

The change in the stress-strain state of polymer concrete occurs in the process of vibration loading. In this case, the nature of the stresses distribution that arise in the compacted polymer concrete by the vibration load will largely determine the modes of vibration exposure, the design and technological features of vibration equipment.

O. Maslov · D. Savielov (✉) · Y. Salenko
Kremenchuk Mykhailo Ostrohradskyi National University, Kremenchuk, Ukraine

M. Javadova
Azerbaijan University of Architecture and Construction, Ayna Sultanova 11, Baku, Azerbaijan

V. Onyshchenko et al. (eds.), *Proceedings of the 3rd International Conference on Building Innovations*, Lecture Notes in Civil Engineering 181,
https://doi.org/10.1007/978-3-030-85043-2_19

To date, considerable experience has been accumulated in scientific research of soils vibration compaction [1–3], cement and asphalt concretes [4–11], composite media [12–21] based on their model representation. A detailed acquaintance with these works allows us to conclude that in order to determine the patterns of change in stresses under various methods of vibration exposure, the dynamic systems "vibrating working body – soil", "vibration platform – compacted medium", etc. were studied. In this case, the choice of rheological models that adequately describe the stress-strain state of such dynamic systems was of great importance. This approach, with a sufficiently high degree of accuracy, made it possible to obtain the regularities of changes in stresses that arise in the medium compacted layer.

It is known that systematic studies of polymer concretes vibration treatment, determination of their physical and mechanical properties and rational areas of application began somewhat later than soils, cement and asphalt concretes.

At the same time, for a scientifically grounded choice of vibration exposure rational modes and equipment parameters for various sizes polymer concrete compaction, it is of great importance to identify the patterns of change in stresses arising from vibration impact on the compacted polymer concrete.

All this suggests that carrying out theoretical studies aimed at studying the nature of the stresses distribution arising in the polymer concrete compacted layer during compaction on a vibrating platform, depending on the product size and the physical and mechanical characteristics of the compacted material, is a very urgent task.

The purpose of this research is a theoretical study of the stresses change nature arising in the polymer concrete compacted layer when compacted on a vibrating platform with vertically directed vibrations.

2 Main Body

In article [22], a design diagram of the dynamic system «vibrating platform – polymer concrete» was presented (Fig. 1).

As a result of the processes theoretical studies occurring in the dynamic system «vibrating platform – polymer concrete», in which polymer concrete is presented as a system with distributed parameters, the law of the vibrating platform movable frame motion was determined, which interacts with polymer concrete in the compaction mode.

The law of motion has the following form [22]:

$$u(x,t) = \frac{A}{\sqrt{(ch\alpha H \cos kH)^2 + (sh\alpha H \sin kH)^2}} \times [ch[\alpha(H-x)]\cos[k(H-x)]\sin(\omega t - \varphi) + sh[\alpha(H-x)]\sin[k(H-x)]\cos(\omega t - \varphi)]. \quad (1)$$

where H – is the height of the compacted layer; k – wave number; α – the vibration load absorption coefficient, which characterizes the decrease in the vibration

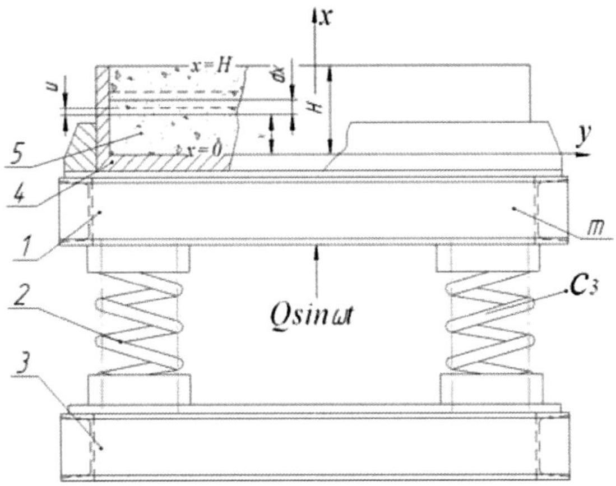

Fig. 1 Design diagram of the dynamic system "vibrating platform – polymer concrete" [22]: 1 – movable frame; 2 – elastic shock absorber; 3 – base; 4 – mold; 5 – polymer concrete

amplitude with distance from the source of disturbance; A – the forced vibrations amplitude of the vibrating platform moving frame and the polymer concrete lower layer; ω – forced angular frequency; ϕ – phase angle; t – current time.

The forced vibrations amplitude of the vibrating platform movable frame and the polymer concrete lower layer is determined from the expression [22]:

$$A = \frac{Q}{\sqrt{\left[c_3 + c_n - (m + m_n)\omega^2\right]^2 + \omega^2 b_n^2}}, \tag{2}$$

where Q – disturbing force amplitude; c_n and b_n – reduced coefficients of stiffness and dissipative resistance of the compacted polymer concrete; m_n – reduced mass of the compacted polymer concrete.

Expressions for the analytical determination of the values k, α, c_n, b_n and m_n are defined in [22].

To theoretically determine the law of change in stresses arising in a layer of polymer concrete compacted on a vibration platform, we substitute expression (1) into the relationship between stress and deformation in polymer concrete [21]:

$$\sigma(x,t) = E_1 \frac{\partial u(x,t)}{\partial x} + \eta \cdot \left(\frac{E_1 + E_2}{E_2}\right) \frac{\partial^2 u(x,t)}{\partial x \partial t} - \left(\frac{\eta \cdot \rho}{E_2}\right) \frac{\partial^3 u(x,t)}{\partial t^3}, \tag{3}$$

where $\sigma(x,t)$ – stresses arising in the compacted layer of polymer concrete; u and x – Euler and Lagrange coordinates; E_1 and E_2 – dynamic moduli of polymer concrete elastic deformation; η – coefficient of dynamic viscosity, taking into account internal friction in polymer concrete.

Substituting expression (1) into dependence (3), it could be defined the law of changes in stresses arising in the polymer concrete compacted layer:

$$\sigma_1(x,t) = \frac{A \cdot \sqrt{E_1^2 + \lambda^2 \cdot \omega^2} \cdot \sqrt{k^2 + \alpha^2}}{\sqrt{(ch\alpha H \cos kH)^2 + (sh\alpha H \sin kH)^2}} \times \{ch\alpha(H-x)\sin k(H-x)\sin(\omega t - \theta_1)$$
$$-sh\alpha(H-x)\cos k(H-x)\cos(\omega t - \theta_1) - \frac{\xi \cdot \omega^3}{\sqrt{E_1^2 + \lambda^2 \cdot \omega^2} \cdot \sqrt{k^2 + \alpha^2}} \quad (4)$$
$$\times[ch\alpha(H-x)\cos k(H-x)\sin(\omega t - \varphi) + sh\alpha(H-x)\sin k(H-x)\cos(\omega t - \varphi)]\},$$

where θ, φ_3 and φ_4 – phase angles

$$\theta = \varphi - \varphi_3 + \varphi_4;\ \varphi_3 = arctg(\lambda\omega/E_1);\ \varphi_4 = arctg(\alpha/k). \quad (5)$$

From the analysis of the obtained expression (4), it can be concluded that the forced vibrations amplitude A of the vibrating platform movable frame, the angular frequency of the forced vibrations ω, the compacted layer height H, the wavenumber k, and the vibration load absorption coefficient α have a significant effect on the magnitude of the arising stresses $\sigma(x,t)$.

From expression (4) it is possible to find the law of stress change at the compacted layer base at $x = 0$:

$$\sigma_1(0,t) = \frac{A \cdot \sqrt{E_1^2 + \lambda^2 \cdot \omega^2} \cdot \sqrt{k^2 + \alpha^2}}{\sqrt{(ch\alpha H \cos kH)^2 + (sh\alpha H \sin kH)^2}} \times \{ch\alpha H \sin kH \sin(\omega t - \theta_1)$$
$$-sh\alpha H \cos kH \cos(\omega t - \theta_1) - \frac{\xi \cdot \omega^3}{\sqrt{E_1^2 + \lambda^2 \cdot \omega^2} \cdot \sqrt{k^2 + \alpha^2}}$$
$$\times[ch\alpha H \cos kH \sin(\omega t - \varphi) + sh\alpha H \sin kH \cos(\omega t - \varphi)]\}, \quad (6)$$

and also on its surface at $x = H$:

$$\sigma_1(H,t) = -\frac{A \cdot \xi \cdot \omega^3}{\sqrt{(ch\alpha H \cos kH)^2 + (sh\alpha H \sin kH)^2}} \cdot \sin(\omega t - \varphi). \quad (7)$$

The stresses that arise in polymer concrete from the action of the polymer concrete layer own gravity can be determined based on the relationship [23]:

$$\sigma_2(x,t) = -\rho g(H-x). \quad (8)$$

The total stress arising in the compacted polymer concrete under vibration action can be determined from the dependence [23]:

$$\sigma(x,t) = \sigma_1(x,t) + \sigma_2(x,t). \quad (9)$$

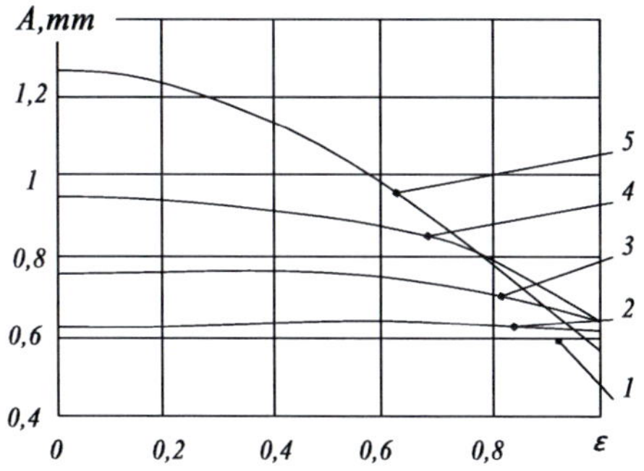

Fig. 2 Change in the vibration amplitude of the vibrating platform moving frame depending on the relative density ε and the compacted layer height H: 1 – at $H = 50$ mm; 2 – at $H = 60$ mm; 3 – at $H = 80$ mm; 4 – at $H = 100$ mm; 5 – at $H = 120$ mm.

The analysis of dependences (4), (8) and (9) shows that the stresses arising in the polymer concrete compacted by vibration load are asymmetric. This means that the magnitude of the compressive stresses is greater than the magnitude of the tensile stresses. Consequently, a more efficient compaction of polymer concrete is provided.

Theoretical studies were tested on a laboratory vibrating platform with the following parameters: the mass of the movable frame $m = 24$ kg; disturbing force amplitude $Q = 981$ N; the angular frequency of the forced vibrations $\omega = 293$ rad/s; stiffness of elastic shock absorbers $c_3 = 235440$ N/m; vibration amplitude of the vibrating platform moving frame in idle mode $A_{im} = 0, 47$ mm. Polymeric concrete in the form of $0, 3 \times 0, 3$ m^2 in plan size of the following structural composition was compacted on a vibrating platform [24]: granite crushed stone of fraction 5–20 (50% of the mixture total volume), river sand with a size modulus $M_k = 1, 8$ (22–27%); marshalite fraction 0.05 mm (10–15%); Filabond 2000 PA polyester resin (5%); hardener MEKPHA-2 (0.5 ... 1%).

The change in the vibration amplitude of the vibrating platform movable frame is shown in Fig. 2.

Amplitude curves are shown in Fig. 2. They show that the physical and mechanical characteristics of the compacted polymer concrete, its relative density ε and the compacted layer height H have a significant effect on the vibration amplitude of the vibrating platform moving frame.

The vibration amplitude A of the vibrating platform movable frame increases with the selected vibration exposure mode with an increase of the compacted layer height H.

In this case, in the vibration compaction process, with an increase in the polymer concrete relative density ε from 0 to 1 at a layer height of $H = 50$ mm, the amplitude of forced vibrations A of the vibrating platform movable frame and the polymer concrete lower layer is equal 0.59 mm and practically does not change.

When compaction of polymer concrete layers with heights H of 60 and 80 mm with an increase in the polymer concrete relative density ε from 0 to 1, the forced vibrations amplitude A of the vibration platform movable frame and the polymer concrete lower layer decreases from 0.63 to 0.61 mm at $H = 60$ mm and from 0.73 to 0.65 mm at $H = 80$ mm.

When compaction of layers with heights H of 100 and 120 mm with an increase in the polymer concrete relative density ε from 0 to 1, there is a significant decrease in the vibration amplitude A of the vibrating platform movable frame, respectively, from 0.93 to 0.65 mm at $H = 100$ mm and 1.26 mm and 0.57 mm at $H = 120$ mm.

The change in the stress amplitude at the base $\sigma(0)$ and on the surface $\sigma(H)$ of the compacted layer depending on the relative density of polymer concrete is shown in Fig. 3 and 4.

The graphical results shown in Fig. 3 and 4 demonstrate that with the selected vibration mode, when compaction of layers with a height H of 50, 60 and 80 mm with increasing relative density ε, the stress amplitude increases both in the base $\sigma(0)$ and on the surface $\sigma(H)$ of the polymer concrete compacted layer.

With a layer thickness of 120 mm, at the final stage of the compaction process, the stress amplitude in the compacted layer decreases from 5689 to 5591 Pa at the compacted layer base (Fig. 3) and from 1962 to 578 Pa on the compacted layer surface (Fig. 4). Consequently, the duration of the vibration compaction process will increase.

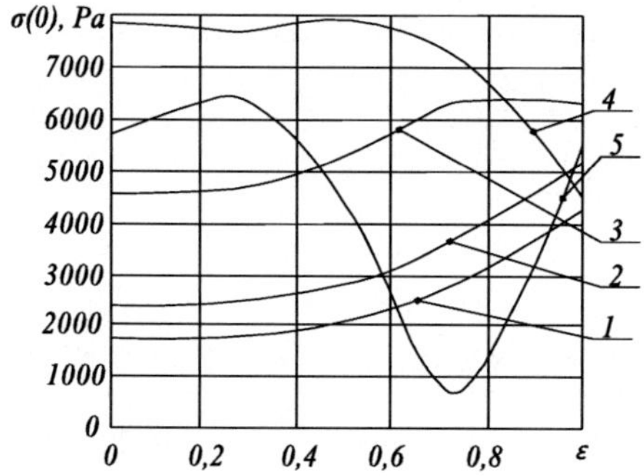

Fig. 3 Changes in stresses at the base $\sigma(0)$ of the polymer concrete compacted layer depending on the relative density ε and the compacted layer height H: 1 – at $H = 50$ mm; 2 – at $H = 60$ mm; 3 – at $H = 80$ mm; 4 – at $H = 100$ mm; 5 – at $H = 120$ mm

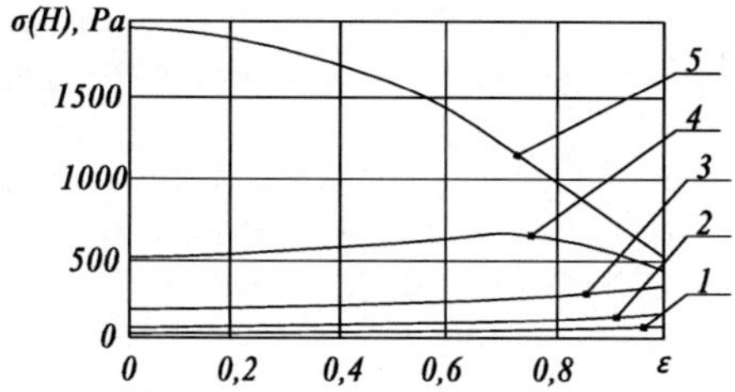

Fig. 4 Changes in stresses on the surface $\sigma(H)$ of the polymer concrete compacted layer depending on the relative density ε and compacted layer height H: 1 – at $H = 50$ mm; 2 – at $H = 60$ mm; 3 – at $H = 80$ mm; 4 – at $H = 100$ mm; 5 – at $H = 120$ mm

One of the main parameters that determine the compacted layers stress–strain state and, as a consequence, the quality of the resulting polymer-concrete compositions, is the nature of the stress distribution along the compacted layer height H.

Graphs of stress changes σ along the compacted layer height H are shown in Fig. 5, 6, 7, 8 and 9. The obtained theoretical expressions (4), (8) and (9) were used to construct these graphs. The compacted layers height H was taken equal to 50, 60, 80, 100 and 120 mm.

In this case, each layer in height H was divided into a number of intermediate values. For example, for the layer height $H = 120$ mm, for each relative deformation ε, the stress values were determined by intermediate values of the height, i.e. at $x =$ 0, 20, 40, 60, 80 and 120 mm.

The values of the relative density ε were taken equal to 0, 0.25, 0.5, 0.75 and 1.

The angular coordinate values ωt were taken in the range from 0° to 270°. Stress values were determined for each value of the intermediate height x and relative density ε every 10°, i.e. for angular coordinates ωt = 0°, 10°, 20°, 30°, 40° … 0.270°.

The largest values of one sign were chosen from the obtained series of numerical values of stresses.

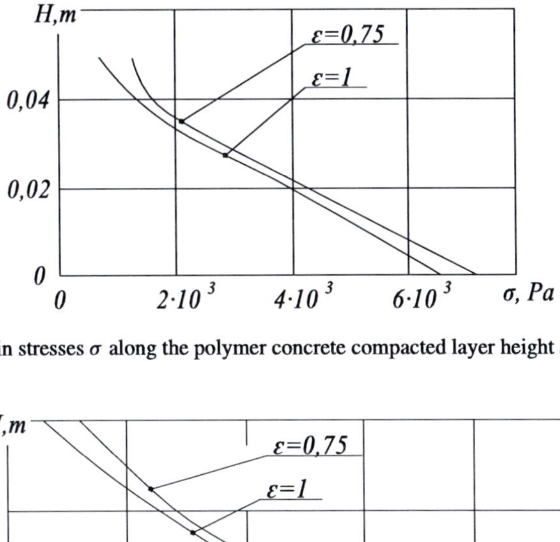

Fig. 5 Change in stresses σ along the polymer concrete compacted layer height at $H = 50$ mm

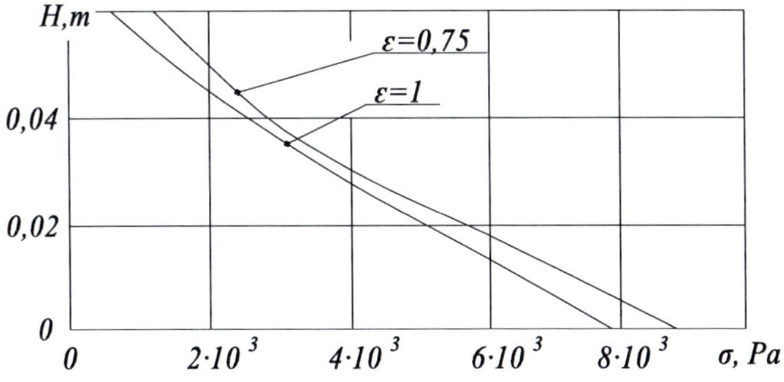

Fig. 6 Change in stresses σ along the polymer concrete compacted layer height at $H = 60$ mm

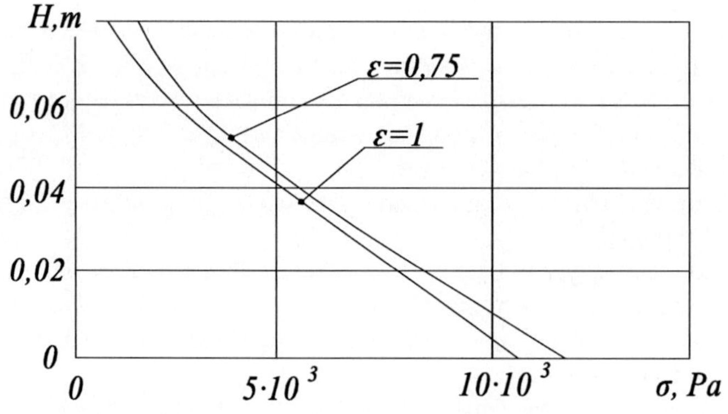

Fig. 7 Change in stresses σ along the polymer concrete compacted layer height at $H = 80$ mm

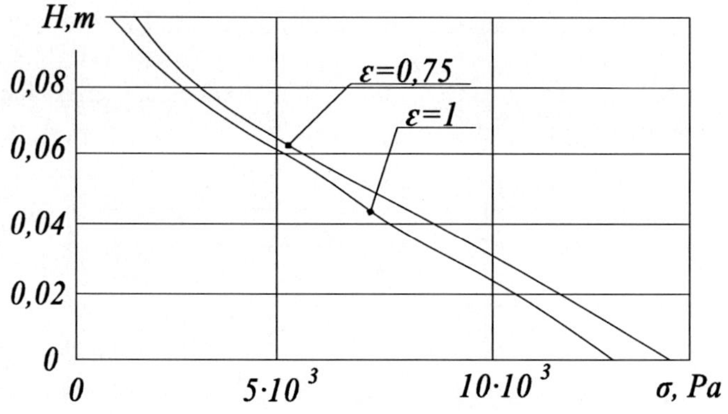

Fig. 8 Change in stresses σ along the polymer concrete compacted layer height at $H = 100$ mm

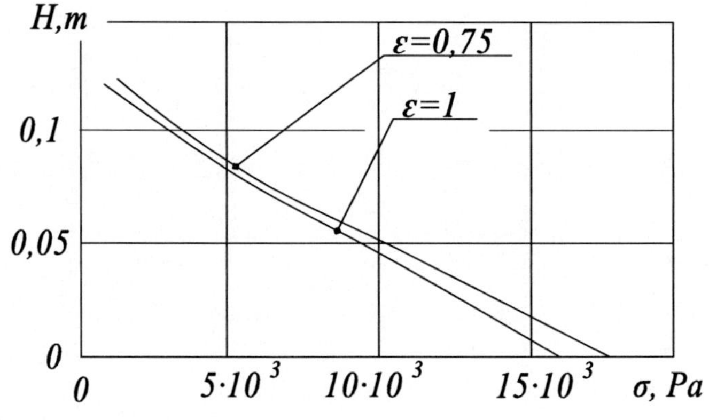

Fig. 9 Change in stresses σ along the polymer concrete compacted layer height at $H = 120$ mm

Analysis of the graphs in Fig. 5 and 9 indicates that the height H and relative density ε of the polymer concrete compacted layer has a significant effect on the value of stresses. So the stress amplitude σ has the greatest values in the polymer concrete lower layers, which interact with the vibrating platform moving frame.

This is due to the fact that the polymer concrete lower layers are located in close proximity to the source of disturbance and at the same time experience additional vertical compressive stresses from the action of the polymer concrete layer gravity forces.

The stress amplitude σ decreases rapidly with an increase in the compacted layer height H. This is because with an increase in the compacted layer height H, there is a simultaneous decrease in the magnitude of vertical compressive stresses from the action of gravity forces, as well as a distance from the disturbance source, as a result of which the amplitude of oscillations A decreases. As a result, the upper layers will be under compacted and, as a result, have lower mechanical strength.

Therefore, to intensify the compaction process at these heights, it is necessary either to change the vibration frequency or to use a cantledge.

3 Conclusion

As a result of the dynamic system "vibrating platform polymer concrete" studies, theoretical expressions were obtained. The expressions make it possible to determine the law of stresses changes arising in the polymer concrete compacted layer during its compaction on the vibratory platform, depending on its physical and mechanical characteristics, the amplitude and frequency of forced vibrations and the compacted layer height.

The found dependences make it possible to determine the nature of stress changes along the compacted layer height of polymer concrete, depending on the relative density. The results obtained are the basis for further theoretical studies on the analytical law definition of motion and changes in stresses arising in the polymer concrete compacted layer under the action of an inertial or vibration cantledge on it.

References

1. Minaev OP (2015) Russian methods and equipment for spatial vibrocompaction foundations and structures, vol 2, no 80, pp 2747–2750. Japanese Geotechnical Society Special Publication. https://doi.org/10.3208/jgssp.TC305-11.
2. Maslov AG, Savelov DV (2019) Vibration puncture of soils. KrNU, Kremenchuk, p 102
3. Maslov AG, Savelov DV (2003) Theoretical studies of interaction process of the soil-piercing installation vibration head with the soil. Acad J Ser Kirovograd State Technical Univ Mach Agric Prod Ind Mach Build Autom 13:26–32

4. Maslov A, Batsaikhan J (2018) The research of the parameters of a vibration machine for composite materials compaction. In: MATEC web of conferences, vol 224, p 02099. https://doi.org/10.1051/matecconf/201822402099
5. Maslov O, Batsaikhan J, Salenko Y (2018) The theory of concrete mixture vibratory compacting. Int J Eng Technol (UAE) 7(3.2):239–244. https://doi.org/10.14419/ijet.v7i3.2.14411
6. Nesterenko M, Maslov A, Salenko J (2018) Investigation of vibration machine interaction with compacted concrete mixture. Int J Eng Technol 7(3.2):260–264. https://doi.org/10.14419/ijet.v7i3.2.14416
7. Maslov A, Batsaikhan J, Puzyr R, Salenko Y (2018) The determination of the parameters of a vibration machinef the internal compaction of concrete mixtures. Int J Eng Technol 7(4.3):12–19. https://doi.org/10.14419/ijet.v7i4.3.19545
8. Batsaikhan J (2019) Development of the theory of a vibrating platform with a concrete mixture interaction. Sci J Trans Kremenchuk Mykhailo Ostrohradskyi Nat Univ 6(119):112–117
9. Bogomolov V, Zhdanyuk VA, Tsynka PG (2016) Viscoelastic structural model of asphalt concrete. Autom Transp 38:117–123
10. Nazarenko II, Baranov YO, Basarab VA (2013) Interaction investigation of the concrete mixture with the working body of the electromagnetic shock-vibration installation. Vibr Eng Technol 3(71):55–60
11. Sviderskii AT, Свідерський AT (2013) Investigation of the vibrating machine dynamics taking into account the stress-strain state of an elastic-plastic medium. Vibr Eng Technol 3(71):41–45
12. Savelov D, Puzyr R, Markevich A (2015) Peculiarities of vibrational press dynamics with hard-elastic restraints in the working regime of metal powders molding. Metall Min Ind 2:67–74
13. Chen X, Wu S, Zhou J (2013) Experimental study and analytical formulation of mechanical behavior of concrete. Constr Build Mater 47:662–670
14. Tattersall GH (1990) Effect of vibration on the rheological properties of fresh cement pastes and concretes. In: Banfill PFG (ed.) Rheology of fresh cement and concrete, proceedings of the international conference, University of Liverpool, UK, 16–29 March. Chapman and Hall, London, pp 323–338
15. Kakuta S, Kojima T (1990) Rheology of fresh concrete under vibration. In: Banfill PFG (ed.) Rheology of fresh cement and concrete, proceedings of the international conference, University of Liverpool, UK, 19 –29 March. Chapman and Hall, London, pp 339–342
16. Banfill PFG et al (2011) Rheology and vibration of fresh concrete: predicting the radius of action of poker vibrators from wave propagation. Cem Concert Res 41(9):932–941
17. Korobko B (2016) Investigation of energy consumption in the course of plastering machine's work. Eastern Eur J Enterp Technol 4(8–82):4–11. https://doi.org/10.15587/1729-4061.2016.73336
18. Korobko B, Vasyliev I (2017) Test method for rheological behavior of mortar for building work. Acta Mechanica Et Automatica 11(3):173–177. https://doi.org/10.1515/ama-2017-0025
19. Korobko B, Virchenko V, Shapoval M (2018). Feed solution in the pipeline aith the compensators mortar pump of various design solutions pressure pulsations degree determination. Int J Eng Technol (UAE) 7(3):195–202. https://doi.org/10.14419/ijet.v7i3.2.14402
20. Korobko B, Khomenko I, Shapoval M, Virchenko, V (2020) Hydraulic single pump with combined higher volume compensator operation analysis.https://doi.org/10.1007/978-3-030-42939-3_12
21. Korobko B, Zadvorkin D, Vasyliev I (2017) Study of the operating element motion law for a hydraulic-driven diaphragm mortar pump. Eastern Eur J Enterp Technol 4(7–88):25–31. https://doi.org/10.15587/1729-4061.2017.106873
22. Maslov AG, Savelov DV (2020) Theoretical definition of vibrating platform movable frame motion law with polymer concrete during its compaction. Sci J Trans Kremenchuk Mykhailo Ostrohradskyi Nat Univ 4(123):84–90

23. Itkin AF (2004) Investigation of a cement-concrete mixture compaction process on a vibration platform with vertically directed vibrations. Sci J Trans Kremenchuk State Polytech Univ 6(29):86–91
24. Maslov AG, Savelov DV (2018) Development of structural composition of polymer concrete. Sci J Trans Kremenchuk Mykhailo Ostrohradskyi Nat Univ 4(111):94–99

Experimental Study of the Efficiency of the Differential Pump of Electromagnetic Action on the Basis of Mathematical Modeling of the Parameters of Its Operation

Bogdan Korobko, Anton Kivshyk, and Dmytro Kulagin

Abstract The operation of a differential pump of electromagnetic action operating on the principle of an inductive electromagnetic accelerator is considered. To improve the accuracy of engineering calculations and analysis of the pump, its mathematical model was developed using Newton's and Hooke's laws, the method of dimensional analysis, the Reynolds and Froud criteria. The analysis of the mathematical model of the pump operation was carried out and the design parameters that most influence the efficiency of its operation were determined. The problem of improving the operation of the differential pump by increasing the speed of its valves is solved. The dynamics of the relative motion of the discharge valve of the plunger of the differential pump is analyzed, using the mathematical model of its operation. A parametric expression is obtained that determines the operation time of the differential pump valve. On this basis, directions for improving the operation of the pump were found and tested experimentally on a prototype. The dependence is obtained, which allows to draw conclusions about which parameters directly affect the value of the valve operation time: spring stiffness, plunger size, location of the ball relative to the coil winding, friction forces of the sleeve seals, energy loss in valve assemblies. It is established that uniform pumping of the material is achieved by reducing the time to open and close the discharge valve, which is achieved by reducing the weight of the shut-off element. The operation of pump valves with shut-off elements made of different materials has been investigated experimentally. The value of the parameter of the mass of the shut-off element at which the optimal mode of operation of the pump is achieved. By setting the minimum operating time of the pump valve, its performance is improved.

B. Korobko (✉) · A. Kivshyk
National University "Yuri Kondratyuk Poltava Polytechnic", Poltava 36011, Ukraine
e-mail: bogdan.korobko@ukr.net

A. Kivshyk
e-mail: anton.kivshik3@ukr.net

D. Kulagin
National University "Zaporozhye Polytechnic", Street Zhukovsky 64, Zaporozhye 69063, Ukraine

V. Onyshchenko et al. (eds.), *Proceedings of the 3rd International Conference on Building Innovations*, Lecture Notes in Civil Engineering 181,
https://doi.org/10.1007/978-3-030-85043-2_20

Keywords Differential pump · Mathematical model · Finishing material · Pumping · Mortar pump

One of the problems in finishing work in construction is the creation of effective designs of pumps for pumping finishing materials. This problem is solved by the use of electromagnetic pumps operating on the principle of inductive electromagnetic accelerator. The undeniable advantage of this design is its relative simplicity and minimal number of moving parts.

However, minimizing the operating time of the working body of the pump will significantly improve its performance and expand its scope.

The problem with improving the design of modern pumps is the relatively low accuracy of engineering methods of their calculation, including calculations of their characteristics in nominal conditions.

The analysis of mathematical model of work of the pump will allow to reveal the most significant parameters variation of which will enable reaching the minimum time of operation of a working body of the pump.

1 Review of the Latest Research Sources and Publications

Electromagnetic pumps for pumping finishing materials are alternate current hydrodynamic machines. The moving part in them is the working body and the mixture that is pumped due to the action on the plunger of the magnetic field created by the alternate current winding [1, 2].

After analyzing the existing models of pumps, it was found that the differential pump of electromagnetic action is not sensitive to external factors, has the highest efficiency among analogues due to the small number of parts that are subject to active wear. The design of the differential pump provides its mobility and convenience in operation. In works [3–8] the analysis of the existing constructive schemes of the equipment, the basic means of small mechanization of manual work in construction, types of working bodies of mortar pumps is carried out, authors [10–16] offered a new design of the small differential pump of electromagnetic action.

The authors of [9] obtained mathematical models in the form of differential equations that reflect the variation in time of the speed of the plunger of the differential pump for finishing materials in the full cycle, presented a graphical interpretation of changes in the speed of the plunger over time. Also in [9] by analyzing the mathematical model, some geometrical parameters of the pump were improved, such as the size of the spring to ensure the preservation of mechanical energy during operation.

2 Problem Statement

Therefore, the purpose of this study is to further improve the design of the differential pump of electromagnetic action based on the analysis of the parameters included in the mathematical model of its work, obtained by the authors in [9]. For this purpose it is necessary: to analyze dynamics of relative movement of the discharge valve of the plunger of the differential pump, using mathematical model of its work. Obtain a parametric expression that determines the operating time of the differential pump valve. On this basis, find ways to improve the operation of the pump and test them experimentally on a prototype.

3 Basic Material and Results

In this research it's investigated the design of the differential pump for injection of finishing material, the structure of which is shown in Fig. 1.

The working element of the pump is a steel plunger, which moves in translational motion. When an electric current is applied to the drive of the differential pump, the force of the magnetic induction moves the plunger to the left.

When the working body moves at different intervals, the discharge and suction valves work alternately (Fig. 2).

Therefore, the drive of the pump 3 is supplied with an electric voltage, which has an alternating nature and forms a magnetic field, the force of which is directed at the plunger, moving it to the left. The plunger begins to retract into the coil, closing the suction valve and opening the discharge. The working chamber of the pump is filled with finishing material. During the transportation process, the pressure in the magnifying fitting begins to rise. The increase in the speed of the working body is due

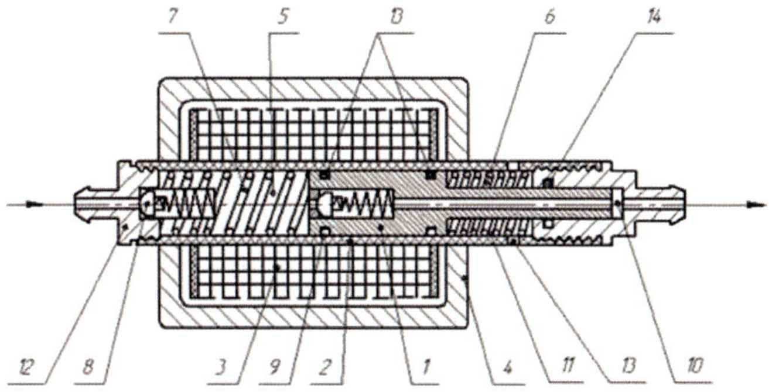

Fig. 1 Differential pump of electromagnetic action: 1—plunger; 2—frame; 3—coil; 4—magnetic coil; 5—suction cavity; 6—compensation spring; 7—working spring; 8—suction valve; 9—discharge valve; 10—compensation chamber; 11, 12—injection and suction fittings; 13, 14—cuff seals

Fig. 2 The main design of the pump

to the increased pressure of the whitewashing material. In the first cycle of pumping begins to compress the working spring 7 and lengthen the compensation spring 6.

As the coil current decreases, the magnetic induction decreases and, at the same time, the speed of the plunger decreases—until the cessation of movement. However, the stop of the working body is carried out a little earlier than the complete reduction of magnetic induction—when the moment of balance of magnetic induction and compression force of the working spring 7. When the sinusoid changes direction, the diode in the power supply removes its lower part and during the movement of working body to the right the magnetic field doesn't act upon the plunger.

The length of the working spring increases by pushing the plunger to the right. The speed of the working body begins to increase. When passing the plunger in the extreme right position, the discharge valve 9 is closed and the pumping cycle is repeated. The increase in transport pressure is proportional to the increase in travel speed.

At the same time, the absorption valve 8 opens and the finishing material is poured into the working volume of the pump. When changing the position of the plunger to the right, the resistance of the working spring 7 decreased. The increase in the length of the spring is limited by the force from the whitewashing material, the force of retraction of the finishing material into the working chamber and the resistance of the compensating spring. As the force of ejection of the spring 7 decreases, the speed of the working body decreases, while the performance of the pump decreases.

By the time the plunger stops, current is supplied to the solenoid and the transport process is repeated.

Thus, the process of pumping the mixture by the pump consists of two cycles, each of which is primarily determined and depends on: the nature of the corresponding movement of the working plunger and those moments t_i when the discharge and suction valves are opened and closed.

Among other factors for improving and harmonizing the operation of the pump is related to the knowledge of which parameters depend on these time points t_i. To establish one of these dependences, we examine the dynamic state of the discharge valve 9 of the plunger 1, considering this valve as a material point whose mass is equal to the mass m_9 of this valve, and entering into consideration the coordinate system OXY and $O_1X_1Y_1$. The beginning of the reference of the Oxy system is connected

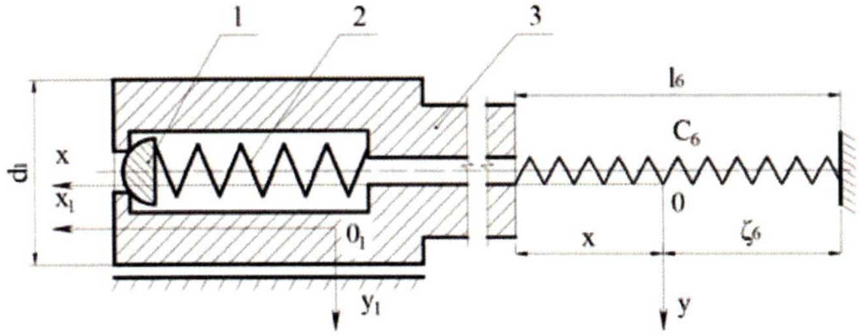

Fig. 3 Kinematic scheme of the differential pump valve due to the influence of bodies on each other, taking into account the force of inertia

with the Earth, which determines its inertia, and the beginning of the reference of the system O1x1y1 is connected with the plunger 1 (Fig. 3).

Since in [9] it was found that this plunger moves translationally, but unevenly, according to [10] the coordinate system O1x1y1 is non-inertial.

The motion of bodies in inertial reference frames can be described by the same equations of motion as in inertial ones, if, along with the forces caused by the influence of bodies on each other, the forces of inertia are taken into account.

Consider the dynamics of the relative motion of the discharge valve of the plunger during the first cycle of pumping. The basic equation of the dynamics of the absolute movement of the discharge valve is as follows.

$$m_9 * \vec{a}_9 = \sum_{i=1}^{\lambda} \vec{F}_i \quad (1)$$

where $\vec{a}_9$—absolute acceleration of the valve; $\sum_{i=1}^{\lambda} \vec{F}_i$—geometric sum of forces acting on the valve; λ—the number of these forces.

Since in this case the portable movement for the valve is the translational movement of the plunger, the absolute acceleration is equal to the geometric sum of two accelerations: relative and portable, in which case due to the translational type of portable movement Coriolis acceleration of the valve is absent.

Given the above, the law of absolute movement of the valve takes the form.

$$m * \vec{a}_9^r + m * \vec{a}_9^e = \sum_{i=1}^{\lambda} \vec{F}_i \quad (2)$$

Determining the product of the mass of the valve on its relative acceleration, we obtain the basic equation of the dynamics of the relative movement of the valve in the form:

$$m_9 * \vec{a}_9^r = \sum_{i=1}^{\lambda} \vec{F}_i + \vec{\Phi}_9^e \tag{3}$$

The last equation shows that the relative movement of the valve can be considered as absolute if to the forces acting on this valve to add its transfer force of inertia. That is, in the inertial reference system OXY the movement of the valve is only the result of forces acting on it (or the result of its mechanical interaction with other material bodies), and in the non-inertial system $O_1X_1Y_1$ the movement of the valve is both the result of forces acting on it and the result of the $O_1X_1Y_1$ If the action of forces is a dynamic factor in the acceleration of the valve, then the movement of the reference system is a kinematic factor in the establishment of this acceleration.

If we replace the portable movement of the valve with the movement of the plunger, replace the acceleration of the plunger according to Newton's second law and project the vector equality on the horizontal axis, we obtain an expression to determine the projection of the acceleration of the plunger on the axis Ox.

$$\begin{aligned} \frac{d_X^2}{dt^2} = \frac{dv}{dt} = \frac{Q_0}{m} * \sin\frac{\pi t}{\tau} - \frac{(c_6 + c_7)}{m} * x + \\ + \frac{c_6(l_{6ned.} - \varsigma_6)}{m} - k\frac{\sqrt{\pi(1-\gamma^2)}}{2*m}\mu d_1 v, \end{aligned} \tag{4}$$

where υ—the speed of the working plunger; x—is the coordinate that determines the position of the plunger in the inertial coordinate system OXY; m—mass of the plunger; Q_0—the maximum value of the modulus of the driving (turbulent) force generated by the magnetic induction of the coil of the differential pump acting on the plunger; τ—the time of movement of the plunger to the left from its initial position to the final; c_6—is the stiffness coefficient of the compensation spring; c_7—is the stiffness coefficient of the working spring; l_{6ned}—the length of the undeformed compensation spring; ς_6—the length of the spring in the extreme right position of the plunger; k—is the dimensionless coefficient; γ—"diameter reduction factor"; μ—coefficient of dynamic viscosity of the finishing material pumped by the pump; d_1—is the inner diameter of the body (cylinder) of the working plunger (or the outer diameter of the plunger, or the diameter of the suction working cavity (chamber).

Having the equation of relative motion of the discharge valve, rejecting the balanced systems of forces, consider the forces acting directly on it:

- the force of elasticity of the spring;
- force of injection pressure, which is a measure of mechanical action on the valve of the part of the finishing material that is being processed by the pump, which is already in the discharge cavity (chamber) and outside the pump in the network (in the pipeline);
- the force of excess pressure, which is a measure of the mechanical action on the valve of the part of the finishing material, which is filled with the suction working cavity (Fig. 4).

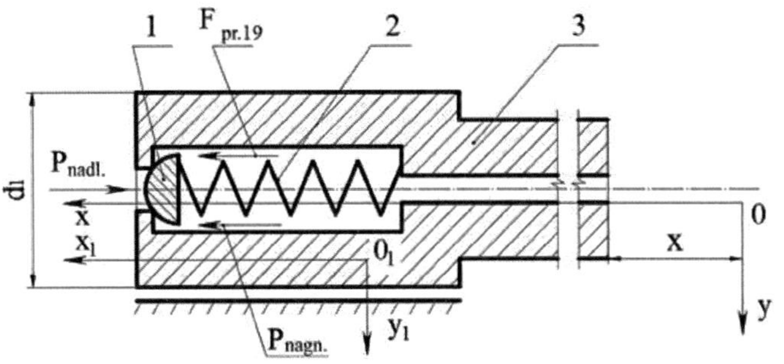

Fig. 4 Kinematic scheme of the pump valve

Let's find out the meaning of these forces. According to Hooke's law, the modulus of elastic force is equal to the product of the stiffness coefficient for spring deformation. The force of injection pressure depends on the specific location in the space of the injection network (pipeline); we assume in this study that in the conditions of each of the possible cases the modulus of this force is a constant.

Excess force is not constant and is quite difficult to depend on several technological and design parameters, but the most influential on the value of the modulus of this force is the value of reducing the volume of the suction working cavity (chamber).

We take into account that the discharge valve carries out its relative movement only along the axis O1 × 1 due to which the vectors of its relative acceleration and relative velocity are projected on this axis only in real (natural) quantities. Substituting all the values and values set in one way or another in the acceleration equation, we obtain the equation of change in the amount of relative movement of the valve:

$$d(m_9 * v_{9x1}^r) = (c_{19} * \xi_{19} + P_{nagn,} - \frac{\pi * d_1^2}{4} * k_5 * f(x) - \frac{m_9 * Q_0}{m} * \sin\frac{\pi t}{\tau}$$
$$- \frac{m_9 * (c_6 + c_7)}{m} * x + \frac{m_9 * c_6 (l_{6ned,} - \varsigma_6)}{m} - k\frac{m_9 * \sqrt{\pi(1-\gamma^2)}}{2 * m} \mu d_1 * v) * dt, \tag{5}$$

As for emergence and growth of speed of movement of the valve it is necessary to fulfill conditions of growth of quantity of movement that mathematical condition of movement of the valve (after integration of both parts within size of time of opening of the injection valve)

$$(c_{19}\xi_{19} + P_{nagn} + \frac{m_9 c_6 (l_{6ned,} - \varsigma_6)}{m}) * t_1 + \frac{m_9 Q_0 \tau}{\pi m} * (\cos(\frac{\pi t_1}{\tau}) - 1)$$
$$> \frac{\pi d_1^2}{4} * \int_0^{t_1} k_5 * f(x) * dt + \frac{m_9 (c_6 + c_7)}{m} * \int_0^{t_1} x * dt + k\frac{m_9 \sqrt{\pi(1-\gamma^2)}}{2 * m} \mu d_1 * \int_0^{t_1} v * dt. \tag{6}$$

The obtained dependence allows us to draw conclusions about which parameters directly affect the value of the valve operation time. It is impossible to obtain its analytical solution due to incoherent integrals in the right part, however, the influence of parameters is obvious, including: spring stiffness, plunger size, location of the ball relative to the coil winding, friction seals, energy loss in valve assemblies.

The greatest influence on energy losses in the valve units of the pump is the ratio of the diameters of the ball and the socket (which is 4/3), the height of the ball above the plane of the socket, the mobility of the mixture and the mass of the locking element.

Thus, it is established that uniform pumping of the material is achieved by reducing the time to open and close the discharge valve, which is respectively achieved by reducing the weight of the shut-off element (Fig. 5).

To study the effect of the mass of the shut-off element on the speed of the valve, an experiment was conducted during which the operation of the pump with shut-off elements made of different materials was investigated.

To determine the speed of operation of the valve with a shut-off element from the above materials, it was first determined the mass of each of them (Table 1):

Acceleration and operating time were determined experimentally, the influence of resistance and density of the pumped medium was neglected because its area of interaction with the shut-off element is insignificant. The authors also suggested that the speed of the shut-off element and the plunger are equal, based on the fact that the discharge valve is mounted in the plunger.

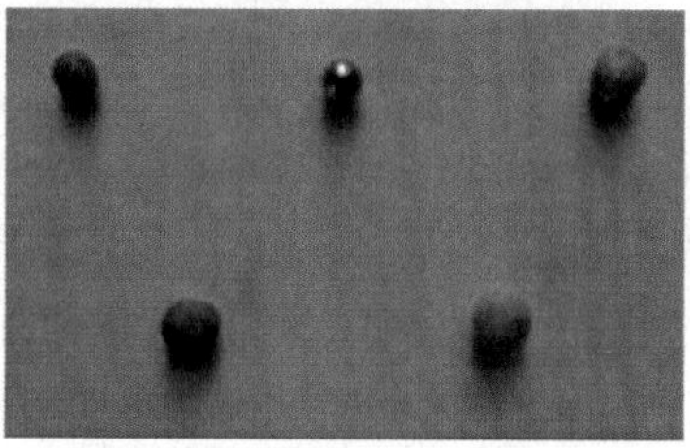

Fig. 5 Locking elements from different material (from left to right): plastic, steel, rubber, cold welding, epoxy glue

Table 1 Masses of locking elements

#	Locking element material	The average value of the mass from a series of 6 measurements, g
1	Polyvinyl chloride (plastic)	0,221
2	Glue on the basis of epoxy resins of cold hardening (epoxy glue)	0,304
3	Rubber amortization, special (rubber)	0,432
4	Crystallized metal welded bath (cold welding)	0,431
5	Low-carbon steel with a density of 7.86 g/cm^3	0,862

Table 2 Dependence of mass and acceleration of the locking element on the material from which it is made

#	Locking element material	Acceleration of the locking element m/s^2
1	Polyvinyl chloride (plastic)	0,04
2	Glue on the basis of epoxy resins of cold hardening (epoxy glue)	0,03
3	Rubber amortization, special (rubber)	0,02
4	Crystallized metal welded bath (cold welding)	0,02
5	Low-carbon steel with a density of 7.86 g/cm^3	0,01

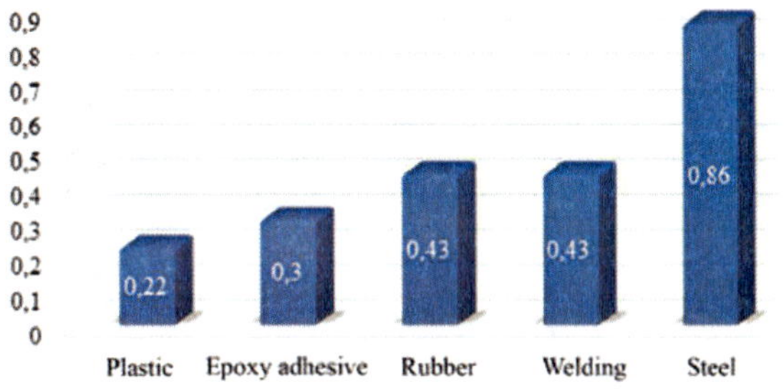

Fig. 6 Diagram of the dependence of the operating time of the discharge valve on the material of the shut-off element

The operating time of the discharge valve was determined by Newton's second law and the dependences of classical dynamics. The obtained data are shown in Table 2.

As you can see from the table, the locking element has the greatest acceleration, made of plastic. Using the described technique, it was determined that for the locking element made of plastic, the valve speed will be 8.63 m/s.

The dependence of the operating time of the discharge valve on the material from which the shut-off element is made, is shown in Fig. 6.

4 Conclusions

To increase the accuracy of engineering calculations and analysis of the pump, its mathematical model was developed.

The analysis of the mathematical model of the pump operation is carried out. The dynamics of the relative movement of the discharge valve of the plunger of the differential pump is analyzed taking into account the forces of inertia. A parametric expression is obtained that determines the operation time of the differential pump valve. The design parameters that determine the effectiveness of its work were determined. These include: spring stiffness, plunger size, location of the ball relative to the coil winding, friction forces of the seals, energy loss in the valve assemblies. The greatest influence on energy losses in the valve units of the pump is the ratio of the diameters of the ball and the socket (which is 4/3), the height of the ball above the plane of the socket and the mobility of the mixture.

The operation of pump valves with shut-off elements made of different materials has been investigated experimentally.

The value of the parameter of the mass of the shut-off element at which the optimal mode of operation of the pump is achieved.

References

1. Parfyonov EP (1972) Determination of the performance of piston mortars. Mekh. tools and finishing machines: information Scientific-Technical Sat, vol 4, pp 12–13. TsNIITEstroymash
2. Korobko B, Vasyliev IE (2017) Test method for rheological behavior of mortar for building work Acta mechanica et automatica 11/3(41):173–177. https://doi.org/10.1515/ama-2017-0025
3. Korobko B, Vasiliev A, Rogozin I (2015) The analysis of mixture kinematics in the mixer body frame with a screw elevator with variable generatrix. Eastern-Eur J Enterp Technol 3(7):48–52. https://doi.org/10.15587/1729-4061.2015.4305348-52
4. Mortelpumpen und ihre Entwicklung (1969) Fordern und Heben, no 15 (Germany)
5. EP 0200026, INT. Cl. 4 F 04 B 43/12, 15/02. Neumuller pumps Walter, Sturmer Gerhard. 10.12.1986. Patentblatt 86/45
6. Rohozin I, Vasyliev O, Pavelieva A (2018) Determination of building mortar mixers effectiveness. Int J Eng Technol 7(3.2) (S.I. 2):360–366. https://doi.org/10.14419/ijet.v7i3.2.14553
7. Shapoval MV, Virchenko VV, Skoryk MO, Shpilka AM (2019) Improving the efficiency of the pump by using a hydraulic actuator. In: Collection of scientific papers of the II international Ukrainian-Azerbaijan conference "building innovations—2019", 23–24 May 2019, pp 202–205. PoltNTU, Poltava
8. Emelyanova IA, Shapoval MV (2017) Determination of productivity and volumetric efficiency of the mortar pump, depending on the geometric parameters of the suction chamber and the compensators of various design solutions. Sci Bull Constr 88(2):195–203
9. Kukoba AT, Vasilyev AV (2000) Investigation of the volumetric efficiency of a hydraulic drive mortar. In: Collection of scientific works (branch mechanical engineering, construction, vol 5, pp 19–24. Polt. state. tech. Yuriy Kondratyuk. PDTU, Poltava
10. Kukoba AT, Vasilyev AV, Yakubtsov OM (2000) The impact of the law on the piston against capacious efficiency pump. In: Collection of scientific works (branch engineering, construction, vol 6, pp 12–17. Polt. state. tech. Yuriy Kondratyuk. Part 1. PDTU, Poltava
11. Kukoba AT, Korobko BO, Vasilyev AV (2000) Changing the volume of the mortar mixture during pumping with a solution pump. Mechanization of Construction, 3
12. Shapoval MV (2013) Analysis of the effect of flow pressure on the volumetric pump efficiency. In: Collection of scientific works sectoral mechanical engineering, construction, vol 1, no 36, pp 19–204. PoltNTU, Poltava
13. Korobko B (2016) Investigation of energy consumption in the course of plastering machine's work. Eastern Eur J Enterp Technol 4(8–82):4–11. https://doi.org/10.15587/1729-4061.2016.73336
14. Korobko B, Virchenko V, Shapoval M (2018) Feed solution in the pipeline aith the compensators mortar pump of various design solutions pressure pulsations degree determination. Int J Eng Technol (UAE) 7(3):195–202. https://doi.org/10.14419/ijet.v7i3.2.14402

15. Korobko B, Khomenko I, Shapoval M, Virchenko V (2020) Hydraulic single pump with combined higher volume compensator operation analysishttps://doi.org/10.1007/978-3-030-42939-3_12
16. Korobko B, Zadvorkin D, Vasyliev I (2017) Study of the operating element motion law for a hydraulic-driven diaphragm mortar pump. Eastern Eur J Enterp Technol 4(7–88):25–31. https://doi.org/10.15587/1729-4061.2017.106873

Air Filter Throughput Impact Experimental Study on Internal Combustion Engine Fuel Efficiency

Bogdan Korobko, Anatolii Kryvorot, Maksym Skoryk, and Viktor Virchenko

Abstract The article discusses the estimated indicators of fuel efficiency of vehicles. It was established that analytical and experimental data on determining the fuel efficiency of vehicles have a significant discrepancy. It was found that well-known mathematical models do not take into account the purity of the air filter element for fuel efficiency. Experimental studies were performed and processed, mathematical dependence between the specified parameters was established: air filter throughput and angular speed of crater shaft of internal combustion engine. The obtained mathematical dependence, in the form of a second degree polyonomy, is suitable for use according to the criteria of Student and Fisher. It is determined that in the future the found regression equation can be used to determine the fuel consumption of cars depending on the pollution of the air inlet system of engines.

Keywords Vehicle · Fuel costs · Planning an experiment

1 Introduction

A decisive trend in the development of modern vehicles is to increase their operational qualities.

Fuel efficiency is a corresponding set of properties that determine fuel consumption when the car performs work in different operating conditions. This property of the car allows the driver to determine the fuel consumption per unit kilometer. The main indicator is fuel consumption per 100 km. At the same time, fuel consumption is distinguished when driving along the urban cycle and outside settlements. The cost of the city cycle will always be greater [1].

The estimated indicators of fuel consumption, that is, the efficiency of the vehicle in general is denoted by such indicators of the engine as hourly fuel consumption G_T, kg/h—the mass of liquid fuel consumed in one hour of engine operation, and the

B. Korobko · A. Kryvorot · M. Skoryk · V. Virchenko (✉)
National University «Yuri Kondratyuk Poltava Polytechnic», 24, Pershotravnevyj Ave., Poltava 36011, Ukraine

V. Onyshchenko et al. (eds.), *Proceedings of the 3rd International Conference on Building Innovations*, Lecture Notes in Civil Engineering 181,
https://doi.org/10.1007/978-3-030-85043-2_21

specific fuel consumption g_e, g/(kW h)—the mass of liquid fuel required per hour per unit of engine power.

In most European Union countries and in our country, the main measure of fuel efficiency of the vehicle is the consumption of liquid fuel in volume (liters) per 100 km of traveled pathway—(road fuel consumption) Q_S, l.

To assess the effective fuel consumption when performing transport work by car, the concept is used as fuel consumption per unit of transport work (100 t km) Q W, l—the ratio of the actual volume of fuel to the work done.

In the United States, along with the track consumption, they use a reverse meter—the length of the way to the volume of fuel consumed.

Consequently, reducing fuel consumption by cars equipped with internal combustion engines is considered one of the most important activities of all interested structures, ranging from designers and vehicle developers, research and testing and repair services and structures, legal entities and individuals who operate them to various enterprises, including vehicles engaged in cargo or passenger transportation, organizations and firms, and ending this list with private individuals.

2 Analysis of the Latest Sources of Research and Publications

Currently, prices for petroleum products are 14-20% of all costs for the cars operation, so it is desirable to use vehicles with the highest efficiency to perform transportation work. With the reduction of fuel consumption by cars, the emissions of carbon monoxide into the environment are also reduced, which leads to the improvement of the ecological environment. Vehicle fuel efficiency determines the rational use of fuel for transportation, both cargo and people, under different operating conditions. To determine the estimated vehicle fuel efficiency, a generalized equation of fuel consumption equation is required. But this is not really real, because the design schemes of cars differ from each other [1–12].

In general, the engine produces N power, kW, and the specific fuel consumption of the engine is equal to g_e, g/kW · h, when driving the car, the hourly fuel consumption will be:

$$Q_{hour} = \frac{q_e \cdot N}{1000}, \quad (1)$$

$$Q_{hour} = \frac{q_e \cdot N}{1000 \cdot \rho_T}, \quad (2)$$

where ρ_T—fuel density, kg/m^3.

To overcome the car's range of 100 km at a speed of V, km/h, the required amount of fuel, l/100 km, will be

$$Q_S = \frac{q_e \cdot N}{10 \cdot V \cdot \rho_T}, \quad (3)$$

where N—total engine power, kW.

The specified power is necessary for the movement of the car, that is, to overcome all resistance forces, in accordance with the powerful balance consists of the following parts:

- loss of power in the transmission; N_{Tp}
- spent power to overcome the rolling resistance of the wheels; N_f
- spent power to overcome the rise of the road; N_h
- spent power to overcome air resistance; N_n
- spent power to overcome the acceleration of the vehicle.N_j

When determining all the specified components, you can get the total engine power required to drive the vehicle into motion

$$N = \frac{M_a \cdot g \cdot (\sin\alpha + f \cdot \cos\alpha) \cdot V + K \cdot F \cdot V^3 + P_j \cdot V}{1000 \cdot \eta_m}. \quad (4)$$

After substitution of the formula in the expression and mathematical conversion, the fuel consumption will take the form (4) (3)

$$Q_S = \frac{q_e}{10 \cdot \rho_T} \cdot \frac{M_a \cdot g \cdot (\sin\alpha + f \cdot \cos\alpha) \cdot V + K \cdot F \cdot V^3 + P_j \cdot V}{1000 \cdot \eta_m}. \quad (5)$$

From the theory of cars [1–7], the force spent on overclocking all the masses of the vehicle, moving rotating and moving in a moving way, has the form

$$P_j = \delta \cdot M_a \cdot \frac{dV}{dt}. \quad (6)$$

Having substituted this force into a formula and turning the speed of the car from m/s to km/h, we get (5)

$$Q_S = \frac{q_e}{10^4 \cdot 3,6 \cdot \eta_m \cdot \rho_T} \times \left(M_a \cdot g \cdot (f \cdot \cos\alpha \pm \sin\alpha) + \frac{1}{3,6^2} K_B \cdot F \cdot V^2 + \delta_{BP} M_a \cdot \frac{dV}{dt} \right). \quad (7)$$

Taking into account the work of well-known scientists in the field of road transport [1–7], we come to the conclusion that in the expression it is advisable to introduce the coefficient of opposition to the road, then we get such an equation of fuel consumption (7) $\psi = f \cdot \cos\alpha \pm \sin\alpha$

$$Q_S = \frac{q_e}{10^4 \cdot 3,6 \cdot \eta_m \cdot \rho_T} \times \left(\begin{array}{l} M_a \cdot g \cdot \psi + \frac{1}{3,6^2} K_B \cdot F \cdot V^2 \\ + \delta_{BP} M_a \cdot \frac{dV}{dt} \end{array} \right). \quad (8)$$

The Eq. (8) makes it possible to assess the quantitative influence of factors that affect economic indicators in different modes of the vehicle.

For a sustainable speed of the car on the horizontal part of the road surface, the force is absent, so fuel costs are reduced to expression P_h.

$$Q_S = \frac{q_e}{10^4 \cdot 3,6 \cdot \eta_m \cdot \rho_T} \times \left(M_a \cdot g \cdot f + \frac{1}{3,6^2} K_B \cdot F \cdot V^2 \right). \quad (9)$$

And when moving the car on the rise

$$Q_S = \frac{q_e}{10^4 \cdot 3,6 \cdot \eta_m \cdot \rho_T} \times \left(M_a \cdot g \cdot \psi + \frac{1}{3,6^2} K_B \cdot F \cdot V^2 \right). \quad (10)$$

All presented expressions do not take into account the technical condition of the car, namely, the technical condition significantly affects unproductive energy costs, causing an increase in fuel consumption. The most significant impact are engine malfunctions, especially—inlet and power supply systems. Well-known mathematical models and automatic control systems for the process of supplying fuel mixture do not take into account the purity of the filter air element.

Setting a Task. The purpose of this article is to highlight the results of experimental studies of the fuel efficiency of the internal combustion engine 4G18, depending on the throughput of the air filter.

Main Material and Results. For the object of research in this work, it was decided to take a four-cylinder atmospheric engine 4G18 with a working volume of 1.6 L and euro-5 toxicity standards, which is massively used by the Mitsubishi Concern.

In order to establish optimal fuel consumption by the engine and obtain a mathematical dependence describing this process, the planning of the experiment was applied. Since the main factors affecting the fuel consumption of the engine, with a serviceable fuel injection system, are the filter element throughput and the angular velocity of the crankshaft, and the appearance of the resulting dependence is unknown decided to adopt a two-factor three-level plan of the experiment [13, 14].

To simulate air filter pollution (Fig. 2), initially the filter rinse was determined, which is 360.75 cm^2, which corresponds to 100% of the patency. Accordingly, 50% of the patency is an area of 180.37 cm^2, and 75% of the throughput is 270.56 cm^2.

After the above data, we perform two stencils made of durable cardboard with a permeability of 50 and 75%, which are shown in Fig. 3 and 4, respectively, and will be installed in front of the filter element (Fig. 1).

To register the results of the study, the ELM 327 canner was used to find and eliminate errors in the electronic systems of the car engine, as well as to visualize various parameters of the engine and car.

Experimental studies were carried out in the laboratory of technical diagnostics of cars of the National University «Yuri Kondratyuk Poltava Polytechnic» (Fig. 5). Initial working conditions: the engine was heated to the operating temperature,

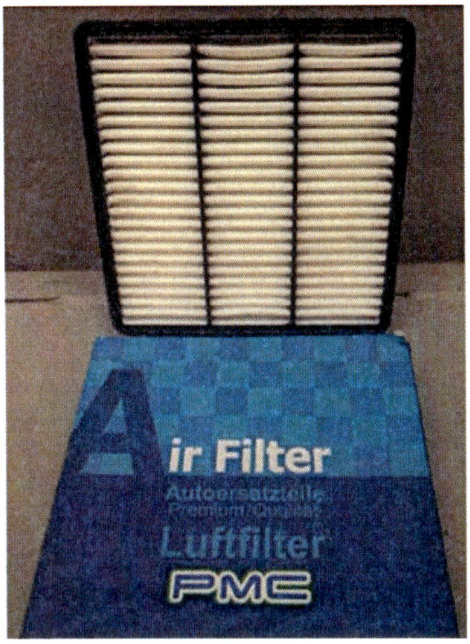

Fig. 1 New filter element Luftfilter PMC AP175

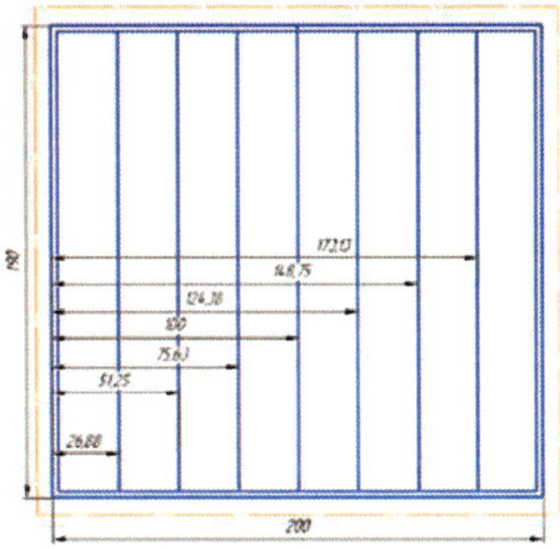

Fig. 2 Scheme of the air filter uniform division.

Fig. 3 Stencil with a patency of 50%

Fig. 4 Stencil with a patency of 75%

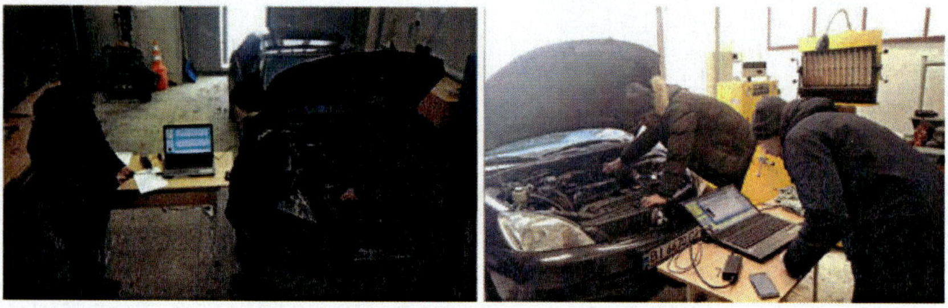

Fig. 5 The process of experimental research

namely 92 °C, the exhaust system is wearing a sleeve for exhaust gases output. The ambient temperature was 10 °C with atmospheric pressure of 745 mm Hg. The time of each of the experiments is 60 s.

During the experiment, the factors varied on three levels—average (main), lower and upper, far from the main one by the same value. The values of the intervals of factors varying and the range of values studied is shown in Table 1.

To simplify records and subsequent calculations, the upper level of factors is indicated by the symbol (+), the middle level—(0), and the lower level—(–). The value of minute fuel consumption at different purity of the filter element and at different angular velocity of the crater shaft is determined according to the plan given in Table 2.

Table 1 Values of intervals of factors varying

Factors	Variation intervals
Cleanliness of the filter element, % (cm^2)	180,37–360,75
Angular velocity of crater shaft and internal combustion engine, rad/s	157–366,33

Table 2 Three-level plan for conducting experiments with the number of factors k=2 ... (N=N₁+Nα+n₀)

No.		Planning matrix (x_i)		Squares of variables (x_i^2)		Interaction $(x_i x_j)$	Surface temperature of the lodge
		x_1	x_2	x_1^2	x_2^2	x_1x_2	y
1		2	3	4	5	6	7
N_1	1	+	+	+	+	+	0,069
	2	+	–	+	+	–	0,028
	3	–	+	+	+	–	0,078
	4	–	–	+	+	+	0,032
N_α	5	+	0	+	0	0	0,051
	6	–	0	+	0	0	0,056
	7	0	+	0	+	0	0,071
	8	0	–	0	+	0	0,03
n_0	9	0	0	0	0	0	0,052
	10	0	0	0	0	0	0,05
	11	0	0	0	0	0	0,051

During the experiment on determining fuel costs, the experiments are divided into groups so that the experiments at zero point are evenly distributed among others. In particular, we accept the following procedure for the implementation of the plan: experiments 1, 2, 9, 3, 7, 4, 5, 10, 6, 7, 8, 11.

The results of experiments are processed using methods of mathematical statistics [13, 14].

For plans of the second order in a two-fact experiment, we obtain an algebraic equation in the form of

$$\bar{y_i} = b_0 + b_1x_1 + b_2x_2 + b_{11}x_1^2 + b_{22}x_2^2 + b_{12}x_1x_2, \tag{11}$$

where—minute fuel costs studied;$\bar{y_i}$.

$x_1,\ x_2$—initial factors;

$b_0,\ b_1,\ b_2,\ b_{11},\ b_{22},\ b_{12}$—coefficients of the equation.

Having determined by the method [15, 16] coefficients and sub stating them in the resulting regression equation in code form (11)

$$\begin{aligned}\bar{y_i} &= 0,05128 - 0,003x_1 + 0,02134_2 \\ &+ 0,00182x_1^2 - 0,00118x_2^2 - 0,00125x_1x_2.\end{aligned} \tag{12}$$

In the future, the Eq. (12) is clarified by checking the difference between coefficients band from zero according to the Standard criterion.

For the convenience of further calculations of the Eq. (12) lead to a natural appearance (13).

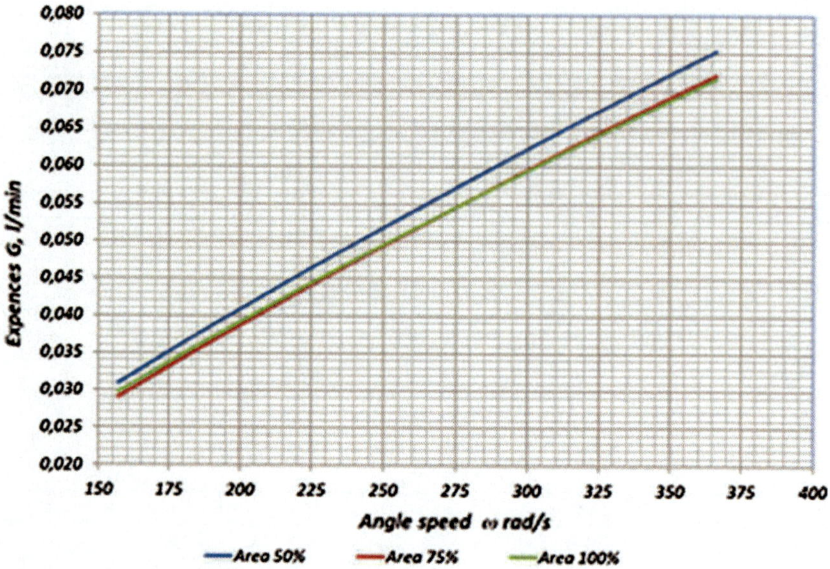

Fig. 6 Graphic dependence of minute fuel consumption on the gradual increase in the angular velocity of the crankshaft and the engine 4G18 with a fixed permeability of the air filter element

$$G = 1,66122 \cdot 10^{-7} \cdot A^2 - 6,62099 \cdot 10^{-8} \cdot A \cdot \omega - 0,0000862 \cdot A \\ -1,07705 \cdot 10^{-7} \cdot \omega^2 + 0,00028 \cdot \omega + 0,00174 \quad (13)$$

Using the Eq. (13) we build graphical dependences of fuel costs on two factors, namely the throughput of the air filter and the angular velocity of the crater shaft of the engine Fig. 6 and 7.

Analyzing the data presented in Fig. 6, we can conclude that even not significant pollution of the air filter leads to an increase in x of the fuel consumption at a rotational speed of the crater shaft of the engine over 2000 rpm.

Analyzing Fig. 7, we conclude that with the increase in the permeability of the air filter element, minute costs decrease, and the greater the speed of the crater shaft of the engine, you are more obvious in the steepness of the fallout of the curves.

Based on the research, the following recommendations can be taken to reduce fuel consumption during the operation of vehicles:

- be sure to observe the frequency of replacement of the air filter;
- every 1000 km of mileage, it is necessary to clean the side house for particulate particles;
- every 1000 km of mileage, blow the air filter in compressed air.

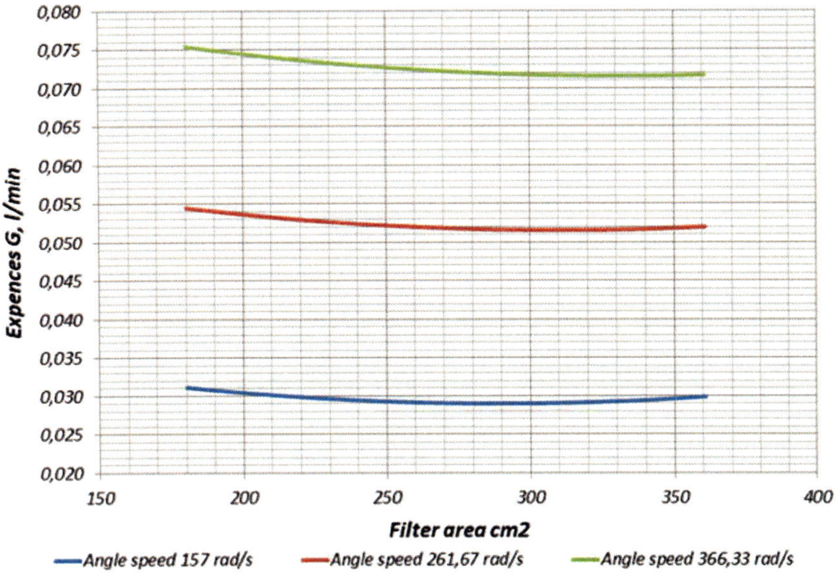

Fig. 7 Graphic dependence of minute fuel consumption on a gradual increase in the area of cross section of the filter element at a fixed angular velocity of the crater shaft of the internal combustion engine 4G18

3 Conclusions

As a result of experimental studies, mathematical dependences were established between the specified parameters: air filter throughput and the angular veloce of the crater shaft of the internal combustion engine.

On the basis of the experimental studies, practical recommendations for improving the fuel efficiency of vehicles have been clarified.

References

1. Sakhno VP, Bezborodova GB, Mayak MM, Sarai SM (2004) Automobiles: Traction-speed properties and fuel efficiency /Textbook/. - K: B-w “KWC”, 2004, p 174. 15. Table 19. Bibliographer. 30. Titles. 30 ISBN 966–7192–55–5
2. Farobin YE, Shuplyakov VS (1983) Otsenko of xpluationnyh voyages for interdunarodnyh-transports. Transport, p 200
3. Lytvynov AS, Farobyn YE Makmobile: Theory of Explutation of Svistystvo
4. Osnovenko MY, Sakhno VP (1992) Automobiles: Training. Guide. NMK NAO, p 344
5. Grishkevych AI (1986) Automolysly: Theory. Student for the esc. - Mn. Vis. Shk, p 208
6. Klympush OD, Rubtsov VA, Gutarevych YF (1988) Ecoconomy drowning on a car transport. Technics, p 144
7. Govorushchenko NY (1990) Ecoconomy drowning toxicity on the car transport/N.Y. Govorushchenko. Transpornt, p 135

8. Korobko B, Zadvorkin D, Vasyliev I (2017) Study of the operating element motion law for a hydraulic-driven diaphragm mortar pump. Eastern-Eur J Enterprise Technol 4(7–88):25–31. https://doi.org/10.15587/1729-4061.2017.106873
9. Korobko B (2016) Investigation of energy consumption in the course of plastering machine's work. Eastern-Eur J Enterprise Technol 4(8–82):4–11. https://doi.org/10.15587/1729-4061.2016.73336
10. Korobko B, Virchenko V, Shapoval M (2018) Feed solution in the pipeline aith the compensators mortar pump of various design solutions pressure pulsations degree determination. Int J Eng Technol (UAE) 7(3):195–202. https://doi.org/10.14419/ijet.v7i3.2.14402
11. Korobko B, Khomenko I, Shapoval M, Virchenko V (2020) Hydraulic single pump with combined higher volume compensator operation analysis. https://doi.org/10.1007/978-3-030-42939-3_12
12. Korobko B, Vasyliev I (2017) Test method for rheological behavior of mortar for building work. Acta Mechanica Et Automatica 11(3):173–177. https://doi.org/10.1515/ama-2017-0025
13. Fleahin VG, Gludkin OP, Gurov AI, Hanyn MA (1997) Soviet experimentation: preparation, conducted, analizov/Pod ed. O.P. Gludkina. Radio and Svyaz
14. Vanhatalo E, Bergquist B, Va¨nnman K (2013) Towards improved analysis methods for twolevel factorial experiments with time series responses. Qual Reliab Eng Int 29(5):725–741
15. Vanhatalo E, Va¨nnman K (2008) Using factorial design and multivariate analysis when experimenting in a continuous process. Qual Reliab Eng Int 24(8):983–995
16. Mukhachov VA (2007) Planning and election of the results of the experiment: Uchebnoe assistance.—Tomsko: Tomsk State University of Management Systems and Radoelectronists, 2007. - 118 s ISBN 966-552-11-7: 325.60. http://www.springer.com/lncs. Accessed 21 Nov 2016

Solution Pressure Pulsations into the Pipeline Size Determination in Dependence on Constructive Parameters of Valve Units of Mortar Pump

Bogdan Korobko, Inna Khomenko, Mykola Shapoval, and Viktor Virchenko

Abstract The article considers a single-piston mortar pump with electromechanical drive, ball suction and spring-loaded discharge valves, a special insert in the suction chamber and a compensator of increased volume.

The analysis of the influence of hydrodynamic pressure on the valve balls is carried out taking into account all the design parameters of the hydraulic part of the mortar pump.

The mechanisms of action of the hydrodynamic pressure force on the ball from the side of its solution flow on the valve ball due to such factors as the cumulative flow of the solution through the hole in the valve seat are established; the influence of normal and tangential stresses that occur in the surface layer on the surface of the valve ball and occur when the solution has a structured viscosity.

Keywords Single-piston mortar pump with combined compensator of increased volume · Suction chamber · Suction and discharge valves · Volumetric efficiency · Mobility of the solution

1 Introduction

The main directions of development of mortar structures development are analyzed and analytical analysis of single-piston mortar pump with combined compensator of increased volume is carried out. Introduction of a mortar in the design, namely the installation of a special insert, which should ensure a decrease in back leaks through the suction valve, which will ensure the growth of the volumetric efficiency of the mortar and reduce the level of the solution pulsation. Also, the use of an elastic element in the supercharger valve should have a positive effect on reducing inverse leaks through the injection valve due to the increase in the rate of lowering of the ball,

B. Korobko · I. Khomenko · M. Shapoval · V. Virchenko (✉)
National University «Yuri Kondratyuk Poltava Polytechnic»,
24, Pershotravnevyj Ave., Poltava 36011, Ukraine

V. Onyshchenko et al. (eds.), *Proceedings of the 3rd International Conference on Building Innovations*, Lecture Notes in Civil Engineering 181,
https://doi.org/10.1007/978-3-030-85043-2_22

especially when pumping solutions of reduced mobility P 7–9 cm. From less to a minimum pulsation of the pressure of the solution during its pumping will positively affect the quality of finishing work, both in the use of m in conjunction with plaster stations or as part of mobile small plaster units.

Therefore, it is necessary for theorists to investigate the process and in the hydraulic part of a single-piston mortar pump with a combined compensator of increased volume.

The analysis of the latest sources of research indicates the inability to investigate the interaction of the pumped medium with valve nodes, taking into account structural changes in the suction chamber of the mortar pump a, which will affect the reduction of the degree of pulsations of the pressure supply of the solution [1–8]. Therefore, the analytical research of the hydrodynamic effect of the cumulative jet of the solution on the elements of the hydraulic part of the mortar pump becomes relevant [9–15].

Highlight parts of a common problem that have not been solved before. It is necessary to analyze the effect and interaction of hydrodynamic pressure force on the ball of valves from the cumulative flow of the solution.

Setting a Task. To increase the level of volumetric efficiency and reduce the level of pressure pulsations of the solution during pumping, it is necessary to use a hydraulic drive in a single-piston mortar pump and with a combined compensator of increased volume.

The Purpose and Objectives of the Study. The purpose of this work is to increase the level of volumetric efficiency and reduce the level of pressure level of the single-piston mortar solution with a combined compensator of the increased volume due to the use of a special insertion in the suction chamber and further reducing its harmful volume.

To achieve a certain goal, the following tasks must be solved:

1. Theoretical investigation the forces in the influence of the structured environment on the ball valve.
2. Install with chem in the effect of hydrodynamic forces on the injection and suction valves of single-piston mortar pump.
3. Carry out the operation of the ball valve of the mortar based on the established dynamic model, which makes it possible to determine the law of movement of the valve ball during operation, depending on the law of movement of the working body of the valve node design, and the properties of the pumped solution

The object of research is the processes occurring in the hydraulic part of the single-piston mortar pump, and with a combined compensator of increased volume during the transportation of mortars through the pipeline.

The subject of research is a single-piston mortar pump with a combined compensator of increased volume.

Research Methods. When conducting researches, I used the basic provisions of hydraulics, hydrodynamics, methods of mathematical physics, physical and mathematical modeling by methods of applied mechanics, computer programming of Microsoft Office, Compass 3D, MathCAD 14.

The results of the study. Theoretically, the influence of the solution pumped into the elements of the hydraulic part of the mortar was investigated on the basis of the laws of hydrodynamics, taking into account the rheological properties of the solution, the pressure of the solution supply and when using a combined compensator of increased volume. The effect of the operation of the mortar valves on the technical characteristics of productivity and volumetric efficiency has been established.

The Main Material. The analysis of the work of the mortar pump a (Fig. 1), which was developed at the National University "Poltava Yuri Kondratyuk Polytechnic" indicates a more thorough study of the interaction of the pumped medium with valve nodes, which directly affects the level of technical characteristics of the mortar pump with electromechanical drive, which has established itself as a reliable and highly efficient volumetric machine. But pumping solutions with a solution of reduced mobility P7-8 cm does not occur at a degree of pressure pulsations near the level $\delta \geq 25\%$.

To determine the number of pulsations of the pressure of the solution into the pipeline from the design parameters of the valve components of the mortar, it is

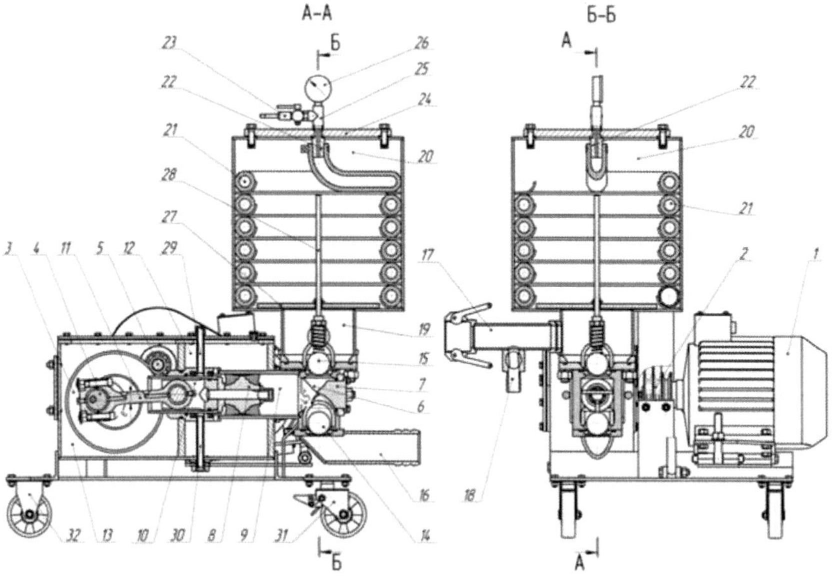

Fig. 1 Single-piston mortar pump with combined compensator of increased volume: 1—electric motor; 2—wedge-pass transmission; 3—gear wheel; 4—crankshaft; 5—shaft-gear; 6—suction chamber; 7—special cylindrical insert; 8—piston; 9—working cylinder; 10—slider; 11—rod; 12—chamber; 13—gearbox body; 14, 15—suction and suction suspended ball valves; 16, 17—suction and injection nozzles; 18—pressure reduction valve; 19—supercharger; 20—cylindrical chamber; 21—locked chamber; 22—fitting of air pumping unit; 23—nipple; 24—cover; 25—crane; 26—pressure gauge; 27—float-limiter; 28—guide rod; 29, 30—channel nozzles; 31, 32—pair of wheels

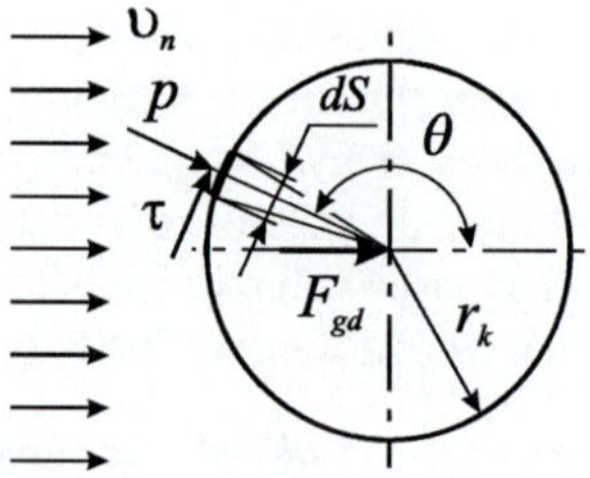

Fig. 2 Design diagram of the influence of the valve ball on the flow of structured liquid

necessary to analyze the effect of hydrodynamic pressure on the valve balls, taking into account all the design parameters of the hydraulic part of the mortar.

The effect of the solution on the valve ball is characterized by the effect of hydrodynamic pressure on the ball from its flow. This effect occurs due to the following factors: the cumulative flow of the solution through a hole in the valve saddle; influence of normal and tangent stresses that occur in the surface layer on the surface of the valve ball and occur when a solution has structured viscosity.

So, to determine the strength of the hydrodynamic effect F_{gd} on the ball, dependence is made [3]

$$F_{gd} = F_s + F_t \tag{1}$$

where F_s—effort of frontal resistance from the cumulative jet of the solution in the hole of the valve saddle; F_t—side friction force that occurs during the flow of the ball by a structured liquid.

At the time of lowering the ball in a statically stationary solution, the force of the frontal pressure on the ball F_s from the side of the cumulative jet and side friction, as well as during the flow of the F_t structured liquid of the valve balls in a limited volume space has the same nature (see Fig. 2) [3, 8].

Therefore, the force of the influence of the structured medium on the ball valve can be presented as a component of the elementary forces of pressures and grated Ion the parts of the surface of the ball [3, 8]

$$\begin{aligned} F_{gd} = & \int_0^{\pi} \left(-\frac{3}{2} \cdot \frac{\upsilon_p \cdot \mu}{r_k} \cos\theta \right) \cdot \cos\theta \cdot 2 \cdot \pi \cdot r_k^2 \cdot \sin\theta d\theta \\ & + \int_0^{\pi} \left(\tau_0 + \frac{3}{2} \cdot \frac{\upsilon_p \cdot \mu}{r_k} \sin\theta \right) \sin\theta \cdot 2 \cdot \pi \cdot r_k^2 \cdot \sin\theta d\theta \\ & = 2 \cdot \pi \cdot \upsilon_p \cdot r_k \cdot \mu + 4 \cdot \pi \cdot \upsilon_p \cdot r_k \cdot \mu + \pi^2 \cdot r_k^2 \cdot \tau_0 \end{aligned} \tag{2}$$

After simplification will look like

$$F_{gd} = 6 \cdot \pi \cdot \upsilon_p \cdot r_k \cdot \mu + \pi^2 \cdot r_k^2 \cdot \tau_0 \tag{3}$$

The frontal resistance of the ball of the solution flow and the resistance of the surface friction of the ball when the mortar stream moves are defined as components of dependence (2). Both components of the ball movement resistance are determined by the flow rate υ_n, the structured viscosity of the liquid μ and the geometric parameter of the ball, namely the radius r_k. From the dependence (2) it can be seen that a much greater effect carries out friction resistance compared to the frontal resistance, and the third addition of the amount means the total force required to shift the valve ball relative to the solution, which in turn creates a static resistance of its movement caused by the presence of tension τ_0, and does not depend on the two components of the structured viscosity μ of the solution and the velocity of relative movement. Addiction is the solution to the inseparability problem of the Naive-Stokes mass when the ball interacts with a structured liquid.

To determine the values of hydrodynamic force, it is necessary to determine the rheological parameters μ and τ_0 of environment. Based on the expressions (2) (3), it is possible to develop analytical dependences to determine the characteristics of the structured solution by studying the lowering of the ball under the action of gravity, as well as to assess the hydrodynamic effect of the solution flow on the valve ball during its operation in the limited round-valve space of the working chamber.

Taking into account the design features of the valve node, namely: the presence of a saddle under the ball, the limitations of the round-valve chamber, the pressure force F_s acting on the ball from the cumulative jet can be considered equivalent to the average velocity of the jet of the solution, which makes the infusion from the saddle hole, and the resistance force on the surface of the ball F_t is equivalent to the average velocity of the solution jet in the medial cross section of the ball (see Fig. 3). The static effect of the resistance of the ball's interaction with the solution (see formulas (3) (4), can be considered independent of the nature of the flow movement.

Provided that at the beginning of the tear of the ball from the saddle its speed υ_k relative to the solution is close to zero, then the expression (3) taking into account $\upsilon_k = \upsilon_p$ will look like [3, 8]

$$F_\tau = \pi^2 \cdot r_k^2 \cdot \tau_0 \tag{4}$$

Where

$$\tau_0 = \frac{F_\tau}{\pi^2 \cdot r_k^2} \tag{5}$$

The value of the force F_τ is defined as the difference between the external force F_0 at which the ball moves and the Archimedes force F_A

$$F_A = \frac{4}{3} \cdot \pi \cdot r_k^3 \cdot g \cdot \rho \tag{6}$$

where ρ—density of the solution under investigation, kg/m^3; g—acceleration of free fall, m/s^2.

The value of the structured viscosity of the soluble mixture μ is determined by the dependence [3, 7], obtained on the basis of the expression (3)

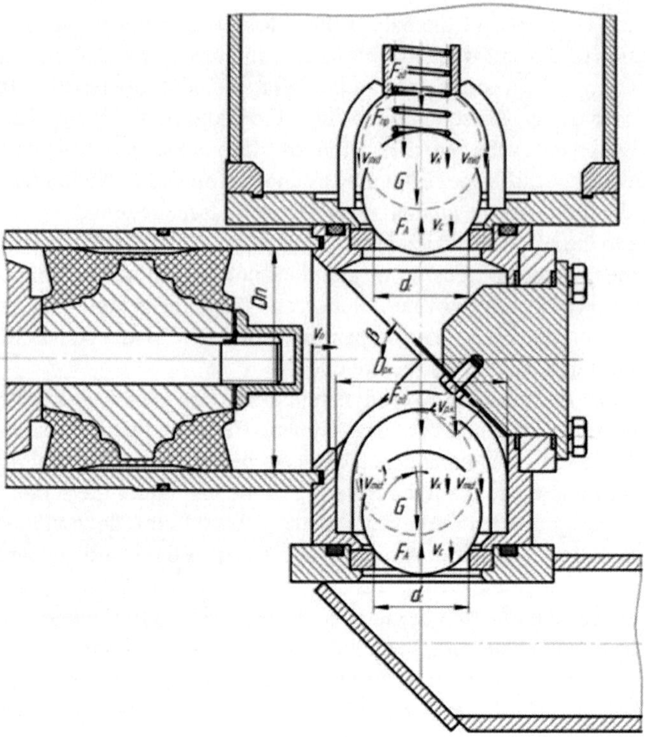

Fig. 3 Scheme of influence of hydrodynamic forces on the injection and suction patch of single-piston mortar

$$\mu = \frac{F_{\eta} - \pi^2 \cdot r_k^2 \cdot \tau_0}{6 \cdot \pi \cdot \upsilon_k \cdot r_k} \tag{7}$$

By dependence, the (6) marginal stress of the shift τ_0, taking into account the influence of the Archimedes force, will take the form

$$\tau_0 = \frac{F_0 - \frac{4}{3} \cdot \pi \cdot r_k^3 \cdot g \cdot \rho}{\pi^2 \cdot r_k^2} \tag{8}$$

where $F_A = \frac{4}{3} \cdot \pi \cdot r_k^3 \cdot g \cdot \rho$—is the component of the force of Archimedes, which has an effect on the ball.

The value of the structured viscosity of the solution, taking into account the effect on the ball of the Archimedes force valve [3, 8] will be determined by dependence (3) will look like

$$\mu = \frac{F_{\mu} - \frac{4}{3} \cdot \pi \cdot r_k^3 \cdot g \cdot \rho - \pi^2 \cdot r_k^2 \cdot \tau_0}{6 \cdot \pi \cdot \upsilon_k \cdot r_k} \tag{9}$$

where F_{μ}—resistance to lowering the ball from the solution, H; υ_k—speed of movement of the ball in a statically stationary solution, m/s; r_k– radius of the ball, m.

Obtained mathematical dependences (8) (9) which allow describing the nature of the interaction of the flow with the elements of the hydraulic part of the mortar (piston working organ, valve nodes, working chamber) is determined on the basis of experimental determination of parameters F_0, F_{μ} and υ_k the movement of the valve ball radius r_k in the solution and calculate the values of the maximum shear stress τ_0 and the coefficient of structured viscosity of the solution μ.

Based on dependence (2), and assumptions given, on the design parameters of the valve nodes and the characteristics of the solution flow, the dependence of the hydrodynamic pressure force will look like

$$F_{gd} = C_r \cdot \pi \cdot \mu \cdot r_{kl} \cdot \upsilon_s + C_{\tau} \cdot \pi \cdot \mu \cdot r_{kl} \cdot \upsilon_{mid} + \pi^2 \cdot r_{kl}^2 \cdot \tau_0 \quad (10)$$

where r_{kl}—the radius of the valve ball, m; υ_s—the speed of the jet of the solution coming out of the hole of the saddle, m/s; υ_{mid}—flow rate of solution in copper cross section, m/s; μ—coefficient characterizing the structured viscosity of the solution, Pa · s;—extreme shear stress of the solution, Pa; C_r, C_{τ}—odds of frontal pressure and grater resistance of the valve ball.

It should be noted that when the ball is fully flown around in the volume of the absorbent chambers, the entered coefficients have a value corresponding to: $C_r = 2C_{\tau} = 4$ (2).

The nature of the change in the pressure force of the cumulative jet on the ball valve is determined by the coefficient C_r and depends on the following parameters: the diameter of the ball, the diameter of the saddle hole, the height of the lifting of the ball. The coefficient C_{τ} of expression (10) characterizes the influence of a limited flow that the surface of the ball and is determined by the design parameters of the valve space. The coefficients C_r and C_{τ} were determined experimentally during the study of the process of interaction of the ball with the solution in a confined space using constructive and physical modeling [4].

During the operation of the valve, the speeds of the jet of the solution are redistributed in the characteristic cross section of the valve node. This process is characterized by the Law of Westphalia [1, 6], according to which, during the movement of the ball in the direction of the jet solution, the flow of the solution through the saddle hole is greater than the consumption of the solution through the valve slit (Fig. 3).

$$Q_s = Q_{shch} + Q_{kl} \quad (11)$$

where Q_s—solution consumption through the hole in the valve saddle, m^3/s; Q_{shch}—solution consumption through the valve slot, m^3/s; $Q_{кл}$—loss of the solution occupying the sub-space, m^3/s.

The law of inseparability of the flow indicates that the consumption of the solution in the medial cross section Q_{mid} is equal to the consumption of the solution through the gap, so the dependence is based on the expression (11)

$$Q_s = Q_{mid} + Q_{kl}. \tag{12}$$

Accordingly, the consumption of the solution of the included nodes that make up, are determined by the dependences of the (11) (12)

$$Q_s = \upsilon_s \cdot S_s;\ Q_{shch} = \upsilon_{shch} \cdot S_{shch};\ Q_{mid} = \upsilon_{mid} \cdot S_{mid}, \tag{13}$$

where S_s, S_{shch}, S_{mid}—area of cross-sectional holes: saddles, crevices and middle, respectively, m^2; υ_s, υ_{shch}, υ_{mid}—average speeds of soluble flow: in the hole of the saddle, crevices and medial cross section, respectively, m/s.

The consumption of the solution covering the sub-space during the lifting of the ball can be approximately determined by dependence

$$Q_{kl} = \upsilon_{kl} \cdot S_s \tag{14}$$

where υ_{kl}—the speed of movement of the valve ball, m/s.

Since the consumption of the solution through the saddle hole is proportional to the consumption of the solution in the suction chamber $Q_{v.k.}$, therefore, dependence is made

$$Q_s = Q_{v.k} = \frac{1}{\cos\beta} \cdot \upsilon_p \cdot \frac{D_p^2}{d_s^2} \cdot S_p \tag{15}$$

where υ_p—piston velocity, m/s; S_p—area of the contact part of the piston with solution, m^2; β—angle of inclination of special insertion.

When opening the valve, the ball moves in the direction of the stream, respectively, the relative flow rate of the ball will decrease by the value of its velocity, that is, the dependence (10) will look like

$$F_{gd} = C_p \cdot \pi \cdot \mu \cdot r_{kl} \cdot (\upsilon_s - \upsilon_{kl}) + C_\tau \cdot \pi \cdot \mu \cdot r_{kl} \cdot (\upsilon_{mid} - \upsilon_{kl}) + \pi^2 \cdot r_{kl}^2 \cdot \tau_0 \tag{16}$$

Given that speeds $\upsilon_s = \upsilon_p \cdot \frac{1}{\cos\beta} \cdot \frac{S_p}{S_s}$, and $\upsilon_{mid} = \frac{\upsilon_p \cdot S_p - \upsilon_{kl} \cdot S_s}{S_{mid}} \cdot \frac{1}{\cos\beta}$ (see (11)–(15)), determine the strength of hydrodynamic pressure can be by dependence

$$\begin{aligned} F_{gd} = {} & C_r \cdot \pi \cdot \mu \cdot r_{kl} \cdot \left(\upsilon_p \cdot \frac{S_p}{S_s} \cdot \frac{1}{\cos\beta} - \upsilon_{kl}\right) + C_\tau \cdot \pi \cdot \mu \cdot r_{kl} \\ & \times \left(\frac{\upsilon_p \cdot S_p - \upsilon_{kl} \cdot S_s}{S_{mid}} \cdot \frac{1}{\cos\beta} - \upsilon_{kl}\right) + \pi^2 \cdot r_{kl}^2 \cdot \tau_0 \end{aligned} \tag{17}$$

By marking the current height of the ball above the saddle with a variable coordinate x, the dependence will look like

$$F_{gd} = C_r \cdot \pi \cdot \mu \cdot r_{kl} \cdot \left(\upsilon_p \cdot \frac{S_p}{S_s} \cdot \frac{1}{\cos\beta} - \dot{x} \right) + C_\tau \cdot \pi \cdot \mu \cdot r_{kl} \times \left(\frac{\upsilon_p \cdot S_p - \dot{x} \cdot S_s}{S_{mid}} \cdot \frac{1}{\cos\beta} - \dot{x} \right) + \pi^2 \cdot r_{kl}^2 \cdot \tau_0 \tag{18}$$

where $\dot{x}$—the first derivative coordinates of the position of the ball x while moving over the saddle, which determines its speed of movement, namely υ_{kl}; S_s—area of the passing hole of the saddle.

Figure 3 shows the scheme of influence of hydrodynamic forces on the injection and suction valves.

It is worth noting that the movement of the valve ball affects not only the level of inverse consumption of the solution, but determines the resistance to the movement of the current solution, the suction ability or level of vacuum in the working chamber and the degree of its filling in the suction tact.

The process of hydraulic interaction and the balance of forces acting on the ball valve during the operation of the suction valve can be described according to Newton's law

$$m_{kl} \cdot \ddot{x} = -F_{gd} - G + FA \tag{19}$$

The direction of force F_{gd} is determined by the vector of the flow rate of the solution υ_r in the near-detached space.

Dependences (18) and (19) indicate that the hydrodynamic force F_{gd} is directly proportional to the flow rate of the solution and is determined depending on the speed $\vec{\upsilon}_p$ of movement of the piston, and depends on the position of the ball, the geometric dimensions of the valve node and the characteristics of the solution. In general, force F_{gd} can be represented as a function from the rheological parameters of the solution and the design parameters of the valve nodes

$$F_{gd} = f\left(x_{kl}, \dot{x}_{kl}, \upsilon_p, \mu, \tau_0, r_{kl}, r_s, D_{nkl}\right) \tag{20}$$

Where x_{kl} is the coordinate of the path that determines the instantaneous position of the ball above the saddle; $\dot{x}_{kl}$—instantaneous speed of the ball (υ_{kl}) in position x_{kl}; υ_p—piston speed; μ, τ_0—extreme shear stress, depending on its mobility; r_{kl}, r_s—radius of the ball and the hole of the saddle; D_{nkl}—diameter of the round-the-circle cavity.

Given dependencies (6) (20), an expression (19) can be represented as a differential dependency

$$m_{kl} \cdot \ddot{x}_{kl} = -F_{gd}\left(x_{kl}, \dot{x}_{kl}, \upsilon_p, \mu, \tau_0, r_{kl}, r_s, D_{nkl}\right) - mkl \cdot g + \frac{4}{3} \cdot \pi \cdot r_{kl}^3 \cdot g \cdot \rho \tag{21}$$

Permanently

$$m_{kl} \cdot \ddot{x}_{kl} = -\begin{pmatrix} C_r \cdot \pi \cdot \mu \cdot r_{kl} \cdot \left(\upsilon_p \cdot \dfrac{S_p}{S_s} \cdot \dfrac{1}{\cos\beta} - \dot{x} \right) + C_\tau \cdot \pi \cdot \mu \cdot r_{kl} \times \\ \times \left(\dfrac{\upsilon_p \cdot S_p - \dot{x} \cdot S_s}{S_{mid}} \cdot \dfrac{1}{\cos\beta} - \dot{x} \right) + \pi^2 \cdot r_{kl}^2 \cdot \tau_0 \end{pmatrix} - m_{kl} \cdot g + \frac{4}{3} \cdot \pi \cdot r_{kl}^3 \cdot g \cdot \rho \tag{22}$$

When considering the opening of the valve, the initial condition is that it corresponds to the lower part of the ball. When closing the valve, the initial condition is (the upper position of the ball) $x_{kl}(t_0) = 0 \ \ (m_{kl} \cdot \ddot{x}_{kl} < 0) \ x_{kl}(t_0) = 0$.

During the operation of the supercharging valve, the process of hydraulic interaction differs from the previous case in that the ball is affected by a column of solution from the combined compensator and the effect of the force of the elastic element, the balance of forces acting on the ball, has the form

$$m_{kl} \cdot \ddot{x} = -F_{gd} - G - F_{pr} + F_A \tag{23}$$

Resistance force of the elastic element acting on the valve ball

$$F_{pr1} = \frac{G_m \cdot d_{pr}^4 \cdot f_1}{8 \cdot D_0^3 \cdot n_{pr}} \cdot \frac{1}{\chi_{op}} = \frac{G_m \cdot d_{pr}^4 \cdot (H_0 - H_1)}{8 \cdot (D_1 - d_{pr})^3 \cdot n_{pr}} \cdot \frac{1}{\chi_{on}} \tag{24}$$

where G_m is the material shear module (pressure that causes deformation by 1 mm), MPa; d_{pr}—spring fiber diameter, mm; D_0—average diameter of the spring, mm; D_1– outer diameter of the spring, mm; H_0—spring length in free condition, mm; H_1—spring stroke length at the length of the pressure valve ball, mm; f_1—change in the length of the spring when compressed by the ball of the supercharger valve; n_{pr}– the number of spring coins; χ_{op}– spring movement resistance coefficient depending on the density of the pumped medium.

Taking into account the law Hutton and Archimedes (6), (20) and (24) the expression (23) can be written in the form of a differential equation

$$m_{kl} \cdot \ddot{x}_{kl} = -F_{gd}(x_{kl}, \ddot{x}_{kl}, \upsilon_n, \mu, \tau_0, r_{kl}, r_c, D_{nkl}) - m_{kl} \cdot g - \frac{G_m \cdot d_{pr}^4 \cdot (H_0 - H_1)}{8 \cdot (D_1 - d_{pr})^3 \cdot n_{pr}} \cdot \frac{1}{\chi_{op}} + \frac{4}{3} \cdot \pi \cdot r_{kl}^3 \cdot g \cdot \rho \tag{25}$$

And finally, in expanded form

$$m_{kl} \cdot \ddot{x}_{kl} = -\begin{pmatrix} C_r \cdot \pi \cdot \mu \cdot r_{kl} \cdot \left(\upsilon_p \cdot \dfrac{S_p}{S_s} - \dot{x} \right) + C_\tau \cdot \pi \cdot \mu \cdot r_{kl} \times \\ \times \left(\dfrac{\upsilon_p \cdot S_p - \dot{x} \cdot S_s}{S_{mid}} - \dot{x} \right) + \pi^2 \cdot r_{kl}^2 \cdot \tau_0 \end{pmatrix} \tag{26}$$

$$- m_{kl} \cdot g - \frac{G_m \cdot d_{pr}^4 \cdot (H_0 - H_1)}{8 \cdot (D_1 - d_{pr})^3 \cdot n_{pr}} \cdot \frac{1}{\chi_{op}} + \frac{4}{3} \cdot \pi \cdot r_{kl}^3 \cdot g \cdot \rho.$$

It can be assumed that the influence of the law of movement of the working body of the pump occurs with a double nature, namely, establishes the value of the piston stroke for the period of time and time of closing the valve, which directly determines the volume of reverse flow of the solution. That is, the establishment of the law of movement of the working body requires rational decisions for the period of time of lowering the valve ball and the speed of movement during this period.

Analysis of the operation of the ball valve of the mortar based on the established dynamic model, which makes it possible to determine the law of movement of the valve ball during operation, depending on the law of movement of the working body of the valve node design, and the properties of the pumped solution. Theoretically, it is necessary to evaluate the interaction of the law of movement of the working body and the development of the suction and injection valves and the consumption of the solution during closing.

The solution of the differential equation presented makes it possible to obtain the law of movement of the valve ball as a function

$$x_{kl}(t) = f\left(m_{kl}, r_{kl}, r_s, D_{nkl}, \upsilon_p(t), \mu, \tau_0, \rho, t\right) \tag{27}$$

which makes it possible to establish the dependence of the nature of the movement of the ball with the time t and design parameters of the valve $(m_{kl}, r_{kl}, r_s, D_{nkl})$, the law of movement of the working body $\left(\upsilon_p(t)\right)$ and the rheological properties of the pumped solution (μ, τ_0, ρ).

The initial movement of the ball with a diameter of 50 mm in a solution with a mobility of P10 cm was carried out with a total weight of 510 g, that is, the force of the initial shift H. $F_0 = 9,81 \cdot 0,51 = 5$.

In Fig. 2 4 the dependences of the movement of the pump and suction valve of the mortar depending on the mobility of the pumped solution and taking into account (Table 1) the rheological parameters of building solutions of varying mobility, the strength of the ball's shift in solutions and its speed of initial movement in solutions are given.

Dependences (Fig. 4) indicate that the triggering of the suction and suction valves occurs faster depending on the mobility of the solution, which is explained by the increasing resistance of movement with a decrease in mobility, as well as the closure

Table 1 Rheological properties and power parameters of sand solutions of different mobility to determine the laws of movement of pump and suction valves

Person-ality solution, cm	Density ρ, kg/m³	Limit (law) displace-ment, Pa τ_0	Factor structured vis-cosity, Pass $\mu \cdot$	Efforts movement, N F_1	Movement speed, m/s υ_0
P8	2100	736,27	35,86	15,3	0,74
P10	2000	602,39	18,58	11,75	1,06
P12	1900	431,24	10,24	7,86	1,2

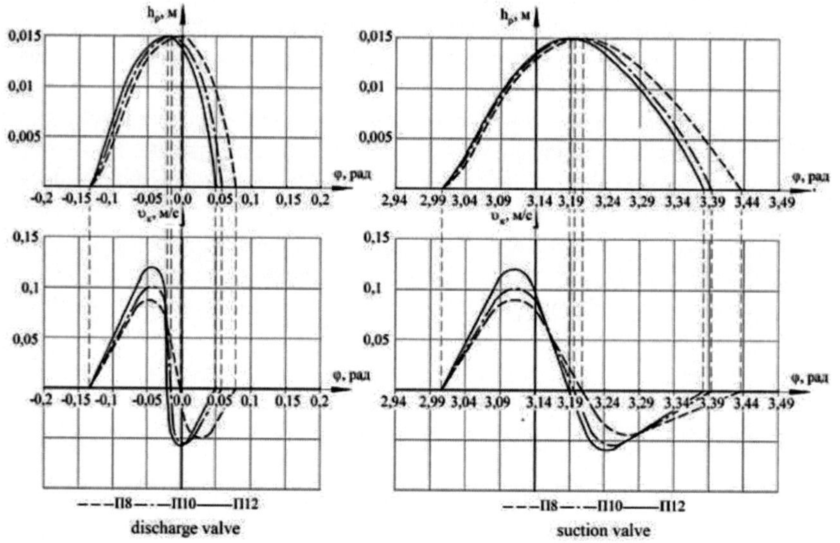

Fig. 4 Dependences of the movement of the pump and suction valve of the single-piston mortar with a combined compensator of increased volume depending on the mobility of the solution:

of valves in the mortar with a combined pressure pulsation compensator, respectively, by more than 1.35° and 0.96° rotation of the crane and filter.

It is worth paying attention to the fact that the faster operation of the valve balls in the mortar with a combined compensator of increased volume affects not only the reduction of the reverse leaks of the solution, but determines the resistance to the promotion of the flow of the solution depending on the installation of a special insertion in the suction chamber, the level of vacuum in it and the degree of its filling in the suction tide, which significantly affects the reduction of the degree of pulsation and increase in volume.

Next, it is necessary to determine the maximum shear τ_0 stress and the value of the structured viscosity of the solution with mobility P10.

$$\tau_0 = \frac{F_0 - \frac{4}{3} \cdot \pi \cdot r_k^3 \cdot g \cdot \rho}{\pi^2 \cdot r_k^2} = \frac{5 - \frac{4}{3} \cdot 3{,}14 \cdot 0{,}025^3 \cdot 9{,}81 \cdot 2000}{3{,}14^2 \cdot 0{,}025^2} = 602{,}39 \text{ Pa}$$

$$\mu = \frac{F_\eta - \frac{4}{3} \cdot \pi \cdot r_k^3 \cdot g \cdot \rho - \pi^2 \cdot r_k^2 \cdot \tau_0}{6 \cdot \pi \cdot \upsilon_k \cdot r_k} =$$

$$= \frac{11{,}75 - \frac{4}{3} \cdot 3{,}14 \cdot 0{,}025^3 \cdot 9{,}81 \cdot 2000 - 3{,}14 \cdot 0{,}025^2 \cdot 602{,}39}{6 \cdot 3{,}14 \cdot 1{,}06 \cdot 0{,}025} = 5{,}92 \text{ Па} \cdot c.$$

Determining the law of movement of the valve ball will determine the time of its movement from height h to saddle and estimate the number of inverse leaks ΔV during the closure of the valve

$$\Delta V = S_b \cdot \frac{1}{\cos \psi} \cdot S_p(\varphi_z) \tag{28}$$

where $S_б$is the size of the area of the side around the valve opening surface depending on the lifting height of the ball $S_б = f(h)$, m^2; ψ—the angle of change in the trajectory of the valve ball, which arises as a result of the tank force of the influence of the working body.

Mass of consumption mortar through the saddle of the suction valve determine based on the preliminary analysis dependence

$$\Delta Q_{vs.kl} = \rho \cdot \upsilon_{r.k.} \cdot S_p - \rho \cdot \upsilon_s \cdot S_b = \rho \cdot \upsilon_p \cdot \frac{1}{cos\beta} \cdot \frac{D_p^2}{d_{r.k}^2} \cdot S_p - \rho \cdot \upsilon_p \cdot \frac{1}{cos\beta} \cdot \frac{D_p^2}{d_s^2} S_b \tag{29}$$

After the transformations, we have

$$\Delta Q_{vs.kl} = \frac{\pi \cdot \rho \cdot D_p^4}{cos\beta} \cdot \left(R \cdot \sin\varphi - \frac{(R \cdot \sin\varphi - e) \cdot R \cdot cos\varphi}{\sqrt{l^2 - (R \cdot \sin\varphi - e)^2}} \right) \cdot \left(\frac{D_p^2}{4 \cdot d_{r.k}^2} - \frac{R_k}{d_s^2} \cdot \frac{h^2 + 2 \cdot h \cdot \sqrt{R_k^2 - r_s^2}}{\sqrt{r_s^2 + \left(h + \sqrt{R_k^2 - r_s^2}\right)^2}} \right), \tag{30}$$

where $S_p(\varphi_z)$—moving the piston during the period of time of lowering the valve ball to the saddle, which is characterized by the law of movement of the working body, m; φ_z—the angle of rotation of the cradle during the t lowering of the valve ball to the saddle.

The dependence (28) can be explained by the fact that during the lowering of the valve ball, the piston will move at a certain distance $S_p(\varphi_z)$ and during this movement will change the volume of the pump suction chamber. It is clear that the simultaneous opening of both valves is impossible, both in the suction and injection tactics of the solution through any valve that is currently closed, the solution flows in the form of reverse leaks and fills the change in volume in the suction chamber.

Under conditions and if the speed of the working body is constant, dependency can be used to determine the magnitude of inverse leaks through the supercharger valve

$$\Delta Q_{nagn.\,kl} = \frac{\pi \cdot d_k^2 \cdot h_r}{4} - V_k + Q_K \tag{31}$$

where d_k is the diameter of the cylindrical part in the suction chamber, m; h_r—lifting path, which the ball passes along with the solution, m; V_k—volume displacing the lower mideal part of the valve ball, $V_k = \pi \cdot h_{k.s.}^2 \cdot \left(R_k - \frac{h_{k.s.}}{3}\right)$, m^3; $h_{k.s.}$—height of ball copper; R_k—flap ball radius; Q_B—a sublimation "dead" volume, which joins the reverse leaks under the Law of Westphalia, m^3, $Q_K = \frac{\pi}{4} \cdot h \cdot \left(D_{kl}^2 - d_s^2\right)$.

Received after conversions

$$\Delta Q_{nagn.\,kl} = \frac{\pi \cdot d_k^2 \cdot h_r}{4} - \pi \cdot h_{k.s.}^2 \cdot \left(R_k - \frac{h_{k.s.}}{3}\right) + \frac{\pi}{4} \cdot h \cdot \left(D_{kl}^2 - d_s^2\right) \tag{32}$$

The established law of movement of the valve ball allows you to determine the time of its lowering in height h to the saddle and set quantitatively the values of the inverse leaks ΔV_{kl} during the closing of the valve at different mobilities of the solution

$$\Delta V_{kl} = F_p \cdot S_p(\varphi_{kl}) \tag{33}$$

where $F_p = \frac{\pi}{2} \cdot D_p^2$—piston area; $S_p(\varphi_{kl})$—part of the movement of the piston for a certain period of time, during which the valve ball is lowered to the saddle, which is characterized by the law of movement of the working body, m; φ_{kl}—the angle of rotation of the cradles during the lowering t of the valve ball.

The path h_r, the valve balls that it passes along with the flow of the solution can be represented by dependence

$$h_r = h - h_t \tag{34}$$

where h—the lifting height of the valve ball to its limiter, m;

h_t—lifting height of the ball, which it passes through the resistance of the solution under the action of gravity, m.

Also, the path h_r can be associated with the diameter and stroke of the piston with this dependence

$$h_r = \frac{D_p^2}{d_k^2} \cdot S_p(\varphi_z) \tag{35}$$

where φ_z is the angle of rotation of the cradle, the value, which coincides with the moment of lowering the valve ball to the saddle, rad.

The lifting height of the ball h_t can be expressed as the product of the speed of lifting at a time

$$h_t = \upsilon_t \cdot t \tag{36}$$

where υ_t—the speed of movement of the valve ball in solution, m/s;

t—the time of the valve ball being sedittered for closing, s.

Lowering the valve ball t can be associated with the angle of rotation φ_z of the cradles with dependence

$$t = \frac{60}{n} \cdot \frac{\varphi_z}{2\pi} = \frac{30 \cdot \varphi_z}{\pi \cdot n} \tag{37}$$

where n—is the number of revolutions of the cradly wasps, s^{-1}.

Then, through expressions (36) and (37), dependency (34) will look like

$$h_r = h - \upsilon_t \cdot \frac{30 \cdot \varphi_z}{\pi \cdot n} \tag{38}$$

To find the closing angle φ_z, you must compare the (35) and (38) expressions. The resulting dependency will look like

$$h - \upsilon_t \cdot \frac{30 \cdot \varphi_z}{\pi \cdot n} - \frac{D_p^2}{d_k^2} \cdot \left(R \cdot (1 - \cos\varphi_z) - \left[l - \sqrt{l^2 - (R \cdot \sin\varphi_z - e)^2} \right] \right) = 0 \tag{39}$$

Conducts direct calculations to determine the value of the angle of rotation of the cradle analytical way in an analytical way φ_z in a cumbersome and inconvenient way (39), so it is more advisable to use the software of computer mathematics Mathcad 15 to get numerical values.

Substantive to the value of the angle φ_z before dependence (38), we obtain the value of the valve ball path h_r, which the valve ball passes along with the flow of the soluble mixture—0.015 m, and substantive finally gives us the value of the volume of leaks $\Delta V_{vs.kl}$ in the injection rate of 0.041 m^3.

Having integrating the dependence of the leakage rate through the valve socket from the beginning of the coordinates to the closing angle of the valve φ_z, you can also find the volume of leaks (excluding values V_{kl} and Q_B).

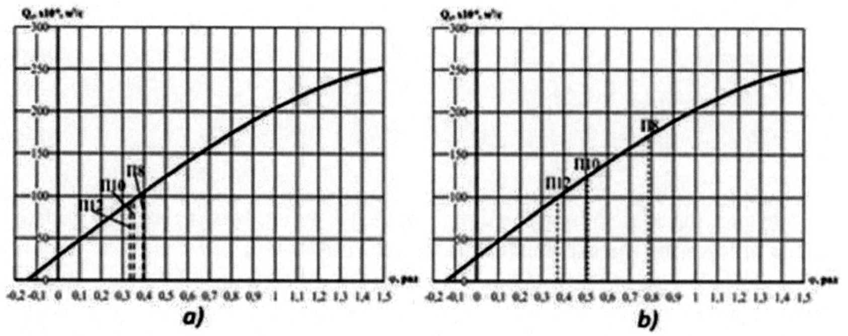

Fig. 5 Dependence of the loss of the soluble mixture through the saddle of the supercharger **a** and the suction **b** valves of the soluble pump with a combined compensator of increased volume depending on the angle of rotation of the cradle when pumping solutions with mobility P8, P10, P12

The loss of the volume of the soluble mixture $\left(\mathrm{m}^3/\mathrm{s}\right)$ through the valve saddle is determined as the product of the speed of the working organ $\upsilon_p(\varphi)$ on the area of the valve saddle passing hole and the counting of piston and saddle areas $\frac{D_p^2}{d_s^2}$, which determines their difference.

$$\Delta V_{kl} = \upsilon_p(\varphi) \cdot \frac{\pi \cdot d_s^2}{4} \cdot \frac{D_p^2}{d_s^2} = \upsilon_p(\varphi) \cdot \frac{\pi \cdot D_p^2}{4} \tag{40}$$

Finally, the volume of leaks through the suction and injection valves in the injection rate is determined depending on the suction and injection rate

$$\begin{aligned} 0 \le \varphi \le \pi,\ \Delta V_{vs.kl} &= \frac{\pi \cdot D_n^2}{4} \cdot \int_0^{\varphi_z} \left(R \cdot \sin\varphi - \frac{(R \cdot \sin\varphi - e) \cdot R \cdot cos\varphi}{\sqrt{l^2 - (R \cdot \sin\varphi - e)^2}} \right) d\varphi - \\ &- \pi\left(R_k - \sqrt{R_k^2 - r_s^2}\right)^2 \cdot \left(R_k - \frac{R_k - \sqrt{R_k^2 - r_s^2}}{3} \right) + Q_B \\ \pi \le \varphi \le 2\pi,\ \Delta V_{ng.kl} &= \frac{\pi \cdot D_p^2}{4} \cdot \int_0^{\varphi_z} \left(R \cdot \sin\varphi - \frac{(R \cdot \sin\varphi - e) \cdot R \cdot cos\varphi}{\sqrt{l^2 - (R \cdot \sin\varphi - e)^2}} \right) d\varphi - \\ &- \pi\left(R_k - \sqrt{R_k^2 - r_s^2}\right)^2 \cdot \left(R_k - \frac{R_k - \sqrt{R_k^2 - r_s^2}}{3} \right) + Q_K - \Delta V_{pr}, \end{aligned} \tag{41}$$

where Q_K—the volume of the solution in the cylindrical compensator chamber, which affects the valve; ΔV_{pr}—reduction of volume from the elastic element of the valve.

Based on dependencies (34), (37), (38), (39), (41) and theoretical studies [5] on determining the rational height of lifting the valve ball in Table 2 and Fig. 6 is given with the allergy of the degree of pressure pulsations when pumping solutions of varying mobility depending on the lifting height of the suction and injection valve.

Dependences (Fig. 6) indicate that the degree of pulsations when pumping solutions with mobility P8-12 cm in the mortar will be reduced to the minimum values when setting the ball lifting limiters in the range of 12.5–17.5 mm, both for the suction and for the injection valves.

This is due to the rational sub-valve space, which is equal to 1260 mm^2 suction and supercharger valves, which provide high suction capacity and filling of the suction chamber and ensuring quick operation.

Table 2 The value of the pulsation level taking into account various factors

Number of double moves of piston, min n^{-1}	Average feed pressure, MPa p_{cp}	Volume of compensator, dm $V_{комп}$ ^{3}	Mobility of solution, P, cm	Lifting height of suction and suction valve, h mm	δ, %
Mortar pump with combined compensator of increased volume					
150	2,0	47	8	5	36,54
				10	18,22
				15	12,81
				20	15,12
				25	25,32
				30	38,09
			10	5	31,03
				10	12,05
				15	6,93
				20	10,04
				25	20,89
				30	34,61
			12	5	29,03
				10	10,04
				15	5,10
				20	8,13
				25	18,22
				30	33,23

*In gray, rational heights of lifting of suction and supercharger valves and minimum levels of pulsation degree are determined

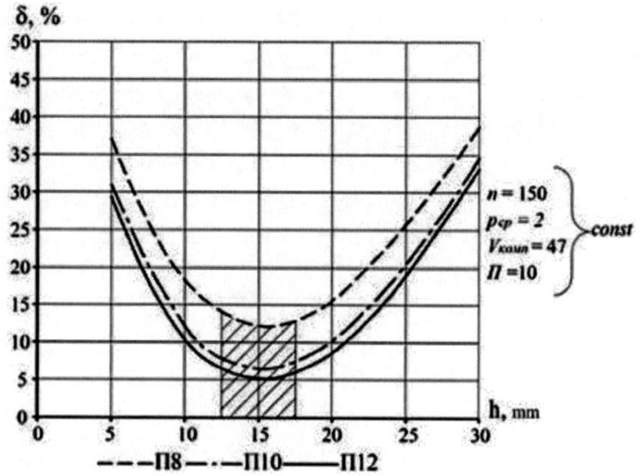

Fig. 6 Dependence of the degree of pulsations of the pressure of the solution and on the lifting height of the suction and suction valves

Increasing the lifting height of the valve ball $h = 15$ mm will lead to an increase in the degree of pressure pulsations as a result of an increase in inverse leaks due to the delay in lowering the valve ball, especially at reduced mobilities of the solution.

2 Conclusions

According to the results of studies, it can be seen that the speed of operation of the suction and supercharger valves is significantly influenced by the mobility of the solution, this is especially noticeable when pumping solutions P8 cm.

The height of the lifting of the valve ball mm according to the results of studies and $h = 15$ the level of pulsation of pressure of the pumped solution is considered rational.

References

1. Cenyaev IA (1983) Piston cradle pumps Machine-state, Leningr. otd-nye, p 156
2. Parfyonov EP (1972) Production definition of piston-like mortar pumps. Meh. tools and otdelochnye machines: Inform. nauchn.-techn. – Vip.4. CNIITEStroymash, pp 12–13
3. Onishchenko OG (2000) Interaction of the mortar valve ball with the flow of pumped solution/ Onishchenko OG, Vasilyev AV, Korobko BO. Car transport: Sat. nauchn. Labor/Harkov state motor vehicle and road technical university. – Release 5/Universal machines for earthlings and roadworks. Harkov: KSTU, pp 116–118
4. Onishchenko OG (2000) Interaction of the mortar valve ball with the flow of pumped solution/Onishchenko OG, Vasilyev AV, Korobko BO. Automo-bull transport: Sat. nauchn. labor/Harkov state motor vehicle and road technical university. Release 5/Universal machines for earthlings and roadworks. Harkov: KSTU, pp 116–118

5. Mortelpumpen und ihre Entwicklung. Fordern und Heben (1969). № 15. (FRG) 8. EP 0200026, INT. Cl. 4 F 04 B 43/12, 15/02. Pumpe neumuller Walter, Sturmer Gerhard. 10.12.1986. Patentblatt 86/45
6. Karaev MA (1978) Reserch of working of drilling pump valves on clay mortar/Karaev MA. Neft and gas 11, pp 31–37
7. Kukoba AT, Onishchenko OG, Nadobko VB (1998) Methodology for calculating reverse losses of the solution through the injection valve of vertical differential mortar. Collection of scientific works (industry engineering, construction)/Polt. State technical. 2. Poltava: PTU, pp 3–7
8. Onishchenko AG The work of the ball valve in the flow plunger of a differential mortar pump. Onishchenko, AV, Vasiliev BO, Korobko // Problems of creating new machines and technologies. Scientific works of the KSPU. Issue 1/2001 (10). Kremenchug: KSPU, 2001. pp 466–471.
9. Korobko B, Zadvorkin D, Vasyliev I (2017) Study of the operating element motion law for a hydraulic-driven diaphragm mortar pump. Eastern-Eur J Enterprise Technol 4(7-88):25–31. https://doi.org/10.15587/1729-4061.2017.106873
10. Korobko B (2016) Investigation of energy consumption in the course of plastering machine's work. Eastern-Eur J Enterprise Technol 4(8–82):4–11. https://doi.org/10.15587/1729-4061.2016.73336
11. Korobko B, Virchenko V, Shapoval M (2018) Feed solution in the pipeline aith the compensators mortar pump of various design solutions pressure pulsations degree determination. Int J Eng Technol (UAE) 7(3):195–202. https://doi.org/10.14419/ijet.v7i3.2.14402
12. Korobko B, Khomenko I, Shapoval M, Virchenko V (2020) Hydraulic single pump with combined higher volume compensator operation analysis. https://doi.org/10.1007/978-3-030-42939-3_12
13. Korobko B, Vasyliev I (2017) Test method for rheological behavior of mortar for building work. Acta Mechanica Et Automatica 11(3):173–177. https://doi.org/10.1515/ama-2017-0025
14. Piskunov VG, Gorik AV, Cherednikov VN (2000). Modeling of transverse shears of piecewise homogeneous composite bars using an iterative process with account of tangential loads 2. Resolving equations and results. Mech Composite Mater 36(6):445–452. https://doi.org/10.1023/A:1006798314569
15. Piskunov VG, Goryk AV, Cherednikov VN (2000) Modeling of transverse shears of piecewise homogeneous composite bars using an iterative process with account of tangential loads. 1. Construction of a model. Mech Composite Mater 36(4):287–296. https://doi.org/10.1007/BF02262807

Approximation Model of the Method of Design Resistance of Reinforced Concrete for Bending Elements

Marta Kosior-Kazberuk, Dmytro Kochkarev, Taliat Azizov, and Tatiana Galinska

Abstract The paper considers approximation models of the method of design resistance of reinforced concrete for calculation of normal sections of bending reinforced concrete elements. It is proposed to use approximation dependences of two types—polynomial and linear. Both methods are based on the method of design resistance of reinforced concrete, which relies on universally accepted theory-based prerequisites and hypotheses. This method is based on the use of nonlinear deformation curves of concrete, Bernoulli hypothesis (the plane-sections hypothesis) is accepted as the correct one, and the extremum criterion for determining the bearing capacity (carrying force) based on a nonlinear deformation calculation model is used. The proposed techniques can massively simplify the calculation of bending reinforced concrete elements. They relieve from the necessity to use tables and perform complicated calculations with iteration methods, as it is intrinsic to the majority of existing methods. The possibility of determining on their basis both the bearing capacity (carrying force) and the relative height of the compressive zone of the concrete is shown in this article. The authors have conducted a check of the obtained approximation models of calculation on experimental samples of well-known researchers. The computational results indicate the computational accuracy, sufficient for practical calculations of the proposed methods. This paper presents the examples of determination of the carrying force and area of the effective reinforcement of normal sections of bending reinforced concrete elements by both offered methods. The proposed methods of calculating bending reinforced concrete elements can be widely used in design practice.

M. Kosior-Kazberuk
Bialystok University of Technology, Bialystok, Poland

D. Kochkarev
National University of Water and Environmental Engineering, Rivne, Ukraine

T. Azizov
Pavlo Tychyna Uman State Pedagogical University, Uman, Ukraine

T. Galinska (✉)
National University "Yuri Kondratyuk Poltava Polytechnic", Poltava, Ukraine
e-mail: galinska@i.ua

V. Onyshchenko et al. (eds.), *Proceedings of the 3rd International Conference on Building Innovations*, Lecture Notes in Civil Engineering 181,
https://doi.org/10.1007/978-3-030-85043-2_23

Keywords Nonlinear strain model · Bending moment · Deformation · Reinforced concrete beam · Design resistance

1 Introduction

The calculation of reinforced concrete elements at different types of force loads is usually performed taking into account nonlinear properties of materials [1–4, 9, 10]. The calculation of bending reinforced concrete elements involves the use of nonlinear or simplified diagrams of deformation of concrete [1–4]. The calculation of bending elements using nonlinear diagrams of deformation of concrete is quite complicated, and is usually performed using specially designed computer programs. Similar calculations on determination of the durability of normal sections of bending reinforced concrete elements by using simplified schemes are also quite complicated and may lead to certain imprecisions and inaccuracies. In most cases, such methods require the use of either pre-designed dependency graphs [1, 2] or the corresponding tables [4]. The simplest method, which involves the use of nonlinear diagrams of deformation of concrete in order to calculate reinforced concrete elements is the method of design resistance of reinforced concrete [4]. Instead, it also involves the use of certain tables. Let us consider an approximation model for calculating reinforced concrete elements based on this method.

The scientific developments of the authors of the article are associated with their preliminary studies, which are set out in the works [9–17], and are also a further development of the research of leading scientists Pavlikov [17, 18], Piskunov [19, 20], Pliukhin [21], Storozhenko [22], Zhuravskyi [23, 24] Semko [25–27].

2 Approximation Model of the Method of Design Resistance of Reinforced Concrete for Bending Elements

According to EN 1992-1-1, the calculation of bending reinforced concrete elements is performed using Bernoulli hypothesis (the plane-sections hypothesis), generally accepted deformation curves of concrete and reinforcement, as well as failure criteria. The main failure criterion is reaching the ultimate deformations of concrete and reinforcement. In the cases of calculation of beams with limited redistribution of forces, the relative height of the compressive zone of the concrete is also limited.

$$x/d \leq (\delta - 0.4)/(0.6 + 0.0014/\varepsilon_{cu}), \tag{1}$$

where x/d—the relative height of the compressive zone of the concrete, δ—the ratio of the distributed moment to the moment determined at the elastic stage, ε_{cu}—the ultimate strains of the compressive zone of the concrete. The remarkable thing is that according to [], a maximum redistribution of forces up to 20% is allowed for reinforcing steel of class A, and up to 30%—for reinforcing steel of class B.

Thus, regardless of the method of calculating bending reinforced concrete elements, it is necessary to check condition (1) in the case of redistribution of forces.

The calculation of bending elements according to the design codes [1] is as follows:

1. The mechanical reinforcement ratio is determined by the following expression:

$$\omega = \frac{\rho_f \cdot f_{yd}}{f_{cd}}, \tag{2}$$

   where ρ_f – the reinforcement ratio of the cross-section with longitudinal reinforcement, f_{yd} – design resistance of the longitudinal reinforcement, f_{cd} – design resistance of compressive strength of concrete.
2. Next, the parameter k_z $k_z = M/(f_{cd}{\cdot}b{\cdot}d^2)$ is determined according to the graph (Fig. 1).
3. The carrying force of the element is set by the expression:

$$M = k_z f_{cd} \cdot b \cdot d^2. \tag{3}$$

The procedure for calculating by the method of design resistance of reinforced concrete is similar to that described above:

1. The mechanical reinforcement ratio is determined by the formula (2).
2. The design resistance of reinforced concrete is set by the expression [1]

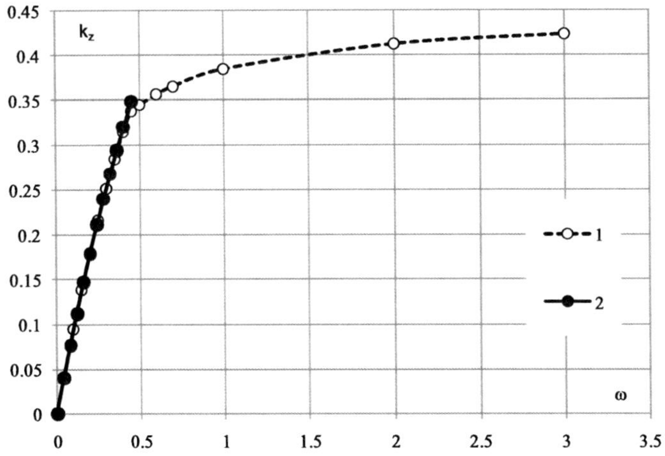

Fig. 1 Functional dependence $k_z = f(\omega)$ by the methods: 1–method of design resistance of reinforced concrete [4]; 2—EN 1992-1-1 [1]

$$f_z = 6 \cdot k_z f_{cd}. \tag{4}$$

3. The bearing capacity (carrying force) of the element is set by the formula:

$$M = f_z W_c, \tag{5}$$

where f_z – the design resistance of reinforced concrete in bending, MPa, W_c – moment resistance of working cross section of concrete, b·h^2/6, m^3.

The functional dependence of the method of design resistance of reinforced concrete, presented in Fig. 1, has a much larger number of values than the similar dependence of the method EN 1992-1-1 [1]. This is primarily due to the fact that the method of design resistance of reinforced concrete uses more accurate diagrams of deformation of concrete and theoretically substantiated failure criteria, in particular the extremum criterion. Therefore, the bearing capacity (carrying force) of the over reinforced elements should be determined more accurately by the method of design resistance of reinforced concrete. Methodology EN 1992-1-1 [1] limits ω to the value 0.45 (at C50 and below) and 0.36 (at C55 and above). This is due to the fact that such elements have brittle failure (fracture failure), which must be avoided in existent building structures. But when testing experimental samples with ω > 0.45, more precise values of the bearing capacity (carrying force) will correspond to the method of design resistance of reinforced concrete. The bearing capacity of the elements by the method EN 1992-1-1 [1] at ω > 0.45 will correspond to the bearing capacity of the elements at ω = 0.45. In the section at ω ≤ 0.45 bearing capacity of the bending elements will be the same for both methods.

The use of the dependency graph, shown in Fig. 1, in the calculation of bending elements somewhat complicates the calculation process itself, so we suggest to perform an approximation of the dependence $k_z = f(\omega)$.

The approximation will be performed in the range ω ≤ 0.45, avoiding the use of elements with brittle failure (fracture failure). We will perform the approximation by linear functions and a polynomial of the second degree (Fig. 1). Linear approximation involves somewhat approximate calculation of bending elements, so in order to simplify the calculation procedure, we will perform the approximation in the form of the following system:

$$k_z = \alpha_1 \cdot \omega,\ 0 < \varpi \leq 0.24; \tag{6}$$

$$k_z = \alpha_2 \cdot \omega,\ 0.24 < \omega \leq 0.45. \tag{7}$$

As a result of the approximation, the coefficients of the system of equations $\alpha_1 = 0.94,\ \alpha_2 = 0.82$. were obtained. The coefficient of variation, determined by the fractional accuracies of this approximating dependence in comparison with the true function, is ν = 4.19%.

It ought to be noted that expressions, obtained by linear approximation, lead to the following expressions of the bearing capacity (carrying force):

$$M_{ED} = 0.94 \cdot A_s \cdot f_{yd} \cdot d,\ 0 < \varpi \leq 0.24; \tag{8}$$

$$M_{ED} = 0.82 \cdot A_s \cdot f_{yd} \cdot d, 0.24 < \omega \leq 0.45. \tag{9}$$

Approximating expressions in the form of a polynomial of the second degree have the following form:

$$k_z = \beta_1 \cdot \omega^2 + \beta_2 \cdot \omega,\ 0 < \varpi \leq 0.45; \tag{10}$$

The approximation coefficients for the expression (10) are $\beta_1 = -0.5,\ \beta_2 = 1.0$, whereby the coefficient of variation is $\nu = 0.82\%$ (Fig. 2).

1—dependence $k_z = f(\omega)$ of EN 1992-1-1 [1]; 2—linear approximation; 3—approximation by a polynomial of the second degree.

There is a relationship between the relative height of the compressive zone of the concrete and the mechanical reinforcement ratio [1]

$$x/d = 1.25 \cdot \omega,\ 0 < \omega \leq 0.45. \tag{11}$$

We shall confirm the validity of the obtained formulas on experimental samples.

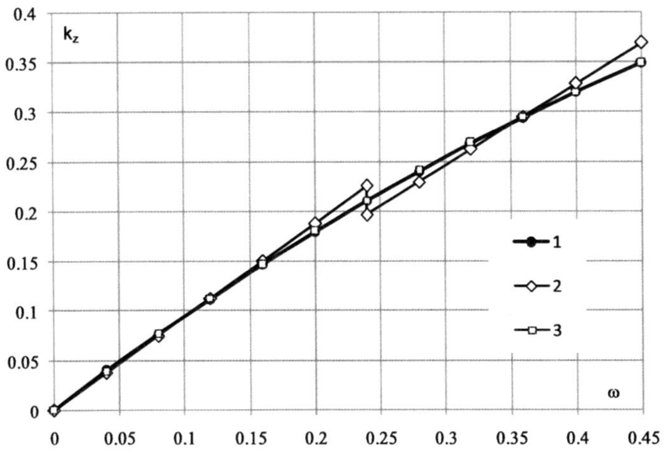

Fig. 2 Approximation of functional dependence $k_z = f(\omega)$:

Table 1 Verification of developed approximation models on experimental samples [5–8]

№.	Author of experiments, number of beams	Experimental data			Statistical value			
		f_c, *MPa*	f_y, *MPa*	ρ_s, %	x_l,	ν_l, %	x_p	ν_P, %
1	Pam et al. (2001) [5], 12 beams	36.4–102.5	520–579	0.76–3.14	1.00	13.17	1.00	13.17
2	Sarkar (1997) [6], 13 beams	77–107	442–470	1.03–4.04	0.95	15.66	0.95	12.66
3	Bernardo, Lopes (2004) [7], 19 beams	62.9–105.2	534–575	1.36–3.61	1.25	6.99	1.24	5.83
4	S.A. Ashour (2000) [8], 9 beams	48.61–102.4	530	1.18–2.37	0.99	4.07	0.99	3.17
	Total amount of samples 53				1.08	16.06	1.07	14.92

3 Verification of the Proposed Design Models on Experimental Samples

In order to check the proposed methods, we will determine the bearing capacity (carrying force) of experimental samples of well-known experiments [5–8]. Table 1 shows fractional accuracies of theoretically determined bearing capacity of the experimentally tested samples by the two abovementioned methods. The conducted calculations confirmed high accuracy of the developed approximating design models. The coefficient of variation of the ratio of theoretically determined and experimentally obtained bearing capacity for the linear dependence is $\nu_l = 16.06\%$, for polynomial dependence it is $\nu_P = 14.92\%$. When performing calculations, the parameter ω was limited to 0.45 (at C50 and below) and to 0.36 (at C55 and above).

4 Examples of Calculating Bending Reinforced Concrete Elements by the Offered Methods

Example 1. The task is to calculate the area of effective reinforcement, required for the reinforcement of the support section of a continuous beam made of concrete grade C20/25, $f_{cd} = 14.5$ MPa, $\varepsilon_{cu} = 350{\cdot}10^{-5}$, and of steel of class A500C, $f_{yd} = 435$ MPa, if the bending moment $M_{Ed} = 120$ кN·m works in the design section. Redistribution of forces in the amount of 15% ($\delta = 0.85$) should be carried out in the support section of the beam. It's advisable to take a rectangular cross-section of a beam with a single reinforcement with dimensions of b × d = 400 × 450 mm. The calculation must be performed by two approximation methods.

The solution. Let us determine the required design resistance of reinforced concrete.

$$f_{zM} = \frac{M_{ED}}{W_c} = \frac{6M_{ED}}{bd^2} = \frac{6 \times 120 \times 10^6}{400 \times 450^2} = 8.88\ MPa;$$

Then we will define the accessory parameter (fault-identifying variable):

$$k_z = \frac{f_{zM}}{6 \cdot f_{cd}} = \frac{8.88}{6 \cdot 14.5} = 0.102.$$

After that we shall determine the mechanical reinforcement ratio:

– by polynomial approximation

$$\omega = 1 - \sqrt{1 - 2 \cdot k_z} = 1 - \sqrt{1 - 2 \cdot 0.102} = 0.108;$$

– by linear approximation

$$\omega = k_z/0.94 = 0.102/0.94 = 0.109;$$

We will determine the limiting value of the height of the compressive zone of the concrete under the condition of redistribution of forces in the amount of 15% (δ = 0.85)

$$(\delta - 0.4)/(0.6 + 0.0014/\varepsilon_{cu}) = (0.85 - 0.4)/(0.6 + 0.0014/350 \cdot 10^{-5}) = 0.45.$$

Let us determine the height of the compressive zone of the concrete

– by polynomial approximation

$$x/d = 1.25 \cdot \omega = 1.25 \cdot 0.108 = 0.135 < 0.45.$$

– by linear approximation

$$x/d = 1.25 \cdot \omega = 1.25 \cdot 0.109 = 0.136 < 0.45.$$

Then we will determine the area of reinforcement and accept the reinforcement:

– by polynomial approximation

$$\rho_f = \frac{\omega \cdot f_{cd}}{f_{yd}} = \frac{0.108 \cdot 14.5}{435} = 0.0036 > \rho_{\min} = 0.0013.$$

$$A_s = \rho_f \cdot b \cdot d = 0.0036 \cdot 400 \cdot 450 = 648\ \text{mm}^2.$$

– by linear approximation

$$\rho_f = \frac{\omega \cdot f_{cd}}{f_{yd}} = \frac{0.109 \cdot 14.5}{435} = 0.00363 > 0.0013.$$

$$A_s = \rho_f \cdot b \cdot d = 0.00363 \cdot 400 \cdot 450 = 654\ \text{mm}^2.$$

As for the range of sizes, we will take 2Ø22 A500C, $A_s = 760\ \text{mm}^2$.

Example 2. The task is to determine the bearing capacity of a concrete beam, made of C20/25 strength grade of concrete, $f_{cd} = 14.5$ MPa, $\varepsilon_{cu} = 350 \cdot 10^{-5}$, and of steel of class A500C, $f_{yd} = 435$ MPa, if the beam is reinforced in the lower zone with reinforcement area $A_s = 760 \cdot \text{mm}^2$. The cross section of the beam is rectangular with single reinforcement with dimensions of b × d = 400 × 450 mm. The calculation must be performed by linear approximation methods.

The solution. Let us determine the mechanical reinforcement ratio:

$$\omega = \frac{A_S \cdot f_{yd}}{b \cdot d \cdot f_{cd}} = \frac{760 \cdot 435}{400 \cdot 450 \cdot 14.5} = 0.127,$$

Then we will determine the bearing capacity of the beam:

$$M_{ED} = 0.94 \cdot A_S \cdot f_{yd} \cdot d = 0.94 \cdot 760 \cdot 435 \cdot 450 \cdot 10^{-6} = 139,84\ kN \cdot m.$$

It is quite interesting to determine the bearing capacity of the beam by area from the previous example at As = 648 mm^2

$$M_{ED2} = 0.94 \cdot A_S \cdot f_{yd} \cdot d = 0.94 \cdot 648 \cdot 435 \cdot 450 \cdot 10^{-6} = 120,34\ kN \cdot m.$$

The fractional accuracy is $\frac{120.34-120}{120} 100\% = 0.28\%$.

5 Conclusions

The considered approximation models of the method of design resistance of reinforced concrete enable to check quickly the durability of normal sections of bending reinforced concrete elements. The obtained dependences were confirmed by testing on a large number of experimental samples of well-known researchers, they fully correspond to the norms of design codes EN 1992-1-1 and can be widely used by engineers of different levels in design practice and in construction of reinforced concrete structures.

References

1. EN 1992–1 (1993) Eurokode- 2. Design of concrete structure. Part 1, General rules and rules for buildings, GEN
2. Bibi EV (2013) Designers' Guide to Eurocode 2: Design of reinforced concrete structures: designers' guide to EN 1992-1-1 and EN 1992-1-2. Eurocode 2: Design of reinforced concrete structures. General rules and regulations for buildings. Fire-fighting civil and structural engineering/Bibi EV, Narayanan RS; series editor Gulvanesyan X; Ministry of Education and Science of the Russian Federation, Federal State-Funded Educational Institution of Higher Professional Education "Moscow State University of Civil Engineering"; science editors Almazov VO, Plotnikov AI. Second edition, Moscow: MSUCE 2013, p 292 (The series "Published in MSUCE: Eurocodes". The science editor of the series –Almazov VO)
3. DBN V.2.6-98:2009. Konstruktsiyi budynkiv i sporud. Betonni ta zalizobetonni konstruktsiyi. Osnovni polozhennya., Minrehionbud, Kyiv
4. Kochkarev D, Azizov T, Galinska T (2018) Bending deflection reinforced concrete elements determination. In: Paper presented at the MATEC Web of Conferences, vol 230.https://doi.org/10.1051/matecconf/201823002012
5. Pam HJ, Kwan AKH, Islam MS (2001) Flexural strength and ductility of reinforced normal- and high-strength concrete beams. Struct Build 146(4):381–389
6. Sarkar S, Adwan O, Munday JGL (1997) High strength concrete: an investigation of the flexural behavior of high strength RC beams. Struct Eng 75(7):115–121
7. Bernardo LFA, Lopes SMR (2004) Neutral axis depth versus flexural ductility in high-strength concrete beams. ASCE J Struct Eng 130(3):425–459
8. Ashour SA (2000) Effect of compressive strength and tensile reinforcement ratio on flexural behavior of high-strength concrete beams. Eng Struct 22:413–423
9. Azizov T, Jurkowska N, Kochkarev D (2019) Basis of calculation on torsion for reinforced concrete structures with normal cracks. In: Proceedings of the fib Symposium 2019: Concrete - Innovations in Materials, Design and Structures 2019, pp 1718–1725
10. Kochkarev D, Azizov T, Galinska T (2020) Design of effective statically indeterminate reinforced concrete beams. https://doi.org/10.1007/978-3-030-42939-3_10
11. Kochkarev D, Azizov T, Azizova A, Galinska T (2021) Designing of standard cross sections of composite bending reinforced concrete elements by the method of design resistance of reinforced concrete. https://doi.org/10.1007/978-3-030-57340-9_25
12. Azizov T, Kochkarev D, Galinska T (2020) Reinforced concrete rod elements stiffness considering concrete nonlinear properties. Lecture Notes in Civil Engineering, vol 47, pp 1–6. https://doi.org/10.1007/978-3-030-27011-7_1
13. Azizov T, Derkowski W, Jurkowska N (2019) Consideration of the torsional stiffness in hollow-core slabs' design. https://doi.org/10.4028/www.scientific.net/MSF.968.330
14. Pavlikov A, Kosior-Kazberuk M, Harkava O (2018) Experimental testing results of reinforced concrete beams under biaxial bending. Int J Eng Technol (UAE) 7(3):299–305. https://doi.org/10.14419/ijet.v7i3.2.14423
15. Krassowska J, Kosior-Kazberuk M (2019) Experimental investigation of shear behavior of two-span fiber reinforced concrete beams. Arch Civil Eng 65(2):35–55. https://doi.org/10.2478/ace-2019-0017
16. Pavlikov A, Kochkarov D, Harkava O (2019) Calculation of reinforced concrete members strength by new concept. In: CONCRETE. Innovations in Materials, Design and Structures: Proceedings of the fib Symposium 2019, pp 820–827, Kraków, Poland, May 27–29 2019. http://reposit.pntu.edu.ua/handle/PoltNTU/6064
17. Pavlikov A, Kochkarev D, Harkava O (2020) Analysis of eccentrically loaded members of circular cross section by nonlinear deformation model. https://doi.org/10.1007/978-3-030-42939-3_19
18. Pavlikov A, Mykytenko S, Hasenko A (2018) Effective structural system for the affordable housing construction. Int J Eng Technol (UAE) 7(3):291–298. https://doi.org/10.14419/ijet.v7i3.2.14422

19. Piskunov VG, Gorik AV, Cherednikov VN (2000) Modeling of transverse shears of piecewise homogeneous composite bars using an iterative process with account of tangential loads 2. resolving equations and results. Mech Composite Mater 36(6):445–452. https://doi.org/10.1023/A:1006798314569
20. Piskunov VG, Goryk AV, Cherednikov VN (2000) Modeling of transverse shears of piecewise homogeneous composite bars using an iterative process with account of tangential loads. 1. construction of a model. Mech Composite Mater 36(4):287–296. https://doi.org/10.1007/BF02262807
21. Cherniha R, Pliukhin O (2013) New conditional symmetries and exact solutions of reaction-diffusion-convection equations with exponential nonlinearities. J Math Anal Appl 403(1):23–37. https://doi.org/10.1016/j.jmaa.2013.02.010
22. Storozhenko L, Yermolenko D, Gasii G (2018) Investigation of the deformation state of a composite cable space frame structures with a photogrammetric method. Int J Eng Technol 7(3.2):442–446. https://doi.org/10.14419/ijet.v7i3.2.14568
23. Panchenko OV, Ivanyts'kyi YL, Kun' PS, Zhuravs'kyi OD (2018) Determination of the durability of ferroconcrete bridge beams reinforced by composite strips. Mater Sci 53(5):660–665. https://doi.org/10.1007/s11003-018-0121-3
24. Kozak O, Zhuravskyi O, Delyavskyy M (2019) Effect of the pre-stressed reinforcement curvature on the bearing capacity of inclined sections of monolithic beams. In: Paper presented at the AIP Conference Proceedings, vol 2077. https://doi.org/10.1063/1.5091888
25. Semko O, Fenko O, Hasenko A, Harkava O, Kyrychenko V (2018) Influence of external and internal cooling at solidification on strength of brittle duralumin in compression. In: Paper presented at the MATEC Web of Conferences, vol 230. https://doi.org/10.1051/matecconf/201823002029
26. Hasenko A, Semko O, Drobotia O, Sirobaba V (2020) Experimental and numerical studies of nodes of light steel-reinforced concrete structures. In: Paper presented at the Proceedings of the 2020 Session of the 13th Fib International PhD Symposium in Civil Engineering, pp 173–178
27. Semko O, Hasenko A, Kyrychenko V, Sirobaba V (2020) The rational parameters of the civil building steel frame with struts. https://doi.org/10.1007/978-3-030-42939-3_25

Analytical Procedure for Design of Centrally Compressed Bars

Anton Makhinko, Nataliia Makhinko, and Oleg Vorontsov

Abstract The article is devoted to the study of stability problems of centrally compressed bars. The difficulties of the classical solution of this problem are highlighted. In analytical form, we propose a simplified dependence for calculating the stress reduction factor. Using the author's approach shows good agreement between the results received and the calculation results according to the normative methodology for a wide range of slenderness. An algorithm for determining the dimensions of the cross sections of steel elements loaded with a central force is also constructed. At the same time, solutions presented through the Lambert transcendental function are used to test rigidity. The convenience and advantages of using this algorithm are indicated. A practical example of the column sizing with the subsequent verification of the results according to current standards is given. The example of calculation shows the simplicity of the calculation according to the proposed algorithm and the full compliance of the result with the requirements of regulatory documents.

Keywords Stability · Buckling length · Centrally compressed bar · Buckling coefficient · Slenderness · Effective length · Lambert function

1 Introduction

The issues of stability theory are very important for displaying the stress–strain state and understanding the actual operation of compressed bars [1–5]. Stability testing is an integral part of real design [6–10]. However, it is important to have a simple engineering algorithm in order to solve practical problems and develop construction

A. Makhinko
ETUAL LLC, Bortnytska Street 1, Boryspil District, Petropavlivske, Kyiv Region, Ukraine

N. Makhinko (✉)
National Aviation University, Liubomyra Huzara 1, Kyiv, Ukraine

O. Vorontsov
National University «Yuri Kondratyuk Poltava Polytechnic», Pershotravnevyj Avenue 24, Poltava, Ukraine

V. Onyshchenko et al. (eds.), *Proceedings of the 3rd International Conference on Building Innovations*, Lecture Notes in Civil Engineering 181,
https://doi.org/10.1007/978-3-030-85043-2_24

solutions. The analytical method presented in regulatory documents for calculating the stress reduction factor of compressed elements complicates this task [11]. First, it is connected with difficulties of a calculated nature. Thus, the aim of this study was to develop a methodology for calculating centrally compressed bars, which gives an accurate result, but is not complicated by unnecessary calculation procedures.

There is a centrally compressed bar with a length ℓ_k. The cross-section of the bar in general is arbitrary, but then the sections that are most popular in the field of metal structures are considered: equal angle, circular hollow section, channel section, composite I-beam. With the known strength characteristics of steel (yield strength R_y) it is necessary to propose an algorithm for determining one of the overall dimensions of the bar cross-section in a closed form from the condition of ensuring its stability.

2 Main Body

As you know, the calculation of centrally compressed elements in all world norms is performed according to the equation [11]

$$K_R = \frac{N}{\varphi R_y A_k} \leq 1.0, \tag{1}$$

where A_k is cross area; φ is stress reduction factor.

According to clause 8.1.3 of DBN «Steel Structures» [11], the stress reduction factor φ is a function of the element's slenderness λ, steel yield strength R_y and type of buckling curve. The main difficulty in selecting the cross sections of the elements is because the form of the function $\varphi(R_y, \lambda)$ is rather bulky and non-linear with respect to R_y

$$\varphi = \frac{E}{2\lambda^2 R_y}\left[\pi^2\left(1-\alpha+\beta\lambda\sqrt{\frac{R_y}{E}}\right)+\lambda^2\frac{R_y}{E}\right] - \frac{E}{2\lambda^2 R_y}\left[\sqrt{\left(\pi^2\left(1-\alpha+\beta\lambda\sqrt{\frac{R_y}{E}}\right)+\lambda^2\frac{R_y}{E}\right)^2+39.48\lambda^2\frac{R_y}{E}}\right], \tag{2}$$

$$\lambda = \frac{\ell_k}{i_{\min}\mu_k}, \tag{3}$$

where α and β are parameters of the buckling curve from Table 1 DBN [11]; ℓ_k is effective length; $i_{\min}$ is minimum section radius of inertia; μ_k is effective length factor, which takes into account the fastening of the ends of the rod.

To simplify the calculations, it was proposed to replace this «uncomfortable» expression with a more «convenient» one, which greatly simplifies the classical calculation of stability within the framework of norms. An exponential relationship of the form is proposed as such a «convenient» expression [12, 13, 15–19]

$$\varphi = \exp\left(-\delta \frac{\lambda^{\varepsilon}}{\pi^2} \frac{R_y}{E}\right), \tag{4}$$

where ε and δ are parameters of the stability curve similar to α and β.

The expression under the sign of the exponent resembles in its structure the well-known Euler formula for the buckling force and differs from it by the introduction of additional parameters ε and δ. In the general case, these parameters depend on the type of cross section of the element and the steel liquid limit. The parameter values can always be successfully chosen in such a way as to describe the curves of the stress reduction factor given in the design standards with a minimum error. However, we propose a simpler way, slightly affecting the accuracy of obtaining results. Its essence is that the degree of slenderness indicator ε is taken equal to 2, and the parameter δ is considered dependent only on the type of buckling curve: «a»: $\delta = 0.4$, «b»: $\delta = 0.5$, «c»: $\delta = 0.6$.

Thus, for the stress reduction factor, we will have

$$\varphi = \exp\left(-\delta \frac{\lambda^{2}}{\pi^2} \frac{R_y}{E}\right). \tag{5}$$

A comparison of the normative expression for the stress reduction factor (2) with the proposed relationship (5) is performed in Fig. 1 for two types of buckling curve («a» and «b»). The figure shows that in the range of element slenderness values from 0 to 100, the consistency of the results is quite acceptable, and for $\lambda > 100$, although the differences are large, firstly, Eq. (5) gives a lower estimate of the stress reduction factor φ, and secondly, by practice rarely has to deal with values $\varphi < 0.3$. Taking into account these considerations and the simplicity of the proposed relationship $\varphi(\lambda)$, we will further consider the application of Eq. (5) justified. Substituting Eq. (5) in Eq. (1), we obtain the following expression for checking the stability of centrally compressed elements

$$K_R = \frac{N}{\exp\left(-\delta \frac{\ell_k^2}{i_k^2 \pi^2} \frac{R_y}{E}\right) R_y A_k} \le 1.0, \tag{6}$$

where slenderness λ is represented through the effective length ℓ_k and radius of inertia i_k relative to the plane of interest.

Further, we will assume that the cross area A_k and the radius of inertia i_k of the element can always be represented as

$$A_k = b_k \cdot t_k \cdot f_1, \tag{7}$$

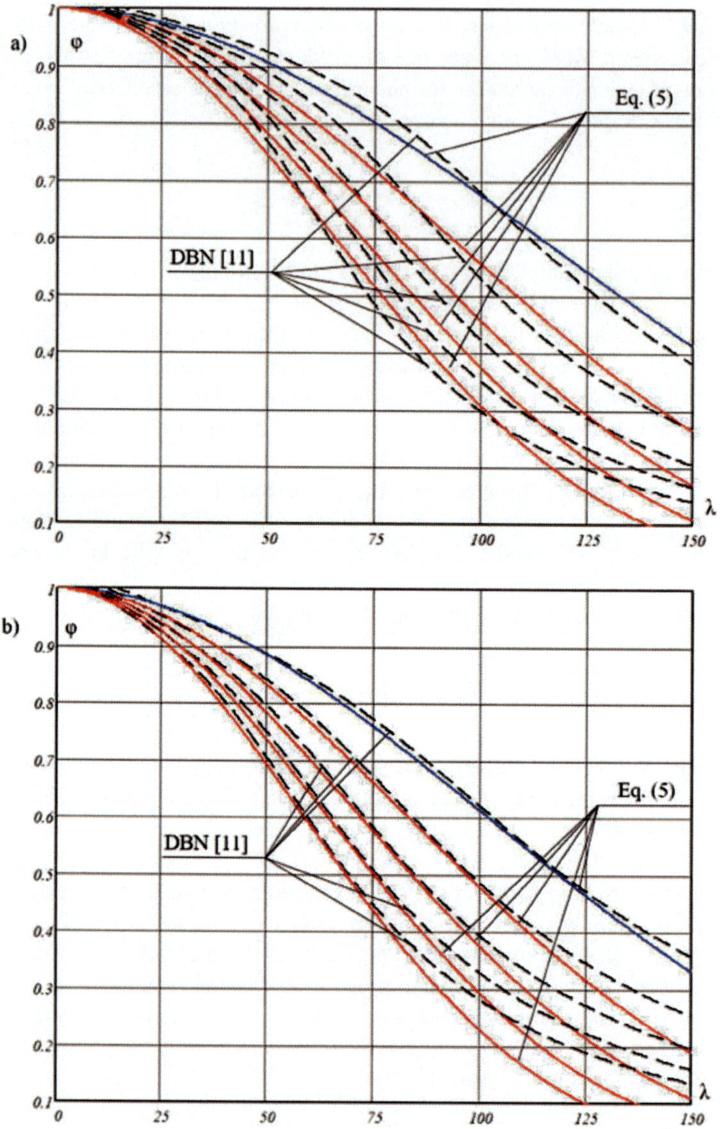

Fig. 1 Approximation buckling curve в DBN [11] according to Eq. (5): **a** - the type of buckling curve «a»; **b** - the type of buckling curve «b»

$$i_k = b_k \cdot f_2, \tag{8}$$

where b_k and t_k are some characteristic overall size and characteristic thickness of the cross section; f_1 and f_2 are some dimensionless coefficients, additionally characterizing other sizes of the section.

Table 1 represents the values for sections in the form of an equal angle, circular hollow section, channel section and I-beam.

Substituting expressions Eqs. (7) and (8) into Eq. (6), we obtain the following equation with respect to the desired cross-sectional characteristic b_k

$$N = K_R \exp\left(-\delta \frac{\ell_k^2}{b_k^2 f_2^2 \pi^2} \frac{R_y}{E}\right) R_y b_k t_k f_1. \tag{9}$$

It is convenient to represent the solution of this equation through the Lambert function. Omitting the intermediate transformations and simplifications, we give only the result

$$b_k = b_{ef} \cdot \Delta_R, \tag{10}$$

$$\Delta_R = \frac{\eta_R}{\sqrt{LambertW(\eta_R^2)}}, \tag{11}$$

Table 1 Characteristics of cross sections in the form of equal angle, circular hollow section, channel section and I-beam

Value	Cross section			
	equal angle	circular hollow section	channel section	I-beam
b_k	width flange	diameter	width flange	width flange
t_k	flange thickness	thickness	flange thickness	flange thickness
h_k	–	–	depth	depth
t_w	–	–	web thickness	web thickness
f_1	2.0	π	$2+\beta_k$	
f_2	0.195	0.35	plane of greatest rigidity $0.5\beta_2\sqrt{(6+\beta_k)/(6+3\beta_k)}$; plane of least rigidity $\sqrt{(1+2\beta_k)/3}/(2+\beta_k)$	$1/\sqrt{6(2+\beta_k)}$
Designations for sections in the form of a channel section and an I-beam $\beta_k = \beta_1 \beta_2,\ \beta_1 = t_w/t_k,\ \beta_2 = h_k/b_k$				

where b_{ef} is a value having a dimension of length, which we will call the effective dimension of the cross section. From the physical point of view, this is the dimension of the cross section calculated at bar in tension.

$$b_{ef} = \frac{N}{K_R \, R_y \, f_1 \, t_k}, \tag{12}$$

where η_R is dimensionless characteristic showing how many times it is necessary to increase the cross-sectional dimension found from tensile analysis.

$$\eta_R = \frac{1}{\pi}\sqrt{2\delta \frac{R_y}{E} \frac{\ell_k}{b_{ef}} \frac{1}{f_2}}. \tag{13}$$

As for the Lambert function, it is defined as a solution of the functional equation

$$Lambert\ W(\eta_R) \exp[Lambert\ W(\eta_R)] = \eta_R. \tag{14}$$

This function is transcendental, i.e. the integer values of the argument correspond to the transcendental values of the function, and vice versa, the integer values of the function correspond to the transcendental values of the argument. The convenience of its use is explained by the fact that the dimensionless quantity Δ_R is enclosed in a very narrow range of values (from 1 to 10) and can be easily tabulated for all design schemes of compressed bars and cross-sectional forms of elements. The table of functional dependency values $\Delta_R(\eta_R)$ is given below.

Thus, in order to select the cross section of a compressed bar, it is necessary to select the required overall size from the conditions of its tensile operation and increase it by the value of Δ_R, found by the dimensionless parameter η_R. We give a numerical example.

3 Practical Calculation

According to the described approach, we perform the selection of a section of an I-beam profile for a 6 m long column with hinged fastening at the ends, with a known flange thickness $t_w = 8.0$ mm, $f_1 = 2.7$ and $f_2 = 0.25$ characteristics. We also set the compressive strength of $N_k = 1600$ kN and the value of the critical factor $K_R = 1.0$. The type of stability curve is taken as «b», which determines the value of the parameter $\delta = 0.5$.

In this case, the dimensionless argument of the Lambert function in the calculation by Eq. (13) is $\eta_R = 1.247$. According to Table 2, we find the value of the function $\Delta_R = 1.45$ and calculate the desired cross-sectional dimension (I-beam flange width) $b_k = 302.9$ mm.

Table 2 Table of values of function $\Delta_R(\eta_R)$

η_R	Δ_R	η_R	Δ_R
0.0	1.0	5.0	3.255
0.5	1.107	5.5	3.482
1.0	1.328	6.0	3.707
1.5	1.574	6.5	3.930
2.0	1.824	7.0	4.50
2.5	2.071	8.0	4.584
3.0	2.315	10.0	5.435
3.5	2.555	12.0	6.264
4.0	2.791	15.0	7.487
4.5	3.025	20.0	9.440

According to the assortment of rolled steel GOST 26,020-83 [14], we accept the 30K1 UC with geometric characteristics $b_k = 300$ mm, $h_k = 296$ mm, $t_w = 9.0$ mm, $A_k = 108$ cm^2, $i_{\min} = 75$ mm.

Let us check the cross section according to DBN [11]. The effective length factor for a bar with fastened ends is $\mu_k = 1$, and the slenderness of the element according to (3) will be $\lambda_k = 80$. Note that for the range of slenderness $\lambda < 100$ there is a high convergence of the results in terms of the stress reduction factor of the author's approach and the normative methodology. Using Eqs. (1) and (2), we calculate the stress reduction factor $\varphi = 0.697$ and the critical factor of the element, which is almost equal to the initial value $K_R^n = 0.996 \approx K_R = 1$.

4 Conclusions

1. To solve the stability problem of centrally compressed bars, an engineering algorithm is formulated. Using the author's approach makes it possible to simplify the procedure for determining the section dimensions. At the same time, the accuracy of the calculation remains quite high and complies with current standards.
2. The algorithm is based on the use of empirical dependence in determining the stress reduction factor. The study shows a fairly accurate correspondence of the results of the author's approach with the relationship given in current standards for a wide range of slenderness.
3. The original solution for the desired characteristics of the sections of centrally compressed elements is presented through the Lambert function. This allows you to choose the desired overall size of the compressed bar from the conditions of its operation in tension.

4. The given practical example of calculation shows the simplicity of the calculation according to the proposed algorithm and the full compliance of the result with the requirements of regulatory documents.

References

1. Streletskii NS (1959) Rabota szhatykh stoek. Gosudarstvennoe izdatelstvo literatury po stroitelstvu, arkhitekture i stroitelnym materialam Moskva
2. Perelmuter AV, Slivker VI (2010) Ustoichivost ravnovesiia konstruktsii i rodstvennye problemy. SKAD SOFT, Moskva
3. Timoshenko SP, Gere JM (2009) Theory of Elastic Stability. Dover Publications, New York (2009)
4. Bazant ZP, Cedolin SL (2010) Stability of Structures: Elastic, Inelastic, Fracture and Damage Theories. World Scientific Publishing Company
5. Alfutov NA (2009) Stability of Elastic Structures. Springer, Berlin
6. Qiusheng L, Hong C, Guiqing L (1995) Stability analysis of bars with varying cross-section. Int J Solids Struct 32(21):3217–3228
7. Nikolic A, Salinic S (2017) Buckling analysis of non-prismatic columns: a rigid multibody approach. Eng Struct 143:511–521
8. Li QS (2009) Exact solutions for the generalized Euler's problem. J Appl Mech 76(4):1015–1024
9. Elishakoff I (2001) Inverse buckling problem for inhomogeneous columns. Int J Solids Struct 38:457–464
10. Otrosh Y, Kovalov A, Semkiv O, Rudeshko I, Diven V (2018) Methodology remaining lifetime determination of the building structures. In: Matec Web of Conferences, vol 230, p 02023
11. DBN V.2.6–198:2014 Stalevi Konstruktsii. Normy Proektuvannia. Minrehionbud, Kyiv
12. Makhinko N (2018) Stress-strain state of the storage silos under the action of the asymmetric load. In: Matec Web of Conference, vol 230, p 02018
13. Makhinko N (2019) Imovirnisnyi rozrakhunok koefitsiientu krytychnoho faktoru dlia tsentralno stysnutykh elementiv. Zbirnyk Naukovykh Prats UkrDUZT 183:80–86
14. GOST 26020–83 Hot-rolled steel I-beam with parallel flange edges. Dimensions. Izdatelstvo standartov, Moskva (1983)
15. Pichugin S (2017) Probabilistic description of ground snow loads for ukraine. Snow engineering 2000: Recent advances and developments, pp 251–256. https://doi.org/10.1201/9780203739532
16. Pichugin S, Severin V (2004) Reliability of structures under snow load in ukraine. Snow engineering V, pp 67–72
17. Pichugin SF, Makhin'Ko AV (2009) Calculation of the reliability of steel underground pipelines. Strength Mater 41(5):541–547.https://doi.org/10.1007/s11223-009-9153-0
18. Pavlenko A, Koshlak H (2015) Design of processes of thermal bloating of silicates. Metallurg Mining Ind 7(1):118–122
19. Pichugin SF (2018) Reliability estimation of industrial building structures. Mag Civil Eng 83(7):24–37.https://doi.org/10.18720/MCE.83.3

Thermal Calculation and Its Optimization

Borys Makovetsky, Petro Sankov, Volodymyr Hilov, and Mykhailo Troshin

Abstract In 2013 significant changes in respect of thermal properties for enclosures of buildings were introduced to the DBN "Thermal building insulation". Requirements to thermal properties of enclosing structures have nearly doubled. Such changes were aimed at reducing heat consumption for buildings that will be built after the release of DBN. The same buildings that were built in the country before are subject to insulation with thermal insulation materials. The authors proposed the idea of simplifying the thermal calculation in order to have an express method of quick results. For this purpose, work was carried out to unify the types of enclosing structures for the period from 1930 to 2000. The analysis of types of heat-insulating materials which were used at that time in construction is made. As a result two classifications were proposed, presented in the form of tables. The first classification was carried out by types of enclosing structures for all residential buildings using the principles of typification and unification. The second classification, with a reduction in the nomenclature, is devoted to thermal insulation materials that serve as auxiliary insulation to enclosing structures (walls). The proposed classifications became the basis for the development of a graphical analytic method of calculation in the form of nomograms. Nomograms accelerate the calculation by obtaining instant results. Graphical analytic method of calculation allows determining the thickness of the shell of the outer walls of buildings that are being insulated.

Keywords Outer walls · Thermal insulation · Nomograms · Thermal resistance · Heat insulation depth

B. Makovetsky · P. Sankov (✉) · V. Hilov · M. Troshin
Pridneprovsk State Academy of Civil Engineering and Architecture, Street Chernyshevsky, 24a, Dnipro 49000, Ukraine

V. Onyshchenko et al. (eds.), *Proceedings of the 3rd International Conference on Building Innovations*, Lecture Notes in Civil Engineering 181,
https://doi.org/10.1007/978-3-030-85043-2_25

1 Introduction

1.1 Analysis of Recent Research on the Insulation of Existing Residential Buildings

Earlier, during the Soviet era, heating cost a penny. From now on, reducing heat consumption is the way to significant savings. And it is possible only with auxiliary heat insulation of enclosing structures. For countries in Europe and Asia, this problem is not relevant, as their regulations on thermal insulation of buildings have long been introduced with regulatory thermal resistance of wall enclosures, which corresponds to the weather conditions of these countries and they always paid for energy products at world prices. Therefore, to make the analysis on heat insulation of houses of existing inhabited building it is possible in a part of existing systems of heat insulation of enclosing structures which there were offered in 70th and in years of the last century. In 2020, the EU introduced strict rules to reduce primary energy consumption in buildings. These standards have also been introduced to reduce carbon dioxide emissions from energy (gas, coal, fuel oil). The strictest rules are in Denmark [1]. It should be noted that a fairly simple thermal calculation by the authors was not found, because in most countries there are computer programs that are distributed and used free of charge [2]. According to the literature analysis [3–5], known systems for external insulation of buildings, which are generally as multi-layered "pies" that perform the functions of insulation and decoration of the facade. There are wet plaster system (light and heavy) and ventilated facades [3, 4]. These systems came to Ukraine in the late twentieth century from Europe, where they were successfully used. So our task is to find out which systems are most attractive for use in Ukraine, and which can be combined to reduce species and nomenclature. Having a smaller number of input data, it is possible to simplify the thermal calculation, where the required value is the thermal resistance of the outer walls.

1.2 Analysis of the Typology of Residential Buildings by Types of Construction of Outer Walls

As have been mentioned earlier, heat consumption in residential buildings is distributed through enclosing structures (walls, windows, attics) [6–8]. We stopped at the walls, where more heat is consumed due to large volumes [9, 10]. The effect on heat transfer mainly depends on the properties of the material, the wall thickness [11, 12].

Theoretical foundations of thermal calculation emerged in the 30 s of the twentieth century [13–16]. Prior to that, the material and thickness for the walls were chosen by experience.

Therefore, on this basis, the analysis and classification of materials [7] and structures of walls for residential buildings since 1930 has been performed. Housing construction from this time on structural systems, schemes is divided into 3 periods in terms of changes in time of structures and materials of enclosing walls. The classification is made in the form of a table. Which became the basis for the share of factors that became a component for obtaining a simplified engineering thermal calculation [17].

1.3 Analysis of the Nomenclature of Thermal Materials

The main feature of heat-insulating materials is their high porosity, small density and low thermal conductivity.

The modern range of thermal materials impresses with a variety of forms, physical factors and raw material base [3, 18]. From all this, attention was focused on the following properties: the lowest thermal conductivity; ecological purity; fire danger; durability; manufacturability in installation; estimated cost.

The analysis gave a classification of materials that met the needs of thermal insulation, which would be mounted on the outer walls from the outside. In such a way the second important component of the simplified technological calculation was received.

1.4 Formulation of the Problem

For Ukraine, the issues of improving people's living conditions and health have been and are one of the main directions of socio-economic policy. There should be a comfortable life in residential buildings with an internal temperature that ensures health with low amounts of energy losses and, as a consequence, the funds of citizens [19]. We also made a small contribution to solving this problem, making the design of thermal protection more accelerated.

2 Goal

Development of the simplified engineering thermotechnical calculation at designing of external thermal insulation on earlier constructed apartment houses.

To achieve the goal the following tasks are set:

- to study the world experience of neglecting thermal insulation in residential buildings;

- to develop a classification of residential buildings and their parts subject to additional insulation in accordance with the rules [19];
- to develop a classification of thermal insulation materials for outer walls;
- using the graphical analytic method to develop nomograms that simplify the thermal calculation.

3 Methodology

In this paper we present the analysis of building systems, constructive schemes of apartment houses of mass building, as well as the analysis of the modern range of heat-insulating materials, the thermal engineering calculation by means of nomograms.

3.1 Structural and Thermal Properties of Outer Walls of Residential Buildings

The analysis covered the construction of Ukraine, which was built from the 30 s of the last century to 2000 years.

Three periods of construction were identified, which influenced the qualitative and quantitative characteristics of buildings. These are the houses of 1930–1960, 1960–1970 and 1970–2000 (Fig. 1a, b, c).

On the basis of the conducted researches of the whole array of residential buildings, the main representatives of which are the houses, which are shown in Fig. 1a–c the classification of the basic inhabited designs of outer walls which are the most widespread and need auxiliary external heat insulation "the Common materials of constructions of outer walls of houses (Table 1)" was made [20, 21].

a) b) c)

Fig. 1 Buildings built in **a** 1930–1960, **b** 1960–1970, **c** 1970–2000

Table 1 Common material of constructions of outer walls of houses

Material	Thickness ρ_o, kg/m^3	Floor thickness δ_1, m	Thermal conductivity λ_p, W/(m*K)
Buildings built in 1930–1950			
Clay brick	1700	0,51–0,64	0,64
Buildings built in 1950–1970			
Silicate brick	1800	0,51	0,76
Slag concrete blocks	1000	0,4	0,33
The panels are single-layer	2500	0,35–0,4	0,50
Buildings built in 1970–2000			
Silicate brick	1800	0,51	0,76
Slag concrete blocks	1000	0,4	0,33
Panels are two-layer and three-layer	2500	0,35–0,4	0,40
Buildings built in 2000 and later			
Two-layer reinforced concrete and lightweight concrete	2500	0,2–0,4	0,40
Aerated and foam concrete block	1000	0,2–0,4	0,30

3.2 *Structural and Physical Properties of Heat-Insulating Materials for Long-Term Heat Insulation of Houses*

When selecting insulation, it is necessary to pay attention not only to its price (in Ukraine, this position is most relevant), but also to the technological characteristics, in addition to the complexity of installation. The application includes: soft and flexible products (mats, cords, harnesses); rigid heaters (plates, shells); liquid spray compositions; granular modifications.

Of the soft, mineral wool insulation is considered to be quite cheap and effective, durable and fireproof, vapour-permeable. Of the rigid insulation, the cheapest is expanded polystyrene, but it is flammable and moisture-proof. Extruded polystyrene foam does not have these defects.

Liquid spray compositions are quite expensive, but not technological in terms of forming the surface of the facade of buildings. Granular modifications for the walls were not considered. Polyurethane foam has the lowest thermal conductivity, does not burn, but is quite expensive. Soft insulation is converted into hardboard, usually impregnated with formaldehyde. But there are those that are environmentally friendly, but of course very expensive. Based on the analysis of most indicators, a classification is made from a small list of insulation materials "Materials for heat insulation of building of buildings" (Table 2).

Table 2 Materials for heat insulation of building of buildings

Name	Thickness ρ_o, kg/m^3	Thermal conductivity λ_p, W/(m*K)	Price, UAH m^2
Plates from mineral wool (from 3,5 to 4,2%)	70	0,039	117–156
Basalt-fiber plates	90	0,050	100–150
Fiberglass plates	45	0,040	100–200
Polystyrene foam plates	50	0,040	10–20
Expanded polystyrene extrusion	39	0,037	30–60
Polystyrene foam plates	40	0,040	150

4 Development of a Graphic Method of Thermal Calculation

4.1 Calculation for Enclosing Structures

Thermal calculation is to determine the minimum sufficient value of heat transfer resistance of the outer enclosing structure. In this case, the calculated value of the heat transfer resistance of the enclosing structure must be not less than the value required for sanitary and hygienic and construction and technical indicators.

The main task of thermal calculation is the selection of the thickness of the insulation for the outer surface of the walls; this task is performed using nomograms.

For external thermally homogeneous enclosing structures of heating buildings and structures it is obligatory to fulfill the condition:

$$R_{\sum} \geq Rq\ min, \tag{1}$$

$$\Delta R = R_{qmin} - R_{\Sigma}, \tag{2}$$

R_{Σ}—reduced total thermal resistance of impervious enclosing structure,

ΔR—the minimum permissible resistance value of heat transmission of impervious enclosing structure.

Using Formulas (1) and (2), we construct nomograms for wall enclosing structures for 1 temperature zone (Fig. 2) and 2 temperature zones (Fig. 3).

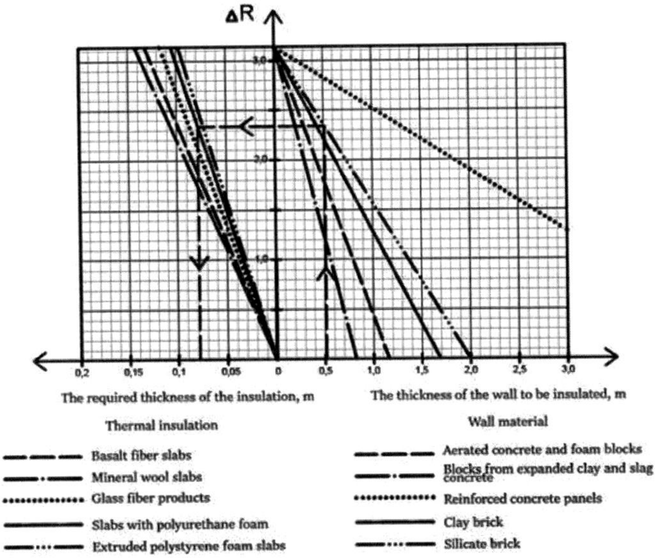

Fig. 2 Nomogram depending on the depth and material of the wall, as well as thermal insulation for the 1 temperature zone of Ukraine (ΔR = 3, 142)

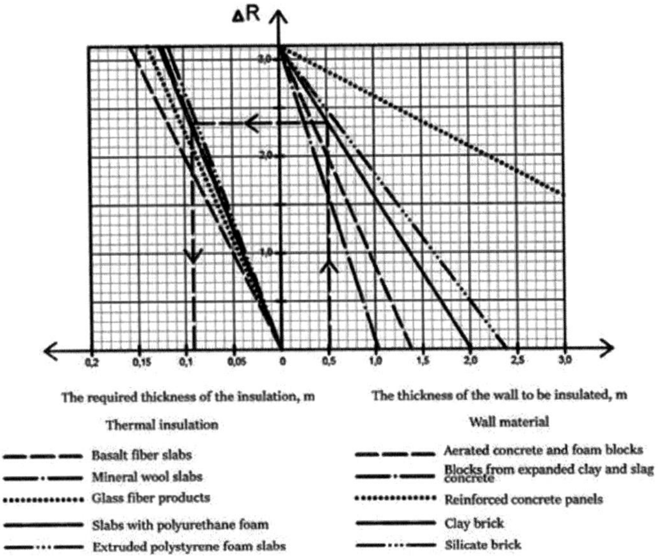

Fig. 3 Nomogram depending on the depth and material of the wall, as well as thermal insulation for the 2 temperature zone of Ukraine (ΔR = 2, 642)

5 Conclusions

To achieve this goal, the work was carried out to survey the types of residential buildings that were built before the release of DBN. The classification on building and constructive systems, on enclosing structures (walls) is presented in a tabular form.

It was established that auxiliary thermal insulation depends on the thermal insulation material (physical properties, thickness, service life, manufacturability and cost). An analytical review was made, which revealed almost the entire range of these products. The classification of heat-insulating materials which are most suitable for application of warming is made in the tabular form. Their list is reduced to 10 items.

Having the material in the form of two classifications, nomograms were developed by graphical analytic method, which greatly simplifies the thermal calculation of enclosing structures.

Using the nomograms (Fig. 2) the following conclusions have been made:

1) as the main thermal insulations for 1 climatic zone it is necessary to use expanded polystyrene extrusion, expanded polyurethane;
2) the thickness of the insulation, depending on the material of the walls must be not less than: a) for walls from a clay brick 510 mm thick, thickness of expanded polystyrene extrusion is equal to 100 mm, b) for walls from large blocks 400 mm thick - the thickness of expanded polystyrene extrusion is 150 mm, c) for aerated and foam concrete blocks with a thickness of 200–150 mm, with a thickness of 400–100 mm.

In the capacity of directions for future research, authors have been started a work on the development of similar nomograms for the roofs of residential buildings.

References

1. September 1993 Limit Carbon Dioxide Emissions bi improving Energy Efficiencu (SAVE), Council Direktive 93/76EEC. Official J L237, 22.09.1993, pp 28–30. https://www.irbnet.de/daten/iconda/CIB2512.pdf
2. Matrosov YA (2015) Moscow, Russia. Comparative analysis of the Russian system of norms and standards for the energy performance of buildings with the requirements of the European Union. Research Institute of Building Physics RAASN
3. Sobinova KS, Ozhishchenko OA, Savitsky NV (2013) Analysis of existing systems of thermal insulation finishing of facades. Dnipro, Ukraine. Bulletin of the Prydniprovska State Academy of Civil Engineering and Architecture
4. System of bonded external thermal insulation of buildings and structures: TUU V.2.7–21685172.002–2001
5. (2004) Ventilated Walls Design principles Building Insulation
6. Pushkareva KK (2012) Modern Ukrainian building materials, products and structures. K: Association “VSVBMV”. 664

7. Minarova M (2006) Boundary value problems in bulding physics – thermal engineering of buildings. Czech Technical University in Prague, pp 14–17
8. Bouache T, Ginectet S, Limam K, Lindner G, Bosschaerts W (2013) Identification of thermal characteristics of a building. Elsevier Ltd.
9. Zakharov Y, Sankov P, Trifonov I, Tkach N, Toshyna L (2019) The essence and features of the reconstruction of residential buildings of various structural systems.https://www.scopus.com/record/display.uri?eid=2-s2.0-85074266669&origin=resultslist
10. Sankov P, Zakharov Y, Zakharov V, Hvadzhaia B (2020) Proceedings of the 2nd International Conference on Building Innovations, pp 215–223. Springer, Berlin, Heidelberg, Germany
11. Kunzel HM (1995) Simultaneous heat and moisture transport in building components. One- and two-dimensional calculation using simple parameters
12. Dikarev K, Kuzmenko O, Petrenko V, Kyslytsia L, Ibadov N (2020) Experimental study of operating indicators of a thermalactic covering panel. Sci Innov 16(2):62–71.https://doi.org/10.15407/scin16.02.062
13. Kalis H, Kangro I (2003) Simple methods of engineering calculation for solving heat transfer problems. Math Model Anal 8(1):33–42. https://doi.org/10.1080/13926292.2003.9637208
14. Yurin O, Galinska T (2017) Study of heat shielding qualities of brick wall angle with additional insulation located on the outside fences. Paper presented at the MATEC Web of Conferences, vol 116.https://doi.org/10.1051/matecconf/201711602039
15. Yurin O, Azizova A, Galinska T (2018) Study of heat shielding qualities of a brick wall corner with additional insulation on the brick. Paper presented at the MATEC Web of Conferences, vol 230https://doi.org/10.1051/matecconf/201823002039
16. Leshchenko MV, Semko V (2015) Thermal characteristics of the external walling made of cold-formed steel studs and polystyrene concrete. Mag Civil Eng 60(8):44–55. https://doi.org/10.5862/MCE.60.6
17. Makovetsky BI, Gilov VV, Shadronova SS (2020) Improved method of calculating the thermal insulation of buildings. In: The 2nd International Scientific and Practical Conference —Modern Science: Problems and Innovations, 3–5 May 2020, Stockholm, Sweden: SSPG Publish
18. Sobinova KS, Ozhyshchenko OA, Savytskiy MV (2013) External thermal insulation composite facade systems for existing constructions. Visnyk Prydniprovsk State Acad Civil Eng Archit 1(2):59–64
19. SBC B.2.6. – 31:2006. Thermal insulation of buildings (2006)
20. EN1745 Masonry and masonry products – Methods for determining thermal properties. CEN TC 125. 2009/
21. Korniyenko S (2015) Evaluation of thermal performance of residential building envelope. Procedia Eng 117(1):191–196. https://doi.org/10.1016/j.proeng.2015.08.140

Control of Thermal Energy Consumption Mode for Multifunctional Buildings

Nurmammad Mammadov and Oleg Yurin

Abstract At present there are many mathematical models of optimal control, planning and distribution of thermal energy, however in real conditions, their use is often difficult. First of all it concerns prompt control and distribution of thermal energy among consumers when there arises a problem of prompt correction of initial information and feedback on actual consumption. At the same time, one has to deal with uncertainties of goals arising on simultaneous execution of various management tasks and providing necessary values under extreme climatic conditions or emergency situations. As a result of impact of disturbing factors (climatic parameters of environment, operating conditions of buildings, organizational factors, emergency situations and so on), and also incompleteness and inaccuracy of initial information, controlled parameters turn out to be fuzzy. Obviously, further improvement of the system of planning, control and distribution of thermal energy among consumers of different categories, improving the efficiency and reliability of the operation of modern buildings especially in extreme cases, are related to transition to a new modeling system, in particular to new information technology and creation on its basis new systems of optimal and prompt control, distribution and exploitation. In the paper we analyze possibility of using expert systems for regulating heat consumption mode in modern buildings. We often a structure of hybrid expert system of thermal energy control for modern multifunctional buildings.

Keywords Modern multifunctional buildings · Expert systems · Thermal energy consumption mode · Database · Operational dispatch control · Emergency situations

N. Mammadov (✉)
Azerbaijan University of Architecture and Construction, Ayna Sultanova 11, Baku, Azerbaijan
e-mail: nurmammad.mammadov@azmiu.edu.az

O. Yurin
National University "Yuri Kondratyuk Poltava Polytechnic", Poltava, Ukraine

V. Onyshchenko et al. (eds.), *Proceedings of the 3rd International Conference on Building Innovations*, Lecture Notes in Civil Engineering 181,
https://doi.org/10.1007/978-3-030-85043-2_26

1 Introduction

Complication of the structure, spread in extension and increase in energy intensity of modern buildings and also intensification of operational modes reduce to the fact that traditional control methods applied in these buildings have ceased to be an effective means of rational distribution of thermal energy among consumers. This led to sharp increase in unproductive costs of material and energy resources, to a decrease in degree of satisfaction of thermal energy consumers.

As management objects, modern residential and public multifunctional buildings belong to the class of multidimensional multiplyconnected nonlinear stochastic systems with distributed parameters whose specific feature is their multilevel structure, high level of uncertainty of the structure, parameters and states of the control object (building), availability of continuous and discrete components in the control vector.

At present there is a considerable experience in modeling and optimization of the mode of distribution and control of thermal energy in modern buildings [1, 2]. However, the worked out optimization methods as a rule are deterministic, ignore real operating conditions of the buildings related to uncertainty both of the control object and environment.

Optimal solution on distribution and control of thermal energy among consumers obtained by means of these methods correspond only to specific boundary conditions and are as a rule are on the border of the feasible area. Practically, this led to the fact that even minor variations of boundary conditions could essentially change not only optimal solution but also to take it out of the area of technological feasible modes: i.e. to lead to an emergency situation. Naturally, such optimal solutions are unacceptable for practice.

At present, there are many mathematical models of optimal control, planning and distribution of thermal energy, however in real conditions, their use is often difficult. First of all it concerns prompt control and distribution of thermal energy among consumers when there arises a problem of prompt correction of initial information and feedback on actual consumption. At the same time, one has to deal with uncertainties of goals arising on simultaneous execution of various management tasks and providing necessary values under extreme climatic conditions or emergency situations. As a result of impact of disturbing factors (climatic parameters of environment, operating conditions of buildings, organizational factors, emergency situations and so on), and also incompleteness and inaccuracy of initial information, controlled parameters turn out to be fuzzy.

Traditional regulation, control and exploitation methods applied in modern centralized heat supply systems ceased to be effective and there was a sharp increase material and thermal power resources, decrease in the quality of uninterrupted provision and degree of satisfaction of consumers.

Obviously, further improvement of the system of planning, control and distribution of thermal energy among consumers of different categories, improving the efficiency and reliability of the operation of modern buildings especially in extreme cases, are related to transition to a new modeling system, in particular to new information technology and creation on its basis new systems of optimal and prompt control, distribution and exploitation.

2 Methodology

The main goal of prompt control of thermal energy distribution for residential buildings is to ensure more complete satisfaction of continuously changing requirements of consumers. As known, the existing mathematical models of optimal planning and control in real conditions are often not workable. Especially it concerns operational scheduling where there arises a problem of feasibility of models and necessity of their prompt correction with regard to feedback information on actual heat consumption. This time, one has to face with uncertainty of goals arising from the desire to perform simultaneously unattainable tasks for providing consumers with necessary amount of thermal energy on one hand, and providing extreme value of the selected economic criteria on the other hand. Furthermore, as a result of impact of disturbing factors and also incompleteness and uncertainty of initial information the basic system parameters (thermal energy consumption, limitations, coefficients) turn to be fuzzy. In this connection, production planning specialists tend to use in practice their own decision rules based on their experience and intuition. Although such heuristic rules do not guarantee mathematical optimality, turn out to be adequate to real conditions.

The structure of the process of prompt control of thermal energy distribution can be represented in the form of two main stage or level of management:

1. Prompt control of thermal energy distribution with regard to prevailing influence of some criteria on the given planning period (minor operating costs, maximum heat energy supply of consumers, etc.).
2. Stabilization of all or at least of some phase coordinates (consumption, temperature) providing minimum variance of these coordinates related to the calculated or planned settings.

The solution of such problems, is usually separated in time and space, requires various volume and character of prompt information, mathematical models describing the control objects (buildings or placements), various methods or criteria.

In this direction we have studied new approaches to the solution of problems of planning and distribution of thermal energy based on the conception of L. Zadeh's fuzzy sets [3, 4]. The offered approaches allow to consider such factors as experience and intuition difficult to formalize.

Further improvement of the system of planning and control, their efficiency and reliability are related to transition to a new information technology and creation on its basis qualitatively new systems based on buildings. As noted above, a modern building with continuous character of exploitation is a complex production system consisting of a dozen of maintenance services. The goal of an operational-dispatcher control of a building with continuous exploitation is to provide trouble-free, rhythmic and uninterrupted functioning of all communal systems. A great majority of specific decisions on the actions of these systems is accepted by the dispatcher on duty. He constantly analyzes the incoming information to reveal deviations from normal functioning of the system. If such a deviation happens, the dispatcher should necessarily find the reason for its occurrence. After identifying the reason, the dispatcher develops a sequence of actions to eliminate the unfavorable situation. Thus, the dispatcher constantly performs the tasks of tracking the progress of operation, analysis of situation and planning of actions. Furthermore, he has periodically deal with calculated computational problems. It should be noted that unlike other modern mathematical tools, the expert systems model human thinking mechanism in relation to the solution of the problem, form certain considerations and conclusions based on the knowledge they have. These systems unlike traditional approaches, offer a set of conceptions fit for solving complex problems in the cases when mathematical solutions are either unknown or ineffective.

The operating experience of automated operational-dispatcher control systems in modern buildings showed that the used traditional "rigid" formal mathematical models are not sufficiently adequate to the control object. Therefore there arises a need to create dispatcher room that controls the expert technological system, that combines in itself such advantages of modern computers as high performance, large memory and huge computing power and also a highly qualified specialist accumulating experience and knowledge of an expert in this field [5–7].

Functionally, dispatching control expert technological system is intended to help the dispatcher in the process of his work and provide him with advices, recommendations on actions in the established situation, with prompt and reliable information and also possibility to perform calculated character tasks. The process of expert system building can be divided into 5 stages (Fig. 1).

Stage 1 is the definition of goals and objectives for which the system is structured. Here, first of all the spectrum of problems and their typical features are established. The precise technical task for the system under development will help in future to outline correctly the area of knowledges of the expert, necessary to define actual energy consumption. Secondly, it is important to define potential users of the system that also influences on necessary level of the work of the expert system and consequently on the level of required knowledges.

Stage 2 is highlighting the main concepts of the subject area that reflects the knowledge of experts. Highlighting of such concepts allows to analyze what type knowledges uses the expert when defining heat energy consumption. This helps engineers to select the formal means of representation of knowledges and procedures for getting decisions that are the most suitable for the decision making process by the expert in the field of exploitation of communal systems of buildings.

Stage 3 is the selection of the language of representation of knowledges and a solver that in our opinion to a large extent causes the success of the creation of the expert system to accurately define heat energy consumption.

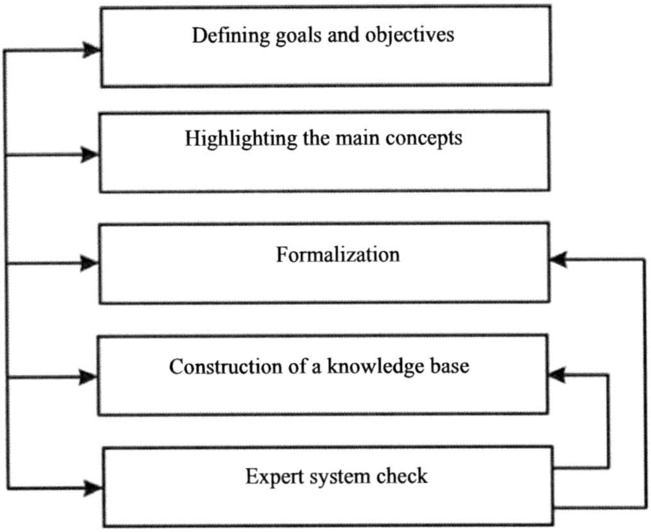

Fig. 1 Expert system building stages

Stage 4 is the direct construction of knowledge base of the expert system. A knowledge engineer being actually a translator between the expert and computer brings the knowledges in the field of exploitation of buildings obtained from the expert and written in the language of representation of knowledges.

Stage 5 consists of checking the work of the expert system. Verification is carried out by solving the benchmarks on control of heat energy consumption mode by the expert system. Evolution of views on the problem of control of complex systems in the context of the use of these or other formalisms to construct adequate models of controlled objects, passing through the stage of creating the need to take into account the human factor in them led the specialists in the field of engineering and construction to formation of the concept of construction of "intellectual" knowledges.

Expert systems, as the most actively developing areas of ideology of classic artificial intelligence, imitating human ways of reasoning are capable to solve control problems no worse than a man, an expert. Moreover, such systems intelligently combining dignity of human – machine dialogue systems with heuristic models get a number of principally new and extremely important properties. This, first all, is the possibility to represent the expert knowledges in the language close to the natural language of human communication, to solve the problems based on these knowledges and perhaps the most important thing to verify the goals of control, i.e. to confirm the correctness of the adopted decision [8]. The fact is that the level of complexity of operation of modern buildings, uncertainty of their functioning parameters, large dimension of their management problems do not allow to create such universal knowledge base of the expert system that could give solutions adequate to frequent emergency situations out of bounds.

It was natural to look for ways out of this situation by imparting some dynamism to the expert system in the sense of the possibility of a more flexible and prompt response to changing comfortable conditions inside the building in real time. And moreover periodically updated knowledge base models should allow based on the analysis of trends in the change in the parameters of the outside air, to predict possible changes in the parameters in the rooms and generate appropriate decisions.

For the purpose of prompt data processing in a real time environment and solutions of the problems of control of thermal mode in modern buildings, the general problem of control was decomposed by us into a number of narrowly focused subproblems as:

- Recognition of all operational events uniquely associated with some unforeseen problems that lead to existing changes in the mode of thermal energy consumption;
- Arbitration of the priorities of these events;
- Analysis of the event for the purpose of recognition and identification of the corresponding unforeseen problem;
- Arbitration of the priorities of the basic problems;
- Coordination of above processes.

Efficiency of thermal energy consumption mode control largely depends on the timeliness and promptness of the recognition of unforeseen operational problems.

Traditionally, to solve the optimal control problem, the accumulated experience of preliminarily prepared plan of exploitation measures and the way of life of people living in this building is used. However, practical implementation of prompt and optimal control of thermal energy consumption mode for modern buildings does not allow to get desired results, rhythmic control and stability of provision of all consumes with necessary amount of thermal energy and this is accompanied with violation of comfort of individual consumers and promptness of the work of the heat supply system of the building as a whole.

The problem of real world, its problematic areas as urban planning, environmental protection, ecology of the city and others are characterized by weak structurenedness and fuzzy formalization, the knowledge in these areas are insufficient. The greatest practical results in the solution of above problems can be achieved by means of expert systems being in essence one of the methods of artificial intelligence.

The offered technique for constructing a mathematical model of an ecologically friendly model of "intellectual" and safe modern building with efficient use of thermal energy based on expert system allows to solve the above problems.

In this connection, we developed a hybrid expert system based on a complex exploitation of a building and that will be implemented on the base of experience and knowledge of experts in the given field and optimization methods.

The given hybrid expert system is intended for solving the problems of control regulation of thermal energy consumption problems based on combination of knowledge of experts and optimization conditions, ensuring the development of practically acceptable control formulations.

The structure of the hybrid expert system includes the following blocks (Fig. 2):

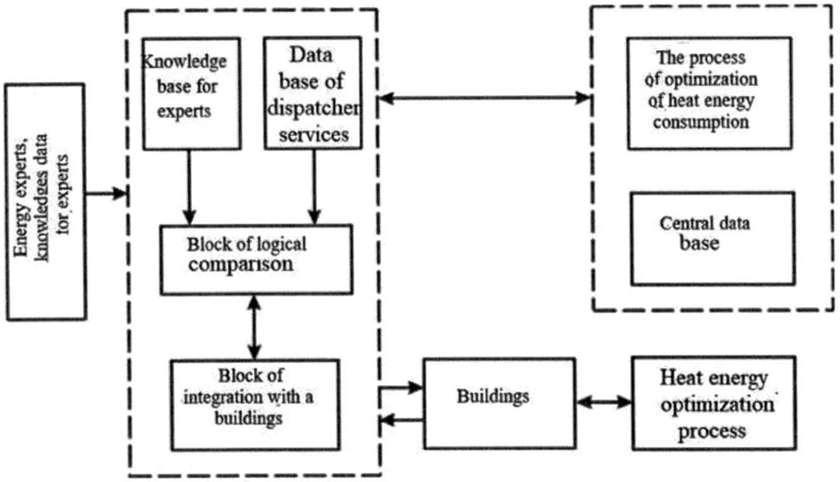

Fig. 2 Approximate scheme of the hybrid expert system of thermal energy control in modern buildings

1. Expert system. It consists of a knowledge base, database, inference block and interface unit with a user.
- Optimization calculation block.
- A dispatcher participates in a hybrid expert system to organize the thermal energy control mode. He performs the following actions:
- Filling the data file containing the current values of indicators for the thermal energy consumption, transfer of the file to the optimization block.
- Activation of optimization and control procedures.
- Interpretation of optimization and control results.
- The knowledge base contains the rules that use experts when composing optimal forecast value of thermal energy consumption.

The database of the system contains current data on thermal energy consumption. The logical comparison block implements the procedure of comparison of expert and archive data. The integration unit with a building includes a dialogue system and a subsystem of explanations on divergence of expert and archive data on thermal energy consumption of the building.

The dialogue subsystem allows to implement input and interpretation of data that can be entered from the central database. The subsystem of explanations on divergence of expert and archive data forms the text of explanation, recommends formulations based on tracing of inference. The block of optimizational calculations allows to solve a optimization problem consisting of determination of such relations between climatic parameters of inner air under which thermal energy consumption for ensuring data of climatic conditions would be optimal.

Working session with a hybrid expert system starts with input of dialogue subsystem of current data on thermal energy consumption, that are send to database. The inference block based on current data forms admissible value of consumption and admissible norms of deviation. If the value given by the system sharply differs, then the subsystem of explanations forms the reason of deviation. Based on the standard optimization package the optimization block performs calculations within admissible norms. The knowledge base contains informal expert calculations based on the construction of "abstract" plan on heat energy consumption.

3 Conclusions

According to this plan, the process of solution of the problem is decomposed into three levels. The principle on the base of which internal parameters components are chosen on each level is that qualitative and quantitative parameters of thermal energy correspond to the values required on this level. Such on approach allows to highlight the purposes in the solution of the problem, that can be implemented in less detail, in a general form. To implement these goals the procedure of formation of forecast value of thermal energy consumption can be decomposed into the following stages:

- analysis of outside air, thermal energy and inside air parameters;
- formation and definition of optimization parameters;
- formation of optimal output values of thermal energy consumption.

The system constantly comparises the output values with the contents of the knowledge base, monitors to have not been violations of the norms of marginal admissible values. When solving the problem one can return to the beginning, to correct the optimal output value and then again to solve the problem. If the obtained value sharply differs from archive data and the explanation does not satisfy the expert, he makes necessary corrections and the solution process starts over.

References

1. Tabunshikov Y, Brodach M (2002) Mathematical Modeling and Optimization of Thermal Efficiency of Buildings. ABOK – Press
2. Bogoslovskiy V (1982) Constructional thermal physics. Vysshaya Shkola
3. Zadeh L (1976) The notion of linguistic variable and its application to the adoption of approximate values. Mir, Translate from English
4. Yager R, Zadeh L (1994) Fuzzy sets, neural networks and Soft Computing (eds.) VAN Nostrand Reinhold, New York
5. Abdikeev N, Yakovenko I (1988) Dispatcher Managirp expert technological system to control continuous production. In: All Union Conference Artificial Intellegence: Abstracts of Papers. v.2. pereslavl-Zalesski, pp 261–266
6. Shostak V (1999) Control of complex objects in real time on the base of dynamical expert systems. In: Aviatsionno-kosmicheskaya tekhnika I tekhnologiya », issue 10, Kharkov, 204–210

7. Leshchenko MV, Semko V (2015) Thermal characteristics of the external walling made of cold-formed steel studs and polystyrene concrete. Mag Civil Eng 60(8):44–55. https://doi.org/10.5862/MCE.60.6
8. Aliyev F, Mehdaliyev A, Mammadov N, Aliyev F (2008) Development of mathematical model of intellectual building in Baku city. In: The Second International Conference Problems of Cybernetics and Informatics. Dedicated to the 50th Anniversary of the ICI in Azerbaijan 1, Baku, 275–278

Production Risks Assessment as a Method of Construction Industry Safety Management

Roman Pahomov, Oleksandr Zyma, Ferahim Veliyev, and Ivan Peleshko

Abstract The dynamics of occupational injuries with fatal consequences in Ukraine was considered in the article. The main reasons that significantly affect the injuries level in the construction industry and in the construction materials industry have been identified and analyzed. The results of the production process factors comprehensive survey, which allow to assess the safety state, provide a basis for the analysis of working conditions and the operational management actions development in the construction industry and in the building materials production industry were considered. The concept of risk, main stages, methods and criteria of risk assessment are defined. The main activity directions of state and branch management bodies on reduction of risks occurrence are also recommended. It is explained in detail why risk assessment as the safety management method of technological processes and production, and as a means of practical measures implementation of industrial hazards prevention or reduction should be part of a systematic approach to the labor process implementation.

Keywords Construction industry · Occupational safety · Production risks · Production traumatism · Risk assessment

1 Introduction

The construction industry is one of the most promising and profitable economic activities, but at the same time – problematic and dangerous. Analyzing the circumstances and causes of the construction site accidents, it was established that the

---

R. Pahomov · O. Zyma (✉)
National Universuty «Poltava Yuri Kondratyuk Polytechnic», Poltava, Ukraine

F. Veliyev
Azerbaijan University of Architecture and Construction, 5 Ayna Sultanova, Baku, Azerbaijan

I. Peleshko
Lviv Polytechnic National University, S. Bandera Street 12, Lviv 79013, Ukraine

V. Onyshchenko et al. (eds.), *Proceedings of the 3rd International Conference on Building Innovations*, Lecture Notes in Civil Engineering 181,
https://doi.org/10.1007/978-3-030-85043-2_27

main accidents causes at work are organizational, technical and psychophysiological causes. Organizational causes predominate among the fatal accidents causes (58, 7%). That is, employees ignore the requirements of labor protection instructions and job descriptions; violate the rules of safety and technological process. The second place in the anti-rating (26.4%) is occupied by psychophysical, man-made, natural, environmental and social causes. In third place – technical causes (14.9%) [1–3]. But it is impossible to improve the activity conditions, to increase security with the help of organizational measures alone. Certain technical measures are required. Mistakes of various kinds can cost lives, deteriorate health, and cause significant material loss. Absolute safety cannot be achieved in practice, there is always some excess risk.

Risk is a measure of expected failure, failure to act, the adverse effects risk on human health; certain phenomena that are accompanied by possible material losses. Risk is characterized by surprise, the sudden onset of a dangerous situation, which involves rapid and decisive action to eliminate or reduce the danger source impact. The risk concept combines two elements: the frequency (probability) with which a dangerous event occurs and the adverse effects consequences.

In the article, the construction industry and the construction materials industry are considered as industries in the traditional definition. The construction industry includes enterprises and companies engaged in activities related to the design, construction and operation of buildings and structures. Production risks inherent in the enterprises activities in the construction industry and the construction materials production, are determined primarily by the content of these activities. Analysis and assessment of occupational injuries are necessary conditions for the successful operation of any system, including – labor protection. Based on the dynamics, it is possible to judge to what extent the state of the system has become better or worse compared to the base period, to assess the implementation of targets and management effectiveness for the possibility of preventive measures planning. This can be done only on the basis of targets and criteria set, their numerical evaluation and comparison with set or baseline values. The determination problem of the objective quantitative indicators that characterize the labor protection state (level), production safety or hazard, the human–environment system reliability, is relevant and specific to each enterprise. However, there must be common approaches to solving it.

Risk assessment as a safety management method of technological processes and production, and as a means of practical measures implementation to prevent or reduce industrial hazards should be part of a systematic approach to the labor process implementation. With regard to practical application, it should be noted that the methodological apparatus of risk analysis and assessment is already used in various areas of economic activity. The idea of the article is to use the developed comprehensive criteria that take into account harmful production factors; justify the funds allocation to reduce occupational injuries to improve formation and evaluation methods of socio-economic measures effectiveness to improve working conditions.

2 Formulation of the Problem

The accident risks analysis and assessment methodology at the objects of the construction industry and the construction materials industry is actively developing. Therefore, the development of new and improvement of existing approaches, models and methods of accidents risk assessment, their computer implementation remains an urgent task for our state. Determination of accident risk assessment should be based on the monitoring results of the potentially dangerous objects technical condition, statistics on man-made nature accidents and emergencies, dangerous geological and hydrometeorological processes comprehensive monitoring, the natural complexes state, as well as relevant dangerous situations modeling and their public health impact.

3 Scientific Novelty

The paper further developed the scientific task of developing an effective method of applying a risk-oriented approach to the industrial hazards assessment in the construction industry and construction materials industry.

4 Analysis of Recent Studies

Analysis of risk assessment and decision-making modern methods, in conditions of uncertainty, showed that different authors consider the term «risk» differently. Most often, the concept of risk is associated with the adverse event probability.

Defining risk as the degree of a certain adverse event probability that may occur at a certain time or under certain circumstances on the high-risk facility territory and / or outside it, is to determine the accident probability at the facility for a certain period of time, as usually for a year. A.O. Imasheva uses a static method to quantify the risk, according to which the risk criteria are the average deviation and the coefficient of variation [4].

In technical systems, reliability is the property of an object to function satisfactorily for a certain time in a given mode and use or maintenance conditions [5].

Analysis of other well-known publications [6–11, 13–17] showed that although today there are some studies in the theory and practice of risk-oriented approach usage to hazard assessment. They are not enough for practical implementation in the current occupational safety management system.

5 The Aim

The aim of this article is to improve the occupational injuries analysis methodology, taking into account the accidents risks, as well as to develop measures to reduce injuries and occupational diseases.

When analyzing the enterprise activities must take into account the risks it faces. Therefore, the main tasks include the following: the study of occupational injuries existing methods; occupational injuries methods selection and improvement; identification of traumatic factors influencing the indicators of occupational injuries; study of the production risks existing methods determination.

6 The Method

The analytical method of researches is used in the work. The methodological basis of the study is a systems approach, which is based on such principles as integrity, structure, the relationship of system and environment.

7 The Main Material and Results Presentation

Construction, in comparison with the coal, chemical, socio-cultural sphere and trade, transport, etc., belongs to one of those industries that are characterized by the risks significant number probability. In addition, an important role is given to financial risks in the construction industry. These risks arise from unforeseen changes in legislation or economy. They can have a negative impact on the results, given the investment duration and the construction process capital intensity.

Currently, risk is increasingly used to assess the negative factors impact in the building materials and products production. This is due to the fact that risk as a hazards realization quantitative characteristic can be used to assess working conditions, economic losses, the number of accidents and diseases at work, as well as to form a system of social policy at the enterprise (compensation, benefits).

Quantitative risk assessment is the values assessing process of the probability and consequences of adverse events. Hazards can be realized in the form of injuries or diseases, only if the hazards formation intersects with the human activity area. In industrial conditions, it is a work area and a source of danger (one of the industrial production elements).

In modern domestic and foreign practice to formalize the risk (R) is widely used model that links the probability (P) of a negative event A (accident, catastrophe) and the probable magnitude of possible consequences (W) as a result of this event, as shown in Eq. (1):

$$R(A) = P(A) \times W(A) \quad (1)$$

The probability P (A) present in this model numerically expresses the degree of negative event A possibility realization connected with an uncertain situation. The probable magnitude of the expected consequences W (A) as a negative event A result implementation depends not only on possible losses (number of dead, injuried, material losses), but also on the object vulnerability degree to event A, determined by Eq. (2):

$$W(A) = V(A) \times U(A) \quad (2)$$

Where W(A)—the probable magnitude of the event A possible consequences implementation; V(A)—the object vulnerability degree for event A; U(A)—conditional total loss as a result of the event A implementation.

Thus, substituting expression (2) in Eq. (1), it could be obtained a model for determining the risk level:

$$R(A) = P(A) \times V(A) \times U(A) \quad (3)$$

Where R(A)—risk (negative event A); P(A)—the probability that there is a certain risk; V(A)—the probability that risk can be avoided; U(A)—a category that determines the risk severity.

Therefore, Eq. (3) is common to all types of risks, characterized by their risk scale. But in its practical use in each case there may be a need for additional research.

The criteria for the risk degree determining in the general case should be: legislation analysis; working hours timing; connection with regulations on dangerous equipment; connection with fire safety regulations; connection with environmental regulations; analysis of injuries and morbidity in the workplace (for the last 5 years); existing risk factors and their measurement (job-places certification data); available employee complaints; job review; poll; documentation (results of various inspections); data on the service life and technological equipment wear degree; staff qualification and motivation data. The values of P and U, according to the proposed Tables 1 and 2, can be quite objectively selected from statistics (Table 3).

Systematizing information about the work process circumstances that affect risk factors by workplaces groups with similar working conditions, it is necessary to begin risk assessment. The risk degree and action in this process determination it could be determined after the risk magnitude [12]:

- more than 100—risk reduction is mandatory. If due to lack of funds it is not possible to take preventive measures, then work in the danger zone is strictly prohibited;
- 85…100—work can be continued until the risk reduce or eliminate measures are taken. If the work cannot be interrupted, then measures (collective) must be taken within 1… 3 months;

Table 1 Failure consequences (category U)

Consequences, P	Description	Rank
No effect	Failure has no serious impact on the production process	0
Not important	Very small failure (damage), significantly does not affect safety and technological process	1
Small	Failure with short-term effect, risk of personal injury and emergencies are absent	2
Medium	Failure, that may pose a risk to personnel, requires safety precautions	4
Serious	Failure creates serious obstacles to work, damage to equipment, requires special protection and safety measures, more time is needed to eliminate	6
Very serious	The failure created a serious danger, possibly a serious injury or death	8
Catastrophic	The failure posed a serious threat to the health of a large number of people	10

Table 2 Possible failure rate (probability P)

Event, A	Description	Rank
Very seldom	Failure is almost impossible	0
Seldom	Failure can occur every 2…3 years	3
Medium frequency	Failure can occur once per year	5
Frequently	Failure can occur 2… 3 times per year	8
Very friquently	Failure can occur frequently, perhaps even 2…3 times per month	10

Table 3 Work process circumstances that affect risk factors (probability V)

Impact rank	Impact description
0,6	These circumstances, which fully affect the risk factor, can lead to a significant reduction
0,8	The impact of these work process circumstances on the risk factor is not as complete as possible and may lead to a partial reduction of risk
1,0	The impact of these work process circumstances on the risk factor is negligible
1,2	The impact of these work process circumstances on the risk factor is not as complete as possible in its intensity and may lead to a partial increase in risk
1,4	These circumstances, which fully affect the risk factor, can lead to a significant increase

- 55…85—risk reduce measures are needed, but they do not have to be implemented immediately, economic considerations must be taken into account. Measures should be taken at least 3… 5 months after the risk assessment;
- 25…55—medium risk, it is necessary to clarify security measures, set priorities;
- 0…25—low risk, the order of labor management organization and labor protection is necessary.

Injuries analysis by the events main type shows that most workers in the construction industry are deceleration injured; when collapsing objects, materials, rocks, soil; actions of objects and parts that move, rotate. Injuries are often caused by unsatisfactory technical condition of industrial facilities, buildings, structures, territory. The analysis allows to determine the main directions for the measures development to improve working conditions and reduce injuries in the construction industry and the construction materials industry. An important factor in reducing injuries is the industrial discipline introduction among workers.

Accidents unfavorable trend continuation at enterprises is an obstacle to the industry effective functioning as a whole. To overcome negative phenomena, it is required high-quality, systematic and purposeful activities of state and sectoral authorities. At the sectoral level, the following areas require implementation:

- professional selection system improvement for work with dangerous and harmful working conditions;
- restoration and ensuring the medical service effective functioning at the enterprise and occupational health supervision;
- system development and implementation for the pathologies prevention after the end of work in hazardous conditions, as well as workers transfer to workplaces that do not have harmful and dangerous production factors.

8 Conclusions

The methodology of accident risks assessment analysis at the construction site is actively developing, so the development of new and improvement of existing approaches, models and methods of accident risk assessment, computer implementation remains an urgent task for our country. Determination of accident risk assessments should be based on the monitoring results of potentially dangerous objects technical condition; statistics on accidents and emergencies of man-made nature; comprehensive monitoring of hazardous production factors; the state of natural complexes; as well as the results of modeling relevant situations and their impact on public health. The use of the risk indicator allows to compare the action of different nature dangerous factors, to determine, taking into account each individual factor contribution, the integrated degree of any industrial facility danger. The application of risk assessment methodology makes it possible to develop mechanisms and strategies for various regulatory measures to improve the safety of the construction industry; set limits on the variability of risk values and uncertainties associated with limited input or unresolved scientific problems.

References

1. Analysis of occupational injuries for the first half of 2020. [Electronic resource] - Access mode: https://www.sop.com.ua/news/2695-analz-virobnichogo-travmatizmu-za-pvrchchya-2020-roku.
2. Pahomov R, Zyma O, Dyachenko E (2018) Analysis and prevention of industrial injury in the construction sector. Int J Eng Technol (UAE) 7(3):285–290. https://doi.org/10.14419/ijet.v7i3.2.14421
3. Polukarov YA (2011) Analysis of construction injuries and recommendations for its reduction. In: Problems of Labor Protection, Industrial and Civil Safety: a Collection of Materials of the Sixth Scientific-Methodical Conference, pp 76–78
4. Imasheva AO (2013) Mathematical modeling in occupational safety management. Sci Vector Togliatti State Univ 2(24):283–287
5. Henley EJ, Kumamoto H (1984) Reliability of Technical Systems and Risk Assessment. 2nd edn. Moscow: Mechanical Engineering, p 528
6. Rumezhak O (2012) From violations finding to risks identifying. Res Prod J Labor Protection No.12:8–9. Kyiv
7. Zyma O, Pahomov R, Dyachenko E (2020) Analysis of emergency management methods in oil and oil-product reservoirs. In: Onyshchenko V, Mammadova G, Sivitska S, Gasimov A (eds) Proceedings of the 2nd International Conference on Building Innovations. ICBI 2019. Lecture Notes in Civil Engineering, vol. 73, pp 335–344. Springer, Cham. https://doi.org/10.1007/978-3-030-42939-3_34
8. Pichugin S, Zyma O, Vynnykov P (2016) Reliability level of the buried main pipelines linear part. Recent Progress in Steel and Composite Structures. In: Proceedings of the 13th International Conference on Metal Structures, ICMS 2016, 551–558. https://doi.org/10.1201/b21417-76
9. Naumenko V (2003) Risk-oriented approach. Emerg Situations 1:18–20
10. Abdrakhmanov NK, Shutov NV et al (2015) Research and analysis of non-stationary occurrence and development of potentially dangerous situations during dangerous production facilities operation. Electron Sci J Oil Gas Bus 1:292–306. http://ogbus.ru/issues/1_2015/ogbus_1_2015_p292-306_AbdrakhmanovNKH_ru.pdf.
11. Artemov VI (2008) To the issue of principled approaches to risk assessment in decision-making. Mod Prob Sci Educ 1:39–47. [Electronic resource] - Access mode: www.science-education.ru/20-671.
12. Kalkis V, Kristinesh I, Roya J The main directions of working environment risk assessment. Riga: SIA «Jelgavas tipogrāfija», p 76
13. Khazeev LF (2015) Assessment of production risks at the enterprise. Innov Sci 3:55–58
14. Pyzhikova NI, Titova EV, Kozlov MA (2015) Methodology for assessing financial, industrial and aggregate risk. Age Sci 4:149
15. Drobyazko S, Bondarevska O, Klymenko D, Pletenetska S, Pylypenko O (2019) Model for forming of optimal credit portfolio of commercial bank. J Manage Inf Dec Sci 22(4):501–506
16. Iasechko S, Haliantych MK, Skomorovskyi VB, Zadorozhnyi V, Obryvkina O, Pohrebniak O (2020) Contractual relations in the information sphere. Syst Rev Pharm 11(8):301–303. https://doi.org/10.31838/srp.2020.8.46
17. Dyachenko E, Zyma O, Pahomov R, Shefer O (2020) Non-crane method of reconstructing buildings with additional storey erection. In: Onyshchenko V, Mammadova G, Sivitska S, Gasimov A (eds) Proceedings of the 2nd International Conference on Building Innovations. ICBI 2019. Lecture Notes in Civil Engineering, vol 73, pp 35–44. Springer, Cham. https://doi.org/10.1007/978-3-030-42939-3_4

Design of Reinforced Concrete Members Taking into Account the Influence of Biaxial Bending

Andrii Pavlikov, Olha Harkava, and Kelvis Atembemoh

Abstract The majority of reinforced concrete members are subjected to biaxial bending due to influence of many factors of constructive, technological and operational character. As a rule, a complex stress–strain state is led to a special case. This approach leads to a distortion of the true picture of the reinforced concrete member in conditions of complex loads. One of the ways to eliminate the existing problem is to introduce into the theory of calculation the complex reinforced concrete deformation model using complete stress–strain diagrams of materials. This will allow to objectively taking into account the full range of physical and mechanical properties of concrete and reinforcement. The development of the deformation model, which would equivalently reflect the patterns of phenomena occurring during the load deformation of reinforced concrete members at the critical capacity of their components depending on various factors, is an urgent problem. Its solution will make it possible to obtain effective design solutions for structural systems of buildings, to solve problems in a closed form.

Keywords Reinforced concrete · Biaxial bending · Design · Minimal amount · Ultimate strain

1 Introduction

It is known that the phenomenon of biaxial bending is quite common in the practice of designing reinforced concrete structures and can be caused by many factors [1–3]. Nevertheless, today in the current regulations [4] its occurrence is considered only in the tasks of the second type—strength analysis. As for taking this phenomenon into account in the problems of the first type—strength design or calculating the required area of the longitudinal reinforcement, the number of studies in this direction is insufficient. Their replenishment at the current level is advisable to carry out on

A. Pavlikov · O. Harkava (✉) · K. Atembemoh
National University "Yuri Kondratyuk Poltava Polytechnic", 24 Pershotravnevyj Avenue, Poltava 36011, Ukraine
e-mail: olga-boiko@ukr.net

V. Onyshchenko et al. (eds.), *Proceedings of the 3rd International Conference on Building Innovations*, Lecture Notes in Civil Engineering 181,
https://doi.org/10.1007/978-3-030-85043-2_28

the basis of the introduction of the deformation model, equations of mechanics of deformable solids and stress–strain diagrams of materials as evidenced by works [5–7], as well as based on extreme principles [8].

In [2, 9–12], reinforced concrete members operating in biaxial bending conditions were experimentally and theoretically investigated, and proposals for determining the strength of such members based on a nonlinear deformation model are presented. Scientists solve [13–18] the problem of calculating the bearing capacity of biaxially bended members in general for sections of arbitrary configuration with and without holes, but this approach is difficult to implement and does not solve the problem of optimal reinforcement of such elements. Therefore, the development of a method for determining the required area of the longitudinal reinforcement in the normal cross section of biaxially bended columns based on the deformation model is an urgent task.

2 Main Material

The problem of taking into account the phenomenon of supercritical deformation of concrete in the calculations of the area of the longitudinal reinforcement of reinforced concrete members, which work in the conditions of biaxial bending, is considered and solved. The main purpose of the problem is to obtain analytical formulas of the main parameters of the stress–strain state of biaxially bended members according to the calculation scheme based on the deformation model. This is dictated by the fact that the presence of such formulas opens the possibility of constructing almost new methods of calculating the bearing capacity of the reinforced concrete members in their design.

In analytical studies of strength, the introduced concept of the deformation criterion of strength is used [5]. When deriving the design dependences for calculating the fiber ultimate values of strains ε_{cu} of concrete under compression both discretely and continuously distributed, a complete diagram of the physical stress–strain state of concrete under compression is used according to [4].

In order to simplify the derivation of design formulas, the problem is solved under the condition that the total area of tensile discrete bars of reinforcement with value $\sigma_s = f_{yd}$ can be reduced to the resultant force.

For the accepted type of force deformation, it is assumed that the strains in the reinforcement-equivalent are equal to ε_s at stresses $\sigma_s = f_{yd}$, and the ultimate values of strains in concrete after the phenomenon of its unstrengthening at stresses $\sigma_{c(1)}$ at the point farthest from the neutral axis are $\varepsilon_{c(1)} = \varepsilon_{cu}$.

Taking into account the accepted preconditions of nonlinear analysis according to [4], the design scheme in the plane coinciding with the Y axis will have the form shown in Fig. 1. The main parameters of deformation mode of reinforced concrete members at the time of reaching the maximum value of bearing capacity will be:

X is the neutral axis depth;

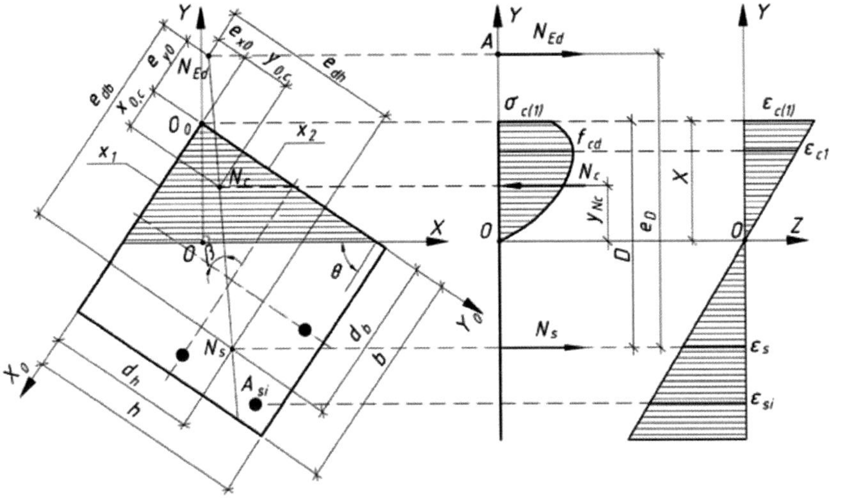

Fig. 1 Design scheme for determining the parameters of deformation mode in the calculations of the strength of biaxially bended members with a triangular form of the compressed zone

ε_{cu} is the ultimate value of strains in concrete of the compressed zone at the level of the point furthest from the neutral axis;

θ is the angle of inclination of the neutral axis to the horizontal axis of the section;

A_s is the calculated value of the area of the longitudinal reinforcement required for the perception of the external force N_{Ed}.

Given the above assumptions and simplifications, the design equilibrium equations are taken in the following general functional form:

$$\sum N_z = 0; \quad N_{Ed} - N_c\left(\eta_{c(1)}\ldots\right) + \sum_{i=0}^{n} \sigma_{si}\left(\eta_{c(1)}\ldots\right)A_{si} = 0; \tag{1}$$

$$\sum M_x = 0; \quad N_{Ed}e(X) - N_c(\eta_{c(1)}\ldots)y_{Nc}(\eta_{c(1)}\ldots) - \sum_{i=1}^{n} \sigma_{si}\left(\eta_{c(1)}\ldots\right)A_{si}y_{si} = 0; \tag{2}$$

$$\sum M_{Y0} = 0; \quad -N_{Ed}e_{y0} + N_c(\eta_{c(1)}\ldots)y_{0,Nc}(\eta_{c(1)}\ldots) - k\sum_{i=1}^{n} \sigma_{si}\left(\eta_{c(1)}\ldots\right)A_{si}\,y_{0,si} = 0; \tag{3}$$

$$\sum M_{X0} = 0; \quad -N_{Ed}e_{x0} + N_c(\eta_{c(1)}\ldots)x_{0,Nc}(\eta_{c(1)}\ldots) - k\sum_{i=1}^{n} \sigma_{si}\left(\eta_{c(1)}\ldots\right)A_{si}x_{0,si} = 0. \tag{4}$$

In Eqs. (1)–(4) $N_c(\eta_{c(1)}\ldots)$ is a function of resultant in concrete of the compressed zone;

$D_{si}(\theta) = x_{0,si}\sin\theta + y_{0,si}\cos\theta$, $y_{Nc}(\eta_{c(1)}\ldots) = y_{Nc}(\eta_{c(1)}, X, \theta)$ respectively, the effective height of the i-th reinforcing bar in the plane orthogonal to the neutral axis and the distance from the neutral axis to the point of N_c application;

$N_{Ed}e(X)$ is the functionally expressed moment from the N_{Ed} force with the arm function $e(X) = X - e_{x0}\sin\theta - e_{y0}\cos\theta$ about the coordinate axis that coincides with the neutral axis;

$\eta_{c(1)} = \varepsilon_{c(1)} / \varepsilon_{c1}$ is the level of strains in the extreme concrete fiber of the compressed zone;

$\sigma_{si}(X, \eta_{c(1)}) = \eta_{c(1)}\varepsilon_{c1}E_{si}(D_{si}-X) / X$ and $k = (D_{si}-X) / |D_{si}-X|$ are the stress function in the i-th reinforcing bar of cross section and the coefficient of the sign of the force N_{si}.

The system of equilibrium Eqs. (1)–(4) is considered together with the following preconditions:

stress distribution in concrete based on [4] is modeled by the function

$$\sigma_c = f_{cd}\left(\frac{k\eta_{c(1)}y}{X} - \frac{\eta_{c(1)}^2 y^2}{X^2}\right) \Big/ \left(1 + (k-2)\frac{\eta_{c(1)}y}{X}\right), \tag{5}$$

in which the level of reduction of elastic properties of concrete in the state of its critical deformation ($\varepsilon_c = \varepsilon_{c1}$);

$$k = E_{cd}\varepsilon_{c1}/f_{cd}, \tag{6}$$

at $k = 2$ the dependence (5) is used in the form of a parabola known as the "Hognested diagram";

the stress–strain relation in the reinforcement is accepted in the form of the two-line diagram;

theorem of the location of internal and external forces in one plane is used to obtain the dependence $\theta = f(\beta)$.

For reinforced concrete member undergoing biaxial bending and axial loading at a triangular form of the compressed area (Fig. 1) expressions for calculating the resultant N_c in (1), as well as the distance y_{Nc} from the X axis to the point of its application in (2) using function (5) at $\eta = \eta_{c(1)}y / X$ are obtained as follows:

$$N_c = \frac{f_{cd}X^2}{\sin 2\theta}\omega_1\left(k, \eta_{c(1)}\right), \tag{7}$$

$$y_{Nc} = S_c/N_c = X\frac{\varphi_1(\eta_{c(1)}, k)}{\omega_1(\eta_{c(1)}, k)}; \tag{8}$$

$$S_{cX} = \frac{f_{cd}X^3}{\sin 2\theta}\varphi_1\left(k, \eta_{c(1)}\right). \tag{9}$$

The coefficients φ_1 $(k, \eta_{c(1)})$ and ω_1 $(k, \eta_{c(1)})$ are obtained by integrating function (5) in (7)–(9) by introducing the notation $c = 1 + k\eta_{c(1)} - 2\eta_{c(1)}$:

$$\omega_1 = \frac{(k-1)^2}{(k-2)\eta_{c(1)}^2}\left[\frac{c(c-2\ln c)-1}{(k-2)^3} - \frac{\eta_{c(1)}^3}{3(k-1)^2}\right], \quad k \neq 2; \tag{10}$$

$$\varphi_1 = \frac{(k-1)^2}{(k-2)^2}\left(\frac{c(c-2\ln c)-1}{(k-2)^3\eta_{c(1)}^3} - \frac{c-2(k-1)^2-1}{6(k-1)^2}\right), \quad k \neq 2. \tag{11}$$

Expressions (7)–(9) taking into account (10)–(11) allow equilibrium Eqs. (1) and (2) for the case of a triangular compressed zone of biaxially bended members in the supercritical stage to lead to a convenient form for practical use:

$$N_{Ed} = \frac{f_{cd}X^2}{\sin 2\theta}\omega_1\left(k, \eta_{c(1)}\right) - A_s f_{yd}, \tag{12}$$

$$N_{Ed}e_D = \left(N_{Ed} + A_s f_{syd}\right)\left(D - \frac{\omega_1 - \varphi_1}{\omega_1\sqrt{\omega_1}} \cdot \sqrt{\frac{\left(N_{Ed} + A_s f_{yd}\right)\sin 2\theta}{f_{cd}}}\right), \tag{13}$$

where $e_D = D - e_{y0}\cos\theta - e_{x0}\sin\theta = e_{dh}\cos\theta - e_{db}\sin\theta$ is the eccentricity of the force N_{Ed} with coordinates e_{x0}, e_{y0} about the point of application of resultant N_s in the tensile reinforcement in the direction of the effective height $D = d_b\sin\theta + d_h\cos\theta$;

e_{dh} is the eccentricity of the force N_{Ed} about the point of application of the resultant N_s in the tensile reinforcement in the direction of the effective height d_h;

e_{db} is the eccentricity of the force N_{Ed} about the point of application of the resultant N_s in the tensile reinforcement in the direction of the effective width d_b.

As the analysis of the last two Eqs. (12) and (13) shows for their solution with respect to the unknowns A_s, X, θ and $\eta_{c(1)}$ it is necessary to receive two additional dependences.

To derive the required dependence $\theta = f(\beta)$, the theorem on the location of resultant internal N_s and N_c and external N_{Ed} forces in one plane is used to calculate the angle θ of inclination of the neutral axis. Applying this condition in the coordinate system $X_0O_0Y_0$ (Fig. 1) a ratio is written

$$\frac{\sum_{i=1}^{n} x_{0,si}A_{si}\sigma_{si} \Big/ \sum_{i=1}^{n} A_{si}\sigma_{si} - x_{0,c}}{\sum_{i=1}^{n} y_{0,si}A_{si}\sigma_{si} \Big/ \sum_{i=1}^{n} A_{si}\sigma_{si} - y_{0,c}} = \tan\beta, \tag{14}$$

where $x_{0,si}$, $y_{0,si}$ are the coordinates of each of the reinforcing bars, respectively, in the direction of the axes X_0 and Y_0;

$x_{0,c}$, $y_{0,c}$ are the coordinates of the point of application of the compressed zone resultant in concrete in the coordinate system $X_0O_0Y_0$ when the biaxially bended reinforced concrete member reaches the supercritical state.

Experimental-theoretical researches of degree of influence of various factors, which are a part of (14), on values θ, were carried out. The values of all parameters of deformation mode of reinforced concrete columns during their operation in the supercritical stage were calculated. The graphs constructed for four series of experimental columns with the same cross-sectional dimensions 350 × 250 mm are shown in Fig. 2. Analysis of the data of tables and graphs presented in Fig. 2 showed that the value of the angle of inclination of the neutral axis is practically not affected by the parameter k, which is the level of reduction of elastic properties of concrete in the state of critical deformation.

Each series included from 9 to 12 columns, equally reinforced in the tensile zone within the series. Each series of columns consisted of 3 groups of columns 3 or 4

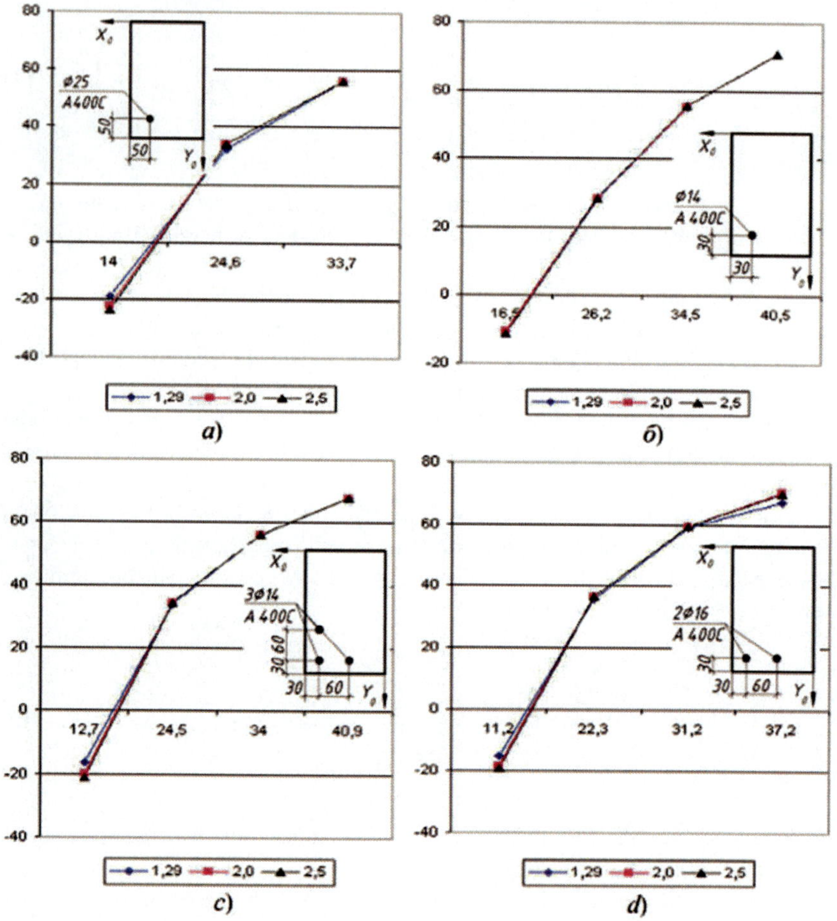

Fig. 2 Graphs θ—β (β—horizontally) for columns with coefficients in each series $k = 1.29$; 2.0; 2.5 and reinforcement: ***a***—1st series; ***b***—2nd series; ***c***—3rd series; ***d***—4th series

columns in a group. The first series included 9 columns reinforced with 1Ø25A400C ($d_h = 300$ mm, $d_b = 200$ mm). The second series included 10 columns reinforced with 1Ø14A400C ($d_h = 320$ mm, $d_b = 220$ mm). The third series included 12 columns reinforced with 3Ø14A400C ($d_{h1} = 260$ mm, $d_{b1} = 220$ mm; $d_{h2} = 320$ mm, $d_{b2} = 220$ mm; $d_{h3} = 320$ mm, $d_{b3} = 160$ mm). The fourth series included 12 columns ($d_h = 320$ mm, $d_b = 220$ mm; $d_{h1} = 320$ mm, $d_{b1} = 160$ mm), reinforced with 2Ø16A400C (Fig. 2).

In each series of columns the physical and mechanical characteristics of concrete are changed corresponding to one (fixed) value of the coefficient of elastic–plastic properties of concrete: $k = 1.29$ ($f_{cd} = 25$ MPa, $E_{cd} = 17.00 \times 10^3$ MPa, expanded clay concrete); $k = 2.0$ ($f_{cd} = 25$ MPa, $E_{cd} = 26.32 \times 10^3$ MPa, fine-grained concrete); $k = 2.5$ ($f_{cd} = 25$ MPa, $E_{cd} = 32.89 \times 10^3$ MPa, heavy concrete). In this case, for all columns $f_{yd} = 420$ MPa, $E_s = 20 \times 10^4$ MPa. The position of the longitudinal force was changed by its application along the contour in the system of coordinate axes Y_0 and X_0 by eccentricities e_{x0} and e_{y0} [5]. From the analysis of graphs (Fig. 2) it follows that for each group of columns within the series the value of the coefficient k has little effect on changing the value of the angle θ (inclination of the neutral axis) depending on the angle β (inclination of the plane of external load).

This conclusion opens the way to quickly obtain a generalized formula for calculating the angle θ taking into account the properties of concrete in the supercritical state. To implement it, the simplest way to perform algebraic transformations over functions without the use of boundary transitions is to take $k = 2$ in the coordinate system $X_0Y_0Z_0$.

It is obtained that the coordinates of the point of application of the force N_c

$$y_{0,c} = \frac{X}{\cos\theta}\frac{\varphi_{1,\text{Yo}}}{\omega_1}, \tag{15}$$

$$x_{0,c} = \frac{X}{\sin\theta}\frac{\varphi_{1,\text{Xo}}}{\omega_1}. \tag{16}$$

To determine the parameters of the position of the neutral axis X and θ from (14) and the equilibrium equations taking into account expressions (7), (15), (16) to solve strength problems, the dependences are obtained

$$X = \frac{a\sin\theta}{(1 - \tan\theta\tan\beta)u_1}, \tag{17}$$

$$\sin 2\theta = \frac{f_{cd}X^2}{A_s f_{yd} e_{dh}}\omega_1 e_{c,Ed}. \tag{18}$$

In Formulas (17) and (18) the notation is accepted: $u_1 = \varphi_{1,X0} / \omega_1$; $\varphi_{1,X0} = \varphi_{1,Y0}$; $a = e_{x0} - e_{y0}\tan\beta = d_b - d_h\tan\beta$; e_{x0}; e_{y0} are taken into account as coordinates; $e_{c,Ed}$ is the eccentricity of the force N_{Ed} about the force N_c in the direction of the Y_0 axis.

When solving problems of the first type (reinforcement selection) to determine the parameter θ, the equation is derived from (17) by replacing the parameter X with an expression through strains according to the hypothesis of plane sections at optimal reinforcement $\varepsilon_s = \varepsilon_{s0} = f_{yd} / E_s$ and $\varepsilon_{c(1)} = \varepsilon_{cu}$:

$$X = \frac{\eta_u \varepsilon_{c1} D}{\varepsilon_{s0} + \eta_u \varepsilon_{c1}}. \tag{19}$$

Such a substitution leads to a quadratic equation, the solution of which with respect to $\tan\theta$ at $a \geq 0$ gives the desired dependence

$$\tan\theta = \frac{a(ku_1 - 1) + \sqrt{a^2(ku_1 - 1)^2 + 4k^2u_1^2 d_b d_h \tan\beta}}{2d_b k u_1 \tan\beta}. \tag{20}$$

In (20) coefficient

$$k = \eta_u \varepsilon_{c1} / (\varepsilon_{s0} + \eta_u \varepsilon_{c1}). \tag{21}$$

In order to obtain the dependences of determining the deformation mode parameters of reinforced concrete members in design problems, it is proposed to use the criterion of the minimum number of reinforcement $A_{s,\min}$ in the cross section at which in the supercritical state the reinforced concrete is able to perceive maximum force. This criterion is written as.

$$A_s(\eta_u) = \min A_s(\eta_{c(1)}), \tag{22}$$

In the expression (22), the value of the level of relative strains η_u must also satisfy the condition of the extreme criterion of strength of the considered section, i.e.

$$N_{Rd}(\eta_u) = \max N_{Rd}(\eta_{c(1)}). \tag{23}$$

Therefore, to obtain the necessary dependences the Eq. (2), after excluding from it $\sigma_c(\eta_{c(1)})$ using (5), is represented as:

$$N_{Ed} / (f_{cd} A_c) = \left(k\eta_{c(1)} - \eta_{c(1)}^2\right) / (1 + (K - 2)\eta_m) + \eta_{c(1)} B \rho_l\left(\eta_{c(1)}\right), \tag{24}$$

where $B = E_s \varepsilon_{c1} / f_{cd}$.

From (24) the functional dependence necessary for researches between quantity of reinforcement A_s (or reinforcement ratio ρ_l) and level of strains $\eta_{c(1)}$, which is unknown value, is obtained for a supercritical condition of section. The desired dependence is convenient for extreme research and has the form:

$$\rho_l\left(\eta_{c(1)}\right) = \frac{1}{B}\left(\frac{N_{Ed}}{A_c f_{cd} \eta_{c(1)}} - \frac{k - \eta_{c(1)}}{1 + (k - 2)\eta_{c(1)}}\right). \tag{25}$$

Using the concept of the minimum number of reinforcement criterion (22) and taking into account that this criterion exists for the values of $\eta_{c(1)} = \eta_u$, which will correspond to the equation $d\rho_l(\eta_{c(1)})/d\eta_{c(1)} = 0$, from (25) after mathematical transformations it is obtained

$$\left((k-2)^2 - C(k-1)^2\right)\eta_u^2 + 2(k-2)\eta_u + 1 = 0, \tag{26}$$

where $\eta_u = \varepsilon_{cu}/\varepsilon_{c1}$, $C = A_c f_{cd}/N_{Ed}$.

The solution of Eq. (26) has the form of a formula

$$\eta_u = \left((k-1)\sqrt{C} - (k-2)\right)^{-1}, \tag{27}$$

which allows in practical calculations to determine the ultimate values of compression strains of concrete in the supercritical state of reinforced concrete members, which are subjected to axial load by N_{Ed}.

The derived Formula (27) can be represented by a diagram of the ultimate values of the levels of fibre strains of concrete in the form of the function $\eta_u = f\ (k, A_c f_{cd}\ /\ N_{Ed})$, which is convenient to implement in a three-dimensional graphic (Fig. 3) image in the practice of design reinforced concrete members under axial load with random eccentricities.

The obtained formulas allow to design a cross section of reinforced concrete members with the minimum amount of reinforcement by substituting the values of η_u calculated by (27) into Formula (25) instead of the value of $\eta_{c(1)}$.

The obtained formula makes to solve problems of the first type, i.e. in the problems of calculating the minimum number of reinforcement A_s when designing reinforced concrete members on the action of compressive force N_{Ed} with any physical and mechanical characteristics of concrete:

$$\rho_{l,\min} = \frac{2\sqrt{C}(k-1) - kC - (k-2)}{BC}, \tag{28}$$

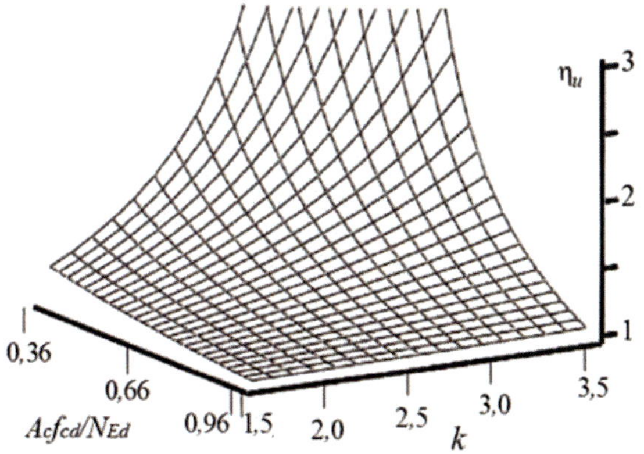

Fig. 3 Diagram of the ultimate values of the levels of strains of concrete $\eta_u = f\ (k, A_c f_{cd}/N_{Ed})$ by (27) under axial load and biaxial bending: $k = E_{cd}\varepsilon_{c1}/f_{cd}, f_{cd} = 8.5\ldots60$ MPa, $\rho_l = 1\ldots3\%; f_{yd} = 225\ldots500$ MPa.

where $C = A_c f_{cd} / N_{Ed}$ is the level of reloading of the concrete section of reinforced concrete members with the longitudinal force N_{Ed}, for the perception of which it is necessary to ensure the minimum amount of reinforcement.

3 Conclusions

Developed on the basis of the model of stress–strain state in the supercritical stage for biaxially bended reinforced concrete members, the method of determining the area of longitudinal reinforcement can be implemented as algorithms in creating optimization programs for the design of building structures. To solve the problem of designing reinforced concrete structures operating in individual cases of biaxial bending, the equation for determining the minimum amount of reinforcement is derived depending on the physical and mechanical characteristics of materials. On the example of some cases of biaxial bending of reinforced concrete members in the supercritical stage, the solution of the equations of the mathematical nonlinear deformation model of the stress–strain state with respect to the basic parameters is obtained in a closed form, which confirms the universality and generalization of the proposed physical model.

References

1. Pavlikov AM, Mykytenko SM, Hasenko AV (2018) Effective structural system for the construction of affordable housing. Int J Eng Technol 7(3.2):291–298
2. Baryliak B, Pavlikov A, Harkava O (2020) General method of structural analysis of reinforced concrete columns under axial load and biaxial bending. In: Proceedings of the 2020 Session of the 13th fib International PhD Symposium in Civil Engineering, pp 179–186, Paris, France
3. Semko OV, Hasenko AV, Kyrychenko VA, Sirobaba VO (2020) The rational parameters of the civil building steel frame with struts. In: The 2nd International Conference on Building Innovations. ICBI 2019. LNCE, vol 73, pp 235–243. Springer, Cham
4. EN 1992-1-1:2004, Eurocode 2 (2006) Design of Concrete Structures – Part 1-1: General rules and rules for buildings. Model code for service life design. Fédération Internationale du Béton (fib), Lausanne, Switzerland
5. Pavlikov AM (2007) Neliniina model napruzheno-deformovanoho stanu kosozavantazhenykh zalizobetonnykh elementiv u za krytychnii stadii. PoltNTU, Poltava, Ukraine
6. Pavlikov A, Harkava O (2020) Structural deformability of concrete. In: Concrete Structures for Resilient Society: Proceedings of the fib Symposium 2020, pp 449–455, China
7. Pavlikov A, Kochkarev D, Harkava O (2020) Analysis of eccentrically loaded members of circular cross section by nonlinear deformation model. In: The 2nd International Conference on Building Innovations. ICBI 2019. LNCE, vol 73, pp 171–181. Springer, Cham
8. Pohribnyi V, Dovzhenko O, Maliovana O (2018) The ideal plasticity theory usage peculiarities to concrete and reinforced concrete. Int J Eng Technol 7(3.2):19–26
9. Bonet JL, Barros MHFM, Romero ML (2006) Comparative study of analytical and numerical algorithms for designing reinforced concrete sections under biaxial bending. Comput Struct 84(31–32), 2184–2193

10. Pavlikov AM, Harkava OV, Prykhodko YO, Baryliak BA (2018) Experimental and theoretical testing results of reinforced concrete columns under biaxial bending. Int J Eng Technol 7(4.8):145–151
11. Pallarés L, Bonet JL, Miguel PF, Prada MAF (2008) Experimental research on high strength concrete slender columns subjected to bending and biaxial bending forces. Eng Struct 30(7):1879–1894
12. Chang S-Y (2010) Experimental studies of reinforced concrete bridge columns under axial load plus biaxial bending. J Struct Eng 136(1(12)):12–18
13. Bonet JL, Romero ML, Miguel PF (2011) Effective flexural stiffness of slender reinforced concrete columns under axial forces and biaxial bending. Eng Struct 33(3):881–893
14. Jin-Keun K, Sang-Soon L (2000) The behavior of reinforced concrete columns subjected to axial force and biaxial bending. Eng Struct 22(11):1518–1528
15. Sfakianakis MG (2002) Biaxial bending with axial force of reinforced, composite and repaired concrete sections of arbitrary shape by fiber model and computer graphics. Adv Eng Softw 33(4):227–242
16. Piskunov VG, Gorik AV, Cherednikov, VN (2000). Modeling of transverse shears of piecewise homogeneous composite bars using an iterative process with account of tangential loads 2. resolving equations and results. Mech Compos Mater 36(6):445–452. https://doi.org/10.1023/A:1006798314569
17. Kochkarev D, Azizov T, Galinska T (2018). Bending deflection reinforced concrete elements determination. Paper presented at the MATEC Web of Conferences, 230. https://doi.org/10.1051/matecconf/201823002012
18. Piskunov VG, Goryk AV, Cherednikov VN (2000). Modeling of transverse shears of piecewise homogeneous composite bars using an iterative process with account of tangential loads. 1. construction of a model. Mech Compos Mater 36(4):287–296. https://doi.org/10.1007/BF02262807

Strength of Masonry Indirectly Reinforced by Expanded Steel Sheets

Andrii Pavlikov, Nataliia Pinchuk, and Taras Kachan

Abstract The results of experimental investigations of the compressed masonry elements strength reinforced by expanded steel sheets are presented. These experiments make it possible to evaluate the efficiency of using this type of indirect reinforcement for brick structures that work on compression. Experimental investigations have shown that the use of high-tech reinforcement products—expanded steel sheets for the reinforcement of masonry is technically possible and economically feasible. With the same power of indirect reinforcement by meshes or sheets, the strength of elements reinforced with expanded steel sheets is on average 40% higher than analogues reinforced with welded mesh of BpI wire. Ease of manufacture and high strength make it possible to predict the widespread use of these reinforcement products in the future.

Keywords Masonry · Compression · Indirect reinforcement · Expanded steel sheets

1 Introduction

Natural stone structures were built in the Stone Age. To this day, partially preserved buildings made of natural stone, built in the Stone Age from large depths and slabs—dolmens. With the development of society and the improvement of the means of production instead of large heavy stones began the use of convenient for manual masonry hewn stones. Later, the material for the stone structures began to be artificial coarsely formed blocks of clay, followed by raw bricks and fired bricks.

A. Pavlikov · N. Pinchuk (✉)
National University "Yuri Kondratyuk Poltava Polytechnic", Pershotravnevyj Avenue 24, Poltava, Ukraine

T. Kachan
Poltava Scientific Research Forensic Center of the Ministry of Internal Affairs of Ukraine, Lane Rybalsky, 8, Poltava, Ukraine

V. Onyshchenko et al. (eds.), *Proceedings of the 3rd International Conference on Building Innovations*, Lecture Notes in Civil Engineering 181,
https://doi.org/10.1007/978-3-030-85043-2_29

Nowadays, there is a steady trend to increase the use of brick structures in construction, due to such characteristics of brick structures as high strength, fire resistance, durability, low operating costs, architectural expressiveness, accessibility for all consumers.

However, despite all the advantages of stone structures, there are situations when they can not meet the requirements that apply to them [1].

So, at construction of apartment houses the question of increase of illumination of the room quite often arises. In such cases, increase the area of window openings, thereby reducing the area of the openings. And if this house is also multi-storey, then there is a problem of increasing the strength of such a structure, because the brickwork of the walls, especially the first floor, will not withstand the actual load.

If we consider public buildings made of brick, then their construction also causes similar problems.

As a rule, on the ground floor of a public building a completely different planning solution is used than in a residential building. Such a building should have a large and spacious lobby or hall. The number of floors of such a building can be significant. To implement such a project, it is necessary that the load instead of load-bearing walls is perceived by separate columns, the cross-sectional area of which is controlled by the requirements for the spatial planning solution. And again there is a question of bricklaying strength increase for a given cross-sectional area of the element.

In such cases (but others are possible) to increase the strength of masonry, it is possible to use its indirect reinforcement [2, 4, 9, 10]. The vast majority of stone structures in operation work on compression—central or off-center. In this type of work, the destruction of the masonry is due to the loss of stability formed after its cracking of the columns, so the strength of the masonry, even with a very strong solution, is always less than the compressive strength of the brick.

Indirect reinforcement is used to increase the load-bearing capacity of compressed masonry elements. In addition, such reinforcement ensures the joint work of individual parts of buildings, significantly increases the seismic resistance of masonry structures [6–8].

1.1 Experimental Research of Masonry Strength Indirectly Reinforced by Expanded Steel Sheets

Experimental studies of the bearing capacity and deformability of indirectly reinforced masonry elements have been carried out at the National University «Yuri Kondratyuk Poltava Polytechnic» for the introduction of expanded steel sheets reinforcement in construction practice.

The experimental part of the work consisted of testing masonry columns reinforced by expanded steel sheets on central compression in order to further comparison the experimental data with the data obtained for analogues (twins), which are reinforced with welded meshes.

1.2 Characteristics of Test Specimens and Test Equipment

In accordance with the purpose and objectives of the study, in the Reinforced Concrete and Masonry Structures laboratory were made 3 groups of test specimens.

The test specimen of the first group (C1) was a prismatic non-reinforced masonry column with cross section $b \times h = 380 \times 380$ mm, working height 770 mm. This sample was made to determine the strength of non-reinforced masonry.

The test specimens of the second group (C2-1, C2-2, C2-3, C2-4, C2-5) were reinforced masonry columns with cross section $b \times h = 380 \times 380$ mm, working height 770–850 mm.

Samples of the second group were reinforced with meshes C1, C2, C3, C4 made from wire BpI Ø 4 and 3 mm (Fig. 1a), the size of the mesh cell was 60 and 120 mm, and the step of the grids in the central part of the samples was $S = 310$ mm or 400 mm.

The coefficient of indirect reinforcement by meshes is determined by the formula

$$\mu_{S,xy} = \frac{n_x \cdot A_{SX} \cdot l_x + n_y \cdot A_{Sy} \cdot l_y}{A \cdot S}, \tag{1}$$

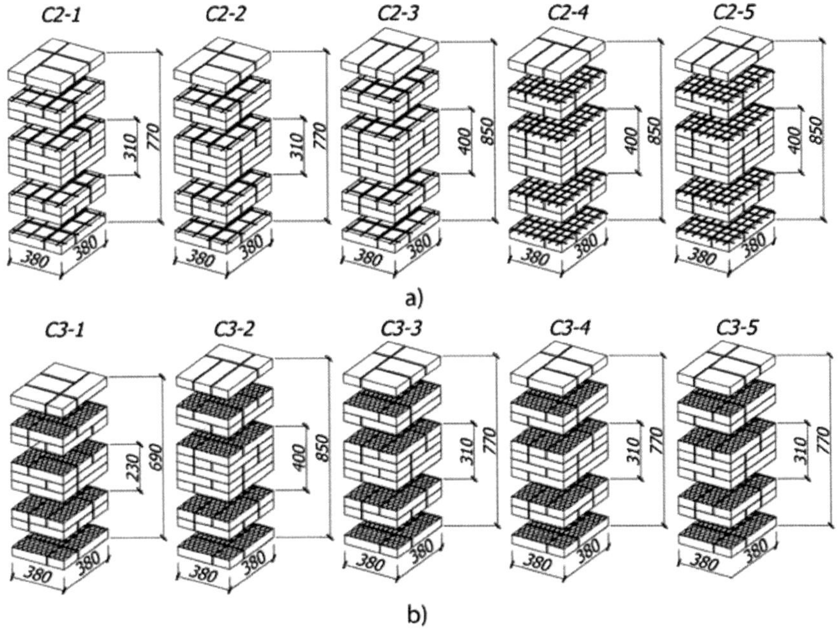

Fig. 1 Schemes of experimental samples reinforcement: **a** samples with traditional reinforcement; **b** samples with expanded steel sheets

where n_x, n_y; A_{sx}, A_{sy}; l_x, l_y – respectively, the number of bars, area and their length in the appropriate directions (within the brickwork);

A—the cross-sectional area of the element;

S—step of meshes.

The test specimens of the third group (C3-1, C3-2, C3-3, C3-4, C3-5) were in the form of reinforced masonry columns with cross section $b \times h = 380 \times 380$ mm, working height 690–850 mm.

Samples of the third group were reinforced by expanded sheets marks PVL 202, PVL 204, PVL 082 (Fig. 1b) according to [3].

The coefficient of expanded steel sheets indirect reinforcement was estimated by expression

$$\mu_{S,xy} = \frac{P_M \cdot \rho_m}{A \cdot S}, \quad (2)$$

where P_M—weight of expanded steel sheet 380 × 380 mm;

$\rho_m = 7,85\ g/\text{mm}^3$—density of steel;

A—the cross-sectional area of the element;

S—step of meshes.

Step of the expanded steel sheets $S = 230, 310, 400$ mm was selected so that the corresponding samples of the 2nd and 3rd groups in terms of indirect reinforcement power $\mu_{S,xy} \cdot \sigma_y$ were about the same (were "twins"). Thus, 5 pairs of prototype twins were created.

For all test specimens, clay brick of plastic molding with a strength of M75 produced by the Poltava plant "Ceramics" was used.

The mortar for masonry was prepared using sand from the Poltava quarry (with a coarse modulus up to 1.0 mm) and Portland cement M400 from the Balaklia Cement Plant. At the time of testing, the strength of the mortar was M80.

Control samples were taken made to control the strength of brick, mortar, BpI wire and steel sheets. Evaluation of the strength of brick, mortar, wire, sheet steel was performed according to standard methods, in accordance with regulatory requirements (Table 1).

The test specimens (brick columns) were subjected to a short-term load on the central compression. The tests were performed on a 500-ton hydraulic press type PG-500. Transportation and installation of experimental samples on the bottom plate of the press was performed by a crane-beam.

The installation of experimental samples in the press was reduced to the coincidence of the dimensional axes with the axis of the hinges and ended with the clamping of the sample between the plates (traverses) of the experimental equipment. Symmetrically located and centered brick columns were subjected to uniform pressure of the slab of the experimental mechanism.

Table 1 Parameters of experimental samples

Sample mark	Dimensions of the cross section, mm $b \times h$	Height of the sample, mm	Type of reinforcement	Reinforcement step in the central part of the sample S, mm	Reinforcement ratio $\mu_{s,xy}$, MPa
1	2	3	4	5	6
C2-1	380 × 380	770	meshes C1	310	0,00,086
C3-1	380 × 380	690	PVL 202	230	0,0013
C2-2	380 × 380	770	meshes C2	310	0,00,048
C3-2	380 × 380	850	PVL 202	400	0,00,075
C2-3	380 × 380	850	meshes C2	400	0,00,037
C3-3	380 × 380	770	PVL 082	310	0,00,088
C2-4	380 × 380	850	meshes C3	400	0,0012
C3-4	380 × 380	770	PVL 204	310	0,00,167
C2-5	380 × 380	850	meshes C4	400	0,00,065
C3-5	380 × 380	770	PVL 202	310	0,00,097

To measure the compression deformation of the brickwork on the two adjacent flat boards of the samples were installed mechanical indicators of the watch type brand 6 PAO-LISI with a division price of 0.01 mm. In addition to this function, the indicators allow you to "center" the research samples so that the line of force application applications with the geometric axis of the sample.

Loading of test specimens was carried out in stages, the value of which in fractions of the expected destructive load N_u was $0,1N_u$. After each stage of the load, the instrument readings were recorded twice: the first time immediately after the application of the load, the second time—after 5–10 min of exposure under load.

During the test process, the log of observations was recorded by devices at each stage of loading, the appearance and nature of the development of cracks, the destructive load were recorded.

To avoid premature destruction of the masonry from crumpling under the supports of the press, the supporting parts of the samples were reinforced with metal clamps.

The criterion for the destruction of the sample was taken as the inability of the test column to perceive the increasing load (Fig. 2).

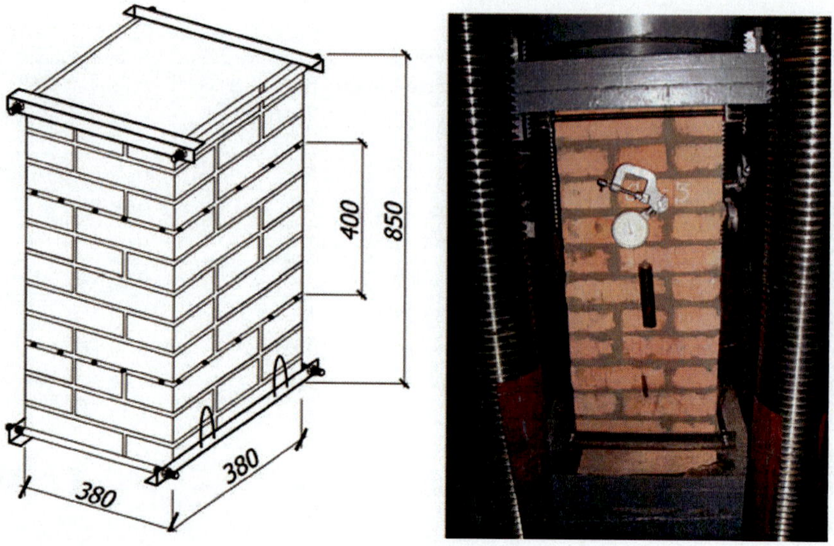

Fig. 2 Design and general view of the sample in the compression machine PG-500

2 Results of Experimental Research

2.1 *The Change in the Stress State and its Effect on the Nature of the Destruction of the Test Specimens*

Experiments have shown that at low load levels ($N/N_u = 0.1$–0.4) the stress state of reinforced masonry is close to non-reinforced. Deformations of a laying are proportional to force N, formation of longitudinal cracks (cracks of separation) is not observed.

As the load increases, structural changes occur in the masonry material (caused by its inhomogeneity), which lead to an increase in transverse deformations of the masonry. The development of these deformations is actively hindered by indirect (transverse) reinforcement, as a result of which tangential stresses τ_{iy} occur on the contact-reinforcement contact planes, which cause tensile deformations of the mesh rods or plates. Given that the perimeter of the solution "mortar—plate" is much larger than the perimeter of the contact "mortar—bar", the expanded steel sheets are much more actively involved in tensile work.

At the load level $N/N_u = 0.4$–0.99, the stress state of the masonry becomes triaxial (ie, $\sigma_1 > \sigma_2 = \sigma_3$ for sheet reinforcement $\sigma_1 > -\sigma_2 = -\sigma_3$).

a b c

Fig. 3 The nature of the samples destruction: **a**—non-reinforced sample (C1); **b**—the sample is reinforced with welded mesh (C2-4); **c**—the sample reinforced by expanded steel sheets (C3-4)

Due to the formation of a three-axis stress state (clamp effect [5, 7, 8]), the stresses in the masonry significantly exceed the corresponding values characteristic of non-reinforced masonry under uniaxial compression.

In the boundary stage $N/N_u = 0.9–0.99$ the process of viscoplastic deformation takes place, the characteristic features of which are the formation of new and opening of existing vertical cracks, the appearance of local places of destruction of brick and mortar (Fig. 3b,c). The fracture process is slow and the ultimate load as a result of stress redistribution is perceived by the sample for 3–8 min.

The process of samples destruction with indirect reinforcement is much slower than samples without reinforcement (where it can be sudden).

Experiments show that even a slight transverse reinforcement of the test specimens leads to a significant increase in the strength of the reinforced masonry.

Comparing the experimental strength of mesh-reinforced elements with analogues where expanded steel sheets was used, it can be argued that the reinforcement of expanded steel sheets masonry is more effective than grids with the same reinforcement strength. For example, the increase in the strength of the masonry of the sample C3-1 is 47% greater than that of the sample C2-1, reinforced with mesh.

We explain the best effect from the use of expanded steel sheets by the significant influence of tangential stresses τ_{ij} acting on the contact "mortar—sheet surface", increasing the tensile deformation of sheet steel and the completeness of the stress diagram (Table 2).

Table 2 Characteristics of test specimens and evaluation of the masonry strength

Sample mark	Type of reinforcement	Reinforcement ratio $\mu_{s,xy}$	Reinforcement power, $\mu_{S,xy} \cdot \sigma_y$, MPa	Strength of reinforced masonry, f_{sk}, MPa	Indicator of reinforcement efficiency for samples with expanded steel sheets
1	2	3	4	5	6
C2-1	meshes C1	0,00,086	0,344	3,15	2,01
C3-1	PVL 202	0,0013	0,325	3,92	
C2-2	meshes C2	0,00,048	0,192	3,74	1,21
C3-2	PVL 202	0,00,075	0,188	4,02	
C2-3	meshes C2	0,00,037	0,148	3,10	1,14
C3-3	PVL 082	0,00,088	0,220	3,67	
C2-4	meshes C3	0,0012	0,480	4,05	1,16
C3-4	PVL 204	0,00,167	0,420	4,08	
C2-5	meshes C4	0,00,065	0,260	4,24	1,51
C3-5	PVL 202	0,00,097	0,240	5,01	
				Average	1,41

3 Conclusions

The experimental investigations showed that due to the formation of the triaxial stress state (clamp effect) the stresses in the masonry significantly exceed the corresponding values characteristic of non-reinforced masonry under uniaxial compression. The process of destruction of samples with indirect reinforcement is much slower than samples without reinforcement (where it can be sudden).

Even a slight transverse reinforcement of the prototypes leads to a significant increase in the strength of the reinforced masonry. Comparing the experimental strength of mesh-reinforced elements with analogues where expanded steel sheets was used, it can be argued that the reinforcement of expanded steel sheets masonry is more effective than meshes with the same reinforcement strength.

The use of high-tech reinforcing products expanded steel sheets for reinforcing masonry is technically possible and economically feasible. This is explained by the fact that with the same power of indirect reinforcement by grids or sheets, the strength of elements reinforced with expanded steel sheets is on average 40% higher than analogues reinforced with welded mesh of BpI wire.

The economic effect of using expanded steel sheets compared to non-reinforced brickwork is more than 10%.

Ease of manufacture and high strength make it possible to predict the widespread use of these reinforcement products in the near future.

References

1. EN 1996–1–1 Eurocode 6 (2005) Design of masonry structures - Part 1–1: General rules for reinforced and unreinforced masonry structures
2. Pinchuk N, Fenko O, Kyrychenko V (2018) The stress-strain state computer modelling of reinforced masonry under local loading. Int J Eng Technol 7(3.2):316–321. https://doi.org/10.14419/ijet.v7i3.2.14427
3. TU 5262–001–23082253–96 Technical conditions. Expanded steel sheets, p 16
4. Pavlikov A, Kochkarev D, Harkava O (2020) Analysis of eccentrically loaded members of circular cross section by nonlinear deformation model. In: Proceedings of the 2nd International Conference on Building Innovations. ICBI 2019. Lecture Notes in Civil Engineering, (73), pp 171–181
5. Pinchuk N, Byba V (2020) Experimental investigation of masonry and reinforced masonry walls under local loading. In: Proceedings of the 2nd International Conference on Building Innovations. ICBI 2019. Lecture Notes in Civil Engineering, (73), pp 205–213 (2020).https://doi.org/10.1007/978-3-030-42939-3_22
6. Sementsov SA (1959) Local and extra-center compression of concrete and masonry. Struct Mech Calculation Struct 1(1):11–19
7. Shapoval SL (2005) Stress-strain state and strength of brickwork at local compression: Dis. … Cand. tech. Sciences: 05.23.01 Poltava National Technical Yuri Kondratyuk University. Poltava: PNTU
8. Baryliak B, Pavlikov A, Harkava O (2020) General method of structural analysis of reinforced concrete columns under axial load and biaxial bending. In: Proceedings of the 2020 Session of the 13th fib International PhD Symposium in Civil Engineering, pp 179–186
9. Pavlikov A, Kosior-Kazberuk M, Harkava O (2018) Experimental testing results of reinforced concrete beams under biaxial bending. Int J Eng Technol 7(3.2):299–305 (2018)
10. Dharek MS, Raghunath S, Ashwin CP (2021) Experimental behaviour of unreinforced and reinforced concrete block masonry walls under uniaxial compression. Materials Today: Proceedings, ISSN 2214-7853 (2021). https://doi.org/10.1016/j.matpr.2021.01.398

Statistical Strength Characteristics of Building Structures Materials

Sergii Pichugin

Abstract The strength of material is an important parameter of the load-bearing capacity of building structures. Therefore, an objective assessment of the strength of building structures materials is of great importance for ensuring and calculating the reliability of structures and the proper justification of design standards. The temporary resistance and other mechanical characteristics of building materials have a statistical variance, which is well described by normal law, which has been repeatedly confirmed by test data of material samples. Relevance of regular statistical studies of material strength is linked to the constant revision of design standards. Factory tests of building material strength are performed for many years on a large scale, creating a significant array of statistical information. However, there is no common information database for these data. Some of them have been published in various scientific and technical journals, collections of articles, conference proceedings. Access to these publications is difficult. The translation into electronic form has taken place only for publications published after 2000. The article contains a systematic review of publications in leading scientific and technical journals on the problem of statistical description of the strength of structural materials such as concrete, brick, mortar, brickwork, wood. The main attention is paid to the selection of statistical strength characteristics of materials of different periods, such as mathematical expectation, standard deviation (standard), coefficient of variation, etc. Statistical approach took into account the statistical variability of material strength, allowed to reasonably control the resistance, identify total reserves and compare them for different structures on a common basis. The data presented in the article are intended for use in numerical calculations of structural reliability. In addition, the evolution of design standards for building structures is analyzed in the sense of justification of normative and design resistances and the involvement of experimental statistics.

Keywords Strength of concrete · Strength of masonry · Strength of wood · Temporary resistance · Normative resistance · Design resistance · Coefficient of homogeneity

S. Pichugin (✉)
National University "Yuri Kondratyuk Poltava Polytechnic", Pershotravnevyj Avenue, 24, Poltava, Ukraine

V. Onyshchenko et al. (eds.), *Proceedings of the 3rd International Conference on Building Innovations*, Lecture Notes in Civil Engineering 181,
https://doi.org/10.1007/978-3-030-85043-2_30

1 Introduction

The strength of materials is a crucial parameter of the load-bearing capacity of building structures. Therefore, an objective assessment of material strength is of great importance for ensuring and calculating the reliability of structures and the proper justification of design standards. In particular, concrete is characterized by heterogeneity and statistical variability of strength and deformation properties. They depend on such random factors as cement activity, aggregate size, manufacturing and curing technology, service life, and so on. The strength of masonry also depends on many factors. They include: strength, size and shape of the stones; strength, ease of laying and elastic–plastic qualities of the solution; masonry quality. Less important factors in the work of masonry for compression are: bandaging masonry, adhesion of the mortar to the stone, filling with mortar vertical joints. Wood is an anisotropic structural material that has a heterogeneous structure, different nature of work in different stress states. Wood of different breeds has different strength. In addition, the strength of wood is affected by the presence of defects, humidity and service life. In the presence of such numerous factors that affect the strength of materials, it is natural that the strength indicators have a certain variability, a clear idea of which is given by statistical distribution curves of different characteristics of materials. The temporary resistance and other mechanical characteristics of modern materials have a statistical variance, which is well described by normal law, which has been repeatedly confirmed by test data of material samples. Therefore, the undoubted relevance of regular statistical studies of material strength is associated with the constant revision of design standards.

The initial data of the mechanical characteristics of building materials are obtained as a result of standard acceptance tests of samples in the laboratories of enterprises of the construction industry. The main purpose of these data is to assess the quality and rejection of substandard materials and structures. In addition, statistical test results of materials are used in the preparation and revision of design standards. Statistical studies of the concrete strength unfolded in the 50s of last century in connection with the adoption of the calculation method of limit states and continued in subsequent years [1–16]. Studies of the masonry strength and its components—brick and mortar—have been conducted since the 30s, and intensified in the 80s of last century [17–20]. Statistical studies of the mechanical characteristics of wood for building structures were constantly carried out [21–28]. This problem is actively discussed by foreign experts [29–34]. Reliable statistical parameters of material strength are especially needed to assess the reliability of building structures. This is emphasized, in particular, in the publications prepared by the scientific school "Reliability of building structures" of the National University "Yuri Kondratyuk Poltava Polytechnic" [35–40].

Factory tests of material strength are performed for many years on a large scale, creating a significant array of statistical information. However, there is no common information database for these data. Some of them have been published in various scientific and technical journals, collections of articles, conference proceedings.

Access to these publications is difficult, especially since some institutions have begun to destroy paper magazines in recent years, citing the transition to electronic publications. However, in reality, the translation into electronic form has so far occurred only for publications published after 2000.

The article contains a systematic review of publications in leading scientific and technical journals on the problem of statistical description of the strength of main construction materials: concrete, brick, mortars, masonry, wood. The principal attention is paid to the selection of statistical strength characteristics of materials of different periods, such as mathematical expectation, standard deviation (standard), coefficient of variation, etc. These data are intended for use in numerical calculations of structure reliability. In addition, the evolution of the design norms of structures is analyzed in terms of changes in the purpose of normative and design resistances and the involvement of experimental statistics.

2 Basic Material and Results

The content of the article is an organized review of publications of such scientific and technical journals as "Industrial and civil construction" (formerly "Construction industry" and "Industrial construction"), "Industrial construction and engineering structures", "Construction mechanics and calculation of structures", "University News. Construction and architecture", "Building materials", "Concrete and Reinforced Concrete", "Concrete and Reinforced Concrete in Ukraine", "Lesnoy Zhurnal", etc. The review is compiled for the period from the 40s of the twentieth century to the present. The paper version was mainly used for journals published before 2000, which were in the scientific and technical library of the National University "Poltava Polytechnic Yuri Kondratyuk", one of the most complete book storages in Ukraine. Information on later editions digitized from electronic libraries and electronic versions of journals.

2.1 *Statistical Characteristics of Concrete Strength*

The initiator of statistical studies of concrete strength was KE Tal, one of the developers of the new method of limit states, an employee of the Research Institute of Reinforced Concrete Structures (NDIZB). In one of the first publications on the parameters justification of the method of limit states, he compared the then average and defective values of concrete strength with the newly introduced design resistances (Table 1) [1].

It was emphasized that the calculation according to the new method is based on the design resistance of concrete, which is equal to the lowest probable value of resistance, and not on the allowable stresses, as it was the case with the previous norms of NiTU 3–49 "Norms and technical conditions of concrete structures". Under

Table 1 Design resistances, average and defective strength of concrete

Concrete	Design resistance		Defective strength		Average strength, MPa
	Designation	Value, MPa	Value, MPa	TU	
Brand 150	R_{cub}	8.5–9.0	11.5–13.0*	TU 200–54	~15.0–16.0
Brand 300		18.0–19.5	22.5–25.5*		~30.0–33.0

Note * Depending on the number of tested samples

production conditions, the strength of concrete was controlled by tests, but taking into account the selectivity of these tests, the defective values of concrete resistance were taken to be much higher than the design resistance (Table 1).

Then KE Tal stated that the strength of concrete has a statistical variance, which is well described by normal law [2]. Tests showed that for concrete brand 200 the average value of compressive strength in bending was 22.3 MPa, the standard 2.7 MPa with "good" quality of work (3700 samples-cubes) and 4.9 MPa with "poor" quality of work (675 samples). The coefficient of variation was 0.121 and 0.219, respectively.

In NiTU 123–55 "Norms and technical conditions for the design of concrete and reinforced concrete structures" for the normative resistance of concrete R^n it was taken as the average (or close to it) value of its compressive strength in bending $\overline{\sigma}_b$. The design resistance was determined by providing 0.999 at a distance of three standards $\hat{\sigma}_b$ from the average value. Accordingly, the theoretical value of the homogeneity coefficient, greater than one, was defined as

$$k_{theor} = \frac{\overline{\sigma}_b - 3\hat{\sigma}_b}{R^n} \tag{1}$$

Statistically substantiated homogeneity coefficients, introduced in NiTU 123–55 after some adjustment and rounding, are given in Table 2.

In Table 2, line A corresponds to the automatic or semi-automatic dosing of concrete components; line B—when there is no automatic dosing.

A similar concept of the purpose of normative and design resistances and homogeneity coefficients of concrete was implemented in the following general norms of SNiP II-A.1062 "Building structures and foundations. Basic design provisions".

Table 2 Homogeneity coefficients for concrete k_b

Stress state	Preparation mode of concrete	Homogeneity coefficients for concrete brand	
		100–200	300–600
Axial compression in bending	A	0.60	0.65
	B	0.55	0.60
Axial tension in bending	A	0.45	0.50
	B	0.40	0.45

Table 3 Statistical characteristics of concrete strength after 28 days of exposure

Plant	Concretebrand	With heat treatment			Without heat treatment		
		$\overline{\sigma}_b$, MPa	$\hat{\sigma}_b$, MPa	V, %	$\overline{\sigma}_b$, MPa	$\hat{\sigma}_b$, MPa	V, %
ZBK №6	300	33.2	2.7	8.1	32.5	3.5	10.8
ZBK №7	200	24.4	5.0	20.0	27.2	5.0	19.0

With the development of statistical substantiation of design standards of reinforced concrete structures, attention was paid not only to the average strength of concrete, but also to its variability, which is estimated by the coefficient of variation V. Statistical evaluation of this coefficient was performed on several ZBK of Moscow [3].

The results of tests on 20 technological lines were considered, 28 representative samples were composed and more than 10 thousand results were processed. The obtained statistical characteristics of concrete are given in Table 3.

The obtained coefficient of variation of concrete strength was in the range of 8.1–20.0%. It was confirmed that it depends mainly on the variability of the properties of the materials and the composition of the mixture. As can be seen from Table 3, the coefficient of variation is practically independent of the heat treatment of concrete at its stable mode. It is shown that the coefficient of variation with the set of concrete strength, as a rule, decreases due to the lag of growth of the strength standard from the increase of the average value.

In 1972, new standards were introduced in the form of SNiP II-A.1071 "Building structures and foundations. Basic design provisions". Designing norms of reinforced concrete structures were developed under the direction of prof. A.A. Gvozdev on their basis. They reflected achievements in the field of the calculation theory of reinforced concrete for the last 12 years and were vigorously discussed in the scientific and technical periodical [4–6]. According to the codes of SNiP II-2175 "Concrete and reinforced concrete structures" the normative resistance of concrete was provided not less than 0.95. Therefore, it was determined by Eq. (2).

$$R^n = \overline{R}(1 - 1.64V), \tag{2}$$

where $\overline{R}$ is the average value of strength; V—coefficient of variation; 1.64—a factor corresponding to the security of 0.95 when applying the normal law for the concrete strength.

The design resistance of concrete on compression is now determined by dividing by the "safety factor for concrete" $R = R^n / k_{b.s.}$. In particular, for heavy concrete, this factor was $k_{b.s.} = 1.3$, the coefficient of variation was $V = 0.135$.

This approach took into account the statistical variability of concrete strength, allowed to reasonably control the strength, identify total reserves and compare them for different structures on a common basis. Now, in industries that produce concrete with high homogeneity (low value of the variation coefficient), the average strength could be lower than the design brand with the corresponding savings in cement. On the contrary, productions with the high coefficient of variation were forced to assign an average concrete strength above the design brand [7].

It should be noted that the variation coefficient of concrete strength is accepted in the design standards equal to $V_b = 0.135$ as the average value for a large number of plants which produce reinforced concrete structures. Meanwhile, studies of technological lines at reinforced concrete plants have shown that the coefficient of variation can deviate significantly in both larger and smaller directions and is $V_b = 0.05\ldots0.25$ (data from AS Lychev [8]). With increasing strength of concrete, its structure is compacted, which leads to increased homogeneity and reduced coefficient of variation.

This is confirmed by the data in Table 4, which shows the results of laboratory tests conducted in ODASA by MM Zastava (462 tests, 14,320 samples), adjusted for the impact of production conditions on the recommendations of AP Kudzis (Vilnius Technical University).

The statistical scatter of concrete strength significantly depends on the level of manufacturing technology and construction of structures (Table 5).

The scatter of concrete strength is significantly affected by the variability and possible inaccuracy of dosing of its components, which may be within the following limits:

- strength of cement ±(6...25)%;
- strength of aggregates ±(15...50)%;
- cement content ±(5...15) t%;
- water content ±(8...23)%;
- content of fillers ±(5...10)%;
- moisture content in the filler ±(1...21)%.

Table 4 Variation of concrete strength depending on its average strength

Type of hardening	Coefficients of variation in % at average strength, MPa							
	10	20	30	40	50	60	70	≥80
Natural	15.9	12.0	10.5	8.8	6.6	5.4	5.1	5.1
Steaming	12.1	11.1	9.4	9.0	7.8	6.6	5.5	5.2

Table 5 Variation coefficients of concrete compressive strength, %

Type of structures	Concrete mixture consistence	Production culture and erection of structures		
		high	medium	low
Monolithic	Plastic	8…11	11…16	16…20
	Tough	10…15	15…20	20…25
Prefabricated	Plastic	6…9	9…14	14…18
	Tough	8…11	11…16	16…22

Mass statistical studies of compressive strength of concrete were performed in 1979–82 by leading research institutes—NDIZB and CNDIPromzdanii [9]. Data from 67 production lines of reinforced concrete plants of different cities were collected. In total, about 17 thousand test results of control series of samples were processed, representing 2.5 million m^3 of concrete after heat treatment at the age of 18 days. This was a large-scale test of the effectiveness of the statistical strength control which was recently introduced according to GOST 18,105–80.

According to the entire study information, it was found that the average strength of concrete in 89% of cases exceeds the design strength and in 11% of cases exceeds the design strength with the provision of 95%. The average variation coefficient of strength was 0.057–0.083, ie it was more than twice less than the value of 0.135, regulated by SNiP II-21-75.

The calculation of the provision of normative concrete resistances $P(R > R_n)$for the studied technological lines was performed. The average value of the safety characteristics was quite high $-\beta = 5.25$, in all cases $\beta > 2$. To identify the possibility of increasing the standard resistance of concrete, a safety estimate $P(R > 1.1R_n)$was calculated, which gave values β greater than 1.64 according to SNiP, for all technological lines. When assessing security $P(R > 1.15R_n)$, it was found that in 17% of production lines the index β is less than regulated. These conclusions clearly confirm the data of Table 6, inwhich all probabilistic indicators (except one) significantly exceed the normative provision of 0.95.

Table 6 Probabilistic provision of normative resistance of concrete

Plant (city)	Wares	Concrete brand	Number of concrete tests	$P(R > R_n)$	$P(R > 1{,}1R_n)$	$P(R > 1{,}15R_n)$
ZBK-18 (Moscow)	Road plates	M400	288	–	0.9965	0.9965
ZBK -2 (Moscow)	Stairs elements	M300	303	0.9961	0.9933	0.9538
	Vent-gutters	M200	455	0.9912	0.9880	0.9868
	Columns	M300	1123	0.9991	0.9973	0.9904
ZBK -22 (Moscow)	Columns	M400	1245	0.9991	0.9927	0.9871
	Stairs elements	M300	265	–	0.9962	0.9924
ZBK -8 (Moscow)	Flooring hollow plates	M300	213	–	0.9859	0.9812
ZBK -5 (Kharkov)	Plates 3×6 м	M300	717	0.9958	0.9888	0.9791
	Piles	M300	898	–	0.9944	0.9781
	Columns	M400	582	0.9931	0.9600	0.9070
ZBK (Yoshkar-Ola)	Beams	M400	418	0,9952	0.9832	0.9832

Although in some cases the distribution curves had a right-hand (positive) asymmetry, the strength of concrete was described by the normal law of distribution, which did not lead to significant errors.

The authors of the article [9] concluded that the obtained results open the possibility of increasing the normative and design resistances of precast concrete by 5–10% in comparison with norms SNiP. However, given that the study was conducted only at the plants of large industrial centers with a fairly high culture of production, this proposal was not included in the design standards of reinforced concrete structures. Perhaps it is now rational to return to these conclusions.

In 1986 SNiP 2.03.01–84 "Building constructions. Concrete and reinforced concrete structures" came into force, which introduced the concept of "concrete strength class "as the main indicator in the design and manufacture of reinforced concrete structures. As always when changing design standards, this transition was actively discussed in the publications of scientific and technical journals [10, 11]. Previously, in SNiP II-21-75, the strength of concrete was characterized by a brand that corresponded to the average value $\overline{R}$ of the strength of the reference concrete cubes. Now the concrete class B was introduced, equal to the standard cube strength $R^n$with a provision of 0.95. It was determined, as in the previous edition of the rules, by Eq. (2), from which the average strength was

$$B = \overline{R} = \frac{R^n}{1 - 1.64V} \tag{3}$$

When using the class as a characteristic of the concrete strength, the normative resistance $R^n$is set, according to which Eq. (3) determines the average value of the cubic strength $\overline{R}$ depending on the actual coefficient of variation V.

In the article [12] it was shown that if in Eq. (2) to substitute the normative coefficient of variation $V = 0.135$ according to SNiP II-21-75 for heavy concrete, we obtain

$$R^n = \overline{R}(1 - 1.64 \cdot 0.135) = 0.78\overline{R}$$

Accordingly, the average strength of concrete is

$$\overline{R} = R^n / 0.78 = 1.28R^n$$

A new situation has arisen when, depending on the value of the coefficient of variation, the average strength of concrete, corresponding to a certain class, can vary widely. For example, according to Eq. (3), with a variation coefficient of concrete strength of 20%, the average strength is 150% of the class. Obviously, it makes no sense to compare the brand and class by the average values of cubic strength.

The authors of the article [12] reasonably believed (and this was confirmed in practice) that the transition to concrete classes will not make significant changes in the design. At the same time for technologists it will lead to fundamental changes in the technique of appointment of standard strength at selection of concrete structure and its control on production. It is no coincidence that numerous articles published over many years have been devoted to the organization and improvement of statistical control of concrete strength in plants of reinforced concrete structures [11, 13, 14].

Publications of the 2000s summarize the long-term evolution of the rationing of the calculation of reinforced concrete structures for 70–80 years [15] and the application of the probabilistic method [16].

2.2 *Statistical Characteristics of Masonry Strength*

As can be seen from publications from the 1930s [17], the greatest influence on the masonry strength has the strength of stones and mortars, which have a statistical scatter.

Strength of Mortars. Statistical studies of the mortars strength on various binders were conducted by CNDIBK in the period 1980–85 at construction companies and construction laboratories in Moscow and the Moscow region [18]. According to the norms, the mortars were distributed by compressive strength by 8 brands, the most common of them are listed in Table 7. The mortar brand $\overline{R}$ corresponded to the average strength of the mortar after curing for 28 days. The number of tests of mortar brands was 500–8000.

Processing of test results of mortars revealed a statistical scatter of results and showed (Table 7) that for brands M50, M75 and M100 the actual strength $\bar{R}_{\phi}$ is almost equal to the standard strength, and for brands M150 and M200—lower by 10%. The strength of mortars with a provision of 0.95 is lower than the normative by 30–40%. The coefficient of variation was obtained in the range of 18–35%. It was found that the actual strength distributions of the mortars are asymmetric and slightly different from the normal distribution. Therefore, an attempt was made to involve Pearson curves of different types. However, the difference in results compared to the normal law was insignificant. Therefore, the authors of the publication [18]

Table 7 Mechanical characteristics of mortars

Mortar brand	Mortar strength, MPa		
	$\overline{R}$	$\bar{R}_{\phi}$	$R_{0,95}$
M50	5.0	5.1	3.5
M75	7.5	7.2	4.5
M100	10.0	9.6	6.0
M150	15.0	14.0	9.0
M200	20.0	18.0	12.5

Designations $\overline{R}$—strength according to norms; $\bar{R}_{\phi}$—actual average strength; $R_{0,95}$—strength with a provision of 0.95

Table 8 Mechanical characteristics of brick

Brick brand	Stress state	Number of brick tests	$\overline{R}$, MPa	$\bar{R}_{\phi}$, MPa	$R_{0,95}$	V, %
M75	Compression	744	7.5	9.33	5.80	33.6
	Tension	595	1.8	1.61	0.691	40.0
M100	Compression	1915	10.0	11.12	7.551	40.4
	Tension	1602	2.2	1.82	0.748	40.1

Designations $\overline{R}$—strength according to norms; $\bar{R}_{\phi}$—actual average strength; $R_{0,95}$—strength with a provision of 0.95; V – coefficient of variation

concluded that it is possible to apply the normal law to describe the strength of the mortars.

The obtained statistics on the mortar strength were used to assess the strength of brickwork [20]. Based on the research, the authors proposed to replace the strength division of mortars by brands with the division by classes. The basic parameter is the strength with a provision of 0.95 (Table 7), and the main statistical characteristics are mathematical expectation and coefficient of variation.

Strength of Brick. Statistical studies of the bricks strength (ordinary clay) were conducted by CNDIBK in the period 1980–85 years according to the data of construction laboratories in Moscow and the Moscow region [19].

According to the codes, bricks were distributed by strength by 9 brands, more common were brands M75 and M100. The brand of brick corresponded to the average compressive and tensile strength at bending (Table 8). The tests number of brick brands was 595–1915.

Processing of brick test results revealed a statistical scatter of results and showed (Table 8) that for brands M75 and M100 the actual compressive and tensile strength of the brick in bending almost meets the requirements of the norms. The strength of bricks with a provision of 0.95 is lower than the norm by 30–40%. The coefficient of variation was obtained in a fairly high range of 33.6–40.4%, which is due to the large diversity of raw materials and the instability of the technological process. It was found that the actual distributions of brick strength are asymmetric with left-handed asymmetry; they therefore differ from the normal distribution. Therefore, Pearson curves of different types were involved.

The obtained statistics on the bricks strength were used to assess the strength of brickwork [20]. On the basis of the conducted researches the author has made the offer to replace division of brick durability on brands on division on classes with maintenance of statistical control of brick durability that was not introduced in norms.

Strength of Masonry. The combination of a large number of random factors that affect the strength of the masonry, give it a random character and the ability to study by statistical methods. The initial data for assessing the strength of masonry were obtained in 1932–1958 in the tests of many thousands of large samples of various masonry from brick, concrete stones, large blocks, rubble, ceramic stones and more. Tests were performed on different types of resistance [17]. Proposed by prof. L.I. Onishchik empirical formula allowed to determine the strength of the masonry for different combinations of its strength components:

$$R = AR_1\left[1 - a/\left(b + R_2/2R_1\right)\right]\eta, \quad (4)$$

where $R,\ R_1,\ R_2$—the compressive strength of masonry, brick (stone) and mortar; a, b—empirical coefficients, which depend on the type of masonry; η—correction factor for masonry with a mortar of low brands.

To assess the accuracy of Eq. (4), the tests results of 747 samples of brickwork were processed. The distribution of deviations of the average breaking strength of experimental samples from the values calculated by Eq. (4) was constructed, which was well described by the normal law. The coefficient of variation of these deviations was equal to 9% taking into account the actual strength of the mortar (class of work A) and 14.7%—if the strength was determined by the composition of the mortar (class of work B) [17].

When determining the homogeneity coefficient of the masonry, the following factors were taken into account:

a) the necessary provision of the design resistance of the masonry (based on the rule of "three sigma"):

- for class of work A $k' = 1 - 3V_A = 1 - 3 \cdot 0.09 = 0.73$;
- for class of works B $k' = 1 - 3V_B = 1 - 3 \cdot 0.147 = 0.56$;

b) possible reduction in the quality of masonry $k'' = 0.9$;
c) possible decrease in the masonry strength $k''' = 0.95$.

The resulting homogeneity coefficient of the masonry was as follows:

- for the case of class of works A $k = 0.73 \cdot 0.9 \cdot 0.95 = 0.6$;
- for the case of class of works B $k = 0.56 \cdot 0.9 \cdot 0.95 = 0.48 \approx 0.5$.

At the suggestion of prof. S.A. Sementsov [17] these coefficients were taken into account in the norms of SNiP II-22-81 "Stone and reinforced stone structures" when assigning the design resistances of masonry. It should be noted that at the stage of implementation of the calculation method of limit states, the system of correction factors for the strength of masonry was considered cumbersome and difficult to use in the calculations. Therefore, the rationing of calculations of masonry, starting from the 40s of the twentieth century, did not fit into the general procedure of the method of limit states, and remained based on the calculated characteristics of strength.

In the late 80's of the twentieth century in CNDIBK under the leadership of prof. V.D. Raizer an attempt was made to translate the calculation of stone structures into the format of the method of limit states. To do this, statistical studies of the masonry strength were realized [20]. They showed that the actual average strength of brickwork is slightly higher than strength calculated by the Eq. (4) of LI Onishchik)

Table 9 Average compressive strength of the brickwork, MPa

Brick brand	Brick average strength (GOST/factual, MPa)	Mortar average strength (GOST/ factual, MPa) for brands				
		M150	M100	M75	M50	M25
		15/13.93	10/9.65	7.5/7.23	5/4.92	2.5/2.38
M75	7.50/9.36	–	2.97/3.37	2.81/3.16	2.57/2.88	2.14/2.38
M100	10/11	3.81/4.01	3.53/3.72	3.31/3.48	3.00/3.15	2.49/2.60

Notes Actual data on the average strength of the mortar are taken from the publication [18], for bricks from the publication [19]. Data on the average strength of the masonry: the numerator - strength calculated by Eq. (4) with the substitution of the strength of mortar and brick according bricks from the publication [19]. Data on the average strength of the masonry: the numerator - to GOST (given values), the denominator - strength calculated by the same formula with the substitution of experimental strength of mortar and brick

(Table 9). This is due to the fact that the average compressive strength of the brick is greater than that required by the standard.

The largest sample of test results of M100 brickwork samples with M25 mortar was considered separately. The normative and design values of the strength of this masonry were evaluated. With a provision of 0.995, the experimental value of the masonry strength of 2.1 MPa was 1.5 times higher than the above 1.42 MPa; with a provision of 0.95, the corresponding values were 2.12 MPa and 1.7 MPa. Taking into account the correction factors included in the homogeneity coefficient, the value of the design resistance of 1.30 MPa contained in the standards was confirmed for the specified sample of material strengths.

To expand the combinations of the original components of the masonry strength VD Raizer applied statistical modeling. The test results obtained in this case differ slightly from the calculations by the Eq. (4) by LI Onishchik. The conclusion about the possibility of increasing the calculated strength of the masonry by reducing the variation coefficient of the mortar is substantiated. The general offer is made: a) to accept values with provision not less than 0.995 for design values of masonry resistance at compression taking into account the gained experience (results are in article [20]); b) to equalize the provisions of resistances and in some cases to increase them essentially (to 20%), c) to enter the normative resistance of masonry with provision of 0.95. To date, these proposals are not included in the design standards. The current norms for stone structures have no normative values of strength characteristics and give only the design values of these values.

2.3 *Statistical Characteristics of Wood Strength*

The strength of wood is evaluated by the results of tests of small pure samples of wood (so-called "pure wood"). Some of the published generalized statistical data on

Table 10 Statistical indicators of softwood strength

Indicators	Strength limit along the fibers			
	Compression MPa	Bending MPa	Tension MPa	Chipping MPa
Average	46.8	80.6	93.7	7.37
Minimum	33.6	43.8	50.5	4.04
Maximum	69.5	117	131	12.2
Variant number	61	61	31	55
Standard	7,82	15.0	22.9	1.5
Coefficient of variation, %	16.7	18.7	24.4	21.0
Kurtosis	0.449	0.318	−0.821	0.90
Asymmetry	0.816	0.273	−0.174	0.45

Table 11 Statistical indicators of hardwood strength

Indicators	Strength limit along the fibers			
	Compression, MPa	Bending, MPa	Tension, MPa	Chipping, MPa
Average	52.6	97.0	113.2	10.3
Minimum	31.4	55.7	63.2	4.67
Maximum	81.7	160.0	212.0	19.0
Variant number	123	115	39	101
Standard	9.86	21.99	32.84	3.19
Coefficient of variation, %	18.7	22.7	29.0	30.9

the wood strength of different species of the former USSR are given in Tables 10 and 11 (humidity 12%) [21, 22].

In Tables 10 and 11 coniferous and deciduous wood species are considered as separate statistical aggregates. This allowed to significantly expand the range of studied characteristics of pure wood. Each variant represents the arithmetic mean for a number of values that characterize the species and growing area. Judging by the values of asymmetry and excess, all data sets, without exception, obey the normal distribution law. The generalized average values of wood strength were higher for deciduous species, the standard and coefficient of variation was lower for conifers that characterize the variability of wood strength. The coefficient of variation is different for different stress states of wood. The tensile strength of pure wood with average values of about 100 MPa and maximum more than 200 MPa, commensurate with some grades of steel. The average compressive strength of wood along the fibers is twice less. The average tensile strength in static bending occupies an intermediate position between tension and compression.

Statistical data of Tables 10 and 11 gave a general probabilistic picture of the wood strength and served as a basis for its rationing. It should be emphasized that the method of substantiation of normative and design resistances of wood is much more complicated than other materials. The average values of the strength of pure softwood from Table 10 were included in the norms of SNiP II-25-80 "Wooden structures" (marked as R_p^u). Normative values of the strength limit of pure wood R_p^n, determined by the standard method at a distance of 1.64 standards from the average value, were also given. In the specified codes the average $R^u$and normative resistances $R^n$of lumber were resulted with essential reduction taking into account influence of wood defects and the sizes of working section of wooden elements.

The fact is that yard lumber is a qualitatively new statistical set, another material, the mechanical properties of which are determined not so much by the numerous elements of the wood microstructure, but by the presence of a large defect such as a knot, all characteristics of which are very different from pure wood. Therefore, the variability of lumber is higher than that of pure samples, as the presence of defects adds additional variance to the results. It is noteworthy that the absolute values of tensile strength of lumber along the fibers are 2–4 times lower than strength of small samples. The compressive strength of pine and spruce lumber along the fibers is approximately the same and twice less than strength of small pure samples [23].

For the transition to the design resistance, the coefficient of long-term resistance of wood m_{mp} and the coefficient of reliability $\gamma = (1 - \beta_n V)/(1 - \beta V)$ were introduced, where $\beta_n = 1.64$ is the coefficient (safety characteristic), which corresponds to the provision of 0.95 of the normative resistance; $\beta = 2,33$—coefficient that took into account the provision of 0.99 of the design resistance. Let us illustrate these transformations with the example of the tensile strength of softwood along the fibers: pure wood—average value $R_p^u = 100$ MPa, normative value $R_p^n = 60$ MPa; 1st grade lumber—average value $R^u = 34$ MPa, normative value $R^n = 20$ MPa; stretched elements of the 1st grade—design resistance $R_p = 10$ MPa. Thus, the design resistance of lumber was much less than the initial strength of small pure samples. In addition, the design resistance is adjusted by a number of lowering coefficients, taking into account the operating conditions and features of the structures.

The strength characteristics of wood are influenced by numerous factors, including growth conditions and the actual age of the trees from which the structural wood is obtained. In this regard, it is interesting to compare the strength characteristics of softwoods belonging to different construction periods—the nineteenth century. ("old" wood) and the late twentieth century ("new" wood)—on the example of pine wood of the Arkhangelsk region, one of the main forestry regions of the European part of Russia [24]. For this purpose, during the restoration of wooden architecture objects, the specialists of the Moscow State University (MSU) selected samples of structural "old" and "new" wood aged 75 and 110 years. The test results of the obtained samples of pure wood, processed by statistical methods, are given in Table 12.

Experimentally obtained average and normative resistances of wood of different ages for compression and bending are illustrated, which are compared with each other and with resistances according to SNiP II-25-80. Note that for the "new" wood,

Table 12 Comparison of wood resistances of different ages with each other and with codes

Wood age	Stress state	Desig-nation	Resistance, MPa		$\frac{OW}{NW}$	Resistance SNiP, MPa
			OW «old» wood	NW «new» wood		
1	2	3	4	5	6	7
75 years	Compression along the fibers	R^u	48.7/1.11	40.3/0.91	1.10	44.0
		R^n	41.3/1.25	35.2/1.07	1.17	33.0
	Bending	R^u	74.7/0.93	72.0/0.90	1.04	80.0
		R^n	57.4/1.01	53.4/0.94	1.00	57.0
110 years	Compression along the fibers	R^u	63.7/1.44	48.8/1.10	1.31	44.0
		R^n	47.8/1.45	43.6/1.32	1.10	33.0
	Bending	R^u	97.4/1.22	96.1/1.20	1.01	80.0
		R^n	81.0/1.42	69.6/1.22	1.16	57.0

Notes: The denominator of the values in columns 4 and 5 shows the ratio of experimental resistances to resistances according to SNiP; column 6 shows the ratios of resistances of "old" and "new" wood; column 7 shows the resistances of pure wood according to SNiP

the coefficients of variation of compressive and bending resistances are 7.7% and 12–14%, respectively.

As can be seen from Table 12, in all cases, the strength of wood of the XIX century which aged 75 and 110 years is more than 1.00–1.31 times the strength of modern wood of the appropriate age. Obtained as a result of tests of small samples of pure wood and subsequent statistical processing, the normative resistance of compression along the fibers is more than established by current standards: for wood under 75 years—by 7 and 25%, for wood under 110 years—by 32 and 45% for modern wood and wood of the XIX century respectively. Normative bending resistances of modern wood and wood of the XIX century are more than the resistances established by norms for wood up to 75% of age—by 1%; for wood up to 110%—by 22 and 42% respectively. This shows that wood belonging to the construction periods of the nineteenth century and present, can be used as a structural building material.

Some authors believe that the parameters of defects do not fully characterize the strength and deformability of structural lumber, and in this aspect a statistical approach may be promising [25]. As an argument for the statistical assessment of the strength of the beams proposed defect index $V\%$—the relative total volume of knots of the fracture site—which is defined as the ratio of the sum of the volumes of knots located on the beam length equal to the width of the beam to the volume of this sample. To obtain the initial data, tests were performed on samples of spruce wood of the 1st grade and pine of the 2nd grade, the results are given in Table 13.

Statistical indicators of wood strength of the tested samples of spruce and pine, as in previous tests (Tables 10, 11 and 12), correspond to the values of normative and temporary resistances of SNiP II-25-80. The values of the new indicator of wood defects are of the following order: 1st grade—average value $\overline{V} \approx 3\%$, standard $\hat{V} \approx 2\%$; 2nd grade—average value $\overline{V} \approx 8\%$, standard $\hat{V} \approx 5\%$. The correlation

Table 13 Test results of wood samples [25]

Wood	Stress state	N	Designation	$\overline{X}$	$\hat{X}$	$X_{0,95}$
Spruce	Bending	54	R, МПа	45.8	8.8	31.3
			V, %	2.7	1.7	–
Pine	Compression	42	R, МПа	31.7	4.9	23.6
			V, %	7.1	4.7	–
	Bending	52	R, МПа	35.9	9.5	20.2
			V, %	7.6	4.8	–

Designations N—number of tested samples; R—resistance of wood; V—relative total volume of knots of the destruction site; $\overline{X}$—average value; $\hat{X}$—standard; $X_{0,95}$—value with provision of 0.95

analysis revealed a linear decrease in wood strength with increasing V at a correlation coefficient $\rho = 0.5–0.7$.

As shown above, in all tests, the strength characteristics of wood were well described by the normal distribution law. But it should be noted that the initial distributions of the limit strength of whip wood deteriorate with prolonged storage of raw materials in stacks in warehouses, in interoperable stocks after dimensional sorting of logs, in drying stacks and stacks of finished products. Therefore, the limit strength distribution of lumber can take the form of the sum of four normal distributions with their arithmetic means and standard deviations [26].

In modern design standards DBN B.2.6–161: 2017 "Wooden structures. Basic provisions" wood is divided into strength classes separately for coniferous and deciduous species depending on the characteristic values of strength, stiffness and density. The designations of the classes coincide with the characteristic bending strength of wood, expressed in N/mm^2.

In parallel with the development of probabilistic substantiation of the parameters of the method of calculation of wooden structures by limit states, the calculation of reliability of such structures was developed, and the lognormal distribution law was used to describe the strength of wood [27, 28].

3 Conclusions

A systematic review of work is carried out on the problem of statistical description of the strength of building materials. Structural materials such as concrete, brick, mortar, brickwork, wood are considered. The main attention is paid to the statistical characteristics of the materials strength of different periods, such as mathematical expectation, standard deviation, coefficient of variation, etc. These data are intended for use in numerical calculations of reliability of structures. The evolution of design norms of building structures is analyzed in the part of changes of purpose and provision of normative and design resistances and utilization of experimental statistical data.

References

1. Tal KE (1956) Some questions of the calculation of reinforced concrete structures according to new standards. Concr Reinforced Concr 4:139–142
2. Tal KE (1961) On the reserve factors in reinforced concrete structures. In: Materials of the International meeting on the calculation of building structures (Moscow, December 1958), pp 60–70. Gosstroyizdat, Moscow
3. Khayutin YuG, Kozlov ED (1970) On the strength of concrete under different modes of hardening. Concr Reinforced Concr 12:20–21
4. Gvozdev AA, Dmitriev SA, Lessig NN (1971) On the structure and general provisions of new standards for design of structures. Concr Reinforced Concr 5:1–4
5. Tal KE (1971) Normative and design characteristics of concrete. Concr Reinforced Concr 5:4–6
6. Sychev VI (1976) On the design standards for concrete and reinforced concrete structures. Concr Reinforced Concr 6:2–5
7. Yashin AV (1976) On new requirements for the standardization of the basic design characteristics of concrete. Concr Reinforced Concr 6:12–14
8. Lychev AS (2001) Statistical information for calculating the reliability of elements of building systems. Univ News Constr 5(509):17–20
9. Sapozhnikov NYa, Berdichevsky GI, Brusser MI (1983) Statistical analysis of the strength of concrete at the age of 28 days (according to factory tests). Concr Reinforced Concr 5:27–29
10. Tal KE (1977) On the improvement of standardized methods for calculating of reinforced concrete structures. Concr Reinforced Concr 5:20–21
11. Gvozdev AA, Krakovsky MB, Brusser MI, Igoshev VL, Dorer VA (1985) Relationship of statistical control of concrete strength with the reliability of reinforced concrete structures. Concr Reinforced Concr 3:37–38
12. Gushcha YuP, Brusser MI, Krakovsky MB, Serykh RL, Skubko VM (1985) On the transition from grades to concrete strength classes. Concr Reinforced Concr 10:39–40
13. Ratz EG (1968) Statistical control of concrete strength in reinforced concrete factories. Concr Reinforced Concr 10:5–10
14. Danilenko EL, Sadovich MA (1975) Statistical control and quality management of concrete. Concr Reinforced Concr 1:35–37
15. Zvezdov AI, Zalesov AS, Mukhamediev TA, Chistyakov EA (2002) 75th Anniversary of the Design Standards of concrete and reinforced concrete structures. Ind Civil Constr 9:10–15
16. Rastorguev BS, Vanus DS (2016) Calculation of reinforced concrete structures for limit states using the probabilistic method. Ind Civil Constr 7:11–15
17. Sementsov SA (1961) Modern methods of stone structures calculating in the USSR. In: Materials of the International meeting on the calculation of building structures (Moscow, December 1958), pp 234–248. Gosstroyizdat, Moscow
18. Raizer VD, Kruchinin NN (1988) On the normalization of the mortars strength. Struct Mech Struct Calc 4:50–53
19. Kruchinin NN (1987) On the statistical assessment of brick strength. Struct Mech Struct Calc 3:59–63
20. Raizer VD, Kruchinin NN (1990) To standardizing of the bearing capacity of masonry. Struct Mech Struct Calc 2:80–86
21. Ryumina EB, Uspenskaya GB (1992) Confidence level of standardized strength indicators of structural sawn timber. Lesnoy Zhurnal (Russ Forestry J) 2:81–86

22. GSSD 69–84 (1985) Standard reference data tables. Wood. Indicators of physical and mechanical properties of small clean samples. Publishing House of Standards, Moscow
23. Tuturin SV (2004) Mechanical strength of wood. Phys Mesomech 7(6):85–88
24. Linkov NV (2018) To the issue of comparing the physico-mechanical characteristics of structural timber of the XIX and XX centuries. Ind Civil Eng 1:58–62
25. Byzov VE (2016) Melekhov VI (2016) Structural sawn timber: resource enhancement. Mag C.E. 65(5):67–76
26. Dounaev VF (2014) Variability of young's modulus and wood tensile strength in the work cycle. Lesnoy Zhurnal (Russ Forestry J) 3:106–113
27. Gromatsky VA, Turkovsky SB, Filimonov MA (2011) On the assessment of the reliability of wooden structure elements. Struct Mech Struct Calc 6:22–25
28. Denesh ND, Sukhov DYu, Andreev OO (1988) On the use of the lognormal load distribution law in assessing the reliability of wooden structures. Struct Mech Struct Calc 4:8–9
29. Mirza SA, Mac Gregor JG, Hatzinikovas M (1979) Statistical description of strength of concrete. J Struct Div 6(105):1021–1037
30. Indelicato F (1999) In-place compressive strength of concrete: statistical methods to evaluate experimental data. Mater Struct 32:394–399
31. Kausay T, Simon T (2007) Acceptance of concrete compressive strength. Concr Struct (Ann J Hung Group ***fib*** Budapest) 8:54–63
32. Kabantsev O (2015) Modeling nonlinear deformation and destruction masonry under biaxial stresses. Part 1. Masonry as simulation object. Appl Mech Mater 725–726:681–696
33. Grishchenko AI, Semenov AS, Semenov SG, Melnikov BE (2014) Influence of structural parameters of the masonry on effective elastic properties and strength. Mag Civil Eng 5(49):95–106
34. Foschi RO, Folz BR, Yao FZ (1989) Reliability-based design of wood structures. Structural research series. Rep No 34 Dep of Civil Eng, Univ of British Columbia, Vancouver, Canada
35. Perelmuter AV, Pichugin SF (2017) On one safety characteristic of buildings. J Civil Eng Arch Res 5(4):2035–2044
36. Pichugin SF (2018) Reliability estimation of industrial building structures. Mag Civil Eng 83(7):24–37
37. Pichugin SF (2019) Scientific school«reliability of building structures»: new results and perspectives. Acad J Ser Ind Mach Build Civil Eng 2(53):5–12
38. Pichugin S (2017) Probabilistic description of ground snow loads for Ukraine. In: Snow engineering 2000: Recent advances and developments, pp 251–256. https://doi.org/10.1201/9780203739532
39. Pichugin SF, Makhin'Ko AV (2009) Calculation of the reliability of steel underground pipelines. Strength Mater 41(5):541–547. https://doi.org/10.1007/s11223-009-9153-0
40. Pichugin S, Severin V (2004). Reliability of structures under snow load in ukraine. In: Snow engineering, pp 67–72

Forecasting the Possible Accident Scenario on the Example of Self-framing Metal Buildings

Sergii Pichugin and Lina Klochko

Abstract The article aims to provide a graphical modeling of the possible accident probability at a construction site. A model feature is the calculation and frame design with a cold-formed steel. The article reveals the essence, features and disadvantages of light steel framing (LSF). The general concept of accidents statistical processing in construction is also considered in the work. For the further possibility of modeling the accident, the main stages have been identified the highest percentage of the building collapse probability. The LSF collapse examples are highlighted and presented. The accidents statistics is considered as a prerequisite for the correct and reasonable modeling of a possible accident. The calculation for modeling the frame of a real construction object, namely a superstructure with LSF, has been performed. The peculiarity of the light thin-walled steel structures calculation is highlighted. Attention is focused on the roof (truss)calculation for this project. With the help of the method of knocking out (withdrawing) individual elements of the frame supporting structures, possible destruction scenarios of the structure are considered. As a result of the work carried out, the most vulnerable area with a dangerous influence as a result of structural element stability loss was determined. Consequently, appropriate conclusions were drawn.

Keywords Cold-formed steel · Light steel framing (LSF) · Accident · Building collapse · Building collapse probability · Modeling · Statistic

1 Introduction

Modern construction industry is at the rapid state of innovation development. The main design goal is the maximum construction site reliability along with its economic feasibility. For these reasons the construction industry is in constant search of new designs that can meet all the customer needs. The main structural element goal is the necessity for bearing capacity and the ability to withstand deformations using

S. Pichugin (✉) · L. Klochko
National University «Yuri Kondratyuk Poltava Polytechnic», Pershotravnevyj Avenue 24., Poltava 36011, Ukraine

V. Onyshchenko et al. (eds.), *Proceedings of the 3rd International Conference on Building Innovations*, Lecture Notes in Civil Engineering 181,
https://doi.org/10.1007/978-3-030-85043-2_31

more economical sections and visually perfect forms to implement the architectural design.

As a result of this development of the construction industry, cold-formed steel, namely LSF (light steel framing), is being actively investigated and introduced into construction. This structure type is a thin steel building structure used for building houses. These structures include profiled sheets and thin-walled galvanized steel profiles.

Light steel framing conquer the world construction market with a number of advantages, namely:

- wide architectural possibilities;
- high rates of construction and all-season installation;
- the possibility of construction in hard-to-reach places;
- light weight and no shrinkage;
- high energy efficiency;
- environmental friendliness;
- absence of harmful effects;
- economic benefit [13].

Taking into account the above structures advantages of this material, it is impossible not to pay attention to the disadvantages as well, especially if this can lead to an accident of a building or structure.

Thus, in the process of researching and improving the design of LSF in the world, including Ukraine, where the next construction type is gaining more and more popularity, the task is to implement a complete design project, including modeling a possible accident at a construction site.

2 Main Body

2.1 The Analysis of Recent Research Sources and Publications

Light steel framing, methods of their improvement and ways of using in civil engineering are actively pursued by scientists around the world, including in their research and the specifics of the country construction study. More detailed information on this issue is given in the works of such scientists as Remo F Pedreschi [5], R.M. Lawson [6], Enrico de Angelisa [2], Cláudio Martins [8]. A significant contribution to the LSF design in Ukraine was made by such scientists as Semko V.O. [11] and Pichugin, S.F. [10, 14–17].

2.2 *The Selection of Previously Unsolved Parts of General Problem*

The issues of modeling a possible accident at a construction site are widely studied and introduced into the world design standards. Such modeling aims to create the most reliable structure frame. In the case of the exclusion of some load-bearing elements, the modeling purpose is to preserve a full-fledged structure frame. This approach allows to limit the negative factors influence at different construction stages.

2.3 *Problem Definition*

The main goal of the work is to study the features of LSF, based on the accidents statistical material in buildings and structures, including this structures type, to develop and simulate a scenario of the accident possibility at a construction site made of light steel framing (knocking out from the work of the frame supporting elements).

2.4 *The Main Material and Results*

2.4.1 Characteristics and Features of LSF Using

Light steel thin-walled structures are gaining wide popularity nowadays both in Ukraine and around the world. If we consider the history of the this frames type introduction, the first thing to mention is the frames development in the 50 s of the twentieth century in Canada.

Light steel framing can be presented in low-rise buildings shape (see Fig. 1) as well as multi-storey (Fig. 2), including multi-occupancy and panelised structural frames [1].

Speaking of light steel framing, it is important to note that they consist of galvanized C-profiles, which have a depth of 70 to 100 mm in wall panels and 150 to 300 mm in depths of C-profiles or lattice beams in the slabs. All this allows for sufficient design flexibility to create individual projects.

Characteristics of prefabricated wall panels: usually floor heights from 2.7 to 3.2 m; length up to 8 m, depending on the conditions of transportation and lifting.

A house can be constructed from as little as 12 wall panels. It is also important to note that about four houses (large apartments) can be transported by truck. The floors can be constructed in two ways: individual joists or prefabricated floor cassettes.

It should be noted the popularity of the composite slabs use supported by light steel walls in special fields. The floor slab usually varies from 150 to 180 mm and has a span of up to 5.5 m when temporarily supported during construction. It is important

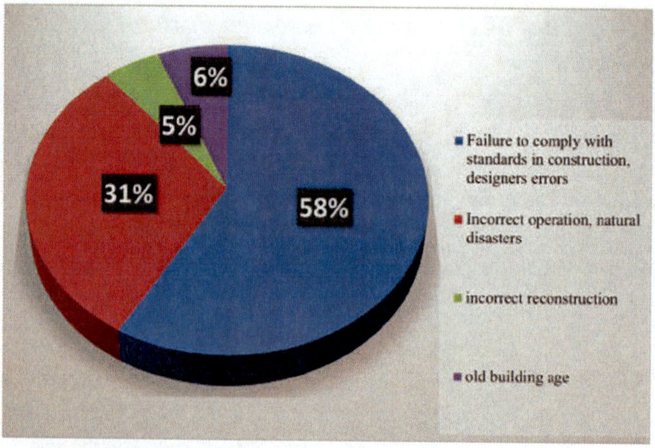

Fig. 1 Accident statistics by construction stages

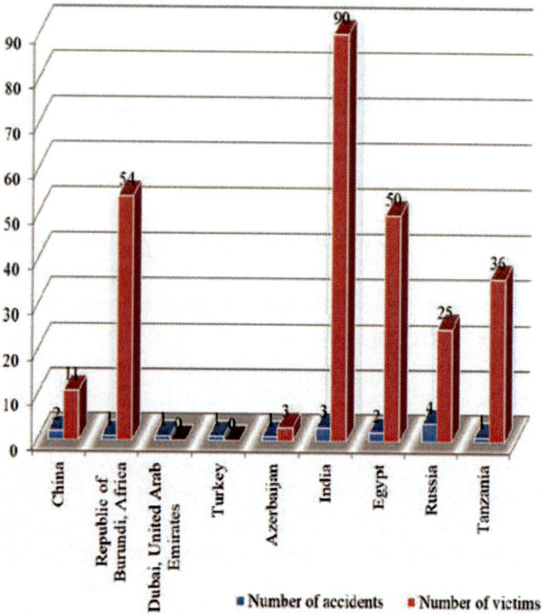

Fig. 2 Accidents statistic during construction, depending on the country where it was occurred

to note that such a slab is shallower and makes it possible to provide routing of ducts suspended from the decking.

Joisted and composite floors often use thin floor beams to create a large enough space and open plan space. An example of such a design can be lobby areas and etc. These beams are integrated into the floor area. Steel and concrete stairs are usually supplied as part of a light steel package and are integrated into the Building Information Model (BIM) provided by the light steel supplier.

2.4.2 Prerequisites for LSF Frame Modeling. Construction Accident Statistics

Having a clear understanding of LSF design and operation and taking into account the advantages and disadvantages of this material, there is certainly a need to exclude the occurrence of a possible accident at this construction site.

Knowing about accidents in construction in general and separately LSF, it is worth noting that the problem arises due to the rapid influx of the latest technologies. Objects are beginning to be created, the analogs of which do not yet exist, which means that they cannot use the trial and error method, referring to the past generations experience. Construction projects are acquiring new forms, using the latest material and technical base, while requiring maximum efficiency, energy efficiency and, at the same time, reliability. These newest technologies include the manufacturing and design technology of LSF.

To successfully fulfill all the above conditions, the scientific sector for the construction industry development is faced with the task of creating a unified algorithm for eliminating a possible event of a building or structure accident.

The most probable and effective way to solve the posed problem is to solve it by contradiction. That is, in the engineering building design, it is necessary to be able to simulate various options for the destruction of a building structure, the exclusion of load-bearing elements from the work to further determine the most vulnerable structural scheme sections, which will strengthen the necessary structures. The carried out work will reduce the likelihood of an accident in general.

Based on the study of the accidents occurrence experience around the world for the period from 2000 to 2020, the construction accidents classification example was created and attempts were made to calculate and simulate the destruction of the industrial building structure. When searching for accidents, scientific works of previous years, Internet resources, news information sources, statistical data from the relevant centers of Ukraine, Russia and the EU were used.

As a result of the carried out work, the following data were obtained. Speaking about the structure operation stage and the accidents occurrence, according (see Fig. 1), it can be concluded that the largest destruction falls occurrence sector on the stage of construction and putting into operation a construction site. A particularly high percentage of such accidents can be observed in the countries with fragile economies and below-average wealth levels. Countries such as India, Egypt and African countries are at the top of this rating [11] (see Fig. 2).

It is obvious from the diagram, 31% are accidents during the buildings or structures operation. Basically, the buildings destruction at this stage is caused by improper construction site operation, the load of the supporting structures is more than the permissible by design, and the failure by repair work and reconstruction on time. But even during reconstruction, serious construction accidents can often occur. Their percentage is 5% of the total and lead to serious economic and non-economic losses. 6% is the sector of accidents due to the old building age. Basically, the object destruction occurs as a result of exceeding the service building life, failure to carry out repair and reconstruction work planned according to the building passport, as well as postponing the necessary dismantling.

When it comes to cold formed steel construction only, attention should be paid to the problem of tracking such accidents. Basically, the media do not pay attention to many of these incidents, since they have little impact and consequences.

But despite the tracking information complexity, there are several examples of an accident with the LSF.

One of the striking examples is the cowshed construction (32 × 150 m) in the Smolensk region (Russia) in 2012. The framework of the light steel structures collapsed in the first winter after construction. The building was partially put into operation. Steel structures were destroyed (span 16 m). During the investigation, obvious errors were found directly in the design itself [3].

Also on January 26, 2014 in the morning in one of the Russian regions a hangar was destroyed. The structure could not withstand the load from the snow and fell like a cards house. The hangar has not been put into operation. Thanks to this, possible human casualties were avoided [7].

LSF accidents caused by their main disadvantages, specifically:

- the LSF supporting structures durability and the building (structure) as a whole strongly depends on the steel profiles production quality and the LSF installation. On average, the LSF load-bearing frames durability in our conditions is 40 years (due to the corrosion of the galvanized profile, the LSF durability is low, compared to stone, concrete and brick buildings;
- low fire resistance of building frame steel structures (it is necessary to increase it by installing fire-resistant protective sheathing.
- after a fire, the LSF supporting structures cannot be restored and require complete replacement [4].

2.4.3 Design of LSF Structures

To appropriately design LSFs, it is important to model the structure in three dimensions and with high accuracy.

After making 3D model of the intended building, appropriate framing would be assigned based on engineering principles and computations. Then, drawings which are corresponding to manufacturing and installation of panels would be prepared. The issued drawings should take into account all the engineering details and considerations in a specific manner.

Drawings are divided into following groups:

1. Drawing of walls, roofs and corresponding connections including size of structural members, thickness of plates, distance of members, location of load

bearing and non-load bearing walls, details of connections and considerations of regulation.
2. Drawings of roof and trusses (if necessary) and relevant connections which include trusses production drawings, braces of roofs, connections and regulation implications [9].

As a result of the analysis of the structure's operation and main disadvantages determination, it becomes necessary to simulate the accident likelihood at a construction facility of LSF at the design stage, in order to enhance the structure reliability and to prematurely prevent the destruction of the most unreinforced building sections as a result of an unforeseen situation during building operation.

This method of modeling the accident likelihood in a building or structure allows predicting and assessing the risk of possible destruction, if necessary, to strengthen the structure. Moreover, the following method allows you to check the calculations correctness, significantly reduces the percentage of the building accident likelihood as a result of engineering errors.

Carrying out modeling on the example of a real engineering project.

This model was created three-dimensional, with a total length of 26.894 m, created to calculate six frames with a step of 1.2 m. The truss height on the support is 300 mm. The length of the columns is 3 m. The spans are 11,144 m. The general view of the building frame is shown in Fig. 3.

Based on the created 3D model of the LSF frame, we proceed to modeling the scenario of a possible accident of this object. To perform the task directly, you must perform the following calculations:

1) perform linear calculation with the determination of the frame deformation (see Figs. 4, 5);

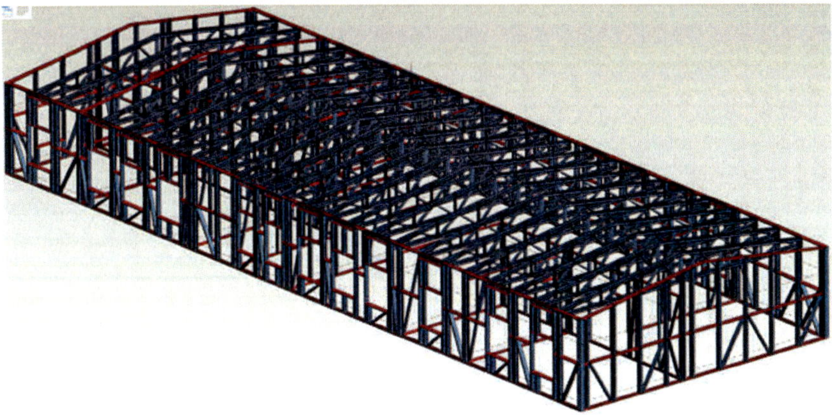

Fig. 3 General view of the structure frame

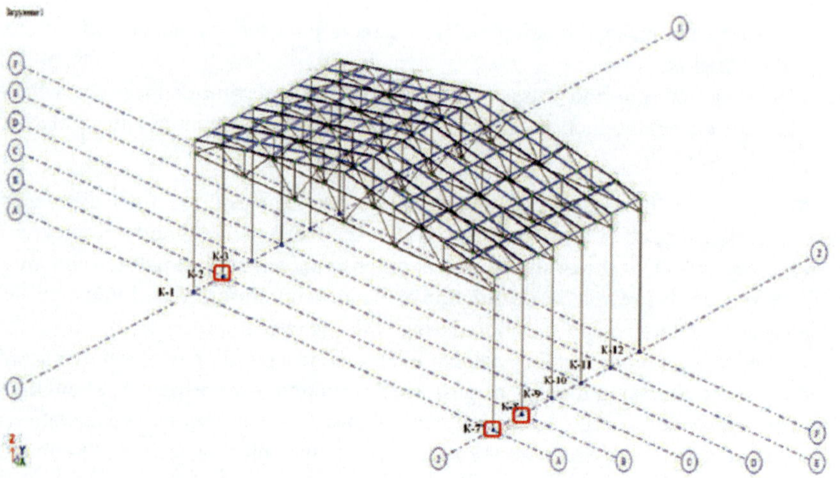

Fig. 4 3D model of the frame with certain removed elements

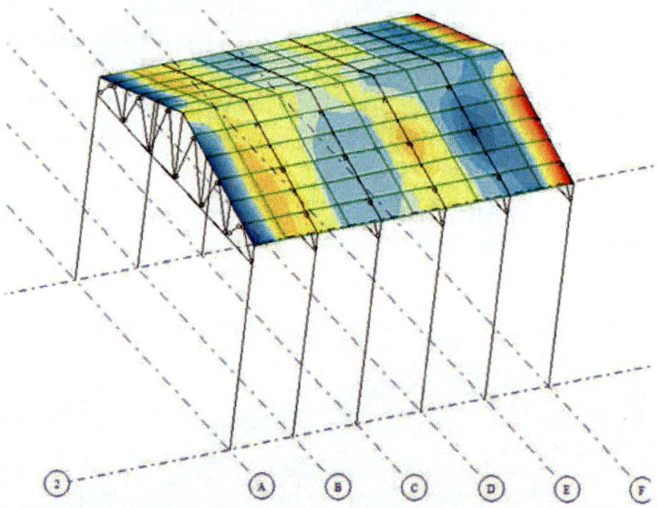

Fig. 5 Linear calculation with the determination of the frame deformation

2) exclusion from work in the structural scheme of individual load-bearing elements (columns 3, 7, 8), the calculation of the scheme, as a result of which reinforcement is assigned for the model calculation in the non-linear phase;
3) the building calculation taking into account physical and geometric nonlinearity and dynamic coefficient.

It should be noted that at third calculation stage, the criteria for the structures destruction are the geometric system variability at the nth step, an avalanche-like growth of deformations and the system displacement.

The next step is to identify the most dangerous areas of the frame. Using the design model, we determine the areas where the greatest efforts occur in the elements. Guided by the previously created 3D model, we remove one after another the bearing elements of this section, namely the columns K3, K7 and K8 (Figs. 6, 7 and 8).

When the K-3 column was removed from the general scheme, the deformation frame calculation was carried out again. The calculation result is shown in Fig. 4.

A similar frame calculation for deformation was performed when the column K-7 (see Fig. 7).

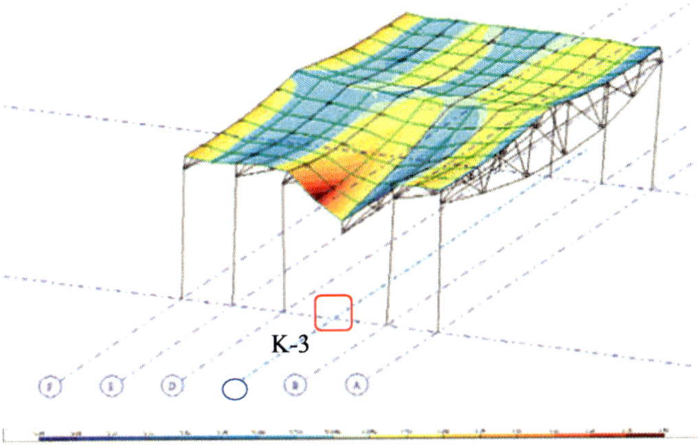

Fig. 6 The frame calculation result on deformation with the decommissioning of column K-3

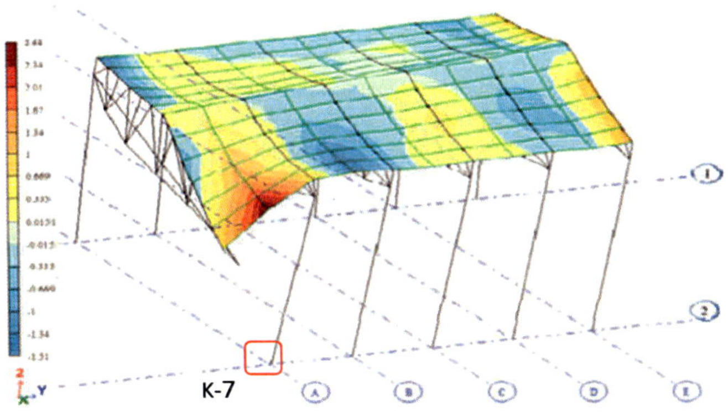

Fig. 7 The result of calculating the frame on deformation with the decommissioning of column K-7

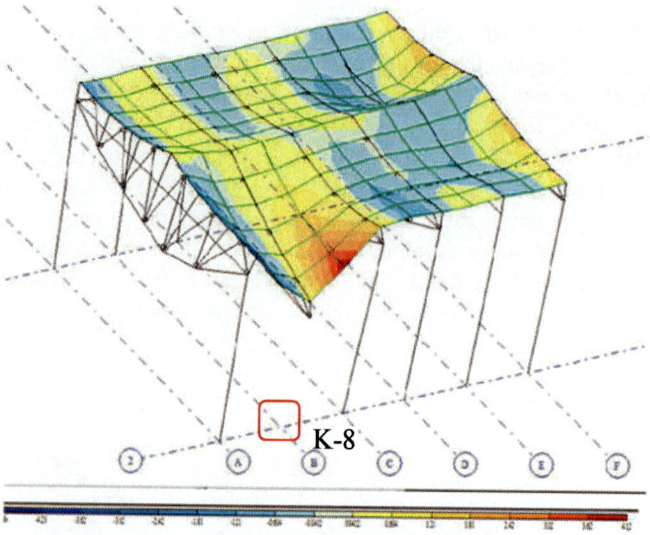

Fig. 8 The results of the design calculation in the accident simulation - removal of the K-8 column

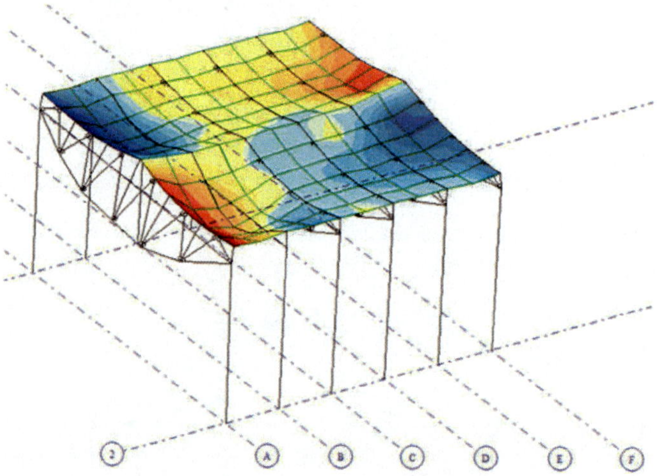

Fig. 9 The result of calculating the frame on deformation with the decommissioning of truss T-2

Also, the calculation of the frame was carried out when the column K-8 was excluded from the work, which made it possible to carry out further analysis of the progressive failure effect on the structure (see Fig. 8).

On the basis of the accidents analysis in construction, in particular the study of LSF accidents statistics, the most vulnerable frame element was determined. For this type of structures, the largest accidents percentage falls on the roof destruction from snow loads. It was simulated a situation in which the snow loads will be higher than the critical ones. It was removed a constructive element from work—a roofing truss F2 (see Fig. 9).

As a result of the analysis of the simulated possible accidents scenarios in Figs. 4, 5, 6, 7, 8, when the load-bearing frame elements are removed from LSF, appropriate conclusions can be drawn. The framework destruction occurs locally, only within the elements limits removed from work and does not proceed to the progressive entire model destruction. So, the structure does not need additional reinforcement.

3 Conclusion

As a result of the work carried out, LSF features material were investigated, based on the accidents statistics material in buildings and structures, including this structures type. And also LSF modeling was carried out on the example of a real construction object. The processed material on the accidents statistics in construction, in particular cold-formed steel structures accidents, made it possible to consider the most probable scenarios for the possible structural failure occurrence.

To create the most reliable building frame, it is necessary to envisage various accident scenarios. Such scenarios can be divided into separate groups as a result of the statistical materials processing on this topic. These include: accidental, reliable and impossible accidents [11]. When modeling LSF accidents, it is worth paying attention to the specifics of the this particular type structures destruction occurrence. The experience and construction incidents statistics at LSF highlight the main reasons for the structures destruction, namely: design errors (in particular, roof destruction due to incorrect calculation of snow loads) and installation errors (poor-quality installation, negligence during work). The above accidents causes arise at the construction stage and facility commissioning and belong to the possible accidents group. Therefore, when modeling the LSF accident scenario, first of all, it is necessary to remove from work the structural roof elements (also allows to check the loads calculation correctness) and key vertical bearing elements (determined by direct frame calculation).

All of the above modeling stages were presented in this work using the example of a real building object of LSF.

The result of this simulation is the determination of the need to strengthen the structure. In this case, amplification is not required, since the object fully complies with the design requirements. But it is worth focusing on the importance of this modeling type for unique architectural structures, large construction projects and increased danger objects. They are the ones that have the highest danger degree in the emergency event an and should be designed with consideration of all possible accident probability scenarios.

Subsequently, it becomes necessary to simulate the accidental collapse possibility, it is especially important to carry out such a calculation for objects of increased danger. Accidental collapce occurrence includes—explosion, fire and other natural factors.

References

1. Andrew Way Benefits of light steel framing Homepage (2019). https://www.pbctoday.co.uk/news/planning-construction-news/light-steel-framing/56740/#:~:text=Light%20steel%20framing%20comprises%20galvanised,flexibility%20in%20internal%20space%20plannin. Accessed 15 Dec 2020
2. de Angelis E, Serra E (2014) Light steel-frame walls: thermal insulation performances and thermal bridges. Energy Procedia 45:362–371. https://doi.org/10.1016/j.egypro.2014.01.039
3. Farm from LSTK Homepage. http://www.steelbuildings.ru/Barenboym. Accessed 11 Jun 2020
4. Gidproekt Homepage. http://gidproekt.com/texnologiya-lstk-stroitelstvo-zdanij-iz-legkix-stalnyx-konstrukcij.html. Accessed 10 Mar 2020
5. Heatley T, Popo-Ola S, Lawson R, Pedreschi R (2010) Durability of light steel framing in residential applications. In: Proceedings of the ice - construction materials, pp 109–121. https://doi.org/10.1680/coma.2010.163.2.109
6. Lawson RM, Kermani A, Stergiopoulos M, Coste G, Way A (2020) Diaphragm action in light steel framing by sheathing boards. Engineering Structures. https://doi.org/10.1016/j.engstruct.2020.110952
7. LSTK (Light steel thin-walled structures) – accidents and safety Homepage. http://stroy-trading.ru/information/article/264. Accessed 11 Jun 2020
8. Martins C, Santos P, Silva L (2013) Lightweight Steel Framed Construction System. Homepage. https://www.researchgate.net/publication/312160591_Lightweight_Steel_Framed_Construction_System. Accessed 25 Sept 2020
9. Mashhadifarahani S (2015) Lightweight steel frames vs. common building structures - structural performance evaluation. Am Sci Res J Eng Technol Sci 12:222–229. https://core.ac.uk/download/pdf/235049745.pdf
10. Pichugin S, Semko V, Prohorenko D (2010) Application in the new construction of lightweight, fast-mounting systems. Acad J Ser Ind Mach Build Civ Eng 2(1(52)):62–67
11. Pichugin S, Dmytrenko L (2018) Building accident causes at a stage of construction and acceptance in operation. Int J Eng Technol (UAE) 7:311–315. https://doi.org/10.14419/ijet.v7i3.2.14426
12. Semko V (2017) Steel cold-formed thin-walled designs. Poltava
13. Steelco Homepage. https://steelco.com.ua/lstk/5. Accessed 15 Dec 2020
14. Pichugin S (2017) Probabilistic description of ground snow loads for Ukraine. In: Snow engineering 2000: recent advances and developments, pp 251–256. https://doi.org/10.1201/9780203739532
15. Pichugin S, Severin V (2004) Reliability of structures under snow load in Ukraine. Snow engineering V, pp 67–72
16. Pichugin SF, Makhin'Ko AV (2009) Calculation of the reliability of steel underground pipelines. Strength Mater 41(5):541–547. https://doi.org/10.1007/s11223-009-9153-0
17. Pichugin SF (2018) Reliability estimation of industrial building structures. Mag Civ Eng 83(7):24–37. https://doi.org/10.18720/MCE.83.3

Plasticity Theory in Strength Calculations Concrete Elements Under Local Compression

Volodymyr Pohribnyi, Oksana Dovzhenko, Oleksiy Fenko, and Dmytro Usenko

Abstract To determine the strength of compressed elements, a variational method was proposed in the theory of concrete plasticity using the principle of virtual velocities. Plastic strain of concrete are considered to be localized in thin layers on the surface of failure (jump of velocities). The given basic dependences of the variational method. The problem of the strength of a concrete base when a rectangular stamp is pressed in by the method of characteristic lines under conditions of a plane stress state and plane strain has been solved and by the proposed method with a comparison of the results obtained. Two cases of destruction of a concrete base under local compression without splitting and during its implementation are considered. Dependencies provided for determining the ultimate load value under the stamp. The boundary between the cases of destruction is established depending on the ratio of the height of the base to the width of the stamp.

Keywords Principle of virtual velocities · Functional of the method · Stationary state · Strength · Concrete base · Indentation of a stamp

1 Introduction

In construction, widespread concrete elements with various shapes, geometric dimensions, the nature of the load application, the specifics of the stress state. As a result of the variety of proposals for assessing their strength [1–7], questions arise regarding the choice of design dependencies. Modern energy efficient structural solutions require refinement of calculations [8–10]. The empirical approach has a narrow field of application, limited by the experimental conditions, and the extension of the obtained formulas to other cases of operation of compressed elements can lead to errors in assessing their strength. Therefore, the creation of a fairly general methodology for calculating the strength of concrete elements in compression on a theoretical basis is an urgent task. As such a theoretical basis, the theory of plasticity can be used,

V. Pohribnyi (✉) · O. Dovzhenko · O. Fenko · D. Usenko
National University «Yuri Kondratyuk Poltava Polytechnic», Poltava, Ukraine

V. Onyshchenko et al. (eds.), *Proceedings of the 3rd International Conference on Building Innovations*, Lecture Notes in Civil Engineering 181,
https://doi.org/10.1007/978-3-030-85043-2_32

the mathematical apparatus of which is widely tested for plastic materials. Scientific research is devoted to this direction [11–17]. At the National University «Yuri Kondratyuk Poltava Polytechnic» a variational method in the theory of plasticity of concrete using the principle of virtual velocities has been developed for calculating the strength of concrete and reinforced concrete elements under the shear [18–20]. The method considers discontinuous solutions and has experimental confirmation on keyed joints, models of a compressed zone above a dangerous inclined crack, samples that were recommended for determining the resistance of concrete with a "pure shear", etc. [21, 22]. There is a possibility of using it to solve problems of the strength of concrete elements under local compression [23]. Taking into account the specifics of their stress–strain state will make it possible to add additions to the basic dependencies of the variational method.

2 The Main Dependencies of the Principle of Virtual Speeds

The limited plastic properties of concretes cause localization of severe strain in the area of uneven compression in thin layers on the failure surface. Therefore, the functional of the principle of virtual velocities, at which only the strain rates vary, can be written in the form

$$J = \int_{S_l} (TH' + \sigma\xi')\Delta n dS - \int_{S_F} F_i \upsilon_i' dS, \tag{1}$$

where T – the shear stress intensity; H' – the intensity of shear strains; σ – the average stress; ξ' – the volumetric strain; S_l – the failure surface area; Δn – the thickness of the plastic layer; F_i – the surface force; υ_i' – the speed of the F_i application point; S_F – the area of action F_i.

The condition of concrete strength is accepted [24], which in the area of triaxial non-uniform compression is considered as a condition for the onset of yield of materials with different resistance to axial tension f_{ct} and compression f_c

$$T^2 + m\sigma - T_{sh}^2 = 0, \tag{2}$$

here $m = f_c - f_{ct}$, $T_{sh}^2 = f_c f_{ct}/3$.

Strength condition (2) at plane stress state and plane strain in coordinates $|\tau_n| - \sigma_n$, respectively, is written

$$|\tau_n| = \varphi(\sigma_n) = \sqrt{d^2 - \frac{1}{4}(\sigma_n - m)^2}, \tag{3}$$

$$|\tau_n| = \varphi(\sigma_n) = \sqrt{m\left(\sigma_n + \frac{1}{4}m + \frac{1}{12}n^2/m\right)}, \tag{4}$$

where $d = \sqrt{(f_c^2 - f_c f_{ct} + f_{ct}^2)/3}$, $n = f_c + f_{ct}$.

The functional of the principle of virtual velocities J [25] is investigated for a stationary state using the equation $\delta J = 0$.

In a plane stress state in the range of available slip planes at $\Delta n \to 0$, using (2) and (3), we have

$$\delta \int\limits_{S_l} \left[d\sqrt{4\Delta \upsilon_n'^2 + \Delta \upsilon_t'^2} - m\Delta \upsilon_n' \right] dS - \int\limits_{S_F} F_i \upsilon_l' dS = 0, \tag{5}$$

here $\Delta \upsilon_n'$ and $\Delta \upsilon_t'$ – jumps of normal and shear velocity components on the failure surface (velocity jump line).

Under conditions of plane strain, taking into account (2) and (4), the functional J is investigated using the equation

$$\delta \int\limits_{S_l} \left[\frac{d^2}{m} + \frac{m}{4} \left(\frac{\Delta \upsilon_t'}{\Delta \upsilon_n'} \right)^2 \right] \Delta \upsilon_n' dS - \int\limits_{S_F} F_i \upsilon_i' dS = 0. \tag{6}$$

Introducing the characteristic of strength $\chi = f_{ct}/f_c$ and parameters k' and $\tan \gamma'$, where υ_1' and υ_2' – the velocity components in the direction of stress σ_1 and σ_2, but γ'– the angle between the failure surface and the direction of action of the principal stresses σ_1, and solving Eq. (5) with respect to $\sigma_1 = F_1/S_F$, the basic dependence for the plane stress state is received

$$\frac{\sigma_1}{f_c} = \frac{2\sqrt{(1-\chi+\chi^2)/3}\sqrt{(k'-\tan\gamma')^2 + 0.25(1+k'\tan\gamma')^2} - (1-\chi)(k'-\tan\gamma') + k\sigma_2/f_c}{\tan\gamma}. \tag{7}$$

The value σ_1/f_c is set by varying the parameters k and γ and corresponds to the minimum power of plastic strain.

According to the obtained parameters k and $\tan\gamma$ on the failure surface, the level of shear stresses

$$\frac{\tau_n}{f_c} = \frac{\sqrt{(1-\chi+\chi^2)/3}(1+k\tan\gamma)}{\sqrt{4(k-\tan\gamma)^2 + (1+k\tan\gamma)^2}}, \tag{8}$$

and normal stresses

$$\frac{\sigma_n}{f_c} = 1 - \chi - \frac{4\sqrt{(1-\chi+\chi^2)/3}(k-\tan\gamma)}{\sqrt{4(k-\tan\gamma)^2 + (1+k\tan\gamma)^2}}, \tag{9}$$

is established.

When used as a vary parameter of the angle $\psi = \pi/2 - 2\gamma$ between the tangent to the strength condition (3) and the direction of normal stresses σ_n, the following expressions were obtained to determine σ_1, τ_n and σ_n

$$\frac{\sigma_1}{f_c} = \frac{2\left[\sqrt{(1-\chi+\chi^2)/3}\sqrt{1+4\tan^2\psi'} - (1-\chi)\tan\psi'\right] + \left(\sqrt{1+\tan^2\psi'} + \tan\psi'\right)\sigma_2/f_c}{\sqrt{1+\tan^2\psi'} - \tan\psi'}, \quad (10)$$

$$\frac{\tau_n}{f_c} = \frac{\sqrt{(1-\chi+\chi^2)/3}}{\sqrt{1+4\tan^2\psi}}, \quad (11)$$

$$\frac{\sigma_n}{f_c} = 1 - \chi - 4\frac{\sqrt{(1-\chi+\chi^2)/3}\tan\psi}{\sqrt{1+4\tan^2\psi}}. \quad (12)$$

Under conditions of plane strain, the equations for determining the stresses σ_1, τ_n, σ_n have the form

$$\frac{\sigma_1}{f_c} = \frac{\left(\frac{1}{3}\frac{1-\chi+\chi^2}{1-\chi} + \frac{1-\chi}{4\tan^2\psi}\right)\psi + \frac{\sigma_2}{f_c}\left(\sqrt{1+\tan^2\psi} + \tan\psi\right)}{\sqrt{1+\tan^2\psi} - \tan\psi}, \quad (13)$$

$$\frac{\tau_n}{f_c} = \frac{1-\chi}{2\tan\psi}, \quad (14)$$

$$\frac{\sigma_n}{f_c} = \frac{1-\chi}{4}\left[\frac{1}{\tan^2\psi} - 1 - \frac{1}{3}\left(\frac{1+\chi}{1-\chi}\right)\right]. \quad (15)$$

3 The Problem of the Action of a Rectangular Stamp on a Concrete Base Under a Plane Stress State and Plane Strain

We consider the problem of determining the limit value of a uniformly distributed load q, which is transmitted to a concrete base through a rectangular stamp (Fig. 1).

To solve it in a plane stress state and plane strain, the well-known method of characteristic lines is used [24–26]. A thin and infinitely long base is adopted. The strains are considered small, so the change in the outline of the free surface can be neglected.

Two families of characteristic lines z and u and three characteristic areas of the stress state are considered: I (ABC) – simple stress state of uniaxial compression in the horizontal direction, which corresponds to a free rectilinear boundary; II (CBO) – a centered arear with an alternating stress state in the direction of the characteristics u with a family of radial straight lines $z = const$ centered at point B; III (BOB')

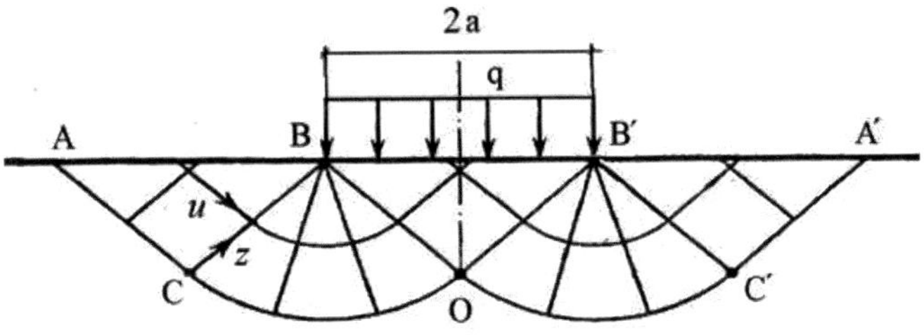

Fig. 1 Calculating the strength of a concrete base when exposed to a rectangular stamp

– biaxial compression with principal normal stresses $\sigma_1 = q$. The location of the OB'C' and C'B'A' arears relative to the central axis is symmetric to the OBC and CBA arears.

The parameters that determine the stress state are first set in the I arear with a sequential transition to the I arear, the value of the $C_{II}(u)$ is set using the properties of the characteristic lines and the boundary conditions on the line BC (B'C').

The calculation is made for concrete with a ratio $\chi = f_{ct}/f_c = 0.1, d = 0.551 f_c$.

In a plane stress state, the angle of inclination of the characteristic lines in the I area to direction of action σ_1 is set by the formula

$$\gamma_I = \frac{1}{2}\text{arc cos}\,\frac{m - 0.5f_c}{1.5f_c} = \frac{1}{2}\text{arc cos}\,\frac{0.5 - \chi}{1.5} \tag{16}$$

and is equal to $\gamma_I = 37.27^{\circ}$.

For the convenience of consideration, the parameter is entered $t = (\sigma_1 - \sigma_2)/2$, which in the area I and when $\sigma_1 = f_c$ and $\sigma_2 = 0$ is equal $t_I = 0.5f_c$; $t_I/d = 0.907$.

The equation on the characteristics $u = const$ is written as

$$\arcsin\left(\frac{5}{3} - \frac{8}{3}\frac{t_{III}^2}{d^2}\right) + \frac{1}{2}\arcsin\left(\frac{5}{3} - \frac{2}{3}\frac{d^2}{t_{III}^2}\right) = \arcsin\left(\frac{5}{3} - \frac{8}{3}\frac{t_I^2}{d^2}\right) + \frac{1}{2}\arcsin\left(\frac{5}{3} - \frac{2}{3}\frac{d^2}{t_I^2}\right) - \pi. \tag{17}$$

Substituting the values of the parameters t_I and d received: $t_{III}/d = 0.82$, $t_{III}/f_c = 0.452$.

The principal stresses in area III are determined from the equations

$$\sigma_1 = m + \sqrt{3\left(d^2 - t_{III}^2\right)} + t_{III} = \left(1 - \chi + \sqrt{1 - \chi + \chi^2 - 3t_{III}^2} + t_{III}\right)f_c, \tag{18}$$

$$\sigma_2 = m + \sqrt{3\left(d^2 - t_{III}^2\right)} - t_{III} = \left(1 - \chi + \sqrt{1 - \chi + \chi^2 - 3t_{III}^2} - t_{III}\right)f_c \tag{19}$$

and are equal to $\sigma_1 = 1.9f_c$, $\sigma_2 = 0.995f_c$.

Limit value of uniform load $q = 1.9f_c$.

The angle between the characteristic u and the direction in area III is determined by the formula

$$\gamma_{\text{III}} = \frac{1}{2}\arccos\frac{m - (\sigma_1 + \sigma_2)/2}{3t_{\text{III}}} = \frac{1}{2}\arccos\frac{1 - \chi - (\sigma_1 + \sigma_2)/2}{3t_{\text{III}}} \quad (20)$$

and is equal to $\gamma_{\text{III}} = 57°$.

The sizes of areas I, II and III are: $BO = a/\sin\gamma_{\text{III}} = 1.19a$; the radius of the u-characteristic at the boundary of arears I and II is defined as

$$r_{BC} = \sqrt[4]{\frac{4t_{\text{III}}^2 - d^2}{4t_{\text{I}}^2 - d^2}}\, r_{BO}, \quad (21)$$

$BC = r_{BC} = 1.1a$, $AB = 2r_{\max}\cos\gamma_1 = 1.75a$, $AA' = 2(AC + a) = 5.5a$.

Considering that $\chi = 0.15$: $\gamma_{\text{I}} = 38.25°$, $\gamma_{\text{III}} = 58.04°$, $t_{\text{III}} = 0.428 f_c$, $\sigma_1 = q = 1.85 f_c$, $\sigma_2 = 0.989 f_c$; $\chi = 0.05$: $\gamma_{\text{I}} = 36.27°$, $t_{\text{III}} = 0.475 f_c$, $\sigma_1 = q = 1.95 f_c$, $\sigma_2 = 0.999 f_c$.

Under conditions of plane strain, the placement of arears I, II, and III is similar to that shown in Fig. 1.

In the arear I a simple stress state ($\sigma_2 = 0$) is realized, here the unknown angle of inclination of the characteristics u to the direction of the stress action σ_1, the value and boundary conditions in the plane of the BC, which is adjacent to the area II. The values of γ_{I} and σ_1 are set from the dependencies

$$\gamma_{\text{I}} = \frac{1}{2}\arccos\frac{m}{m + 2d} = \frac{1}{2}\arccos\frac{1 - \chi}{1 - \chi + 2\sqrt{(1 - \chi + \chi^2)/3}}, \quad (22)$$

$$\sigma_{\text{I}} = 2t_{\text{I}} = m + 2d = \left(1 - \chi + 2\sqrt{(1 - \chi + \chi^2)/3}\right) f_c \quad (23)$$

and are equal to $\gamma_{\text{I}} = 31.63°$, $\sigma_{\text{I}} = 2 f_c$, $t_{\text{I}} = f_c$.

The condition on the characteristics $u = const$, which makes it possible to access the parameters of arear III through a centered arear, with plane strain is written in the form

$$\tan 2\gamma_{\text{III}} - 2\gamma_{\text{III}} = \tan 2\gamma_{\text{I}} - 2\gamma_{\text{I}} + \pi. \quad (24)$$

After substituting the angle values γ_{I} and d are installed: $2\gamma_{\text{III}} = 79.53°$, $\gamma_{\text{III}} = 39.77°$.

Parameter is equal to $t_{\text{III}} = 0.5m/\cos 2\gamma_{\text{III}} = 0.5(1 - \chi)/\cos 2\gamma_{\text{III}} = 2.47 f_c$.

The principal stresses in area III are determined from the equations

$$\sigma_1 = \frac{t_{\text{III}}^2}{1 - \chi} + t_{\text{III}} - \frac{(1 + \chi)^2}{3(1 - \chi)}, \quad (25)$$

$$\sigma_2 = \frac{t_{\mathrm{III}}^2}{1-\chi} - t_{\mathrm{III}} - \frac{(1+\chi)^2}{3(1-\chi)}, \tag{26}$$

$$\sigma_3 = \frac{\sigma_1+\sigma_2}{2} + \frac{m}{2} = \frac{\sigma_1+\sigma_2}{2} + \frac{1-\chi}{2} f_c \tag{27}$$

and are equal to $\sigma_1 = 9.18 f_c$, $\sigma_2 = 4.22 f_c$, $\sigma_3 = 7.15 f_c$.

The sizes of areas I, II and III are: $BO = a/\sin\gamma_{\mathrm{III}} = 1.56a$; the radius of the u-characteristics at the boundary areas I and II is defined as

$$r_{BC} = \sqrt{1 + \frac{2\gamma_{\mathrm{III}} + \pi - 2\gamma_{\mathrm{I}}}{\tan 2\gamma_{\mathrm{I}}}} r_{BO}, \tag{28}$$

$BC = r_{BC} = 2.58a$; $AB = 2r_{BC}\cos\gamma_{\mathrm{I}} = 4.39a$, $AA' = 2(AC + a) = 10.78a$.

Having that $\chi = 0.15$: $\gamma_{\mathrm{I}} = 31.92°$, $\gamma_{\mathrm{III}} = 39.80°$, $t_{\mathrm{III}} = 2.35 f_c$, $\sigma_1 = 8.75 f_c$, $\sigma_2 = 4.04 f_c$, $\sigma_3 = 6.82 f_c$; $\chi = 0.05$: $\gamma_{\mathrm{I}} = 31.39°$, $\gamma_{\mathrm{III}} = 39.75°$, $t_{\mathrm{III}} = 2.61 f_c$, $\sigma_1 = 9.66 f_c$, $\sigma_2 = 4.45 f_c$, $\sigma_3 = 7.53 f_c$.

According to [26], the distribution of velocities is as follows: the triangular area I moves downward with a speed V relative to areas II and III, which move away from the central axis and up. Velocity jumps take place along the boundaries of the arears on the CO and C’O lines. An analysis of the distribution of strains in this problem is given in [27].

In a simplified version for applying the dependences of the variational method, the influence of areas I and II on the strength of the concrete base is proposed to be taken into account by lateral compression.

Under the conditions of a plane stress state at $\sigma_2 = f_c$, the stress value σ_1 and the geometrical dimensions of the III area (angle γ_{III}) correspond to the characteristics obtained by the method above.

To determine the value of the ultimate load, the dependence is proposed

$$q = \sigma_1 = (2 - \chi) f_c. \tag{29}$$

Under conditions of plane strain at $\sigma_2 = 4.22 f_c$ the stress value σ_1 and geometrical dimensions of the III area, the characteristics are similar to those obtained by the method. To determine the ultimate load and principal stresses when a rectangular stamp is applied to a concrete base under plane strain conditions, the following dependencies are proposed

$$q = \sigma_1 = (10 - 8\chi) f_c, \tag{30}$$

$$\sigma_2 = 4.7(1 - \chi) f_c. \tag{31}$$

Consider the case of destruction of the concrete base when pressing the stamp with splitting in the tensile zone (Fig. 2).

The values of stresses in the tensile zone are penetrated by equal to the value of the resistance of concrete to axial tension f_{ct}.

Dependence for determining the value of the ultimate load in the kinematic scheme shown in Fig. 2, under conditions of plane stress has the form

$$q = \sigma_1 = \frac{f_c}{\tan\gamma'}\left[\sqrt{(1-\chi+\chi^2)/3}\sqrt{4(k'-\tan\gamma')^2+(1+k'\tan\gamma')^2} - (1-\chi)(k'-\tan\gamma')\right] + k' f_{ct}\left(\frac{h}{a} - \frac{1}{\tan\gamma'}\right). \quad (32)$$

The strength of the concrete base is influenced by the ratio of its height to the width of the stamp h/a.

For $\chi = 0.1$ a minimum power of plastic strain with varying parameters k and $\tan\gamma$ in relation to a ratio $\frac{h}{2a} = 6.32$ is equal to $q = \sigma_1 = 1.9 f_c$, the value of the load when concrete is destroyed only in the compressed zone.

The shear and normal stresses on the failure surface in the compressed zone are established from Eqs. (8), (9), (11) and (12) and are equal to $\tau_n = 0.55 f_c, \sigma_n = 0.9 f_c$.

The calculation results are given in Table 1.

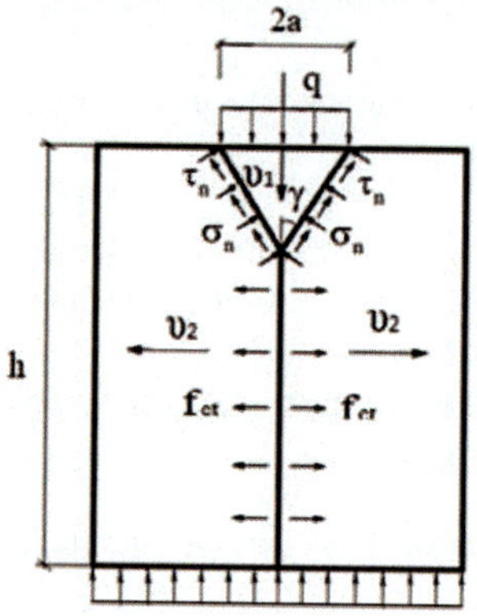

Fig. 2 Kinematic scheme of destruction of the concrete base when pressing a rectangular stamp with splitting

Table 1 The results of the calculation of the strength of the concrete base with simultaneous failure in compressed and tensile zones at $\chi = 0.1$

$\frac{h}{2a}$	$\frac{\sigma_1}{f_c}$	$\frac{\tau_n}{f_c}$	$\frac{\sigma_n}{f_c}$	k	γ,°	ψ,°
4	1.61	0.535	0.639	0.724	28.95	6.96
5	1.74	0.546	0.749	0.643	28.77	3.97
6	1.86	0.550	0.862	0,572	28.79	0.99
7	1.97	0.549	0.983	0.507	29.05	−2.16

For concretes with the ratio $\chi = 0.15$ and $\chi = 0.05$, the boundary between the cases of destruction only in the compressed zone and simultaneously in the compression and tension zones, respectively, is $\frac{h}{2a} = 4.38$ and $\frac{h}{2a} = 12.1$.

To establish the boundary between the cases of destruction, the dependence

$$\frac{h}{2a} = \frac{1}{3} + \frac{0.6}{\chi}. \tag{33}$$

For a larger value determined according to (33), the value $\frac{h}{2a}$ the I case of destruction is realized (Fig. 1), for a smaller value, the II case of destruction is realized (Fig. 2).

4 Conclusions

1. When solving the problem of the strength of a concrete base for indentation of a rectangular stamp by the method of characteristic lines and dependencies of the proposed variational method using the principle of virtual velocities and the criterion for the minimum power of plastic strain, the same values of the ultimate uniform load and geometric parameters of the area under the stamp were obtained.
2. The size of the principal stress σ_2 in the horizontal direction in the triangular arear of biaxial compression under the stamp at the plane stress state is equal to the principal stress $\sigma_1 = f_c$ in the triangular arear of uniaxial compression, which is adjacent to the free surface of the concrete base. Analysis of the stress state of the area under the stamp made it possible to propose for determining the values of large principal stresses σ_1 and the value of the ultimate load in it the dependence $q = 2f_c - f_{ct}$.
3. In conditions of plane strain, the value of the lower principal stresses in the area of biaxial compression under the stamp is equal to $\sigma_2 = 4.7(f_c - f_{ct})$ which correspond to the value of the principal stresses and the ultimate load on the concrete base $q = \sigma_1 = 10f_c - 8f_{ct}$.
4. The arear of realization of destruction only in the compressed zone near the stamp (case I) and simultaneous destruction in the zone of compression and tension (case II) at a plane stress state was determined. The boundary of cases of destruction corresponds to the ratio of the height of the concrete element to the width of the stamp $\frac{h}{2a} = \frac{1}{3} + 0.6\frac{f_c}{f_{ct}}$. If the ratio is greater than or equal to the specified value, the case I is realized, otherwise, the case of II.
5. The results obtained by the authors indicate that the theory of concrete plasticity is promising for solving problems of the strength of concrete elements under local compression.

References

1. Ince R, Arici E (2004) Size Effect in Bearing Strength of Concrete Cubes. Constr Build Mater 18:603–609
2. Yahya NA, Dhanasekar M, Zain M, Oh CL, Lee SW (2018) Bearing strength of concrete for difference heights of concrete blocks. In: IOP conference series: materials science and engineering, vol 431, p 072011
3. Gasii G, Hasii O, Klimenko V (2020) Testing of the combined structural elements of support of a mine opening. In: Web of conferences, vol 168, p 00028
4. Gasii GM (2014) Technological and design features of flat-rod elements with usage of composite reinforced concrete. Metall Min Ind 6(4):23–25
5. Venckevicius V (2005) About the calculation of concrete elements subjected to local compression. J Civ Eng Manag 11(3):243–248
6. Roberts-Wolimann CL, Banta T, Bonetti R, Charney F (2006) Bearing strength of light-weight concrete. ACI Mater J 103(6):459–466
7. Pavlikov A, Kochkarev D, Harkava O (2020) Analysis of eccentrically loaded members of circular cross section by nonlinear deformation model. In: Proceedings of the 2nd international conference on building innovations. ICBI 2019. Lecture Notes in Civil Engineering, vol 73, pp 171–181. Springer, Cham
8. Filonenko O, Maliushytskyi O, Mahas N (2018) Design features of thermomodernization of combined roofs. Int J Eng Technol 7(3.2):111–115
9. Filonenko O, Yurin O, Kodak O (2018) Thermal modernization of the panel buildings external walls. Int J Eng Technol 7(3.2):116–122
10. Dovzhenko O, Pohribnyi V, Klymenko Y, Oreškovič M (2020) Use of extreme properties of deformation for estimation of strength of constructive concrete and reinforced concrete. News Natl Acad Sci Repub Kaz Ser Geol Tech Sci 3(441): 32–39. https://publons.com/publon/33368786/
11. Yahya NA, Manicka D (2017) Strategies for mitigation of the failure of concrete pedestals supporting bridge girder bearings. Queensland University of Technology
12. Nielsen MP, Hoang LC (2011) Limit analysis and concrete plasticity. CRC Press, Taylor & Francis Group
13. Sorensen JH, Hoang LC, Olesen JF, Fischer G (2017) Test and analysis of a new ductile shear connection design for RC shear walls. Struct Concr 18:189–204
14. Yu MH (2002) Advances in strength theories for materials under complex stress state in the 20th century. Mech Rev 55(3):169–218
15. Sokolov BS (2014) Sozdaniye. Primeneniye i Razvitiye Teorii Silovogo Soprotivleniya Anizotropnykh Materialov Szhatiyu Stroitelnyye konstruktsii. Zdaniya i sooruzheniya 3(20):68–74. (in Russian)
16. Karpov VV, Salnikov AY (2009) Variatsionnyye Metody i Variatsionnyye Printsipy Mekhaniki pri Raschete Stroitelnykh Konstruktsiy. SPbGASU. (in Russian)
17. Morar GA (1990) Metod Razryvnykh Resheniy v Mekhanike Deformiruyemykh Tel. Shtiinia, Kishinev. (in Russian)
18. Mitrofanov V (2006) Teoriya Ideal'noy Plastichnosti kak Elementarnaya Mekhanika Psevdoplasticheskogo Predel'nogo Sostoyaniya Betona: Osnov y, Ogranicheniya, Prakticheskiye Aspekty Sovershenstvovaniye. City utilities 72:6–26. (in Russian)
19. Pohribnyi V, Dovzhenko O, Maliovana O (2018) The ideal plasticity theory usage peculiarities to concrete and reinforced concrete. Int J Eng Technol 7(3.2):19–26
20. Dovzhenko O, Pohribnyi V, Pents V, Mariukha D (2018) Bearing capacity calculation of reinforced concrete corbels under the shear action. In: MATEC web conferences, vol 230, p 02005
21. Dovzhenko O, Pohribnyi V, Yurko I (2018) Concrete and reinforced concrete strength under action of shear, crushing and punching shear. In: IOP conference series: materials science and engineering, vol 463, p 022026

22. Dovzhenko O, Pohribnyi V, Yurko I, Shostak I (2017) The bearing capacity experimental determination of the keyed joints models in the transport construction. In: MATEC web of conferences, vol 116, p 02011
23. Pohribnyi V, Dovzhenko O, Kuznietsova I, Usenko D (2018) The improved technique for calculating the concrete elements strength under local compression. In: MATEC web conferences, vol 230, p 02025
24. Geniyev GA, Kissyuk VN, Tyupin GA (1974) Teoriya Plastichnosti Betona i Zhelezobetona. Strojizdat, Moscow (in Russian)
25. Kolmogorov VL (1986) Mekhanika Obrabotki Metallov Davleniyem. Metallurhyia, Moscow (in Russian)
26. Kachanov LM (2013) Osnovy Teorii Plastichnosti. Ripol Klassik, Moscow (in Russian)
27. Bukhanko AA, Khromov AI (2002) Polya Deformatsiy pri Vnedrenii Klinoobraznykh i Ploskikh Shtampov. Dalnevostochnyy matematicheskiy zhurnal 2(3):311–319 (in Russian)

Construction Companies Development and Competitiveness Ensuring Based on Project Management Standards Implementation

Oleksandr Redkin, Oleksandr Zyma, Sahib Farzaliyev, and Olha Tsvihunenko

Abstract The state of industry and construction in Ukraine, the world experience of effective project management application in construction and investment projects and enterprise development programs are studied. The need for construction of new or reconstruction, modernization and technical re-equipment of existing buildings and structures for industrial and civil purposes, including infrastructure, was identified. Specific requirements, procedures and tools were defined. They allow to successfully manage the implementation of "turnkey" construction projects. Recommendations and procedures for project managers and project teams, as well as project management standards to be implemented by Ukrainian construction companies to ensure their continued development and competitiveness were proposed. The recommendations given in the article concern: the project manager; the project initiating procedure; the order and principles of communications construction; project or program staff; features of report preparation and their analytical processing; work with contracts and contractors; implementation of the project engineering main functions; their software and hardware; activities and interactions of leading managers, program managers and senior management; development of basic business processes regulations; project planning, budgeting and evaluation; decision-making. The application of the proposals and recommendations presented in the work will facilitate the rapid transition of Ukrainian construction companies to world standards of construction management, as well as ensure their development and competitiveness.

Keywords Construction · Construction company · Program · Project · Project management

O. Redkin · O. Zyma (✉) · O. Tsvihunenko
National University «Poltava Yuri Kondratyuk Polytechnic», Poltava, Ukraine

S. Farzaliyev
Azerbaijan University of Architecture and Construction, 5 Ayna Sultanova, Baku, Azerbaijan

V. Onyshchenko et al. (eds.), *Proceedings of the 3rd International Conference on Building Innovations*, Lecture Notes in Civil Engineering 181,
https://doi.org/10.1007/978-3-030-85043-2_33

1 Introduction

Today the depreciation of fixed assets of enterprises in Ukraine reaches 85–90% [1]. That is why there is a need to build new or reconstruct, modernize and technically re-equip existing buildings and structures for industrial and civil purposes, including infrastructure. According to experts, to achieve the European standard of living and development, Ukraine needs a total investment of 3.0–3.5 trillion USD., including in the field of capital construction - 1.5–1.6 [2]. This ratio of investment needs and the experience of the world's leading countries shows that domestic construction companies must become the locomotives of modern development. At the same time, they must be the most advanced and competitive enterprises. One of the ways to implement this task (along with the introduction of advances in science and technology) is the application of standards of strategic and project management, development and engineering.

2 The Main Material

The traditional form of separate objects construction organization (buildings and constructions), and especially their large and unique complexes (industrial enterprises, objects of oil and gas complex and transport, other difficult systems) which was applied earlier and extended in the present Ukraine construction complex, outdated and has a number of disadvantages:

- the presence of construction participants large number who have different purely corporate interests, which often bring contradictions between these participants, significantly slow down and worsen the entire construction process;
- the diversity of participants not only does not ensure the integrity of the process of design, construction and operation of the facility, "breaks" this process into separate, fairly isolated phases, stages and complexes, but also leads to significant economic overruns, unreasonable increase in cost and construction time, significant decline in its quality;
- poor economic conditions, lack of cheap and long-term investment in construction, crises in all management areas have led to the decline of construction companies, falling to zero levels of their opportunities for innovative development and technological renewal: complete aging and wear of construction equipment, lack of renewal during recent decades, scientific and technological lag behind the world's leading construction companies, falling skills of workers and other personnel, etc.;
- today, the construction industry, like other sectors of the economy, have worn-out fixed assets, use materials, technologies, scientific, technical and design developments, standards of production, construction and quality, which correspond to 3–4 technological way. At the same time, the world's leading construction and manufacturing companies use science-intensive products and advanced technologies, production and construction organization forms, which correspond to 6–7

technological way. Thus, today the level of scientific and technical (innovative), technological and organizational development of the construction industry and the main sectors of the Ukraine economy lag behind world management standards and competitiveness by 60–70 years [1, 2, 13–22];

These and other problems of socio-economic development and innovative progress must be addressed in an integrated manner at all levels of government, in various spheres of life and management by developing and implementing appropriate strategic target programs at the state, regions, enterprises and business levels. The construction industry has a leading role in the practical implementation of Ukraine modern economic reforms, in ensuring the economic entities development and increasing the population's welfare, solving their everyday problems.

The world practice of effective application of the project management modern standards in the construction investment projects development and implementation indicates ways and mechanisms to eliminate the above shortcomings of traditional for today's Ukraine construction organization and management forms [1, 3–16]. At the same time program-target management (project approach) allows to ensure real progress, construction efficiency and competitiveness by significantly reducing the construction duration and cost while increasing its quality and effectiveness. This is possible under the conditions of advanced technical, technological, organizational and other innovations systematic implementation. Today in the capital construction and organization field of enterprise and other real estate development, it is necessary to implement the realization of "turnkey" relevant unique projects and programs that provide:

effective development of design solutions, solving all construction tasks and obtaining the necessary competitive results under the project team unified leadership based on project management standards;

creation of such teams on the basis of project-oriented, construction-investment, engineering and development companies activity.

During the construction investment projects implementation and the project-oriented construction organizations operation the following must be performed:

- marketing and other research that contributes to the work growth within the enterprise national and international projects;
- scientific and technical research and design works to ensure continuous improvement of applied engineering and technology in the implementation, operation and further development of projects and their products;
- pre-project research and feasibility study of the main goals and directions of enterprise various future projects and programs; their investment support expediency and efficiency; technical tasks preparation, projects and programs feasibility studies and business plans, etc.;
- project activities planning;
- innovation, investment and other projects implementation;
- preparation and execution of source and permit documentation;
- providing a full range of engineering and other services;

- schemes selection and finance forms, their implementation organization and control, including financial flows control;
- preparation of agreements with design, survey and other organizations for their involvement in the project documentation development and adjustment, as well as for the author's supervision implementation;
- projects development and provision with the necessary design and estimate documentation;
- organization of production and technical equipment for projects;
- formation of subject or branch databases;
- selection of project participants on a competitive basis with the conclusion of the contracts most effective types;
- construction and reconstruction organization, including preparation of sites, manufacture, supply, installation, commissioning and facilities testing;
- production and transport logistics (material flow management with minimal costs);
- work coordination of all project participants in a uniform technological mode for the purpose of maximum final results achievement;
- providing project technical supervision and system quality management on the basis of total quality management methods and the requirements of ISO 9001: 2000;
- project activities management during the projects and programs implementation;
- projects legal support;
- organization of delivery-acceptance and complex maintenance of project results (construction objects);
- project products sales and other marketing services;
- solving current tasks with the authorities and the public;
- issues solution related to the company operation and further improvement, including its production, economic, financial, commercial, innovative, foreign economic and other areas;
- branded and lifelong service implementation of project results, participation in their operation and improvement, etc.;
- leasing and other types of rent;
- modern standards development and implementation of professional activity and effective project management, as well as other areas of enterprise operation.
- It is necessary to know and be able to apply the key principles of successful project and program management and work organization of different categories of staff. These principles should be grouped and considered in the following groups:

1. *Project manager:*

    - must have information on all stakeholders;
    - must have all the tools to motivate project participants;
    - the main task of the head is to find contractors and determine work execution ways;
    - must be honest and objective;

- positive experience of cooperation with the same specialists should be used in the implementation of future projects and programs;
- must be responsible, decisive and proactive;
- the the project manager work is temporary and limited by the project requirements;
- must be able to identify the causes of the problem and quickly propose a solution;
- project activities, as a rule, require skills to quickly make an effective (rational) decision (manage change). However, in any case, time should be allocated to identify risks and to develop a management plan;
- should be result-oriented, including when he does not know how to achieve it;
- key qualities in project management are competence, diligence and intuition. However, it should be borne in mind that not all successful managers are competent, just as not all managers who have had a bad work experience are incompetent;
- should take into account the views of all stakeholders and the project team, and if necessary - bring controversial views for general discussion;
- the project manager exercises his powers and performs his / her functional duties personally and, if necessary, delegates them to the team members;
- the project success mainly depends on the professionalism, effective interaction and mutual trust of team members.

2. *Project initiation is one of the defining processes at the first stage of pre-project research and justification, so it must first perform a preliminary planning of all actions and activities, which will avoid many problems in the future.*

3. *Communications construction in the project should take into account the following principles for their implementation:*

   - project development in an undesirable direction can be prevented if manager interacts effectively with all stakeholders (even with those whose role in decision-making is insignificant). Therefore, a successful joint activity requires an effective communication system and a system of anti-risk measures. While in the communications center, the project manager must notify all stakeholders of changes in the project (or program);
   - negotiations – one of the most effective tools to identify the wishes and requirements of stakeholders, analyze and resolve the problem in a timely manner, successfully, with minimal damage to any interests;
   - in order to achieve maximum mutual understanding, it is necessary to ensure a common vision of the problem by all project participants and create a single terminology base;
   - a modern manager must have the basic computer-integrated tools, methods and tools of project management, be a highly qualified and competent specialist.

4. *Project (or program) staff:*

- it is expedient for the manager to control the staff duties performance;
- to achieve successful project "turnkey" implementation is possible only if the unity of the leader, staff, the project purpose;
- the role of project documentation should not be underestimated and overestimated: highly qualified specialists who are able to increase the efficiency of project implementation are the most valuable human resource;
- the manager needs to know the staff advantages and disadvantages, as the human factor significantly affects the implementation process and results;
- if staff overloaded, then in the future the work efficiency, as a rule, decreases rapidly. Therefore, care must be taken to ensure that project team staff have sufficient time to rest;
- it is necessary to control the works performance order: if to devote most of the time to secondary tasks, the main ("critical") works and tasks will not be performed in due time;
- the manager can discuss the problems that arise during the project with the team members directly performing the work - in order not only to find the optimal solution to the problem, but also to achieve a deeper understanding;
- all team members should participate in the discussion of situations;
- staff working time should be valued and therefore the project team should not be distracted from performing optional tasks.

5. *Preparation of reports and their analytical processing:*

- any project usually includes a list of standard reports and a list of officials reviewing them;
- the responsibilities of the project manager include the preparation of analytical information on the presentation and project reports;
- the project manager must respect the team members;
- the reporting system and form should be developed in advance, taking into account the requirements of all project stakeholders;
- when preparing reports and analytical data, the project manager should take into account not only statistical information and project indicators, but also assess the possible development of certain trends in the course of the project;
- The project manager must constantly receive updated information on the results of the work and the progress of the project;
- in order to save time and resources, it is advisable to reduce the number of documents to a minimum.

6. *Contracts and contractors:*

- before hiring a contractor, the project manager must make sure that he is able to perform the work efficiently, on time and within the agreed budget;
- contracts payment - a tool that allows you to discipline both the contractor (subcontractor) and the customer (government and other authorities,

investors, business, owner, etc.). This tool is based on the charter and status of the project, and also takes into account the qualifications of its managers. To assess the status of the contract it is necessary to use a system of evaluation indicators of its implementation within the project. When the system of contracts performance evaluation does not correspond to the project evaluation system, senior management is obliged to find out why this happens;

- one of the main manager tasks is to motivate staff to create a high quality product - the result of the project;
- the project manager should be extremely objective. If stakeholders express opposing views, only one that will help improve project implementation should be taken;
- keep in mind that the contractor may put pressure on the team;
- in order for the interaction with the contractor to be effective, it is necessary to control its activities and results.

7. *Project manager:*

- the first signs of problems in the field of engineering are non-compliance with deadlines and excess costs. Engineering (design, research) unit, as a rule, seeks to constantly improve the product (project results), which can increase the time and cost of its creation, while the project manager aims to perform it within the specified time limits within the criteria specified by the customer quality. This leads to permanent conflicts that need to be resolved by the project manager;
- if changes in project areas pose risks to research, research units may slow down project progress, require additional funding, and generate conflicts in the project. The manager should take these factors into account.

8. *Hardware:*

- if the functional and technological requirements, as well as the external project environments differ from the previous ones, it is necessary to analyze the need to replace or update hardware or software;
- at the design stage, the project manager should provide additional time and budget to resolve inconsistencies related to the placement, installation, and operation of the hardware and software.

9. *Computer and software:*

- today a prerequisite for the project successful implementation is the use of modern technologies, primarily - information;
- it is necessary to understand that a new, more diverse version of a product is not always more reliable, and therefore should return to the previous version in case of failure;

- new knowledge in the project may be limited or based on erroneous initial data. Therefore, when developing a strategy, an important place is occupied by intuition and professional knowledge of the project manager.

10. *Leading managers, program managers and senior management:*
    - the team and the project manager must be informed in time about the reasons for making certain decisions;
    - the manager makes decisions and is responsible for them, so this decision should be clearly implemented, even when subordinates consider them wrong. Everyone must make an effort to achieve a positive result;
    - project and program managers must act as one team. The program manager should interact with the senior management of the decision-making organization and lobby the interests of each project included in the program.

11. *Project planning, budgeting and evaluation:*
    - because the decision maker at the program level and above can influence external project stakeholders, the project manager must have an effective relationship with him;
    - during the project implementation it is necessary to manage resources, not to exceed the budget, to implement the work schedule in a timely and high-quality manner, to control the main indicators, financing plans, volume development curve;
    - when developing a work schedule, the project manager must provide sufficient time in case of unforeseen problems;
    - the project manager and his team must understand that when the project has a fixed price, it is undesirable to change the completion date;
    - The capabilities of the project team as well as stakeholder teams need to be realistically assessed. In case of unforeseen situations, the project manager can ask for help from any team of stakeholders (or vice versa, they must offer their help);
    - The project manager and other stakeholders should not hide information about the results of the project. The stakeholder, as well as the project manager, will be able to make a rational decision if he is aware of the whole real situation.

12. *Regulations of business processes:*
    - in order to develop business process regulations (work structure, calendar, resource and organizational plans, etc.) for each project participant, it is necessary to analyze their activities and participation in project management, as well as reflect changes in the structure and composition of functional responsibilities in accordance with project plans and job descriptions (performance standards).

13. *Decision making:*
    - the manager must understand that sometimes it is better to make any decision, even a wrong one, but timely, than absolutely correct, but too late;
    - it should be borne in mind that in some cases the inaction of one of the stakeholders may contribute to the resolution of situations by other parties;
    - the project manager should analyze the problems and situations that arise, eliminate the consequences and prevent the causes of their occurrence in the future.
14. *Professional ethics and behavior:*
    - the project manager must approve ethical norms and codes of conduct at the level of the project team, and project team must implement and develop them;
    - the project management standards in force in the project must be met in a quality, timely and complete manner by all team members and project participants.
15. *Project management and working group:*
    - project manager delegates to his subordinates to perform certain powers and tasks. It should be understood that the quality of the final result has a higher priority than the unquestioning implementation of the project schedule;
    - The project manager must take into account that the team is able to draw the right conclusions, even if the input parameters of the project were incomplete.
16. *Discussion and prevention of failures:*
    - in case of negative trends, the project manager should determine at what stages mistakes were made, establish the preconditions and causes of the problem to avoid similar situations in the future;
    - project failures should be seen as new experience from which conclusions can be drawn for the successful implementation of subsequent projects;
    - the manager should take into account previous mistakes and develop risk response scenarios and alternative solutions to possible problems;
    - the project manager should improve the team's skills in project management in the use of knowledge and experience of previous projects, risk management and communication with stakeholders;
    - there is no single unified solution to the problem: the project is considered as a unique process, and therefore the solution of the problem in different projects is also unique;
    - Instead of looking for excuses for problems, the project manager and his team should develop a plan to solve them.

The above principles of successful project and program management, based on many years of global experience in project management, should be applied in project and program management processes implemented by a construction company that seeks to continuously develop and gain competitive advantage.

3 Conclusions

The experience of the leading project-oriented construction and investment enterprises of the world, which in the construction of buildings, structures, various unique objects use the standards of project management, shows that they can get: business project total duration reduction by 10–15%, including number in its active investment part by 15–25%; labor intensity reduction by 10–25%; operating costs reduction by 20–25%; whole project total cost reduction by 10–15% [1, 3, 8–12]. Therefore, the application of the proposals and recommendations presented in this paper should not only facilitate the transition of Ukrainian construction companies to construction management world standards, but also ensure their development and competitiveness.

References

1. Redkin OV (2019) Strategic and project management of national economy innovative development, its high-tech research and production systems and enterprises: [monograph], p 334. PoltNTU, Poltava
2. Bardish GO (2008) Project financing: a textbook. 3rd edn., p 464. Hi-Tech Press, Kyiv
3. Kaplan R (2004) Strategy-oriented organization. How organizations with a balanced scorecard thrive in the new business environment, p 380
4. Mazur II (2010) Project management: a textbook, p 960. Omega-L, Minsk
5. Redkin OV (2019) Theory and practice of business project management and development programs (in three parts): a textbook, p 874. PoltNTU, Poltava
6. Redkin OV (2013) Construction organization. Modern forms and mechanisms of organization and management of construction of oil and gas complex facilities. Part 1. The essence of modern forms of design, organization and management of construction projects and programs for the development of oil and gas: textbook. Manual, p 214. PoltNTU, Poltava
7. Redkin OV (2013) Construction organization. Modern forms and mechanisms of organization and management of construction of oil and gas complex facilities. Part 2. Organizational and technological planning and ensuring the implementation of construction investment projects and enterprise development programs: textbook. Manual, p 196. PoltNTU, Poltava
8. A Guide to the Project Management of Knowledge (2017) Sixth Edition (PMBoK Guide) Newtown Square, PA: Project Management Institute
9. P2M (2005) A Guidebook to the Project & Program Management. For Enterprise Innovation. Japan, PMAJ
10. Cleland D (2016) Project management strategic desing and implementation, p 440. McGraw-Hill Inc., New York
11. Kendail G (2008) Advanced project portfolio management and the PMO: multiplying ROI at warp speed. International Institute for Learning, Inc. and J Ross Publishing, Inc. Boca Raton, Florida, p 428

12. Pinto J (2014) Project management handbook, p 464. Bass Publishers, San Francisco
13. Redkin OV, Tolkachiv DM (2015) World standards of Ukraine economy innovative development management. Economics and management organization. Collection of scientific works of Donetsk National University, 1(17):215–223. Vinnytsia
14. Redkin OV, Pahomov RI, Zyma OE (2018) Now forms and world experience of organization and management of business processes and build-investment projects in the field of the complex objects development of Ukraine. Acad J Ser Ind Mach Build Civ Eng, 1(50):238–245. https://doi.org/10.26906/znp.2018.50.1081
15. Redkin OV, Chaikina AO (2019) Multi-project management of innovation and high-tech development of national economy. Acad Bull Econ Reg, 3(74):50–56. https://doi.org/10.26906/eip.2019.3(74).1762
16. Redkin O (2019) Organizational innovations in the activities of construction companies in Ukraine in the transition to world standards of management. Acad J Ser Ind Mach Build Civ Eng, 2(53):151–156. https://doi.org/10.26906/znp.2019.53.190
17. Zyma OE, Dyachenko EV, Pahomov RI (2018) Works execution organization at reconstruction and renovation of buildings after the fire with usage of slabs lifting method. Int J Eng Technol 7(2.23):242–246. https://doi.org/10.14419/ijet.v7i2.23.11951
18. Pahomov R, Zyma O, Dyachenko E (2018) Analysis and prevention of industrial injury in the construction sector. Int J Eng Technol (UAE), 7(3):285–290. https://doi.org/10.14419/ijet.v7i3.2.14421
19. Dyachenko E, Zyma O, Pahomov R, Shefer O (2020) Non-crane method of reconstructing buildings with additional storey erection. In: Onyshchenko V, Mammadova G, Sivitska S, Gasimov A (eds) Proceedings of the 2nd international conference on building innovations. ICBI 2019. Lecture Notes in Civil Engineering, vol 73. Springer, Cham. https://doi.org/10.1007/978-3-030-42939-3_4
20. Mykytas M, Terenchuk S, Zhuravska N (2018) Models, methods and tools of optimizing costs for development of clusterized organizational structures in construction industry. Int J Eng Technol (UAE), 7(3):250–254. https://doi.org/10.14419/ijet.v7i3.2.14414
21. Iasechko S, Haliantych MK, Skomorovskyi VB, Zadorozhnyi V, Obryvkina O, Pohrebniak O (2020) Contractual relations in the information sphere. Syst Rev Pharm 11(8):301–303. https://doi.org/10.31838/srp.2020.8.46
22. Drobyazko S, Bondarevska O, Klymenko D, Pletenetska S, Pylypenko O (2019) Model for forming of optimal credit portfolio of commercial bank. J Manag Inf Decis Sci 22(4):501–506

Classification of Self-stressed Steel-Concrete Composite Structures

Oleksandr Semko and Anton Hasenko

Abstract The essence of building structures preliminary stressing is the creation of internal stresses in them, opposite to those that arise during operation. The creating of initial stresses in building structures from their own weight greatly simplifies the pre-stressing process due to the unnecessary spending on additional measures and adaptations. Preliminary stresses in this case can be created due to the well-chosen nodes sizes and the development of technology for building structures preliminary assembly. In this sense, the term "self-stress" is used in this work. As is known, steel—concrete composite structures are made of successfully combined steel mainly rod elements and monolithic concrete. The creation of pre-compression of concrete in steel—concrete composite structures will have a positive effect on increasing their overall load-bearing capacity. In this work the classification of self-stressed steel—concrete composite structures depending on the method of pre-stresses creating in them and the type of deformed state on which these structures will work in the frame of the building. Preliminary self-stresses in steel—concrete composite structures can be created by the following measures: using the stressing cement for concrete mortar preparation, prediction of constructive measures for additional compaction of fresh concrete mortar (centrifugation) or for the hardened concrete pre-compression, constructive tension of the external reinforcement (sprung), changing the cross section geometric characteristics or the design scheme of the elements work in the manufacturing process, specially developed step-by-step technology of constructions manufacturing.

Keywords Steel-reinforced concrete structures · Self- stressed

1 Introduction

Steel-reinforced concrete is one of the effective types of complex structural elements [1]. It is known that it combines rolled steel profiles with reinforced by rods concrete

O. Semko · A. Hasenko (✉)
National University «Yuri Kondratyuk Poltava Polytechnic», Poltava 36011, Ukraine

V. Onyshchenko et al. (eds.), *Proceedings of the 3rd International Conference on Building Innovations*, Lecture Notes in Civil Engineering 181,
https://doi.org/10.1007/978-3-030-85043-2_34

[2]. Such structures are created both during the new design and during the reinforced concrete structures strengthening with rolled steel and metal structures - with concreting [3]. However, during strengthening, the issue of the existing structure and reinforcement elements joint operation ensuring, that is the creation of prestresses in the reinforcement elements is relevant. Increasing the crack resistance of concrete in such composite structures is also an urgent issue [4, 5]. The creation of pre-compression of concrete in steel—concrete composite structures will have a positive effect on increasing their overall load-bearing capacity. One of the reinforced concrete using advantages is the possibility of creating inseparable statically indeterminate design schemes [6]. This advantage allows to regulate their stress—strain state, to increase load-bearing capacity and rigidity and, as a result, to save materials.

2 Analysis of Recent Research and Publications

The essence of building structures preliminary stressing is the creation of internal stresses in them, opposite to those that arise during operation [7]. By means of pre-stressing constructions rigidity is reached which, in turn, increases the crack resistance of reinforced concrete structures. Due to this, it is possible to use composite structures materials work more efficiently and to reduce their material consumption. The creating of initial stresses in building structures from their own weight greatly simplifies the pre-stressing process due to the unnecessary spending on additional measures and adaptations. Preliminary stresses in this case can be created due to the well-chosen nodes sizes and the development of technology for building structures preliminary assembly. In this sense, the term "self-stress" is used in this work.

Constructive schemes of buildings and structures are optimized in order to create a planar (not spatial) scheme of its elements work [8, 9]. However, due to uneven loading and random eccentricities, it is not always possible to avoid the structural elements work on complex types of deformations. Therefore, quite often rod structures count on such combined types of deformations as bending compression, oblique bending, torsional bending, and so on [10–13]. With the pre-stress help it is possible to minimize the number of types for which the designed structure will work or at least to reduce the impact of a particular type of deformation on its overall load-bearing capacity. The scientific developments of the authors of the article are also a further development of the research of leading scientists Storozhenko L. [2], Pavlikov A. [8–10], Kochkarev D. [5, 6, 10], Azizov T. [4–6], Piskunov V. [14, 15] and other [16].

3 Formulating of the Article's Goals

The purpose of the work is to create a general classification of self-stressed steel—concrete composite structures depending on the used constructive and technological manufacture measures and the order of the composite structures components stress.

4 The Main Material with the Justification of New Scientific Results

Based on the analysis of existing experience in the design, construction and operation of self-stressed steel-reinforced concrete structures, Fig. 1. Self-stressed steel—concrete composite structures classification scheme. shows the classification depending on the type of their deformed state.

Figure 2 shows the classification of the self-stressed structures shown in the previous figure only depending on the used constructive and technological manufacture measures and the order of the composite structures components stress. These include using the stressing cement for concrete mortar preparation, prediction of constructive measures for additional compaction of fresh concrete mortar (centrifugation) or for the hardened concrete pre-compression, constructive tension of the external reinforcement (sprung), changing the cross section geometric characteristics or the design scheme of the elements work in the manufacturing process, specially developed step-by-step technology of constructions manufacturing.

Described constructive and technological manufacture measures and steel—concrete composite structures components pre-stressing order allow to eliminate certain shortcomings of such constructions. In particular, the concrete compaction in the pipe by centrifugation, which is widely used in manufacture of steel—concrete composite structures with round cross section, allows to simplify the technology of laying concrete in a long pipe, to increase the concrete strength compared to conventional vibrating concrete, to create a cavity in the concrete core, which can be in the center of the section or shifted in any direction which is of great importance when elements work on the eccentric compression or bending, and to reduce the construction weight [2].

Pre-stressing of steel structures concrete core or steel holder of reinforced concrete strengthening structures allows immediately include the strengthening elements in the work of damaged structure. That is, the formed steel—concrete composite structures have pre-stresses in the strengthening elements, which reduce the stresses in the existing part of the building element.

In order to ensure rigidity in its own plane of the buildings transverse frames arrange struts between the racks and overlapping crossbars. In this case, the overlapping can be arranged as a monolithic reinforced concrete over the steel overlapping crossbars. Regulating the struts rigidity and the scheme of their establishment it can be easily regulated the meaning and sign of tension at the overlapping crossbar. This is

SELF-STRESSED STEEL-REINFORCED CONCRETE STRUCTURES:

1) *working on compression:*
   - a) pipe concrete with a concrete core on stressing cement:
     - – with shell from steel pipes [17 – 19];
     - – with shell from fiberglass pipes [17];
   - b) centrifuged pipe concrete [20 – 21];
   - c) pipe concrete with compressed core and cut by deformation seams into separate sections with an outer steel shell [22];
   - d) pipe concrete with compressed by jacks concrete core, formed during the strengthening of steel structures and the inclusion of the core in the work [23];
   - e) pipe concrete with strained shell (double shell);
2) *working on bend:*
   - 2.1) solid cross section:
     - a) steel-reinforced concrete beams with external tape fittings [24];
     - b) reinforced concrete beams reinforced with a lever-rod system [25];
     - c) steel-reinforced concrete beams with local tension:
       - – by sprengel [26];
       - – of support sticks (construction of continuous beams) [27];
       - – as a result of strengthening [28];
     - d) precast-monolithic steel-reinforced concrete beams loaded through the stages of production (concreting) [29];
   - 2.2) through cross section (spatial):
     - a) structural cable-stayed steel-reinforced concrete constructions [30 – 33];
     - b) combined rod constructions on intermediate elastic supports [34];
     - c) transport construction objects [35 – 36];
     - d) steel-reinforced concrete structures of a covering with pre-stressing from own weight [37 – 38].

Fig. 1 Self-stressed steel—concrete composite structures classification scheme

SELF-STRESSED STEEL-REINFORCED CONCRETE STRUCTURES:

1) *creation of preliminary stresses in concrete:*
   - a) self-straining concrete (cement):
     - – with shell from steel pipes [17 – 19];
     - – with shell from fiberglass pipes [17];
     - – centrifuged pipe concrete [20 – 21];
   - b) artificial (mechanical) stress of concrete:
     - – pipe concrete with compressed core and cut by deformation seams into separate sections with an outer steel shell [22];
     - – pipe concrete with compressed by jacks concrete core, formed during the strengthening of steel structures and the inclusion of the core in the work [23];
2) *creation of preliminary stresses by steel reinforcement:*
   - a) pipe concrete with strained shell (double shell);
   - b) reinforced concrete beams reinforced with a lever-rod system [25];
   - c) steel-reinforced concrete beams with local tension by sprengel [26];
   - d) structural cable-stayed steel-reinforced concrete constructions [30 – 33];
   - e) transport construction objects [35 – 36];
3) *creation of preliminary stresses due to change of geometrical characteristics of the cross-section:*
   - a) steel-reinforced concrete beams with external tape fittings [24];
   - b) steel-reinforced concrete beams with local tension as a result of strengthening [28];
4) *creation of preliminary stresses due to manufacturing technology:*
   - a) precast-monolithic steel-reinforced concrete beams loaded through the stages of production (concreting) [29];
   - b) steel-reinforced concrete beams with local tension of support sticks (construction of continuous beams) [27];
5) *stress regulation by changing the design scheme of the element:*
   - a) combined rod constructions on intermediate elastic supports [34];
   - b) steel-reinforced concrete structures of a covering with pre-stressing from own weight [37 – 38];
   - c) adjustment of efforts on length of an element [39].

Fig. 2 Classification of the self-stressed steel—concrete composite structures depending on the used constructive and technological manufacture measures and the order of components stress the skill of obtaining a cost-effective construction. In [39] it is proved that the optimal angle of the struts inclination relative to the vertical in terms of internal forces in the strut is equal to 40…50°. That is, the design moments on the middle supports of the floor beam can be reduced by using a frame-struts scheme of the transverse building frame. In this case, it is possible to achieve a diagram of along the beam length with four calculated sections instead of two. That is to align the values of the span and support moments by aligning the steps of the crossbar supports "rack-strut-rack" installation.

The change of internal efforts in cross sections of constructions depends on supporting and span moments due to the support nodes design (height). As the shoulder of the inner pair of forces of the support nodes increases, the supporting moment increases, and the span moment decreases.

With the help of the possible self-stress measures of the components of steel—concrete composite structures presented in Fig. 2, it is possible to optimize the scheme of their joint work, increase the bearing capacity and reduce material costs for the manufacture of steel—concrete composite structures.

5 Conclusion

The redistribution of forces in self-stressed steel—concrete composite structures, which occurs at the stages of their manufacture, installation and operation, allows you to adjust within the specified allowable limits of internal forces in the cross sections at all specified stages of these structures work. However, appropriate constructive measures should be taken to ensure the adequacy of the design schemes adopted during the planning to the real operating conditions of the designed structures. Rational constructions of nodes have a special influence on fulfillment of this condition.

References

1. Pichugin SF (2018) Reliability estimation of industrial building structures. Mag Civil Eng 83(7):24–37. https://doi.org/10.18720/MCE.83.3
2. Storozhenko LI, Surdin VM, Yefimenko VI, Verbyts'kyy VI (2007) Stalezalizobetonni konstruktsiyi: doslidzhennya, proektuvannya, budivnytstvo, ekspluatatsiya: monohrafiya. Kryvyy Rih
3. Dmytrenko T, Semko O, Dmytrenko A, Derkach T, Voskobiinyk O (2018) Development and implementation of algorithms of building structure engineering calculations for shear fraction under pressing-through. In: MATEC Web of Conferences, vol 230
4. Azizov T, Jurkowska N, Kochkarev D (2019) Basis of calculation on torsion for reinforced concrete structures with normal cracks. In: Paper presented at the Proceedings of the Fib Symposium 2019: Concrete - Innovations in Materials, Design and Structures, pp 1718–1725

5. Kochkarev D, Azizov T, Galinska T (2018). Bending deflection reinforced concrete elements determination. In: Paper presented at the MATEC Web of Conferences, p 230. https://doi.org/10.1051/matecconf/201823002012
6. Kochkarev D, Azizov T, Galinska T (2020). Design of effective statically indeterminate reinforced concrete beams. https://doi.org/10.1007/978-3-030-42939-3_10
7. Semko O, Fenko O, Hasenko A, Harkava O, Kyrychenko V (2018) Influence of external and internal cooling at solidification on strength of brittle duralumin in compression. In:Paper presented at the MATEC Web of Conferences, p 230. https://doi.org/10.1051/matecconf/201823002029
8. Pavlikov A, Harkava O, Prykhodko Y, Baryliak B (2019) Highly constructed precast flat slab frame structural system of buildings and research of its slabs. In: Proceedings of the International fib Symposium on Conceptual Design of Structures. Torroja Institute, Madrid, Spain
9. Pavlikov A, Mykytenko S, Hasenko A (2018). Effective structural system for the affordable housing construction. Int J Eng Technol (UAE) 7(3):291–298. https://doi.org/10.14419/ijet.v7i3.2.14422
10. Pavlikov A, Kochkarov D, Harkava O (2019) Calculation of reinforced concrete members strength by new concept. In: CONCRETE. Innovations in Materials, Design and Structures: Proceedings of the fib Symposium, pp 820–827
11. Pavlenko A, Koshlak H (2015) Design of processes of thermal bloating of silicates. Metall Mining Ind 7(1):118–122
12. Hudz S, Gasii G, Pents V (2018) The problem of consideration torsion emergence in beams. Int J Eng Technol 7(3.2), 141–148
13. Semko VO, Hasenko AV, Mahas NM, Fenko OG, Sirobaba VO (2019) Stability of light steel thin-walled structures filled with lightweight concrete. In: IOP Conference Series: Materials Science and Engineering, vol 708, no 1, p 012071
14. Piskunov VG, Gorik AV, Cherednikov VN (2000) Modeling of transverse shears of piecewise homogeneous composite bars using an iterative process with account of tangential loads 2. Resolving equations and results. Mech Comp Mater 36(6):445–452. https://doi.org/10.1023/A:1006798314569
15. Piskunov VG, Goryk AV, Cherednikov VN (2000) Modeling of transverse shears of piecewise homogeneous composite bars using an iterative process with account of tangential loads. 1. Construction of a model. Mech Comp Mater 36(4):287–296. https://doi.org/10.1007/BF02262807
16. Loburets AT, Naumovets AG, Senenko NB, Vedula YS (1997) Surface diffusion and phase transitions in strontium overlayers on W(112). Zeitschrift Fur Physikalische Chemie 202(1–2):75–85. https://doi.org/10.1524/zpch.1997.202.part_1_2.075
17. Shakhvorostov AI (2000) Issledovaniye napryazhonno-deformirovannogo sostoyaniya trubobetona na napryagayushchem tsemente: dis. kand. tekhn. nauk: 05.23.01. NIIZHB
18. Bondarenko VM (2010) Soprotivleniye osevomu szhatiyu staletrubobetonnykh elementov kruglogo secheniya s yadrom iz napryagayushchego betona: dis. kand. tekhn. nauk: 05.23.01. Belorusskiy natsional'nyy tekhnicheskiy universitet, Brest
19. Rezvan IV (2012) Trubobetonnyye kolonny na vysokoprochnom samouplotnyayushchemsya napryagayushchem betone: dis. kand. tekhn. nauk: 05.23.05:05.23.01. Rostovskiy gosudarstvennyy stroitel'nyy universitet, Rostov-na-Donu
20. Taranovs'kiy OL (1993) Napryazhenno-deformirovannoye sostoyaniye gibkikh szhatykh elementov iz tsentrifugirovannogo trubobetona: avtoref. dis. kand. tekhn. nauk: 05.23.01. Poltavskiy inzhenerno-stroitel'nyy institut, Poltava
21. Yefimenko VI (2009) Nesuchi konstruktsiyi zi stalevykh trub, zapovnenykh tsentryfuhovanym betonom: avtoref. dys. dokt. tekhn. nauk: 05.23.01. PDABA, Dnipropetrovs′k
22. Chekanovych MH (2014) Trubobetonni konstruktsiyi, zmitsneni pozdovzhnim poperednim obtyskom. Zb. nauk. pr. Poltav. nats. tekhn. un-t im. Yuriya Kondratyuka. Haluzeve mashynobuduvannya, budivnytstvo. Vyp 3(1):221–226

23. Semko OV, Hasenko AV (2006) Pat. 26462 Ukrayina, MPK. E 04 G 23/00. Sposib pidsylennya stalevykh stysnutykh elementiv obetonyuvannyam iz nastupnym obtysnennyam betonu. zayavnyky vlasnyk PoltNTU imeni Yuriya Kondratyuka. - № u 2007 04651.
24. Kinash RI (1993) Vytryvalist′ stalebetonnykh balok iz zovnishn′oyu strichkovoyu armaturoyu: avtoref. dys. kand. tekhn. nauk: 05.23.01. NU "L′vivs′ka politekhnika"
25. Chekanovych OM (2013) Napruzheno-deformovanyy stan zalizobetonnykh z·hynal′nykh elementiv, pidsylenykh vazhil′no-stryzhnevoyu systemoyu: dys. kand. tekhn. nauk: 05.23.01. Nats. akad. pryrod. ta kurort. bud-va. Simferopol
26. Shahyn AL, Yzbash MYU, Shemet RN (2004) Otsenka nesushchey sposobnosty dvukhproletnykh stalezhelezobetonnykh lokal′no predvarytel′no napryazhennykh balok. Naukovyy visnyk budivnytstva. Vyp. 38:81–89
27. Hnidets′ VH (2008) Zalizobetonni konstruktsiyi z napruzhuvanymy stykamy i rehulyuvannyam zusyl′: monohrafiya. Vydavnytstvo NU "L′vivs′ka politekhnika", L′viv
28. Plevkov VS, Mal'ganov AI, Baldin IV (2014) Otsenka tekhnicheskogo sostoyaniya, vosstanovleniye i usileniye stroitel'nykh konstruktsiy inzhenernykh sooruzheniy: uchebnoye posobiye. - 2-ye izd., pererab. i dop. Izdatel'stvo ASV, Moskva
29. Bibik DV (2011) Napruzheno-deformovanyy stan zbirno-monolitnykh stalezalizobetonnykh balok z urakhuvannyam fizychnoyi neliniynosti: avtoref. dys. kand. tekhn. nauk: 05.23.01. PoltNTU imeni Yuriya Kondratyuka, Poltava
30. Abovskiy NP (1998) Upravlyayemyye konstruktsii: ucheb. posobiye. KrasGASA, Krasnoyarsk
31. Ivanyk YUI (2017) Mitsnist′ ta deformatyvnist′ kombinovanykh stalezalizobetonnykh poperedn′o napruzhenykh konstruktsiy: dys. kand. tekhn. nauk: 05.23.01. Natsional′nyy universytet "L′vivs′ka politekhnika", L′viv
32. Gasii G, Hasii O, Klimenko V (2020) Testing of the combined structural elements of support of a mine opening. In: E3S Web of Conferences, vol 168, p 00028
33. Storozhenko LI, Gasii GM (2014) Experimental research of strain-stress state of ferrocement slabs of composite reinforced concrete structure elements. Metall Mining Ind 6(6):40–42
34. Myron G, Zygun A, Maksiuta N (2018) New effective combined steel structures. Int J Eng Technol 7(3.2):343– 348
35. Babayev V, Shmukler V (2017) New constructive solutions for building of transport construction facilities. In: MATEC Web of Conferences
36. Vatulia GL, Lobiak OV, Deryzemlia SV, Verevicheva MA, Orel YF (2019) Rationalization of cross-sections of the composite reinforced concrete span structure of bridges with a monolithic reinforced concrete roadway slab. In: IOP Conference Series: Materials Science and Engineering
37. Seríya 7011 "Zhelezobetonnyye sbornyye bystromontiruyemyye zdaniya s proletami 18 m"
38. Semko OV, Hasenko AV, Krutybich OV (2020) Pererozpodil vnutrishnikh zusyl′ pid chas ukrupnyuval′noyi zbirky zalizobetonnoyi arky iz zatyazhkoyu. Tezy 72-yi nauk. konf. univ-tu. Tom 1:106–107
39. Semko OV, Hasenko AV, Kyrychenko VA, Sirobaba VO (2020) The rational parameters of the civil building steel frame with struts. In: Proceedings of the 2nd International Conference on Building Innovations. Part of the Lecture Notes in Civil Engineering book series, LNCE, vol 73, pp 235–243 (2020)

Study of Structural-Sorption Properties of Materials Made on the Basis of Clay-Ultradispers Additional Mixture

Irada Shirinzade, Bulud Baghirov, and Olena Mykhailovska

Abstract The article is devoted to the research of structural and sorption properties of the material obtained on the basis of a mixture of clay and ultradispersed additives. For this purpose, one of the most common methods of regulating hydrophilicity was considered to be the use of surface-active additives. To determine the sorption capacity of the ceramic mass, both hydraulic clays, which are widely used in the building ceramics industry, and bentonite clays with high sorption capacity were used. According to the experiments, the use of surface-active additives led to an increase in the adsorption of water by the ceramic mass. The influence of increased water adsorption on the properties of the ceramic material was determined experimentally. Since the properties of ceramic materials, such as absorption and density, reflect their structure, the effect of surface-active additives on these properties was studied, and it was determined that the density of the ceramic material increased significantly. Surface-active additives increase the plasticity of the ceramic mass and, consequently, improve the structural and technical properties of the material.

Keywords Clay · Surface active substance · Sorption · Ceramic mass · Ultradispersed additives · Water adsorption

1 Introduction

Ceramic materials with high ecological and physicomechanical properties have been widely used in construction at all times and are still being used. At present, one of the main conditions for the production of durable ceramic material is the ability to make scientific predictions through an in-depth study of the physical and chemical processes occurring in the production of local raw materials. The experience

I. Shirinzade (✉) · B. Baghirov
Azerbaijan University of Architecture and Construction, Ayna Sultanova -11, Baku, Azerbaijan
e-mail: i.shirinzade@azmiu.edu.az

O. Mykhailovska
National University «Yuri Kondratyuk Poltava Polytechnic», Pershotravnevyj Avenue, 24, Poltava, Ukraine

V. Onyshchenko et al. (eds.), *Proceedings of the 3rd International Conference on Building Innovations*, Lecture Notes in Civil Engineering 181,
https://doi.org/10.1007/978-3-030-85043-2_35

of foreign manufacturers with extensive experience in the production of building ceramics shows that it is impossible to organize production at a high level without studying the composition and properties of polymineral raw materials, physical and chemical processes occurring during the preparation of raw materials [1–5].

The structuring of ceramic materials obtained from low-melting clay, widespread in nature, can be influenced by various factors. In such systems, learning the factors affecting structuring is always a challenge and one of the areas that still need research.

In the production of building ceramics, mainly hydraulic clays (Fig. 1) are used, in some cases (in the production of ceramic slabs with internal cladding) kaolins. In the production of building ceramics, mainly hydraulic clays are used as raw materials, and in some cases (in the production of inner lining ceramic tiles) kaolin. Both types of clay are known to swell very slightly. Water in clay minerals is of three types [6, 7]: physically-bound water, physicochemically bound water, and chemically bound water. Physicochemically bound water is also called adsorption water. This water differs from free water for its viscosity and structure. Water around clay minerals is always in the form of charged water molecules, which is adsorption water. The maximum plasticity of the ceramic mass is provided when the hydrated layers are composed of particles (clay particles and water molecules) through a sufficiently strong bond [6].

One of the most common methods of regulating hydrophilicity in such systems is the use of surface active additives. Surface-active additives also significantly improve the molding properties of the ceramic mass. The use of surface-active additives improves the deposition of water molecules of solid particles on their surface, which increases the sorption capacity of clays.

Fig. 1 General view of the clay sample

2 Main Part

2.1 *Purpose of the Article*

The main purpose of the study is to investigate the improvement of the adsorption capacity of ceramic masses based on light clays and ultradispersed additives using surface-active additives.

2.2 *Methods*

In the course of the study, the adsorption capacity of the ceramic mass was evaluated by determining the number of adsorbed water molecules. This was done by the calorimetric method [9]. The effect of surface-active substances species and additives on the hydrophilicity of monomineral clays, as well as polymineral pres-powders has been studied by determining the heat spent on wetting of mineral powder amount [10, 11].

The amount of heat used for wetting was studied by differential microcalorimetry using a Beckmann thermometer. The thermometer records the temperature change in the system when the powder-like materials are soaked with water. The amount of heat used for wetting is determined by the following formula:

$$\Delta H = (t_k - t_h) \cdot K/m$$

here, ΔH - the amount of heat spent on wetting, kC/kg;

t_k and t_h - temperature of the Beckmann thermometer when wetting water with dust, °C;

K - coefficient of Beckmann thermometer (285,44);

m - weight of tested material in dry state, g.

Influence of the amount of heat consumed for the specified wetting on the amount of chemically bound water (A,%) is calculated by the following formula:

$$A = (Q_{is.m} \cdot \rho \cdot h/q) \cdot 100\%$$

ρhere, $Q_{is.m}$ - Amount of heat used to wet 1 g of solid phase, C/g;

ρ - density of chemically bound water, $(1{,}2 - 1{,}3) \cdot 10^6$ g/m^3;

h - chemically bound water thickness, $2{,}7 \cdot 10^{-10}$ m;

q - full surface energy of "water–vapor", 0,116 C/m^2.

3 Results

To determine the adsorption capacity of the ceramic mass used a mixture based on the clay Ashaghy Guzdek and ultradispersed metakaolin. This type of clay is the most widely used in the production of ceramic materials. Dash-Salahly bentonite clays with high sorption capacity were also used for comparison. The results are presented in Table 1, Table 2, and Table 3. From the results it can be seen that the amount of adsorption water of ordinary clays is 2 times lower than the amount of adsorption water of bentonite clays.

The next step was to study the effect of the type and amount of additives on the sorption properties of the mass prepared on the basis of a mixture of clay and clay-ultradisperse mixtures. Test results were presented in Table 1, Table 2 and Table 3. During the study of the sorption properties of these systems as chemical additives were used POZZOLITH 42 CF and C-3.

The technological properties of the Ashaghy Guzdek clay and ceramic mass consisting of a mixture of this clay and metakaolin, with the addition of surface active additives and without adding surfactant, were studied and it was found that the surface-active additive improves the properties of the mixture (Table 4).

As can be seen from Table 1 and Table 3, the highest value of water adsorption in the mixture of ceramic mass prepared using a mixture of clay and clay-ultradisperse additives and surface-active additives was obtained in the amount of 0.5% surfactant (surface-active additive). Therefore, 0.5% surface-active additive was used to determine the technological properties of the ceramic mass.

Table 1 Influence of type and quantity of surface-active additives on sorption properties of clay

Composition of samples (amount of additive by mass, %)	The amount of heat released during wetting, ΔH, C/q	Amount of adsorption water, A,% (by mass)
Clay Ashaghy Guzdek + water	21.7	6.6
Clay Ashaghy Guzdek + water + POZZOLITH 42 CF (0.3)	34.7	10.5
Clay Ashaghy Guzdek + water + POZZOLITH 42 CF (0.5)	38.9	11.8
Clay Ashaghy Guzdek + water + POZZOLITH 42 CF (0.7)	33.6	10.2
Clay Ashaghy Guzdek + water + C-3 (0.3)	30.6	9.2
Clay Ashaghy Guzdek + water + C-3 (0.5)	34.2	10.3
Clay Ashaghy Guzdek + water + C-3 (0.7)	29.1	8.8

Table 2 The influence of the type and amount of surface-active additives on the sorption properties of bentonite clay Dash-Salahly

Composition of samples (amount of additive by mass, %)	The amount of heat released during wetting, ΔH, C/q	Amount of adsorption water, A,% (by mass)
Bentonite clay + water	38.5	11.6
Bentonite clay + water + POZZOLITH 42 CF (0.3)	35.0	10.6
Bentonite clay + water + POZZOLITH 42 CF (0.5)	36.2	10.9
Bentonite clay + water + POZZOLITH 42 CF (0.7)	33.4	10.1
Bentonite clay + water + C-3 (0.3)	30.9	9.3
Bentonite clay + water + C-3 (0.5)	29.5	8.9
Bentonite clay + water + C-3 (0.7)	29.5	8.9

Table 3 Influence of the type and amount of surface-active additives on the sorption properties of the clay-ultradisperse mixture

Composition of samples (amount of additive by mass, %)	The amount of heat released during wetting, ΔH, C/q	Amount of adsorption water, A,% (by mass)
Clay-ultradisperse additive mixture + water	18.9	5.7
Clay-ultradisperse additive mixture + water + POZZOLITH 42 CF (0.3)	25.3	7.6
Clay-ultradisperse additive mixture + water + POZZOLITH 42 CF (0.5)	27.8	8.4
Clay-ultradisperse additive mixture + water + POZZOLITH 42 CF (0.7)	25.6	7.7
Clay-ultradisperse additive mixture + water + C-3 (0.3)	24.9	7.5
Clay-ultradisperse additive mixture + water + C-3 (0.5)	26.8	8.1
Clay-ultradisperse additive mixture + water + C-3 (0.7)	24.5	7.4

Table 4 Ceramic indicators of clay

Combustion temperature, ^{0}C		Shrinkage during burning, %		Moisture absorption, %		Medium density, g/sm^3	
Clay	Clay + metakaolin + POZZOLITH 42 CF (0.5)						
950	950	3.7	3.3	15.8	15.5	1.70	1.73
1000	1000	4.3	3.7	12.5	10.8	1.74	1.80
1050	1050	5.4	4.2	8.8	6.7	1.80	1,88
1100	1100	5,1	4,1	5,6	5,1	1,86	1,92

4 Discussion

As is known from the results of experiments, the use of surface active additive increases the plasticity of clay minerals, which leads to an increase in density. Due to the high mechanical strength and low water absorption of densely structured material, its frost resistance will be even higher, which is one of the important factors in the production of ceramic materials for exterior cladding.

As can be seen from Table 1, the water absorption of the samples decreases with increasing temperature. When the temperature is increased by 150 ^{0}C, the water absorption of samples consisting only of clay decreases by 13%, and the water absorption decreases by 23% when using metakaolin additive. This indicates that the softening additive material had a positive effect on the compaction of the additive material.

The effective special surface is the surface taken up by adsorption water. The results of the study show that the effective specific surface area of clay samples of the Bina deposit increases by 0.5% of the surface active additive (from 10.5 to 11.8 cm^2/g), and at higher values of the additive the effective specific surface area begins to decrease.

In samples of Dash-Salahly bentonite clay, the effective specific surface area changed very little, that is, from 10.6 to 10.9 cm^2/g. As can be seen, in ordinary kaolin-hydraulic clays this amount is greater. This is due to their activity. Hydraulic clays are considered more active clays, and this property is very important in the presented work. This is because the more active clay minerals are, the easier they will interact with the oxides in the clay and lead to the formation of the structure of ceramic materials.

5 Conclusions

Therefore, the experiments allow to conclude that the use of surface active additive in the preparation of ceramic material with the addition of clay-ultradisperse materials can improve the efficiency of adjusting of the molding process and improve the properties of the resulting material.

References

1. Weiwei X, Zengbin Y, Juntang Y, Zenhua W, Yue L (2017) Preparation and characterization of Si3N4-based composite ceramic tool materials by microwave sintering. Ceram Int 43:16248–16257. https://doi.org/10.1016/j.ceramint.2017.08.209
2. Abdellahi N, Bahmanpour M, Bahmanpour M (2014) Optimization of process parameters to maximize hardness of metal/ceramic nanocomposites produced by high energy ball milling. Ceram Int 40:16259–16272. https://doi.org/10.1016/j.ceramint.2014.07.063
3. Yan Y, Mingsheng M, Fagiang Z, Feng L, Guanyu C, Zhifu L, Yongxiang L (2020) Low-temperature sintering of Al2O3 ceramics doped with 4CuO-TiO2–2Nb2O5 composite oxide sintering aid. J Eur Ceramic Soc 40:5504–5510. https://doi.org/10.1016/j.jeurceramsoc.2020.06.068
4. Moshnyakov I, Orlova T (2017) Study of the features of firing ceramic granite from an experimental mass with a partial replacement of imported clays with domestic ones. News of Universities, Investments, Construction, Real Estate, 7(3):95–106
5. Verchenko A (2014) Influence of X-ray amorphous phases of a ceramic granite shard on the strength properties of products. In: Don's Engineering Bulletin, vol 4
6. Pavlov V (1977) Physicochemical fundamentals of firing building ceramics. M: "Stroyizdat", p 272
7. Dyatlova E (2012) Chemical technology of ceramics and refractories. Minsk, p 46
8. Shirinzade I, Mamedova I (2015) Investigation of the physical and mechanical properties of a ceramic material based on clay from the Umbakinsky deposit of Azerbaijan. In: International Conference. Kiev, pp 31–32
9. Dubinin M (1970) Basic problems of the theory of physical adsorption, p 269. Nauka, Moscow
10. Zavadsky V et al (1987) Changes in the nature of surfactant adsorption on the surface of clay minerals in the presence of water. Construction and architecture. Izv. Universities. 8:74–77
11. Loburets AT, Naumovets AG, Senenko NB, Vedula YS (1997) Surface diffusion and phase transitions in strontium overlayers on W(112). Zeitschrift Fur Physikalische Chemie 202(1–2):75–85. https://doi.org/10.1524/zpch.1997.202.part_1_2.075

Automated Measuring Station for Road Structures Stamp Tests

Ivan Shuliak, Nadiya Pavlenko, Volodymyr Ilchenko, and Akif Gasimov

Abstract A research prototype of a measuring station was developed and created. It is a set of test equipment integrated into a single automated complex for the rapid determination of road structures and their layers deformation characteristics during stamp tests. This excludes the operator's manual work and increases tests productivity. At the same time, any truck or road vehicle may be used to create a load. A measuring unit with installed specific software is used for measurement results registration and processing.

Keywords Road construction · Elastic deflection · Static stamp tests · Automated measuring station · Modulus of elasticity

1 Introduction

One of the most objective road structures condition and their layers assessment methods is stamp tests, in which the load is transmitted through a rigid stamp, and the vertical deformation (full or elastic) is measured before or after the load is removed. The deformation modulus or the elastic modulus, which are the main indicators of the layer or road structure deformability is located at a certain full or elastic deflection.

At the operation stage, it is advisable to use stamp tests in layer-by-layer tests in order to determine the destruction causes of the road structure. The reason may

I. Shuliak (✉)
National Transport University, Omelianovycha-Pavlenka Str 1, Kyiv, Ukraine

N. Pavlenko
Representative of «Credo-Dialogue» in Ukraine, Kharkiv, Ukraine

V. Ilchenko
National University «Yuri Kondratyuk Poltava Polytechnic», Pershotravnevyj Ave. 24, Poltava, Ukraine

A. Gasimov
Azerbaijan University of Architecture and Construction, Ayna Sultanova 11, Baku, Azerbaijan

V. Onyshchenko et al. (eds.), *Proceedings of the 3rd International Conference on Building Innovations*, Lecture Notes in Civil Engineering 181,
https://doi.org/10.1007/978-3-030-85043-2_36

be insufficient bearing capacity of the soil base or road surface structural layers, including pavement layers.

During the overhaul and roads reconstruction, it is also necessary to perform stamp tests to make reasoned project decisions.

However, in Ukraine these tests are rarely used in practice, as they require significant time and are associated with manual labor when creating a load. At the same time, the tools and devices development level for such tests lags significantly behind the level of similar devices developed abroad.

2 Defining the Problem

Static stamping tests consist in loading the stamp mounted on the road structure surface and measuring the surface coating vertical deformation depending on the applied load magnitude.

These tests give the most objective results in determining the road structures deformation characteristics, as the load scheme complies with the calculation scheme used by J. Businescu in solving the problem of stress–strain state of the half-space under the action of force applied to its surface and evenly distributed on the pressure stamp surface [1].

In most cases, a round rigid stamp and a hydraulic jack are used for stamp tests, and the stamp settling is measured by displacement indicators mounted on the deflectometer (Fig. 1 a, b) [2, 3].

The disadvantages of this measuring equipment are low tests performance due to the need to use the operator's manual labor when creating a load on the stamp with a manual hydraulic jack.

In addition, the considerable weight and bulkiness of the equipment create additional inconveniences while moving the equipment elements from one test site to another and during their installation.

a)  b)

Fig. 1 Measuring equipment for static stamp tests: **a** NTU stamp equipment; **b** SNPTs «ROSDORTECH» press-stamp PSH-050

Fig. 2 NTU hinged equipment on a motor grader (photo Ivan Shuliak 2016): 1 – bracket; 2 – hydraulic cylinder; 3 – grubber frame; 4 – hydraulic distributor; 5 – control lever; 6 – hydraulic hoses; 7 – rigid stamp; 8 – base of the hydraulic load meter; 9 – manometer; 10 – throttle valve; 11 – mounting handles; 12 – rod tip of the hydraulic cylinder

During the test specifically on the road, there is a risk that the operator gets under the wheels of road construction equipment or moving transport.

NTU has developed a hinged equipment for a motor grader (Fig. 2), designed to create a calculated load during stamp tests of road surface layers and soil bases, which is devoid of many these shortcomings [4, 5].

But at the same time it requires a connection to the motor grader hydraulic system, i.e. it is focused on the use of road machinery certain types.

The aim of this study is to improve the method of road structures static stamp tests in the roads diagnosis by increasing its productivity, accuracy, informativeness and safety.

3 Research Results

The road structures and their layers stamp tests measuring station has been developed in the National Transport University to achieve the goal [6–13].

A station is a complicated measuring instrument, which is a set of test equipment combined into a single automated complex for the rapid determination of the road structures and their layers deformation characteristics during stamp tests.

The station includes: pump station; pump station control panel; stamp with hydraulic cylinder and pressure sensor; lever-holding mechanism; lever-holding mechanism control unit; deflectometer; measuring unit.

A general view of the measuring station mounted on the basic car "GAZelle" is shown in Fig. 3.

The measuring station also performs:

- an ambient temperature determination;
- a determination of the surface temperature of the structural layer under test;
- determination of measurements place current geographical coordinates;
- geospatial binding of the obtained results.

Technical characteristics of the station are given in Table 1.

A force created by the oil pressure in a hydraulic cylinder connected to a pump station, an autonomous controlled source of oil pressure, is applied to the pavement

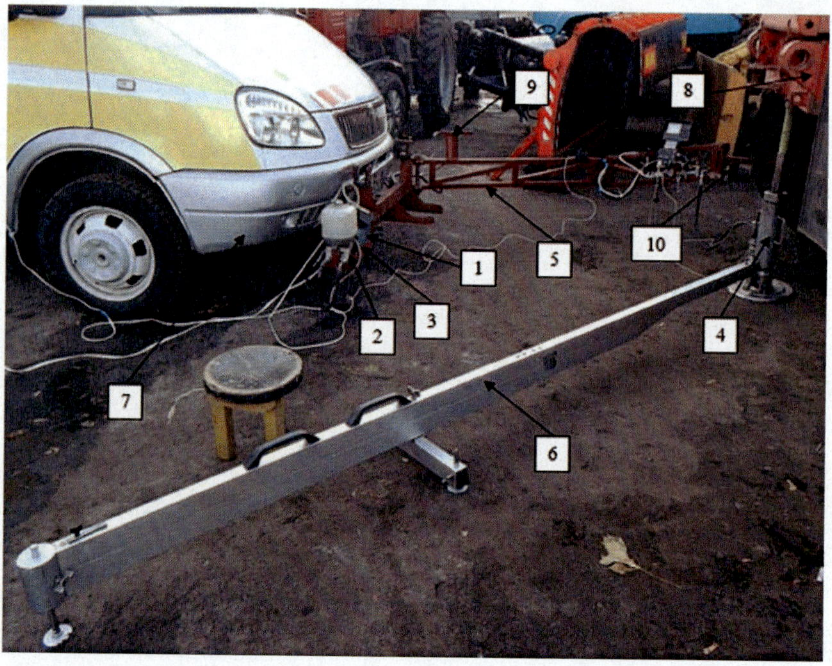

Fig. 3 General form of the measuring station mounted on the basic car "GAZelle" (photo Ivan Shuliak 2016) 1 – pumping station; 2 – hydraulic pump; 3 – engine; 4 – a stamp with the hydraulic cylinder; 5 – lever-holding mechanism; 6 – deflectometer; 7 – basic car; 8 – loading means; 9 – mounting bracket; 10 – grip of the lever-holding mechanism

Table 1 Technical characteristics of the measuring station for road structures and their layers stamp tests

Parameter	Value
Weight, kg	80 ± 0,5
Overall dimensions:	
- height, mm	500
- width, mm	400
- length, mm	1600
Stamp diameter, mm	300
Absolute load measurement error, %	5
Maximum load on stamp, kN	100
Displacement measurement error, mm	0,0025
Operating temperature, °C	From + 5 to + 40
Maximum relative humidity, %	95
Required power supplies:	
- alternating current (power plant), V/kW	220/1,0
- direct current, V/A	12/0,5
Storage temperature, °C (PC is stored at a temperature of +10 to +35 °C)	From –10 to + 60
Service personnel, pers	2

Fig. 4 A pump station remote control (photo Ivan Shuliak 2016) (1 – force creation button on stamp)

through a rigid stamp. Force measurement is carried out by registering the fluid pressure in the power hydraulic system using a pressure sensor. This sensor is connected to the hydraulic system in close proximity to the hydraulic cylinder, thus ensuring high accuracy and stability of measurements.

The pump station is managed by the operator remotely, using a remote control (Fig. 4), by pressing the function button which creates a force on the stamp, and adjusts the load speed as well.

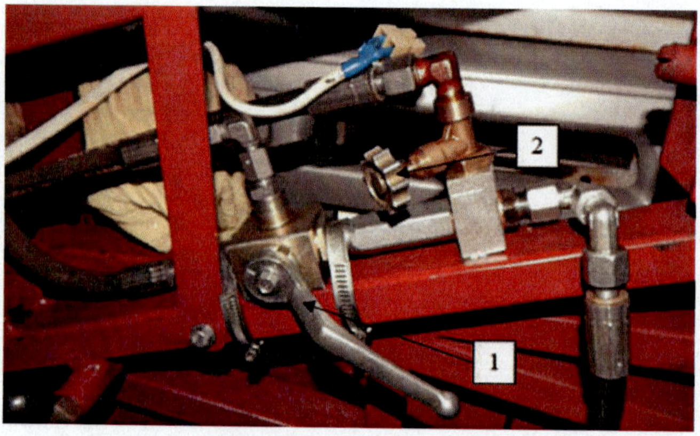

Fig. 5 Control unit VUM (photo Ivan Shuliak 2016) (1 – tree-way crane; 2 – oil drane tap)

Fig. 6 Measuring unit (photo Ivan Shuliak 2016) 1 – graphic display; 2 – keyboard; 3 – connector for an external data carrier; 4 – power switch; 5 – system connector

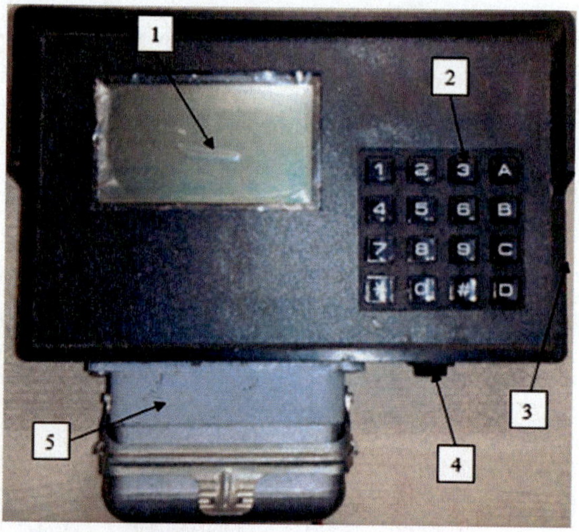

Meanwhile, to create a load, it it possible to use any truck or road vehicle that could provide a load on the stamp of at least 7.5 Tf. Thus, dependence on a certain type of road machinery is excluded.

The equipment is mounted on a vehicle. Transportation of equipment elements from one test site to another and installation of a stamp in the test point is carried out by means of the lever-holding mechanism. The presence in the latter of three lever links with a lifting cylinder allows the operator to move and install at the measuring point a heavy set of hydraulic cylinder with a stamp without much effort.

The control of the lever-holding mechanism and work of the lifting hydraulic cylinder is carried out by means of the VUM control unit of (Fig. 5). It consists of the three-way crane and the oil drane tap for effort removal on a stamp.

The measuring unit (Fig. 6) is used for registration and processing of measurements results. A special software is installed, and its operation algorithm is resulted in Figs. 7, 8 and 9.

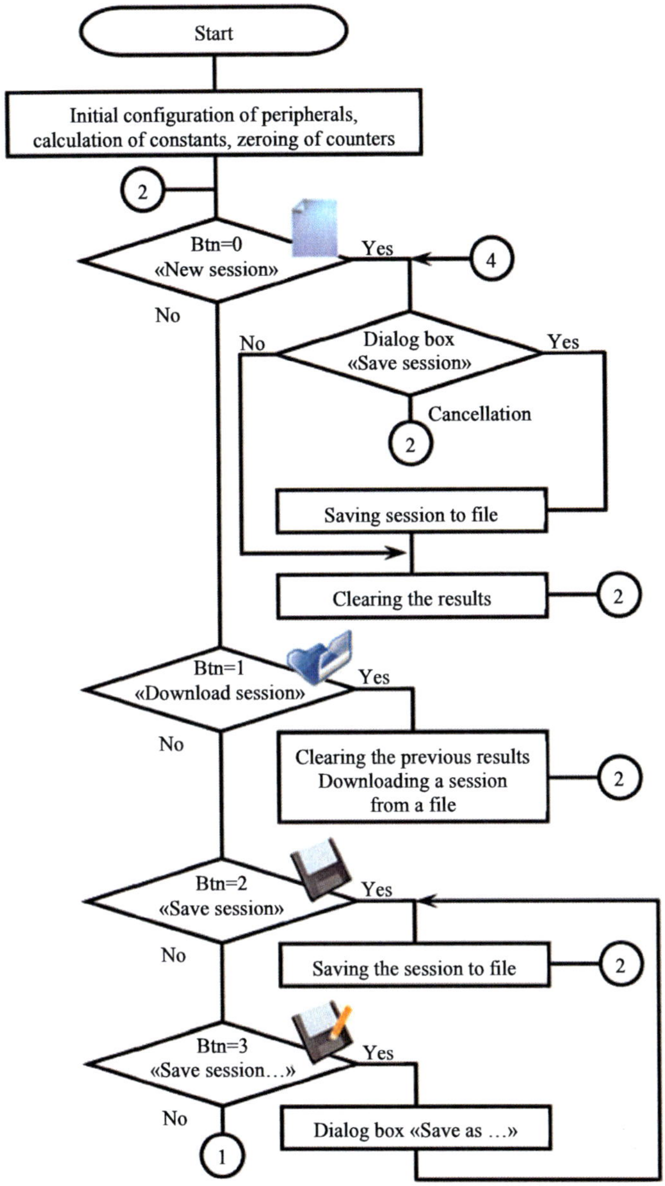

Fig. 7 Operation algorithm of automated control system of measuring processes at the station

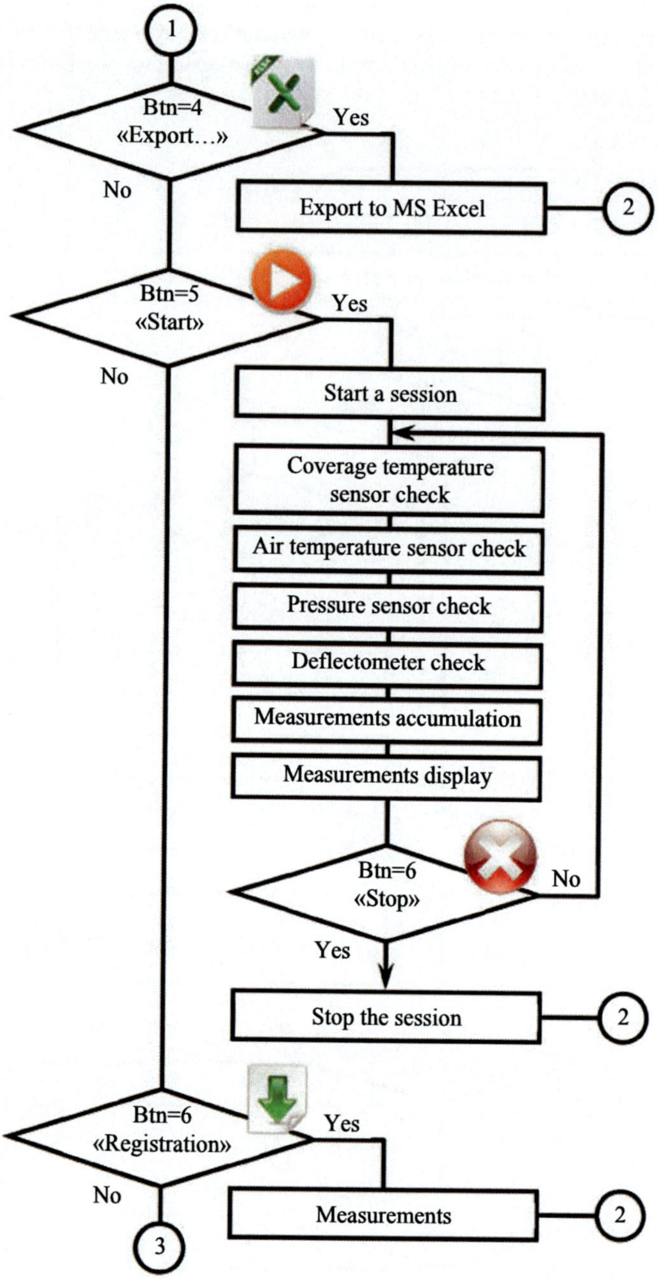

Fig. 8 Operation algorithm of automated control system of measuring processes at the station (continuation)

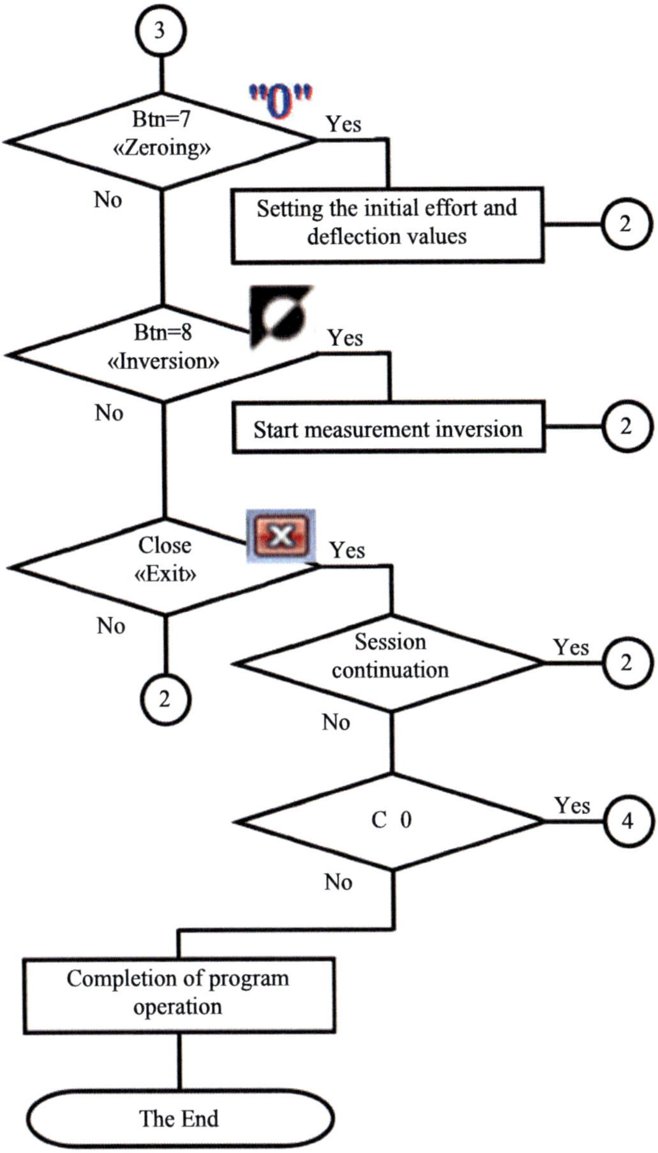

Fig. 9 Operation algorithm of automated control system of measuring processes at the station (the end)

One of the main measuring unit features is the ability to display the measurement results in text and graphical form on the display, with subsequent storage and export of these results to an external non-volatile data carrier.

The deflectometer is a metal telescopic reference beam on three supports, the rear of which has a vertically movable base and the other two on wheels. This allows the operator to quickly and easily configure it for measurements.

4 Conclusions

The static stamp tests method has been improved due to the development of an automated measuring station, which combines test equipment into a single complex mounted on a basic car and eliminates the need for operator manual operation, provides multiple loads, and increases test performance by 5 times.

The research results are protected by a patent for the utility model in Ukraine [7], and an improved method of road structures testing when diagnosing roads is reflected in the state standard DSTU B V.2.3 42:2016 [2].

References

1. Tymoshenko SP, Hudier J (1975) Theory of Elasticity. Nauka, Moscow
2. DSTU B V.2.3-42:2016 (2016) Highways. Methods for Determining the Deformation Characteristics of the Road Bed and Surface
3. JSC SNPTs ROSDORTECH (2013). Product Catalog
4. Bulakh EO (2010) Improvement of Methods of Road Structures Deformability Assessment (PhD thesis). National Transport University, Kyiv
5. Pavliuk DO, Pavliuk VV, Lebediev OS, Bulakh YeO, Peristyi OO (2008) NTU Hinged Equipment for Assessment of Strength and Deformability of Road Structures and Soil Bases. Motorways of Ukraine (Avtoshliakhovyk Ukrainy) 3:33–36
6. Shulyak IS, Pavliuk DO, Pavliuk VV, Pavliuk VV, Lebediev OS, Havryshchuk VV, Ivashchenko AP, Shuriakov MV (2013) Acceptance tests results of a station for road structures and their layers stamp tests. Roads and Road Construction (Avtomobilni dorohy i dorozhnie budivnytstvo) 90:56–62
7. Pavliuk DO, Pavliuk VV, Lebediev OS, Shulyak IS, Hladun SA (2016) Patent of Ukraine 109061. Device for road structures and their layers stamp tests. Ukrpatent, Kyiv
8. Talakh S, Dubyk O, Bashynska O, Ilchenko V (2020) Some technical solutions for the use of aerodrome pavements in the soft soil conditions. Proceedings of the 2nd international conference on building innovations (ICBI 2019). Lecture Notes in Civil Engineering 73:303–311. https://doi.org/10.1007/978-3-030-42939-3_31
9. Korobko B (2016) Investigation of energy consumption in the course of plastering machine's work. East Eur J Enterp Technol 4(8–82):4–11. https://doi.org/10.15587/1729-4061.2016.73336
10. Shulyak IS (2018) Improvement of Road Structures Tests Methods at Diagnosing of Highways (PhD thesis). National Transport University, Kyiv. http://diser.ntu.edu.ua/Shuliak_dis.pdf

11. Piskunov VG, Gorik AV, Cherednikov VN (2000) Modeling of transverse shears of piecewise homogeneous composite bars using an iterative process with account of tangential loads 2. resolving equations and results. Mechanics of Composite Materials 36(6):445–452. https://doi.org/10.1023/A:1006798314569
12. Piskunov VG, Goryk AV, Cherednikov VN (2000) Modeling of transverse shears of piecewise homogeneous composite bars using an iterative process with account of tangential loads. 1. Construction of a model. Mechanics of Composite Materials 36(4):287–296. https://doi.org/10.1007/BF02262807
13. Pichugin SF (2018) Reliability estimation of industrial building structures. Mag Civ Eng 83(7):24–37. https://doi.org/10.18720/MCE.83.3

Research of Risks of Depressurization of Steel Oil Pipelines

Olena Stepova, Elena Hanoshenko, Gulnar Feyziyeva, and Yuliia Chukhlib

Abstract Modern development of safety theory justifies the introduction into practice of providing the necessary parameters of the state of technical facilities, namely pipelines and the environment, including operational conditions, climatic features of areas, standardized parameters of risk and safety, based on reliability, strength, resource, survivability. Based on mathematical model of electrochemical corrosion of oil pipeline in the coating cavity under the aggressive affect of electrolyte environment on the metal of the oil pipelines, the correlation has been found and studied, which enables counting of the remaining depth pit of the corrosive oil pipeline wall of macrogalvanic corrosive couples on the condition of aggressive liquid being in the affected area. The advantage of the given model is an opportunity to predict the development of corrosion progression over time irrespective of aggressive electrolyte chemical composition, opportunity of the necessary design parameters for the constructions that are operated. Correlation of section the assessment of the remaining corrosion cavity of an oil pipeline has been developed which gives an opportunity to rationally plan the repairs, predict the real terms of construction operating, reconsider the operation conditions etc. The given results enable valid assessment of the loadbearing capacity of the construction that operate in the aggressive environment with the defects.

Keywords Ecological safety · Oil pipeline · Corrosion · Pipe depressurization · Risk

1 Introduction

Ukraine has an extensive network of steel pipelines with a total length of almost 5,000 km, which are objects of increased danger in terms of modern environmental

O. Stepova (✉) · E. Hanoshenko · Y. Chukhlib
National University "Yuri Kondratyuk Poltava Polytechnic", Poltava, Ukraine

G. Feyziyeva
Azerbaijan University of Architecture and Construction, Ayna Sultanova 11, Baku, Azerbaijan

V. Onyshchenko et al. (eds.), *Proceedings of the 3rd International Conference on Building Innovations*, Lecture Notes in Civil Engineering 181,
https://doi.org/10.1007/978-3-030-85043-2_37

requirements. In case of their depressurization there are ecological risks of environmental pollution due to leakage of oil products, possible fires, explosions, etc. One of the negative factors that increase the environmental risks of emergencies related to soil, water and air pollution is the corrosion of steel oil pipelines. Understanding the patterns of such processes and taking them into account is a scientific basis for determining the residual life of steel oil pipelines, as well as developing measures to prevent increased risks of environmental pollution during their operation.

Modern development of safety theory justifies the introduction into practice of providing the necessary parameters of the state of technical facilities, namely oil pipelines and the environment, including taking into account operating conditions, climatic features of areas, normalized risk and safety parameters, based on reliability, strength, resource, survivability. A key factor in solving this problem should be the use of the concept of risk monitoring, based on continuous or periodic information on the diagnosis of conditions and basic parameters of hazards in the operation of these facilities. One of the conditions for safe operation, is the interconnected development and use of a comprehensive system for diagnosing and monitoring the condition of materials and structural elements in normal and emergencies, monitoring the risks of their operation at all stages of the life cycle, and operation of protection systems for accidents and disasters, as the risks go beyond acceptable approach them to the limit [1–5].

Risks as an interdisciplinary scientific basis for assessing integrated safety, including environmental, are based on laws, methods, equations and criteria obtained in fundamental fields of knowledge—mathematics, physics, chemistry, mechanics, computer science, mechanical engineering, biology, physiology, geology, geophysics. atmospheric and ocean physics, geography, philosophy, sociology, psychology, economics, law [3–5].

Generalized for the analysis of integrated risks are the developed theories of systems analysis, control theories, theories of catastrophes and protection constructions, methods of mathematical and simulation modeling, forecasting, mathematical statistics, methods and systems of diagnostics and monitoring.

Therefore, the development of scientific bases for environmental safety of existing steel oil pipelines, which take into account the features and patterns of their electrochemical corrosion as a source of environmental pollution is an urgent problem, the solution of which creates preconditions for reducing the risk of environmental pollution during operation.

2 Aim and Objectives of the Research

The aim of the work is to develop a methodology for assessing the residual life of steel pipelines and the risks of their depressurization under the influence of a corrosive environment in compliance with the requirements of safe operation on the basis of real characteristics. The following tasks are set to solve the goal:

based on a mathematical model of local electrochemical corrosion of oil pipeline steel and dependence, which allows to calculate the residual life of the pipeline in the crack of the insulating coating during macro galvanic corrosion pairs to develop a method for assessing the risks of depressurization of the pipeline;

to carry out verification calculations of strength characteristics and residual resource of the section of the operated oil pipeline.

3 Materials and Methods of the Research

Risks (R) in safety theory are understood as such combinations of probabilities P (t) of occurrence of adverse events in time (dangerous and crisis phenomena, catastrophic and emergency situations), on the one hand, and mathematical expectation of the losses caused by it U (t)—on the other hand, which determine the change in the level of safety and the state of human protection systems, infrastructure and the environment from threats and dangers of internal and external nature—man-made, natural, anthropogenic.

$$R(t) = F_R\{P(t), U(t)\}$$

As is known [4, 5], during the operation of the pipeline there is an accumulation of damage along with a certain trajectory D (N, t, σ), which is determined by the load parameters—the number of cycles N, the intensity σ, the defect l (Fig. 1).

To ensure safe operation of the structure instead of critical damages Dc, corresponding to the achievement of limit states, using a system of reserve ratios in the calculation can be introduced allowable damage [D]. Levels D_c and [D] divide the

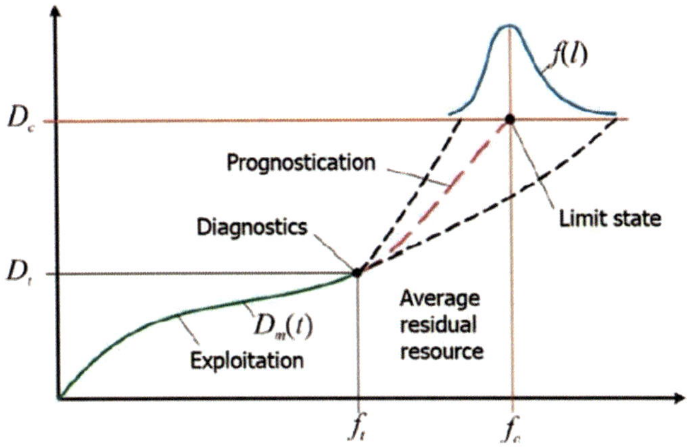

Fig. 1 Trajectories of accumulation of damages in the course of operation of the oil pipeline

area of safe operation and the area of limited safety and danger (risk). Monitoring the parameters of the state of the object in these areas and is the basis for analyzing the risks of the object in a particular state and the conditions of its transition between them. The results of assessing the state of the object as shown in Fig. 1 scheme has the form of a statistical function f and is not the final solution of the problem, which also includes determining the time interval Δt before the next examination of the state of the object.

In the probabilistic estimation of the interval Δt, the risk R_f of the possibility of reaching the limit state should be accepted as a safety criterion. The assigned interval Δt must ensure the probability of possible destruction not higher than the specified level [Rf] of risk. The magnitude of this risk should be determined taking into account the nature (class) of the potential danger of the object. If we use recurrent relations for the probability of transition of the object to the limit state $R_f(t)$, we can obtain an expression to estimate the optimal time until the next moment of the control inspection of the object.

In the general case, there are two main types of risk change scenarios R(t) in time t (Fig. 2). The first includes scenarios for managing the safety of the analyzed facilities in the normal operation of the oil pipeline section as a whole with a monotonous increase in risks R(t) to acceptable levels [R(t)] at time [t] at point A on line 1. The critical risks $R_i(t)$ are not achieved. At this time, special measures are required to reduce the current risks R(t) along with line 1 “to point C and further along line 1”, when the risks for this system remain acceptable levels.

The second type includes scenarios in which there may be points of instability A and B with a dangerous increase in risks on lines 2 and 3 to the critical point K at time $t = t_i$, and the value of R (t) in this case is equal to R_i (t). Points of instability in the system may be the appearance of areas of local damage in the area of the pipe,

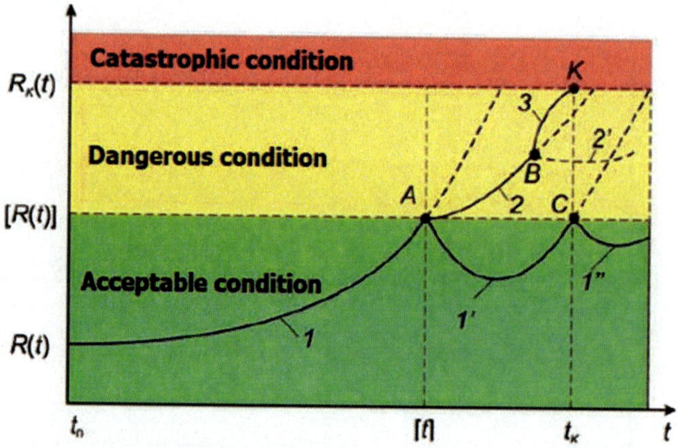

Fig. 2 Scenarios of change of risks R(t) in time t

the emergence of external threats to its normal operation, unauthorized impact on the object, etc.

The wall thickness of the pipeline is a determining parameter that characterizes its strength. Although each pipe has its own certificate, it must be calculated on strength, for which it is necessary to know the actual wall thickness, taking into account the working pressure and the maximum allowable wall thickness at which there is no depressurization of the pipeline with subsequent spillage of oil or oil products.

The allowable residual wall thickness of the oil pipeline corresponds to the complete depletion of the resource of the structure. Radial, longitudinal and annular strain act in the pipeline. Radial strain are much smaller in value than the ring and longitudinal tension, so the strength calculations are not taken into account in the test calculations.

Checking the strength of the pipeline is carried out by the known method of limit states. Considered so strain state of the pipeline is considered, at which its further operation is impossible. The first limit state is the bearing capacity (destruction of the pipeline under the influence of internal pressure), the second is the maximum allowable deformation. The characteristic of the bearing capacity of pipelines is the temporary resistance of the metal of the pipes or the tensile strength.

For failure of the steel oil pipeline on bearing capacity the condition is accepted when the strain from the design loads and influences on the investigated area in the crack, exceeds the yield strength of tubular steel

$$\sigma > R \tag{1}$$

where σ—longitudinal axial strain from design loads and influences, MPa;
R_s—the calculated resistance of the pipe material (yield strength).

The strength of the pipeline is ensured by calculating the strain arising in it during operation and comparing it with the resistance of the pipe material R.

When determining the stress state of the pipeline to check the first limit state, the strain that affect the destructive pressure are taken into account.

Check on the durability of underground pipelines for the purpose of an exception of inadmissible deformations carry out proceeding from conditions.

$$[\sigma_{npN}] \leq \phi_2 R_1, \tag{2}$$

$$\sigma_{ann} \leq \frac{m}{0{,}9k_n} R_2^n \tag{3}$$

where $[\sigma_{npN}]$—longitudinal axial strain from design loads and influences, MPa;
ϕ_2—coefficient taking into account the biaxial strain state of the metal of the pipe (at tensile strain is taken equal to 1);
R_1, R_2—calculated tensile strength (compression), MPa.

$$R_1 = \frac{R_1^n m}{k_1 k_n},\ R_2 = \frac{R_2^n m}{k_2 k_n}, \tag{4}$$

m—coefficient of working conditions of the pipeline;
k_1, k_2—reliability coefficients for the pipeline material;
k_n—reliability factor for the purpose of the pipeline.

The longitudinal axial stress measured at design load and impacts, taking into account of resilient metal. For rectilinear sections of underground pipelines in the absence of longitudinal and transverse displacements and subsidence of the soil, the longitudinal axial strain from the influence of internal pressure, temperature drop and elastic bending, MPa, are determined by the formula.

$$\sigma_{npN} = \frac{0{,}15 p D_{in}}{\delta} - \alpha E \Delta t \pm \frac{E D_{ext}}{2\rho}, \tag{5}$$

where p—working pressure, MPa;
$D_{вн}$—the inner diameter of the pipeline section, cm;

$$\sigma_{кц} = \frac{p D_{in}}{2\delta}. \tag{6}$$

де δ—the nominal wall thickness of the pipeline section, cm;
α—coefficient of linear expansion of metal pipes, deg^{-1};
E—variable modulus of elasticity of the pipe material, MPa;
Δt—calculated temperature difference, °C;
ρ—the minimum radius of elastic bending of the axis of the pipeline, cm.

The difference between the ultimate bearing capacity at the time of inspection and the design force acting on the structure during operation creates a margin for bearing capacity, which can be taken into account when calculating the residual life of the structure, with the affected corrosion of the pipeline section, in cracks in the insulation coating.

The calculation of the actual stresses occurring in the pipeline at the time of the survey is performed by taking into account the reduction of the wall thickness of the pipeline, which is entered into the calculation

$$\Delta\delta = \delta - h, \tag{7}$$

where $\Delta\delta$—the residual wall thickness of the oil pipeline on the site affected by corrosion, mm;

h—depth of corrosion, mm;

The level of annular strain in the pipeline having corrosion lesions (Fig. 2) must meet the condition [9]:

$$\frac{p(D_{in}+2h)}{2(\delta-h)} \leq [\sigma_{ann}] \quad (8)$$

$D_{вн}$—the inner diameter of the pipe, mm;
$[\sigma_{кц}]$—allowable ring tension.
The allowable depth of corrosion of the pipe wall [h] is calculated by the formula

$$[h] = \delta - \frac{pD_з}{2([\sigma_{ann}]+p)} \quad (9)$$

where $D_з$—outer diameter of the pipeline, mm.

Since for pipes $\delta \ll D$, formula (9) can be applied to cases of internal and external corrosion. Formula (9) can be written as:

$$[\varepsilon] = (1 - \frac{p \times D_з}{2\delta([\sigma_{ann}]+p)} 100\% \quad (10)$$

where $[\varepsilon] = \frac{[h]}{\delta}$—permissible relative thinning of the pipeline wall.

The actual absolute h (or relative ε) thinning of the wall must be less than the allowable: $h \leq [h]$ (or $\varepsilon \leq [\varepsilon]$).

Similarly, the test is performed on the longitudinal strain and the allowable depth of corrosion is calculated.

According to the requirements for sections of oil pipelines having corrosive thinning of pipe walls within certain limits, the calculation of operating pressure is carried out according to the formula:

$$[p] = \frac{2[\sigma_{ann} \times (\delta - h)]}{D_{ext} - 2(\delta - h)} \quad (11)$$

where h—depth of corrosion of the oil pipeline wall, mm.

In addition, having the value of the allowable depth of pipeline corrosion section of the oil pipeline and knowing the speed of the corrosion process, it is possible to determine the residual life of the oil pipeline [6–8]

$$T = \frac{[h]}{i} - t_1 \quad (12)$$

where *[h]*—allowable depth of corrosion of the pipeline section, mm, *i*—corrosion rate on the investigated section of the pipeline, mm/year, $t_в$—time of the pipeline in these conditions, years.

Main oil pipelines are used in natural conditions, mainly, below the ground. For protection of steel oil pipelines from aggressive attack of the environment, different coatings are used as a rule. But in the process of operating wholeness of coatings is broken and the aggressive environment affects the steel of an oil pipeline. In such a case the question of the remaining lifetime of the oil pipeline is very urgent. In the below ground with the sectors, where the coating is broken, anode and cathode polar characteristics of steel essentially change, and as a result, the potentials are steady in these places. Taking into account that operating of an oil pipeline with sectors where broken coating is connected with electrochemical corrosion of the metal of a pipeline, when examining an oil pipeline, attention should be paid to determining of characteristics of corrosive process. Electric potential of the given galvanic couples is a universal indicator for calculating the expenses for metal damage in the cavity.

Coating, as capillary-porous material, is a conductor of 2-d type, that's why the process of steel corrosion in it is possible to consider from ordinary electrochemical corrosion of metals in the electrolytes. The authors [10–15] consider the electrical field near the heterogeneous electrode, model of which consists of 2 sectors of arbitrary width, which differ in steady potentials. Local corrosive element is in the sector of oil pipeline under the coating (cathode) and sector of the pipeline in the cavity under the electrolyte.

Electrical boundary of the electrical field potential is determined in this case on the basis of the two-dimensional solution of Laplace's equation.

$$\begin{aligned}\phi(x,y) &= \frac{a(E_a - E_K) + E_K}{c} + \sum_{K=1}^{\infty} \frac{2(E_a - E_K)}{\pi k\left(1 + \frac{\pi k}{c} L\right)} \sin\frac{\pi k}{c} a \cos\frac{\pi k}{c} x e^{-\frac{\pi k}{c} y} \\ &= \frac{a(E_a - E_K) + E_K}{c} + \frac{2(E_a - E_K)}{\pi} \sum_{K=1}^{\infty} \frac{\sin\frac{\pi k}{c} a}{\left(1 + \frac{\pi k}{c} L\right) k} \cos\frac{\pi k}{c} x e^{-\frac{\pi k}{c} y},\end{aligned} \tag{13}$$

where φ—potential; a—width of anode selector, м; E_a, E_κ—currentless potentials of anode and cathode, мВ; c—width of cathode selector, м; x, y—flow coordinates; $k = 1, 2, 3$;L— coefficient that depends on electroproductivity of electrolyte and coefficient of polarization, L—coefficient depending on the specific electrical conductivity of the electrolyte and the polarization coefficient, $L = \gamma \cdot b$; γ—electric conductivity of electrolyte; b—polarization efficiency.

Taking into account Ohm's law in the differential form in the Eq. (1) expression for determining current density on the surface of one local element is found out.

$$i(x) = \frac{2(E_a - E_k)\gamma}{c} \sum_{k=1}^{\infty} \frac{\sin\frac{\pi k a}{c} \cos\frac{\pi k x}{c}}{k\left(1 + \frac{\pi k L}{c}\right)} \tag{14}$$

The proposed method of calculating the residual allowable wall thickness of the oil pipeline as a parameter that determines the life of the pipe and its safe operation is tested in assessing the condition of the section of the current state of pipeline under the conditions listed in Table 1.

Thus, the operating pressure on the investigated section of the oil pipeline should not exceed 8,17 MPa.The results of calculations of the residual life of the oil pipeline

Table 1 The results of the calculation of the residual resource

Parameter	Marking	Units of measurement	Value
Output data			
Outer diameter of the pipeline	D_3	mm	530
The wall thickness of the oil pipeline	δ	mm	9
The strength limit of the pipe material	σ	MPa	
Working pressure in the oil pipeline	p	MPa	5,04
The potential difference of electroplating	ΔE	мB	0,06
The service life of the oil pipeline	t_e	year	15
The area of the corroded section	a	cm^2	0,0024
Calculated values			
Permissible ring tension	σ	MPa	363
Corrosion current of electroplating	I	A/cm^2	$0{,}88 \times 10^{-4}$
The corrosion rate of electroplating	$i_в$	mm/hour	$3{,}17 \times 10^{-5}$
The corrosion rate of electroplating	$i_в$	mm/year	0,27
The estimated value of corrosion depth at the time of inspection	h	mm	4,05
The value of the allowable pressure at the defect at the time of inspection	$[p]$	MPa	8,17
The value of the allowable wall thickness	$[h]$	mm	5,37
Residual resource-1	T	year	20,94–15 = 5,94
Residual resource-2	T	year	19,88−15 = 4,88

according to the dependences proposed by the author coincide with the calculations according to existing methods with a relative error of 17%.

The novelty of the research is the development of a calculation method for determining the depth and allowable depth of corrosion of a steel oil pipeline during macro galvanic corrosion vapour under the influence of aggressive electrolytic solution, as well as a method for calculating the residual life of its environmentally safe operation, which will allow predicting the development of corrosion processes on the steel oil pipeline, to plan the necessary measures to prevent environmental pollution.

4 Conclusion

Based on mathematical model of electrochemical pipeline corrosion in the coating cavity correlation is found out and researched which enables counting the remaining lifetime of the oil pipeline of operating macrogalvanic corrosive couples in the conditions of aggressive liquid in the damaged coating. The developed correlation enables possibility of rational planning of the repairs, to predict the real terms of construction operation, to reconsider operating regime etc. The results enable valid assessment of the loadbearing capacity and the remaining lifetime of the constructions with the cavities in the coating that operate in the aggressive environment.

The results of calculations of the residual life of the oil pipeline according to the dependences proposed by the author coincide with the calculations according to existing methods with a relative error of 17%.

In order to reduce the likelihood of abnormal situations (with a corresponding increase in risks of their implementation) and reduce possible losses from their manifestation for potentially dangerous objects of the technosphere should be implemented a set of measures that take into account the nature of sources of danger and peculiarities of their manifestation, permissible modes of operation for each phase of risk increase, the possibility of using means of reflecting threats based on the results of comprehensive diagnostics and monitoring of the object.

To solve the problem of ensuring the safe operation of main oil pipelines, it is advisable first of all to use a set of modern methods and means of monitoring the parameters of these facilities and the environment in the considered conditions with the widest possible range of changes (including outside regular modes), application of monitoring systems and analysis of environmental data and possible external influences on the studied system, use of data banks with sources of danger and scenarios of emergency situations, criteria for their evaluation and methods of preventive action to reflect emergencies according to the programmed teams of the risk monitoring system.

References

1. Makhutov NA, Lebedev MP, Hadenyn MM, Bolshakov AM (2013) The scientific basis for the analysis and reduction of the risks of emergency situations in the regions of Siberia and the North. Ecol Econ 4(12):4–15
2. Lebedev MP (ed) Proceedings of the VI Eurasian Symposium on the Strength of Materials and Machines for Cold Climate Regions. Yakutsk, 24–29 June 2013, vol 2, pp 256–332. Akhsaan, Yakutsk
3. Makhutova NA (1998/2013) Security of Russia: legal, socio-economic and scientific-technical aspects. In:MGOF Knowledge, Under scientific. Hands, pp 1–40
4. Makhutov NA (2005) Structural strength, resource and industrial safety, pp 494–610. Nauka, Novosibirsk
5. Makhutov NA (2008) Strength and safety: basic and applied research, p 528. Nauka, Novosibirsk
6. Stepova O, Paraschienko I (2016) Mathematical modeling of local corrosion element in pipelines at galvanic couple's operation in soil conditions. Geotech Mech 127:49–55
7. Stepova O, Paraschienko I (2017) Modeling of the corrosion process in steel oil pipelines in order to improve environmental safety. East-Eur J Enterp Technol Ind Technol Syst 2(1(86)):15–20
8. Stepovaja E, Holik Yu, Fraňa K (2018) Methods for precautionary management of environmental safety at energy enterprises. Sci Bull Natl Mining Univ Sci Tech J 6(168):173–177. https://doi.org/10.29202/nvngu/2018/21
9. Stepova O, Paraschienko I, Lartseva I (2018) Calculation of steel pipeline corrosion depth at the work of galvanic corrosive element operating. Int J Eng Technol 7(3.2):431–435. https://doi.org/10.14419/ijet.v7i3.2.14566

10. Golik Y, Illiash O, Chuhlib Y, Maksiuta N (2020). Environmental Areas of poltava planning development. https://doi.org/10.1007/978-3-030-42939-3_38
11. Onischenko VA, Soloviev VV, Chernenko LA, Malyshev VV, Bondus SN (2014) Acidic-basic interactions in tungstate melts based on tungsten electroplating out of them. Materialwiss Werkstofftech 45(11):1030–1038. https://doi.org/10.1002/mawe.201400222
12. Loburets AT, Naumovets AG, Senenko NB, Vedula YS (1997) Surface diffusion and phase transitions in strontium overlayers on W(112). Zeitschrift Fur Physikalische Chemie 202(1–2):75–85. https://doi.org/10.1524/zpch.1997.202.part_1_2.075
13. Cherniha R, Serov M (2003) Nonlinear systems of the burgers-type equations: Lie and Q-conditional symmetries, ansätze and solutions. J Math Anal Appl 282(1):305–328. https://doi.org/10.1016/S0022-247X(03)00155-0
14. Cherniha R, Pliukhin O (2013) New conditional symmetries and exact solutions of reaction-diffusion-convection equations with exponential nonlinearities. J Math Anal Appl 403(1):23–37. https://doi.org/10.1016/j.jmaa.2013.02.010
15. Pavlenko A, Koshlak H (2015) Design of processes of thermal bloating of silicates. Metall Mining Ind 7(1):118–122

The Main Directions of the Solution of the Problems of Providing the Houses at Affordable Prices in the Azerbaijan Republic by Effective Project Management of Housing Construction

Ferahim Veliyev and Oleksandr Redkin

Abstract It is very important to Providing of housing to the population at affordable prices in the developing countries. It should be noted that the concept of "reasonable price" is differently stated in the experience of different countries of the world. However, the fact that the cost of obtaining housing should not exceed 40–50% of family income is a standard adopted in most countries. The Studies show that today about 330 million families in developing and also in developed countries are faced with the problem of obtaining quality housing at affordable prices, and this figure is gradually increasing. Despite the fact that the number of apartments built in the Azerbaijan Republic, which is rapidly developing in the residential real estate market, is quite large, it is impossible to purchase housing for the low-income population, since the prices are very high. It also plays a role the high price of land in cities and low mortgage lending. In investments in housing construction, the proportion invested in land purchases is very high.

Taking into account the above, the Strategic Road Map was developed and adopted to increase the affordability of housing in the Republic of Azerbaijan. To achieve the nominated goals established the State Agency for Housing Construction (abbreviation in Azeri-MIDA) and started construction of two large residential complexes. This article touches upon the task of providing the population of Azerbaijan with housing at affordable prices.

Keywords Construction · Road map · Affordable housing · Project management · The plot of land · Mortgage lending · Population income · Cost of housing · Residential complexes

F. Veliyev (✉)
Azerbaijan University of Architecture and Construction, 5 Ayna Sultanova, Baku, Azerbaijan

O. Redkin
National University «Poltava Yuri Kondratyuk Polytechnic», Poltava, Ukraine

V. Onyshchenko et al. (eds.), *Proceedings of the 3rd International Conference on Building Innovations*, Lecture Notes in Civil Engineering 181,
https://doi.org/10.1007/978-3-030-85043-2_38

1 Introduction

Providing the population with affordable housing is one of the key issues for any country. All countries are trying to resolve this issue in accordance with the socio-economic situation in the country. Some countries are trying to solve this problem through low-interest mortgage lending. World practice shows that one of the most effective ways to do this is to apply the project management method to the organization of work on the implementation of housing construction projects.

It should be noted that the project is characterized by limitations and requirements for its duration, cost and quality. Analysis of international experience shows that effective project management reduces the duration of their implementation by an average of 20–30% and costs 10–15% [2]. In this article, we will examine the issues of providing the population with "affordable" housing through the creation of an effective organizational mechanism for the management of housing construction projects in Azerbaijan. These houses are called social houses because they are accessible to low-income families in Azerbaijan.

2 Main Part

2.1 *Purpose of the Article*

The main purpose of the article is to study the activities of the State Agency for Housing Construction (Azeri MIDA), established by the relevant decree of the President of the Republic of Azerbaijan, to provide the population with "affordable" housing and to make recommendations for further expansion.

2.2 *Research Methodology*

Methodology is a system of principles and methods of organization and construction of theoretical and practical activities, as well as the study of this system [4].

During the research, I am successfully implementing "affordable" housing construction projects in Azerbaijan. Actual materials of the State Agency for Housing Construction, as well as a number of construction contractors performing large-scale monolithic reinforced concrete works were used.

The research is based on a systematic approach, statistical methods, statistical grouping, comparative analysis, business process theory, principles and design methods.

2.3 Results

It is very important Providing affordable housing to the population in developing countries at affordable prices. It should be noted that the term "affordable prices" is interpreted differently in the experience of different countries of the world. However, the fact that the cost of obtaining housing should not exceed 40–50% of family income is a standard adopted in most countries. According to statistics, today about 330 million families in developing and developed countries of the world are faced with the problem of obtaining quality housing at affordable prices and this figure is gradually increasing.

To obtain affordable prices for apartments under construction, it is especially important to use the mechanism of rational management at all stages, starting with the purchase of land, the design and implementation of construction. This fact was taken into consideration when was established the State Agency for Housing Construction under the President of the Republic of Azerbaijan by the order No. 858 dated April 11, 2016. The State agency for housing Construction for the development of the housing fund of Azerbaijan is designing housing supply projects at affordable prices, organizing the formation and implementation of the main mechanisms that create conditions for the efficient operation of the housing market (state support, pooling project resources, participation of private sectors and ensuring mortgage market productivity).

In the international practice, the following four areas are taken into account to reduce the cost of housing and provide the population with housing that meets the standards:

- Identification and allocation of suitable land for housing at an affordable price;
- The application of the mechanism of state support in the provision of engineering and communication systems and lines;
- Ensuring efficiency in the construction process;
- Selection of financing mechanisms and beneficiaries.

Since the cost of land constitutes a significant part of the cost of housing, and the territory where the housing is located are the most influential factors for the quality of housing and the standard of living, for housing agencies, the acquisition of land in the territories at an affordable price is of exceptional importance. This problem is solved by the executive on the most favorable terms. For the residential complex in the Yasamal district, whose construction was completed, a land plot was allocated in the most picturesque part of the city.

The new Residential Complex consists of 40 buildings, including 22 nine-storey and 18 twelve-storey. In total, these buildings will consist of 2202 apartments, of which 160 are one-room, 114 are studio-type, 674 are two-room, 1128 are three-room, 126 are four-room. The area of the apartments will be 33–96 m^2. All apartments will be offered to residents completely renovated and equipped with kitchen furniture. [5]. Prices for these apartments are shown below in the table (Table 1).

Table 1 Prices for fully renovated and furnished apartments in a residential complex in the Yasamal district of Baku

Number of rooms in the apartment	Type of repair	The area around the outer perimeter (м^2)	The area around the inner perimeter (м^2)	Selling price, AZN
1	1	40,52	33,1	31455
2	1	66,92	54,8	52060
3	1	81,64	69,51	66082
1	2	40,52	33,1	31455
2	2	66,92	54,65	51918
3	2	81,64	69,51	66035

The residential complex in the Hovsan settlement of Surakhani region of the Baku, also under construction, is located in the rapidly developing eastern part of Baku, just 2 km from the coast of the Caspian Sea, in a very convenient place for living and recreation. In the western part of the complex with an area of 20 ha there is a large green area with pines and olive trees [5].

These zones have been allocated by the executive authority. It should be noted that for this purpose, special attention should be paid to vacant and improperly used land plots.

One of the essential factors necessary for determining the areas for construction is the quality of the infrastructure in these areas, especially the transport lines. I should note that both of the above complexes are very close to the main communications and transport lines. Basically, in developed countries due to energy savings, housing costs are reduced. In the United States and in the UK, 20–30% of energy costs were saved due to housing requirements [2].

Since the cost of land is a significant part of the cost of housing and the area in which it is located is the most influential factor for housing quality and living standards, land acquisition in areas at a reasonable price is of exceptional importance for housing agencies. This problem is solved by the executive power on the most favorable terms.

One of the important aspects of the management of housing construction projects in Azerbaijan is characterized by the near and far external environment of projects. Thus, the country has achieved full macroeconomic stability. There is financial stability and very low inflation.

By the rational management of housing projects and optimization of the construction process, significant cost reduction can be achieved. The most important innovation proposed in this direction is the industrialization of the construction process. According to international experience, it is possible to reduce costs up to 30%, and the project completion time - up to 50% [2]. For this purpose was established entity LLC State Agency for Housing Construction (abbreviation in Azeri - MİDA) under the State Agency for Housing Construction. The entity State Agency for Housing Construction (abbr. In Azeri - MIDA) LLC has created an extensive material base, a transparent mechanism for selecting subcontractors, design and procurement organizations.

The houses built by the MIDA LLC are designed for socially poorly endowed unprotected groups of the population - workers of science and education, young scientists, young families. Sale of apartments for the specified layer of the population is made online. The focus should be on families in need of housing. The main goal here is to eliminate by 2025 the shortcomings in the provision of housing at an affordable price through the measures taken by MİDA LLC. [6]

Affordable housing projects and apartments under construction each year will serve to meet the housing needs of young and low-income families and certain social groups.

It should be noted that «MİDA» realizes its activities with the Mortgage Fund of Azerbaijan, through which it can issue mortgage loans with very low and long-term interest rates. This will serve the development of mortgage lending in the country. It is worth noting that over the past 15 years, the average living area in Azerbaijan has gradually increased from 9.3 to 12 m^2. This indicator is a criterion for the MİDA when developing a project [6–8].

Project monitoring is given special attention and is carried out at all levels of project management, where independent experts can also be involved. Processes of procurement and spending of funds and the compliance of the planned goals of the project with the current situation are subject to especially strict control. Various criteria are used during the assessment. For example, EU organizations use criteria such as adequacy, cost-effectiveness, productivity, efficiency, impact, economic and financial viability [4].

In determining the pricing policy MIDA takes as a basis the average monthly income of the population (Table 2).

The financial stability of MİDA is ensured by the state budget, the State Oil Fund and the Azerbaijan Mortgage Fund.

In the future, MIDA will be able to rationalize its activities in the regions and provide affordable housing for residents of the regions. Thus, in 2020, MIDA is

Table 2 Population size, living space per person, average monthly salary by region [6]

Cities and regions with high density	Number of the population (thousand)	Living space per person (m^2) (2015)	Average monthly salary (2015)
Baku	2225,8	11,6	666,8
Absheron	168,7	14,5	336,0
Sumgayit	336,2	9,3	378,5
Ganja	330,1	11,8	321,0
Sheki	67,3	11,8	255,6
Khachmaz	87,2	14,2	267,2
Lenkoran	68,0	12,1	266
Nakhichevan	80,9	20,8	424,6

building a new residential complex in Ganja, the second largest city in the country. The residential complex consists of 18 residential buildings with 12 nine-storey and 6 twelve-storey buildings. Also, as in other MIDA projects, large yards, green areas and children's playgrounds will be put into operation in the territory of the Complex, which will be built in Ganja [5].

According to estimates, as a result of the implementation of housing affordability in Azerbaijan, according to forecasts, in 2021 Azerbaijan's real GDP will reach a total of 83 billion 224 million and create 18,000 jobs in the construction sector [7].

MIDA LLC will provide land plots, infrastructure and other privileges to private construction companies to encourage their participation in this sector. On the other hand, private contractors can transfer their resources and methods to MIDA and assume some of the risks associated with housing construction.

The Distribution of projects among private contractors, as well as tenders will lead to the provision of the highest quality to the beneficiaries.

It should be noted that the sale of apartments is carried out electronically.

2.4 Scientific Novelty

For the first time, the article comprehensively examines the activities of the State Agency for Housing, which was established by the state by adopting a Roadmap to increase the affordability of the housing market in a market economy, confirms the effectiveness of the organization of the implementation of housing construction projects in providing the socially vulnerable with affordable housing by applying the project method, also proves that such state intervention in the housing market has led to a significant drop in prices in the housing market.

2.5 Practical Importance

The results of the study and MIDA's experience can be used to provide affordable housing for vulnerable groups in developing countries, as well as in the Commonwealth of Independent States.

3 Conclusions

The study of the activities of the State Agency for Housing of the Republic of Azerbaijan and its results shows that by developing and implementing the right state policy, it can be used to provide socially vulnerable groups with housing at affordable prices.

References

1. Strategic Development Roadmap for affordable housing in the Republic of Azerbaijan, 2016, 44 p
2. Kovalskiy MI (2014) Upravlenie stroitelstvom: opyit SShA, Yaponii, Velikobritanii, FRG, Kanadyi. M: Stroyizdat, 427
3. Upravlenie proektami (2013) fundamentalnyiy kurs. Kollektiv avtorov: Vyisshaya shkola ekonomiki, Moskva, 529
4. https://mida.gov.az/
5. https://www.stat.gov.az
6. www.economy.gov.az
7. Drobyazko S, Bondarevska O, Klymenko D, Pletenetska S, Pylypenko O (2019) Model for forming of optimal credit portfolio of commercial bank. J Manag Inf Decis Sci 22(4):501–506
8. Iasechko S, Haliantych MK, Skomorovskyi VB, Zadorozhnyi V, Obryvkina O, Pohrebniak O (2020) Contractual relations in the information sphere. Syst Rev Pharm 11(8):301–303. https://doi.org/10.31838/srp.2020.8.46

Influence of the Rear Verge Configuration of the Retaining Wall and Surface of Backfill on Active Pressure of Heterogeneous Anisotropic Soil

Yuriy Vynnykov and **Inha Voitenko**

Abstract Active pressure of heterogeneous anisotropic backfill on a massive retaining wall with different rear verges orientations was determined with various options for flat boundaries of soil layers. Numerical researches were carried out for loose soil, anisotropy of strength properties of which is presented in the form of piecewise linear hodographs of the angle of internal friction. Various combinations of anisotropic strength characteristics of contact soil layers were considered, which made it possible to evaluate the effect of anisotropy on lateral soil pressure. The results of numerical simulation are presented in tabular form. Comparative analysis indicates a significant effect of the configuration of the rear verges of the wall and the features of the stratification of contact soil layers on lateral pressure.

Keywords Retaining wall · Hodographs of internal friction angle · Lateral pressure · Strength anisotropy

1 Introduction

Retaining walls are widely used in different branch of building. Massive constructions, taking big lateral soil pressure are popular enough. Strength anisotropy, specific for contact soil environment, hence lithogenesis or artificial backfilling formation was estimated by the researches [1, 2]. The question of lowering of soil pressure on retaining structures for their cost lowering is actual. Therefore, the question of economic wall configuration, and the influence of backfilling surface orientation on the value of anisotropic soil's lateral pressure is of interest. In particular, such an approach was used for the retaining system of the excavation of the complex

Y. Vynnykov
National University «Yuri Kondratyuk Poltava Polytechnic», Pershotravnevyj Ave. 24., Poltava 36011, Ukraine

I. Voitenko (✉)
Odessa State Academy of Civil Engineering and Architecture, Didrikhsona Street, 4, Odessa 65029, Ukraine
e-mail: voytinna@ukr.net

V. Onyshchenko et al. (eds.), *Proceedings of the 3rd International Conference on Building Innovations*, Lecture Notes in Civil Engineering 181,
https://doi.org/10.1007/978-3-030-85043-2_39

"Panorama" in Frankfurt-am-Main. [3]. For the cost optimization, a supporting wall was erected from the anchored bored piles, combined with jet grouting columns, that significantly speeded the construction process in cramped conditions, and provided existing buildings and underground communications safety.

A calculating-theoretical device of basements and foundations design on the base of phenomenological orthotropic model of soil's compaction and the decision of the axis symmetrical task by the method of finite elements in non-linear setting was described in [4]. The chute researches of anisotropic environment's lateral pressure on the rigid walls, closely located at a different angle of chute filling, are also looked through. The pressure was measured by the strain gauges in different levels, the testimony of which testify the results' differences, and the curves of the soil lateral pressure depend on the angle of backfill laying. The task for the homogeneous, anisotropic in shear resistance, environment on the base of classical Kulon's theory and the solution of R. Nedderman was solved.

The authors [5–8] conducted numeral and physical modeling of strength condition of anisotropic soils of different mineralized composition. In [9] the solution of spatial task of stability of a non-drained slopes, folded by layered anisotropic soils, from the positions of probabilistic approach was proposed. The influence of strength anisotropy on lateral pressure of non-homogeneous soil was investigated [10]. Taking it into account allows lowering the value of active pressure.

By numerous investigations of a thin wall [11], it was established that passive pressure grows with the increase of the soil's bulk density and the depth of embedment of the wall, but doesn't depend on wall's thickness. In its turn, total active pressure increases with depth of embedment (or total depth) of the wall and decreases with the increase in bulk density of sand. The combination of Yansen's equations with the Kulon's theory about the slip surface [12–17] demonstrated the character of pressure distribution along the rear surface of different rigidity, and the maximal meanings of contact stresses are identified somewhere higher a wall's sole.

2 Defining the Problem

However, the described investigations don't contain the results of influence of wall and backfill parameters on active pressure of non-homogeneous anisotropic soil. For identification of more rational and economic configuration of a wall's contact edge, the numerous researches of active pressure of anisotropic non-homogeneous soil on a massive retaining structure was being conducted. The task of the investigation was the definition of active pressure of anisotropic soil under different orientation of a rear edge and the surfaces of contact soil layers for the searching of their correspondence to minimal pressure.

3 Research Results

The numerous experiments were being conducted for the retaining wall with smooth rear surface of a wall, oriented at an angle β_3, take equal 270°, 285° and 255° (Fig. 1a). The contact soil environment was two-layered loose backfill, which surfaces are oriented at the angles $\beta_{1,1}$ and $\beta_{1,2}$.

Soil's active pressure was identified, which anisotropy of strength properties is presented by the hodographs' internal friction angle, in the view of piece-linear graphics (Fig. 2), satisfying the condition:

$$\varphi_n \cdot (\beta) = \varphi_n \cdot (\beta + \pi);\ c_n(\beta) = c_n(\beta + \pi). \tag{1}$$

Active pressure of an arbitrary n-layer for loose soil (c = 0) and at the absence of a surface load (q = 0) is defined according to the formula:

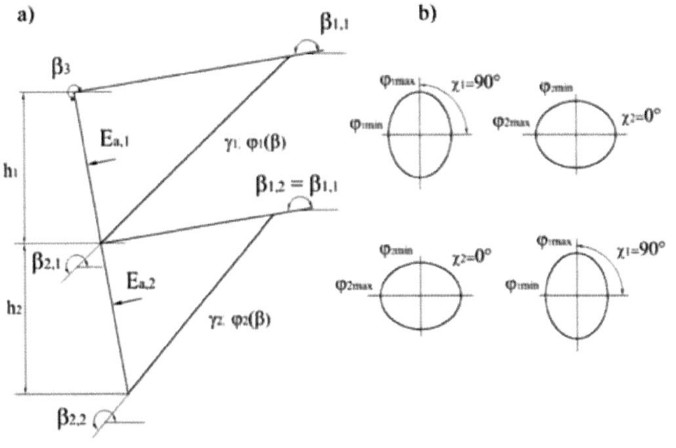

Fig. 1 The calculating scheme **a** and the variants of hodograph's combinations of an angle of internal friction φ (β) at different orientation of an axis of maximum strength of n-layer, χ_n **b**

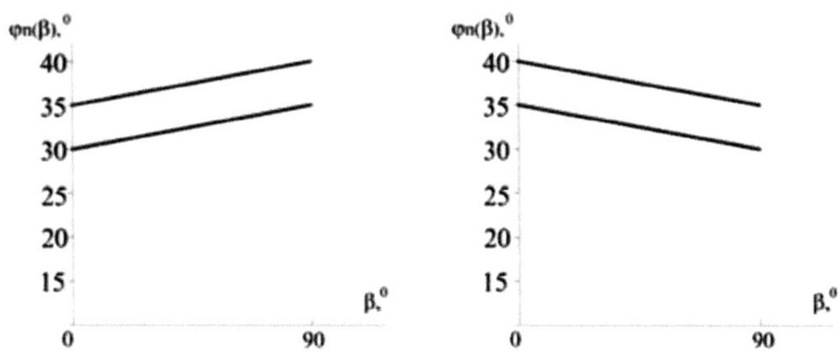

Fig. 2 The piece-linear hodographs of an internal friction angle, taken in the investigation

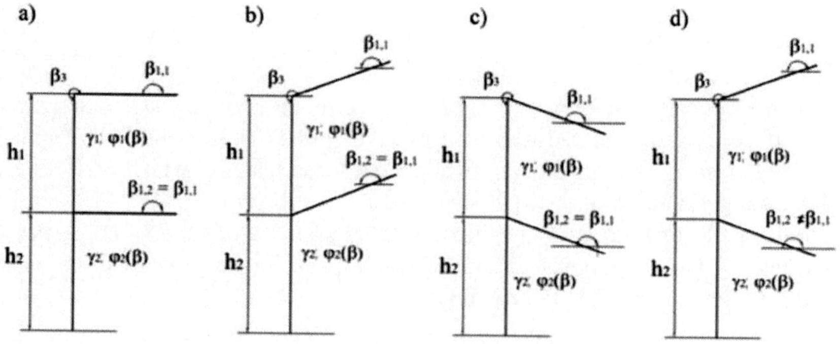

Fig. 3 The calculating schemes for a retaining wall with the orientation of a rear verge $\beta_3 = 270°$ at the orientation of surface layers of backfill: **a** $\beta_{1,1} = \beta_{1,2} = 180°$; **b** $\beta_{1,1} = \beta_{1,2} > 180° = 190°$; **c** $\beta_{1,1} = \beta_{1,2} < 180° = 170°$; **d** $\beta_{1,1} = 190° > \beta_{1,2} = 170°$

$$E_{a,n} = \gamma_n h_n^2 N_{\gamma,n}\left(1 + N_{cor,n}\right), \tag{2}$$

where $N_{\gamma,n}$, $N_{cor,n}$ – the coefficients, received from the expressions (3) and (4):

$$N_{\gamma,n} = \frac{1}{2}\frac{sin(\beta_3 - \beta_{1,1})\,sin(\beta_3 - \beta_{2,n})\,sin(\phi_n(\beta_{2,n}) - \beta_{2,n})}{sin^2\beta_3\,sin(\beta_{2,n} - \beta_{1,1})\,sin(\beta_3 - \beta_{2,n} + \phi_n(\beta_{2,n})}; \tag{3}$$

$$N_{cor,n} = \frac{sin(\beta_3 - \beta_{2,n})\,sin(\beta_{1,n} - \beta_{1,1})}{sin(\beta_3 - \beta_{1,1})\,sin(\beta_{2,n} - \beta_{1,n})}, \tag{4}$$

where β_3 – the angle of wall's orientation;
$\beta_{1,1}$ – the angle of backfill surface orientation;
$\beta_{1,n}$ – the angle of surface soil of anarbitrary n-layer orientation;
$\beta_{2,n}$ – the angle of failure surface of n-layer;
γ_n – a soil's bulk density of n-layer;
h_n – the thickness of n-layer of soil on the wall's vertical projection.

When investigating, different variants of orientation of a wall's rear verge were being looked through $\beta_3 = 270°$ (Fig. 3), $\beta_3 > 270° = 285°$ (Fig. 4), $\beta_3 < 270° = 255°$ (Fig. 5), for each of them the calculations were made when differently oriented surface layers of backfill: $\beta_{1,1} = \beta_{1,2} = 180°$, $\beta_{1,1} = \beta_{1,2} > 180° = 190°$, $\beta_{1,1} = \beta_{1,2} < 180° = 170°$, $\beta_{1,1} = 190° > \beta_{1,2} = 170°$ (Figs. 3, 4 and 5).

Strength anisotropy of soil layers was modeled by hodographs of an internal friction angle $\varphi = 30–35°$ and $\varphi = 35–40°$ with variations in the relative position of axes of maximal hodograph's strength χ_n horizontal relatively (Fig. 1b).

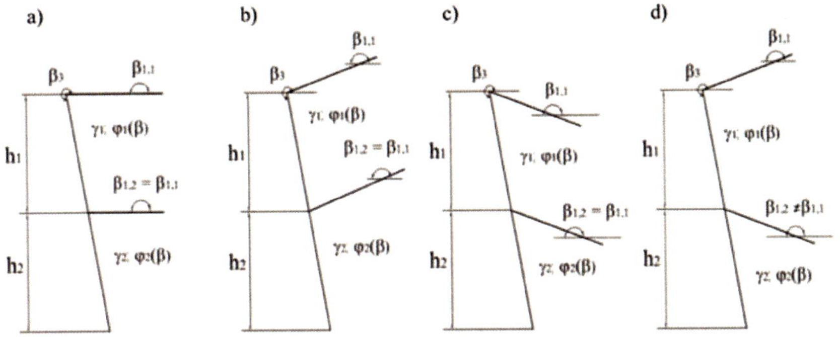

Fig. 4 The calculating schemes for a retaining wall with the orientation of a rear verge $\beta_3 = 285°$ at the orientation of surface layers of backfill: **a** $\beta_{1,1} = \beta_{1,2} = 180°$; **b** $\beta_{1,1} = \beta_{1,2} > 180° = 190°$; **c** $\beta_{1,1} = \beta_{1,2} < 180° = 170°$; **d** $\beta_{1,1} = 190° > \beta_{1,2} = 170°$

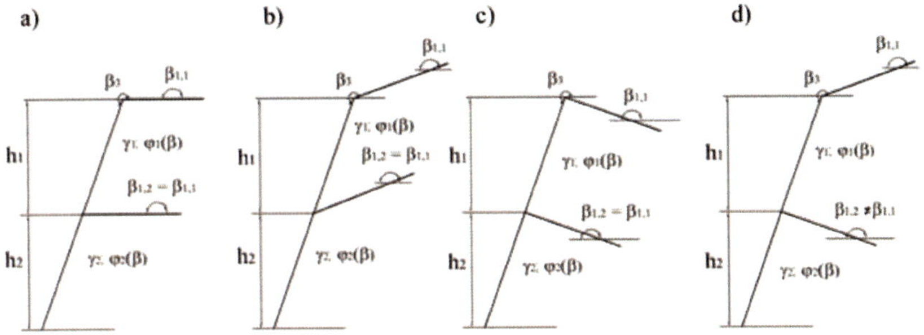

Fig. 5 The calculating schemes for a retaining wall with the orientation of a rear verge $\beta_3 = 255°$: $\beta_3 = 285°$ at the orientation of surface layers of backfill: **a** $\beta_{1,1} = \beta_{1,2} = 180°$; **b** $\beta_{1,1} = \beta_{1,2} > 180° = 190°$; **c** $\beta_{1,1} = \beta_{1,2} < 180° = 170°$; **d** $\beta_{1,1} = 190° > \beta_{1,2} = 170°$

The results of calculations are presented in the Table 1, Table 2, Table 3, Table 4, Table 5 and Table 6.

In the result of calculation the following indexes were identified:

– the component of layers' active pressure $E_{a,n}$;

– the angles of orientation of sliding surfaces $\beta_{2,n}$, corresponding to extreme pressure.

– the coefficients of anisotropy of n-layer, defined according to the formula:

$$k_{a,n} = \frac{E_{a,n}(anis.)}{E_{a,n}(is.)}, \tag{5}$$

where $E_{a,n}(anis.)$ – active soil pressure of n-layer, defined, taking into account strength anisotropy;

$E_{a,n}(is.)$ – active soil pressure of n-layer, defined in the conditions of isotropic soil environment at $\varphi_n = \varphi_{n,min} = \text{const}$.

Table 1 Parameters $E_{a,1}$, $E_{a,2}$, $\beta_{2,1}$, $\beta_{2,2}$, $k_{a,1}$, $k_{a,2}$ depending on the orientation of contact soil layers surfaces at their minimal and maximal internal friction angle and the orientation of a wall's rear verge $\beta_3 = 270°$

The orientation of backfill soils surfaces, $\beta_{1,1}$, $\beta_{1,2}$	$E_{a,1}$	$E_{a,2}$	$\beta_{2,1}$	$\beta_{2,2}$	$k_{a,1}$	$k_{a,2}$
$\beta_{1,1} = \beta_{1,2} = 180°$	$\varphi = \varphi_{min} = 30°$					
	11,33,333	34,00,000	240	240	1	1
	$\varphi = \varphi_{max} = 40°$					
	8,07,942	24,23,827	243	243	1	1
$\beta_{1,1} = \beta_{1,2} = 190°$	$\varphi = \varphi_{min} = 30°$					
	12,70,509	38,11,525	237	237	1	1
	$\varphi = \varphi_{max} = 40°$					
	8,07,942	24,23,827	243	243	1	1
$\beta_{1,1} = \beta_{1,2} = 170°$	$\varphi = \varphi_{min} = 30°$					
	10,32,815	30,98,445	242	242	1	1
	$\varphi = \varphi_{max} = 40°$					
	6,84,575	20,53,726	246	246	1	1
$\beta_{1,1} = 190°$ $\beta_{1,2} = 170°$	$\varphi = \varphi_{min} = 30°$					
	12,70,509	37,99,565	237	237	1	1
	$\varphi = \varphi_{max} = 40°$					
	8,07,942	24,17,724	243	243	1	1

Table 2 Parameters $E_{a,1}$, $E_{a,2}$, $\beta_{2,1}$, $\beta_{2,2}$, $k_{a,1}$, $k_{a,2}$ depending on axes orientation of hodographs' maximal strength, χ_n and the given hodographs of an internal friction angle of soil layers at the orientation of a wall's rear verge $\beta_3 = 270°$

The orientation of backfill soils surfaces, $\beta_{1.1}$, $\beta_{1.2}$	Axes orientation of maximal strength of hodographs of the internal friction angle, χ_1, χ_2	$E_{a,1}$	$E_{a,2}$	$\beta_{2,1}$	$\beta_{2,2}$	$k_{a,1}$	$k_{a,2}$
$\beta_{1,1} = \beta_{1,2} = 180°$	$\varphi = 30$–$35°$						
	$\chi_1 = 0°$, $\chi_2 = 90°$	10,82,523	29,57,552	242	240	0,93,752	0,86,987
	$\chi_1 = 90°$, $\chi_2 = 0°$	9,85,851	31,87,568	240	242	0,86,987	0,93,752
	$\varphi = 35$–$40°$						
	$\chi_1 = 0°$, $\chi_2 = 90°$	8,65,296	23,68,448	244	243	0,93,939	0,85,710
	$\chi_1 = 90°$, $\chi_2 = 0°$	7,89,494	25,95,889	243	244	0,85,709	0,93,938
$\beta_{1,1} = \beta_{1,2} = 190°$	$\varphi = 30$–$35°$						
	$\chi_1 = 0°$, $\chi_2 = 90°$	11,78,399	33,06,345	240	238	0,92,750	0,86,746
	$\chi_1 = 90°$, $\chi_2 = 0°$	11,02,115	35,35,198	238	240	0,86,746	0,92,750
	$\varphi = 35$–$40°$						
	$\chi_1 = 0°$, $\chi_2 = 90°$	9,49,115	26,09,889	242	241	0,93,132	0,85,365
	$\chi_1 = 90°$, $\chi_2 = 0°$	8,69,963	28,47,346	241	242	0,85,365	0,93,132
$\beta_{1,1} = \beta_{1,2} = 170°$	$\varphi = 30$–$35°$						
	$\chi_1 = 0°$, $\chi_2 = 90°$	9,74,870	26,98,500	244	242	0,94,390	0,87,092
	$\chi_1 = 90°$, $\chi_2 = 0°$	8,99,500	29,24,611	242	244	0,87,092	0,94,390
	$\varphi = 35$–$40°$						
	$\chi_1 = 0°$, $\chi_2 = 90°$	7,99,511	21,80,621	246	245	0,94,459	0,85,877
	$\chi_1 = 90°$, $\chi_2 = 0°$	7,26,874	23,98,532	245	246	0,85,877	0,94,459
$\beta_{1,1} = 190°$; $\beta_{1,2} = 170°$	$\varphi = 30$–$35°$						
	$\chi_1 = 0°$, $\chi_2 = 90°$	11,78,399	32,96,323	240	238	0,92,750	0,86,755
	$\chi_1 = 90°$, $\chi_2 = 0°$	11,02,115	35,25,222	238	240	0,86,746	0,92,780
	$\varphi = 35$–$40°$						
	$\chi_1 = 0°$, $\chi_2 = 90°$	9,49,115	26,02,792	242	241	0,93,132	0,85,374
	$\chi_1 = 90°$, $\chi_2 = 0°$	8,69,963	28,39,891	241	242	0,85,337	0,93,151

Table 3 Parameters $E_{a,1}$, $E_{a,2}$, $\beta_{2,1}$, $\beta_{2,2}$, $k_{a,1}$, $k_{a,}$ depending on the axes orientation of hodographs' maximal strength, χ_n and the given hodographs of the internal friction angle of soil layers at the orientation of a wall's rear verge $\beta_3 = 285°$

The orientation of backfill soils surfaces, $\beta_{1.1}$, $\beta_{1.2}$	Axes orientation of maximal strength of hodographs of the internal friction angle, χ_1, χ_2	$E_{a,1}$	$E_{a,2}$	$\beta_{2,1}$	$\beta_{2,2}$	$k_{a,1}$	$k_{a,2}$
$\beta_{1,1} = \beta_{1,2} = 180°$	$\varphi = 30–35°$						
	$\chi_1 = 0°$, $\chi_2 = 90°$	14,75,252	40,82,400	250	248	0,96,542	0,89,053
	$\chi_1 = 90°$, $\chi_2 = 0°$	13,60,800	44,25,755	248	250	0,89,053	0,96,542
	$\varphi = 35–40°$						
	$\chi_1 = 0°$, $\chi_2 = 90°$	12,69,126	34,67,077	252	251	0,96,774	0,88,121
	$\chi_1 = 90°$, $\chi_2 = 0°$	11,55,692	38,07,376	251	252	0,88,124	0,96,774
$\beta_{1,1} = \beta_{1,2} = 190°$	$\varphi = 30–35°$						
	$\chi_1 = 0°$, $\chi_2 = 90°$	16,64,549	46,42,673	246	246	0,95,564	0,88,848
	$\chi_1 = 90°$, $\chi_2 = 0°$	15,47,558	49,93,647	244	246	0,88,848	0,95,565
	$\varphi = 35–40°$						
	$\chi_1 = 0°$, $\chi_2 = 90°$	14,17,852	38,92,376	249	248	0,96,013	0,87,861
	$\chi_1 = 90°$, $\chi_2 = 0°$	12,97,459	42,53,557	248	249	0,87,861	0,96,013
$\beta_{1,1} = \beta_{1,2} = 170°$	$\varphi = 30–35°$						
	$\chi_1 = 0°$, $\chi_2 = 90°$	13,26,289	36,47,168	253	251	0,97,173	0,89,072
	$\chi_1 = 90°$, $\chi_2 = 0°$	12,15,723	39,78,866	251	253	0,89,072	0,97,173
	$\varphi = 35–40°$						
	$\chi_1 = 0°$, $\chi_2 = 90°$	11,47,736	31,22,980	254	253	0,97,319	0,88,268
	$\chi_1 = 90°$, $\chi_2 = 0°$	10,40,993	34,43,209	253	254	0,88,269	0,97,320
$\beta_{1,1} = 190°$; $\beta_{1,2} = 170°$	$\varphi = 30–35°$						
	$\chi_1 = 0°$, $\chi_2 = 90°$	16,64,549	42,26,061	246	244	0,95,564	0,98,856
	$\chi_1 = 90°$, $\chi_2 = 0°$	15,47,558	49,76,666	244	246	0,88,848	0,95,590
	$\varphi = 35–40°$						
	$\chi_1 = 0°$, $\chi_2 = 90°$	14,17,852	38,79,820	249	248	0,96,013	0,87,869
	$\chi_1 = 90°$, $\chi_2 = 0°$	12,97,459	42,40,204	248	249	0,87,860	0,96,030

Table 4 Parameters $E_{a,1}$, $E_{a,2}$, $\beta_{2,1}$, $\beta_{2,2}$, $k_{a,1}$, $k_{a,2}$ depending on the orientation of contact soils layers' surface at their minimal and maximal internal friction angle and the orientation of a wall's rear verge $\beta_3 = 285°$

The orientation of backfill soils surfaces, $\beta_{1,1}$, $\beta_{1,2}$	$E_{a,1}$	$E_{a,2}$	$\beta_{2,1}$	$\beta_{2,2}$	$k_{a,1}$	$k_{a,2}$
$\beta_{1,1} = \beta_{1,2} = 180°$	$\varphi = \varphi_{min} = 30°$					
	15,28,087	45,84,262	247	247	1	1
	$\varphi = \varphi_{max} = 40°$					
	11,16,994	33,50,983	252	252	1	1
$\beta_{1,1} = \beta_{1,2} = 190°$	$\varphi = \varphi_{min} = 30°$					
	17,41,808	52,25,423	243	243	1	1
	$\varphi = \varphi_{max} = 40°$					
	12,44,579	37,33,736	250	250	1	1
$\beta_{1,1} = \beta_{1,2} = 170°$	$\varphi = \varphi_{min} = 30°$					
	13,64,871	40,94,612	251	251	1	1
	$\varphi = \varphi_{max} = 40°$					
	10,11,808	30,35,425	255	255	1	1
$\beta_{1,1} = 190°$ $\beta_{1,2} = 170°$	$\varphi = \varphi_{min} = 30°$					
	17,41,808	52,06,254	243	243	1	1
	$\varphi = \varphi_{max} = 40°$					
	12,44,579	37,22,336	250	250	1	1

Table 5 Parameters $E_{a,1}$, $E_{a,2}$, $\beta_{2,1}$, $\beta_{2,2}$, $k_{a,1}$, $k_{a,2}$ depending on the surface orientation of contact soil layers at their minimal and maximal internal friction angle and the orientation of a wall's rear verge $\beta_3 = 255°$

The orientation of backfill soils surfaces, $\beta_{1,1}$, $\beta_{1,2}$	$E_{a,1}$	$E_{a,2}$	$\beta_{2,1}$	$\beta_{2,2}$	$k_{a,1}$	$k_{a,2}$
$\beta_{1,1} = \beta_{1,2} = 180°$	$\varphi = \varphi_{min} = 30°$					
	8,18,668	24,56,004	232	232	1	1
	$\varphi = \varphi_{max} = 40°$					
	4,47,138	13,41,412	237	237	1	1
$\beta_{1,1} = \beta_{1,2} = 190°$	$\varphi = \varphi_{min} = 30°$					
	9,07,003	27,21,009	231	231	1	1
	$\varphi = \varphi_{max} = 40°$					
	4,81,102	14,43,305	237	237	1	1
$\beta_{1,1} = \beta_{1,2} = 170°$	$\varphi = \varphi_{min} = 30°$					
	7,57,557	22,72,671	234	234	1	1
	$\varphi = \varphi_{max} = 40°$					
	4,21,790	12,65,369	238	238	1	1
$\beta_{1,1} = 190°$; $\beta_{1,2} = 170°$	$\varphi = \varphi_{min} = 30°$					
	9,07,003	27,13,717	231	231	1	1
	$\varphi = \varphi_{max} = 40°$					
	4,81,102	14,40,513	237	237	1	1

Table 6 Parameters $E_{a,1}, E_{a,2}, \beta_{2,1}, \beta_{2,2}, k_{a,1}, k_{a,2}$ depending on the axes orientation of hodographs' maximal strength, χ_n and the given hodographs of the internal friction angle of soil layers at the orientation of a wall's rear verge $\beta_3 = 255°$

The orientation of backfill soils surfaces, $\beta_{1.1}, \beta_{1.2}$	Axes orientation of maximal strength of hodographs of the internal friction angle, χ_1, χ_2	$E_{a,1}$	$E_{a,2}$	$\beta_{2,1}$	$\beta_{2,2}$	$k_{a,1}$	$k_{a,2}$
$\beta_{1,1} = \beta_{1,2} = 180°$	$\varphi = 30–35°$						
	$\chi_1 = 0°, \chi_2 = 90°$	7,31,364	20,75,013	234	233	0,89,335	0,84,487
	$\chi_1 = 90°, \chi_2 = 0°$	6,91,670	21,94,091	233	234	0,84,487	0,89,335
	$\varphi = 35–40°$						
	$\chi_1 = 0°, \chi_2 = 90°$	5,46,961	15,16,940	237	236	0,89,134	0,82,402
	$\chi_1 = 90°, \chi_2 = 0°$	5,05,647	16,40,881	236	237	0,82,401	0,89,134
$\beta_{1,1} = \beta_{1,2} = 190°$	$\varphi = 30–35°$						
	$\chi_1 = 0°, \chi_2 = 90°$	8,00,989	22,87,924	233	231	0,88,311	0,84,083
	$\chi_1 = 90°, \chi_2 = 0°$	7,62,642	24,02,965	231	233	0,84,083	0,88,311
	$\varphi = 35–40°$						
	$\chi_1 = 0°, \chi_2 = 90°$	5,91,085	16,45,990	236	235	0,88,339	0,81,999
	$\chi_1 = 90°, \chi_2 = 0°$	5,48,664	17,12,833	235	236	0,81,999	0,88,339
$\beta_{1,1} = \beta_{1,2} = 170°$	$\varphi = 30–35°$						
	$\chi_1 = 0°, \chi_2 = 90°$	6,81,361	19,23,522	235	234	0,89,942	0,84,637
	$\chi_1 = 90°, \chi_2 = 0°$	6,41,174	20,44,083	234	235	0,84,637	0,89,941
	$\varphi = 35–40°$						
	$\chi_1 = 0°, \chi_2 = 90°$	5,13,951	14,20,840	237	237	0,89,681	0,82,642
	$\chi_1 = 90°, \chi_2 = 0°$	4,73,613	15,41,853	237	237	0,82,642	0,89,681
$\beta_{1,1} = 190°; \beta_{1,2} = 170°$	$\varphi = 30–35°$						
	$\chi_1 = 0°, \chi_2 = 90°$	8,00,989	22,82,031	233	232	0,88,311	0,84,092
	$\chi_1 = 90°, \chi_2 = 0°$	7,62,642	23,97,143	231	233	0,84,083	0,88,334
	$\varphi = 35–40°$						
	$\chi_1 = 0°, \chi_2 = 90°$	5,91,085	15,86,447	236	235	0,88,339	0,82,009
	$\chi_1 = 90°, \chi_2 = 0°$	5,48,664	17,69,613	235	236	0,81,999	0,88,361

4 Conclusions

At the orientation of the hodograph's maximal strength axis $\chi_n = 0°$ active pressure of anisotropic soil increases for both layers. Maximal pressures of soil layers at any orientation of a wall's rear verge correspond the orientation of their surfaces $\beta_{1,1} = \beta_{1,2} > 180° = 190°$, here in 1,17–1,23 times exceeding minimal pressure, received at $\beta_{1,1} = \beta_{1,2} < 180° = 170°$. The orientation of a wall's rear verge $\beta_3 = 285°$ corresponds maximal soil pressure, that in 1,96–2,32 times more that minimal lateral pressure, got for $\beta_3 = 255°$, moreover such results are characteristic for both anisotropic and isotropic soil environment. The influence of strength anisotropy is mostly characteristic at the orientation of a hodograph's maximal strength axis $\chi_n = 90°$. Maximal influence of strength anisotropy corresponds the orientation of a wall's rear verge $\beta_3 = 255°$ ($\beta_3 < 270°$), minimum influence appears at $\beta_3 = 285°$ ($\beta_3 > 270°$). So, the orientation of a wall's rear verge and the surface of contact layers significantly effect on the value of soil's lateral pressure, in some cases allowing to lower it more than in 2 times. The results can be used in retaining structures' design for the choice of optimal solutions of wall's configuration and contact environment, which anisotropic properties can be formed by means of sand washing or a technology of dumping and consolidation.

References

1. Amadei B (1983) Rock anisotropy and the theory of stress measurements. Springer, Berlin
2. Oda M, Koishikawa I, Niguchi T (1978) Experimental study of anisotropic shear strength of sand by plan strain test. Soil and Found 18(1):25–38
3. Katzenbach R, Leppla S, Seip M, Kurze S (2015) Value engineering as a basis for safe, optimized and sustainable design of geotechnical structures. In: XVI ECSMGE geotechnical engineering for infrastructure and development, pp 601–606. ICE Publishing, Edinburg. https://doi.org/10.1680/ecsmge.60678
4. Vynnykov Yu, Aniskin A (2019) Practical problems of anisotropic soil mechanics: monograph. University North, Croatia, Varazdin
5. Cudny M, Partyka E (2017) Influence of anisotropic stiffness in numerical analyses of tunneling and excavation problems in stiff soils. In: Proceedings of 19th international conference on soil mechanics and geotechnical engineering, pp 707–710. ICE Publishing, Seoul. https://doi.org/10.1680/geot.1997.47.3.693
6. Tong Z, Zhang L, Zhou M (2013) DEM simulation of biaxial compression experiments of inherently anisotropic granular materials and the boundary effects. J Appl Math, 2013:1–13. Article ID 394372, Hindawi. https://doi.org/10.1155/2013/394372
7. Tong Z (2014) Experimental investigation of shear strength of sands with inherent fabric anisotropy. Acta Geotechnica 9:257–275. https://doi.org/10.1007/s11440-014-0303-6
8. Zhang K, Zhu J, Charkley F (2017) Recent developments in geomaterial's stress-induced anisotropy research at Hohai university: testing, modeling and application. In: Proceedings of 19th international conference on soil mechanics and geotechnical engineering, pp 997–1000. ICE Publishing, Seoul. https://doi.org/10.1680/geot.1997.47.3.693
9. Allahverdizadeh P, Griffiths D, Fenton G (2015) Influence of highly anisotropic properties on probabilistic slope stability. In: Proceedings of XVI ECSMGE geotechnical engineering for infrastructure and development, pp 1555–1559. Edinburg. https://doi.org/10.1680/ecsmge.60678
10. Shkola A (2012) Lateral pressure of anisotropic soils on buildings. International association of hydrotechnical engineers of marine and river transport, Odessa
11. Chowdhury SS (2017) A study on lateral earth pressure against strutted retaining wall in cohesionless soil deposit. Int J Geotech Eng, 122–138. https://doi.org/10.1080/19386362.2017.1326683
12. Khosravia MH, Pipatpongsa T, Takemura J (2016) Theoretical analysis of earth pressure against rigid retaining walls under translation mode. Soils Found 56(4):664–675. ScienceDirect. https://doi.org/10.1016/j.sandf.2016.07.007

13. Zotsenko M, Vynnykov Y, Lartseva I, Sivitska S (2018) Ground base deformation by circular plate peculiarities. In: Paper presented at the MATEC web of conferences, vol 230. https://doi.org/10.1051/matecconf/201823002040
14. Zotsenko ML, Mykhailovska OV, Sivitska SP (2020) Construction features durable storage of toxic waste in boreholes. https://doi.org/10.1007/978-3-030-42939-3_33
15. Loburets AT, Naumovets AG, Senenko NB, Vedula YS (1997) Surface diffusion and phase transitions in strontium overlayers on W(112). Zeitschrift Fur Physikalische Chemie 202(1–2):75–85. https://doi.org/10.1524/zpch.1997.202.part_1_2.075
16. Kochkarev D, Azizov T, Galinska T (2018) Bending deflection reinforced concrete elements determination. In: Paper presented at the MATEC web of conferences, vol 230. https://doi.org/10.1051/matecconf/201823002012
17. Pavlenko A, Koshlak H (2015) Design of processes of thermal bloating of silicates. Metall Min Ind 7(1):118–122

Constructive Features of the Device "Trombe Wall" During Thermal Modernization of the Existing Building

Dmytro Yermolenko, Grigoriy Sharyi, Oleksandr Lapenko, and Oleksandr Palyvoda

Abstract The article presents a constructive solution for the modernization of the enclosure structures of the operated building. Measures were taken to reduce the annual power delivery for heating. A comprehensive approach has been implemented by combining measures to increase the heat transfer resistance of enclosure structures and the introduction of solar heating systems. In particular, it is proposed: replacement of translucent parts with energy efficient ones; heat insulation of opaque parts of the walls; arrangement of two partitions of the southern wall in the form of elements of passive solar heating - "Trombe Wall". The constructive decision of an element of passive heating does not demand installation of the additional equipment and saves the useful area of the room. To preserve the carrying function of the partition when installing thermal insulation from another wall array, it is proposed to use the method of perforation. Experimental design of complex modernization of the shell and heating system of the existing residential building has been performed. The comparison of initial and design heat needs confirmed the feasibility of introducing a set of energy efficient measures, in particular, the integration into the existing heating system of the house of a passive heating element - "Trombe Walls". The introduction of elements of the passive heating system, which works on the principle of utilization and use of solar radiation on the scale of a single house has reduced the total cost of heat generation during the heating season by 6%.

Keywords Energy efficiency · Renewable energy · Heat input · Carrying capacity

D. Yermolenko (✉) · G. Sharyi
National University «Yuri Kondratyuk Poltava Polytechnic», Poltava 37011, Ukraine
e-mail: yeda@ukr.net

O. Lapenko
National Aviation University, Kyiv 03058, Ukraine

O. Palyvoda
Kryvyi Rih National University, Kryvyi Rih 50027, Ukraine

V. Onyshchenko et al. (eds.), *Proceedings of the 3rd International Conference on Building Innovations*, Lecture Notes in Civil Engineering 181,
https://doi.org/10.1007/978-3-030-85043-2_40

1 Introduction

Requirements for the technical condition of buildings and structures of different functional purposes are changing over time. The change in the regulatory framework in construction in recent years is mainly in the direction of improving energy efficiency. It should be noted that now the economic component includes not only the cost of design and construction, but also added the cost of ensuring safe operation during the estimated service life of buildings or structures. As practice shows, the material costs of maintaining the normal operation of buildings can be many times higher than the cost of their construction.

Among the operational ones, a significant share is occupied by the costs of ensuring the appropriate sanitary and hygienic parameters of the microclimate of the premises. Among the buildings and structures that have already been put into operation, the vast majority are facilities that are designed in accordance with outdated regulatory requirements. The introduction of energy efficiency methods is mandatory, both in new construction and in the reconstruction of existing buildings and structures. From the point of view of a design engineer, this is the most difficult to do for the last category of objects. It is necessary to take into account the architectural, planning and design solutions of the existing building and modernize them to new requirements, in particular, for thermal insulation of buildings. Therefore, the development of rational design solutions for modernization of enclosing structures of buildings and structures operated in order to reduce the annual power delivery for heating by combining measures to increase the heat transfer resistance of enclosing structures and the introduction of solar heating systems is relevant.

2 Analysis of the State of Research

Of course, thermal modernization of the building is a valuable measure. However, these funds have a clear tendency to return in a relatively short time. The process of thermal modernization of an apartment building is mainly carried out in three directions: insulation of enclosing building structures [1]; modernization of heating and ventilation systems [2]; extensive use of renewable energy resources [3]. The greatest effect is observed in the case of an integrated approach by combining these areas in one project.

Solar heating systems are used to reduce the consumption of traditional energy resources. Currently, the most developed are power and heat supply systems that use solar energy. In our case, the advantages of solar energy include [4]: environmental friendliness—does not create hazardous waste; great energy potential—solar energy is inexhaustible and constant; accessibility; low operating costs. Significant disadvantages are [5]: high initial cost; unstable energy capacity; large land consumption.

But the disposal, accumulation and subsequent use of solar radiation is possible by other methods. The most interesting are passive systems that do not require expensive equipment. The most well-known example of a passive system of providing additional heat to buildings and structures is the "Trombe Wall". This is a massive stone structure, which is installed on the south side of the building behind the facade glass fence [6–10]. The idea was developed by Edward Morse [11] in 1881, and French professor Felix Tromb revived it in 1960. This wall device allows you to collect and store solar energy for the whole sunny day, and then give this heat to the room after a certain time (usually the return time falls at night). Depending on the thickness of the structure, a longer delay in heat transfer to the room is provided: at a wall thickness of 20 cm—the delay occurs for about 5 h;—at a wall thickness of 40 cm—the delay occurs at about 10.12 h [11, 12].

Traditionally, the "Trombe Wall" is a separately erected massive, usually flat structure [13], located on some distance from the inner surface of the translucent enclosure structure within the heating room. A significant disadvantage of this design solution is the loss of part of the usable area of the premises and the complexity of the planning solution of the latter. And in the conditions of reconstruction of the existing building erection of such design can demand measures for strengthening of bearing designs.

3 The Goal of the Work

As the practice of designing multilayer enclosing structures has shown, the thickness of an effective insulation with a finishing layer can reach several tens of centimeters (Fig. 1a). This means that in the place of installation of the "Trombe Wall" layer

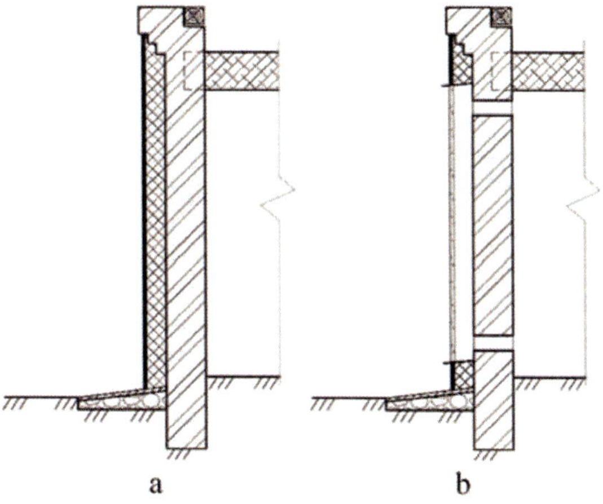

Fig. 1 Example of modernization of a fencing structure: **a** opaque part of the wall; **b** "Trombe wall"

of insulation can be replaced by an air-ventilated layer, and the finishing layer—a translucent element (Fig. 1b).

The paper considers the design features of the "Trombe Wall" in the form of a load-bearing structural element of the building, which receives the vertical compressive load from the above elements of the building. And also in addition I carry out function of an element of system of heating. In this case, for the possibility of heat accumulation, such a structure is located as close as possible to the shell of the building opposite the translucent element (window).

4 The Main Material

The efficiency of solar heating systems directly depends on the amount of solar radiation that can be collected by the receiving surface. The amount of solar radiation is characteristic of the area and is determined by the parameters of the receiving surfaces: size and spatial orientation; geographical coordinates of the location; reflection and absorption of solar energy; heat transfer characteristics. Evaluation of the feasibility of using elements of passive solar heating was carried out for a single-storey residential building (Fig. 2), which was built in 1960 near Poltava.

The purpose of experimental design is to try to identify the design features of the technical part of the design in the implementation of measures to attract renewable energy technologies by partial or complete replacement of traditional means. Among these, it is proposed to apply a partial replacement of heat supply needs—the arrangement of part of the load-bearing wall in the form of "Trombe Wall".

Fig. 2 South facade

Table 1 Estimated indicators of the months of the heating period [15]

Indicator	Oct.	Nov.	Dec.	Jan.	Feb.	Mat.	Apr.
The average monthly out-door temperature is θ_c,°C	7,7	1,3	– 3,4	– 5,6	– 4,7	0,3	9,0
Duration of the month *t*, h	408	720	744	744	672	744	264
Total daily receipts of solar radiation, W/m^2	3179	2352	2281	2828	3308	3035	1898

4.1 Climatic Data

The calculated indoor air temperature [14] in the premises is taken t_B = 20 °C, the estimated value of the relative humidity of the premises is 55%. The calculated outside air temperature [15] is t_3 = –23 °C. The average temperature of the coldest month is –5.6 °C, the relative humidity of the coldest month is 83%.

The duration of the heating period is 179 days [16]. The number of degree-days of the heating period is ≈ 3813 °C-days (Table 1).

The average monthly averages of solar radiation are given for radiation in the vertical plane.

4.2 Description of the Existing House

The total area of the house (Fig. 2) is 74.2 m^2, the height of the room is 2.7 m. The main facade of the house is oriented to the south with a deviation of 2.5° to the west—A = 182.5°). Entrance to the building from the west. The yard is located on the north side of the house.

The external walls are made of double-height silicate brick 380 mm thick. The area of opaque external fences was 117.3 m^2. Glazing is made in two rows, in the form of separate window frames made of wood. Area of translucent structures: north—5.0 m^2; east—1.1 m^2; south—3.4 m^2; west—5.2 m^2.

Power Delivery for Heating an Existing Building. Heat balance is calculated under the condition of quasi-stationary physical processes for a period of one month. It consists of: transmission heat losses of heated zones through external enclosing structures, unheated attic and floor construction on the ground; ventilation heat loss to heat the air; possible heat input from solar radiation. The monthly calculation is performed in general for the heating period. The air-conditioned area is A_f = 74.2 m^2.

Prior to the reconstruction, the structural solution of the building envelope and the heating system led to high energy consumption for heating. Based on the calculations of the internal heat capacity of the building and revenues from solar radiation, the value of specific energy consumption was 193.4 kWh/m^3. The difference between specific and limit power delivery was +61%. Which indicates a low efficiency of the building envelope. And the building as a whole has an energy efficiency class for specific power delivery—"F" [17].

4.3 Energy Efficient Measures

Three constructive decisions of modernization of an external cover of the building are projected: replacement of translucent parts with energy efficient ones; insulation of opaque parts of the walls; arrangement of two partitions of the southern wall in the form of elements of passive solar heating—"Trombe Wall".

The implementation of the first measure involves the replacement of obsolete wooden bindings with energy-efficient double-glazed windows. An effective mineral wool insulation with a density of 80 kg/m^3 and a thickness of 130 mm was chosen for wall insulation.

The implementation of the third measure increases the heat flow to the house.

The Design of the Proposed "Trombe Wall". Elements of the "Trombe Wall" are designed from two partitions (Fig. 3a) on the south side of the house (the area along the facade is 5.4 m^2). Structurally, it is part of a monolithic wall, which is made in the form of masonry of solid silicate brick, thickness $\delta_{cT} = 380$ mm. On the "inside" is applied a layer of lime finishing plaster of dark gray color, thickness $\delta_{\text{вш}}$

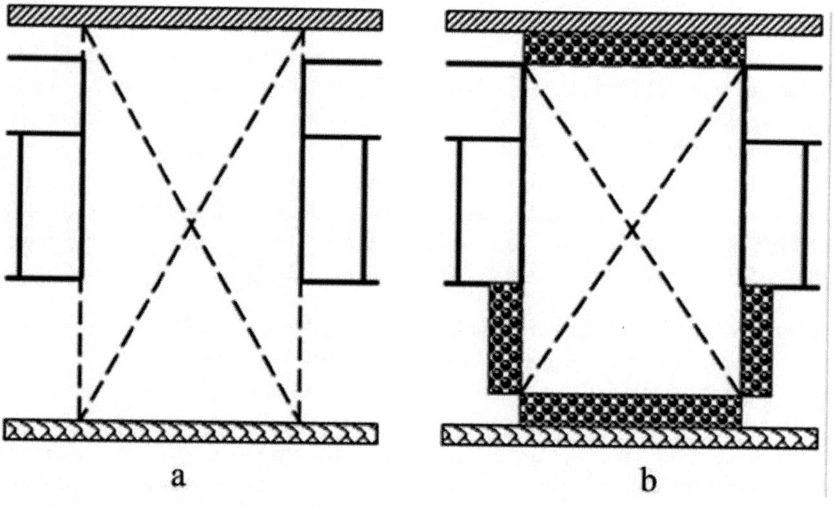

Fig. 3 The section of the enclosing structure, which has been transformed into the "Trombe Wall": **a** the selected area; **b** perforation zones

= 30 mm. With characteristics [18]: coefficient of absorption of solar radiation—$\alpha_{S,c}$ = 0,6; thermal radiation of the outer surface—ε = 0,90. The sun-absorbing surface is formed by a layer of cement plaster of dark green color, thickness $\delta_{3\text{пт}}$ = 30 mm. With characteristics: absorption coefficient of solar radiation—$\alpha_{S,c}$ = 0,6; thermal radiation of the outer surface—ε = 0,93.

Upon completion of the external insulation, the "Trombe Wall" will be located in the interior of the heating room. Separated from the translucent part of the shell by a ventilated air layer 130 mm thick. Thus, part of the outer wall will perform the functions of: heating element; load-bearing structure. For effective performance of the first function it is necessary to carry out constructive measures which will allow to reduce losses of thermal energy from other elements of a wall of the building.

The most effective of such measures will be the arrangement of temperature seams between the wall massif and the sections of the "Trombe Wall". Which are filled with effective insulation. But such a solution has limited application. A factor that significantly affects this is the vertical compressive force on the structure of the "Trombe Wall" from the higher structural elements. The temperature seams that propagate along this force will not be affected by the force factor. And horizontal seams—will be in a state of compression. The magnitude of the components of the stress state exceeds the compressive resistance of the material of such joints, which will lead to its destruction. And, as a consequence, the vertical movement (subsidence) of the "Trombe Wall". This condition is unacceptable.

To preserve the load-bearing function of the partition when installing thermal insulation from another wall array, it is proposed to use the method of perforation (Fig. 3b). Horizontal perforation zones will perform the function of supply (above the floor) and outlet (under the ceiling) ventilation of the working surface of the "Trombe Wall". Vertical perforation zones are arranged in the window sill area.

Vertical perforation does not affect the bearing function of the partition, so the volume of perforation (weakening of the cross section) of the partition is selected under conditions of minimizing thermal energy consumption by heat transfer to adjacent wall arrays. Horizontal sections of the perforation of the sheet are involved in the resistance of the vertical load from the above structural elements of the building. Therefore, for such areas, the second condition when choosing the volume of perforation is to ensure the load-bearing capacity of the partition to the action of external compressive load.

Energy Consumption for Heating a Modernized Building. Developed modernization measures reduce the cost of heating the house. But it is necessary to check to which class of energy efficiency the reconstructed building will be carried. And it is also important to establish how much the result will improve in the rooms, provided that the additional element of the heating system - the "Trombe wall" (Table 2).

Table 2 Energy parameters of the heating season total heat transfer by transmission

Indicator	Oct.	Nov.	Dec.	Jan.	Feb.	Mat.	Apr.
Total heat transfer by transmission, kWh	730	1958	2520	2769	2413	2131	645
Q_{tr} for the heating season, kWh	12224						
Ventilation heat losses for air heating Q_{ve}, kWh	39,6	106,3	137,0	150,5	131,4	115,8	35,1
Q_{ve} for the heating season, kWh	708						
Solar thermal overcurrents Q_{sol}, kWh	115,2	51,6	38,1	61,1	111,3	161,0	195,6
Q_{sol} for the heating season, kWh	733						
Heat flow from the "Thrombus Wall", Q_{wT} kWh	98,1	109,7	100,8	125,0	120,0	146,3	27,1
Q_{wT} for the heating season, kWh	727						

The specific energy consumption for heating during the heating period was 60.1 kWh/m^3. And the share of heat revenues from the "Trombe Wall"—5.9%.

The difference of –50% indicates that the building as a whole has an energy efficiency class for specific energy consumption—"A" [17].

5 Summary

Productivity of internal engineering systems of heat generation of separate buildings can be increased by integration into their structure of the devices working on principles of renewable energy. The analysis of the conducted research indicates that together with the modernization of the outer shell it allows to improve the energy efficiency of the building as a whole. The introduction of elements of the passive heating system, which works on the principle of utilization and use of solar radiation on the scale of the house allowed to reduce the total heat consumption during the heating season by 5.9%.

References

1. Farenyuk G, Kolesnyk E, Farenyuk E, Ralchuk V (2014) Examples of calculation to DSTU B B.2.6-189: 2013 "methods of selection of thermal insulation material for building insulation": a guide for design. DP NDIBK, Kyiv
2. Recommendations for the design of heating systems. https://alterair.ua/uk/articles/rekomendatsii-po-proektirovaniyu-sistem-otopleniya/. Accessed 22 Dec 2020
3. Basok BI, Deshko VI, Goncharuk SM, Bilous IY Influence of solar radiation on the thermal condition of the building. http://ittf.kiev.ua/wp-content/uploads/2017/06/s2_9_tezi.pdf. Accessed 27 Dec 2020
4. Solar energy and myths about it. https://setech.in.ua/sonjachna-energetika-ta-mifi-pro-nei/. Accessed 22 Dec 2020

5. Gifford D 5 ECO friendly home building ideas. https://www.smallfootprintfamily.com/eco-friendly-home-building#ixzz6hmetabv8. Accessed 22 Dec 2020
6. Konev A Stena Tromba v dome—kak ispolzovat passivnoe solnechnoe teplo? http://gidproekt.com/stena-tromba-v-dome-kak-ispolzovat-passivnoe-solnechnoe-teplo.html. Accessed 10 Dec 2020
7. Bisbano L, di Trombe M origini e applicazioni più innovative. https://www.architetturaecosostenibile.it/architettura/criteri-progettuali/muro-trombe-applicazioni-784?utm_content=buffer24abe&utm_medium=social&utm_source=plus.google.com&utm_campaign=buffer. Accessed 21 Dec 2020
8. Wilson A Tromble walls. https://www.buildinggreen.com/news-article/trombe-walls/. Accessed 21 Dec 2020
9. Poindexter J 23 Awesome DIY Rainwater Harvesting Systems You Can Build at Home. https://morningchores.com/rainwater-harvesting/. Accessed 21 Dec 2020
10. Shwarz BS Conqna enepgi. https://www.firstinarchitecture.co.uk/solar-energy/. Accessed 22 Dec 2020
11. Wenz PSOld idea, warm wall. https://www.sfgate.com/homeandgarden/article/Old-idea-warm-wall-3222133.php. Accessed 21 Dec 2020
12. Muro parietodinámico, climatizaciǿn bioclimática. https://rubenllera.wordpress.com/2014/09/20/muro-parietodinamico/. Accessed 22 Dec 2020
13. Leshchenko MV, Semko V (2015) Thermal characteristics of the external walling made of cold-formed steel studs and polystyrene concrete. Mag Civ Eng 60(8):44–55. https://doi.org/10.5862/MCE.60
14. DBN B.2.5-67 (2013) Heating, ventilation and air conditioning. State Enterprise "Ukrarchbudinform", Kyiv
15. DSTU-N B B.1.1-27: 2010 (2011) Construction climatology. State Enterprise "Ukrarchbudinform", Kyiv
16. DSTU-N B A.2.2-5: 2007 (2008) Guidelines for the development and compilation of energy passports and buildings for new construction and reconstruction. State Enterprise "Ukrarchbudinform", Kyiv
17. DBN B.2.6-31: 2016 (2016) Thermal insulation of buildings. State Enterprise "Ukrarchbudinform", Kyiv
18. DSTU B A.2.2-12: 2015 (2015) Energy efficiency of buildings. Method of calculating energy consumption for heating, cooling, ventilation, lighting and hot water supply. State Enterprise "Ukrarchbudinform", Kyiv

Analysis of the Humidity Condition of Wall Enclosing Structures of Cooling Warehouses and Possible Ways to Improve It

Oleg Yurin, Nurmammad Mammadov, Pavlo Semko, and Nataliia Mahas

Abstract The paper deals with the wall enclosing structures of cooling warehouses, erected recently (which mainly use metal sandwich panels) and erected in the second half of the 20th century (using bricks, expanded clay and reinforced concrete panels and effective insulation). The analysis of a humidity condition of wall enclosing structures of cooling warehouses which showed their unsatisfactory condition is executed. The analysis of possible ways to improve the humidity condition of wall enclosing structures of existing cooling warehouses, such as: reducing the thickness of the outer layer from the cooled premises, increasing the resistance to vapor penetration of the vapor barrier layer, using of ventilated air layer between the inner outer and insulation. The analysis showed that it is possible to fulfill the norms of the humidity state by reducing the thickness of the inner outer layer to 5 mm or by using a ventilated air layer which ventilated by air from the cooled premises.

Keywords Humidity condition · Insulation · Enclosing walls of cooling warehouses

1 Introduction

The operations main problem of cooling warehouses' buildings is maintenance of heat-protective qualities of enclosing designs at the norm recommended level which to a large extent depend on their humidity condition.

Recently, the sandwich panels, which consist of effective insulation and metal cladding layers are mainly used in the construction of cooling warehouses. The humidity condition of such fencing structures largely depends on the quality of the joints of sandwich panels and the quality of work performed during the construction of fences. At the same time, there are a large number of cooling warehouses erected

O. Yurin · P. Semko · N. Mahas (✉)
National University "Yuri Kondratyuk Poltava Polytechnic", Poltava, Ukraine

N. Mammadov
Azerbaijan University of Architecture and Construction, Ayna Sultanova 11, Baku, Azerbaijan

V. Onyshchenko et al. (eds.), *Proceedings of the 3rd International Conference on Building Innovations*, Lecture Notes in Civil Engineering 181,
https://doi.org/10.1007/978-3-030-85043-2_41

in the second half of the 20th century, in which bricks, expanded clay concrete and reinforced concrete panels with a layer of effective insulation were used as enclosing structures. As the experience of operation of such enclosing structures has shown, the humidity condition of the insulation in them and, as a consequence, the heat-protective properties are in unsatisfactory condition. During the service life of these buildings, the humidity of the insulation becomes significant. In some cases, there are formation of ice on the surfaces of the structural layers adjacent to the surface of the insulation from the cooled premises.

The humidity condition of the enclosing structures of cooling buildings is significantly different from the humidity condition of the enclosing structures of residential and public buildings. This is due to the fact that the flow of vaporous moisture in the cooling warehouses throughout the year moves in one direction—in the middle of the premises. While in the fences of residential and public buildings, this flow changes its direction during the year. In winter it moves from the premises to the outside, and in summer to the premises. That is, the period of humidity accumulation in the enclosing structures in winter changes to the period of humidity return in summer. And in protections of cooling warehouses there is a constant moisture accumulation. The fact that the structural layer in these structures is sometimes located on the inside also contributes to the accumulation of moisture. It slows down the release of moisture from the enclosing structures into the middle of the premises. The greater the density of this layer, the more humidity condenses in the enclosure. The search for ways to improve the humidity condition of enclosing structures was studied by the authors in [1–3], the thermal condition of enclosing structures improvement was also studied in [4–13]. Therefore, the study of the humidity condition of wall enclosing structures, search and analysis of possible ways to improve the moisture condition of wall enclosing structures of existing cooling warehouses is an urgent task of the study.

2 Analysis of the Humidity State of the Enclosing Wall Structures of Existing Cooling Warehouses

According to the existing standards for thermal protection and humidity [14–17] of enclosing structures for cooling warehouses, the required resistance to heat transfer of the outer walls of refrigerated premises for summer conditions for different climatic regions should be taken according to Table 1, the required resistance to vapor barrier in enclosing structures according to the Table 2.

Estimated values of temperature and relative humidity in refrigerated warehouses depend on the type of products stored in them. The values of these parameters for the most common products are given in the Table 3.

Assessment of the humidity condition of the enclosing structures was performed according to the method given in [9] for the climatic conditions of Poltava region. The humidity condition of the enclosing structures is the worst in July at the lowest

Table 1 The required heat transfer resistance depending on the air temperature in the cooled premises, m^2 °C/Watt

Average annual outdoor air temperature in the construction area, °C	−30 °C	−20 °C	−10 °C	−5 °C	0 °C	5 °C	12 °C
−2 °C and lower	4,8	3,9	3,1	2,6	2,4	2,1	1,9
Upper −2 °C and lower	5,1	4,3	3,6	2,8	2,4	2,1	1,9
7–2 °C and upper	5,4	4,8	4,3	3,7	3,3	2,8	2,2

Table 2 Necessary resistance to vapor penetration of vapor barrier depending on air temperature in the cooled premises, m^2 °C/Watt

Estimated humidity of the outside air in the construction area, gPa	−10 °C and lower	From −9 °C to 1 °C
To 14	6,6	2,7
From 14 to 18	9,3	4,6
More 18	13,3	6,6

Table 3 Temperature-humidity regime of cooled premises

Type of premises	Estimated temperature, °C	Estimated relative humidity, %
Long-term storage of ice cream, pork, poultry, fatty fish	−30	95
Storage of frozen products (meat, butter, fish)	−25	95
Storage of fats, mélange	−20	95
Long-term storage of frozen cheese	−20	85
Storage of herring and salted fish	−10	95
Short-term oil storage	−5	85
Storage of chilled eggs	−2	90
Storage of ghee	0	85
Storage of culinary products	2	80

indoor air temperature. Therefore, Table 3 adopted the calculated temperature of the indoor air $t_в = -30$ °C, and its relative humidity 95%.

According to the accepted calculation schemes (Fig. 1) for enclosing wall structures made of brick, expanded clay concrete panels and reinforced concrete panels, the characteristics of the enclosing structure layers the minimum insulation thicknesses and vapor permeability resistance of the vapor barrier layers were determined.

Calculations of humidity accumulation and humidity return were performed for all months of the year, conditional cross sections of enclosing structures with obtained

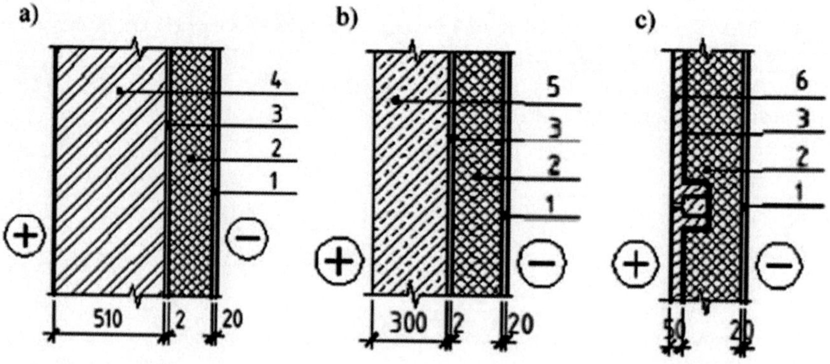

Fig. 1 Calculation scheme of the enclosing structure: **a** enclosing structure made of bricks, **b** enclosing structure made of expanded clay concrete panels, **c** enclosing structure made of reinforced concrete panels; 1—outer layer, 2—insulation, 3—vapor barrier, 4—brick, 5—expanded clay concrete panel, 6—reinforced concrete panel

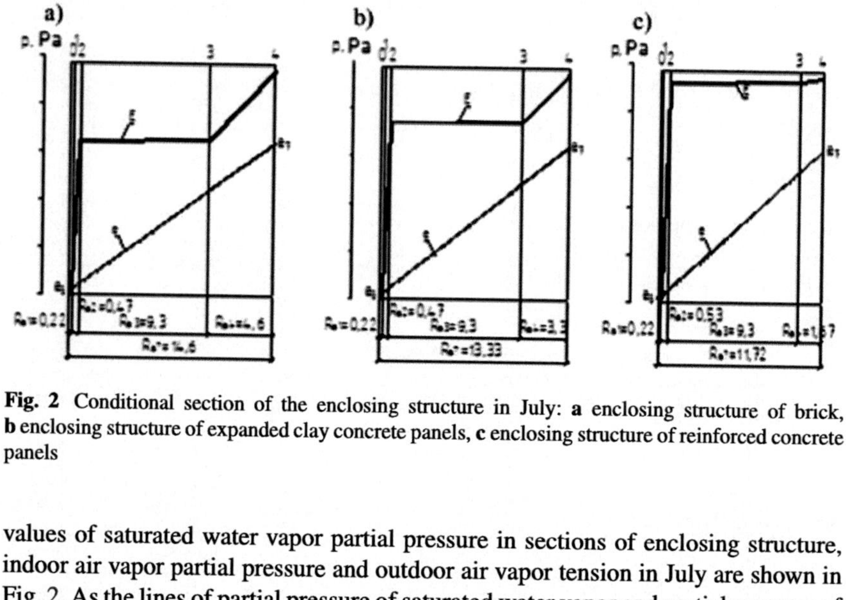

Fig. 2 Conditional section of the enclosing structure in July: **a** enclosing structure of brick, **b** enclosing structure of expanded clay concrete panels, **c** enclosing structure of reinforced concrete panels

values of saturated water vapor partial pressure in sections of enclosing structure, indoor air vapor partial pressure and outdoor air vapor tension in July are shown in Fig. 2. As the lines of partial pressure of saturated water vapor and partial pressure of water vapor of indoor air intersect, condensation of water vapor occurs in the fence.

Analysis of humidity accumulation and return during the year showed that there is a constant accumulation of humidity in enclosing structures.

Increasing the humidity of the insulation in the enclosing structures leads to the loss of their heat-protective properties. This, in turn, leads to increased heat loss through the structure, increased load on refrigeration equipment, lower internal temperature and loss of stored products quality. The greatest increase in the humidity

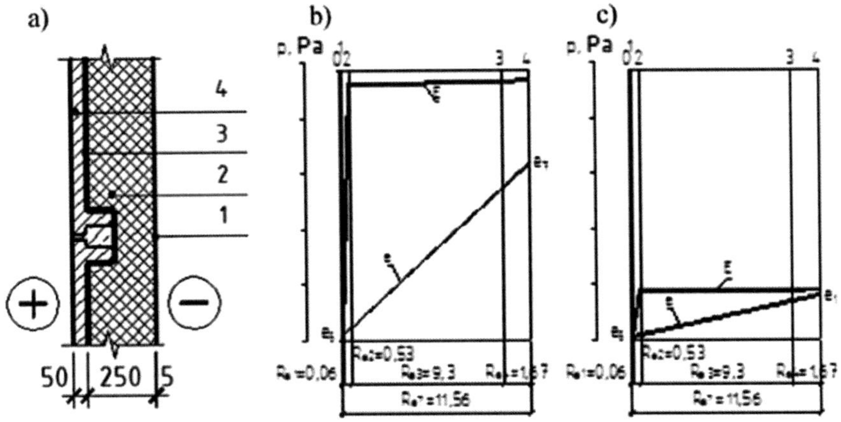

Fig. 3 **a** The design scheme of the enclosing structure of reinforced concrete panels with a reduced outer layer: 1—outer layer, 2—insulation, 3—vapor barrier, 4—reinforced concrete panel; **b** conditional cross-section of the enclosing structure made of reinforced concrete panels with a reduced outer layer in July; **c** conditional section of the enclosing structure of reinforced concrete panels with a reduced outer layer in January

of the insulation is observed in the walls with reinforced concrete panels. Therefore, this design was adopted for further research.

3 Analysis of Possible Ways to Improve the Humidity Condition of Existing Cooling Warehouses Wall Enclosing Structures

3.1 Reducing the Thickness of the Outer Layer in the Wall Structure

Vapor like moisture enters the insulation from the outside air, which has a much higher partial pressure of water vapor, due to the higher temperature than the air in the building. Moisture is removed from the insulation in the direction of the indoor air. The amount of moisture removed depends largely on the resistance to vapor penetration of the outer layer, which in turn depends on the thickness of this layer. The outer layer was usually made of lime-sand mortar with a thickness of 20 mm. Modern materials and technologies enable to reduce it to 5 mm. The calculation of humidity accumulation in the wall with an outer layer 5 mm thick was performed. The design scheme of the enclosing structure is given in Fig. 3a.

Calculations of humidity accumulation and return for all months of the year were performed, conditional sections of enclosing structures made of reinforced concrete panels with a reduced outer layer in July and January are shown in Fig. 3b, c. Analysis of humidity accumulation and return during the year showed that the enclosing structure is accumulating moisture $W_{sp} = 0{,}0186$ kg/m^2, however, more evaporates $W_{wp} = 0{,}2478$ kg/m^2.

3.2 *Increasing the Resistance to Vapor Penetration of Vapor Barrier in the Wall Structure*

As mentioned earlier, vapor like moisture enters the insulation from the outside air. The amount of moisture entering the insulation depends largely on the resistance to vapor penetration of the layers located between the insulation and the outside air. This is a layer of reinforced concrete panel and a layer of vapor barrier. By increasing the resistance to vapor permeability of the vapor barrier, can be reduced the amount of moisture coming from the outside air to the insulation. It is possible to determine the resistance to vapor penetration of the vapor barrier layer in the enclosing structure by both theoretical and graph-analytical methods. The graph-analytical method is simpler and clearer.

The required resistance to vapor penetration of vapor barrier in the wall structure was determined. The obtained partial pressure values of indoor airs water vapor, the average for the years water vapor partial pressure of outdoor air, the saturated water vapors partial pressure in the cross sections of the enclosing structure are plotted on the conditional section of the enclosing structure (Fig. 4). The horizontal section from the outer surface of the enclosing structure to the point of intersection of the lines is a necessary resistance to vapor penetration of the vapor barrier layer, which ensures the absence of humidity accumulation in the insulation during the year. General resistance to vapor penetration of vapor barrier $R_{e1} = 28{,}9$ (m^2 hour Pa)/mg, that is, to prevent moisture accumulation in the wall of the cooling warehouses, it is necessary to increase the resistance to vapor penetration of vapor barrier by 3 times.

Calculations of moisture accumulation and return for all months of the year were performed, conditional sections of enclosing structures made of reinforced concrete panels in July and January are shown in Fig. 5a, b. Analysis of moisture accumulation and return during the year showed that the amount of accumulated moisture in the enclosing structure $W_{sp} = 0{,}0379$ kg/m^2 equal to the amount of evaporating humidity $W_{wp} = 0{,}0379$ kg/m^2, that is, the accumulation of moisture does not occur. The vapor barrier layer with a vapor permeability resistance of at least $R_{e1} = 28{,}9$ (m^2 hour Pa)/mg can be made using the materials listed in Table 4.

It is quite difficult to perform vapor barrier with such resistance to vapor penetration, from the materials given in Table 4.

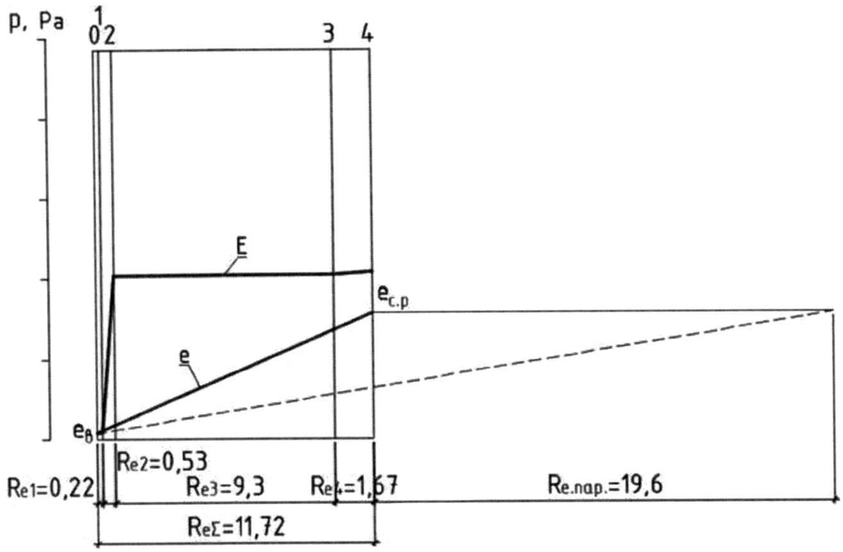

Fig. 4 Determination of resistance to vapor penetration of an additional layer of vapor barrier by graph-analytical method

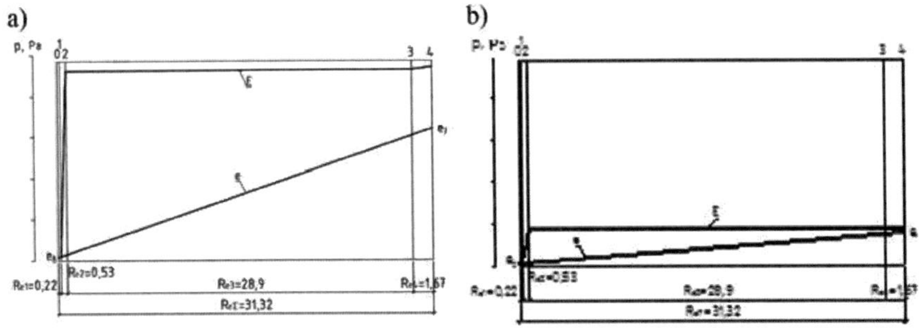

Fig. 5 Conditional section of the enclosing structure of reinforced concrete panels: **a** in July, **b** in January

3.3 *The Use of a Ventilated Air Layer in the Wall Structure*

Air layers, ventilated by air from the cooled rooms, located between the insulation and the inner layer, accelerate the release of humidity from the insulation. This is explained by the fact that in this case, between the insulation and the air in the cooled room there is no outer layer, which prevents the release of moisture from the insulation.

Table 4 Options for vapor barrier with resistance to vapor penetration is not less $R_{e1} = 28{,}9$ (m^2 hour Pa)/mg

Type of vapor barrier	Number of layers	General resistance to vapor penetration, (m^2 hour Pa)/mg
Painting with hot bitumen	97	29,1
Painting with enamel paint	61	29,28
Coating with insulating mastic	49	29,4
Coating with bituminous-kukersol mastic	28	29,44
Roofing glassine	88	29,04
Polyethylene film	4	29,2
Ruberoid	27	29,7

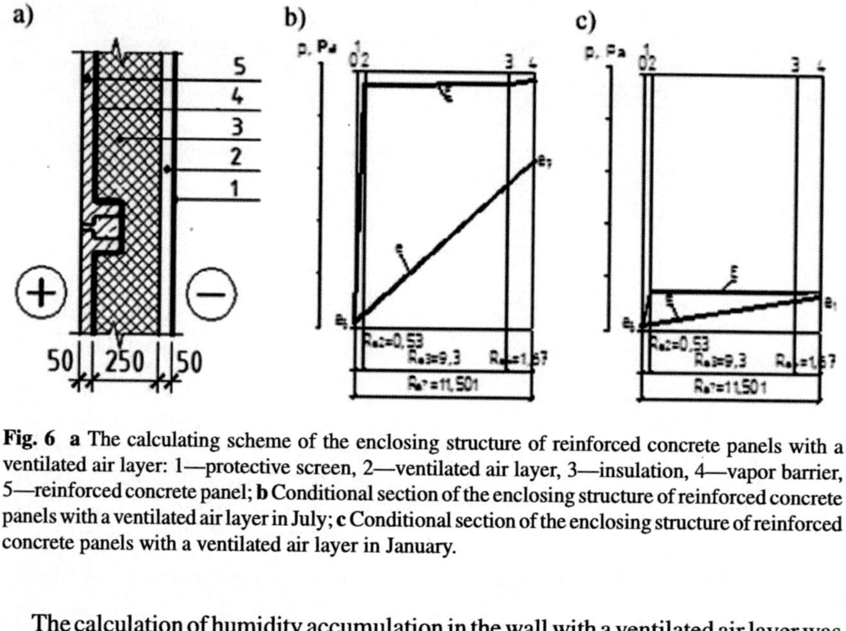

Fig. 6 **a** The calculating scheme of the enclosing structure of reinforced concrete panels with a ventilated air layer: 1—protective screen, 2—ventilated air layer, 3—insulation, 4—vapor barrier, 5—reinforced concrete panel; **b** Conditional section of the enclosing structure of reinforced concrete panels with a ventilated air layer in July; **c** Conditional section of the enclosing structure of reinforced concrete panels with a ventilated air layer in January.

The calculation of humidity accumulation in the wall with a ventilated air layer was performed, the calculation scheme of the enclosing structure of reinforced concrete panels with a ventilated air layer is shown in Fig. 6a.

Calculations of humidity accumulation and return were performed for all months of the year, conditional sections of fencing structures made of reinforced concrete panels in July and January are shown in Fig. 6b, c. Analysis of moisture accumulation and moisture removal showed that when using a ventilated air layer in the wall of moisture accumulation in the insulation during the year does not occur.

4 Conclusions

The analysis of the wall enclosing structures of recently erected cooling warehouses (mainly using metal sandwich panels) and erected in the second half of the 20th century (using bricks, expanded clay concrete and reinforced concrete panels and effective insulation) has been made. The analysis of the humidity condition of the wall enclosing structures of the cooling warehouses has been performed, and their unsatisfactory condition has been revealed. The possible ways to improve the humidity condition of the wall enclosing structures of existing cooling warehouses has been analyzed, which revealed that it is possible to meet the norms by reducing the thickness of the outer layer located on the side of the cooled room to 5 mm or using a ventilated air layer situated between inner layer and insulator which ventilated by air from cooling premises.

References

1. Semko OV, Yurin OI, Filonenko OI, Mahas NM (2020) Investigation of the temperature–humidity state of a tent-covered attic. In: Onyshchenko V, Mammadova G, Sivitska S, Gasimov A (eds) Proceedings of the 2nd international conference on building innovations. ICBI 2019. Lecture notes in civil engineering, vol 73. Springer, Cham, pp 245–252. https://doi.org/10.1007/978-3-030-42939-3_26
2. Filonenko O, Yurin O, Kodak O (2018) Thermal modernization of the panel buildings external walls. Int J Eng Technol 7(32):116–122. https://doi.org/10.14419/ijet.v7i3.2.14386
3. Yurin OI, Galinska TA, Pashchenko AM, Kaminska LS, Tverdokhlib VS (2014) Analiz norm oporu paropronyknenniu sharu paroizoliatsii v pokrytti budivel kholodylnykiv. Budivelni konstruktsii (80):223–230 . http://nbuv.gov.ua/UJRN/buko_2014_80_42
4. Kariuk A, Rubel V, Pashynskyi V, Dzhyrma S (2020) Improvement of residential buildings walls operation thermal mode. In: Onyshchenko V, Mammadova G, Sivitska S, Gasimov A (eds) Proceedings of the 2nd international conference on building innovations. ICBI 2019. Lecture notes in civil engineering, vol 73. Springer, Cham, pp 75–81. https://doi.org/10.1007/978-3-030-42939-3_9
5. Lapenko O, Skrebnieva S, Omelchenko K, Mashkov I (2020) Modern systems of heat insulation buildings. Key Eng Mater 864:128–133. https://doi.org/10.4028/www.scientific.net/KEM.864.128
6. Leshchenko MV, Semko V (2015) Thermal characteristics of the external walling made of cold-formed steel studs and polystyrene concrete. Mag Civ Eng 60(8):44–55. https://doi.org/10.5862/MCE.60
7. Yurin O, Azizova A, Galinska T (2018) Study of heat shielding qualities of a brick wall corner with additional insulation on the brick. Paper presented at the MATEC web of conferences, vol 230. https://doi.org/10.1051/matecconf/201823002039
8. Yurin O, Galinska T (2017) Study of heat shielding qualities of brick wall angle with additional insulation located on the outside fences. Paper presented at the MATEC web of conferences, vol 116. https://doi.org/10.1051/matecconf/201711602039
9. Mammadov N, Akbarova S (2018) Multi-disciplinary energy auditing of educational buildings in Azerbaijan: case study at a university campus. In: IFAC, International federation of automatic control. International conference. https://www.sciencedirect.com/journal/ifac-papersonline, 51(30), 311–315. https://doi.org/10.1016/j.ifacol.2018.11.308. Accessed 16 Jan 2021

10. Mammadova GH, Mammadov NY, Akbarova SM, Feyzieva GH (2021) Expert recommendations for energy improvements in educational facilities: case study-school buildings in azerbaijan. Paper presented at the IOP conference series: materials science and engineering, vol 1030, no 1. https://doi.org/10.1088/1757-899X/1030/1/012063
11. Akbarova S (2018) Trends of energy performance certification of buildings in azerbaijan. Int J Eng Technol (UAE) 7(3):563–566. https://doi.org/10.14419/ijet.v7i3.2.14590
12. Mammadova G, Sharifov A, Akbarova S (2021) Estudio experimental del rendimiento térmico de la cavidad de aire de fachadas ventiladas opacas en condiciones de viento ex-tremo: Estudio de caso bakú. Informes De La Construccion 73(561):1–8. https://doi.org/10.3989/IC.74247
13. Akbarova SM, Gahramanov SH, Mirzayeva DM (2020) Study of the thermophys-ical kinetics of additional concrete samples. Modern Phys Lett B 34(24). https://doi.org/10.1142/S0217984920502528
14. Teplova izoliatsiia budivel: DBN V.2.6-31:2016 (2016) [Chynni vid 2016–08–07]. K.: Minrehion Ukrainy. 30 s.
15. DSTU-N B V.1.1-27:2010 (2011) Zakhyst vid nebezpechnykh heolohichnykh protsesiv shkidlyvykh ekspluatatsiinykh vplyviv, vid pozhezhi. Budivelna klimatolohiia. K.: Minrehionbud Ukrainy. 123 s.
16. DSTU B V.2.6-189:2013 (2013) Metody vyboru teploizoliatsiinoho materialu dlia uteplennia budivel. K.: Minrehionbud Ukrainy. 51 s.
17. DSTU-N B V.2.6-192:2013 (2014) Nastanova z rozrakhunkovoi otsinky teplovolohisnoho stanu ohorodzhuvalnykh konstruktsii. K.: Minrehion Ukrainy. 37 s.

Influence of Strengthening Flat Slab by External Prestressed Reinforcement on Deformation Characteristics of the Slab

Oleksandr Zhuravskyi, Vladyslav Tymoshchuk, Nataliia Zhuravska, and Mukhlis Hajiyev

Abstract When designing and erecting monolithic frame buildings it can happen that at one stage of construction an error was made. Or during the reconstruction of the building, the operational load was increased.

In this case, the structure may become unusable in the conditions for which it was designed.

In the event of such situations, it is necessary to perform reinforcement of structures that do not meet the operating conditions.

This article presents the results of experimental studies of a model of a flat reinforced concrete slab reinforced with external prestressed reinforcement.

A comparison of the results of the plate under load without reinforcement, and with load and reinforcement.

According to the results of the experiment, it was found that this method of strengthening flat plates gives a positive result. That is why this method can be used for strengthening flat slabs.

Keywords Strengthening · External reinforcement · Flat slab

1 Preparing for Laboratory Test

Laboratory test was prepared and conducted on the basis of the department reinforced and stone constructions of Kyiv National University of Construction and Architecture.

At the laboratory was created reinforced concrete slab with scale to original 1:6 (Fig. 2) with dimensions 2500 × 2500 mm with step of columns 1000 mm, and cantilevers 250 mm on the perimeter.

O. Zhuravskyi (✉) · V. Tymoshchuk · N. Zhuravska
Kyiv National University of Construction and Architecture, Povitroflotskyi Avenue, 31, Kyiv, Ukraine
e-mail: zhuravskyi.od@knuba.edu.ua

M. Hajiyev
Azerbaijan University of Architecture and Construction, 5 Ayna Sultanova, Baku, Azerbaijan

V. Onyshchenko et al. (eds.), *Proceedings of the 3rd International Conference on Building Innovations*, Lecture Notes in Civil Engineering 181,
https://doi.org/10.1007/978-3-030-85043-2_42

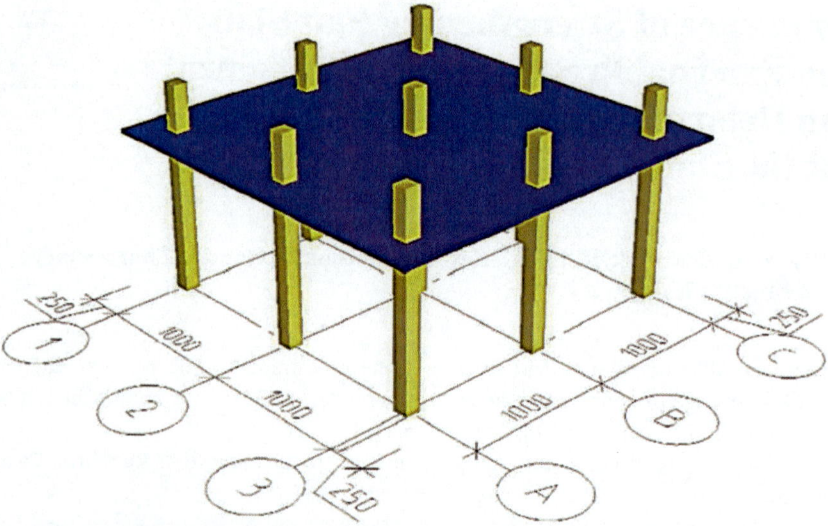

Fig. 1 Computer's model of reinforced flat slab

Class of concrete C35/40, reinforcement—1 layer mesh with Ø3 50 × 50 mm.

Thickness of slab 25 mm.

External reinforcement—Ø5 Bp-2 (Ukrainian number).

Before started with laboratory test of flat concrete slab, was made theoretical calculation of experimental model of slab (Fig. 1).

For these purposes was created four calculation model which correspond for stages of experiment.

1st stage—sum load on the slab was 251 kg
2nd stage—sum load on the slab was 581 kg
3rd stage—sum load on the slab was 677 kg
4th stage—sum load on the slab was 773 kg

2 Operation Test of Flat Slab with Strengthening By External Reinforcement

As was mentioned above all loading steps was broke for four stages. It gives the opportunity to control deformations of concrete slab.

Experiment had three parts:

1. Full load with four steps without strengthening by external reinforcement;
2. Only prestressed external reinforcement without full load—this part of experiment had only two stages of loading—0.5 N and N, what means tensile forces in external reinforcement.

Fig. 2 Research model of reinforcred flat slab

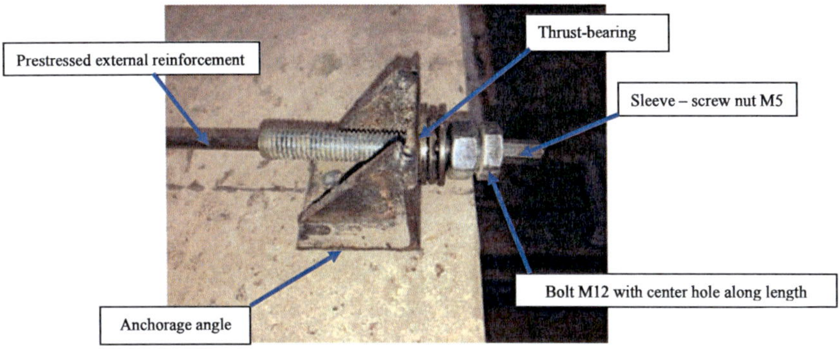

Fig. 3 Anchorage of external reinforcement on the flat slab

3. Prestressed external reinforcement and full load with four steps.

For anchorage external reinforcement on slab was used steel angles with stiffening ribs. For generation tensions on reinforcement was used bolt M12 with center hole along length (Fig. 3), for control of tension was used torque wrench.

For calculation torque was used formula:

$$T = 0.5 \times F_0 \times d_2 \frac{P/(\pi \times d_2) + f_p}{1 - f_p \times P/(\pi \times d_2)} \tag{1}$$

For measurement was used analog clock-type indicators with tolerance 0.001 mm for deformations (Fig. 4), and 0.01 mm for vertical deflections (Fig. 5).

For all model was used 16 indicators with tolerance 0.001 mm, and 4 indicators with tolerance 0.01 mm.

Loading for slab created by steel springs with weight from 15 to 19 kg, and bags with dry sand with weight 3 kg (Fig. 6).

Fig. 4 Clock-type indicator—tolerance 0.001 mm on slab

Fig. 5 Clock-type indicator—tolerance 0.01 mm on slab

Fig. 6 Loading for experimental model. Second stage—sum loading 581 kg

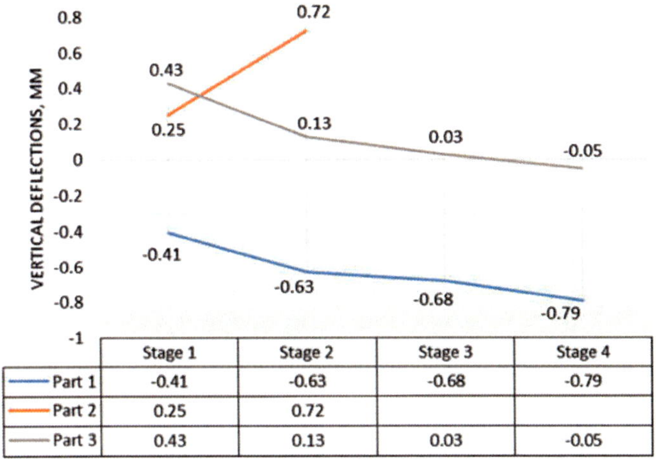

	Stage 1	Stage 2	Stage 3	Stage 4
Part 1	-0.41	-0.63	-0.68	-0.79
Part 2	0.25	0.72		
Part 3	0.43	0.13	0.03	-0.05

Fig. 7 The value of vertical deflections for indicator № 1

3 Processing the Results

After laboratory test of reinforced concrete slab obtained values of vertical deflections for each of three parts of loadings.

As can be seen from the graphics (Figs. 7, 8, 9 and 10) after strengthening with external reinforcement vertical deflections reduced. Irregularity of values could arised, because perfect tolerance of model slab in scale 1:6 is hard to achieve.

The scientific developments of the authors of the article are associated with their preliminary studies, which are set out in the works [4–6], and are also a further development of the research of leading scientists Piskunov [7, 8], Pavlikov and

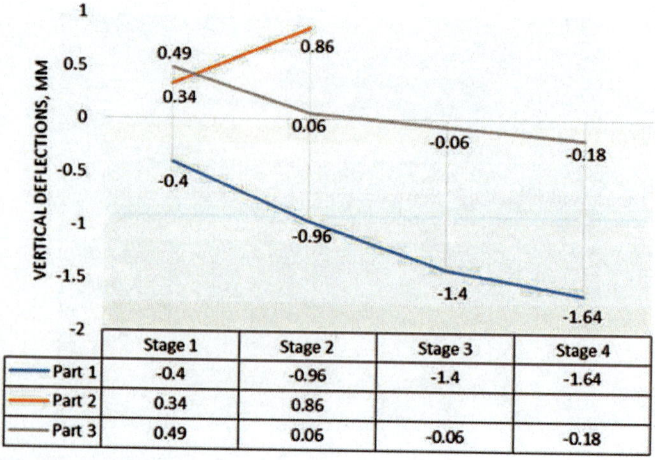

Fig. 8 The value of vertical deflections for indicator № 2

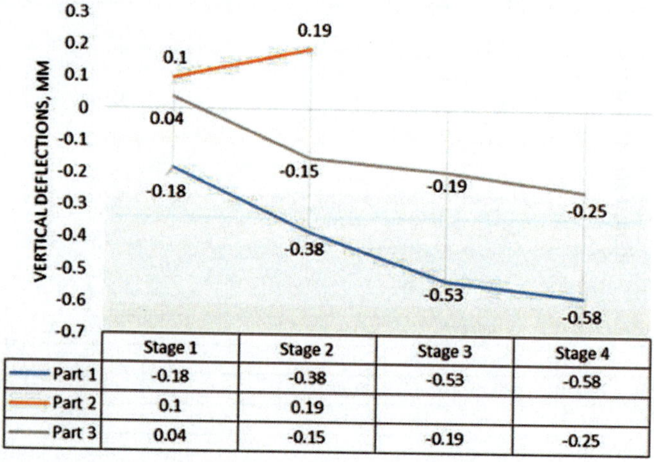

Fig. 9 The value of vertical deflections for indicator № 3

Kochkarev [9, 10], Azizov [11, 12], Semko [13], Pichugin [14], Loburets [15] with their co-authors.

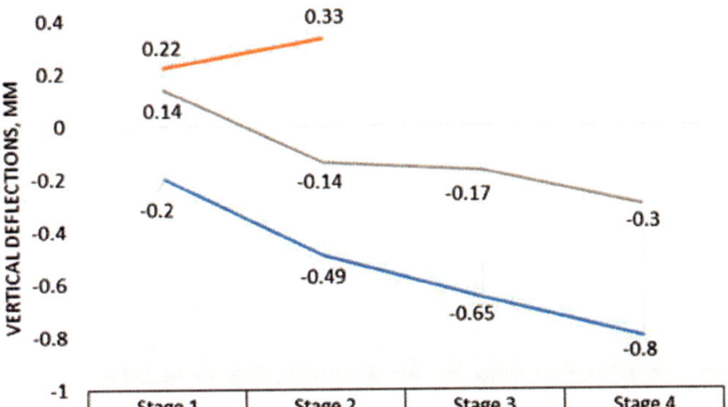

	Stage 1	Stage 2	Stage 3	Stage 4
Part 1	-0.2	-0.49	-0.65	-0.8
Part 2	0.22	0.33		
Part 3	0.14	-0.14	-0.17	-0.3

Fig. 10 The value of vertical deflections for indicator № 4

4 Conclusions

As can be seen from the laboratory test of model of concrete slab, the method of strengthening by external reinforcement by external reinforcement has a positive effect on the performance of reinforced concrete slabs.

This method of reinforcement allows you to reduce vertical deflections in the slab and increase operational loading to the structure.

In further studies of flat reinforced concrete slabs, it is necessary to study and analyze the operation of the support sections of the anchors, as well as their impact on the reinforced concrete slab under prolonged exposure.

References

1. Zhuravskyi OD, Tymoshchuk VA (2017) Calculation of flat reinforced concrete slabs reinforced with external prestressed reinforcement. Building structures. Theory and practice. KNUCA, Kyiv. Issue 1, pp 193–198
2. DSTU B V.3.1-2:2016 (2017) Repair and reinforcement of load-bearing and enclosing building structures and foundations of buildings and structures. K.: DP «UkrNDNTs». 68s.
3. Khan S, Williams M (1995) Post-tensioned concrete floors, 312 p.
4. Hasenko A, Semko O, Drobotia O, Sirobaba V (2020) Experimental and numerical studies of nodes of light steel-reinforced concrete structures. Paper presented at the proceedings of the 2020 session of the 13th fib international PhD symposium in civil engineering, pp 173–178
5. Semko O, Hasenko A, Kyrychenko V, Sirobaba V (2020) The rational parameters of the civil building steel frame with struts. https://doi.org/10.1007/978-3-030-42939-3_25
6. Kochkarev D, Azizov T, Galinska T (2018) Bending deflection reinforced concrete elements determination. Paper presented at the MATEC Web of Conferences, vol 230. https://doi.org/10.1051/matecconf/201823002012

7. Piskunov VG, Gorik AV, Cherednikov VN (2000) Modeling of transverse shears of piecewise homogeneous composite bars using an iterative process with account of tangential loads 2. Resolving equations and results. Mech Compos Mater 36(6):445–452. https://doi.org/10.1023/A:1006798314569
8. Piskunov VG, Goryk AV, Cherednikov VN (2000) Modeling of transverse shears of piecewise homogeneous composite bars using an iterative process with account of tangential loads. 1. Construction of a model. Mech Compos Mater 36(4):287–296. https://doi.org/10.1007/BF02262807
9. Pavlikov A, Kochkarev D, Harkava O (2020) Analysis of eccentrically loaded members of circular cross section by nonlinear deformation model. https://doi.org/10.1007/978-3-030-42939-3_19
10. Kochkarev D, Azizov T, Galinska T (2020) Design of effective statically indeterminate reinforced concrete beams. https://doi.org/10.1007/978-3-030-42939-3_10
11. Azizov T, Jurkowska N, Kochkarev D (2019) Basis of calculation on torsion for reinforced concrete structures with normal cracks. Paper presented at the proceedings of the fib symposium 2019: concrete - innovations in materials, design and structures, pp 1718–1725
12. Kochkarev D, Azizov T, Azizova A, Galinska T (2021) Designing of standard cross sections of composite bending reinforced concrete elements by the method of design resistance of reinforced concrete. https://doi.org/10.1007/978-3-030-57340-9_25
13. Semko O, Fenko O, Hasenko A, Harkava O, Kyrychenko V (2018) Influence of external and internal cooling at solidification on strength of brittle duralumin in compression. Paper presented at the MATEC Web of Conferences, vol 230. https://doi.org/10.1051/matecconf/201823002029
14. Pichugin SF (2018) Reliability estimation of industrial building structures. Mag Civ Eng 83(7):24–37. https://doi.org/10.18720/MCE.83.3
15. Loburets AT, Naumovets AG, Senenko NB, Vedula YS (1997) Surface diffusion and phase transitions in strontium overlayers on W(112). Zeitschrift Fur Physikalische Chemie 202(1–2):75–85. https://doi.org/10.1524/zpch.1997.202.part_1_2.075

Influence of Fly Ash Additives on Strength Characteristics of Soil–Cement as a Material for Waste Storage Construction

Mykola Zotsenko, Olena Mykhailovska, Irada Shirinzade, and Iryna Lartseva

Abstract Innovative ways of using fly ash to improve the strength of soil–cement are analyzed. It is proposed to solve the environmental problem of utilization of this material, when using the addition of fly ash during the construction of soil–cement elements. The strength of soil–cement samples with the addition of fly ash (with the main content of fly ash with a percentage of 5, 10 and 15% by weight of cement) at the age of 28 and 90 days was studied. It is established that the average compressive strength of the samples with the addition of Mikolaev CHP fly ash with a fraction of inclusions up to 4 mm in the amount of 5% increases the compressive strength by 30% (at the age 28 days). The expediency of the application of fly ash for soil–cement elements in the construction of waste storage is analyzed.

Keywords Soil–cement · Fly ash · Waste storage · Compressive strength

1 Introduction

Mixing of soils with cement allows to receive quite strong and environmentally friendly material soil–cement which can be used both for preparation of the basis under the foundations, and for construction of the foundations, and also at the decision of various geotechnical problems.

In recent years, the use of resource-efficient through the development of new methods of arranging soil–cement structures is increasingly used in the construction of foundations: as bored piles, in the arrangement of retaining walls of pits and soil–cement anchors. Building envelopes are designed for the action of horizontal loading from soil pressure and groundwater, as well as on a vertical loading to the surface of the soil near the envelope. Special requirements for the strength of soil–cement

---

M. Zotsenko · O. Mykhailovska (✉) · I. Lartseva
National University «Yuri Kondratyuk Poltava Polytechnic», Pershotravnevyj Ave. 24, Poltava, Ukraine

I. Shirinzade
Azerbaijan University of Architecture and Construction, Ayna Sultanova -11, Baku, Azerbaijan

V. Onyshchenko et al. (eds.), *Proceedings of the 3rd International Conference on Building Innovations*, Lecture Notes in Civil Engineering 181,
https://doi.org/10.1007/978-3-030-85043-2_43

elements and the tightness of their styling are determined for building envelopes [1]. The use of soil–cement structures for the construction of waste storage is promising.

In this case, soil–cement anti-filtration curtains are used, which can be vertical and horizontal. The thickness of anti-filtration curtains is in the range of 0.5 … 2 m and depends on the filtration capacity, strength of soil–cement, pressure gradient [1]. Special requirements are set for anti-filtration curtains to ensure a low filtration coefficient and resistance to groundwater aggressiveness. Soil–cement is used in the installation of drilling mud, artificial reservoirs, pools, reservoirs, fountains [2].

From the point of view of manufacturability soil–cement piles (elements) have advantages over all other driven piles that at their manufacturing there is no need for additional actions for the maintenance of walls of wells (casings, drilling muds, special shells, etc.).

The most common in our country is the drilling method due to the simplicity of mechanisms for its implementation. The essence of this method is to mechanically develop and mix the soil with a binding material, which is fed in the form of a solution. Mixing of soil with binding material, as a rule, occurs in a working well (without excavation of soil), but also application of technology with partial excavation of soil or in special bunkers mixers is possible. The advantage of this method is the economy of the binding material, a constant cross-section of the soil–cement element in accordance with the diameter of the applied working auger. One of the disadvantages of soil–cement elements is their relatively low strength of the material.

On the other hand, the problem of accumulation of waste from enterprises, namely ash, ash and slag waste from thermal power plants (CHP), which occupy large areas, is relevant in Ukraine. For example, Trypillya CHP has accumulated 28 million ash slags, 140 ha. The result of waste storage is large-scale dust storms that cover nearby settlements. All this is complicated by the fact that there is no sanitary protection zone between the ash dump.

In Western Europe and Japan, ash dumps have been virtually eliminated at thermal power plants. In 1990, the European Coal Combustion Products Association was established by European electricity producers to ensure the profitable and high-quality use of ash. At modern thermal power plants, coal is burned in a dusty state. Slag is formed by the adhesion of softened ash particles. The maximum grain size of slag in the ash-slag mixture is not more than 20 mm.

Ash is removed from the furnace with flue gases (fly ash). Most ash fragments have a vitreous surface texture. The size of spherical particles varies from a few microns to 50–60 μm [3].

Ash from thermal power plants have different chemical composition depending on the type of coal burned. Ash from oil shale coal is less acidic than ash from the combustion of brown or hard coal [3]. During the combustion of coal in thermal power plants and slag, 7–9 million tons (50–200 g of ash per 1 kW-gram of produced electricity).

Kovalsky V.P., Sidlak O.S. (2014) note that the use of ash and slag waste from thermal power plants in the production of concrete and reinforced concrete products allows you to: reduce cement consumption by 10–20%; to improve physical and mechanical properties of concrete; reduce the cost of creating and operating dumps;

to release the lands occupied by dumps; to exclude pollution of air and water basins. The use of Portland cement hydration processes with the addition of 40% of CHP fly ash has proved the feasibility of replacing a significant part of the clinker component in concrete [3, 4].

Malyovaniy M.S. (2017) proposes to use CHP waste for the purpose of making additives by mixing a dispersed mineral additive—CHP fly ash [5].

Blaschuk N.V. and Maevska I.V. (2017) propose to add fly ash in order to improve the properties of soil–cement elements [6]. As a result of experiments, they found that the addition of fly ash in the amount of 3.4 to 15% by weight of the soil in the manufacture of soil–cement samples in the laboratory gives a positive effect.

2 Purpose of the Work. Methodology and Research

The purpose of the article is to study the influence of the Mykolayiv CHP fly ash on the strength characteristics of soil–cement.

Soil cement is a building material formed by mixing soil with cement mortar. It has a fairly high shear strength, but is brittle and has a low tensile strength. This leads to the formation of cracks.

Under natural conditions, the soil in most cases is a three-phase system consisting of solid, liquid and gaseous phases. Depending on the value of the exchange capacity and the composition of the exchange cations, different soils, even with the same particle size distribution, have quite different physical and mechanical properties. This circumstance significantly affects the stability of soils in engineering structures [7].

Clay soils are suitable for the production of soil–cement. Of these, loess and loess loams and sands meet the requirements the most.

When soil–cement is produced directly in the soil of the construction site, the quality of mixing of the mixture is reduced by boring-mixing technology in comparison with the production of such elements in the enterprise. In the field, additional compaction of the mixture is almost impossible. In flooded soils, especially below its groundwater level, the water-cement ratio of the mixture increases significantly. The consequence of this will be a decrease in the density of soil–cement and its mechanical characteristics. The disadvantages of soil–cement piles include their low strength of the material.

Recently, a significant amount of waste in Ukraine is incineration waste (ash).

Annually in Ukraine thermal power plants generate about 6.5 million tons of ash and slag waste, a significant part of this waste is fly ash, that is a fine material consisting of particles up to 0.14 mm in size.

The mineralogical composition of ash depends on the type of fuel used at the CHP and the conditions of its combustion. It contains metakaolin, quartz of various modifications, γ-alumina, mullite, various iron compounds, sulfur compounds and other substances, but in very small quantities. The mineralogical composition of the ash determines their activity, chemical composition and some other properties.

The true density of the individual ash fractions in most fuels can range from 2 to 5 g/cm^3. The high density of the individual components of the ash is due to the fact that it is dominated by particles consisting mainly of iron compounds or minerals, with a significant density. Different actual density of ash particles obtained by burning the same fuel is associated with different chemical composition of individual fractions [8].

Storozhuk N.A. (2017) found that during the combustion of the organic part of the fuel, clay impurities that are part of the mineral part of the fuel are dehydrated and amorphized. Temperature intervals of dehydration of clay matter depend on its mineralogical composition. First, the main part of the water is released without destroying the crystal lattice of the mineral. After that, the obtained material differs in physical properties from the original mineral, has a distorted crystal structure and increased reactivity [8].

Most fragments of ash have a spherical shape, smooth glassy surface texture. The size of spherical particles ranges from a few microns to 50–60 μ. The particle size of the furnace dust ranges from 3–5 to 100–150 mkm [6]. The number of large particles usually does not exceed 10–15%.

The Mikolaev CHP fly ash before carrying out experiment was sifted on a sieve of 4 mm (Fig. 1). The content of spherical inclusions from 1 to 4 mm was up to 40%. The average humidity of the removal ash was 0,6%.

The starting materials for the preparation of soil–cement in both laboratory and field conditions are clay-sandy soil, water, Portland cement.

For the production of the studied soil–cement samples it was used loess, yellow–brown, hard loam. The soil had the following characteristics: soil moisture 14%, plasticity limit, W_p—0.25; yield strength W_L—0.37; the density $\rho = 1.72$ g/cm^3.

For the preparation of soil–cement samples it was used hydrocarbonate-calcium water, slightly mineralized, slightly alkaline, pH = 8, which does not contain harmful impurities and microorganisms, which interferes with the normal hardening of cement.

The experiment is carried out as follows: cement and water mixed to obtain "cement milk". The amount of cement was taken as 20% by weight of dry soil for all series of tests. The W/C was taken as 1. Then to the cement mortar was added fly ash, then obtained mixture was blended with soil. The study is carried out for the

Fig. 1 General view of the components of the mixture: 1—loam loess disturbed structure of natural humidity; 2—fly ash of the Mikolaev CHP

amount of fly ash of 5, 10 and 15% of the amount of cement. The prepared soil–cement mixture was mixed to a homogeneous mass for at least 5 min. After mixing, the soil–cement mixture is inserted into cylindrical molds with a diameter of 2,8 cm and a height of 3,5—4 cm.

Prior to testing, the samples were stored immersed in water for 28 and 90 days to gain strength. The tests were carried out in accordance with DSTU B B.2.7-214:2009 as for concrete, taking into account DSTU B B.2.1-4-96. The tests were performed for 6 samples in each series.

The test results of the compression samples at the age of 28 and 90 days are shown in Table 1.

The cracks that preceded the final destruction of the samples were mainly vertical in nature (Fig. 2). The samples studied are subject to brittle fracture. Sample cracks when applying a vertical load usually occurred in local areas of high stress concentration. The decisive role in the destruction process was played by microstresses that occurred in the soil–cement element during the application of step load. The primary cause of microstresses is microhomogeneity and microanisotropy of the structure. Plastic deformation developed mainly in the sliding planes, while most of the body volume continued to work elastically beyond the yield strength. Uneven plastic deformation led to the development of uneven fields of microdeformations and microstresses of turbulent nature. The presence of such strong energy concentrators contributes to the formation of microcracks, the number and size of which increase as plastic deformation continues. This process of accumulation of damage—the process of plastic loosening—is the first stage of destruction. Further growth of

Table 1 Characteristics of compressive strength of soil–cement samples with the addition of furnace dust (at the age of 28 and 90 days) with coefficient of variation υ

Average compressive strength, MPa	Without additive	With the addition of fly ash		
		5%	10%	15%
R_{28}/υ	2,78/0,30	3,64/0,11	3,17/0,11	3,13/0,09
R_{90}/υ	3,06/0,28	3,91/0,13	3,40/0,08	3,33/0,08

Fig. 2 General view of the soil–cement sample, which is soaked in water for 28 days with the addition of 5% fly ash after application of the load

the crack during compression of the element is explained by the fact that it creates a concentration of stresses at its edge. The development of cracks leads to the destruction of the sample. Fragile fracture is characterized by a sharp, often branched crack. The rate of propagation of a brittle crack is significant and sudden.

As a result of the research it was found that the average compressive strength of the samples without any additive is 2,78 MPa at the age of 28 days and 3,06 MPa at the age of 90 days; with the addition of furnace dust in the amount of 5%—3,64 and 3,91 MPa, respectively. Blaschuk N.V. and Maevska I.V. (2017) also noted an improvement in the compressive strength of soil–cement with the addition of a small amount of Ladyzhinskaya CHP fly ash (3,4—15% of soil weight) [6].

As we can see from Table 1, the test results of soil–cement samples for compression at the amount of furnace dust additive 5, 10 and 15% are almost the same. Therefore, it is necessary to increase the range of additives and to study soil–cement samples.

According to the results of research it is established that with increasing of the hardening period of samples in water up to 90 days increases the average compressive strength of soil–cement samples without additives and with the addition of the appropriate percentage of removal ash by 6–9%. For example, the average compressive strength of samples at the age of 90 days is 7.5% higher than samples at the age of 28 days.

Research is of great importance in the construction of structures made of soil–cement elements, waste storage, protective screens. One of the main indicators that must have the material used in the construction of waste storage—water resistance, sufficient strength and durability. Soil–cement is a waterproof material, its porosity is mostly closed, and the filtration coefficient compared to natural soil is reduced by 10,000–100,000 times [9]. According to the results of research, an increase in the average compressive strength of soil–cement elements over time in an aqueous medium, which increases durability. That is, soil–cement with additive of fly ash is recommended for the construction of protective screens and waste storage in flood conditions.

The following construction of waste storage is offered. On perimeters of the planned storage the monolithic vertical watertight diaphragms like "wall in the ground" from soil–cement elements is constructed (Fig. 3). It is offered to arrange soil–cement elements with addition of 5% by weight of ash of removal of the Mikolaev thermal power plant. This improves the compressive strength and durability of the structure and it will allow to dispose of waste from fuel combustion CHP. It is recommended to build soil–cement elements according to boring-mixing technology. The distance between the elements in the axes should be equal to 0.8d (d is the diameter of the soil–cement elements) [10–15]. This method allows you to loosen the soil without removing it. At the same time, a cement slurry is injected into the loose soil with the addition of fly ash in the amount of 5 wt.%. After that, mixing and compaction of the soil–cement mixture is performed. An important factor in the design of the storage is the choice of the location of the storage, provided that there is a waterproof layer at the optimal depth from the surface (8–20 m). Water-permeable layers can be located above it.

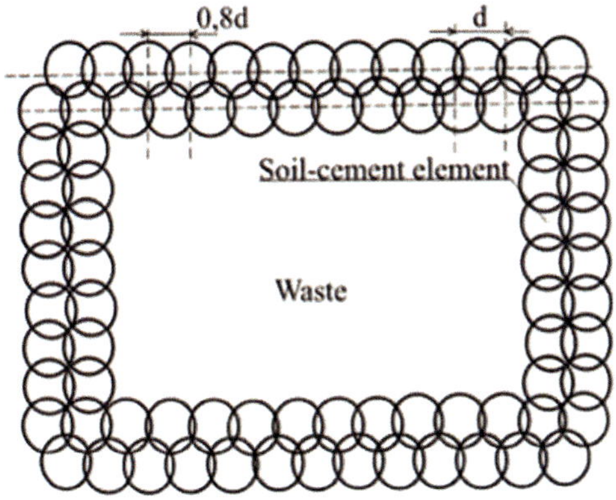

Fig. 3 Waste storage scheme

3 Conclusions

As a result of the researches it is established that the average compressive strength of the samples that were kept in water increases with time. In particular, the compressive strength of samples at the age of 90 days is 7.5% higher than samples at the age of 28 days.

The average compressive strength of the soil–cement samples with the addition of Nikolaev CHP fly ash with a fraction of inclusions up to 4 mm in the amount of 5% increases by 30%.

The use of Mikolaev CHP fly ash as a mineral additive in the soil–cement elements production in the amount of 5% by weight of cement increases the compressive strength. This expands the range of applications of soil–cement elements and increases their efficiency. Soil–cement with additive of removal ash is recommended for the construction of protective screens and waste storage in flood conditions, because the average compressive strength increases over time.

However, in the production of soil–cement, the rheological properties of fly ash are important, especially the high fineness of grinding, favorable fractionation and the shape of its particles.

References

1. Kirichek YuO, Komissarov GV (2016) Analysis of the field of application of soil-cement in geotechnics. Build Constr 83(1):529–538. http://nbuv.gov.ua/UJRN/buko_2016_83%281%2958
2. Zotsenko ML (2013) Soil-cement piles made by boring-mixing method. Acad J Ind Mach Build Civ Eng 3(38):110–122
3. Kovalsky VP, Sidlak OS (2014) The use of ash removal of thermal power plants in building materials. Mod Technol Mater Struct Constr 1:35–40

4. Ivakhnenko AS, Khlopitsky OO (2018) Formation, classification and prospects of ash and slag waste utilization in Ukraine. In: Theory and practice of modern science, Odessa, 23–24 November 2018, pp 111–112. http://molodyvcheny.in.ua/files/conf/other/31nov2018/96.pdf
5. Maliovanyy MS, Bondar MO, Kanda MI (2018) A method of manufacturing a modifier for landslide structures. №126483 IPC C04B 28/14 (2006.01) Ukraine. № u201713144; Application 29/12/2017; Publ. 25.06.2018, 12, 4 p.
6. Blaschuk NV, Maevskaya IV The use of removal ash in the composition of soil cement. https://conferences.vntu.edu.ua/index.php/all-fbtegp/all-fbtegp-2020/paper/download/9154/7828
7. Zotsenko ML, Nesterenko TM (2012) Comparative analysis of the effect of vibration on the mechanical characteristics of soil cement from fine homogeneous sand and clay soil. Acad J Ind Mach Build Civ Eng 4(34):123–130
8. Storozhuk NA, Pavlenko TM, Abbasova AR (2017) Features of thermal power plant ash as aggregate for concrete. Science and progress of transport. Bull Dnipropetrovsk Natl Univ Railway Transp 5(71):149–157. https://doi.org/10.15802/stp2017/113200
9. Zotsenko ML, Vynnykov YL, Zotsenko VM (2016) Drilling soil-cement piles, which are made by drilling method: Monograph. Madrid Printing House, Kharkiv
10. Zotsenko NL, Vynnykov YL (2016) Long-term settlement of buildings erected on driven cast-in-situ piles in loess soil. Soil Mech Found Eng 3:189–195
11. Zotsenko ML, Mykhailovska OV, Sivitska SP (2020) Construction features durable storage of toxic waste in boreholes. Lecture notes in civil engineering, vol 73, pp 325–334 (2020). https://doi.org/10.1007/978-3-030-42939-3_33
12. Kochkarev D, Azizov T, Galinska T (2018) Bending deflection reinforced concrete elements determination. Paper presented at the MATEC web of conferences, vol 230. https://doi.org/10.1051/matecconf/201823002012
13. Zotsenko M, Vynnykov Y, Lartseva I, Sivitska S (2018) Ground base deformation by circular plate peculiarities. Paper presented at the MATEC web of conferences, vol 230. https://doi.org/10.1051/matecconf/201823002040
14. Vynnykov Y, Voskobiinyk O, Kharchenko M, Marchenko V (2017) Probabilistic analysis of deformed mode of engineering constructions' soil-cement grounds. Paper presented at the MATEC web of conferences, vol 116. https://doi.org/10.1051/matecconf/201711602038
15. Vynnykov Y, Manhura A, Zimin O, Matviienko A (2019) Use of thermal and magnetic devices for prevention of asphaltene, resin, and wax deposits on oil equipment surfaces. Min Miner Depos 13(2):34–40. https://doi.org/10.33271/mining13.02.034

Planning of Cities and Building Economics

Logistic Component of Regional Waste Management Systems Efficiency Improvement

Viktor Bredun, Natalia Smoliar, and Ahmad Sarkarli

Abstract The article is dedicated to development of current comparative analysis methods of different types of garbage trucks using effectiveness in logistics schemes of solid waste collection during their planning, adapted to regional conditions. The priority of regional approach to planning of solid waste management systems within the framework of wide implementation of regional waste management programs in Ukraine, which provide organization of solid waste collection in both urban and rural settlements, is substantiated. Factors influencing the choice of a garbage truck are considered, the role of the limiting logistical indicator is noted. A number of methods for determining the effectiveness of planning regional logistics schemes for solid waste collection are analyzed. The requirements to the criteria of specialized equipment use efficiency are given. The system of indicators is proposed for current analysis of logistics scheme technological efficiency of solid waste collection during planning, adapted to regional conditions and based on the use of indicators calculated directly in the planning process, characterize the scheme technological rationality, are easy to be calculated and can be applied under conditions of existing regional design specifics.

Keywords Solid household waste · Waste management · Regional logistics systems · Choice of garbage truck · System of indicators · Efficiency · Ecological safety

1 Introduction

In the current conditions of growing anthropogenic load on the environment, the problem of the mankind ecological safety is quite acute, thus one of the areas of its

V. Bredun (✉) · N. Smoliar
National University «Poltava Yuri Kondratyuk Politechnic», Pershotravnevyj Ave. 24., Poltava 36011, Ukraine

A. Sarkarli
Azerbaijan State Oil and Industry University, 34 Azadliq Avenue, Baku, Azerbaijan

V. Onyshchenko et al. (eds.), *Proceedings of the 3rd International Conference on Building Innovations*, Lecture Notes in Civil Engineering 181,
https://doi.org/10.1007/978-3-030-85043-2_44

solving is developing an appropriate waste management systems. Many countries in the world have gained experience in this field, yet in Ukraine the most common way of waste management is their storage in open areas, which causes environmental pollution and biota destruction. According to the unanimous opinion of scientists dealing with waste management, the only and most acceptable way of waste management is their processing at the factories and use as secondary raw material in the future [1].

Issues related to collection, disposal and burial of production and consumption waste are relevant for almost all regions of Ukraine. Collection, transportation and burial are interrelated stages of the process of settlements sanitation together with mechanisms of rational use of nature and environmental protection constitute an urban system ecological safety system. From economic and environmental point of view organization of valuable raw materials extraction from solid waste (SHW) is of no less importance. This is evidenced by experience of advanced countries around the world, where separate collection of SHW has long become the norm.

The efficiency of SHW collection systems, in particular separate one, is largely ensured by the use of modern technological equipment within implementation of environmentally, technologically and economically sound logistics approaches. Regional waste management programs have recently been widely implemented in Ukraine. However, today there is almost no separate collection of SHW in rural areas of Poltava region, and most of the garbage trucks fleet has long been morally and physically obsolete. Therefore, efficiency improving of regional waste management systems through rational planning of the logistics component using modern technological means is an urgent task.

The purpose and objectives of the study is to increase the efficiency of regional waste management systems through logistics component rational planning using different types of garbage trucks during their planning.

Object of Research: logistical and technological aspects of rational SHW collection planning and transportation systems.

Subject of Research: the role of the garbage truck fleet species composition in ensuring logistical and technological efficiency of the processes of SHW collection and transportation, ways of multi-section garbage trucks possible use for the needs of amalgamated territorial communities of Poltava oblast.

Methods of Research. When reviewing existing analysis methods of the use of garbage trucks technological and logistical efficiency, the methods of system analysis were used. Determining the factors influencing technological efficiency of the use of garbage trucks in specific territorial conditions was based on the use of factor analysis methods. Development of a system of indicators for t current analysis of logistics scheme technological efficiency of SHW collection during planning, adapted to regional conditions, methods of structural-logical synthesis are used.

Formulation of the Problem in General. The structure of specialized vehicles fleet according to capacity and method of loading must best meet the distribution of requirements for the disposal of solid household waste (SHW), ensuring quality and timely work performance. Also, in terms of necessity of re-equipping an enterprise with other brands of garbage trucks, it is required to determine the most effective area of existing fleet of new specialized vehicles application.

Discussion of Research Results. The work of garbage trucks can be considered highly efficient provided it completely and in a timely manner in respective conditions ensures implementation of tasks with the minimum possible resources spent. Thus, efficiency criteria must contain parameters and characteristics that take into account the purpose, results of the use of and the system operating costs. The main purpose of garbage trucks efficiency indicators is that they should be critical as for their parameters, be quite simple to quantify, be universal and enable performing comparative analysis.

Most often, as criteria for solving the problem of SHW transportation efficiency improving, are used indicators that characterize special equipment transport efficiency. Significant fluctuations in the volume of SHW generation are important in the process of garbage trucks operation. The choice of garbage trucks is influenced by technical and operational properties of trucks. For the optimal choice of garbage trucks it is expedient to substantiate a number of indicators and perform a comprehensive comparative analysis of the project options for SHW collection system.

Thus, for example, the term of solid waste disposal is defined by local sanitary-epidemiological service authorities of Ukraine and can range from 1 to 3 days. Working time of the garbage truck driver should not exceed 12 h, the distance of places or objects of SHW disposal from the places of their generation should not exceed 50 km. Despite these limitations, it may be affirmed regarding possibility and feasibility of using garbage trucks of such a capacity at which transport work will be performed with the lowest operating costs. In cases of restrictions on the condition of disposal, time of work or anything else, it is logical to use garbage trucks with the highest hourly productivity. If for quality and timely waste disposal there is a need to use garbage trucks with the highest hourly productivity along with garbage trucks with the lowest cost of transportation.

In the research dedicated to an issue of improving the garbage trucks work organization, the number of indicators characterizing various aspects of its functioning are noted [2, 3]. Most often, as criteria for solving the problems of transportation, indicators that characterize sectorial or economic efficiency of the transport system are used. All these indicators are quantified, and according to their nature can be attributed to either natural or value ones. Natural indicators are recognized as derivatives for the calculation of cost, i.e. cost indicators are more general estimates of the transport work organization.

Important factor in the process of garbage trucks work is a significant fluctuation in the volume of traffic on one route during a week and months of the year. At the same time, the daily volume of traffic is detected only during the movement of the garbage truck on the route, which virtually eliminates the possibility of operational routing of traffic. A small amount of garbage arrivals leads to inefficient use of the vehicle, while excessive arrival leads to excessive hours of the garbage truck operation on the route. This requires solving two interrelated problems of transportation routing and justification of the choice of garbage truck for transportation. The choice of a garbage truck is influenced by technical and operational properties of the car, town-planning (dimensions of passages, location of sites with garbage etc.) and technological (coordination of technical means, multiplicity of the sizes, arrangement of containers etc.) factors. At this regard at the initial stage of designing, classification of the city areas on such indicators as population and features of housing is provided.

Characteristics of garbage truck operation on the route are determined by the vehicle operation modes. Traffic on the route involves performing of the following operations at each race: acceleration, movement at a constant speed, braking, maneuvering, performing cargo operations.

Due to the closure of landfills near cities, there is a problem of using two-stage technology for their removal, which involves use of transshipment and sorting stations and transport garbage trucks. In the scientific literature, the transport of such technology is practically not covered. Therefore, the problem of determining the scope of this technology needs to be solved. Its solution is possible through comparing the cost of SHW collection with garbage trucks and the cost of removal, taking into account the costs associated with reloading from garbage truck to transport one. In addition, for this technology it is necessary to calculate the optimal number and location of transshipment stations in the city as well as requirements for transport garbage trucks.

One of the components of garbage trucks efficiency is quality of the territory cleaning. There is a system for assessing the level of garbage trucks quality, which included application of cumulative scoring for each crew of garbage trucks and cleaners.

The choice of vehicles is made on the basis of criteria substantiation: cost of the garbage truck, fuel consumption, bulk capacity, type of loading, existence of the pressing installation and type of fuel.

Based on the practical experience of designing local and regional schemes of the territories sanitary cleaning [4–6], we have established a number of general and regionally determined principles of garbage trucks choice, providing construction of technologically and ecologically efficient logistic component of regional systems SHW management:

1. Adequate choice of specialized transport plays a primary role in creation of rational systems of SHW collection within individual settlements, amalgamated territorial communities or districts of the region.
2. The choice of a garbage truck is influenced by technical and operational properties of the car, town-planning (dimensions of passages, location of platforms with garbage etc.) and technological (coordination of technical means, multiplicity of the sizes, arrangement of containers, etc.) factors.
3. Under availability of different types and brands of cars for performing certain types of work, it is necessary to comprehensively consider the use of the most efficient vehicles according to certain criteria, including their type, which will

most completely meet the operating conditions. One of the factors influencing the change in the efficiency of garbage trucks is the speed of movement, which is significantly affected by road conditions.

4. Different possibilities of the main types of waste processing, generated in the field of municipal responsibility should be organizationally and technologically connected so that as a result the integrated concept of of all spectrum of waste processing taking into account local features and requirements within the corresponding territorial structure was realized.
5. An important role in choosing the type of garbage trucks has a limiting logistical and technological factor, which is largely determined by the adopted system of SHW collection.
6. In areas of private construction with installed private containers, the limiting factor in route planning is a number of SHW loading operations from containers into the hopper of the truck. The volume of waste on the route is equal to 4–11 m^3. This necessitates the use of low-volume garbage trucks. In apartment buildings areas, the limiting factor in route planning is the amount of SHW removed on the route. At the same time, it is rational to use garbage trucks with large bunkers. Such circumstances require a high-quality garbage trucks fleet. In terms of route sections combination it is possible to achieve a certain level of its unification.
7. One of the most important technological factors is also the efficiency of the garbage truck compaction system, bulk capacity and fuel consumption.
8. Often, the most effective is use of garbage trucks with universal gripping devices to work with containers of all types.
9. When designing logistics schemes for SHW collection for amalgamated territorial communities of Poltava oblast, there is a possibility, as well as technological, logistical and economic feasibility of using multi-section garbage trucks to organize separate SHW collection.

One of the most important conditions for ensuring the logistics processes efficiency in SHW management systems is analysis of technical and logistical system parameters at the design stage. There are a number of methods for determining the effectiveness of planning logistics schemes for SHW collection, each of which has its own scope, advantages and disadvantages.

Thus, methods described in [7–9] are designed to analyze the efficiency of garbage trucks in a large city. The main criterion is the maximum distance of transportation when comparing two vehicles.

In modern conditions of Ukraine, large cities mostly have sanitation schemes. The most urgent task nowadays is development of these schemes for small towns, urban-type settlements and villages. The current trend in the development of the territories administrative structure is unification of small settlements into amalgamated territorial communities. Often in such territorial structures the linear dimensions of settlements are smaller than distances between these settlements. Thus, they cannot be considered as a single urbanized area. Logistic system of SHW collection in this case is a complex of separate low-productivity (with insignificant daily volumes of waste generation) collection sites, united into a single network. In such cases, the structure of route planning within the settlements has little effect on overall length of the route. Correspondingly, the maximum distance of transportation, considering the degree of the garbage truck loading and the duration of the work shift may not always be a criterion for optimizing the choice of vehicle.

The method of mathematical modeling of the logistics scheme individual elements [10] is based on the analysis of a significant amount of statistical and design information. This is both advantage and disadvantage of this method. On the one hand, this allows taking into account all the technological and economic aspects of design, on the other—requires significant computing resources and time as well as availability of an extensive database. In terms of logistics schemes practical design, considering regional conditions, this can significantly slow down the design process and affect its rationality.

The energy model of optimizing the use of garbage trucks [11] is based on the principle of minimizing energy consumption for route maintenance. This technique enables to quite meaningfully analyze technological aspects of design. However, this principle cannot always be fully applied due to the regional specifics of architectural and planning constraints (availability and quality of the road network, the possibility of transport, such as in sanitation projects [5], features of economic administration of sanitation scheme [6].

Economic approach to solving the problem of choosing technical means [12–18] is quite compact, which allows quickly use it as a test at the planning stage of logistics scheme every route. Still it is of generalizing nature and does not detail the influence of each technological parameter of the scheme on its final efficiency, which is necessary in the design process.

Therefore, we have proposed a system of indicators for the current analysis of logistics scheme technological efficiency of SHW collection while planning, adapted to regional conditions.

Ideology of this system of indicators development is use of those parameters that are calculated directly in the planning process, characterize the scheme technological rationality, are easy to calculate and can be applieed in the existing regional specifics of design.

Under such approach, the criteria for specialized equipment use effectiveness should include parameters and characteristics that take into account purpose, results of use and costs on system operation. Natural indicators are more specific, cost ones are more general.

The efficiency of logistics scheme organization is determined by two main components: efficiency of the use of technological equipment and efficiency of movement along the route. The main technological equipment in the logistics scheme are garbage containers and garbage trucks. Efficiency of container fleet use can be expressed in terms the average containers filling coefficient:

$$C_{cf} = V_{SHW}/V_{cont} \quad (1)$$

where V_{SHW} is volume of SHW to be taken out on the route at one run;

V_{cont} is a total volume of containers set along the route.

The closer the coefficient value to 1, the higher the efficiency.

Freight efficiency of the truck fleet can be expressed in terms of average hopper filling coefficient

$$C_{fil}=C_{pr}/C_{pm}, \quad (2)$$

where C_{pr} is a real coefficient of SHW pressing in the truck hopper (obtained through calculation);

C_{pm} is a maximum coefficient of SHW pressing in the truck hopper (according to the truck passport data).

The closer the coefficient value to 1, the higher the efficiency.

Rationality of route planning according to the number of loading operations characterizes the efficiency of loading operations coefficient

$$C_{lo} = V_{SHW}/Q_{lo} \quad (3)$$

where Q_{lo}—quantity of loading along the route.

The higher the indicator value, the higher the efficiency.

Efficiency of the garbage truck planned stops using along the route to perform loading operations is characterized by the stops using efficiency coefficient

$$C_{su} = 1 - \left(Q_{stop}/Q_{lo}\right), \quad (4)$$

Where Q_{stop} is a number of stops along the route.

The closer the coefficient value to 1, the higher the efficiency.

Efficiency of the road network and the optimum speed mode of the garbage truck movement is characterized by coefficient of mainstream fuel consumption

$$C_{mfc} = B_{fuepas}/B_{fuereal}, \quad (5)$$

where B_{fuepas}—linear fuel consumption are passport data for the vehicle serving the route;

$B_{fuereal}$—linear fuel consumption are real for the vehicle serving the route.

The closer the coefficient value to 1, the higher the efficiency.

The relative technical and economic indicators of logistics scheme planning efficiency include:

- specific distance coefficient

$$C_{epd} = V_{SHW}/S_r, \quad (6)$$

where S_r is an extension of the route;

- specific fuel consumption coefficient

$$C_{fpe} = V_{SHW}/B_{totfuelreal} \quad (7)$$

where $B_{totfuelreal}$—total fuel consumption along the route;

- specific time factor coefficient

$$C_{rtpe} = V_{SHW}/t_{rt} \quad (8)$$

where t_{rt}—route transition time.

The higher this group value, the higher the efficiency.

This system of indicators for the current analysis of the SHW collection logistics scheme technological efficiency while planning is adapted to regional conditions.

2 Conclusions

The regional waste management system must have the following main properties: economic and technological efficiency and high level of environmental safety of technological processes. One of the ways to ensure these characteristics is optimal organization of the logistics process, based on a scientifically sound choice of specialized transport and routing of SHW flows, taking into account regional conditions. Under current conditions of urbanization, demographic, socio-economic and administrative-organizational structure of the regions of Ukraine, the process of planning logistics schemes for SHW collection as structural components of regional waste management systems, currently has no perfect methodological mechanisms. An important aspect of ensuring technological efficiency and environmental safety of SHW collection systems is forecasting system parameters at the planning stage, which is based on the principles of inseparable unity of all elements of technological efficiency and environmental safety mechanism, as well as the primary logistics component.

References

1. Popovych NP, Malòvanyj MS, Popovych VV (2017) Pidvyshchennia regionalnoyi ekologichnoyi bezpeky shlyakhom udoskonalennia logistychnoyi systemy povodzhennia z vidkhodamy. Ekologichni nauky T.2. 1(20):11–14. [Попович Н.П., Мальований М.С., Попович В.В. Підвищення регіональної екологічної безпеки шляхом удосконалення логістичної системи поводження з відходами. Екологічні науки Т.2. 1(20), 11–14 (2017)]
2. Protsyk OP, Kotsiuk OYa (2006) Analiz efektyvnosti roboty smittievoziv. Visnyk KDPU 6(41), 52–54. [Процик О.П., Коцюк О.Я. Аналіз ефективності роботи сміттєвозів. Вісник КДПУ 6(41), С. 52–54 (2006)]
3. Shmarin AA, Shmarin AP (2013) Metodika ekonomycheskogo obosnovaniya perekhoda na novyy tip transporta dlya vivoza munitsipalnykh otxodov. Problemy ekonomiki i menedzhmenta 6(22):69–72. [Шмарин А.А., Шмарин А.П. Методика экономического

обоснования перехода на новый тип транспорта для вывоза муниципальных отходов. Проблемы экономики и менеджмента 6 (22), 69–72 (2013)]
4. Skhema sanitarnoyi ochystky smt Semenivka Semenivskogo raionu Poltavskoyi oblasti. [Схема санітарної очистки смт Семенівка Семенівського району Полтавської області]
5. Skhema sanitarnoyi ochystky dlya m. Lokhvytsa ta sil Krynytsa, Vasyl'ky, Xrystanivka, Gayivshchyna Lokhvytskogo raionu Poltavskoyi oblasti (dogovir № 0150/19 vid 02.05.2019). [Схема санітарної очистки для м. Лохвиця та сіл Криниця, Васильки, Христанівка, Гаївщина Лохвицького району Полтавської області» (договір № 0150/19 від 02.05.2019)]
6. Skhem' sanitarnogo ochyshchennia dlya Gadyatskoyi miskoyi terytorial'noyi gromady Gadyatskogo raionu Poltavskoyi oblasti (dogovir zaklyucheno 15.04.2020). [Схеми санітарного очищення для Гадяцької міської територіальної громади Гадяцького району Полтавської області» (договір заключено 15.04.2020)]
7. Illyash OE, Bredun VI, Rad'ko VI, Bylym LR, Gubar' OV (2019) Osoblyvosti logistychnogo planuvannia system povodzhennia z TPV na prykladi mista Lokhvytsa. Ekologichna bezpeka 2(28):12–17. [Ілляш О.Е., Бредун В. І., Радько В.І., Билим Л.Р., Губарь О.В. Особливості логістичного планування систем поводження з ТПВ на прикладі міста Лохвиця. Екологічна безпека 2(28). 12– 17 (2019)]
8. Popovych VV, Prydatko OV, Sychevs'kyi MI, Popovych NP, Panasiuk MA (2017) Efektyvnist' ekspluataciyi smittievoziv u seredovyshchi «misto – smittiezvalyshche». Naukovzi visnyk NLTU Ukrainy 27(10):73–76. [Попович В.В., Придатко О.В., Сичевський М.І., Попович Н.П., Панасюк М.А. Ефективність експлуатації сміттєвозів у середовищі «місто – сміттєзвалище». Науковий вісник НЛТУ України, Т. 27 (10), 73–76 (2017)]
9. Gerelytsia RO, Gerelytsia NC (2010) Upravlinnia transportnoiu logistykoiu v sil's'kogospodars'ky'xkh pidpryiemstvakh. Innovatsiina ekonomika. 292–298. [Герелиця Р.О., Герелиця Н.С. Управління транспортною логістикою в сільськогосподарських підприємствах. Інноваційна економіка. 292–298 (2010)]
10. Tevyashev A, Matviienko O, Shiyan O (2014) Geoinformatical analytic control system of the collection of municipal solid waste. Econtechmod. Int Q J 3(3):77–89
11. Bányai T, Tamás P, Illés B, Stankevičiute Ž, Bányai Á (2019) Optimization of municipal waste collection routing: impact of industry 4.0 technologies on environmental awareness and sustainability. Int J Environ Res Public Health 16:634. https://doi.org/10.3390/ijerph16040634
12. Verbztszkyi GM (2006) Osnovy optimalnogo isspol'zovaniya mashyn v stroitel'stve i gornom dele. Khabarovsk. [Вербицкий Г.М. Основы оптимального использования машин в строительстве и горном деле. Хабаровск. 105 p. (2006)]
13. Stepovaja E, Holik Y, Fraňa K (2018) Methods for precautionary management for environmental safety at energy facilities. Naukovyi Visnyk Natsionalnoho Hirnychoho Universytetu (6):173–177. https://doi.org/10.29202/nvngu/2018/21
14. Golik Y, Illiash O, Chuhlib Y, Maksiuta N (2020) Environmental areas of Poltava planning development. https://doi.org/10.1007/978-3-030-42939-3_38
15. Stepova O, Parashchiienko I, Lartseva I (2018) Calculation of steel pipeline corrosion depth at the galvanic corrosive element operation. Int J Eng Technol (UAE) 7(3):431–435. https://doi.org/10.14419/ijet.v7i3.2.14566
16. Loburets AT, Naumovets AG, Senenko NB, Vedula YS (1997) Surface diffusion and phase transitions in strontium overlayers on W(112). Zeitschrift Fur Physikalische Chemie 202(1–2):75–85. https://doi.org/10.1524/zpch.1997.202.part_1_2.075
17. Pavlenko A, Koshlak H (2015) Design of processes of thermal bloating of silicates. Metall Min Ind 7(1):118–122
18. Onyshchenko VO, Filonych OM, Bunyakina NV, Senenko NB (2020) A new agent for removing concrete residues from the surfaces of polypropylene molds in the manufacture of paving slabs and its advantages. https://doi.org/10.1007/978-3-030-42939-3_67

Lexical and Semantic Features of Foreign Building Terms

Liudmyla Cherednyk, Olena Balatska, Sadagat Valiyeva, and Oksana Vorobiova

Abstract The article investigates current problems of lexical and semantic features of some terms in the field of building and architecture. As a result of the analysis carried out, it has been established that in Ukrainian terminology of building and architecture there are many words of foreign origin, which were borrowed into Ukrainian due to intercultural communication and commodity exchange. External sources of enriching building terminology include borrowings from Latin, German, French, and Italian. It should be noted that the balance between their Ukrainian names, conceptual content and etymological reference is observed in terms. The standardization of term nomination is determined on the basis of its compliance with productive word-forming, morphological models of the modern Ukrainian language.

Keywords Term · Terminological system · Terminology · Etymology · Semantics

1 Introduction

At the present stage of development it is impossible to do without terminology in any field of human activity. Term nomination is a purposeful creative process.

Modern Ukrainian terminology is a relatively stable, traditionally established lexico-semantic system which can be said to be in a state of continuous development and gradual improvement. The development of terminology is caused by factors of socio-political, professional and linguistic nature.

It should be noted that the process of branch term system formation is quite long and complex. Leading domestic and foreign scholars devote their research studies to the problems of clarifying not only the emergence of the term and its development, but also to its place in the language system. Due to rapid development of various fields of science and technology, there is a need to organize and unify terms for ensuring more effective communication between professionals from different countries.

L. Cherednyk (✉) · O. Balatska · S. Valiyeva · O. Vorobiova
National University "Yuri Kondratyuk Poltava Polytechnic", Poltava, Ukraine

Azerbaijan University of Architecture and Construction, 5 Ayna Sultanova, Baku, Azerbaijan

V. Onyshchenko et al. (eds.), *Proceedings of the 3rd International Conference on Building Innovations*, Lecture Notes in Civil Engineering 181,
https://doi.org/10.1007/978-3-030-85043-2_45

In addition, the issue of typology of branch term systems, their classification and characteristics is also important. Research papers by many famous scientists are devoted to the study of these issues, among which are the works of Yu. Safonova, I. Myronets, O. Kvasnitska, Ye. Komar, V. Oliukha, A. Matviichuk, V. Yevteiev, and others.

Empirical material source basis of the research is formed by terminological corpus taken basically from specialized dictionaries of building and architectural terms, definition and etymological dictionaries [3, 5, 7, 13–16] as well as from the works by scientists whose research interests focus on various current construction and architecture issues [8–10, 12, 19–24].

The article was written with the purpose of analyzing lexical and semantic features of foreign terms in building and architecture, as well as their etymology and structure.

2 Results

Structural and semantic studies of lexical taxonomy, namely terminological systems, including the one of building and architecture, has a long and though-provoking history [3–5, 14, 16–18].

It is certain that scientists date back the origins of building and the onset of architecture to the days of the primitive communal system (about 10 thousand years BC), when the first man-made dwellings and settlements were made. This type of human activity is closely connected with the history of human society. Ancient master builders left behind magnificent buildings that still amaze mankind. Among them are the Egyptian pyramids, the monumental and cult buildings of the Aztecs, the temples of ancient Greece and Rome, and more. In his novel "The hunchback of Notre-Dame", the famous French writer V. Hugo wrote that architecture is a stone chronicle of mankind [6, p. 165]. One cannot disagree with it.

In general, construction professions (stonemasons, stone setters, masons, carpenters, stove-makers) were among the first in the world. Accordingly, building vocabulary is considered to be one of the oldest. Scientists believe that there are many words of foreign origin among Ukrainian building terms. The lexical and semantic aspect of the study of this problem indicates that the ways of entry of terminological units into the Ukrainian language were different, the most often it was carried out due to industrial and trade relations.

Needless to say, many words changed their meaning in the process of development. For example, in our opinion, the word *архітектура* "architecture" is interesting. It is well-known to be of Greek origin (*αρχιτεκτονικη*). It meant originally "building" and "construction" [16, p. 58]. Morpheme analysis of the word helps to understand how these words differ. Thus, the term *архітектура* consists of two morphemes: *архі-* "main", "higher degree" and *тектура* "construction" [16, p. 59]. It is now clear why architecture is considered to be building art, which began with the construction of ancient repositories, religious buildings, and then developed into construction of amazing cultural monuments. It also started to denote artistic decoration of the building, its character or style. As we can see, these two words acquired a separate meaning in the process of their existence. However, in the modern Ukrainian language the synonyms of the term *архітектура* "architecture" are the words *будівництво* "construction", *будування* "building" and *зодчество* "architectonics" (marked "learned word").

In construction of different periods, the word *галерея* "gallery" is widely used with the meaning of communication space, which has the form of a covered passage, arch, colonnade and so on. It came into the Ukrainian language from French (French *galerie*), where, in turn, it came from Italian (Italian *galleria*) [15, p. 344].

The curved stucture that spans the space between two supports (such as pylons, pillars, columns) or the aperture in the wall and transmits not only the load but also the strut, is called *арка* "arch" (from Latin *arcus* "arc") [13, p. 10]. The term is also used to denote any curved arch-shaped structure. An arch can be used as a structural architectural element, e.g., above the door opening or gate, or as a purely decorative element.

In addition, in the Ukrainian terminology of building and architecture, there are terms with the headword *арка*, which consist of three or four components. These are, e.g., attributive components that denote various features or properties. They can specify the meaning in terms of type of shape (*арка подвійна гостроконечна* "double pointed arch"); shape of the axis (*арка симетрична кругова* "symmetrical circular arch"); material and type of construction (*арка дерев'яна* "wooden arch" + *арка клеєна* "glued arch" → *арка дерев'яна клеєна* "glued wooden arch"); material and static scheme (*арка дерев'яна* "wooden arch" + *арка тришарнірна* "three-hinged arch" → *арка древ'яна тришарнірна* "three-hinged wooden arch"), etc. According to Ukrainian linguist O. Petukhova, the formation of such compound terms is mainly because of the fact that "new industry concepts appear that are difficult to terminologize with compact phrases" [11, p. 45] with the development of a science. It is a coumpound term, which is able to most completely reflect the necessary characteristic features of the notion being named [1].

Builders know that a part of the building that rises above the ground floor and protrudes beyond the main part in the form of a glazed balcony of various shapes (semicircular, triangular, quadrangular, polyhedral) is called *еркер* "bay window". This is a German term (German *erker*) and it means "lantern" [3, p. 75].

Recently, the word *лоджія* "loggia" has become popular in everyday life. It means the room covered and enclosed in plan on three sides and open to the outside. This word is of Italian origin (Italian *loggia*) [3, p. 187].

The word *консоль* "console" went a long way until it reached Ukrainian builders. Its roots go back to the German language (German *Auslegerm*, *Konsolef*). Later it appeared in English (English *console*, *cantilever*), from there it went to the Russian language (Russian *консоль*) and only after a long time it appeared in the Ukrainian language [15, p. 148].

The term *фальц* or *фалець* "lock joint", i.e. the type of seam used to connect the sheets of metal roof, is of German origin (German *Falz* "fillitser", "groove", "lock joint", "gutter") [7, p. 139].

The history of the word *мансарда* "mansard" is also interesting. It is known to be the name of the floor in the upper space, the facade of which is completely or partially formed by the surfaces of a sloping or gambrel roof. This place was a favorite of creative people, including writers and artists. In the world literature there are even works with such a name, namely: one of the works by a modern Serbian writer Danilo Kish is called "Mansarda: satirična poema". The term owes its origin to the French architect François Mansart, who proposed to use attic rooms for housing in the XVII century. Accordingly, the concept "mansard" has French roots as it came from the French word *mansarde* [7, p. 168].

A niche in the wall of the bedroom, separated from the rest of it and designed to accommodate the bed, was also popular in the houses of the nobles of the XIX century. It was called *альков* "alcove". At present, alcoves are often used in habitable rooms. Today, this word can mean a small dead-end room (usually without windows) adjacent to the living room, or somehow enclosed small recessed parts of such a room.

The term emerged due to the Arabs. It is well-known that in Arabic the word "al-kubbah" meant a tent. Later, through the Moors, meaning a bedroom or side room, it came first in Spanish (*alcoba*), and then in other Romance languages, e.g. Italian *alcóve*; French *alcôve* [15, p. 10]. Wolfram von Eschenbach, one of the greatest epic poets of German medieval literature, borrowed it from Old French, where it took the form of *aucube*, into Middle German, whence it spread among European builders.

In high-rise buildings constructed in the XVIII century as well as in the first half of the XIX century, the upper part of the high room was divided horizontally into two floors. Thus was formed *антресоль* "mezzanine', "entresol" (French *entresol* "mezzanine") [3, p. 28]. Sometimes they are made in modern homes and then they become shelves used for storage under the ceiling.

In Baroque and Renaissance architecture, architects used *абак* "abacus" (from Latin *abacus* "board"). This was the name of the upper plate of the capital of the column, semi-column, pilaster. As a rule, in plan it has always had a simple quadrangular shape (Doric, Ionic, Tuscan orders) or a quadrangular shape with truncated corners and concave sides (Corinthian and composite orders). As an element of the capital, it can also be seen in semi-columns and pilasters. The abacus is known to occupy an honorable place in the folk wooden architecture of Poltavshchyna and Slobozhanshchina [13, p. 14].

In the architecture of late Renaissance and Baroque, architectural decoration became widely used in the form of a spiral curl with a circle in the center. In the Ukrainian language it was called *волюта* "volute" (Latin *voluta*, Greek κάλχη "curl", "spiral"). From ancient times, the circle in the middle of the curl was called the "eye of the volute". In the Corinthian order, the volute took on a slightly different appearance. It started to resemble a branch of a plant that rises from the capital and spirals when it encounters an obstacle near the abacus. Four large volutes (Latin *helices majores*) support the abacus at its four corners. Two smaller ones (Latin *helices minores*) get together on each side of the capital under the rosette, which adorns the middle of the abacus. This architectural motif was a part of the Ionic capital. It was also a part of the composition of the Corinthian and composite capitals. In addition, sometimes architectural details that served to connect parts of the building were shaped as volutes, as well as the cornice consoles, portals, doors, and window framing. In ancient Greek pottery, especially in the region of Puglia, there were popular craters with volutes, which decorated high handles of the vessel.

The middle part of the capital was called *ехін* "echinus", which if translated from the Greek word "echnnos" means "sea urchin" or "shell". In the classical Greco-Doric and Tuscan orders, this part was made in the form of a circular block that bulges outward. In Roman Doric and Ionic orders, it was decorated with ionics, or ova (from the Latin word *ovum* "egg"), i.e. ornamental ribbons depicting ovoid bulges (ellipsoids) framed by fascias, and alternating with leaves and arrows. They can usually be seen on capitals and cornices of classical Ionic and Corinthian orders, etc.

As far back as in the architecture of the Ancient East and antiquity, *колони* "columns" (French *colonne*, Latin *columna* "pillar") were used in the composition of both the building facades and their interior space. A column is a vertical support processed architecturally, which is as a rule round in cross section, the core element of the building, load-bearing structure", etc. [13, p. 110]. They often used columns as supports for vaults and arches. Together with sculptures, the columns are built in the form of separate architectural elements as monuments. An example you can see on Maidan Nezalezhnosti in Kyiv, Ukraine.

Everyone knows that no building is possible without *каркас* "frame", which is the supporting base, a kind of skeleton of houses, buildings, manufactured objects or structures. Interestingly, the etymology of this term is closely related to the French word *carcasse* "skeleton".

Since ancient Greece, a building called *базиліка* "basilica" (Latin *basilica* from the Greek word *βασιλικη* "royal house") is well-known [16, p. 24]. Later, it became the name of a public house intended for the court, with court hearings taking place in the middle nave and trading being conducted in the side aisles. In ancient Rome, basilicas remained public buildings (for court and stock exchange transactions). In some time, the basilica turned into one of the main types of Christian church. The first known basilica dates back to the II century BC (in Pompeii).

When the Roman Empire became Christian officially, this term also came to mean an important Catholic church, which was granted this status by the Pope. So, today the word has two meanings: architectural meaning and Catholic meaning.

As a rule, as far as architecture is concerned, basilica is a rectangular structure in plan, divided inside into 3–5 parts called naves by rows of columns or pillars. It should be noted that the middle nave, the highest one, is illuminated through the windows above the roofs of the side naves and usually ends with a semicircular protrusion—the apse (from the Greek word *άψίς*, Genitive *άψῖδος* "vault"; Latin *absis* < άψίς) [15, p. 12]. The apses first appeared in ancient Roman basilicas. In Christian churches, the word "apse" is used to denote the altar, which is usually oriented to the east.

In the Catholic meaning of the word *basilica*, until recently there was only one basilica in Ukraine, that is the Archcathedral Basilica of the Assumption of the Blessed Virgin Mary (Lviv, Ukraine).

In the construction of palaces, the ancient Assyrians used such a decoration element of stairs, balconies, terraces, and gazebos, as *балюстрада* "balustrade" (Italian *balaustrata*, French *balustrade*, from *baluster* "colonnette", "fencing", "railing") [15, p. 8; 3, p. 16]. In fact, it was an enclosure made of figured columns, which served as decoration at the same time. Due to their attractive appearance and functionality, balustrades were introduced into architecture of Medieval Europe. Later they were used to decorate furniture, including the backs of chairs and beds.

Traditionally, balustrades were made of wood, plaster, stone, and concrete. One of the most popular modern materials for the production of balustrades is polyurethane, which is a practical and durable material that ensures sophistication and beauty of the form. It is worth noting that the balustrade is an all-purpose decoration element, which effectively accentuates any architectural style. So we can say with confidence that it will never go out of fashion. For instance, a baluster in the Baroque style will make the overall design classic, and balusters in the style of minimalism will add a touch of high-tech style to the overall design.

In the 16th century, green tunnels of late medieval and Renaissance gardens appeared in France, which were often made of resilient willow twigs. For that purpose, young willow or hazel shoots were intertwined around their tops, thus creating a series of arches. Then they were connected to long planks along which creepers grew. These were the so-called *берсо* "berceaux" (French *berceau*) [13, p. 9; 3, p. 26]. They could have started to use this word for this purpose because of similarity of appearance between the arches of the tunnel and the cradles (*berceau*). In French, there is also a phrase *voûte en berceau* ("cradle vault"), which is also used as an architectural term. It could also have been influenced by the consonant word *arceau* ("small arch"), which comes from the word *arc* ("arch", "bow", "arc"). It should be mentioned that the berceau can also both be an independent element of the landscape, and serve as a transfer tunnel that connects individual gazebos, and park buildings. The material for the frame of the berceau is usually metal or wood, but sometimes they use brick pillars and even stone columns instead of typical supporting structure. Placing a statue or flower garden at the exit of the shady berceau has a special effect.

Пергола "pergola" has a certain resemblance to berceau. It is a garden covering, a structure made of many repeating sections of arches connected by crossbars to protect the passage from the scorching sun. It can be both a separate building, and a part of the building that covers outdoor terraces. In addition, pergolas are more permanent architectural structures than berceaux. According to linguists, the Italian word *pergola* comes from the Latin word *pergula* which means "roof", "lean-to" [3, p. 489]. It is well-known from the history of architecture that the pergola originated in the south as a utilitarian house structure. It was built to support the vines and it was a place for day rest simultaneously. The architecture of the pergola varies from simple straight wooden structures of medieval monastery gardens to florid marble columns of the XIX century and modern pergolas made of reinforced concrete. One of the most beautiful elements of the pergola is the shadow pattern inside it.

As far as garden and park buildings are concerned, *альтанка* "gazebo" is worth mentioning. At first, this word was used to denote large balconies, then they started to use it for platforms, ledges and pergolas, where you can enjoy the view and the landscape. As a rule, gazebos are open on all sides, but they can also be closed (similar to pavilions). This word has Italian roots and originates from the word *altana* which means "terrace on the roof" or "small shed" sometimes. The gazebo has become a symbol of many cities of Ukraine, their landmarks, e.g.: Sumy gazebo (Sumy), gazebo at Volodymyrska Hill (Kyiv), Mirror Stream—a gazebo and fountain in Kharkiv.

In palaces of Baroque and Classicism and in houses of the nobles in the XIX–XX centuries they often used *анфілади* "enfilades" (French *enfilade*, from *enfiler* "to thread"). The enfilade is the name of a suite of rooms aligned with each other, whose doorways are located on one axis, due to which a through perspective of all interiors is created when the door is open [13, p. 11; 3, p. 9]. In addition, the enfilade may be formed by several courtyards located one after the other on the same axis and connected by driveways or passages, or several squares connected by streets or alleys. The enfilade can also be in the form of a circle or a curved line.

In medieval Western European architecture *вимперги* "gables" were popular (German *Wimperg* from *Windberg* "protection from wind") [13, p. 17; 5, p. 14]. A gable is a high triangular pediment above the portal or window opening. A gable originated from the abutting ends of houses with high double-pitch roofs. In Gothic-style buildings, they accentuated the vertical orientation of the composition, i.e. that of pointed arches and portals. Such forms first appeared in French architecture in the late XII century and early XIII century as a result of using a temporary wooden roof in the reconstruction of old Romanesque cathedrals. They covered the abutting end of the building that helped to protect the interior from rains and winds. But at the end of the XIII century, the gable turned into a decorative element. It was florid with openwork carvings and sculptural details, steep profiled edges, and its top was decorated with a finial. Sometimes a quatrefoil or "blind rose" of stained glass was placed in the center of the tympan.

Besides, *квадрифолій* "quadrifolium" (Latin *quadrifolium* "four-leaf clover"), a decorative motif combining a square and a flower with four symmetrical petals, is also interesting [16, p. 78]. It was widely used in European architecture, book miniatures and sculptures belonging to different time periods and styles (Middle Ages, early Italian Renaissance, Gothic, etc.). In order to decorate literary monuments, it was used to separate episodes of the stories from each other.

From time immemorial, people have always been attracted by castles shrouded in legends and mysteries. The origin of the word *замок* "castle" draws attention as well. It came from the Polish word *zamek*, where it, like in Czech *zámek*, is a calque from the Middle High German *slōȥ* ("castle", "bolt", "fortress"), and, in its turn, the latter word is a calque from Latin *clūsa* ("castle", "lock", "fort", "fortification") [15, p. 59]. That was the name of the building (or complex of buildings), which performed both housing, and defense and fortification functions. In the most general meaning of the word *замок* "castle" is a fortified housing of the feudal lord in medieval Europe. Many castles have survived to the present day and are recognized monuments of history and culture.

It is impossible to imagine ancient castles without expensive furniture, an integral part of whose decoration was *інтарсія* "intarsia" (from Italian *intarsio*), that is a woodworking decoration technique; patterns of wood plates different in texture, color, and cut into the wooden surface. Intarsia technique, like inlay, originated from the Ancient East. In particular, it was widespread in Egypt. Craftsmen from ancient Greece and Rome achieved incredible success in this field, making ornaments of maple, boxwood, holly, ironwood and dogwood. It is interesting that this technique is used in almost all the regions of Ukraine, it is especially widely used by amateur craftsmen, especially in Prykarpattia and Zacarpattia.

Today, the word *шале* "chalet" has become commonplace for people of different nationalities. It is known to be of French origin, although it has Latin roots (French *chalet* from Latin *cala* "protected place") [16, p. 186, 19, 20]. In ancient times, it was the name of shelters for shepherds who herded flocks of sheep in the Swiss Alps. In romantic parks of the XVIII century, the chalet is a garden pavilion in the form of a country house, which brought a pastoral shade to the landscape. In the XIX century, the interest in traditional wooden houses in the Swiss Alps grew among the European nobility and bourgeoisie. Now, the chalet is perceived as a type of house typical for folk architecture of the mountainous areas of the Alps, a kind of a small cottage. With the development of tourism, chalets began to be perceived as holiday homes, especially in Austria, the Alpine region (South Tyrol).

From long ago, people have built amazing structures that made it easier for them to cross rivers. Of course, these were bridges. Initially, a bridge was just a log thrown over the stream (such bridges are called footbridges). Then they started to use stone as a material for the bridge. The first bridges of this type began to be built in the era of slave-owning society. The ancient Romans achieved great success in stone bridge construction, using vaulted structures as supports and cement, the secret of whose manufacture was lost in the Middle Ages, but then rediscovered. Bridges (to be more precise, aqueducts) were used to provide cities with water. It should be mentioned that some ancient Roman bridges still function to this day. It is of interest that this architectural structure has taken its rightful place in literature. The most famous bridges in literary works are: "Le Pont Mirabeau" by G. Apollinaire, "The Bridge of San Luis Rey" by T. Wilder, "The Problem of Thor Bridge" by A.K. Doyle, "The Bridge Across Forever" by R. Bach, "The Bridges of Madison County" by R.J. Waller, and others. The term *міст* "bridge" has Proto-Slavic roots: **mostъ*, from **mesti*, **metati* "to throw". By the way, scientists believe that the bridge is one of the oldest engineering structures of mankind.

3 Conclusions

The brief analysis of building and architectural terminology shows that the creation of Ukrainian building terminology is subject to general word-forming language laws.

A significant number of building and architectural terms belongs to the vocabulary of foreign origin. Many modern terms have a long history and were borrowed into

Ukrainian from German, French, Italian and Latin as a result of intercultural and commodity exchange.

Building and architectural terminology provides significant material concerning modifications in semantic volume of borrowed terms.

References

1. Ageicheva A, Bolotnikova A, Hunchenko Y, Perederii I (2020) English compound construction economic terminology: current aspects of professional text cohesiveness. In: Onyshchenko V, Mammadova G, Sivitska S, Gasimov A (eds) Proceedings of the 2nd international conference on building innovations. ICBI 2019. Lecture notes in civil engineering, vol 73. Springer, Cham, pp 517–523. https://doi.org/10.1007/978-3-030-42939-3_51
2. Balatska O, Fedorenko Yu (2020) Linguocultural Specificity of Explicit Lexical and Grammatical Axiological Means in Criticism in the English-language Research Paper. Linhvistychni Studiyi Linguistic Studies: collection of scientific papers. Vasyl' Stus Donetsk National University. In: Krasnobaieva-Chorna Zh (ed) Vasyl' Stus Donetsk National University, Vinnytsia, vol 40. In 2 p. p 2. pp 7–14. https://doi.org/10.31558/1815-3070.2020.40.2.1
3. Bybyk SP, Siuta HM (2006) Dictionary of foreign words: interpretation, word formation and word usage. In: Yermolenko SY (ed) Folio, Kharkiv. 623 p.
4. Cherednyk L, Hunchenko Yu (2019) Translation as integrative component of intercultural communication. New stages of development of modern science in Ukraine and EU countries: monograph/edited by authors, 6th edn. "Baltija Publishing", Riga, pp 232–252. https://doi.org/10.30525/978-9934-588-15-0-116
5. Cresswell J (2010) The Oxford Dictionary of Word Origins (2 ed.) OUP Oxford. 502 p. https://doi.org/10.1093/acref/9780199547920.001.0001
6. Hugo V (1988) The hunchback of Notre-Dame. Collected works in 6 volumes, vol 1. Pravda, Moscow. 652 p. (in Russian)
7. Klymeniuk TM, Proskuriakov VI, Kovalchuk HI (2014) Dictionary of architectural terms. Lviv Polytechnic Publishing House, Lviv, p 180
8. Kochkarev D, Azizov T, Azizova A, Galinska T (2021) Designing of standard cross sections of composite bending reinforced concrete elements by the method of design resistance of reinforced concrete. https://doi.org/10.1007/978-3-030-57340-9_25
9. Korobko B, Khomenko I, Shapoval M, Virchenko V (2020) Hydraulic single pump with combined higher volume compensator operation analysis. In: Onyshchenko V, Mammadova G, Sivitska S, Gasimov A (eds) Proceedings of the 2nd international conference on building innovations. ICBI 2019. Lecture notes in civil engineering, vol 73. Springer, Cham, pp 103–114. https://doi.org/10.1007/978-3-030-42939-3_12
10. Onyshchenko V, Sivitska S, Cherviak A (2020) Buildings reconstruction within the new educational space project in Ukraine. In: Onyshchenko V, Mammadova G, Sivitska S, Gasimov A (eds) Proceedings of the 2nd international conference on building innovations. ICBI 2019. Lecture notes in civil engineering, vol 73. Springer, Cham, pp 669–676. https://doi.org/10.1007/978-3-030-42939-3_66
11. Petukhova O (2011) Structural modeling of Ukrainian terms of metallurgy. Bulletin of the National University "Lviv Polytechnic": Series "Problems of Ukrainian Terminology", № 709, pp 43–46
12. Semko O, Yurin O, Filonenko O, Mahas N (2020) Investigation of the temperature–humidity state of a tent-covered attic. In: Onyshchenko V, Mammadova G, Sivitska S, Gasimov A (eds) Proceedings of the 2nd international conference on building innovations. ICBI 2019. Lecture notes in civil engineering, vol 73. Springer, Cham, pp 245–252. https://doi.org/10.1007/978-3-030-42939-3_26

13. Shmyh RA, Boiarchuk VM, Dobrianskyi IM, Barabash VM (2011) Terminological reference dictionary of building and architecture. Shmyh RA (ed) Lviv. 222 p.
14. Hoad TF (ed) (2003) The concise Oxford dictionary of English etymology. Oxford University Press, Oxford, New York. 570 p. https://doi.org/10.1093/acref/9780192830982.001.0001
15. Tymofiienko VI (2002) Architecture and monumental art: Terms and concepts. IPSM AUC, Holovkyivarkhitektura, Kyiv. 472 p.
16. Vlasov VH (2004) Dictionary of terms. Drofa, Moscow, p 191
17. Vorobiova O, Saienko L English phrasal verbs: derivation and semantics. New stages of development of modern science in Ukraine and EU countries: monograph/edited by authors, 6th edn. "Baltija Publishing", Riga, pp 57–72. https://doi.org/10.30525/978-9934-588-15-0-107
18. Vorobiova O, Lazebnyk A Telescopic neologisms of modern English: structure and semantics. Linhvistychni Studiyi Linguistic Studies: collection of scientific papers/Vasyl' Stus Donetsk National University. In: Krasnobaieva-Chorna Zh (ed) Vasyl' Stus Donetsk National University, Vinnytsia, vol 40. In 2 p. p 2. pp 33–40. https://doi.org/10.31558/1815-3070.2020.40.2.4
19. Zotsenko M, Vynnykov Y, Lartseva I, Sivitska S (2018) Ground base deformation by circular plate peculiarities. Paper presented at the MATEC web of conferences, vol 230. https://doi.org/10.1051/matecconf/201823002040
20. Zotsenko ML, Mykhailovska OV, Sivitska SP (2020) Construction features durable storage of toxic waste in boreholes. https://doi.org/10.1007/978-3-030-42939-3_33
21. Kochkarev D, Azizov T, Galinska T (2018) Bending deflection reinforced concrete elements determination. Paper presented at the MATEC web of conferences, vol 230. https://doi.org/10.1051/matecconf/201823002012
22. Pichugin SF (2018) Reliability estimation of industrial building structures. Mag Civ Eng 83(7):24–37. https://doi.org/10.18720/MCE.83.3
23. Sivitska S, Vartsaba V, Filonych O (2018) Buildings energy-efficient renovation investment. Int J Eng Technol (UAE) 7(3):408–412. https://doi.org/10.14419/ijet.v7i3.2.14562
24. Leshchenko MV, Semko V (2015) Thermal characteristics of the external walling made of cold-formed steel studs and polystyrene concrete. Mag Civ Eng 60(8):44–55. https://doi.org/10.5862/MCE.60.6

Development of Creative Economy Objects as a Means of Industrial Territories Revitalization

Andrii Dmytrenko, Oleksandr Ivashko, and Yulia Ivashko

Abstract The article considers the possibilities of using the creative economy objects (i.e. refers to a range of economic activities which are concerned with the generation or exploitation of knowledge and information) as a means of industrial territories revitalization in the modern conditions of Ukraine. Based on a comparative analysis of the development of the creative economy in Ukraine and other European countries, it is determined that despite the insignificant role of the creative economy in Ukraine now there are certain prerequisites for its further dynamic development. The world and Ukrainian experience of revitalization of industrial territories with placement of objects of creative economy, and also existing conceptual offers concerning development of objects of creative economy in Ukraine is analyzed. It is stressed that solving the problem of industrial areas revitalization is impossible without the development of creative economy organization new forms, including art clusters. The list of urban planning, organizational and financial, spatial planning and engineering and design criteria for the selection of industrial facilities for conversion into creative economy objects in the modern socio-economic conditions of Ukraine is determined.

Keywords Industrial territories · Re-profiling · Creative economy · Art cluster · Art centre

1 Introduction

In recent decades, in many cities we can see the emergence of industrial areas that are not used for their intended purpose and need revitalization, due to a range of factors,

A. Dmytrenko (✉)
National University "Yuri Kondratyuk
Poltava Polytechnic", 24 Pershotravnevyi Avenue, Poltava, Ukraine
e-mail: metr5555@ukr.net

O. Ivashko · Y. Ivashko
Kyiv National University of Construction and Architecture,
31 Povitroflotskyi Avenue, Kyiv, Ukraine

V. Onyshchenko et al. (eds.), *Proceedings of the 3rd International Conference on Building Innovations*, Lecture Notes in Civil Engineering 181,
https://doi.org/10.1007/978-3-030-85043-2_46

including socio-economic, scientific and technological, urban, environmental and others. The closure of unprofitable industrial enterprises is an objective pattern for countries that have entered the so-called post-industrial era, i.e. primarily for the most developed countries in the world.

There are different approaches to the use of such industrial areas, some of which involve the complete demolition of existing buildings and structures and the use of areas for another function (usually residential), others—the preservation of at least part of existing buildings and its adaptation to other functions. The second approach is used in particular when the buildings and structures of non-operating enterprises are monuments of industrial architecture, when the areas to be revitalized are located near the historic city centre, and so on. One possible means of revitalizing industrial areas with preservation of existing buildings is the deployment of objects of so-called "creative economy" (i.e. refers to a range of economic activities which are concerned with the generation or exploitation of knowledge and information). According to Howkins [1, pp. 88–117] "creative economy" includes art, film industry, music industry, performing arts, fashion, advertising, publishing, TV and radio, toys and games, videogames, craft, design, architecture, research and development, software. Some researchers not only note the importance of partnership and collaboration across higher education institutions and the creative and cultural industries but also consider the education industry as a part of the creative economy [2].

In general, the problems of preserving monuments of industrial architecture, revitalization of industrial areas (especially located near the central parts of historic cities) and the development of new sectors of the economy, typical of the post-industrial period are closely linked and need comprehensive research.

General problems of degradation of cultural heritage (including due to industrial and domestic pollution, which is a significant problem for monuments of industrial architecture) are studied in the work of Spiridon, Sandu, Stratulat [3]. The development of creative industries in European countries was analyzed by Boix, Capone, de Propris [4], Selada, da Cunha, Tomaz [5], in Ukraine—Skavronska [6]. Current trends in the revitalization of the urban environment and the role of art in this process have been studied by Kashchenko, Kovalska, Gnatiuk [7]. The problems of revitalization of industrial enterprises are directly investigated in the works of Kulikov, Dyomin, Chernyshev, Kuśnierz-Krupa, Krupa, Orlenko, Ivashko, Kobylarczyk, Stefański, Gryglewski, Dmytrenko, Ivashko, Leshchenko, Tovbych [8–12]. Examples of revitalization of former industrial facilities in Lodz, Poland by placing objects of the creative industry are covered in the works of Ugorowicz [13] Rawicka [14] and Cysek-Pawlak [15]. General and specific problems of restoration of monuments of industrial architecture are investigated in the works of Orlenko and Buzin [16, 17].

At the same time, the issue of using the objects of creative economy as a means of revitalization of industrial areas in Ukraine remains insufficiently studied.

2 Main Part

2.1 *Purpose of the Article*

The purpose of the article is to formulate the criteria for the preservation of individual objects of industrial architecture with the subsequent re-profiling of the objects of the creative economy in the modern socio-economic conditions of Ukraine.

2.2 *Research Methodology*

Existing scientific sources, regulations and materials of their own research were used in preparing the article. Methods of historical analysis and comparative analysis were chosen as the main ones. The method of historical analysis allowed analyzing the origins of problems related to industrial areas in the structure of cities and their specifics in different cities, the method of comparative analysis allowed to analyze the effectiveness of revitalization measures in different countries and new concepts and draw conclusions.

2.3 *Results*

Socio-economic Preconditions. In European countries, the role of the creative economy is constantly, albeit rather slowly, growing [18]. Thus, in 2018, 3.7% of the EU population was employed in the creative economy [19]. The level of the creative economy development varies greatly in different European countries. The undisputed leader here is the United Kingdom, where in 2019 there were 5.3 million jobs in the DCMS (Digital, Culture, Media & Sport) sector, which was 15.7% of the total number of jobs [20]. Ukraine, along with Latvia and Bulgaria, are among the countries with a low percentage of employees in the creative economy and a small contribution of the creative economy to GDP. This situation is due to a number of factors, but it should be noted that Ukraine has certain prerequisites for the development of a creative economy, one of which is a high level of education, including a significant percentage of people with higher education: 50.3% of Ukrainians under 30–34 have higher educational attainment rates (compared to 36.9% of EU's citizens) [6, p. 99]. This allows us to consider the location of creative economy objects as one of the possible promising means of industrial areas revitalization in Ukraine.

Analysis of the World Experience of Industrial Territories Revitalization. The study of world experience proves that with a sufficient level of investment, no design and spatial planning features of revitalized industrial buildings can be an obstacle to the re-profiling of the industrial enterprise under the objects of the creative economy.

For example, in the re-profiling of industrial enterprises, perhaps one of the most difficult tasks is the revitalization of old silos, as round shapes are the most difficult to saturate with a new function. However, this did not prevent the conversion of silos of the former malt production of the nineteenth century near Antwerp in Belgium into a residential and public complex with a combination of residential apartments, offices and exhibitions. Preserving the original form, the authors modernized it by replacing it with glass parts. Another example of the conversion of grain silos to a cultural object is the Zeitz Museum of Contemporary Art Africa in Cape Town. Decommissioned in 1990, the silo is a monument of industrial architecture and was once the tallest building in South Africa. After the revitalization, the area of the museum with the atrium, galleries, roof garden, reading and exhibition halls and maintenance is 9.5 thousand m^2.

Given the decisive role of socio-economic factors, the analysis of the experience of Poland, where the level of economic development in the 1990s was comparable to the Ukrainian one, seems more relevant for Ukraine. The most striking examples of the industrial areas revitalization are concentrated in the former centre of the textile industry—Lodz, where almost all specialized enterprises were closed in the 1990s. In this city, the monuments of industrial architecture are the main part of the historical and architectural heritage, so there are special requirements for their preservation. A certain contrast between the image of the object and its function did not repel investors and today in Lodz in the former factory buildings of red brick are higher education institutions (Politechnika Łódzka), hotels (Andel), public and shopping centres (Manufaktura), luxury hosing (Księży Młyn) and objects of the creative industry, the most famous of which are Art_Inkubator, "OFF Piotrkowska" and EC1 [10].

The Art_Inkubator complex provides a space for presenting educational initiatives, exhibitions and for activities that combine art and entrepreneurship. In 2012–2013 three buildings placed in the area of the former textile warehouses in Wincentego Tymienieckiego Street were adapted for this complex. The project was financed half from the city budget and half from the EU funds [13].

"OFF Piotrkowska" is a multifunctional complex located in the city centre at 138/148 Piotrkowska Street, on the premises of the former spinning and weaving mill of Franciszek Ramisch. Adaptation works in this area began in 2006. Now the complex houses not only music clubs, restaurants, cafés and exhibition halls, but also studios of architects, fashion and art designers [14].

The former power plant known as EC1 was renovated in 2010–2018. The oldest power plant building (so-called EC1 East), built in 1906–1907, was turned into the "City of Culture" complex, housing the National Centre for Film Culture, Planetarium and exhibition spaces. EC1 West, consists of buildings from the interwar period, They was adapted as an Interactive Science and Technology Centre—a multifunctional museum-educational-scientific facility [15].

Analysis of the Experience of Industrial Territories Revitalization in Ukraine. In Ukraine, old unprofitable enterprises are most often demolished in order to free up land for modern housing, office or shopping centres. Existing industrial facilities, some of which belong to the listed objects, no longer meet modern requirements—their territory is too large, the buildings are not designed for modern equipment; in addition, the increase in traffic load in recent decades will lead to intersections industrial and passenger traffic. Recently, experts pay attention to the rationality of the approach to the preservation of existing industrial facilities, which in accordance with their purpose were built of durable materials and structures and are able to withstand significant loads [8, 9, 12]. In addition, there were entrances and approaches to these facilities, and appropriate engineering networks were installed. Such objects have large areas and many floors of significant height. Many industrial facilities in Kyiv have been demolished in recent decades (Kyiv Yeast Factory, Kyiv Refrigeration Plant, etc.), some are leased to various tenants (Bakery and Confectionery, Sausage Factory on Pavlivska Street), some are abandoned (separate buildings of Motorcycle plant, industrial shops on Telychka, former Richert plant on Kyrylivska Street, elevator on Naberezhno-Khreschatytska, industrial shops on Rybalskyi Peninsula).

Preservation of historic industrial enterprises with their subsequent revitalization has not yet become widespread in Kyiv. Among the implemented examples, the most famous is the revitalization of the Darnytsia Silk Factory on the left bank of the Dnipro River, which was built in 1947. The territory of the enterprise has been in an abandoned state since the 1990s, until in 2001 the Darynok shopping complex appeared in the former industrial buildings. Measures to revitalize the industrial area continued in 2014 with the formation of exhibition and entertainment facilities "Art Factory Platform" on the basis of buildings, which became the venue for festivals of street food, concerts and "Courage Bazaar"—the first charity "flea market" in Kyiv. In 2018, part of the buildings was repurposed as an office and business centre.

Another successful Kyiv example of revitalization is the UnitCity created in 2017—a campus for business development in the field of IT and creative industries on the basis of the buildings of the Kyiv Motorcycle Plant.

The need for new forms of public leisure requires owners to create institutions focused on the needs of the modern consumer. Terminological usage includes new terms—"food hall", "food market". The food hall of Kyiv, Kyiv Food Market, had opened in September 2019 and united more than two dozen Kyiv establishments in the former premises of the Arsenal plant. The owners and organizers have linked the name "food market" with the specifics of the newly created atmosphere of the former industrial facility. Despite the fact that the exterior of the building has the features of industrial architecture, the interior has been redesigned and renovated with three levels (the first two floors—more than twenty different bistros and 550 seats for visitors, the third—for DJs). The peculiarity of such a food market (co-founded by the owners of many restaurants) is that food is cheaper here, because the food market combines all the services: from waiters and washing dishes to security and the Internet and restaurants need less staff.

In addition to Kyiv, measures to revitalize former industrial enterprises were carried out in other cities—in Lviv on the basis of the Plant of Radioelectronic Medical Equipment (REMA), whose history begins with the factory of the Winter brothers ("ReZavod"), in Ivano-Frankivsk on the basis of the plant of industrial equipment ("Promprylad. Renovatsiya", with an area of 37 thousand m^2), in Kharkiv on the basis of the plant for sorting and storage of breeding crops "Soyuzsortnasinneovoch" closed in the 1990s—a modern IT hub with coworking and event spaces, service sector "Fabryka.space") and in Poltava ("Art Platform 11"). "Art Platform 11"is located in the former shop of the Poltava Turbomechanical Plant, areas of activity are typical for the creative industry: educational, cultural, musical, entertainment, art, creative, information (round tables, press conferences), space for work, meetings.

Concepts of Using Creative Economy Objects as Means of Industrial Territories Revitalization in Modern Conditions of Ukraine. Despite the significant difference in the scale of former industrial buildings use to accommodate the creative economy objects in Poland and Ukraine, there is something in common: the adaptation of industrial buildings (especially monuments of industrial architecture) requires large investments and large companies can be considered as investors (especially foreign), municipal authorities or the state. One of the most famous projects of transforming the Soviet-era defence enterprise located on the territory of the Old Pechersk Fortress in a complex of historic buildings (1784–1803) into a cultural, artistic and museum complex "Mystetskyi Arsenal" ("Art Arsenal") was implemented primarily through funding from the state budget. The vast majority of creative economy actors both in Ukraine and in the world are small and medium-sized enterprises, which, taken separately, do not require large areas for accommodation and for which large investments in the redevelopment of existing buildings for their needs are irrational.

One of the ways to solve this problem in Ukraine can be proposed by O. Ivashko to create the so-called "art clusters" and "art centres", where there is a synergistic combination of the potential of many small and medium-sized creative industry objects and as a result of joint action of art clusters (where the art product is produced) and art centres (where the product receives public expression) there is a holistic art formation. An example of the application of this concept is the experimental project developed by O. Ivashko for the art-formation of the former Richert brewery in Podil at 35 Kyrylivska Street in Kyiv. The functional filling of the facility after revitalization was proposed on the basis of such creative objects of Lodz as "Off Piotrowska" and "Art_ incubator". The same applies to a certain change in the appearance of the historic building, which contrasts with the modern added volume.

Among the experimental projects of re-profiling of Ukrainian industrial facilities, we can also single out the revitalization project of the former Kyiv grain elevator complex of 1950 on the 10 Naberezhno-Khreschatytska Street, near the Rybalskyi Peninsula in Kyiv (author S. Reshetnyk, head of the project part J. Vig, head of the scientific part Y. Ivashko). The need to preserve this complex with a change in the original function is justified by the accent location in the structure of the coastal part of the city near the Dnipro River and the structure of the building, which consists of a reinforced concrete working tower 66.9 m high, two silos, connecting bridges and gallery, receiving and pneumatic device for receiving grain. The author proposed to repurpose the elevator complex just for the cultural and public centre with the presence of an exhibition area, catering area, educational area, recreation, office and business area, sports and entertainment area. The modernization of the former elevator buildings appearance with a contrasting combination of modernized authentic buildings and the emphasized modern new buildings is provided.

Thus, summarizing the foreign and Ukrainian experience of industrial areas revitalization with the placement of creative economy objects, we can identify certain criteria for the selection of revitalization.

1. Industrial areas located near the central parts of cities, highways, on the compositionally significant parts of the urban landscape are the most suitable for revitalization with re-profiling as objects of creative industry. The most justified is the re-profiling of industrial facilities in large cities, which provides an influx of visitors and tenants.
2. Industrial buildings recognized as architectural monuments usually need restoration in addition to changes in use. This significantly increases the cost of redevelopment and requires the involvement of investors or large companies (usually, in this case, in addition to the creative industry, the complex may also include housing, shopping and entertainment centre or other highly profitable facilities), or use the funds local and state budgets, grant funds of international organizations, etc. The experience of Lodz shows that from the point of view of the local community long-term interests, the revitalization of the industrial architecture monument with the placement of creative economy objects is a profitable investment that helps increase the tourist attractiveness of the city, create new jobs and renew urban infrastructure.
3. In the absence of large investors and the conversion of industrial facilities into the creative industry objects by means of formation on their basis art clusters and art centres the buildings of former food and light industry enterprises with a floor height of not more than 4.8 m, not entered in the state register of architectural monuments, are the most suitable. This will minimize the buildings redevelopment cost.
4. An important issue is to minimize operating costs when using repurposed facilities. In Ukraine, it is especially important to reduce heating costs. A combined approach to the installation of heating systems is appropriate, when buildings with a large internal space, periodically or occasionally used for short-term activities, can be heated only during these activities (air heating, use of infrared emitters, etc.). The constantly heated internal volume should be minimized.

2.4 Scientific Novelty

The scientific novelty is to determine the list of urban planning, organizational and financial, spatial planning and engineering and design criteria for the selection of industrial facilities for conversion into creative economy objects in the modern socio-economic conditions of Ukraine. The important role of a fundamentally new direction of revitalization based on the interaction of creative industries with the emergence of art education based on the addition of the functions of the art cluster (production of creative product) functions of the art centre (public presentation of creative product) is emphasized. This is especially important, given that the scientific concept of "art cluster" and its content has not yet been formed.

2.5 *Practical Importance*

The defined selection criteria can be used by local governments, government agencies of cultural heritage protection, potential investors and design organizations in creating projects for industrial facilities revitalization by re-profiling them as objects of creative economy, and understanding art education as a harmonious combination of two components—the one which produces the product and the one which publicly presents it will allow to bring the theoretical basis under the term "art cluster" and to ensure the economically justified functioning of such art formations of a new type.

3 Conclusions

Converting non-functioning industrial facilities into creative economy facilities can be considered as one of the promising ways to revitalize industrial areas in Ukraine. The main factors influencing the choice of revitalization objects are: the location of territories in the city structure; the list of potential investors and users; the monument protection status of objects located in the territories proposed for revitalization; spatial planning and engineering features of industrial buildings and structures to be repurposed. Solving the problem of industrial areas revitalization is impossible without the development of creative economy organization new forms, including art clusters.

References

1. Howkins J (2001) The creative economy: how people make money from ideas, 1st edn. Penguin, London
2. Comunian R, Gilmor A (eds) (2018) Higher education and the creative economy: beyond the campus, 1st edn. Routledge, London
3. Spiridon P, Sandu I, Stratulat L (2017) The conscious deterioration and degradation of the cultural heritage. Int J Conserv Sci 8(1):81–88
4. Boix R, Capone F, de Propris L (2016) Comparing creative industries in Europe. Eur Urban Reg Stud 4:935–940
5. Selada C, Vilhena da Cunha I, Tomaz E (2011) Creative clusters in low density areas: a case-study approach. In: ERSA conference papers ersa10p1366, European Regional Science Association

6. Skavronska IV (2017) Creative industries in Ukraine: analysis and prospects of the development. Econ Sociol 10(2):87–106
7. Kashchenko O, Kovalska G, Gnatiuk L (2020) Revitalization of the urban environment and contemporary trends of its humanization via the means of art. Wiadomości Konserwatorskie – J Heritage Conserv 61:31–34
8. Kulikov P, Dyomin M, Chernyshev D, Kuśnierz-Krupa D, Krupa M (2019) The issues of preservation and revitalization of residential, public and industrial buildings from the second half of the 19th and early 20th centuries in Kyiv and Krakow. Wiadomości Konserwatorskie – J Heritage Conserv 60:140–146
9. Orlenko M, Ivashko Y, Kobylarczyk J, Kusnierz-Krupa D (2020) Ways of revitalization with the restoration of historical industrial facilities in large cities. The experience of Ukraine and Poland. Int J Conserv Sci 11(2):433–450
10. Stefański K, Gryglewski P, Ivashko Y, Dmytrenko A, Ivashko O (2020) Revitalization specifics of industrial enterprises made of brick and concrete. Examples of Lodz, Kyiv and Poltava. Int J Conserv Sci 11(3):715–730
11. Ivashko O (2019) The issues of conservation and revitalization of the monuments of industrial architecture. Wiadomości Konserwatorskie – J Heritage Conserv 58:113–117
12. Leshchenko N, Tovbych V (2019) Modern approaches to the revitalization of historical ex-industrial architecture. Wiadomości Konserwatorskie – J Heritage Conserv 60:51–58
13. Ugorowicz P (2016) Art_Inkubator. Przykład modelowej adaptacji zabytku poprzemysłowego/Art_Inkubator – an example of post-industrial monument model adaptation. Renowacje i zabytki 1:122–133
14. Rawicka A (2016) OFF Piotrkowska. Zagospodarowanie zabudowań fabryki Franciszka Ramischa wobec zachowania podstawowych wartości zabytkowych zespołu/OFF Piotrkowska street. Development of Franiszek ramisch's factory in the face of preserving the basic historical values of the complex. Renowacje i zabytki 1:134–141
15. Cysek-Pawlak MM (2016) Adaptacja zespołu dawnej elektrociepłowni EC1/Former EC1 heat and power station complex adaptation. Renowacje i zabytki 1:114–121
16. Orlenko M (2019) The system approach as a means of restoration activity effectiveness. Wiadomości Konserwatorskie – J Heritage Conserv 57:96–105
17. Orlenko M, Buzin I (2019) Stylistic peculiarities, features of the use of building materials and structures in buildings of the latter half of the nineteenth century and the early twentieth century, and methods for their restoration (evidence from Ukraine). Wiadomości Konserwatorskie – J Heritage Conserv 59:97–105
18. Dent T, Comunian R, Conor B, Pica V, Wilson N, Burlina C (2020) Creative and cultural workforce in Europe statistics report. DISCE Publications. https://disce.eu/wp-content/uploads/2020/05/DISCE-Report-D3.2.b.pdf
19. Eurostat Cultural employment by sex. https://appsso.eurostat.ec.europa.eu/nui/submitViewTableAction.do
20. National Statistics DCMS Sectors Economic Estimates 2019: Employment. https://www.gov.uk/government/statistics/dcms-sectors-economic-estimates-2019-employment

Cleanroom Air Control

Ganna Khoroshun, Halyna Tatarchenko, Mykola Diomin, and Zakhar Tatarchenko

Abstract Cleanrooms are the necessary requirement for the further development of mankind in such areas as microelectronics, the pharmaceutical industry, biotechnology, medical device manufacturing and the space industry. The technique of clean rooms in solving these problems should provide control and reduction of particles and microorganisms and gas inclusions in the air. Construction and finishing materials must meet the requirements for clean rooms of the appropriate class. Concentration rates of microparticles for different classes of clean rooms are presented. The specific physical, chemical and structural properties of cleanroom envelope elements must meet four main criteria: functionality, durability, and ability to be cleaned, repair ability. We have presented the visual method of microparticle counting by the optical setup. The registration of particles in them occurs due to the overlap of the light flux from the laser when passing the sample through the measuring cell. The microparticles obtained from a machine with the HEPA filter are used for investigation with digital and optical microscope systems. High resolution optical microscope performs the duster and microorganism structure with the size 1–20 μm. The method of describing laser field pattern and model of service providing on the use of optical laboratory for visual-optical control are used for development of the optical air control system. The movement of microparticles and counting their number in the area of observation for processing video files by the software application is described. The method of optical image compressing is increased the speed of image data processing and database management system. Decision-making support methods for definition optimal construction criteria taking into accounts its cleanliness level, lifetime and profitability by fuzzy logic are started to develop.

Keywords Clean room · GMP standard · Microparticle counting · Fuzzy set

G. Khoroshun · H. Tatarchenko (✉) · Z. Tatarchenko
Volodymyr Dahl East Ukrainian National University, pr. Central 59-a, Severodonetsk 93400, Ukraine

M. Diomin
Kyiv National University of Construction and Architecture, Kyiv, Ukraine

V. Onyshchenko et al. (eds.), *Proceedings of the 3rd International Conference on Building Innovations*, Lecture Notes in Civil Engineering 181,
https://doi.org/10.1007/978-3-030-85043-2_47

1 Introduction

Clean rooms are the necessary requirement for the further development of humankind in such areas as microelectronics, pharmaceuticals, biotechnology, medical devices and the space industry [1]. Cleanroom technology in solving the problems has to provide the control and reduction of particles and microorganisms and gas inclusions in the air.

Cleanroom construction is carried out in accordance with the relevant cleanroom design standards [2–6]. Construction and surfacing materials must meet the requirements established for clean rooms of the corresponding class [7].

Air conditioning systems for clean rooms must supply a certain amount of clean air in order to maintain a given level of cleanliness of the room. The air is supplied to the clean rooms in such a way as to prevent the formation of stagnant zones where dust particles can settle and accumulate. To provide air conditioning for pharmacological, medical or other purposes, balanced ventilation is the most often used.

2 The Problem Statement

Along with increases in the volume of construction in the world of healthcare facilities, laboratories and enterprises for the production of microelectronics, drugs, etc., the demand for ventilation systems for "cleanrooms" has sharply increased. Large-sized flat and spatial rod systems are widely used in industrial and civil construction. For such rod systems, it is necessary to carry out the calculation according to the deformed scheme taking into account their geometrically nonlinear behavior [8, 9].

Cleanroom walls are built either by construction on site in the usual way (the same as used in civil or industrial construction), or by assembling prefabricated elements are connected to each other in place. The materials used in the construction of cleanrooms should be selected by taking into account that they will not generate airborne contaminations to keep the purity of the product.

Decision-making support system allows to define criteria for optimal parameters of the construction taking into account its cleanliness level, lifetime and profitability [10]. So, the development of such a system is an urgent task for the cleanroom air control.

2.1 Cleanroom Standards

In Ukraine since 2009 the requirements for clean rooms are generally described in the standard DSTU [2–5]. These standards are specified air cleanliness classes, requirements for control and monitoring to confirm continued compliance, associated

Table 1 The values of the maximum particle concentration and viable microorganisms number for 4 classes of cleanrooms in accordance with GMP

Cleanroom class by GMP	Maximum concentration of particles, particles/cubic meter, with sizes equal or larger following values, μm				Maximum quantity viable microorganisms per m^3
	Equipped cleanrooms		Operating cleanrooms		
	0,5	5,0	0,5	5,0	
Class A	3500	0	3500	0	Less than 1
Class B	3500	0	350,000	2000	10
Class C	350,000	2000	3,500,000	20,000	100
Class D	3,500,000	20,000	–	–	200

controlled environments, test methods, design, construction and operation of the cleanrooms.

For the purpose of classification, only particle sets with a cumulative distribution of particle concentration are considered, the size of which is in the range of 0.1–5.0 μm. Particular requirements for the cleanroom are expressed in terms of concentration and for aerosol particle counters. The concentration of aerosol particles in the air determines the class of clean rooms. The requirements for clean rooms in the production of pharmaceuticals and microbiological products are described in the GMP (Good Manufacturing Practice) standard, adopted in the world in 1999. GMP standard contains 4 classes with the maximum allowable particle concentration and the number of viable microorganisms in 1 m^3 (Table 1).

The maximum accumulation of aerosol particles in different areas of the room, the number and position of measuring points are determined according to ISO 14,644-2 with taking into account the risk analysis. According to the technical interpretation of Annex 1 (PIC/S PI 032-2), continuous particle monitoring is required for class A and indirectly for class B.

To use cleanrooms, it is necessary to have devices for monitoring and filtering air, as well as the establishment of criteria for the level of cleanliness of the room. Method of establishing such a criterion to analyze the statistical data and taking into account their own subjective feelings, a priori knowledge and personal experience is discussed.

3 The Purpose of the Work

The purpose of the work is to analyze the requirements for cleanrooms, describe the methods for designing cleanrooms and the requirements for materials that are used as protective coatings and in finishing. To show microparticles formed in cleanrooms, to present a modern method of optical control of their concentration. To suggest a mechanism for assessing the level of room cleanliness by fuzzy logic methods.

4 Results and Discussions

Information model of requirements for structural materials used in cleanrooms is shown on pyramid (Fig. 1) The correct choice of construction materials and protective coatings for the cleanroom is determined by building standards and guidelines. The specific physical, chemical and structural properties of cleanroom envelope elements must meet four main criteria:

- Functionality;
- Durability;
- Ability to be cleaned;
- Repairability.

Additional protective measures are needed to clean the air and reduce the concentration of contaminants by the special equipment [11]. To ensure that there is no accumulation of contaminants and to ensure that surfaces are easy to clean, the presence of horizontal surfaces and corners in the cleanroom should be minimized. Thus, when designing clean rooms, the use of:

- flush-glazed windows;
- door blocks mounted flush with the walls;
- flush-mounted luminaire and switch housings;
- hidden channels for engineering communications.

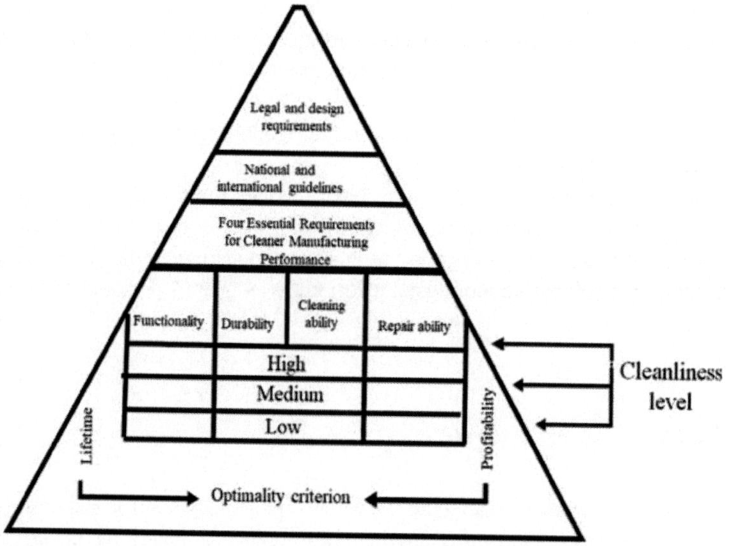

Fig. 1 Hierarchy of requirements for structural materials used in cleanrooms

Materials that do not generate particles and are therefore widely used in cleanroom construction include:

- stainless steel;
- powder-coated metal sheet (or anodized aluminum sheet);
- concrete with a sealed surface;
- hot-welded plastic sheets;
- non-shrinking coatings made of polymeric materials;
- ceramic materials;
- glass.

The filter is the most important component of the climate system for "clean" rooms, as it finally establishes the required degree of cleanliness of the room. A popular cleaning system is the installation of three groups of filter elements after the injection fan:

- coarse filters for mechanical impurities;
- fine filtering elements and antibacterial filters;
- HEPA micro-filters with absolute purification of the supply air.

One of the main problems is the correct organization of the air mixture flows in the design ventilation for cleanrooms. The air supply method and the air distribution device are chosen with according to the purposes of the room using. There are two main types of air movement in a room: laminar and turbulent flows. It becomes possible to ensure high values of air exchange at a low speed - no more than 0.3 m/s across the entire zone when air is supplied with laminar flows.

The recommended air exchange for rooms with an average level of cleanliness is between 30 and 60/h, while for a moderate level, air exchange can be reduced to 20/h. The designer chooses the air exchange rate based on his experience and understanding of the dust emission in the production process. Recently, there has been a tendency to accept lower air exchange values; leading design and construction firms and prudent customers have a successful experience with these parameters.

High-performance efficient technologies of air filtration have been developed [12–14], which become receivers of traditional filters of classical design. Most air analyzers are optical particle counters. The registration of particles in them occurs due to the overlap of the light flux from the diode laser when passing the sample through the measuring cell. Simultaneously, the number of particles in several channels is counted; each of them corresponds to a certain particle size. The microparticles obtained from a machine with the HEPA filter in a digital microscope with a magnification of 1600 times in a view of thread skein are shown in Fig. 2a. High resolution optical microscope performs the duster and microorganism structure with the size 1–20 μm (Fig. 2b).

The method of describing laser field pattern [15] and model of service providing on the use of optical laboratory for visual-optical control [16] are important stages for development of the optical air control system. The movement of microparticles and counting their number in the area of observation for processing video files by a

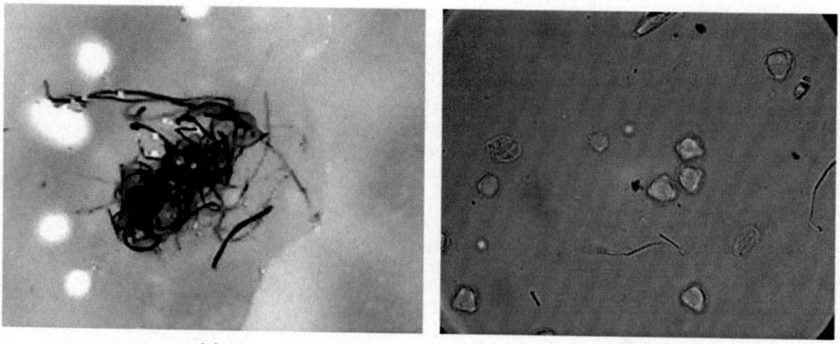

(a) (b)

Fig. 2 View of the microparticles, obtained from the machine with high-efficiency air filter HEPA in a digital microscope with an increase in 1600 (a), in the optical microscope with objects size is about 1–20 μ (b)

software application is described in [17]. The method of optical image compressing [18] is increased the speed of image data processing and database management system.

According to Fig. 1 practical problems of cleanroom development is accompanied by the solution of the optimization problem by three criteria: service life—efficiency—level of cleanliness. Decision support methods are developed for the problems in applied systems [19] and its compliance with the requirements that are determined by its further use. Development of tools for assessing the decision on the level of cleanliness is suggested by fuzzy logic. It allows correctly transforming numerical values into linguistic variables: low, medium, high.

To determine the level of purity we will use fuzzy logic, then the concept of "Cleanliness level" will become a linguistic variable. The basic term-set consists of three fuzzy variables: "High", "Medium", "Low" and the area of reasoning has the form $X = [N_{min}; N_{max}]$, where N_{min} is the threshold value for this class, and N_{max} is the maximum allowable number of microparticles specified in Table 1. Membership function $\mu(x)$ for each linguistic term from the base term set has the form of trapezoids, is shown in Fig. 3.

Definition of the attribute level value of the cleanliness is represented by a real number in the range [0; 1] so that the sum of the attribute estimates is equal to 1. The assessment of the indicator cleanliness level = {High, Medium, Low} = {0.8, 0.2, 0.0} is interpreted as a score of High with 80% confidence and Medium with 20% confidence (Fig. 3).

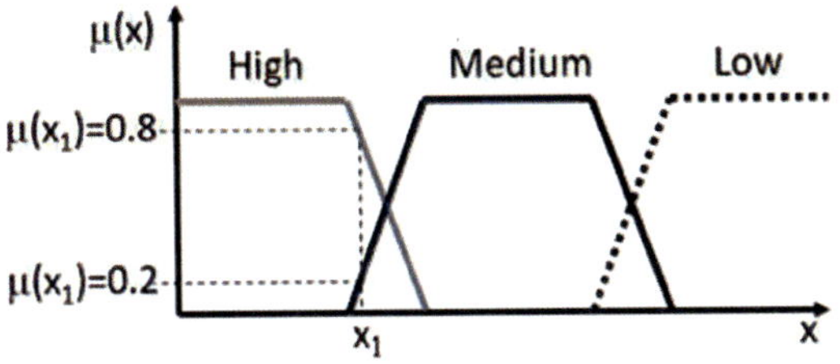

Fig. 3 Membership function μ(x) of a fuzzy set. The real number of undesirable microparticles x_1 in the air corresponds to the value of the linguistic variable High with the Confidence 0.8 and Medium with the Confidence 0.2

5 Conclusions

An analysis of the requirements for cleanrooms and their classification is presented.

Methods of cleanrooms construction and materials requirements that are used in the finishing of cleanrooms as protective coatings are shown.

The view of microparticles formed in cleanrooms is presented. The modern method of optical control of the concentration of microparticles of dust, impurities and microorganisms in the air is discussed.

The mechanism for assessing the level of room cleanliness with some confidence by fuzzy logic is proposed.

Acknowledgements The authors thank to AG&K company from Rubizhne, Ukraine for the providing of dust samples and optical laboratory headed by prof. Jan Masajada of Wroclaw University for the opportunity to study the microobjects by high-resolution microscope.

References

1. Lewis T et al (2016) Clean room technology and IVF. In: Esteves S et al (eds) Clean Room Technology in ART Clinics. CRC Press, United States, pp 361–370. https://doi.org/10.1201/9781315372464-26
2. ISOTechnical committees (2015) Part 1. Classification of air clianliness. In: ISOTechnical committees (ed) ISO 14644-1-2000 Clean rooms and associated controlledenvironments. pp 1–37
3. ISO Technical committees (2015) Part 2. Requirements for control and monitoring to confirm continued compliance. In: ISO 14644-2-2001 Clean rooms and associated controlled environments
4. ISO Technical committees (2015) Part 3. Test methods. In: ISO 14644-3-2007 Clean rooms and associated controlled environments
5. ISO Technical committees (2015) Part 4. Design, construction and commissioning. In: ISO 14644-4-2005 Clean rooms and associated controlled environments
6. ISO Technical committees (2015) Part 5. Operation. In: ISO 14644-5-2005 Clean rooms and associated controlled environments
7. Whyte W (1999) Cleanroom design. Wiley, New York

8. Cherniha R, Pliukhin O (2013) New conditional symmetries and exact solutions of reaction-diffusion-convection equations with exponential nonlinearities. J Math Anal Appl 403(1):23–37
9. Cherniha R, Serov M (2003) Nonlinear systems of the burgers-type equations: lie and Q-conditional symmetries, ansätze and solutions. J Math Anal Appl 282(1):305–328
10. Levashenko V, Liashenko O, Kuchuk H (2020) Building decision support systems based on fuzzy data. Adv Inf Syst 4(4):48–56
11. Tatarchenko GO, Kravchenko IV, Pisarenko NV, Porkuyan SL (2019) Study of atmospheric air pollution by exhaust gases in an urban environment. Visnik Volodymyr Dahl East Ukrainian Natl Univ 256(8):99–104
12. Tsygankov AV, Lysev VI, Dolgovskaia OV, Hildayati A (2019) The selection of a filter for cleanrooms. Vestnik Mezhdunarodnoi akademii kholoda 1:11–16
13. Kim SJ, Chase G, Jana SC (2016) The role of mesopores in achieving high efficiency airborne nanoparticle filtration using aerogel monoliths. Sep Purif Technol 166:48–54
14. Alexandrescu L, Syverud K, Nicosia A, Santachiara G, Fabrizi A, Belosi F (2015) Airborne nanoparticles filtration by means of cellulose nanofibril based materials. J Biomater Nanobiotechnol 7:29–36
15. Khoroshun GM (2020) Diffraction phenomenon description method based on by the topological objects set and the algorithm for distinguishing the minimum from zero intensity. Adv Inf Syst 4(3):17–21
16. Khoroshun GM (2019) Model of service providing on the use of optical laboratory in conditions of individual customer needs. Visnik Volodymyr Dahl East Ukrainian Natl Univ 8(256):118–122
17. Khoroshun G et al (2020) The development of an application for microparticle counting using a neural network. In: Proceedings of the 4th international conference on computational linguistics and intelligent systems (COLINS) vol 1(Ukraine), pp 1186–1195. http://ceur-ws.org/Vol-2604/paper76.pdf
18. Khoroshun GM (2020) The method of diffraction and interference images compression. Open Inf Comput Integr Technol 88:134–140
19. Khoroshun GM (2020) Decision support methods in optical research. Adv Inf Syst 4(4):1–5

The Main Trends in the Development of Production Specialization Rural Settlements in the New Socio-economic Conditions of Ukraine

Tetiana Kuzmenko, Andrii Dmytrenko, Javid Teymurov, and Vasyl Liakh

Abstract The article considers the main trends in the development of rural settlements of production specialization, located in the suburban areas of large and most important cities of Ukraine. Since the early 1990s, significant economic, social and demographic transformations have taken place in rural settlements of Ukraine, which affect the functional and planning structure of rural settlements, covering all its elements: housing, services, social infrastructure, and mainly—production areas and facilities. The preconditions and formation factors of new production facilities and production rural settlements are considered. The reform of the production environment and the transformation of production facilities, as well as the formation and development of new types of production and new types of settlements have been analyzed. New types of rural settlements of industrial purpose are considered. For the first time the factors determining the reconstruction of rural settlements of industrial direction are determined; the typology of production settlements in large cities suburban areas has been improved; proposals for improving industrial settlements depending on the quantitative characteristics of the defining parameters of the territory are developed; recommendations for the location of production of cultivation in the relevant area of the rural settlement depending on its specialization are improved.

Keywords Socio-economic structure · Production environment · Transformation of production facilities · City-centre · Typology of production settlements · New types of production settlements and production facilities

T. Kuzmenko · A. Dmytrenko (✉) · V. Liakh
National University "Yuri Kondratyuk Poltava Polytechnic", Poltava, Ukraine
e-mail: metr5555@ukr.net

J. Teymurov
Azerbaijan University of Architecture and Construction, Ayna Sultanova 11, Baku, Azerbaijan

V. Onyshchenko et al. (eds.), *Proceedings of the 3rd International Conference on Building Innovations*, Lecture Notes in Civil Engineering 181,
https://doi.org/10.1007/978-3-030-85043-2_48

1 Introduction

Since the early 1990s, significant economic, social and demographic transformations have taken place in rural settlements of Ukraine, which affect the functional and planning structure of rural settlements, covering all its elements: housing, services, social infrastructure, and mainly—production areas and facilities. Privatization of agro-industrial complex enterprises, market relations, a significant weakening of state regulation have led to ambiguous changes in the formation of rural settlements, which are most characteristic within the suburban areas.

It is in suburban areas that such changes began much earlier, due to the influence of the city-centre. Structural and morphological changes in suburban rural settlements have manifested themselves in functional changes within production areas and facilities, partial changes in functional zoning and the formation of new types of settlements or the transformation of existing villages into another functional type. It so happened that significant agricultural production began to lose its urban significance. In suburban rural settlements, it is a kind of industrial facility of the city in the countryside (business facilities, high-tech associations, service, tourism, recreation) and the development of new housing on their basis.

The first studies on the rural settlements typology were conducted by scientists in the field of economic and social geography Kovalyov [1] and Zaslavskaya [2], they identified the suburban type of settlement as special, which features belonging to the zone of influence of large and most significant cities. Moiseenko in the study [3] determined the difference between the suburban settlement of Ukraine from Western counterparts, which is a combination of agricultural, medical, recreational and tourist and other functions. In such settlements only 10% of the able-bodied population is employed in agriculture. Of the production types of rural settlements, the main ones were considered to be agrarian and agro-industrial, on the principles of which domestic research was aimed: Vinshu [4], Krysanov [5], Musatov [6], Illiash, Tkach [7] and others. New types of settlements at the present stage have been studied by Antipova [8], Bański [9], Gorbenkova and Shcherbina [10], Kolodin [11], Novikov [12], Pustovetov [13], Zębik [14] and others [15, 16]. However, it should be noted that a significant part of these works does not sufficiently cover industrial settlements and the transformation processes that take place in them. Studies conducted earlier [17–19] have identified production settlements as one of the modern types. Therefore, this study is extremely relevant.

Further study of the production specialized rural settlements typology, preconditions and factors that lead to the formation and development of such settlements is relevant for solving the problems of rural settlement, environmental protection, and even the existence of the rural settlement network, culture and identity of the Ukrainian nation.

2 Main Part

2.1 *Purpose of the Article*

Identify the main directions of development of production specialization rural settlements in large cities suburban areas, improve their typological classification and develop proposals to improve their functional and planning structure.

2.2 *Research Methodology*

The research was conducted using the analytical method in the generalization and analysis of theoretical and practical works, as well as statistical research on the functional-planning organization of rural settlements; field survey and comparative analysis, the method of complex research of literary sources and scientific researches is applied. These methods are based on the main provisions of an integrated approach and include general theoretical and empirical research methods.

2.3 *Results*

During the study, statistical, cartographic analyzes and field surveys of 148 villages of Poltava district of Poltava region (i.e. the vast majority of villages in the large city suburban area) have been conducted. Comparative analysis of trends in the development of existing settlements, as well as the experience of formation and transformation of functional specialization in the new socio-economic conditions proves the need for a typological approach to their development strategy. The nature of the rural settlements functional specialization is determined by their population, territorial location, availability of transport infrastructure and natural and recreational resources. At the same time, the following factors are favourable for the formation of industrial settlements [16]:

- free labour resources;
- availability of vacant land plots for new construction;
- the presence of production facilities that have ceased to function;
- availability of free land for pastures within the village;
- population of the settlement.

Favourable factors for the formation of a particular settlement are identified in the study as determinants, they are in the group of accepted statistical indicators. Depending on their quantitative value, it is possible to provide recommendations for the formation of the type of industrial settlement. For example, a settlement with a population of up to 100 people, in the presence of other favourable factors, may be recommended as a farm, with a population of 1,000 people or more may be agro-industrial or industrial. From 148 rural settlements in the suburban zone of Poltava, 52 (35%) are defined as production. Functional specialization of the production type covers all groups of villages by population, respectively, they are defined for the largest number of settlements, and almost a third of them (27%) are settlements with a population of 300–1000 people and 25%—settlements with a population of more than 1000 people. At the same time, most of these settlements, 65%, are located within 20 km of the city.

Previous studies [16] have developed a method for determining the functional specialization of rural settlements as a strategy for their sustainable development. Accordingly, a modern typology of suburban rural settlements is proposed, among the main types of settlements are identified production type [17, 18]. The production type of rural settlements is typical not only for the suburban area, but also for other rural areas. And the principles and methods of their formation and recommendations for improving the functional-planning structure can also be applied to production settlements located within other territories.

Production types of settlements depending on external and internal factors and other factors have their subtypes: 1—farmers', 2—agrarian, 3—industrial, 4—agrarian-industrial. Table 1 shows proposals for improving the functional structure of production settlements depending on the quantitative characteristics of the territory's defining parameters.

Table 1 Proposals to improve the functional and planning structure of production settlements

Type	Population, people	Free labour resources, people	Undistributed lands, ha	Number of production facilities	Land for new construction, ha	Proposals
1	≤100	4–10	2–5	1–2	2–5	Farmers' settlement in non-functioning production areas
2	100–500	15–80	5–10	1–2	≥1	Re-profiling of an agricultural enterprise into 2–3 farms
3	200–500	15–250	15–30	4–6	10–15	Agro-industrial production, housing and production formations
4	≥1000	50–250	40–70	≥10	≥50	New industrial facilities instead of non-functioning production ones

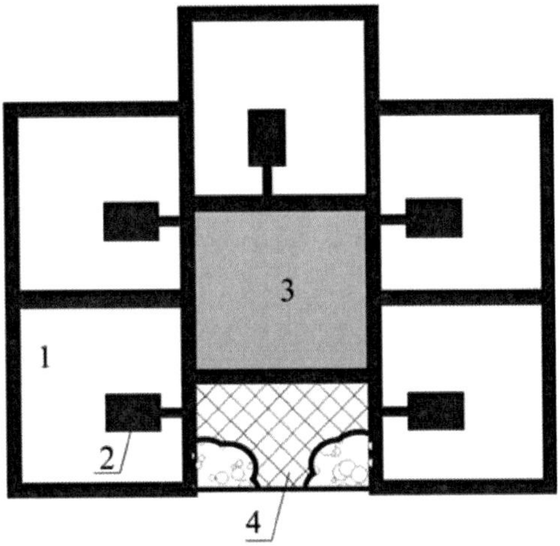

Fig. 1 Scheme of production and residential formation (packing shops, tailoring and repair of clothes, production of footwear, canning of vegetables, etc.); 1—residential land plot; 2—individual house; 3—production unit; 4—entrance zone with elements of public service and greening

Traditional settlements of production type, adopted in domestic practice: workers', agrarian, and agro-industrial. The production zones of these settlements have been extensively studied by domestic and foreign authors. Therefore, the formation of the production function in rural settlements is recommended on the basis of reconstruction, the directions of which are the following factors:

- unbundling of existing production units, zones and enterprises;
- creation in settlements of production and residential formations (modules) on the basis of medium, small and micro-productions (Fig. 1);
- the emergence of new city-forming factors (recreation, tourism, crafts, sports, trade, high-tech science-intensive industries, etc.) [3];
- an important element of the production environment of the village—farms.

The state construction norms of Ukraine do not recommend placing small agricultural enterprises and livestock farms in the suburban area. Therefore, we can recommend the following areas of production: grain, vegetables, greenhouses, fruit and berry, viticulture, milk production and processing, greenhouses and champignon farms. Depending on the specialization of the farm, it is recommended to place them in the appropriate area [3, 12]:

- livestock and poultry farmsteads—in the structure of the settlement, on its border, near the production area;
- creation of new city-forming enterprises of cultivation direction (greenhouse, champignon, etc.) on the territory of the former production zones, which have ceased to function, especially those located within the residential areas (Fig. 2);
- crop direction—outside the rural settlement;
- service—in the structure of the settlement, in the production area, near the highway;

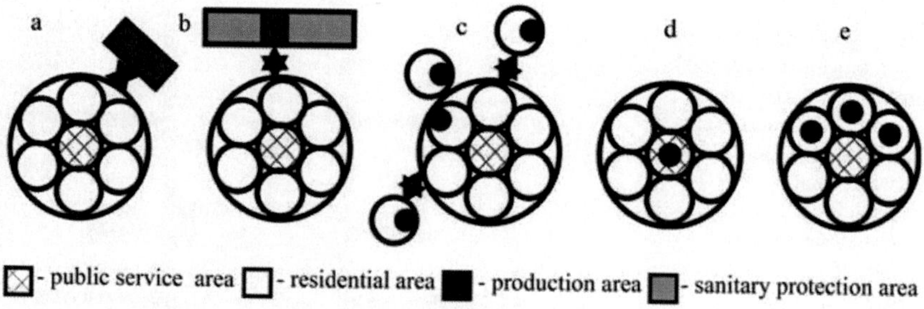

Fig. 2 Placement of cultivation production facility: **a**—outside the village; **b**—in the production area; **c**—in farms; **d**—in the public centre; **e**—in residential quarters

- environmental protection—near forests in need of preservation or rehabilitation or on the border of the nature reserve fund.

2.4 *Scientific Novelty*

The scientific novelty of the obtained results is that for the first time the factors determining the reconstruction of rural settlements of industrial direction are determined; the typology of production settlements in large cities suburban areas has been improved; proposals for improving industrial settlements depending on the quantitative characteristics of the defining parameters of the territory are developed; recommendations for the location of production of cultivation in the relevant area of the rural settlement depending on its specialization are improved.

2.5 *Practical Importance*

The results of the study can be used by design organizations in developing project proposals for the reconstruction of rural settlements in areas of large cities influence, as well as local authorities at various levels and local communities to predict the development strategy of suburban rural settlements and tasks for design and reconstruction of villages.

3 Conclusions

Based on the results of the study, the following conclusions can be drawn.

Current trends in changes in the functional and planning organization of villages are identified: functional restructuring of territories and production zones of the village, redistribution of functions in the balance of the settlement and between neighbouring settlements and the city-centre, which often leads to differentiation of villages by functional specialization and typology. The main types of production rural settlements are determined: farmers', agricultural, agro-industrial, industrial.

Quantitative characteristics for the formation of a certain rural settlement type as a development strategy are determined. Alternative development strategies based on a typological approach can be used to preserve rural settlements or redevelop them. The choice of a promising settlement type for its gradual re-profiling is based on specific conditions: settlement resources, opportunities and priorities of territorial communities, historical longevity, environmental conditions, and location relative to the city.

Therefore, the functional specialization of settlements is an important prerequisite for sustainable development and self-sufficiency of suburban rural settlements and the suburban area as a whole.

References

1. Kovalyov S (1963) Selskoye rasselenie (geograficheskie issledovaniya). Moskovskiy universitet, Moscow
2. Zaslavskaya T (ed) (1970) Migratsiya selskogo naseleniya. Mysl, Moscow
3. Moiseenko Z (1984) Voprosy soverhenstvovaniya metodologii nauchnykh issledovaniy po pereustroystvu sel. Planirovka zastroyka i blagoustroystvo sel Ukrainskoy SSR 7:3–5
4. Vinshu I (1986) Arkhitekturno-planirovochnaya organizatsiya selskikh naselennykh punktov. Stroyizdat, Moscow
5. Krisanov D (1988) Selskoe rasselenie; sotsialno-demograficheskiy aspect. Nauka, Kiev
6. Musatov V (ed) (1990) Arkhitekturnoe proektirovanie agroindustrialnykh kompleksov. Agropromizdat, Moscow
7. Illiash I, Tkach D (2013) Sotsialno-ekonomichnyi rozvytok poselen ahropromyslovoho rehionu v umovakh transformatsii suspilstva. Aston, Ternopil
8. Antipova E (2013) Rural settlement pattern in Belarus. Bull Geogr Socio-Econ Ser 19(19):7–18
9. Bański J (2005) Suburban and peripheral rural areas in Poland: the balance of development in the transformation period. Geografický časopis 57(2):117–130
10. Gorbenkova E, Shcherbina E (2020) Historical-genetic features in rural settlement system: a case study from Mogilev District (Mogilev Oblast, Belarus). Land 9(5):165–182
11. Kolodin K (2004) Formoobrazovanie objektov zagorodnoy sredy. ASS, Moscow
12. Novikov V (2006) Arkhitekturnaya organizatsiya selskoy sredy. ASS, Moscow
13. Pustovetov G (2003) Arkhitektura selskikh poseleniy v novykh sotsialno-ekonomicheskikh usloviyakh (zhilye I proizvodstvennye zdaniya I sooruzheniya) (Sc.D. dissertation) Moscow
14. Zębik G (2011) Typology of suburban communities in Poland. Bull Geogr Socio-Econ Ser 16:173–188
15. Technical handbook on the monitoring and evaluation framework of the common agricultural policy 2014 – 2020. https://ec.europa.eu/info/sites/info/files/food-farming-fisheries/key_policies/documents/technical-handbook-monitoring-evaluation-framework_june17_en.pdf. Accessed 23 Jan 2021
16. Li G, Jiang C, Du J, Jia Y, Bai J (2020) Spatial differentiation characteristics of internal ecological land structure in rural settlements and its response to natural and socio-economic conditions in the Central Plains, China. Sci Total Environ 709:135932. https://pubmed.ncbi.nlm.nih.gov/31884288/. Accessed 25 Jan 2021

17. Kuzmenko T (2017) Pryntsypy funktsionalno-planuvalnoii orhanizatsii prymiskykh silskykh poselen (na prykladi Pivnichno-Skhidnoho rehionu Ukrainy). PhD dissertation, PoltNTU, Poltava
18. Kuzmenko T, Liakh V, Dmytrenko A (2018) Improvement of functional planning organization of rural settlements in suburban zones. Int J Eng Technol (UAE) 7(3):626–630
19. Kuzmenko T (2019) Prymiski silski poselennia: suchasna typolohiia. In: Mykhailyshyn O, Lushnikova N, Potapchuk I, Serhiiuk I (eds) ARKHITEKTURNA OSVITA I NAUKA V UKRAINI TA SVITI: DOSVID PERSPEKTYVY I ROZVYTKU 2019, NUWEE, pp 149–155. Format-A, Rivne

International Cooperation in the Protection of Historical and Architectural Heritage of Azerbaijan

Gulchohra Mammadova

Abstract Active cooperation with international organizations, primarily with UNESCO, contributed to the introduction of a new expanded concept of heritage and the creation of a new heritage management system. Currently, there are 3 monuments of architecture, 9 monuments of intangible culture included in UNESCO lists. The inclusion of Icheri Sheher in the World Heritage List in danger (2003–2009) contributed to the adoption of a whole range of economic, social, scientific and technical measures, the creation of a modern management system for this complex of monuments. Restoration work involving of foreign specialists, the use of new technologies, new approaches to the adaptation of monuments have become good experience for the development of a system of monuments' protection. This and other examples of beneficial international cooperation have contributed to the formation of a new attitude towards the conservation and sustainable development of heritage. At the same time, there are problems that cannot be solved even with the help of international organizations. This is the salvation of monuments in the zone of armed conflict. Monuments become victims not only of military operations, but also of deliberate, systematic destruction and falsification. The article considers various aspects of the history and preservation of the architectural heritage of Azerbaijan.

Keywords Architectural heritage · Monuments of Azerbaijan in UNESCO list · Preservation of tangible and intangible heritage · International cooperation

1 Introduction

Republic of Azerbaijan became a member of many international organizations after the restoration of the independence. UNESCO has a special place among them. Azerbaijan had ratified in the 1993 the Convention on the Protection of the World Cultural and Natural Heritage. In subsequent years, the country had joined other

G. Mammadova (✉)
Azerbaijan University of Architecture and Construction, 11 Ayna Sultanova, Baku, Azerbaijan
e-mail: rector@azmiu.edu.az

V. Onyshchenko et al. (eds.), *Proceedings of the 3rd International Conference on Building Innovations*, Lecture Notes in Civil Engineering 181,
https://doi.org/10.1007/978-3-030-85043-2_49

major UNESCO Cultural Conventions, such as "Fighting against the illicit trafficking of cultural property", "Protection of Cultural Property in the Event of Armed Conflict" and others. Cooperation with UNESCO has made it possible to enrich the experience of protection, restoration and use of architectural monuments, to study and put into practice modern requirements and approaches to heritage management. UNESCO support was also expressed as technical assistance and the provision of specialists, whose participation in specific projects was of great practical importance for the development of the restoration.

2 Main Part

2.1 Historical Background

Preservation of cultural and architectural heritage became a live political issue in Azerbaijani society since the late 19th century marked by acceleration of expansion scale in the development of oilfields together with the development of industry. The beginning of the 20th century saw first societies to study and preserve the cultural heritage. It was gaining of independence and establishment of Azerbaijani Democratic Republic in 1918 that bolstered the rise of public spirit, growth of interest in the national history, culture and heritage. However, establishment of the Soviet power hindered the progress on this track. Even worse, numerous monuments of cult and civil architecture were destroyed; particularly, severe damage was inflicted to the Islamic architecture under the slogan of struggle with religious ideology. Many medieval mosques were destroyed; extant ones were used for economic needs and went to ruin gradually. If at the beginning of the 20th century there were about 2,000 mosques in Azerbaijan, in the 80s of the 20th century there were only 25 religious buildings used for religious purpose.

The attitude to the historical and architectural heritage of past eras starts to change closer to the middle of the 20th century. After the World War II, a horrifying picture of destroyed towns, razed monuments of architecture enhanced population's interests in the cultural heritage. Since 1950s Azerbaijani researchers got down to exploring architectural monuments. Thus, numerous monuments were discovered and taken under state's protection. This process, as well as archaeological explorations and restoration work strongly intensified in the 1960–1980s when the economy and culture of Azerbaijan flourished, self-consciousness and sense of national pride for historical past and cultural heritage intensified. Gaining of independence in 1991 gave a new impetus to the study of the history and cultural heritage of the Azerbaijani people.

Liberation from ideological bans and stereotypes made it possible to take a new look at the architectural heritage; assess its historical and artistic value regardless of class and religious prejudices. These issues were of particular actuality due to the war unleashed by Armenia with the aim to annex Nagorno- Karabakh, and the occupation of 20% of Azerbaijan's territory. It should be noted that thousands of specimens of cultural heritage and hundreds of architectural monuments have remained on the occupied lands. Numerous architectural monuments were destroyed, and whole settlements were leveled.

Despite the economic and social problems of early years of independence, the issues of heritage, its study and protection remained at the focus of attention of Azerbaijan's scientists and community. Integrated into the international community and cooperation with international organizations enabled the young democratic state to study the international experience and modern standards of managing heritage and forming a legal field in order to protect heritage effectively.

2.2 *Azerbaijan Heritage in UNESCO*

In December 1993, Azerbaijan was accepted into the membership of UNESCO; in 1994, there was established a national committee for UNESCO. In first years since regaining its independence, Azerbaijan joined many UNESCO Conventions, including the Convention on the Protection of the World Cultural and Natural Heritage (1972), Protection of Cultural Property in the Event of Armed Conflict" (1954), and the Second Protocol of the Convention for Protection of Immaterial Patrimony (2003), etc. The ratifications of these Conventions were followed by changes of the internal legal provisions that were put in accordance with international standards.

Adopted in 1998 the Law of the Azerbaijani Republic on Preservation of Monuments of History and Culture, other legislative acts were based on international standards of heritage control as set forth in UNESCO documents. It was active collaboration with international organizations contributed to the introduction of a new concept of a new advanced concept of heritage and the creation of a new heritage control system. The cooperation with the UNESCO continues in many respects.

One of the aims of UNESCO operation is to unite all nations for the reasons of saving and protecting monuments, which are of worldwide importance and develop legal bases for the protection of heritage subjects in peace and particularly war time. The study and preservation of world heritage is a universal humanitarian task. This is the heritage that should be protected for the next generations. Objects that are included in the list of World Heritage belong to not only to countries where they are located but also to the whole mankind.

It has to be kept in mind that addition of heritage objects to the World Heritage List is evaluated as recognition of this country's contribution to the culture of mankind and is a mark of pride of the nation. At present, the List includes the most important monuments both of material and non-material culture of Azerbaijan. Note that three objects of material culture have been included in the World Heritage List: Icheri Sheher with Maiden' Tower and Palace of Shirvanshahs (2000); cultural landscape of Gobustan cave paintings (2007) and historical center of Sheki with Palace of Sheki Khans (2019) [2].

It should be noted that a preliminary World Heritage List includes as follows: Ateshgah—a temple of fire in Surakhany (1998), Nakhichivan mausoleums of the 12–13 centuries (1998), historical-architectural site in Ordubad (2001), other cultural and natural objects (in total 9). Representative list of the UNESCO Immaterial Patrimony includes 7 culture specimens of Azerbaijan:

- Azerbaijani mugham (2003).
- Azerbaijan's ashug art (2009).
- The spring Novruz Holiday.
- Azerbaijan's traditional carpet-weaving art (2010).
- Chovghan, the traditional Karabakh horse game of Azerbaijan (2013), etc.

The inclusion of an object into the World Heritage List means that the World Heritage Committee considered this object to be a great universal value. In turn, this imposes on a country additional management commitments aiming to provide high security standards and correspondence to the demands of processes in the system of world heritage. The prestige of the status of a world heritage object substantiates a state's great interest in this object, as countries that have joined the Convention try to use such objects as frontline platforms to improve the general process of cultural heritage management [1].

It ought to be noted that as a party to the World Heritage Protection Convention Azerbaijan fulfills its obligations on discovery, protection and popularization of the cultural heritage through the maximum use of its own resources and international scientific-technological cooperation. Typical in this respect is an experience of Icheri Sheher, a historical center of Baku that has miraculously been escaped from modern development onslaught. It is commonly known that architectural heritage of historical towns is faced with the most serious challenges. Suffice it to remind that the development of towns and rates of present-day building give rise to numerous issues, including modern standard architecture, destruction of historical context of architectural monuments, violation of silhouette and scope of historical towns [3].

It is worth pointing out that economic and financial factors prove to be more effective than curb efforts of guarding stations and public structures. Icheri Sheher' example is proves that government's responsible approach to the heritage protection and compliance with its commitments in line with the World Heritage Protection Convention makes it possible to take effective measures for resolution of managerial, legal and institutional issues and preservation of the historical past in impetuously developing urban construction.

Icheri Sheher is Baku City's heart, a concentration of invaluable monuments of history and architecture that has preserved its historic town-planning structure and composition. Located in an area of 22 ha, the historic town was formed in the medieval times; however, numerous written sources and archaeological excavations prove that it existed yet in the ancient period as a center of fire worshipping with a large number of fire temples. Note that the town's spatial composition was formed and the bigger part of its monuments was built in medieval centuries. Noteworthy is the fact that the 13–16th centuries—a period of prosperity of the Shirvanshahs State—were the most significant period for the town's development. The town's planning structure consisting of different-size and different-configuration mahallas (blocks) separated by narrow streets one from another has mainly been preserved; however, the residential area has endured great changes.

On the territory of Icheri Sheher, there has been preserved a large number of monuments, including the most ancient one—the Maiden's Tower—whose date of origin and functional designation are subject to different opinions. Some researchers tend to consider it to be an ancient temple built in the 8th century B.C. [3]; others believe that it was a fire temple going back to the 5–6th centuries; third opine that it was a defensive fortification. The oldest extant parts of fortress walls go back to the 12th century. Located on the highest part of the relief, ensemble of the Palace of Shirvanshahs came to fruition in the 12–15th centuries. It consists of 9 monuments—Palace of Shirvanshahs, divankhaneh, two mausoleums, Shah Mosque and Keyqubad mosque, a bathhouse, ovdan and oriental portal (Fig. 1). Outside the walls of the palace ensemble there are numerous mosques, bathhouses, caravanserais built in the 11–19th centuries. Apartment blocks of the town have tens of residential houses of the 18–20th centuries of high historical, architectural and art values [4].

Icheri Sheher was included in the List according to the 4th criterion as an outstanding specimen of a type of building, architectural ensemble and landscape illustrating an important stage in the history of mankind. A powerful earthquake that occurred in Baku in November 2000, new construction that threatens the heritage and shortcomings in Icheri Sheher management system caused the Committee's decision to include Icheri Sheher in the List of Endangered World Heritage in 2003. It must be

Fig. 1 Palace of Shirvanshahs

Fig. 2 a, b, c. Gobustan petroglifs

Fig. 3 a, b. Sheki fortress

Fig. 4 a, b. Palace of Sheki khans.

borne in mind that the government of Azerbaijan received DSOC recommendations aimed at safekeeping and preservation of the monument above, specific measures to attain this goal. These recommendations played a great role in modernizing the Old City control system. In addition, the Administration of the State Historical-Architectural Conservation was set up in 2005 under the Cabinet of Ministers. Under the UNESCO documents, the Administration was responsible for the construction of modern control system over the monument of the world heritage.

The monument's peculiarity and complexity is that it was a historic town living a modern life with inhabiting population, its facilities and its own infrastructure. Icheri Sheher's population and its problems were an important factor for the arrangement of the monument's management. It has to be noted that local and foreign specialists and UNESCO experts were involved in the work to define a concept of the town's protection, use and development. The Master-plan they developed made it possible to answer the question: what will the future of Icheri Sheher be? This was forerun by survey and diagnostic of the territory, monuments, common buildings and the infrastructure. All the monuments were stock taken, and opportunities of their future use were determined. The master plan envisioned the conduct of different types of restoration works within the Icheri Sheher territory. It had been planned to carry out restoration works on 31 monuments within 2010–2016, and this project was implemented. The restoration works on the preservation of Maiden's Tower had been carried out nonstop from 2009 to 2013 [3].

Under recommendations of the UNESCO World Heritage Committee, a buffer zone was set up in 2009 around the monument, and the Icheri Sheher Administration was authorized to control architecture conditions in the zone. Taking into consideration the positive changes in the hazard control and preservation of this unique monument, the World Heritage Committee adopted the decision to delete Icheri Sheher from the «List of the monuments in danger» in June 2009. It is worth citing that the system of city management, methods of conservation and restoration of monuments, life conditions of the Old City residents and its involvement in the control system are positively viewed by international experts. Even better, this experience was recommended to put into practice as applied to other historical towns. At present, Icheri Sheher is an example to be initiated in other historical towns of Azerbaijan.

The inclusion of state-owned historical-artistic reserve Gobustan in the World Heritage List in 2007 was also a great event for Azerbaijan. It should be noted that the Gobustan Cultural Landscape with Rocky Paintings was recognized an object with a high universal value and meeting the requirements of criterion III as a unique evidence of cultural tradition and civilization. This invaluable history and culture monument is located 60 km of Baku and represents a 4,000-ha mountainous area where there have been preserved 20 ancient human sites, caves, metallic constructions, about 40 mounds, and more than 100,000 items of tangible culture and some 6,000 rocky paintings [5]. They cover a huge historical period stretching from the Upper Paleolithic to the medieval centuries. Noteworthy is the fact that Gobustan's petroglyphs reflect the everyday life of primitive people, scenes of agriculture, fishing, hunting, shipping, military collisions, and ritual dances (Fig. 2a, b, c).

In terms of the number of rocky paintings, their themes and diversity of their creation technique, Gobustan is comparable to a large picture gallery. They are an invaluable material for study of life, spiritual culture and beliefs of ancient tribes inhabiting not only Azerbaijan but also the whole Caucasus [5]. In 1939, well-known archeologist I. Jafarzadeh was the first person who registered the rocky paintings and introduced Gobustan as a monument in science. In 1966, it was decided to establish a Gobustan historical-artistic reserve. Following the inclusion into the World Heritage List in 2013, Gobustan was granted a status of "Enforced Protection" monument by UNESCO. This has helped substantially increase attention to the monument, tighten protection measures, build a new building of the museum, and enhance the management system.

Sheki is the third monument included in the World Heritage List. Sheki is one of Azerbaijan's historical towns with a substantial well-preserved part of planning structure and ancient monuments. As a capital of the Sheki Khanate, it played an important role in the history of the country and development of culture of Azerbaijan. In 1762, the medieval Sheki was ruined by a disastrous landslide, so construction of a new town on a new site on the opposite bank of the river began later the same year.

The direction of new Sheki streets was determined by the Greater Caucasian Mountain spurs where the town was located, and by the direction of rivers flowing via it. This was one of the factors that substantiated the town's great merger with nature [7]. A type of settlement attributable to a feudal town was reflected in the town planning of Sheki as well (Fig. 3 a, b). It should be noted that the old names of blocks—mahalles—have been preserved up to now. Among these blocks, Yukhary-bash is distinguished for its beneficial location, historical planning and a large number of monuments of the 18–19th centuries. Sheki fortress, located in the town's upper northeastern part, represents a khan's citadel. The fortress walls and towers have been preserved in a semi-destroyed state, so the fortress was restored in 1958–1963. As for constructions inside the fortress, there has been preserved the building of a palace that is considered a summer residence of a khan. The Palace of Sheki Khans is a prominent work of Azerbaijan's medieval architecture. The unique palace was built in 1762 (Fig. 4 a, b).

The Palace of Sheki Khans represents a two-storey building, with its main façade turned southward. The palace's mostly beautiful spaces were halls with deep niches located in the center of the ground and second floors. The composition of the palace's main façade is characterized by the combination of three elements: ornamented wall, open-work shebeke, and spatial portals. On the whole, the façade's architecture represents a highly artistic expression of inner planning structure and constructive solution of the palace. As for the palace's interior, paintings that drew many researchers' attention are of great interest. The paintings cover all the area of walls, niches, stalactite transitions from walls to lamps, as well as lamps in the halls on the both floors and second floor rooms.

The inclusion of Sheki historical part in the UNESCO World Heritage List has increased interest to historical and architectural monuments of that town. Currently, the entire architectural heritage of the town is being studied, including the mosques of the 19th—early 20th centuries, that have traditional architectural and decorative features that distinguish them from similar structures in other regions. These are large prayer halls, well lit by large windows, with arched galleries-eyvans adjoining the façade. Mehrabs with their interesting decorative design play an important role in the interiors. The facades of mosques are faced with decorative brickwork, often combined with river stone masonry. According to their architectural design, Sheki mosques are more like civic buildings than religious ones.

Another prominent history and architecture monument is located 8 km far from of Sheki. This is a church in village of Kish—a mother of Albanian churches, as called by historian Moses Kalankatuklu [6]. This church is of huge historical importance as the first Christian temple of Caucasian Albania-Azerbaijan, with its base laid by Apostle Elysium in the 1 century. This interested multi-layer monument of Christian architecture enriches the historical-architectural image of the Sheki region as a valuable specimen of early medieval period. Also, the inclusion of the Sheki historical part in the World Heritage List should help develop and enhance the management system of these monuments.

The prestige that is attached due to a status of a world heritage object significantly increases interest in these monuments from the side of either the community or state bodies. At the same time, high international appraisal contributes to formation of a more careless attitude to heritage on the whole and recognition of its important role. The system of management of world heritage objects that is formed in accordance with the UNESCO requirements and under the participation of international experts is an important specimen for organization of protection of the country's monuments. World heritage objects become a frontline platform for the improvement of cultural heritage management system [1]. As for Azerbaijan, this platform is Icheri Sheher whose experience is supposed to be used for other historical towns, primarily, Sheki that was included in the list last year.

One of the most important paradigms of today's world—sustainable development—forms new approaches and sets new tasks in all vital life spheres of the world community. Heritage is primarily a cultural resource that should be protected and passed to the next generation. At the same time, heritage and its protection may contribute to a sustainable development of society, by helping it socially and economically. This aspect of monuments' protection and management is particularly important for such monuments as Icheri Sheher, a vivid, developing object of world heritage, as well as Sheki, with its preserved area containing mahalles with residential houses and economic stores.

The management of these monuments should be based on the consideration of interests of local population and carried out with its participation. Local population can take part in all components of the management system, i.e. formation of resources, planning and implementation, as well as observation. Taking into consideration that there are objects with different forms of property, an interactive approach to management becomes essential. The example of Sheki, a provincial town, a part of whose population resides in the reserve area requires particular attention to the aspect of sustainable development. A world heritage object can and should benefit local economy and town residents, contribute to development and increase of the living standards. This may include development of tourism, creation of appropriate infrastructure, motivation of local traditional handicraft, and creation of new jobs, including ones in the monument management system.

It is extremely important to follow the World Heritage Potential Boost Strategy adopted in 2011. It is important not only to develop the knowledge and capabilities of people managing this monument but also of a whole community inhabiting this area because no efficient protection of heritage may be ensured if there is no understanding and support from the part of the population. The protection of heritage may also contribute to the preservation of authenticity and peculiarity of Sheki population with its rich history and traditional culture.

There are many problems in the protection of World Heritage. The most irreparable damage to an architectural monument is still caused by wars and armed conflicts. During military conflicts and invasions of Iraq and Syria the most valuable monuments of ancient cultures were lost. The ruins of Babylon housed the military bases of the multinational forces in Iraq. The ancient city of Palmyra also suffered from the war and was included in the list of endangered world heritage sites. The war destroyed mosques in Timbuktu (Mali), giant Buddha statues in Afghanistan. Culture has moved to the front lines of war and conflict, being both a collateral and deliberate victim of attacks. Heritage is intentionally destroyed or vandalized. After all, the cultural heritage of a nation symbolizes its history and cultural value, and that is why aggressor strives to strike it.

It should be noted that the Azerbaijani people's historical and architectural heritage located in Nagorno-Karabakh and its bordering regions (20% of Azerbaijan territory) occupied by the Armenian troops has been in such a sad state for 30 years already. Unhidden, large-scale terror has been committed in the occupied areas. Half of the medieval town of Shusha, the historical and architectural reserve with its peculiar town-planning structure, that used to preserve its ancient blocks and numerous historical monuments before the war, has been destroyed. At the time, the reserve had 140 architectural monuments protected by state.

As for the issues of world heritage protection, lots of problems exist. It should be noted that wars and armed conflicts damage monuments of architecture most of all. It appears that culture has shifted to the frontline of war and conflicts being both an indirect and intended victim of an attack. The heritage is being destroyed purposefully or vandalized. Note that the cultural wealth of a nation symbolizes its cultural value that is targeted by an aggressor first of all (Fig. 5).

Prominent monuments of civic architecture—eleven-span and fifteen-span bridges at the Khudaferin bridge across River Arax which in ancient times were used as a crossing at a busy caravan route—have now been left in the area under occupation. The war irreparably damaged monuments located in the territories of Karabakh-surrounding regions; these objects include numerous mosques, bridges, bathhouses and memorials. In addition, cemeteries of Azerbaijanis have also been ruined.

While Islamic monuments are being terrorized, the other part of the Azerbaijani cultural heritage—the monuments of Christian architecture of Caucasian Albania—are being "Armenized" by force. Working on these monuments, Armenian researchers carry out the so-called restoration works with the aim to "Armenize"

a) b) c)

Fig. 5 a, b, c. Monuments' damage in war conflict

them. These are illegal works, for they are being done on an occupied territory, at alien monuments and without the participation of Azerbaijan's scholars. As far as we know, traces of their belonging to the Albanian culture are being erased. Under the cover of restoration works, falsification and destruction of the specific features of the Albanian Karabakh architecture were being implemented.

Thus, the principles and specific requirements of the Hague Conference have been violated roughly for 27 years. Regretfully, international organizations, which are designated to contribute to the protection of historical heritage, display no willingness to save the historical and architectural monuments in Karabakh. Despite numerous appeals by Azerbaijan's monument protection bodies as well as scientists, violation against monuments, destructions and falsifications continued during all period of occupation. This violence and destruction of the monuments of Azerbaijan was both destroyed part of the world heritage and a gross violation of the requirements of documents adopted by UNESCO. However, the international community did not take measures to stop it and punish the aggressor.

After the liberation of the occupied territories, a complete picture of destruction and cultural terror was revealed to us. All cities and settlements have turned into ruins, the absolute majority of Islamic religious monuments have been destroyed, irreparable damage has been caused to the entire cultural and architectural heritage of Karabakh and surrounding territories.

3 Conclusions

International cooperation has a positive impact on the spread of modern attitudes and the introduction of modern standards for the identification, study, protection and restoration of architectural heritage. On the example of Azerbaijan, one can see great positive shifts in the protection of monuments and the improvement of the architectural heritage management system. This is facilitated by cooperation with UNESCO, studying the experience of other organizations and attracting international experts.

At the same time, the many years of systematic destruction by Armenia of the cultural and architectural heritage of Azerbaijan in the occupied territories and other examples of vandalism and violence against monuments occurring in the world indicate that international organizations do not take or cannot take measures to prevent them stop. It is time to form a more effective mechanism for monitoring the state of cultural heritage, especially in the territories where wars and armed conflicts occur. We should develop a policy of sanctions in relation to countries that allow the destruction of the cultural heritage of other countries.

References

1. Feilden BM, Jokilehto J (1998) Management Guidelines for World Cultural Heritage Sites, ICCROM Second Edition. OGRARO, Rome, Italy, ISBN 92-9077-150-X
2. Azerbaijan-Unesco. https://whc.unesco.org/en/statesparties/az. Accessed 27 Jan 2021
3. Alekperov A (2019) Bakinskaya krepost Icheri sheher/Baku fortress Icheri sheher. Petropolis, Baku, Azerbaijan. ISBN 978-5-9676-0995-4
4. Akhundov D (1986) Arkhitektura drevnego i rannesrednevekovogo Azerbaijana/Architecture of the ancient and early medieval Azerbaijan. State Publishing House, Baku
5. (2014) Gobustan- a book of millennia. IRS Publishinq House, Dubay
6. Kalankatuyskiy M (1861) Istoriya alban/History of alban (translated by Patkanov)
7. Useynov M, Bretanitskiy L, Salamzadeh A (1963) Istoriya arxitekturi Azerbaidjana/History of Azerbaijan architecture. Iskusstvo, Moscow, USSR

Ways of Integration of the Landform Architecture Buildings with Landscape

Natalia Novoselchuk, Liudmyla Shevchenko, and Mohammad Arif Kamal

Abstract Presently, in the context of ecological crisis, the modern architecture wends the way of the sustained development conception. There are rethinking of approaches and planning methods, including traditional attitude of architecture toward a landscape, influence of artificial, anthropogenic environment on nature and man. It resulted in becoming of landform architecture as a particular trend at the end of XX century. This article is aimed at exposure and systematization of ways of integration of buildings with landscape on the basis of the landform approach. The evidential analysis, empiric and system analyses were involved to achieve this objective. A scientific novelty consists in the exposure of ways, systematization and generalization of possible types of integration of building with landscape. The research resulted in emphasizing of three types of such integration: above-ground, deepened in relief and underground buildings. Building synergy with landscape is considered on the example of different types of relief: plain, hill, mountain, cavity. Based on analysis the composition ways of a building form synergy with natural surroundings are distinguished. Practical value of this research consists in possibility to use the distinguished ways while planning and building the landform architecture objects.

Keywords Landform architecture · Building · Landscape · Terrain · Integration · Landform approach

1 Introduction

Relation of architecture, nature and human being always was indissoluble. The interactive effect of facilities with terrain begins upon since before Christ at creation of caves, mud huts, burial mounds, at the turn of XX–XXI centuries the interrelation of nature and architecture appears.

N. Novoselchuk (✉) · L. Shevchenko
National University "Yuri Kondratyuk Poltava Polytechnic", Poltava, Ukraine

M. A. Kamal
Aligarh Muslim University, Aligarh 202002, India

V. Onyshchenko et al. (eds.), *Proceedings of the 3rd International Conference on Building Innovations*, Lecture Notes in Civil Engineering 181,
https://doi.org/10.1007/978-3-030-85043-2_50

It becomes as never actual—basing on the ecological theory the architecture acquires the further development. New tendencies such as organic, "green", nature integrated, bioclimatic, landform architecture and other appear. Natural materials and components are used as "construction materials". They have a number of positive qualities: from the improvement of the sanitary-hygienic condition of municipal environment to the improvement of physical and psychical health of human being.

Subjects such as stability and maintenance of environment by creation of more ecological structures that prevent negative influence are actual in architecture. New methods of planning, NT, requirement for improvement of the environmental attributes, priority of the organic approaches, ecological materials and constructions caused rethinking of traditional attitude of architecture to the terrain and assisted becoming of the landform architecture as a separate direction at the end of XX century. In modern architecture, conception of «building according to the form of relief» is far more than a formal strategy. Today integration of a building with landscape is the modern and perspective approach that changes the traditional architectural form and creates new character of modern architecture.

Presently the architectural objects based on the landform approach are built in the whole world. Different after their form they are based on a general feature—blurring of limits between the natural and artificial. Conception of landform architecture offers the challenge for synergy between nature, individuals and artificial environment. Such conception allows to architects to design building and environment, that evolves with the course of time and complies with natural, climatic changes and needs of users. Landform architecture, as a result of integration of artificial object with natural terrene, conflict-free interaction becomes a part of environment. Landform architecture adopts the laws of morphogenesis of earth and comes forward as the articulated landscape [1].

Works of architectural theorists and practicians are dedicated to the study of land-form architecture. A number of scientific publications, articles, abstracts of thesis have been analyzed during examination of this subject. Examination of the field experience is based on analyses of the implemented projects of the landform architecture and design documents from every corner of the world.

Research of land-form architecture is represented in work by S. Allen and M. McQuade "Landform Building: Architecture's New Terrain" [3], genesis of new formal strategies is determined, the existent projects are shown with illustrations. Shown here are works of modern architects T. Dean, J. Ishigami, T. Ito, T. Kimura, A.P. Kis, M. Maitzan, R. Nishizawa, C. Taylor, D. Perrault, P. Rahm and other.

First empiric and theoretical researches of landform architecture are presented in works of the practician scientists: Sten Allen [3], Kennet Frempton [4], Diana Balmori [5], Michael Manfred [6]. This conception was examined afterwards as possibility to create ecological spaces and new green belts in the volume of building.

Afterwards this trend was considered as possibility to create the ecological spaces and new green zones in cubic content of the building. Forming of landform architecture as a variety of ecological trend that enables the theory of never-ceasing settlement gardening is considered in works by Karpo [7], Toren [8], Tabb [9], Genks [2], Corner [10].

This issue is examined in works of Zaslkavskaya [11], Shevchenko [12], Denisenko [13], Smirnova [14].

The empiric base to study this theme is presented by works of architects that use the landform approach and architectural planning: M. Sorkin, K. Yeang, E. Ambasz, Sh. Endo, Fr.R.D. Hundertwasser, K. Kum, P. Eisenman and other. Also to study the practical experience the projects of architectural groups—MVRDV (Netherlands), Deca Architecture (Greece), Bercy Chen Studio (THE USA), Chyutin Architects (Israel), Chartier—Dalix Architects (France), GLUCK+ (THE USA), BjarkeIngels Group (BIG) (THE USA), ChartierDalix (France), Signum Architecture (THE USA), Reardon Smith Architects (Great Britain), Grant Associates (Great Britain), Aedas Architects Ltd (China), Studio Odile Decq (France), Jean—Philippe Pargade (France), OFF Architecture (France), Jade + QA (Great Britain), SAALS Architect (Latvia) et al. have been considered.

1.1. *The study is aimed at* exposure and systematization of the ways of building integration with landscape basing on the landform approach.
1.2. *The methodology of the study*. Methods of scientific research used in-process are as follows: evidential analysis that allows to study accessible scientific, literary sources; empiric method for study the practical experience of the investigated objects and exposure of integration ways of objects with relief; the system method by means of which the complex investigation of this issue is possible.
1.3. *Scientific novelty and practical importance*. A scientific novelty consists in an exposure and systematization of integration ways of building with landscapes. As a result of research three types of such cooperation are distinguished: surface, embedded and underground buildings. Building landscape synergy is considered on the example of different types of relief: plain, slope, mountain, cavity. On the basis of analysis the varieties of such integration are distinguished. The obtained results can be of practical use which consists in possibility to use the established ways when planning the landform architecture objects.

2 Results

The beginning of 1990 is considered as initial point of development of the landform architecture, when a nonlinear paradigm and theory of complication were evolved. Ch. Janks, theorist and landscape designer in 2011 in his book "The Universe in the Landscape" put into practice the term "landforms" [2]. His understanding is based on architecture and landscape design, where landform is created in accordance with natural processes and includes the elements of modern sciences and technological development.

The architecture landform of is a trend in architecture, that realizes the organic approach to morphogeny and the spatial organization on the basis of integration of an object with earth [2].

Landform architecture is examined as a new transformation approach of integration of natural surroundings with artificial environment. Practician and theorist of the landscape design Jenkson [2], in his research indicated that landform architecture united the architectural form with natural topography in the single integral system. Thus soil is used as finishing and heat-insulating material.

The landform architecture, creating the modern ways of form making, combines architecture, landscape architecture, landscape design, town-planning, building and ecology. Prerequisites of the landform approach development are the number of factors comprising several principals as follows:

(1) Ecological factors that is maximal maintenance of nature; implementation of principle of compensation for losses that damaged the natural landscape by building of a new object; ecological safety of building and integration with nature
(2) Economic factors that is use of energy effective technologies on principle of "clever" architecture, buildings, that have a zero power balance and which can operate autonomously from engineering structures;
(3) Bioclimatic that is consideration of climatic, geographical, seismic, hydrogeological conditions, landscape environment, land forms of an area;
(4) Psychological that is alliance of human being with nature as a factor of psychological renewal and health improvement.
(5) aesthetic this is the harmonious confluence of architecture with the natural and artificial.

Ecological. The ecology problem appears as one of the basic unsolved tasks for development of humanity. In every case contributory influence on the improvement of ecological situation demands individual efforts of every person and development of conscious strategy of construction of a synergy with natural environment. Ecological personality envisages functional interaction of people in the system "man-environment". Principle of indemnification of loss that was inflicted to the natural landscape by building of some construction is offered in the modern creative conceptions. Architecture creates the "second nature" with principle of geo-equivalence. The cultural landscape created by people shall compensate lost of nature existed at its place.

Economic. Use of the energy-effective technologies based on principle of "clever" architecture whilst building of premises with a zero power balance and circulation that can independently function from engineering structures is important. This is aimed at use of electric power, illumination, alternative sources of heat, ventilation that results in the economy of resources while object functioning.

Bioclimatic pre-conditions effect the direct and substantial influence on formation of the landform architecture. The bioclimatic terms include the following:

(1) land morphology—bioclimatic structure responds to character of relief, adheres to the natural bends or deepens under earth melting into it. When necessary the water aquatorium of locality that refers to the special category of landscape shall be considered;

(2) geology of locality—geological structure of terrain, nature of its massive material, hydrological conditions and terms for the greater part determines character of structural decision of a building;
(3) thermal climate as assembly of natural characteristics of radiation, temperature-moist mode and aeration state of environment (intensity of solar radiation, temperature, humidity, speed and direction of air motion, type and amount of precipitations), that determines normative values and bench-mark data to create the supportive microclimate. Bioclimatic approach also represents the factor of use of biological resources of locality: flora, fauna, mineral minerals, water.

It should be noted, that consideration of the separately taken factors does not provide complete environmental friendliness, comfort and resource preservation of the residential area. Therefore the basic criterion of the bioclimatic approach to the planning is a complex system estimation of the terrain factors.

Psychological. Effect of man-nature synergy is well-known and indissoluble. Man and environment present the single social and nature system. Today scientists distinguish a few stages of synergy of nature and man: (1) dependence of man on nature; (2) dependence of nature on man; (3) unities of man and nature [15]. It should be noted, that today humanity is on the third stage—stage of unity with nature. As of today it is already impermissible to examine the nature dissociated from a human being, it is impermissible to consider it only as an environment for existence and source for satisfaction of pragmatic aims. The human consciousness level must be higher.

Aesthetic. An object fits into a natural environment and has an unusual, attractive look; building ensues the natural forms of the earth surface; supports the tuneable confluence of a summary architecture with nature as well link-up of the internal space with the external one.

Several ways are to be noted during implementation of the landform approach in architecture: (1) by the nature of the building - earth surface synergy; (2) by the nature of location of the building reference area in relation to the level of terrene or soil; (3) for morphologic type of a terrain; (4) by the type of composition synergy of building form with relief and natural environment.

While such building the character of the construction earth surface synergy is essential and allows to emphasize three approaches: (1) polarization; (2) hyperbolization; (3) integration (see Fig. 1).

The prime example of the polarization way and minimum affecting on the environment is the house of Albert in California built by the architect in 1955. The house is the result of careful estimation of lithoidal topography of the area and motion of the sun. The primary objective of architect was the careful inclusion of the building in the landscape using local stones and modern materials, such as glass and metal (see Fig. 1).

Building of university library (by architect Denton Corker Marshall, Indonesia, 2011) is the interesting example of earthen architecture based on extravagance where

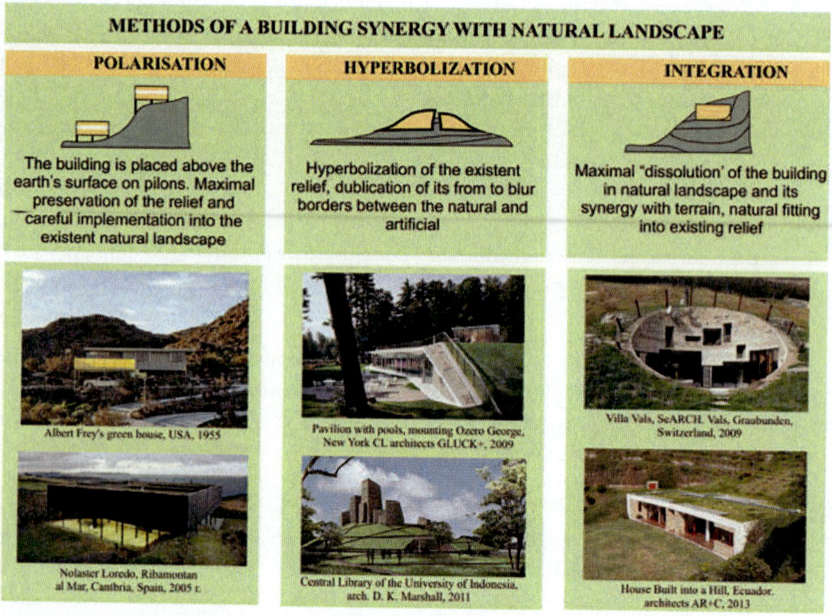

Fig. 1 Methods of a building synergy with natural landscape

successful synergy of natural and artificial environment is shown. The consumption of energy in building is kept to the minimum, collection of sewage and rain waters is envisaged. Green roofs and existent trees provide maintenance of landscape area (see Fig. 1).

Relief is a means of the landform form making. Depending on the location of reference area of building to the level of terrene or soil and deepening in the "body" of landscape the landform objects can be: (1) above-ground (ground, located by a their reference area on the terrene; aligned with rising ground); (2) embedded; (3) underground (see Fig. 2).

Types of half-embedded buildings are appertained to the most widespread in world building practice. They have a number of competitive advantages: the rational use of territory due to the spatial resource, creation of microclimate, wind shield, solar proofing, visual comfort, energy-savings, ecological compatibility etc. As a rule, the half-embedded buildings are erected on areas with flat slopes, if the large volume of soil excavation is impossible (owing to hydrogeological or geological terms, and also economic reasons). The building can rise above the earth surface at 30–40%, and can be embedded at 60–70% [14].

Except it forming of the landform architecture depends on topographical terms of locality—morphological type of the landscape. When the landform architecture is introduced into design the most widespread types of natural relief are flat relief, hill, mountain. Also there are examples of building on the other types of relief: water gap, cavity, hollow, crater.

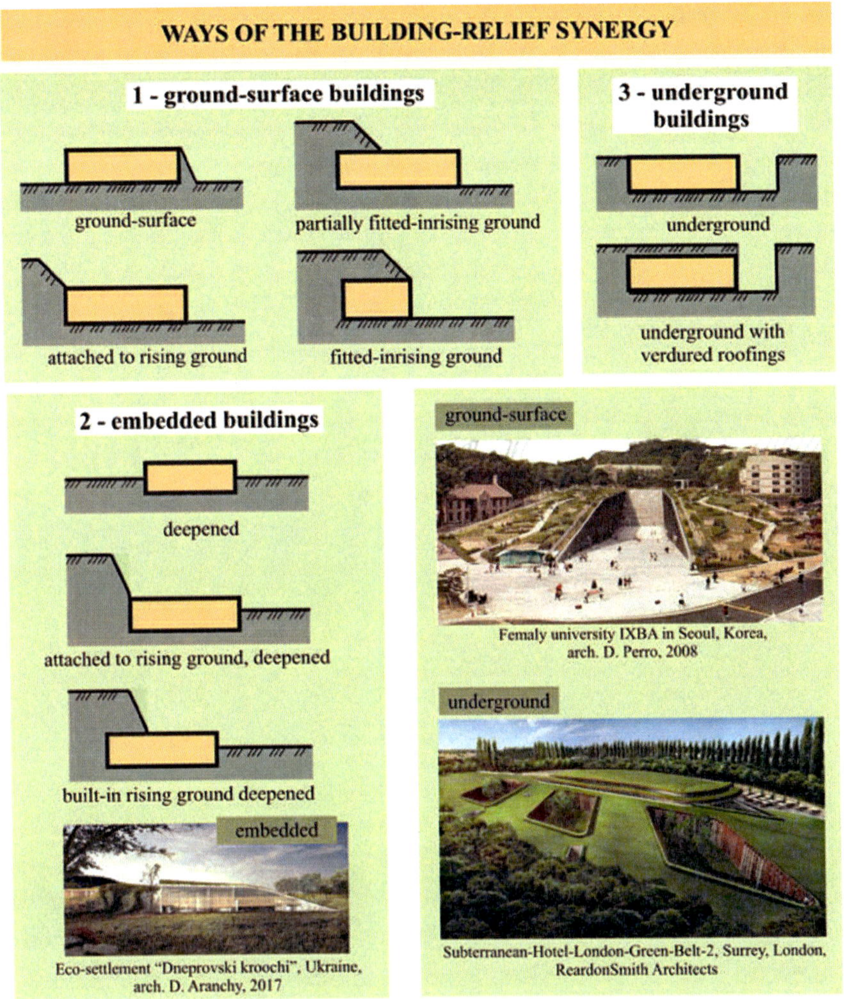

Fig. 2 Ways of the building-relief synergy

When the object form interacts with natural environment and land surface depending on the relief morphology it is possible to emphasize the following compositional ways (see Fig. 3):

(1) *inheritance of the relief form*—where it is possible to singularize the following: nuance—as reiteration of the relief contours by building form and natural landscape; identity—as a complete inheritance of the natural landscape form in silhouette of building, that allows to imbed a building into natural environments softly and without conflicts;

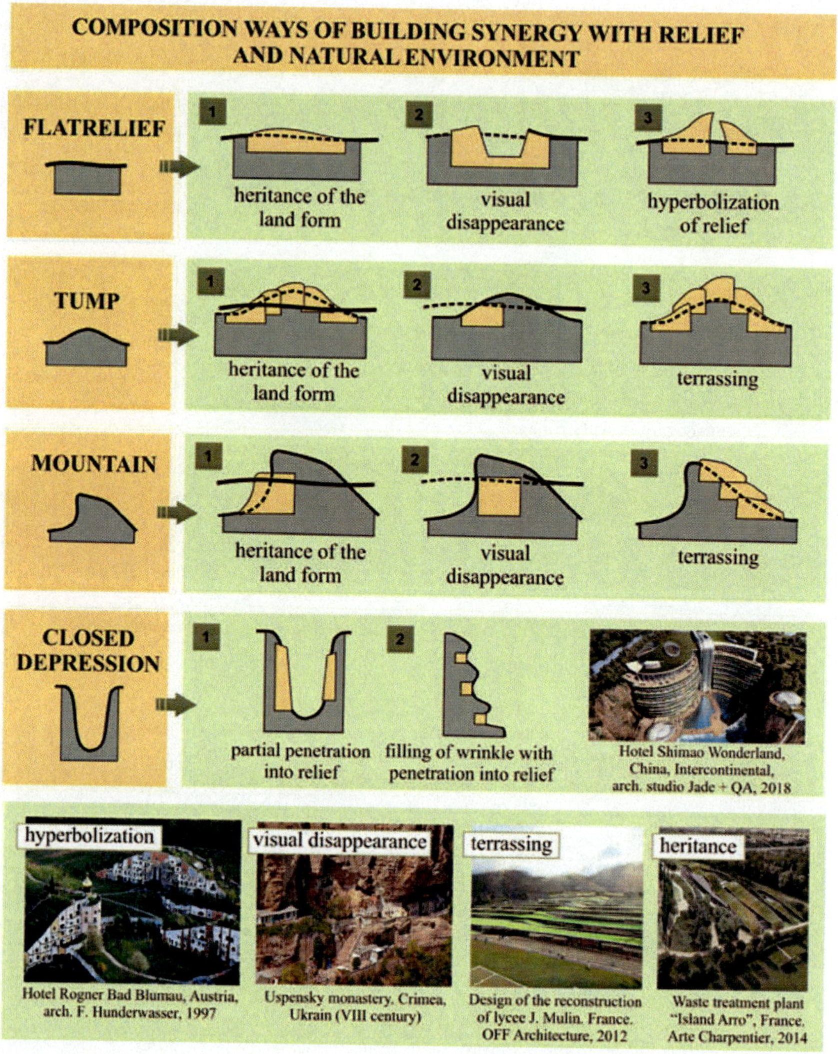

Fig. 3 Composition ways of building synergy with relief and natural environment

(2) *«visual disappearance»*—that is absorption of the architectural object by natural surroundings and its assimilation with ecological structure. Effect of "dissipation", visual "disappearance" of building in natural environment can appear in varying degrees, it is maximal due to deepening, imitation of natural landscape and greenery of facades.

(3) hyperbolization—this is a building which creates the new artificial relief due to exaggeration of the prototype form, use of the planted trees and inclined roofs with greenery. This way more often is used on flat relief. This way is

implemented through the use of geo-plastic as creation of the artificial relief following form of the initial prototype. The building not simply takes place on an area, but constructs area and relief allotted to it;

(4) terracing—this is arranging terraces as wide steps on hills, rising grounds of artificial surfaces with greenery.

The examples of such embedded structures can be the «earthen» houses by the Swiss architect Peter Vetsch. His houses are energy-efficient and environmentally compatible, without square corners.

The extravagance method is typical for landform architecture, whereby building, due to the overstatement of prototype form, use the verdured pitched roofs creates the new artificial relief. This method is more often used on flat relief. It is implemented by geo-plastic way as a creation of the artificial relief following the form of initial prototype.

Building is not simply disposed on an area, but itself constructs the allocated area and relief. One of the form making methods of the landform architecture also is the method of «natural mimetism», when the architectural object maximally imitates, copies natural form. As a result there is visual disappearance or dematerialization of building under the layer of soil and vegetation. The main form making ways here is the geo-plastic and use material, color, textures indicative for this landscape whilst building.

Transformation of traditional architectural form is indicative for the landform approach. Because at the partial or complete deepening in relief, building disappears visually and accordingly loses facades usual for architectural form making. Such building can have one open facade only. Therefore a main form making value in the landform architecture has a natural isomorphism of the architectural object with natural surroundings and relief. It is also necessary to note that the modern energy effective technologies such as geothermal and solar power, wind currents for natural ventilation etc. are used in buildings based on the landform approach [16, 17].

The prime example of steady architecture based on the landform approach is building of Ewha Campus Complex in Seoul, 2008 (South Korea). The building is located in a natural canyon and it masterly entered in the surrounding municipal landscape. The garden that protects internal space from heat and cold is placed on the campus roof. Architect elaborated the so-called "thermal labyrinth", that includes the great number of the metallic batteries, placed on ceiling and connected with pipes going out outside. It provides circulation of crisp air into campus. A rain-water is saved in special reservoirs and then spreads for functioning of lavatories [18].

3 Conclusions

Thus, finally it should be noticed that a landform building, when using the organic approach, is not situated on the design area but itself builds area and its relief basing on integration with the terrene. There is partial or complete visual "disappearance" of architectural form caused by synergy of the building with relief. The basic task of the landform architecture is a submission of building volume to the relief form and integrity of its comprehension in the context of natural relief, character individuality, creation of the comfort biopositive structure.

Today landform architecture doing its contribution to the decision of ecological problems of the earth is one of the effective instruments of humanization of municipal environment and careful attitude toward nature. In the modern architectural practice the landform approach is to investigate the enormous amount of displays of landscape and ecology not as the summary phenomenon, but on basis of use of new methods of planning, new formal strategies and technical issues in architecture.

References

1. Dobritsyna IA (2013) From postmodernism to the nonlinear architecture: architecture within the context of the modern philosophy. Direct media, Moscow
2. Jencks C (2011) The universe in the landscape: landforms. Frances Lincoln, London
3. Allen S, McQuade M (2011) Landform building: architecture's new terrain. Schirmer-Mosel, Baden
4. Frampton K (2010) Megaform as Urban Landscape. In: Hubeli R, Larsen J (eds) University of Illinois, School of Architecture, Urbana, IL. 25 p.
5. Balmori D, Sanders J (2011) Groundwork: between landscape and architecture. Monacelli Press, New York
6. Manfredi MA (2008) Surface/subsurface. Princeton Architectural Press, New York
7. Carpo M (2013) The digital turn in architecture 1992–2012. Wiley, Chichester
8. Thoren R (2014) Landscapes of change: innovative designs and reinvented sites. Timber Press, Portland
9. Tabb P, Senem Deviren A, James P (2013) The greening of architecture: a critical history and survey of contemporary sustainable architecture and urban design. VT: Ashgate Pub. Co, Burlington
10. Corner J, Hirsch A (2014) The landscape imagination: collected essays of James Corner 1990–2010. Princeton Architectural Press, New York
11. Zaslavskaya AY (2008) Peculiarities of the organic approach in architecture at the end XX–beginning XXI centuries. LLC "Publishing house SNTs", Nizhny Novgorod
12. Chebina O, Shevchenko L (2015) The boulevard as a type of urban linear space the historical boulevards of Poltava (Ukraine) and Mons (Belgium). [Bulvar ako typ lineameho urbanneho priestory historicke bulvare Poltavy (Ukrajina) a Monsu (Belgicko)] Architectura Urbanismus 49(3–4):199–215
13. Denisenko EV (2013) Principles of the architectural space development based on bio-approaches. KGASU, Nizhny Novgorod
14. Kryzhanovskaya NYa, Smirnova OV (2016) Genesis of development of the innovative buildings and facilities in urban environment: monograph. KhNUGH them. A.N. Beketova Kharkov
15. Shahiakhmetov MR (2007) Man and nature: system-oriented analysis of synergy. Bull Sanct-Peterburg Univ Ministry Interior Russia 4(36):231–238
16. Pichugin S (2017) Probabilistic description of ground snow loads for ukraine. In: Snow engineering 2000: recent advances and developments, pp 251–256. https://doi.org/10.1201/9780203739532
17. Pavlikov A, Mykytenko S, Hasenko A (2018) Effective structural system for the affordable housing construction. Int J Eng Technol (UAE) 7(3):291–298. https://doi.org/10.14419/ijet.v7i3.2.14422
18. Energy-efficient technologies in architecture. https://alternativenergy.ru/tehnologii/934-energoeffektivnye-tehnologii-v-arhitekture.html. Accessed 13 Aug 2020

Design Principles for Inclusive Environment of Urban Agrorecreational Eco-complexes

Tetiana Pavlenko, Tetyana Lytvynenko, Viktoriia Ivasenko, and Alina Zyhun

Abstract The problem of creating an inclusive environment in urban agrorecreational eco-complexes are considered. The peculiarities of the criteria of accessibility, informativeness and comfort have been determined in the article. There were analyzed and classified requirements for the design of urban agrorecreational eco-complexes taking into account the needs of people with limited mobility. The main means of inclusive environment in urban agrorecreational eco-complexes have also been determined. There have been analyzed the following main means of barrier-free accessibility for the elements of the external space of urban agrorecreational eco-complexes: external ramp, exit ramp, raised pedestrian crossing, lifting device, external stairs, parking space for a person with disabilities. The authors have considered the main elements of universal (inclusive) design and barrier-free environment in interior space of urban agrorecreational eco-complexes: internal ramp, internal stairs, approach ramp, elevator, internal lifting device, escalator, ambulatory compartment for people with disabilities. The main means of universal (inclusive) design for the elements of external and internal space of urban agrorecreational eco-complexes have been outlined in this scientific survey. The following tactile elements of accessibility have been considered: tactile guiding strip, warning tactile strip, informational tactile strip, tactile resource book. Visual accessibility elements (signs in contrast colors, information plates and informers) and auditory accessibility elements (external detectors or speech systems), have also been identified and analyzed.

Keywords Inclusive environment · Inclusive space · People with disabilities · Universal design · Urban agrorecreational eco-complex · Vertical farm

T. Pavlenko (✉) · V. Ivasenko
O.M. Beketov National University of Urban Economy in Kharkiv, Kharkiv, Ukraine

T. Lytvynenko · A. Zyhun
National University «Yu. Kondratyuk Poltava Polytechnic», Poltava, Ukraine

V. Onyshchenko et al. (eds.), *Proceedings of the 3rd International Conference on Building Innovations*, Lecture Notes in Civil Engineering 181,
https://doi.org/10.1007/978-3-030-85043-2_51

1 Introduction

Acceleration of urbanization leads to the situation when high-rise buildings and municipal facilities are extensively constructed on croppable lands suitable for cultivating farm produce. Therefore, vegetables, greens and fruits for the cities with a million-plus population have to be transported from afar off, which has an impact on their freshness. It encourages the emergence of new directions in urban planning, when a considerable part of the afore-referenced products can be produced within the limits of the cities, and moreover–all the year round.

Due to the shortage of sizeable white lands, space in urban agrorecreational eco-complexes [8] is organized vertically for the purpose of compactness; greens and vegetables are grown for local shops, and the orchards growing on the roofs of the houses may help to control air pollution. But this requires special approaches to the creation and improvement of such space, which should be created on the principles of universal design and be inclusive.

In most of developed countries professionals are actively working on the creation of safe, comfortable, accessible, informative environment [1–7, 9–23], including the space for workers and visitors to urban agrorecreational eco-complexes for people with limited mobility, including people with disabilities.

2 Materials and Methods

When conducting the research, basics of the systems approach, the simulation mode, theoretical methods of analysis, synthesis, generalization and abstraction technique were used.

3 Results

3.1 *Defining the Inclusive Environment of Agrorecreational Eco-complexes as a System*

Vertical farm is the generic name for a highly automated agroindustrial complex located in a specially designed high-rise building. The main difference between vertical farms and traditional hothouse facilities and livestock farms is vertical multi-tiered arrangement of plantations and intensive approach to land use [8]. Pertinently, a vertical farm is a multitiered greenhouse. From the very beginning, vertical farms are always planned as a new element of the urban environment, which must comply with the terms of inclusivity, be accessible to all, and have inclusive space for people with limited mobility, including people with disabilities.

People with limited mobility are people who have difficulty with ambulation and orientation in space, in receiving services and necessary information. People with limited mobility are: people with disabilities, people with temporary health problems, pregnant women, the aged (the elderly, people with baby buggies, etc.) [9].

There is a tendency to the usage of such terms as barrier free environment for physically challenged people, adapted environment, **universal design (inclusive design)** [9]. These terms describe the elements of the architectural space taking into consideration the needs of people with disabilities (Fig. 1). The term **«barrier-free environment»** is in most cases used to refer to people with disabilities. Barrier-free environment provides for the installation of approach ramps and sidewalks with high quality roadway paving, fairly wide passages, lanes, drive-throughs and driveways, doors and other elements of the architectural environment that make it easier for people with limited mobility to move [9]. For people with limited mobility, the presence of barrier-free environment is a factor that significantly affects the quality of life [1].

Inclusiveness of buildings and related structures is a comprehensive set of architectural, planning, engineering and technical, ergonomic, structural and organizational measures to ensure the accessibility of buildings and related structures in

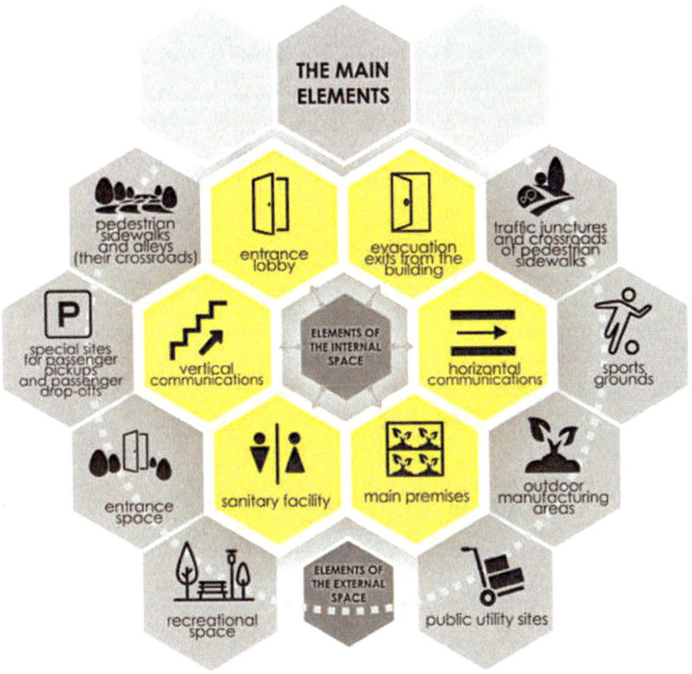

Fig. 1 Basic architectural-planning, engineering-technical and design elements providing for the inclusiveness of the environment

which any person, regardless of age, gender, disability, functional diseases, level of communication skills, can in all circumstances feel safe and comfortable without physical assistance and outside help to the best of their abilities [10].

In accordance with world statistics, from a quarter to the third part of the population are the users of the barrier-free environment at one time or another [9].

Accessibility in urban agrorecreational eco-complexes should ensure easy movement and safe passage in space and the possibility of using public property and public benefits.

Many countries have already developed their accessibility standards, taking into account best practices. As a result, uniform standards for civilized countries have been formed and improved.

In general, **the criteria for accessibility** in urban agrorecreational eco-complexes should have the following requirements for the accessibility provisions:

- unobstructed movement along the sidewalks and surmounting vertical drops (level differences);
- possibility of convenient use of holding points, recreational facilities and additional service points;
- passage and vehicular access to the equipment for different purposes;
- access to public transport stops;
- road crossings;
- access to the information supply;
- availability and equipment of parking spaces;
- availability of means of surmounting pedestrian crossings (overground pedestrian crossings and pedestrian undercrossings);
- availability of means to cross over interchange ramps;
- unobstructed vertical and horizontal movement during technological (agro-industrial) processes in urban agrorecreational eco-complexes;
- unobstructed vertical and horizontal movement of visitors during recreational processes in urban agrorecreational eco-complexes.

Safety should be understood as the possibility of overcoming the obstacles safely, visiting urban agrorecreational eco-complexes without the concern for being injured in any way or causing damage to property or equipment. It also applies to the installation of fences, doors and etc., to the location of public utility sites of support service vehicles, platform jetties and landing stages, steps and lifting devices, their protection from precipitation; routes of people with limited mobility in the middle of pedestrian crossings (both overground pedestrian crossings and pedestrian under-crossings).

It is recommended to include into the requirements of **the informativeness criterion** for urban agrorecreational eco-complexes the following:

- timely identification of landmarks on the territory of urban agrorecreational eco-complexes;
- accurate identification of your location and the places that are the purpose of your visit;

- the use of information distribution media that meet the needs of different user groups;
- the possibility of effective orientation of visitors, both during daylight hours and at night time;
- shortening of time and contraction of efforts to obtain the necessary information;
- warning people of potential threat zones on the road and of the streets that may be possible hazards;
- the ability to have continuous information support while moving along the street.

Landscape comfort includes adaptation of the environment of urban agrorecreational eco-complexes, in which a person with disabilities appears to be, to his/her needs and capabilities.

Universal design (inclusive design) of urban agrorecreational eco-complexes should be suitable for the vast majority of people, as well as for a wide range of people with such disabilities, as trouble seeing, hearing and perception defects, taking into ac-count psycho-emotional and intellectual characteristics. Universal design (inclusive design) of urban agrorecreational eco-complexes addresses the problems of accessibility and gives an opportunity to make all the elements of its environment accessible. This is achieved by detailed planning at all design stages [9].

3.2 *The Main Means of Inclusive Environment of Agrorecreational Eco-complexes*

The peculiarity of creating an inclusive environment in urban agrorecreational eco-complexes is identification of two main groups of elements that must be provided by **means of universal (inclusive) design and barrier-free environment** (Fig. 1):

- *elements of the external space* of urban agrorecreational eco-complexes (pedestrian sidewalks, malls and alleys (their crossroads), traffic junctures and crossroads of pedestrian sidewalks, special sites for passenger pickups and passenger drop-offs [5], entrance space of an urban agrorecreational eco-complex, recreational space, sports grounds, outdoor manufacturing areas, public utility sites, etc.);
- *elements of the internal space* of urban agrorecreational eco-complexes (entrance lobby of an urban agrorecreational eco-complex, vertical communications, horizontal communications, public bathroom and laundry, main premises, etc.).

Taking into account the considered features of creating inclusive (barrier-free) space in urban agrorecreational eco-complexes, we have identified the main design aids for such construction entities.

Means of inclusivity (barrier-free space) for the elements of the external space of urban agrorecreational eco-complexes (Fig. 2):

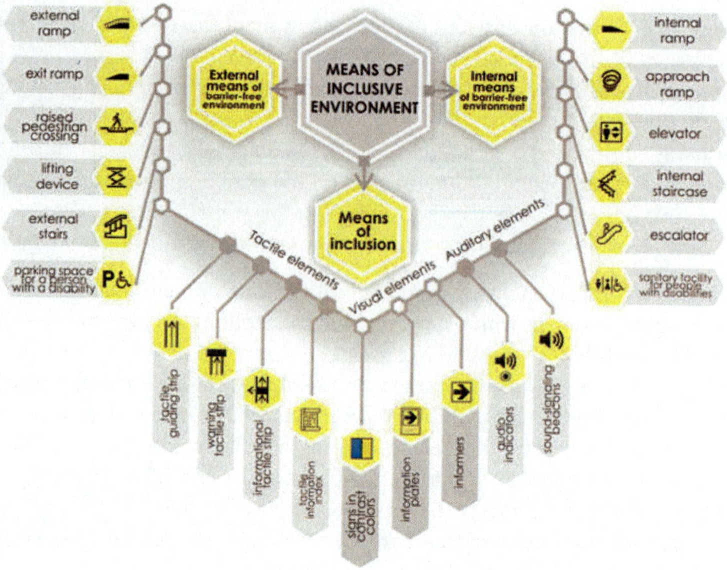

Fig. 2 Means of inclusive environment in urban agrorecreational eco-complexes

- *external ramp* (Fig. 3)–external continuous inclined pane (structure), which connects two uneven horizontal surfaces and is arranged for barrier-free movement of wheelchairs, baby buggies, other various wheeled vehicles and people from one flat surface to another [5];
- *exit ramp* (Fig. 4) is an inclined pane (structure), which is located at all the crossings of pedestrian paths/footwalks/sidewalks with carriage ways of different movement directions (parking lot exits, public transport stops) for barrier-free movement of people and wheelchairs, baby buggies and other various wheeled vehicles [9];
- *raised pedestrian crossing* (Fig. 5) is the means for more convenient barrier-free movement of people and wheelchairs, baby buggies, other various wheeled

Fig. 3 External ramp sample

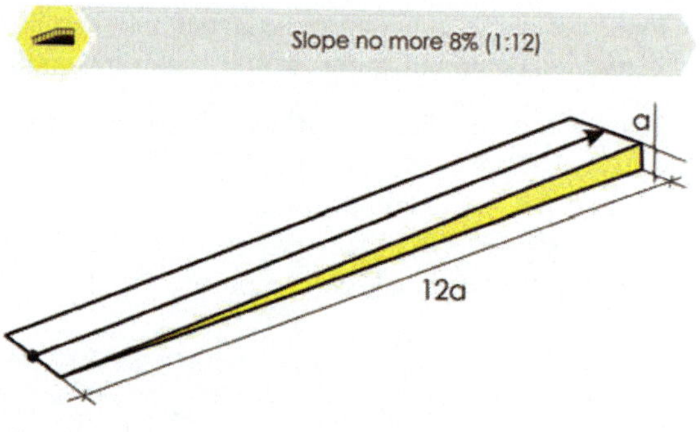

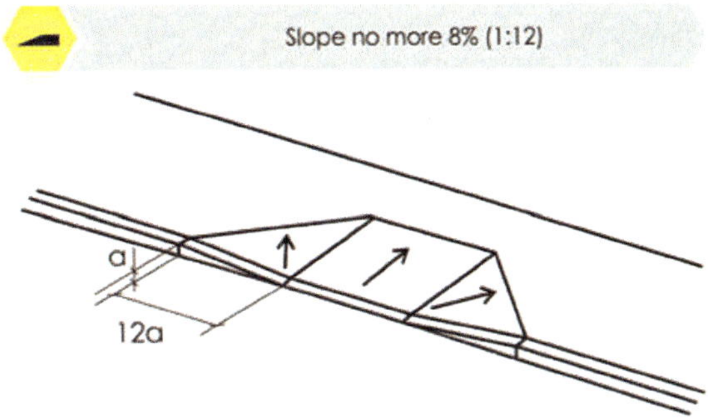

Fig. 4 Exit ramp sample

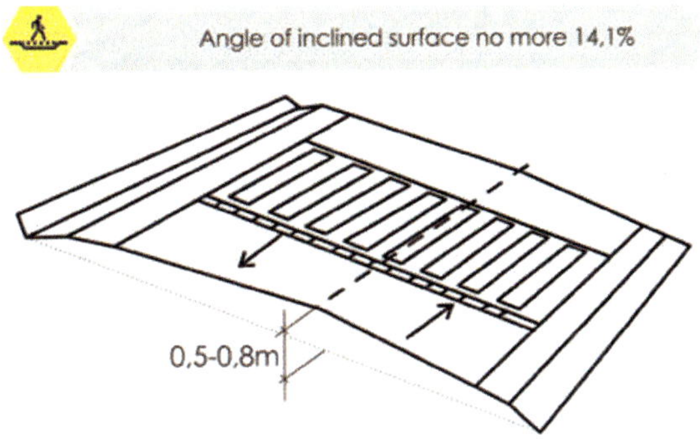

Fig. 5 An example of a raised pedestrian crossing

vehicles across the carriage way at the pedestrian crossing by mechanical speed restriction of vehicles [9];

- *lifting device* (Fig. 6) is the means used to get over the significant difference in levels between the nearest horizontal faces (vertical and sloping direction) [5];
- *external stairs* (Fig. 7)–the stairs on footpaths/footways/sidewalks, planned for the cases when, in a certain place, there is a slope of the ground more than 10%, should be duplicated with ramps, and if necessary–with other lifting devices with vertical movement or moving in parallel with the slope of the stairs and meet the basic requirements for the arrangement of street and road network [9];
- *parking space* (Fig. 8) for a person with a disability (a place for parking private vehicles belonging to people with disabilities or the vehicles used for transporting people with disabilities) is recommended to be placed at the entrance to the buildings and structures, but not further than 50 m in compliance with standards specified [3].

Barrier-free devices for the elements of the internal space of urban agrorecreational eco-complexes (Fig. 2):

- *internal ramp* (Fig. 9) is internal continuous inclined pane (structure), which connects two uneven horizontal surfaces and is arranged for barrier-free movement

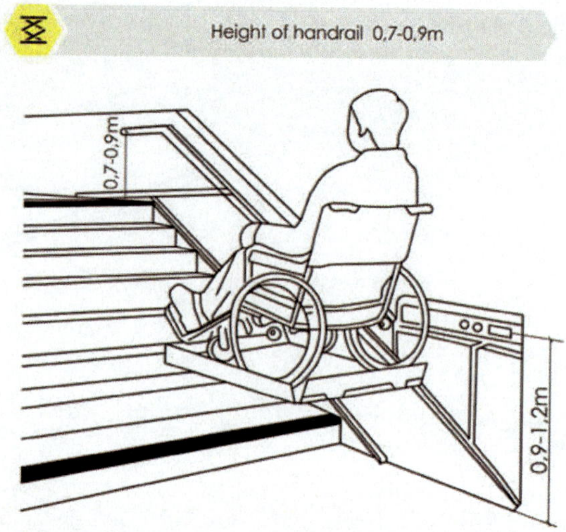

Fig. 6 An example of a lifting device

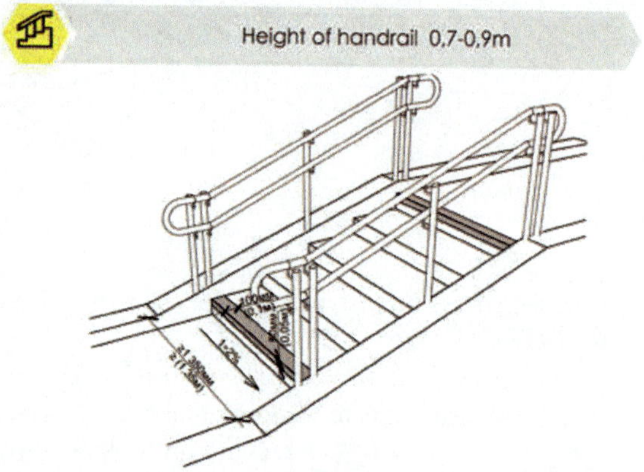

Fig. 7 An example of external stairs

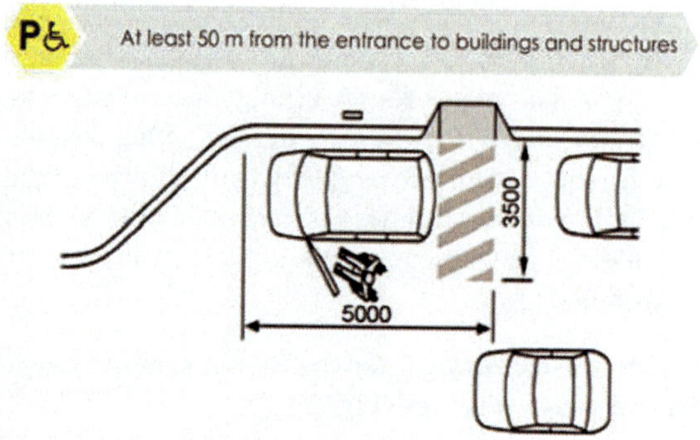

Fig. 8 An example of a parking space for a disabled person

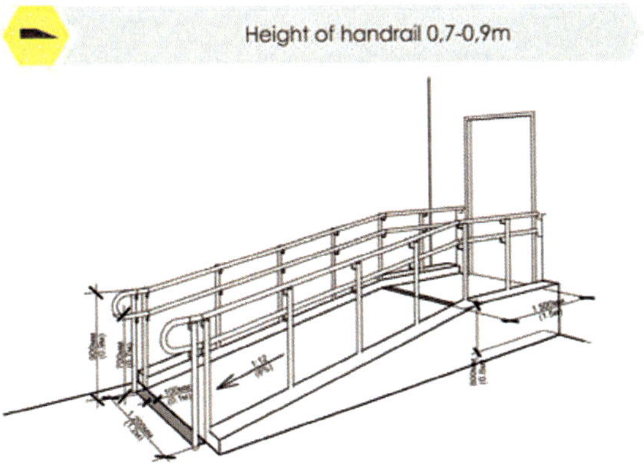

Fig. 9 An example of an internal ramp

of wheelchairs, baby buggies, other various wheeled vehicles and people from one flat surface to another [9];

- *internal stairs* (Fig. 10)–the structure of horizontal ledges or stairs, which is used for equipment maintenance and in order to connect the floors, rooms, roofs of the buildings and structures [10];
- *approach ramp* (Fig. 11)–spiral horizontally curved ramp [9];
- *elevator* (Fig. 12)–an engineering structure with a special cabin for the vertical movement of people or goods [10];

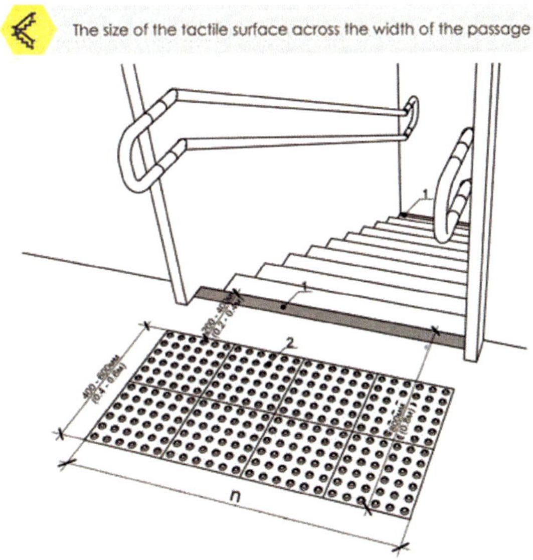

Fig. 10 An example of an internal staircase

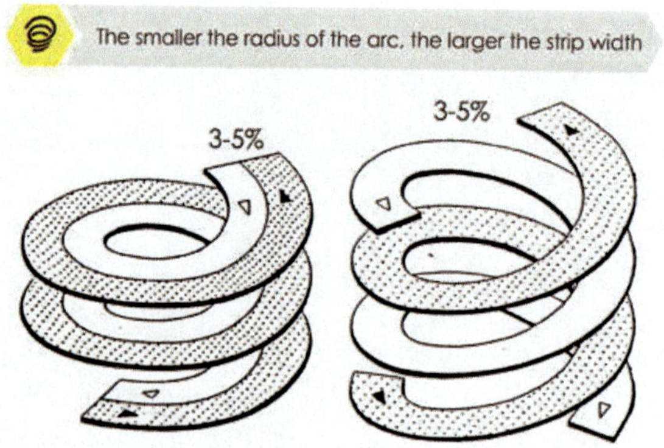

Fig. 11 Examples of ramps

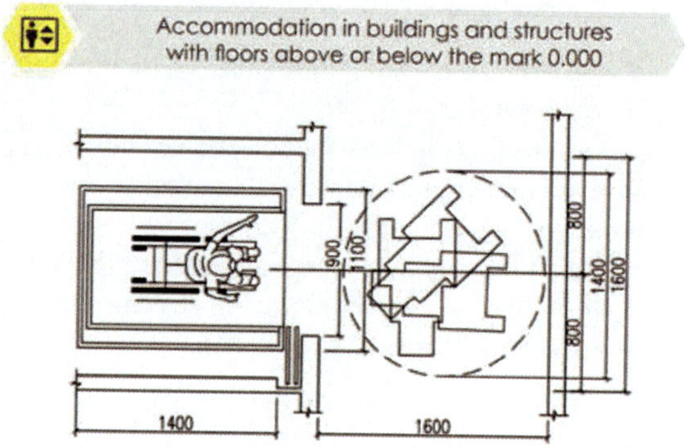

Fig. 12 An example of an elevator

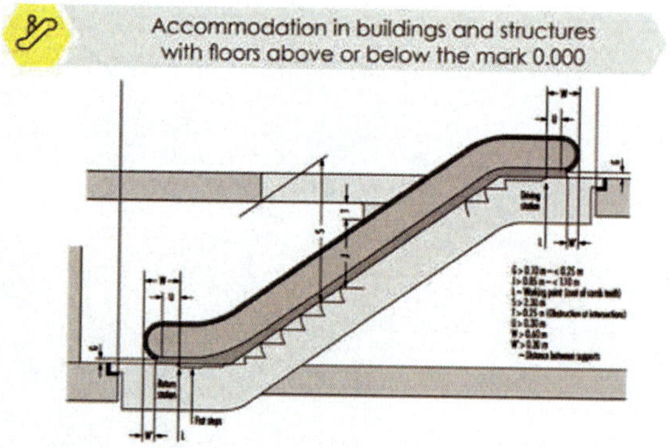

Fig. 13 An example of an escalator

- *internal lifting device* (Fig. 6) is a means used to get over the significant difference in levels between the nearest horizontal surfaces (vertical and sloping direction) [9];
- *escalator* (Fig. 13) is a device, a moving lane in the form of a moving sidewalk

or a sloping staircase with movable horizontal steps for moving horizontally or between different levels of the surface (vertical and sloping direction) [9];

- *sanitary facility for people with disabilities* (Fig. 14)—ablution facility, which is intended for use by all social categories (including people with disabilities) and provides for the installation, in addition to basic accessories (a flush toilet and a sink) of handrails, bars, swiveling seats or fold-back seats, etc. [1, 9, 10].

Means of universal (inclusive) design for the elements of external and internal space of urban agrorecreational eco-complexes (Fig. 2):

Tactile elements of accessibility (TEA)–the system of safety and orientation facilities, means of obtaining information for people with sensory impairments (Fig. 15) [1–7, 9, 10]:

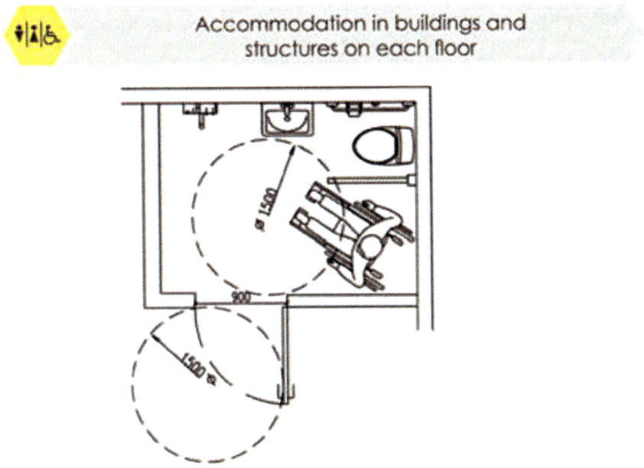

Fig. 14 An example of a sanitary facility for people with disabilities

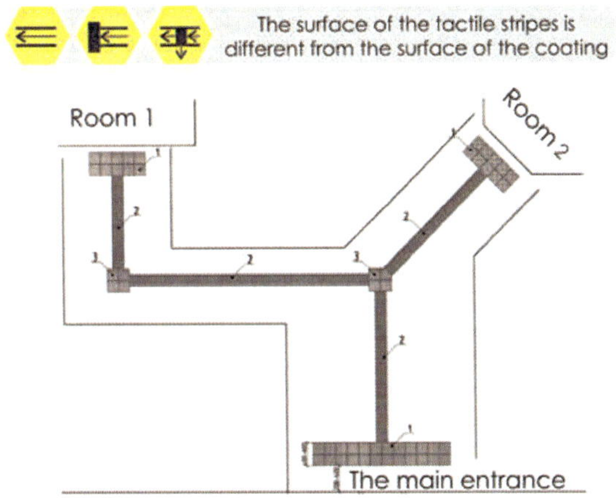

Fig. 15 An example of tactile strips arrangement: 1–warning tactile strip; 2–tactile guiding strip; 3–informational tactile strip; which indicates the place of turning (divergence) of the tactile guiding strip

- *tactile guiding strip* (Fig. 15)–a means of warning or orientation for people with vision impairment and for other users, which warns of the direction of movement to the objects of social, engineering and transport infrastructure, on footpaths (pedestrian routes), on the premises of the objects and inside them [9];
- *warning tactile strip* (Fig. 15)–a means of warning or orientation for people with vision impairment and for other users, which warns of potential danger or hindrances while moving to the objects of social, engineering and transport infrastructure, on footpaths (pedestrian routes), on the premises of the objects and inside them [9];
- *informational tactile strip* (Fig. 15)–a means of warning or orientation for people with vision impairment and for other users, which informs of direction change while moving to the objects of social, engineering and transport infrastructure, on footpaths (pedestrian routes), on the premises of the objects and inside them [9];
- *tactile information index* (Fig. 16)–a tactile element of accessibility that duplicates flatly printed text or graphic information in tactile form and in Braille type [9].

Visual accessibility elements (VAE)–tools that ensure safety and provide for orientation, obtaining information by all users, including people with vision impairments, with the help of color layouts, information plates, informants, markers and signs [9]:

- *signs in contrast colors* (Fig. 17)–visual accessibility elements that use contrast color relation in order to ensure the structuredness of the space (for free orientation, receiving information and safety when moving on the way to the objects of social, engineering and transport infrastructure, on footpaths (pedestrian routes), on the premises of the objects and inside them [9];

Fig. 16 An example of a tactile information index: 1–information guide; 2–informational tactile marking; 3–informational tactile sign plate that duplicates textual information in the tactile form of flatly printed text and in Braille type

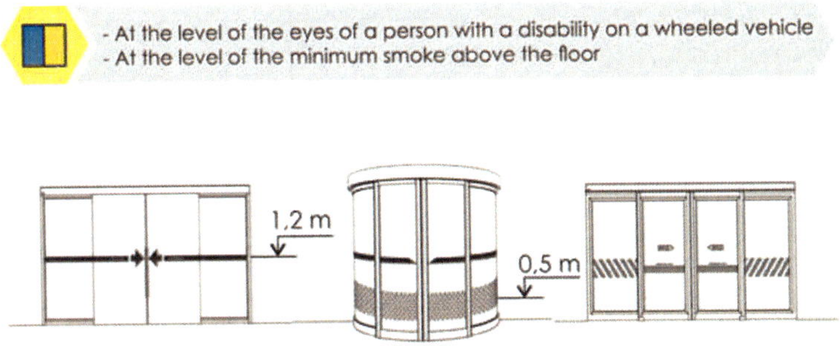

Fig. 17 Usage samples of signs in contrast colors (contrasting markings on transparent doors)

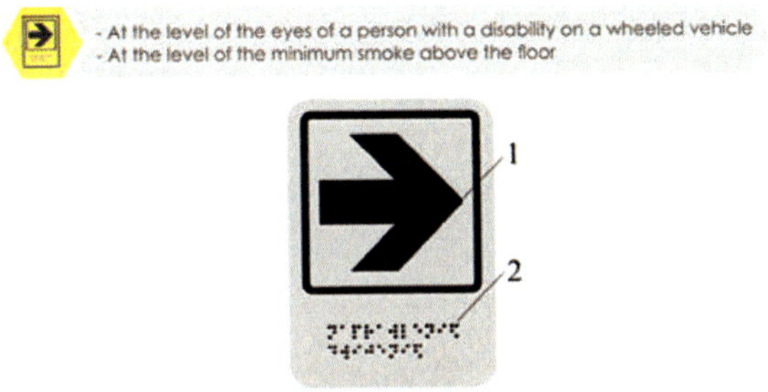

Fig. 18 Examples of information plates combined with a tactile resource book: 1–flatly printed graphic information, 2–duplication of flatly printed graphic information in Braille type

– *information plates and informers* (Fig. 18)–visual accessibility elements that are clear and accessible to all categories of people with limited mobility, made in enlarged font and in a contrasting color relation of fonts to the background of the plates. They can be combined with a tactile resource book [9];

Auditory Accessibility Elements (AAE):

– *audio indicators* (Fig. 19)–auditory elements of accessibility that help people with vision impairments obtain information about the object and the services provided in it [9];

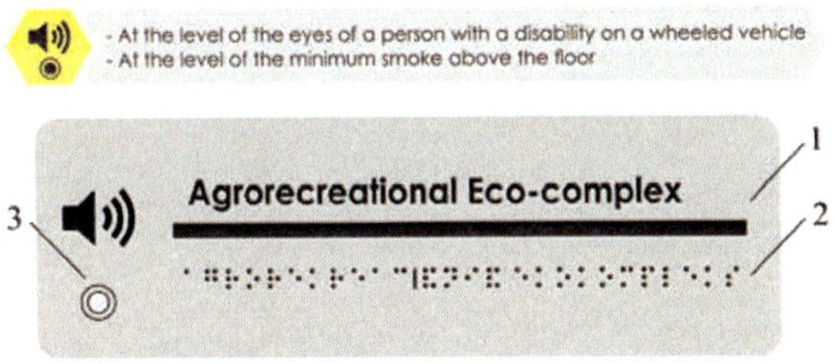

Fig. 19 Examples of an audio indicators combined with an information plate and a tactile resource book: 1–flatly printed textual information, 2–duplication of flatly printed textual in-formation in Braille type, 3–play button for reproducing audio information

Fig. 20 Examples of a sound-signaling beacons combined with an information plate and a tactile resource book: 1–flatly printed graphic information, 2–duplication of flatly printed graphic information in Braille type, 3–play button for reproducing audio information

– *sound-signaling* (Fig. 20) are auditory elements of accessibility, located at the entrance lobbies of the objects that point to the required direction of movement [9].

4 Conclusions

The article studies and classifies the requirements to the space of urban agrorecreational eco-complexes (Fig. 21) taking into account the needs of people with disabilities, which should have the following priority sequence: accessibility, safety, in formativeness, comfort, universal (inclusive) design. The main requirements to the environment of urban agrorecreational eco-complexes have been analyzed and systematized, taking into account the needs of people with limited mobility. The main elements of barrier-free and universal (inclusive) design have been considered and analyzed in this article in order to determine the main design parameters in further studies of urban agrorecreational eco-complexes. It is determined that inclusive design changes the requirements not only to the design of territories, but to civil and structural design as well. It will affect layout concepts, construction solutions and operational characteristics of agrorecreational eco-complexes, and also require further rigorous research.

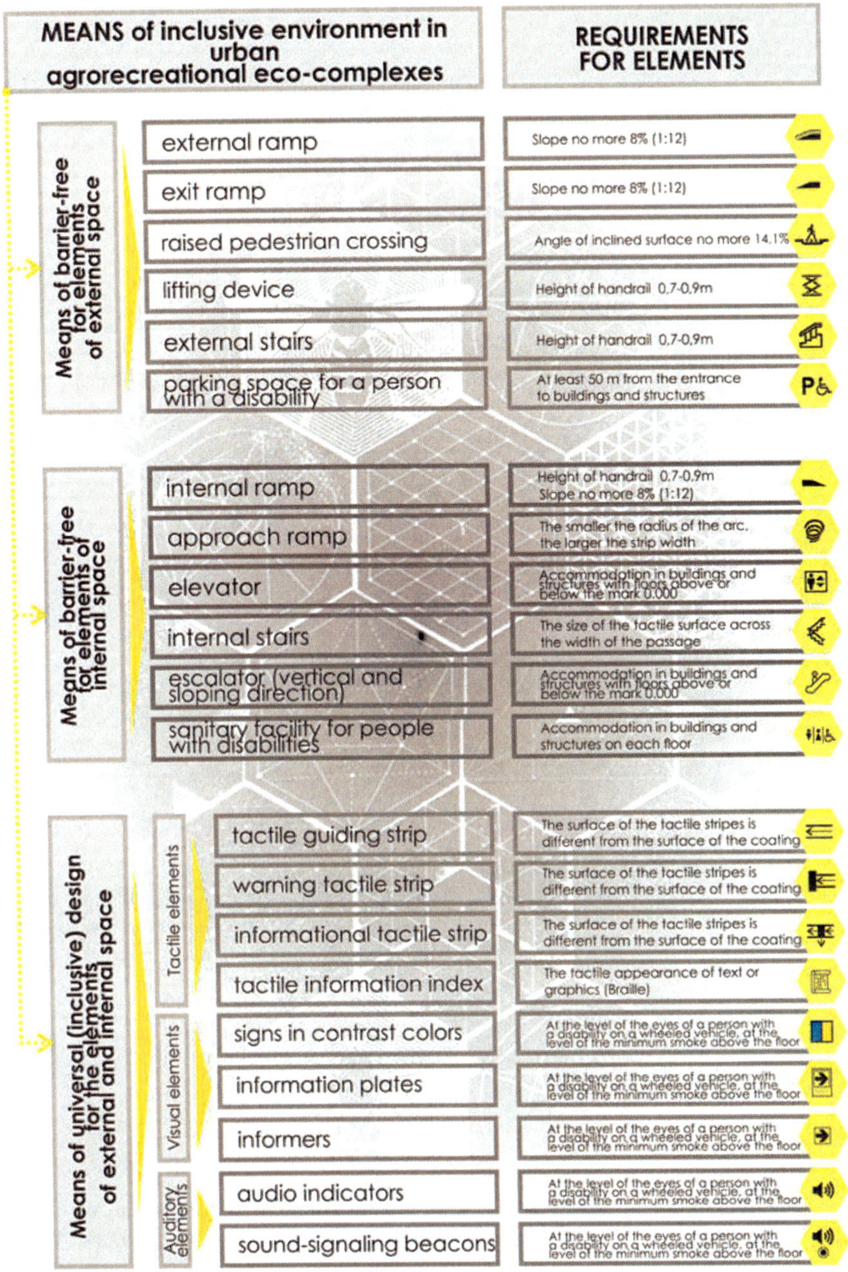

Fig. 21 Defining inclusive environment in urban agrorecreational eco-complexes

References

1. Jones P (2014) Situating universal design architecture: designing with whom? Disabil Rehabil 36:1–6. https://doi.org/10.3109/09638288.2014.944274
2. Khalil M (2016) An assessment of street design with universal design principles: case in Aswan/As-Souq. Megaron/Yıldız Tech Univ Fac Archit E-J 11(4):616–628. https://doi.org/10.5505/megaron.2016.98704
3. Lytvynenko T, Tkachenko I, Ivasenko V, Lvovska T (2020) Application of the universal design principles in the improvement of street and urban road environment. In: Onyshchenko V, Mammadova G, Sivitska S, Gasimov A (eds) ICBI 2019, vol 73. LNCE. Springer, Cham, pp 143–150. https://doi.org/10.1007/978-3-030-42939-3_16
4. Lytvynenko T, Gasenko L, Ivasenko V (2019) Design principles of inclusive street-road environment. In: Mechanisms for ensuring sustainable development of society: collective scientific monograph, pp 203–217. Katowice School of Technology, Katowice. http://reposit.nupp.edu.ua/handle/PoltNTU/6545
5. Lytvynenko T, Ivasenko V (2013) Classification of elements of barrier-free space road network and their requirements. Collection of scientific works [Poltava National Technical Yu. Kondratyuk University]. Ind Eng Const 4(2):66–73. http://nbuv.gov.ua/UJRN/Znpgmb_2013_4(2)__11
6. Lytvynenko T, Tkachenko I, Gasenko L (2017) Principles for road beautification elements placing. Period Polytech Transp Eng 45(2):94–100. https://doi.org/10.3311/PPtr.8592
7. Lytvynenko T, Tkachenko I, Ilchenko V (2018) Principles of street and urban road space formation in modern cities. Int J Eng Technol 7(3.2):642–648. https://doi.org/10.14419/ijet.v7i3.2.14606
8. Mukha T, Demyanchuk A, Savchenko O (2016) Analysis of experience designing vertical agricultural and recreational complexes. Res Prod Collect Archit J KNUBA 10:312–315. http://repositary.knuba.edu.ua/handle/987654321/5073
9. Pavlenko T, Ivasenko V (2020). Basic means of barrier free space in urban agrorecreational ecocomplexes. Municipal Econ Cities, 4(157):54–60. https://khg.kname.edu.ua/index.php/khg/article/view/5633
10. Ward M, Bringolf J (2019) Universal design in housing in Australia: getting to yes. In: Transforming our world through design, diversity and education. Open access by IOS Press, pp 209–355. https://doi.org/10.3233/978-1-61499-923-2-299
11. Rimshin V, Borkovskaya V, Degaev E, Shubin I (2020). Barrier-free urban environment and risks of project solutions implementation. In: Paper presented at the E3S web of conferences, vol 164. https://doi.org/10.1051/e3sconf/202016410019
12. Bascom GW, Christensen KM (2017) The impacts of limited transportation access on persons with disabilities' social participation. J Transp Health, 7:227–234. https://doi.org/10.1016/j.jth.2017.10.002
13. Melnikova E, Lepert M, Popov A, Sorokoymova T (2019) Adapting urban areas for people with limited mobility. In: Paper presented at the E3S web of conferences, vol 97. https://doi.org/10.1051/e3sconf/20199701033
14. Asfaw B, Azage M, Gebregergs GB (2016) Latrine access and utilization among people with limited mobility: a cross sectional study. Arch Public Health 74(1):1–8. https://doi.org/10.1186/s13690-016-0120-5
15. Bekk NV, Taube MV, Boychenko IA, Cherkasova MA (2020) Ergonomic features and social demand for means of transportation for people with limited mobility. Paper Presented IOP Conf Ser Mater Sci Eng 944(1):012002. https://doi.org/10.1088/1757-899X/944/1/012002
16. Bezyak JL, Sabella SA, Gattis RH (2017) Public transportation: An investigation of barriers for people with disabilities. J Disability Policy Stud 28(1):52–60. https://doi.org/10.1177/1044207317702070
17. Aldred R, Woodcock J (2008) Transport: Challenging disabling environments. Local Environ 13(6):485–496. https://doi.org/10.1080/13549830802259847

18. Gamache S, Routhier F, Morales E, Vandersmissen M, Boucher N (2019) Mapping review of accessible pedestrian infrastructures for individuals with physical disabilities. Disability Rehabil Assisti Technol 14(4):410–422. https://doi.org/10.1080/17483107.2018.1449018
19. Gamache S et al (2020) Methodological insights into the scientific development of design guidelines for accessible urban pedestrian infrastructure. J Urban Technol 27(1):87–105. https://doi.org/10.1080/10630732.2019.1632677
20. Al Taweel Z, Challagundla L, Pagan A, Abuzneid A-S (2020) Smart parking for disabled parking improvement using RFID and database authentication. In: 2020 IEEE 6th world forum on Internet of Things (WF-IoT), New Orleans, LA, USA, 2020, pp 1–5. https://doi.org/10.1109/WF-IoT48130.2020
21. Carr K, Weir L, Azar D, Azar N (2013) Universal design: a step toward successful aging. J Aging Res 3:324624. https://doi.org/10.1155/2013/324624
22. Harsritanto B, Wijayanti A (2017) Universal design characteristic on themed streets. IOP Conf Ser Earth Environ Sci 99(1):012025. https://doi.org/10.1088/1755-1315/99/1/012025
23. Harsritanto B (2018) Urban environment development based on universal design Principles. E3S Web Conf. 31:09010. https://doi.org/10.1051/e3sconf/20183109010

Ukrainian Post-socialist Cities and Integrated Development

Vadim Vadimov and Anna Vadimova

Abstract Modern Ukrainian cities in the post-socialist time of their development found themselves in a situation of the lack of an appropriate urbanistic methodology. Socialist urban development doctrines of urban development no longer work in the new socio-political and economic realities and new methodologies are required. One of these methodologies is fractal urbanism. One of the postulates of this methodology for the post-socialist city is the UN Sustainable Development Initiative (SCI) and the Integrated Development Concept (ISEK).

Keywords Information society · Fractal urbanism · Integrated development architectural composition

1 Introduction

Modern Ukrainian cities in the post-socialist period of their development found themselves in a situation where there was no corresponding urban theory and methodology. The demographic situation in most Ukrainian cities remains tense [1].

The international project GIZ "Integrated Urban Development in Ukraine" (ISEK), which is supported in Germany by BMZ, and in Switzerland SECO is implemented in a number of Ukrainian cities, including Poltava [4]. The decrease in the population in Ukraine (41,588,000 people by 2020) is taking place against the background of an increase in the percentage of the urban population. In 1959, the percentage of urban population was 45.7% of the total population of Ukraine, in 1970 it was already 54, 5%, and in 2020 the percentage of urban population was 69.2%. The total number of urban settlements was—1345.

The high dynamics of urbanization in Ukraine did not in itself solve the problem of development of post-socialist Ukrainian cities, but on the contrary exacerbated the existing problems of spatial development of territories and cities, especially given the processes of decentralization and territorial administration in the country.

V. Vadimov · A. Vadimova (✉)
National University «Yuri Kondratyuk Poltava Polytechnic», Pershotravnevyj Avenue 24, Poltava, Ukraine

V. Onyshchenko et al. (eds.), *Proceedings of the 3rd International Conference on Building Innovations*, Lecture Notes in Civil Engineering 181,
https://doi.org/10.1007/978-3-030-85043-2_52

More cities in Ukraine have signs of a monofunctional industrial structure, because the enterprises are mostly one or two profiling sectors of the socialist economy, formulating the revenue side of the city budget, thus ensuring the functioning of social infrastructure. Such times are over. Today, against the background of rapid population decline, the issues of creating new jobs and forming the appropriate social infrastructure of cities have become more acute.

The process of urbanization in Ukraine continues, despite the economic, demographic and social challenges facing post-socialist cities, so cities should be given more attention. Today, cities are becoming points of growth and future development.

The implementation of new approaches in the development of Ukrainian post-socialist cities on the basis of the Concept of Integrated Urban Development (ISEK) requires the application of new methodological foundations of urban planning [2].

The aim of the work is to formulate relevant approaches, based on fractal urbanism and sustainable development methodology, for the development of the Concept of the integrated development of the Ukrainian post-socialist city.

Presentation of the main material. The theory and practice of urban planning in Ukraine during the period of socialist development was closely connected with the socio-economic planning of a directive nature, the central government agencies. The conditions of the centralized planned process (five and seven-year development plans of the state, individual economic districts, regions and cities) contributed to the complexity of urban development. Urban planning theory and practice of Ukraine has a glorious history of its development since the middle of the twentieth century, which was due to the large-scale construction of new industrial facilities and new settlements, reconstruction and development of existing cities.

At the same time, during the construction of Ukrainian cities of the socialist era, certain problems arose in spatial development. The priority was the organization of the location of productive forces, production facilities, and the social sphere was perceived as a supplement to the "city-forming" factors. Large industrial zones in the planning structure of Ukrainian cities after the decline of the economy, which was focused on energy-intensive, resource-intensive production, became "industrial deserts".

During the years of Ukraine's independence, fundamentally new economic, environmental and social challenges arose for the implementation of world and European experience of urban development for Ukrainian cities. Today, the processes of urbanization in Ukraine have their own characteristics. The processes of urbanization occur under the influence of external and internal factors of different territorial levels - from national, regional and local, which are interrelated. Among the external factors of the spatial organization of urbanization in Ukraine are urban agglomerations.

It is the analysis of the spatial characteristics of urban location in Ukraine that indicates the presence of urban agglomerations. Urban agglomerations of Ukraine have different levels of development, population and are morphologically specific to monocentric and polycentric (bicentric). About 19 million people or 36% of the country's population live in the 19 largest urban agglomerations of Ukraine.

Agglomeration forms of settlement must be taken into account in the formation of new territorial units in the process of decentralization.

The ecological component and the preservation of the natural environment remain relevant for the development of Ukrainian cities. The natural landscape component is an important element of spatial planning of territories at different territorial levels from national, regional to local, at the level of individual cities. In addition, for Ukrainian post-socialist cities, it becomes important to take into account the global characteristics of urbanization. At the beginning of the XXI century in the information society, there was a phenomenon of "global city", with the effect of implosion—explosive compression of space and time, removal of urban functions in the information space, social networks—they are objects of fractal urbanism. Such changes in global processes of urbanization are due to the formation of transnational corporations in the fields of industrial production and finance, the development of IT technologies and blockchain in the formation of a digital society with the provision of the latest digital services.

In addition to taking into account external factors on the spatial development of cities and surrounding areas in the process of developing the Concept of Integrated Urban Development, the state of use of internal resources is carefully considered. Such resources include rational land use, optimization of functional zoning and engineering and transport infrastructure. For example, abandoned Ukrainian territories are a significant problem for Ukrainian post-socialist cities. Built on the calculations of economic growth and full employment of the working age population, such industrial zones in cities occupy a large area (from 12–18 to 23–27%). At one time, working enterprises were the main lever for economic growth in cities.

To solve these topical issues of spatial development of Ukrainian post-socialist cities, modern geoinformation technologies (GIS) are needed, on the basis of which a model of permanent project process is implemented.

In Ukraine, since 2005, various attempts have been made to create a national infrastructure of geospatial data—NIGD. steps to build a similar system NSDI, INSPIRE in our country. In 2020, the Law of Ukraine "On the National Infrastructure of Geospatial Data" was adopted, which should summarize the existing scientific and practical experience in the formation of geographic information systems. Urban geographic information systems of urban management are a new tool in the implementation of the Concept of Integrated Urban Development.

Urban geoinformation systems, based on which the cadastral information resources are combined (land cadastre, town-planning cadastre, register of addresses, register of objects of town-planning activity, etc.) become the basis for the formation of a "smart city". The processes of urbanization have the features of fractality—in different areas, in different climatic and socio-economic conditions there is a spatial compaction with different morphology, which should be monitored using GIS technologies.

Therefore, spatial planning in information systems acquires the features of a permanent spatial (urban planning) design process, the essence of which is that the existing hierarchical system of urban planning [3], implemented at three interconnected territorial levels (national, regional and local), in information systems is transformed into a rotational development system. urban planning documentation. At the same time, the subjects of the permanent design process, objects, processes and norms are determined (Fig. 1). The basis of this process are urban development programs of dynamic spatial objects [5].

Urban planning programs are a new type of spatial documentation where urban planning strategy and its constructive development are concentrated. Urban planning programs are not rigidly fixed in their indicators planning documentation, when the conditions change, the recommendations on possible implementation options will change. Spatial objects of different urban territorial levels (national, regional, local) are complex systems. An integral part of the management of complex systems is the methodology of nonlinear equations, on which the theory of dynamical systems is based and which form the methodological foundation for the study of complex systems. Complex systems are first of all dynamic, from the point of view of the available potential of self-development, systems. Urban development programs establish, on the basis of nonlinear characteristics for spatial objects, the anticipated course of events and rules that ensure the implementation of the planned.

Urban development programs are problem-oriented, ie the formulated problem field at a particular point in time is key. This can be ensuring economic stability, environmental security, protection of historical and cultural heritage, elimination of the consequences of man-made disasters, and others. There are options for combining and correlating problems depending on the territorial level for which the program is being developed. In this case, there is a need for subject-oriented and object-oriented programs that can (depending on the problem) be integrated into integrated urban planning programs. Urban development programs include analytical and constructive parts [5].

The analytical part contains: 1) study of the current state of objects and their constituent subsystems—social, economic, environmental (establishment of patterns, trends, formulation of major problems, followed by the allocation of priorities); 2) study of the conditions and characteristics of the factors that determine the functioning and development of objects; 3) development of a hypothesis of development of situations, processes or structural states according to those areas of programming which are recognized as key, formulation of target area and development of variants of its achievement; 4) analysis of resource opportunities, management tools, legal framework, regulatory framework, time constraints, social requirements, political institutions, research and design support and other factors and conditions that potentially must have a society and territory to achieve the results contained in the target area of the hypothesis; 5) development of the concept and relevant urban models, which means the formulation of specific goals (objectives) in the form of a set of desired and potentially achievable results in a generally parameterized form (program-target model) aimed at achieving goals, taking into account resource capabilities. The concept means an algorithm for solving a specific problem (problems), which is a set of the most important elements of the theory (theories) regarding the problem (problems) under consideration, given in a constructive and acceptable for practice form.

The constructive part forms, analyzes and evaluates the range of possible solutions, one way or another focused on achieving the goal and selects those that are most able to change the course of processes, problem solving or change the structural state of

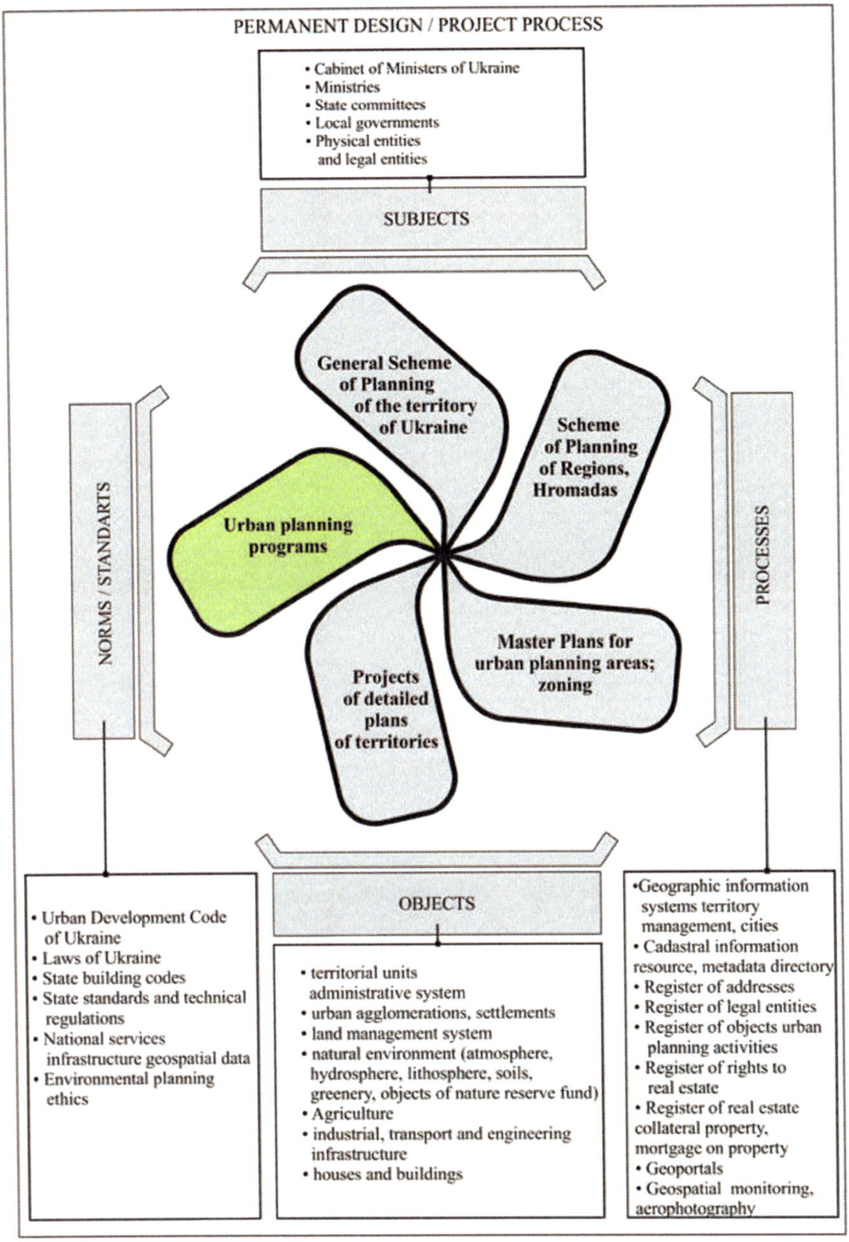

Fig. 1 Permanent design process in conditions information systems for integrated development management

the object in the desired direction. These decisions must be twofold. This means that normative-target models must be resource-based, ie the measures outlined in them, the amount of investment must be invested in real opportunities, take into account real trends in the formation of natural, labor, energy, water, territorial and other resources, investment growth, change their structure, resources of the construction industry, building materials industry and other and the whole system of external and internal constraints.

Moreover, the adopted normative-target models for the near future should be resource-provided, ie all measures should be based on actually proven material, financial, labor and other resources. In addition, in the constructive part it is necessary to: - first, take into account the range of main consequences of decisions affecting the interests of various actors in regional activities; - secondly, based on the target interests of this program, take into account the necessary and sufficient set of territorially related factors that can influence the decision taken and reflect them in it; thirdly, to create the necessary, sufficiently complete, but not excessive information base, able to ensure the implementation of the above two conditions. Implementation of urban development programs should involve the creation of significant information arrays and conduct certain operations with them. There is a need for a database with information about the area, which allows you to track the status of design objects, make the necessary adjustments [5]. The international project GIZ "Integrated Urban Development in Ukraine II" (ISEK) continues for the next three years the implementation of modern models of project processes in Ukrainian post-socialist cities, among which will be urban programs of dynamic spatial objects.

2 Conclusions

The theory and practice of urban development of Ukrainian cities of the post-socialist period, as a result of the implementation (2015–2020) of the international project GIZ "Integrated Urban Development in Ukraine" (ISEK) gained useful experience and legal basis by adopting the Law of Ukraine № 711-IX (17.06.2020). Comprehensive spatial development plans of territorial communities and cities, based on modern methodological approaches, include Concepts of integrated development, thus ensuring the sustainable development of territories and settlements in the future.

References

1. World Bank Statistics [Electronic resource]. Access mode: wbannualreport@worldbank.org
2. White Paper (2017) Proposals for legislative and administrative-legal reform of urban development in Ukraine. Strong cities - strong Ukraine. The first edition. Deutsche Gesellschaft für Internationale Zusammenarbeit (**GIZ**) GmbH – K, 135 p

3. Vadimov V (2018) The latest methodological approaches inthespatial development of Ukrainian cities/Smartproject, building and city. Cracow Univ Technol 23:156–160 Chair of housing environment
4. The concept of integrated development of the city of Poltava. [Electronic resource]. Mode of access to the resource: http://www.2030.poltava.ua/ua/kontseptsiya-intehrovanoho-rozvytku-mista-2030
5. Vadimov V (2019) Features of spatial planning in the conditions of integrated urban development in Ukraine. Practical comment, p 132. Divosvit, Poltava

Lighting Means as Factors Influencing the Formation of Architectural Environment

Aleksandr Vasilenko, Andrii Koniuk, and Kateryna Palii

Abstract Architecture provides comfortable environment, adapts to regional peculiarities of natural environment. Going through the atmosphere, radiant energy of the Sun, is partially scattered, absorbed and reaches the Earth's surface in the form of direct radiation and radiation scattered from the sky. Part of the radiant energy is reflected from the underlying surface, surrounding objects, and creates reflected radiation. Elements of the radiation mode are necessary to assess natural lighting of the premises. As the height of the Sun above the horizon rises, there's observed a rise in both direct and scattered components of radiation. Cloudiness affects the intensity of scattered radiation. Transparent atmosphere increases direct and reduces scattered components of radiation. Among the various concepts of rural houses in the middle of the XXth century the most widespread appeared to be five types of planning (three of them being typical for a large territory of Ukraine), which turned out to be the result of further development and improvement of the previous planning techniques.

Keywords Comfort of architectural environment · Climatological factors · Urban development · Natural light

1 Introduction

Within the concept of urban and ecological approach to the formation of the human life environment one of the main priorities is the research and design direction of creating bioclimatic and ecological architecture, which enables to multilaterally use

A. Vasilenko
Odessa State Academy of Civil Engineering and Architecture, Odessa, Ukraine

A. Koniuk (✉) · K. Palii
National University "Yuri Kondratyuk Poltava Polytechnic", Poltava, Ukraine

V. Onyshchenko et al. (eds.), *Proceedings of the 3rd International Conference on Building Innovations*, Lecture Notes in Civil Engineering 181,
https://doi.org/10.1007/978-3-030-85043-2_53

solar energy and innovative lighting technologies. Under the influence of anthropogenic factors, reduced transparency of the atmosphere due to urban development, there occur significant changes in climatic and insolation resources. Especially it concerns urban development. Light is a natural substance. Natural and artificial lighting is a means of creating an expressive architectural composition which provides favorable conditions for interaction between a human and architectural environment. Daylight reveals the shape of an architectural object and creates a "light image". At night, architecture becomes a source of artificial lighting using technical design tools. Architectural environment creates an "aesthetic image".

The problem of determining energy parameters of the environment requires: -continuous improvement;—implementation of European norms in Ukraine;—harmonization of the national regulatory framework with EU norms. The problem of forming a complex of lighting means in architecture is in the focus of social attention and is a central one in the field of urban planning, architecture and ecology.

The analysis of scientific papers singled out the following unresolved issues. 1) In the process of designing facades of buildings not enough importance is given to the role of the form-creating function of natural and artificial light; 2) The problem of the influence of natural light on comfortable conditions for people has not been studied.

Topicality of the issue is accounted for by the need to improve scientific and well-grounded methods of designing architectural objects, taking into account modern lighting tools. The technique suggested is not fully considered though.

2 Formation of Lighting Architectural Environment

Examples of the first urban planning experience prove that the city plan was oriented on compass points. In northern countries it was sought to get more sunlight. In hot countries it was sought to get protected from warm rays. Residents of southern states when choosing territories for the construction of their cities were looking for favorable natural climatic conditions. This is evidenced by the location of Priena (the fourth century BC), Assos, Cnidus cities and others. Each of them has a pronounced southern orientation. Cities opened up in the direction of the sea and took advantage of the sea breezes. Warehouses, theaters and residential buildings with courtyards were oriented to the south [1].

Ukrainian cities Kyiv, Poltava, Kharkiv, Dnieper are located in the 4-latitudinal belt. Current scientific research has proven the need to take into account the possibility of solar radiation which has a great impact on the light climate of the territory [2].

Aesthetic qualities of natural lighting are revealed: 1. The upper part of the sky has the greatest brightness. Objects on the horizon are characterized by medium brightness. 2. Even distribution of light–on flat surfaces. Uneven distribution of light–on curved surfaces. 3. Natural lighting is created directed, diffused and reflected. 4. Architectural rhythm is accompanied by the rhythm of light and shadow. 5. Lighting contrast (light and shadow–aesthetic category).

Folk housing in the plan has an elongated composition. Large scale type of peasant housing in Ukraine in the nineteenth century was represented by different variants of houses. Such housing traditionally had one room.

The most common are proportions of doorways for these types of housing. Door of rectangular form is widespread. In Poltava region at the end of the nineteenth century hexagonal doorways were used. Greater width of powerful bars endowed the opening with monumentality. Massive plastics of exterior door framing becomes more yielding. Additional elements gradually appear in the solution of the window, creating a cornice profile, eaves are getting larger. In the eastern part of Ukraine, the greatest attention is paid to the treatment of windows. In the western part, attention is given to the design of doorways. In the 50s of the twentieth century, galleries were made glazed. The rest of the gallery was open. In the western regions, the overhang of the roof in the galleries rests on columns. These architectural elements have massive plastic decorative carvings. The entrance to gallery in Lviv region was made in the form of a wooden arch. In Zakarpattia (Transcarpathia region) galleries were made in the form of an arcade [3].

The climate of southern, eastern, coastal regions of Ukraine is characterized by strong, dry winds. The porch began to be built glazed. Due to a large number of wet days with prolonged rains and fogs, porches began to be designed in the western regions of Ukraine (see Fig. 1).

Fig. 1 In the middle of the twentieth century, a porch covered with a gabled roof with a pediment became widespread. The porch (a small canopy over the entrance) was supported by light wooden supports

In the 70–80s of the twentieth century in European countries (Hungary, Czechoslovakia, Poland, Bulgaria), the tendency to provide land plots got widespread. Typical projects become available for individual construction, which include one-storey residential buildings and multi-apartment residential buildings with apartments on two levels. The best individual projects are used for implementation. National and climatic features of the region are taken into account in projects. Traditions of folk architecture of the Czech Republic and Slovakia are used in the solution of the main facade of a one-apartment house of the mansard type aimed for one family living. The project has the main features of the twentieth century housing laid down. The main facade of the house is recommended to be oriented to the north or west, the attic being equipped with a loggia [4].

3 Architectural Form and Natural Light

Vitruvius analyzed the motion of the celestial planets and proposed "analemma of the Sun", which is considered to be the main method guide to calculate insolation. Architects of Vitruvius times called the scientist the first specialist in urban planning. The review of historical samples has showed that in the formation of the form architects solved the problem of harmony of natural light and architecture. Vitruvius emphasized that– "... one part of the Earth lies directly under the path of the Sun, another one–is far away from it, the third one–is in the middle between them. Therefore, due to the inclination of the zodiacal circle (celestial equator) and the path of the Sun, different parts of the Earth receive different amounts of solar energy; it becomes clear that climatic conditions of different countries must be taken into account to the same extent when arranging houses. ...".

According to Vitruvius (see Fig. 2), "in the north buildings should be closed. In southern countries, buildings should be open to the north or northeast. Thus, we can repair the damage caused by nature through art" [5].

Fig. 2 Marcus Vitruvius Pollio. An outstanding Roman architect, engineer, theorist of architecture of the second half of the first century BC. He was born a free Roman citizen in Campania. Got an architectural education. During the Civil War, he participated under the leadership of Julius Caesar in the construction of military equipment [10]

Fig. 3 All the premises of a Greek house were located around the courtyard, in which a room dedicated to god Zeus—the patron saint of the family—was located. Covered passages around the perimeter of the yard turned into a colonnade. This type of Greek interior is called peristyle [11]

At high standing of the Sun, the irradiation of the hemispherical surface will be much weaker. The temperature of surface heating will also be lower. Vaulted shape creates conditions for effective night thermal radiation of the coating surface into space. This promotes rapid cooling [6, p. 38].

To create a comfortable lighting environment, a special design solution was made: the use of natural material; interior design development; the use of wooden sun protection devices. In the Sahara Desert, designing patios is a must in traditional architecture. The courtyard is a protection against excess light and heat. In Egypt, Greece, Italy, Persia, and throughout Islamic architecture, high fences were traditionally designed to protect against sunlight and sand. These methods were most actively used in the area between 15° and 37° north latitude (Fig. 3).

In countries with mild and warm climates (USA, China, Iran, Japan), a residential house is a transformer. And such type of housing has become national. In houses, outer walls are movable elements which are made of lightweight materials (wood, bamboo, reeds, paper). Movable partitions can be used. The system of sliding partitions enables to open walls towards the garden, turns a part of the house into a terrace. In China, a window frame with a paper coating is used. Depending on the position of the Sun, the house is opened in different directions. On hot days, raised walls of the houses form a common roof over the houses. Residential areas are transformed into a single interior. Later on transformers could increase or decrease in volume. Three sections could be shifted and turned into one. Architects around the world design a variety of houses (transformers) which are different from each other and are designed using the latest achievements of science and technology (see Figs. 4 and 5).

The blocks are mounted on special rotating platforms. They are used on theatrical stages and car shows. The architects modified this system and eliminated the possibility of vibration of the walls when the rooms are being rotated. In summer, the blocks get extended from the plane of the building and provide the room with sunlight and ventilation. In the cold season, they are closed. The house consists of six floors. In the basement there is a gym and relaxation areas. On the first and the second floors

Fig. 4 Houses (transformers). Architect–Adam Culkin (USA). Created a structure based on containers. The project direction is popular. Here can be found compact buildings with a minimum of amenities for a dozen of square meters and comfortable apartments, consisting of several modules [12]

Fig. 5 House (transformer–1400 m^2). Tehran, (Iran), 2019. Project of architectural bureau Nextoffice. Residential house. Changes its shape depending on the wishes of the owners. Specialists of Iranian architectural bureau Nextoffice designed this original Sharifiha house. The width of the building is quite narrow. Designers have found an unusual solution to this problem. The house has three huge rooms. Reminiscent of wooden cubes. Depending on weather conditions and the desire of the owners, they can go outside at 90°. From two-dimensional the facade becomes three-dimensional [13]

there is a living room, a kitchen, and a cinema. The other floors have bedrooms, bathrooms and additional guest rooms. The house has a built-in elevator.

Lifting shutters give a possibility to vary the lighting of the room in many ways. Lifting shutters, panel and lattice, common to the east of China. Each wall that is transformed is divided in height into four parts–fixed and movable transoms. Natural light makes it possible to clearly see all the components of nature and architecture.

Light and shadow is the distribution of light gradations on the surface of three-dimensional shapes. Knowing the laws of chiaroscuro, one can use them to create a proper artistic image of an architectural work.

The degree of illumination of the surface of an architectural object can vary within different limits, depending on 1) the brightness of the light source; 2) the size of the angle at which the rays of light fall; 3) the distance of the work surface from the light source; 4) the texture of the illuminated body and color (see Fig. 6).

The intensity of lighting depends on the incidence angle of light rays. The brightest illumination will always be at the point where the beam of light forms the largest right angle with the illuminated surface. If a ray reflected at any point on the surface of an illuminated three-dimensional body enters the eye of the beholder, that point is visible as the brightest point on the illuminated surface and is called a brilliant point or glare. When there is an edge on the form, all points of which approximately

Fig. 6 Some rays of direct light are reflected and scattered. Architectural objects are illuminated by combined light. Such light consists of a complex mixing of streams of direct, reflected and scattered light [14]

correspond to this condition, it appears to be lighter than the adjacent surfaces. The gloss strength of a point and an edge depends on the degree of the reflecting surface.

The texture affects the lighting. The intensity of the illumination of the object decreases with the increase of the roughness and unevenness of the body surfaces. Sometimes irregularities on the surface of the form reach such dimensions that their illumination drops sharply due to the appearance of their own shadows. The luminance of the surface depends on its color. White reflects light, and black absorbs light.

Shadows can be falling and own (proper). Protruding element of the facade is an obstacle in the way of direct light rays. Behind each illuminated object there is a space where direct light does not get. The shadow of this origin is called falling. Own shadow is formed on the object itself, where rays of direct light can not fall. This surface is turned away from direct light rays. If the rays of direct light are parallel to the plane of any of the faces of the body, then such a face will be in the shadow. The plane of such a face is conventionally called the plane of the tangent shadow (see Figs. 7 and 8).

The boundaries of the falling shadow lose their clarity as they move away from the body. The boundaries of the shadow in this case increasingly blur and form a gradual transition from shadow to light. This phenomenon is observed in shadows formed by the action of reflected and scattered light. The reason for this phenomenon is that a semi-shadow zone is formed on the border of the falling shadow. A penumbra on the boundaries of a falling shadow always appears when the light source is not an ideal point but a surface.

The degree of intensity of falling and own (proper) shadows is various and depends on–1) the power and position of the light source; 2) the distance of an object or a

Fig. 7 Residential house. Fragment of the facade. The texture affects the lighting. The intensity of the illumination of the object decreases with the increase of the roughness and unevenness of the body surfaces. Sometimes the irregularities on the surface of the form reach such dimensions that their illumination drops sharply due to the appearance of their own shadows and those that fall on the surfaces themselves. The luminance of the surface depends on its color. White strongly reflects light, and black, on the contrary, actively absorbs it [15]

Fig. 8 Residential house. Fragment of the facade. Falling shadow and shadow proper differ in the strength of the tone. Unilluminated space behind the shadow-giving body absorbs some of the rays of light scattered in the atmosphere, while the areas of its own shadow (shadow proper) have the possibility to receive more scattered and reflected light. Falling shadow becomes weaker as you move away from the body that forms it. It happens for two reasons: due to the action of rays of light scattered in the atmosphere and due to the influence of the air environment (aerial perspective). Both of these factors are the reason why the shadows, respectively, appear to be more pronounced in the close layout [15]

group of objects from the viewer; 3) the shape and the nature of the body surfaces; 4) the presence of illuminated neighboring objects; 5) the state of the atmosphere.

Rays of light scattered in the surrounding atmosphere, as well as those reflected from the surfaces of illuminated objects nearby, can illuminate and make visible the relief and the shape of the shadowed parts of objects. Rays of the reflected light, falling on the shadow area, cause the appearance of lighter spots in the shadow (reflexes). Reflexes are observed mainly in shady or dimly lit parts of the object. Their intensity can be more or less significant, which mainly depends on the degree of illumination of the surface that reflects light, its color (white, as has been repeatedly noted, most strongly reflects light) and the proximity of the surface which reflects light to the surface, on which the reflected light falls.

Particularly noticeable reflexes appear when the rays of reflected light fall on the given surface at right angles. In the falling shadow, reflexes are much weaker.

Light and shadow gradations help to create the impression of volume and spatial depth of the facades of architectural objects. The main stages of chiaroscuro gradations: light that falls at right angle to the surface of the object, which determines the most illuminated place, which is second only to the glare; a halftone that appears when the surface is illuminated by rays of light falling at an acute angle to it; falling shadow; own shadow; reflex; reflection.

Chiaroscuro is a significant means of expression of an architectural form: it participates in light "sculpting", modeling of architectural forms, is a means that characterizes space and depth.

Ukrainian scientists I.N. Skryl and S.I. Skryl were dwelling in their research on the superbly versatile influence of insolation on the environment, urban development processes, architectural and aesthetic role of insolation. Insolation actively affects the contrast of architectural forms, their color perception, the depth of the shadow. Architectural objects are perceived differently in different climatic regions. For instance, small protrusions from the vertical plane on the facades in the southern regions of our hemisphere look contrasting and having a significant depth of shadow at noon and, conversely, in southern latitudes, they remain invisible or blurred due to the reduced contrast and slight depth of shadow [7, pp. 90–93].

The most complete and correct idea of the color of the surface is obtained during sunlight at noon, because sunlight includes the entire monochromatic composition of the visible part of the spectrum with a more or less uniform distribution across the width of the spectral range. In the case of illuminating surfaces with diffused light from the sky, warm colors (red, orange) will be slightly dimmer, because diffused light from the sky contains a smaller part of these rays than the direct one. With the change in the spectral composition of natural light as for the latitude of the area similarly changes the perception of the color of surfaces from south to north.

In southern latitudes, midday colors will be rich, bright and clean, and in the north they will have pastel, calm tones. Herewith, perception of color is greatly influenced by the brightness of lighting. As the brightness of natural light decreases in our hemisphere from south to north, saturation and coloristics of the surfaces decreases accordingly. With the change of the cloudiness of the skies and even the time of the day, the level of illumination of the surfaces also changes, i.e. decreases compared to noon lighting and cloudless sky.

As the brightness of the sky decreases, the level of lighting and color perception decreases too. It shifts towards short waves. Purkine effect comes into action, in which a different darkening of colored surfaces with the same brightness is observed, while their brightness is being reduced. Moreover, using warm (yellowish) colors, thanks to this effect, it's possible to achieve the impression of significant light saturation, and using cold (bluish) tones for decoration, we get the impression of low lighting. With the onset of twilight, the human eye switches to peripheral vision with a high level of sensitivity. The level of color perception by the eye decreases sharply or disappears altogether, because a person is able to distinguish color only due to the central vision and the action of light reflected from the surfaces on the cones. As the light level decreases to a certain value, objects lose color.

At very high brightness of objects, the eye, protecting itself from excessive brightness, reduces the angle of visual perception of objects. As a result, color perception seems to fade, halftones are distinguished poorly. Here appears the so-called phenomenon of colors "absorption" by light [7, p. 92].

Insolation has not only positive but also negative properties, for example, sunlight affects the "fading" of facades, i.e. under the action of ultraviolet rays exothermic, and with some dyes–endothermic reactions occur. Dyes vary in chemical composition and color [8, p. 92].

4 Conclusions

1. As a result of the conducted analysis traditional architectural and compositional methods of creating forms, which promote the formation of the comfortable light architecture in natural and climatic conditions of Ukraine and European countries, are revealed and systematized. By moving from a cold to a warmer climate zone, the degree of compactness of the real estate development changes.

2. It has been established that these are the climatic conditions and the historical and cultural development of the region that served the basis for the emergence of the architectural concepts for the formation of a complex of lighting means.

3. The principles of traditional (invariant) architectural and compositional methods of forming optimally comfortable light architecture of Ukraine include: morphological; 2–two-line; 3–single-line; 4–L-shaped; 5–U-shaped; 6–closed.

4. The study identified the stimuli of the psychological aspect of insolation. The main factor that determines a person's psychological reactions is the time of insolation and the direction of sunlight and the visual "awareness" of a person of the presence of insolation not only indoors but also in the outside environment.

References

1. Gusev NM, Yu AV (1975) On the design of relief finishing of industrial elements of buildings of mass construction. In: Gusev NM (ed) Proceedings of the institute. Construction lighting. Issue 13(XXVII)/Nauch. NIISF, pp 102–115
2. Obolensky NV (1988) Architecture and the sun. Stroyizdat, p 207. ill
3. Samoilovich VP (1989) Folk architectural creativity of Ukraine/VP Samoilovich, 2nd edn, Reworked and ext., 344 p, Budivelnyk
4. Code of rules of JV 52.13330.2016 (2016) "Natural and artificial lighting" updated version of SNiP 23–05–95 * (approved by the order of the Ministry of Construction and Housing of the Russian Federation of November 7, 2016 N 777/pr). Daylighting and artificial lighting
5. Bunin AV, Savarenskaya TF (1979) History of urban planning art. T.2, 486 p
6. Firsanov VM (1982) Architecture of civil buildings in hotcClimates: a textbook for students of architectural and construction specialties of universities, 2nd edn, Reworked and ext., 248 p. Higher school, ill
7. Skril IN, Skril SI (2006) Fundamentals of architectural light science (calculation and design of natural, artificial and combined lighting and insolation): Textbook–К: Вища школа, 214 p, ill
8. Hraska J (1988) Doba insolacie okien tienenych zastavbou. In: Zbornik vedeckych prac Stavebnej fakulty, SVST 1985. Alfa, Bratislava
9. State Building Standards of Ukraine. DBN B.2.5–28: 2018 Engineering equipment of buildings and structures "Natural and artificial lighting". (In force since 28 February 2019)
10. https://www.thefamouspeople.com/profiles/vitruvius-34320.php
11. Internet source: https://kulturologia.ru/blogs/091218/41565/

12. Internet source: https://dom.ria.com/uk/news/interesnye-doma-transformery-176438
13. Internet source: https://novate.ru/blogs/010814/27179/
14. Internet source: https://roomester.ru/dekor/svet/idei-razmeshheniya-sveta-v-interere.html
15. Internet source: https://artfasad.com/fasad/

The Changing Role of Managers of Different Levels in the Increasing Employee Loyalty in the Construction Organizations

Konul Aghayeva and Iryna Krekoten

Abstract The role of managers has been considered in the motivational processes in construction organizations. It has been indicated that modern attitude to the increasing of employee loyalty requires effective communication of all level managers with employees. It makes employees feel valued and important for the organization. In this direction the role of lower-level managers should be increased, and it is necessary to provide their engagement in the decision-making process especially regarding setting motivational policies.

Keywords Motivation · Loyalty · Construction company · Top-level managers · First-level managers

1 Introduction

Obviously, it is a fact that the construction industry in Azerbaijan is one of the fastest growing industries, but there have been a few studies related to human resources management in construction sphere of the country. One of these studies examined the influence of cultural diversity on innovation in the construction industry, another study analyzed how human resource management practices of Azerbaijan companies in construction filed could improve final performance. However, comparative studies assessing the difference in the expectations and performance of employees at different levels must be conducted and completed. When top-level managers design a universal corporate motivation system, they do not take into account the difference between attitude of employees to results of company that exist at different organizational levels. This mistake leads to failure to achieve the maximum productivity.

K. Aghayeva (✉)
Azerbaijan University of Architecture and Construction, 5 Ayna Sultanova, Baku, Azerbaijan

I. Krekoten
National University "Yuri Kondratyuk Poltava Polytechnic",
Pershotravnevyj Avenue 24, Poltava 36011, Ukraine

V. Onyshchenko et al. (eds.), *Proceedings of the 3rd International Conference on Building Innovations*, Lecture Notes in Civil Engineering 181,
https://doi.org/10.1007/978-3-030-85043-2_54

Understanding this gap will lead to increasing of corporate efficiency and provide high competitiveness of construction companies in Azerbaijan.

2 Main Part

2.1 Purpose of the Article

The aim of this study is to identify the optimum system of motivation and encouragement in human resource management in modern building organizations by considering roles of different levels' managers in order to achieve high productivity.

The study draws on the works of world scientists devoted to the problems of forming and developing a system of human resource management of construction companies in a competitive environment, the management and reproduction of their labor potential, the principles of comprehensive solutions of management problems.

2.2 Research Methodology

The solution to the tasks set in the study was carried out using a systematic approach (when studying the motivational potential, the specifics of the activities of construction companies and their influence on the process of human resource management in these companies) and an abstract-logical approach examination (in the analysis of terminology).

2.3 Results

The paper is aimed at improving research and methodological foundations of human resource management in construction companies by means of motivation and incentives.

Attainment of the set goal presupposes the fulfillment of the following tasks:

- analyzing role of managers in motivation processes of employees.
- conducting survey for identifying levels of employee loyalty as an attitude of employees to existing motivational policy in construction organization.
- perfecting the mechanism of incentives for the personnel of construction companies by changing role of managers in this process and thus increasing loyalty of personal.

The basic of employee's loyalty is in their motivational level. Employee motivation is one of main ways to increase labor productivity, which, is a key element of company's human resource policy. The most popular system for boosting employee motivation is a system that provide attaining by employees individual goals, to grow as a professional, and to reach new achievements. The effectiveness of this policy can be checked through the KPI, key performance indicators.

Most theorists, who explored motivation systems, claimed that only the specified motivation system is perfect, as it justifies remuneration and helps employees to earn more and increase their income by spending greater efforts. The system proves flexible solution; each employee has own reason to be motivated in the workplace, and gain a result that he/she desires.

However, some part of motivational policy in the company should be based on the common principles and rules that allow determining which factors can be considered as motivational or demotivational when applied to different teams. The understanding of this process can be based on the traditional theories of motivation (i.e., Maslow's theory, Adam's Equity theory, Herzberg's theory, Vroom's expectancy theory).

In the market economy, all enterprises particularly construction enterprises must continue to operate successfully, and for this purpose, it is necessary to provide correct personnel policy that is important for the sustainable development in this field. Construction companies invest a lot of energy and effort into increasing customer loyalty, but it is necessary to mention that one of the most important keys to a company's success is the level of employee loyalty. Different studies found that a loyal employee is an enthusiastic, interesting, ambitious person who wants continuously to work and develop with the company. In addition, the manager, being a factor that directly affects the employee's performance, is an important factor in the formation of employees' loyalty in the construction enterprises.

At the same time, according to different motivational theories, other factors of working conditions also affect loyalty. Among them, we can mention salary level, relationships with colleagues, management styles, job content, career development, managers' activities, workplace design, team, etc. It has been proved that managers at different levels influence the performance and loyalty of employees.

Our study, based on previous survey [1], identified motivational and demotivating factors that impact the level of employee loyalty in construction companies and thus the impact of it on a company's outcome.

The results are given in Table 1.

2.4 Scientific Novelty

The study provides an opportunity to present its results as a theoretical generalization of managers' role in the motivation process of construction companies. It suggests the need to consider motivation and employee recognition for labor activity in the construction company as the most important factor in work performance improvement. It is important to consider the role of different levels managers, which cause the specifics of human resource management in modern building companies that are considered in the article.

Table 1 List of factors motivating and demotivating construction workers

Motivational factors	Demotivational factors
Compensation	Underpayment
Job security	Poor supervising
Quality of site management	Unsafe work conditions
Timeliness of wage payment	Changing of workmates
Bonuses and benefits	Aggressive and autocratic management style
Good relationship with coworkers	Conflicts
Job satisfaction	Prevalence of individual interests over organizational interests
Appraisal by managers	Long working hours
Good safety program	Incompetency of coworkers
Strict job schedule	Rework
Recognition policy	Lack of recognition
Flexible work	Chaos/adhocracy
Duration of break-time	Overcrowded work areas
Engagement in decision-making process	Communication problems
Complexity of work	Poor organizational ethics
Status of work	Lack of cooperation
Career growth	Underutilization of skill
Availability of contract	Lack of participation in decision making
Job responsibilities	Poor inspection
Training opportunities	Not enough responsibility
Location of work	Not enough challenge
Cultural compatibility	

2.5 *Practical Importance*

However, the research in the field of human resource management in construction organizations in the majority of cases are descriptive in nature; some issues of motivation and encouragement are light poorly or are not considered at all. The lack of science-based approach that takes into account the specific features of building sector is one of the reasons for this kind of research.

The guidance notes for practical application of theoretical leadership developments in building companies also need improvement. The modern leadership methods do not meet the requirements of practical activity. With this aim the results of the study can be applied in the work of construction organizations by making motivation more systematic and theoretically based.

3 Conclusion

Thus, it is important to have systematic basis for making competent managerial decisions:

1. Investigate the level of employee satisfaction and loyalty through a questionnaire.
2. Improve existing programs for increasing staff loyalty based on the information received as a result of the questionnaire.
3. It is important to form loyal staff at the beginning step of work when employees are newcomers, because there is high possibility of conflict of interests at the initial stage, so it is necessary to determine the expectations of the employees. However, the managers of construction companies must keep in mind that satisfying all employees is unrealistic, always due to natural causes, such as movement, illness, childcare, etc. At the same time, the retention of low-productive employees, on the contrary, will have the negative economic impact on the company. Therefore, being aware of the cost of employee resignation and the level of turnover in the company will allow managers to provide the cost-effectiveness of the loyalty programs [3, p. 14].
4. The results of the study concluded that top managers have a big impact on the performance of employees and their loyalty. In this regard, top managers in construction enterprises should improve their communication with employees. It makes them feel valued and a part of the company.
5. In addition, the role of lower-level managers should be increased. They act as a kind of link or communication tool between top management and employees. For this reason, they are more likely to interact with and influence employees in the work process. Therefore, encouraging lower-level managers to involve in decision-making process can increase their self-confidence and, as a result, create more productive work process [5, p. 26].
6. Employee performance appraisals should be conducted periodically. No matter how loyal an employee is to the company, working in the same position for a long time and not being valued by managers can lead to a decline in productivity after a certain period. In this regard, an impartial and objective manager plays an important role.
7. When determining the motivation policy applied to employees, as well as the reward system, the attitude of managers at different levels should be considered and a joint decision should be made.
8. Resolving communication problems between previous and new generation managers can be an important factor in achieving positive results. Business owners try to take their companies to a new level, combining the theoretical base of young

people and the real experience of older leaders and applying modern management systems, this will help construction companies to achieve their goals. In many cases, the old workers do not welcome young people, consider them as a threat to their careers and well-being, and for this reason they openly or covertly resist changes and innovations, and sometimes even openly sabotage them.

9. Injustice in the work environment is unacceptable. If the employee sees equality in the work environment, he will consider the reward system adequate and will show a high level of motivation. If he observes inequality, his level of motivation will decrease. For this reason, managers must be impartial and completely objective in assessing the performance of employees. Managers must be able to distinguish personal relationships from business relationships. It is important to understand that an employee is hired for his/her knowledge and fired for indiscipline. The manager must be able to set tasks correctly and distinguish between responsibilities and obligations.

10. It is mandatory for managers to join self-development programs.

11. Increasing employees' loyalty using only financial tools is not the right approach. It has effect only for a limited time. However, non-financial motivational factors must be used to provide long-term and effective loyalty. Among these factors, there are achievement, recognition, the work itself, responsibility, advancement [7, 8, p. 55].

References

1. Aghayeva K, Ślusarczyk B (2019) Analytic hierarchy of motivating and demotivating factors affecting labor productivity in the construction industry: the case of Azerbaijan. Sustainability 11(21):59–75. https://doi.org/10.3390/su11215975
2. Brown S, McHardy J, McNabb R, Taylor K (2011) Workplace performance, worker commitment, and loyalty. J Econ Manage Strateg 20(3):434
3. Blazovich JL (2013) Team identity and performance-based compensation effects on performance. Team performance management, 98 p
4. Mir-Babayev R, Gulaliyev M, Shikhaliyeva S, Azizova R (2017) The impact of cultural diversity on innovation performance: evidence from construction industry of Azerbaijan. June 2017 Econ Sociol, 10(1):78–93. https://doi.org/10.14254/2071-789X.2017/10-1/6
5. Nguyen L (2017) The impact of employees motivation on organizational effectiveness. Business economics, 51 p
6. Tarvonen J (2018) The role of corporate culture in company performance. Business economics, 37 p
7. Iasechko S, Haliantych MK, Skomorovskyi VB, Zadorozhnyi V, Obryvkina O, Pohrebniak O (2020) Contractual relations in the information sphere. Syst Rev Pharm 11(8):301–303. https://doi.org/10.31838/srp.2020.8.46
8. Heather JD, Younger R (2017) The 7 intuitive laws of employee loyalty, 110 p

The Main Vectors of Labor and Education Transformation of Modern Workforce

Alla Bielova, Svitlana Koval, Nataliia Zhuravska, and Asaf Agayev

Abstract The article reveals that in the long run social and labor relations may undergo radical changes in three main directions or vectors.

First, the boundaries of the traditional division of labor will change, the boundaries of professions will be erased, the rate of "extinction" of traditional professions will accelerate, and new, previously unpredictable ones will emerge.

Second, forms of employment are changing. Along with the traditional contractual forms of labor relations, employment in the form of freelance, crowdsourcing, insourcing, flexible forms of inclusion of professionals in labor activity, remote employment, project form of employment, etc. are actively developing.

Third, human mobility increases throughout employment. This is due to the intensification of migration processes (more than 60% of those who change their place of residence do so for reasons of work and employment), as well as inter-professional, intersectoral, intra-firm mobility.

"Lifelong learning" as a principle and concept is increasingly developed and implemented in practice. The trend is becoming more stable when a person changes his professional affiliation 3…4 times during his working life. In the workplace, traditional barriers to the professional division of labor are increasingly being overcome. Jobs are formed for the tasks, competencies of the employee, the client, technology and more. Therefore, a person needs to be guided in professions not only during their choice, but also throughout their working life.

Keywords Labor · Education · Transformation · Workforce · Creative business · Digital technologies

A. Bielova · N. Zhuravska (✉)
Kyiv National University of Construction and Architecture, Kyiv, Ukraine
e-mail: nzhur@ua.fm

S. Koval
Hryhorii Skovoroda University in Pereiaslav, Pereiaslav, Ukraine

A. Agayev
Azerbaijan University of Architecture and Construction, Baku, Azerbaijan

V. Onyshchenko et al. (eds.), *Proceedings of the 3rd International Conference on Building Innovations*, Lecture Notes in Civil Engineering 181,
https://doi.org/10.1007/978-3-030-85043-2_55

1 Topicality

During the presentation of the Economic Audit of the country and Vectors of economic development until 2030, the Ministry of Economic Development, Trade and Agriculture of Ukraine identified economic vectors that will be able to ensure the recovery and future growth of the domestic economy, including [2]:

- Reduction of the state's share in the economy (continuation of privatization).
- Development of the financial sector.
- Reducing the regulatory burden on business.
- Infrastructure development in the agricultural sector.
- Favorable tax and legal regime for innovative and research companies.
- Digital economy.
- Ukraine's reputation in the world and strengthening trade diplomacy.
- Creating an economic basis for the development of education, medicine, culture and social services.

In turn, the above and other processes change the labor market, defining its main features and vectors of development. In order to predict the probable image of the labor market and the workplace in the conditions of economic transformations, first of all, it is necessary to clearly outline the main processes and megatrends that determine the future of the market and cooperation in the construction sector with innovative development. It is also necessary to determine how the content of the "work" of the workforce will change and what will be the "labor activity". Insufficient scientific substantiation of the regulation of innovative development of construction and setting trends allows to determine the idea of the need and directions for these processes and justifies the choice of research topic.

2 Presenting Main Material

Summarizing the existing numerous studies on the representation of future employment, researchers identify a number of megatrends that will determine the development of the labor market in the near future. Automation and robotization of production processes, the introduction of digital technologies and artificial intelligence is a far from complete list of global trends that will significantly affect the workplace of the future labor market. A global trend is an objective trajectory that affects and changes society as a whole. Along with it, the market, the environment and working conditions are changing. The main current global trends affecting the transformation of labor and workforce skills are [4–10]:

1. ***Creativity of business.*** Modern progressive economy in terms of economic theory is characterized as "knowledge economy" or "information economy", the main features of which are the intellectualization of labor, generation of new ideas, business creativity, digitalization of technology and more.

The new economy is formed on the basis of the use of imagination, creative resources and knowledge, which are transformed in the form of new ideas and values. Accordingly, the success of economic development is determined by the efficiency of the synthesis of creativity, entrepreneurship and technological innovation. Creative industries, operating at the intersection of art, culture, business and new technologies, today form the basis of a developed world economy.

In Ukrainian society, creative industries and creative entrepreneurship are relatively new concepts, but they are well known in Europe, the United States and other developed countries due to the possibility of practical adaptation. Innovation and creativity are one of the priorities of the European Union's socio-economic development program "Europe 2020", the main purpose of which is to promote creativity through lifelong learning as a driver of innovative development and a key factor in human development.

Intensive development of the modern knowledge-intensive economy determines the formation and implementation of the concept of creativity, which provides innovative progress of socio-economic development. It should be noted that the creative economy significantly expands the possibilities of traditional production of goods and services. Creative industries are effectively developing and interacting with sectors of the traditional economy. They now account for 7% of world GDP, growing by 10% annually. The sphere of creative entrepreneurship is developing more rapidly, even than the production and services market, providing jobs for about 8.3 million EU citizens, which gives it the status of an effective economic model [1].

The impact of creativity on the economy is determined by such indicators as job creation, the ability to attract investment and use it effectively, the creation of value chains in which representatives of "traditional" and creative industries are combined and interact. By definition, creative industries are understood as a set of economic activities that are the basis for the formation of added value and the creation of new jobs through cultural (artistic) and creative expression. Goods and services that are the result of such activities are the result of creativity, skills and talent of the individual.

Therefore, to assess the effectiveness of creative industries, the whole chain should be explored: from the generation of a creative idea through the formation of an innovative product to the generation of added value. In the context of growing globalization and the growing needs of companies to ensure sustainable success in the market, there is a need to increase the efficiency of operations. This task should be addressed on a par with others, such as the ability to think innovatively, providing a dynamic strategy for the interaction of resources and competencies in accordance with the requirements of the environment. The emergence of innovative enterprises generates the ability and motivation to find new solutions, concepts and ideas aimed at developing and improving the efficiency of technologies and activities. By increasing its own competitiveness, the company affects the state of the economy not only nationally but also European. The key resources of companies are creative workers with unique competencies, a lively mind, a desire for excellence in business and ensuring effective growth through innovation. In order to initiate innovation, many employers have adopted a strategy of competency management by encouraging and motivating employees through self-realization and promoting the realization of their creative potential and creative abilities.

Because creativity-based production is creative, the extraordinary development of digital technologies and the globalization of information exchange networks have made the creative sector one of the most dynamic in the world. Creative industries create new jobs, diversify the economy and create a comfortable urban environment. In developed countries, the creative sector of the economy generates up to 10% of national GDP [1].

Promising research of the present is the identification of factors influencing the formation of creative industries in Ukraine, the actualization and justification of the implementation of the concept of creativity of the economy of Ukraine. Thus, the identified indicators of creativity should include increasing the competitiveness of existing businesses, increasing local budget revenues, improving the quality of life, reducing the outflow of talent, increasing the investment attractiveness of the territory, increasing tourist activity, improving the quality of cultural life. Some studies have proved the impact on the GDP of Ukraine of innovative development of creative sector enterprises in the chain "generation of creative ideas → formation of innovative products → generation of value added." Numerical measurement of the effectiveness of creativity through correlation-regression analysis proves the existence of a close relationship between GDP, the number of enterprises producing innovative products, the volume of sales of innovative products and the volume of innovative products exported. Such a multifactor model can be used to assess and forecast the impact on GDP of the development of innovation in Ukraine, where a significant share are enterprises belonging to the creative sector.

Thus, the unique properties of human capital, talent, knowledge can be a stimulus for the development of the domestic economy, realized in the form of creative industries and act as a tool for generating value added nationwide, providing consumers with innovative products. Therefore, it is important for Ukraine to adopt a strategic program document, within which the state will carry out targeted work for the formation and development of creative potential. At the state level, it is necessary to address the issue of supporting creative industries, which consist in the simultaneous development of such areas as, for example, the introduction of special education and lifelong learning.

The transformation of the education system requires a radical reform aimed at bringing the quality of training in line with the requirements of the creative economy. A modern professional must not only have certain knowledge, skills and competencies, but also be able to generate them throughout his life. It is also necessary to introduce programs to support creative entrepreneurship, work on grants, introduce educational programs in this area, promote the opening of new enterprises in the creative economy sector, provide them with soft loans. We also need a progressive legal framework, namely a law on support for startups, simplified tax conditions, tax rebates for those who invest in innovative industries, and affordable business conditions. Successful creativity in the form of creating a favorable intellectual environment and developing a system of socio-economic interaction will allow employees to realize their own creative abilities by generating new ideas and implementing them in the form of innovative products that will ensure sustainable development of the domestic economy.

2. ***Digitalization of technologies.*** Given the modern signs of development of society, it can be argued that the modern information society of the world has reached the next stage of developmenthas become digital. This term directly indicates the way all public institutions operate: the activity is realized through digital visualization of objects. Therefore, today the most commonly used terms are: in Europe-digital economy (Digital Economy), in the US-API-technology (Application Programming Interface technology). That is, the digital economy is an economy based on digital computer technology. The digital economy is also sometimes referred to as the Internet economy, the new economy, or the web economy.

According to research by McKinsey Global Institute, cross-border data flows have increased 45-fold over the past decade, and continue to grow. Given today's challenges and responding to rapid change, the EU has embarked on a new common market-digital, which should strengthen existing economic freedoms and foundations throughout the community and develop digital technologies—take full advantage of risks, minimize risks, respond quickly to challenges, including security. The role of digital technologies in EU trade policy is also significantly increasing, which Ukraine must take into account [3].

The digital economy is based on the widespread and widespread use of automatic systems, devices and equipment using computing and control units and devices capable of functioning without human intervention. When equipped with executive bodies—mechanisms, they turn into works that can directly replace human labor or a number of its functions and actions. The modern digital economy is artificial intelligence, robotics, electronic money, industrial biology, large data processing, unmanned vehicles, and so on. For the general public, the "digital economy" means a new level of digital service, when most purchases and services go online.

The peculiarity of Ukrainian digital development is that the users of online technologies are small and medium-sized businesses that somehow work on the Internet and mostly use digital methods to promote their services.

The digital economy is the economy of virtual worlds. Its main space is the Internet. But it should be understood that not only the Internet determines the direction of the digital economy—it is about technology in general: smart apartments, smart cities, jobs that completely replace people, online stores (Amazon, Aliexpress), Internet banking (Privat 24), messengers (Facebook, Telegram), etc. This is exactly what distinguishes the digital economy from the traditional one—it is consumer-oriented. Therefore, you can often hear instead of "digital economy"-"economy on demand" (on-demand economy).

The global digital space is developing dynamically under the influence of active investment activities of the world's leading countries and the influential policies of the world's largest IT companies. There are structural shifts in the capital market, which, first, form a growing trend of investment in global projects through the formation of consortia and integration groups with the participation of leading countries and intensively developing countries. Secondly, global investment flows are directed both in the technology of "mass demand" (Internet games, e-commerce) and in the technology of storing arrays of databases, which leads to the monopolization of global companies on intellectual capital and digital information space. Third, the development of global and local digital markets creates favorable conditions for countries with a high level of education and the level of informatization of national economies [3].

Thus, the digital economy is not only the digitalization of production, distribution, exchange and consumption of goods, but a system of more general processes, in relation to which the economy is one of the applications. The digital economy as a global concept of modern economy puts forward new standards of quality of life, work and communication between people. The ability to adapt quickly to change and optimize in the shortest possible time, adapting to global trends—the main challenges posed by today's digital economy.

No less important are the socio-economic consequences of digitalization, because the displacement of man from the production process is a global social problem. The modern global labor market is a complex multi-component and dynamic system that is permanently affected by information technology, which entails changes in the content of the labor process, its organization, employment structure, social and labor relations. Objective automation processes, even if restrained by governments and society, will accelerate and possibly reach the point where only a few million highly qualified professionals will be enough to support the entire global production and logistics system. In this sense, the reduction of available jobs in the world economy, the emergence of a whole class of "extra" people, total retraining of staff, the destruction of the usual mechanisms of "guarantees of the future" (decent pension, guaranteed social protection, etc.)-quite possible realities.

In addition, the impact on labor resources of technology development and automation is unpredictable. This process is so complex that no country reportedly has a formal strategy for adapting to future change, and governments face ever-changing priorities in trying to develop effective strategies for the digital economy. An analysis of adaptation strategies shows that today we can observe certain types of actions that are to some extent applied or discussed by governments to mitigate future shock. Such actions include, in particular, stimulating the creation of new types of jobs. In today's economy, it is very difficult to create conditions that are guaranteed to create jobs in the long run. Companies can maintain jobs as long as there are subsidies, but abandon them in favor of more efficient solutions when subsidies end. A possible solution would be to support areas of the economy where human action is virtually impossible to replace with a robot or neural system. This leads to a change in the interpretation of the very concept of the workplace, when instead of work a person will be engaged in the realization of their creative potential, which will correspond to the trends of the network society.

At present, digitization technologies are penetrating all areas of human activity. Unlike analog data, digital data is discrete, it can be stored, copied, analyzed and transmitted virtually without restrictions. Another important aspect of digitalization is the gradual "replacement" of familiar reality with digital, augmented or virtual reality. Virtual reality technologies amplify the digital world, and augmented reality technologies blur the boundaries between worlds. Augmented reality is used in the workplace in complex industries, forming new ways of working, communicating and collaborating across the enterprise.

The digital labor market also contributes to the formation of innovative employment. As a result of structural changes in the economy, the share of industrial production is declining, which shaped the demand for standard employment. The service sector, which operates under flexible working hours, is growing with a longer or shorter working day than the current legislation. There is a growing need for greater mobility of labor resources, which leads to a stronger role of fixed-term employment contracts.

Thus, digitalization, like any change, has its pros and cons. The main risk of digital transformation of the economy is a possible rise in unemployment. According to the analytical materials of the World Economic Forum in Davos, each industrial revolution did lead to unemployment at the initial stage, but after a short period of time (1 to 5 years) there were new needs and demands from the market, which led to new professions. In fact, each industrial revolution created new types of economic activity, new approaches, models, and this in turn changed the labor market [4].

McKinsey & Co. estimates that by 2030, 400 to 800 million people (15 to 30% of the world's workforce) will be out of work through the development of artificial intelligence and automation. Many operations performed by employees today have the potential for automation. According to McKinsey, about 60% of all occupations have at least 30% of activities that can be automated using modern technology.

There are other estimates, for example, according to the European Commission, about 50% of current jobs in the world can theoretically be automated. This means that human labor in some processes will not be needed [4].

Another example—according to the analysis of Visual Capitalist, in the US by 2030 will be digitized [4]:

- 50% of jobs in trade;
- 57% of jobs in the field of transport;
- 60% of jobs in agriculture;
- 60% in production;
- more than 70% in the field of accommodation and food.

According to the Ukrainian Institute of the Future, the situation on the labor market in Ukraine in the perspective of 10 years may be less dramatic than in developed countries. In the next 3–5 years, digitalization and automation, on the contrary, will solve the problem of labor shortages [4].

This is due to the fact that digitalization not only leads to the disappearance of professions and reduced use of human labor, but also creates new industries (sectors, professions), provides unlimited opportunities for the realization of human skills and talents. For example, according to a study by McKinsey's Paris office, the Internet has eliminated 500,000 jobs in France over the past 15 years, but at the same time created 1.2 million more. That is, instead of one liquidated job, 2.4 new ones were created [4].

Thus, total digitalization and the development of the digital economy in Ukraine will somehow be accompanied at the first stage by negative processes of job loss, but digitalization will create new directions, which in a few years will lead to new demand, which, as the experience of industrial revolutions shows. much greater than the demand of the past period.

Thus, professions that require social and emotional skills, special cognitive abilities, such as logical thinking and creativity will be in demand. In order to be in demand in an extremely changing world, a person must constantly learn. It is here that the principle of "knowing how to learn throughout life and become self-fulfilling and competitive" is critical.

Ukraine needs to move forward with a modern national curriculum for general and professional digital competences and skills as key components of the digital economy. The priority of the state should be to create and coordinate relevant initiatives and provide resources. The state strategy to prevent the risks of job losses has several levels of implementation:

- educational, the creation of academic programs and educational infrastructure for the training of specialists in new specialties;
- retraining programs for citizens and adaptation programs for citizens;
- structural initiatives: updating the state classifier of professions, is development and approval of the list of digital professions (based on labor market requirements, digital trends), their introduction in specialized educational institutions.

Thus, it can be argued that the world has entered a new era in which the impact of digital technologies is increasingly being felt in all sectors of the economy [4]:

1. Digitalization is radically changing traditional industries and sectors. There is a change of classic business models, analog processes and operations flow into the Internet, it is possible to form personal proposals for each individual client. Automation and robotics minimize the need for human resources, rapidly increasing efficiency and productivity. Radical changes are taking place in those industries that are considered basic for Ukrainian industry, metallurgy, oil and gas production, energy, construction, etc.

2. Digitalization creates new sectors and segments, as well as new professions. According to the Ukrainian Institute of the Future, up to 60% of value added in Ukraine in 2030 will be created in new high-tech sectors of the economy, such as artificial intelligence, robotics, bioengineering, 3D printing, nanomedicine and others. In the future, 65% of today's children will engage in activities that do not yet exist today.

Therefore, the digital economy is not a separate sector of GDP, such as the ICT industry or industry, it is a platform that, like the human circulatory system, permeates all sectors of the economy, radically changes them, changes the structure of the Ukrainian economy, creates new segments and even industries, uniting them into a single whole. The impact of digitalization is determined by the added value it creates for each sector of the economy or sphere of life at the macroeconomic level, or for a specific product or service at the micro level. And the main effect of digitalization is to change the value chains.

However, technology and digitalization will displace people from their usual processes-production, services, entertainment, trade, education and even medicine. First, the work will be in the nature of numerous but short-term projects. Previously, it was believed that having a permanent job is a guarantee of success and security. However, against the background of recent changes in the world economy, permanent employment does not guarantee permanent employment. By 2040, the labor market will mainly consist of people working simultaneously in several places in small firms, working part-time. Daily monotonous work, consisting of the same responsibilities, will be replaced by a business that will provide many short-term projects, the diversity of which a person will control independently throughout life. Thus, people will carry out small projects lasting from a few days to several years, which will create the conditions for an increase in the number of entrepreneurs who work exclusively for themselves.

This will be the locomotive of the bulk of jobs by 2040. Secondly, a large number of different employment agencies will be involved to find highly qualified and creative professionals. If now recruitment agencies are looking for specialists of the general public, in the future it will be a search exclusively for creative people with special abilities, and in the future such recruitment companies and organizations will play a more prominent role in public life. Third, the responsibility for one's own success will increase over time. Professionals will have to constantly develop future projects, as well as permanently develop special skills and obtain the education necessary for the successful implementation of these tasks.

3 Conclusions

Thus, in the long run, social and labor relations may undergo radical changes in three main directions or vectors.

First, the boundaries of the traditional division of labor will change, the boundaries of professions will be erased, the rate of "extinction" of traditional professions will accelerate, and new, previously unpredictable ones will emerge. Experts note that in the coming years, up to 10% of existing professions may disappear. From this point of view, the principles of division and cooperation of labor are changing. Rigid consolidation of functions in a particular profession, including in professional standards, contradicts the dynamics and flexibility of the social and labor sphere. Robotization and automation radically change the content of work in all industries and types of employment, which, accordingly, changes the requirements for employee competencies. Thus, narrow professional training contradicts the need for the formation of cross-cutting, supra-professional competencies (sociability, non-standard and critical thinking, the ability to work with modern digital technologies, etc.), the importance of which is growing every year [5]. Both at the beginning of the career and throughout the work a person needs to navigate in the world of professions, in the dynamics and prospects of their changes in order to timely form and develop their competencies, increase their competitiveness in the labor market taking into account the prospects of its transformation all life.

Second, forms of employment are changing. Along with the traditional contractual forms of labor relations, employment in the form of freelance, crowdsourcing, insourcing, flexible forms of inclusion of professionals in labor activity, remote employment, project form of employment, etc. are actively developing. At the same time, new forms are becoming more and more acceptable for young people and highly competitive staff. Therefore, during professional self-determination and career guidance it is necessary to take into account not only transformations in the content and division of labor, but also opportunities, prospects for the use of flexible (new) forms of social and labor relations.

Third, human mobility increases throughout employment. This is due to the intensification of migration processes (more than 60% of those who change their place of residence do so for reasons of work and employment), as well as inter-professional, intersectoral, intra-firm mobility.

"Lifelong learning" as a principle and concept is increasingly developed and implemented in practice. The tendency becomes more stable when a person changes his professional affiliation 3…4 times during his working life. In the workplace, traditional barriers to the professional division of labor are increasingly being overcome. Jobs are formed for the tasks, competencies of the employee, the client, technology and more. Therefore, a person needs to be guided in professions not only during their choice, but also throughout working life. As a result, objectively there is a need for such a mechanism that would allow the workforce in the face of exponential changes in advance to communicate between technological transformations, transformations in the division of labor, in the "world" of professions, education, labor market, development. Therefore, the main effect of regulation will be to increase the level of innovative development for many industries, including construction-increasing the profitability of its members, competitiveness, as well as increasing tax revenues to budgets at all levels.

References

1. Leonova S (2019) Creativity as a vector of sustainable development of the modern economy. Ser Econ Manage Econ Bus Manage 30(69):133–138 (2019)
2. The Economic Audit of the country and vectors of economic development until 2030 are presented. https://www.kmu.gov.ua/news/predstavleno-ekonomichnij-audit-krayini-ta-vektori-ekonomichnogo-rozvitku-do-2030-roku. Accessed 9 Nov 2020
3. Putzenteilo P (2018) Digital economy as the newest vector of reconstruction of traditional economy. Innov Econ 5–6:131–143
4. Ukraine 2030E is a country with a developed digital economy. Ukrainian Institute of the Future. https://strategy.uifuture.org/kraina-z-rozvinutoyu-cifrovoyu-ekonomikoyu.html. Accessed 17 Dec 2020
5. Digital economy: trends, risks and social determinants. https://razumkov.org.ua/uploads. Accessed 21 Dec 2020

6. Mykytas M, Terenchuk S, Zhuravska N (2018) Models, methods and tools of optimizing costs for development of clusterized organizational structures in construction industry. Int J Eng Technol (UAE), 7(3):250–254. https://doi.org/10.14419/ijet.v7i3.2.14414
7. Chevhanova V, Chichulina K, Skryl V (2018) Effective cluster model of thermal power company management. Int J Eng Technol (UAE) 7(3):65–70. https://doi.org/10.14419/ijet.v7i3.2.14377
8. Chichulina K, Skryl V (2020) Energy service companies are an effective tool for financing energy-efficient projects. Int J Manag 11(5):252–266. https://doi.org/10.34218/IJM.11.5.2020.025
9. Sivitska S, Vartsaba V, Filonych O (2018) Buildings energy-efficient renovation investment. Int J Eng Technol (UAE), 7(3):408–412. https://doi.org/10.14419/ijet.v7i3.2.14562
10. Drobyazko S, Bondarevska O, Klymenko D, Pletenetska S, Pylypenko O (2019) Model for forming of optimal credit portfolio of commercial bank. J Manage Inf Decis Sci 22(4):501–506

Strategies of the New Time in the Perspective Development of Innovation-Oriented Economy of Ukraine

Tetiana Bochulia and Svitlana Sivitska

Abstract The current tasks of ensuring the innovative development of the economy are considered in the article, given the development of new projects and strategies that are innovative and cover the development of each country, taking into account the importance of economic, social and environmental vectors. Redesign of social, economic and environmental vectors is substantiated with the use of intelligent solutions with appropriate technical and technological provision and professional provision. New prototypes of the way of life of society are described, in which new models of economic and social institutions are shown, which have a digital platform and are focused on new values, which are characterized by intangible character. Collectively, such changes are manifested in the development of effective mechanisms for the development of urban areas, based on the efficient use of natural resources, smart urban infrastructure, focus on quality man-made changes, ensuring the basic needs of the population and provision of sustainable development. The innovative dimension of constructing the city of the future is determined, which is manifested in the multistructuring of urban infrastructure and the offer of new services, which are an element of new values.

Keywords Innovative cities · The metropolis of the future · Sustainable development · Digitalization · Strategy

---

T. Bochulia (✉)
Kharkiv State University of Food Technology and Trade,
333 Klochkivska, Kharkiv 61051, Ukraine

S. Sivitska
National University «Yuri Kondratyuk Poltava Polytechnic»,
Pershotravnevyj Ave. 24, Poltava, Ukraine

V. Onyshchenko et al. (eds.), *Proceedings of the 3rd International Conference on Building Innovations*, Lecture Notes in Civil Engineering 181,
https://doi.org/10.1007/978-3-030-85043-2_56

1 Innovative Dimension of Modern Economy Development

1.1 The Concept of Continuous Improvement as an Innovative Transformation

The world is changing every day, which is a logical process of transformation. Any changes can be assessed in different ways–the main thing is the result, which becomes the starting point for further development. The basis of modern economic development is a fundamental rethinking and general redesign of information, technological, economic, socio-cultural processes to ensure the balanced development of social, economic and environmental vectors that collectively characterize the sustainability of the national economy development. Balance, temperance of transformations as a logical process of business development was substantiated by M. Imai, which owns the development of the management concept «Kaizen» [1] (continuous improvement)–Japanese business philosophy, the main idea of which was sustainable gradual development as continuous progress, which does not involve large-scale (radical) transformations, but has a positive effect on the process itself. The importance of the result in this concept is lower than the development of the process itself, in the improvement and modernization of which there is a long-term effect with its further increase. Kaizen's management concept belongs to the idea of innovative development, opening new opportunities for the use of unlimited human potential, thus expanding key competencies as an unconditional competitive advantage of the enterprise. The idea of the management concept of «Kaizen» is implemented in the concepts of Lean Production and Total Quality Control, as the basis of innovative management, namely a modern approach to management to ensure sustainable development. A key aspect of this innovative concept is to achieve results through the use of internal mostly intellectual resources of the enterprise.

In modern business conditions, companies do not compete for supremacy in motion (which is inefficient), but try to achieve leadership positions in the speed of reaction with the definition of the right direction for further development. A new direction of development, which covered all areas of business in the world, is the application of the Lean management concept, which is characterized by the priority of experiments over planning, the predominance of associative thinking over rational, continuous adjustment of management models, production, economic and information relations [2, 3].

This concept has already attracted the attention of large companies, for which it has become a tool to increase business competitiveness. Significant efficiency from innovative transformations can be obtained by implementing the concept of «Lean» management in the strategy of sustainable economic development (Fig. 1).

Lean-transformation is a reaction to the need to form a reference version of the generalization and synchronization of changes in the new rules of formal and informal institutions, adapted to the actual conditions of sustainable economic development. Lean-development of innovative economy is the result of intensification of rational and irrational human activity in the global multimedia environment, which

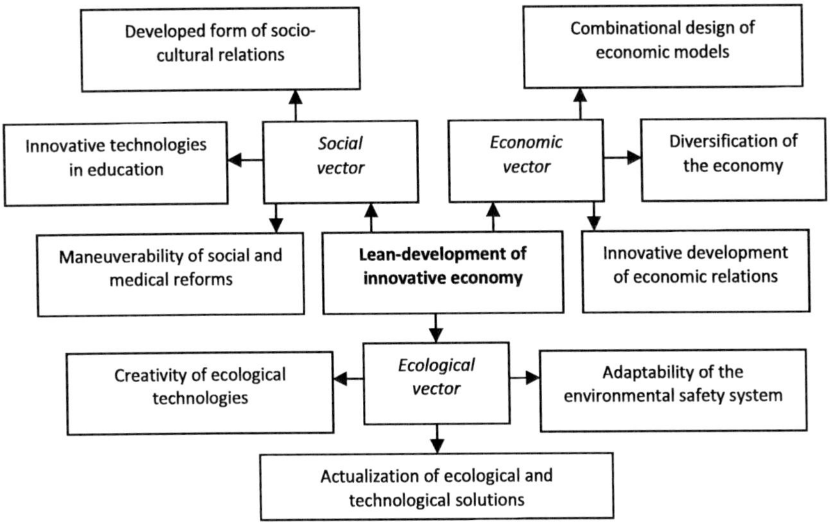

Fig. 1 The lean management concept in the innovative development of the economy

creates better conditions for the development of vectors of sustainable development, increasing the efficiency of economic, environmental and social processes in forming a new economic model. This is to promote greater efficiency of the management process through the formation of such a foundation that best meets the requirements of the current generation and takes into account the expectations and prospects for future generations.

Lean-concept seems simple, but at the implementation stage there are difficulties associated with obtaining the result is not complete, which states a violation of the algorithm for implementing this concept.

The main markers of successful implementation of this concept are:

1. The focus on the human factor–the implementation of innovative transformation should begin with a change in traditional thinking and culture that has developed at the enterprise. This is a transition to new values that should be accepted by staff as the main idea of further development with a new type of behavior and relationships that is implemented in all business processes. This should be a fundamental abandonment of the old templates in favor of a new management model, familiarizing each specialist with the new conditions, technologies and tools that are adopted with an understanding of their prospects as opposed to outdated methods.

2. Transformation as an element of the business system–Lean-concept should be a development strategy, not a program of action. These are much broader and fundamental changes that cover all levels of enterprise activity with the transformation of mission, strategy, values, culture, management and business processes. Changes solely in the level of behavior and decision-making will not give the expected result. This should be the value level at which a new hypothesis of the existence and development of business with a focus on new goals that meet the accepted concept.

3. The culture of continuous improvement–the introduction of a new motivational culture of the desire for constant renewal as in line with modern trends. This is the construction of a new management system with the establishment of ambitious goals and motivational tools that come from each employee with an understanding of the importance of change and continuous improvement, which should be implemented at all levels of the enterprise.

According to the Lean concept, business organization is focused on transferring the idea of step-by-step changes in modifying the development strategy with the priority of sustainability as the basic mission of the enterprise with achieving a balance in the interests of present and future generations.

1.2 Intelligent Solutions in Ensuring the Innovative Development of the Economy

The introduction of new management technologies should be carried out in parallel with the implementation of intellectual innovations, which are the basis for achieving a balance in ensuring sustainable economic development. Today, intelligent solutions are especially in demand for the development and implementation of innovative energy technologies that provide a new dimension of the development of environmental, social and economic vectors.

Intellectualization of technologies has become especially relevant today, because the management systems have the functionality of intelligent data processing, but it is underdeveloped due to the inconsistency of intellectual and technological processes. This encourages the promotion of innovative processes in the development of information technology support management system for well-established business services in accordance with the multi-purpose guidelines for its development. It is necessary to mobilize the intellectual potential to ensure a qualitatively new level of business processes, thus laying the foundation for its continuous coordination with the requirements and requests of the management process for decision-making and development of enterprise development strategy. There is a clear change in the technical side of management to intellectual, creative, analytical, which contributes to the formation of new business properties.

The modern economy is developing in accordance with the main goal of the energy strategy, namely: the creation of an innovation-oriented energy center, which implements the latest technological and intellectual projects designed to ensure moderate use of energy resources in accordance with the internal and foreign economic interests of the country with the provision of social development. Achieving this goal is achieved by gradually solving a set of tasks, including: improving energy efficiency of business, further integration of the national energy complex into the global energy system, taking into account the latest innovative developments and borrowing world experience, the formation of the energy ecosystem in accordance with the concept of sustainable development.

At present, the global scale of energy supply should shift towards the use of innovative energy-saving technologies that use alternative energy sources and renewable energy sources. In particular:

- wind energy–the use of kinetic energy of air masses to form an alternative energy source used for the needs of society and the economy;
- geothermal energy–obtaining an alternative source of heat and/or electricity due to the heat of the earth's depths;
- solar energy systems - the use of collectors and solar systems, as well as solar kits to convert solar energy;
- biofuel energy–the use of wood waste, industrial and household waste, high-yielding plants to obtain an alternative energy source;
- hydropower–systems for converting water flow energy into electricity, etc.

New opportunities in the use of alternative energy sources are based on innovative solutions based on ingenuity, intelligence and savvy. Innovative solutions are based on ideocognitive methods that combine creative approaches with a priori knowledge, forming new knowledge that becomes a unique resource and provides a competitive position of the economy.

New projects are subject not so much to the patterns of formal institutions, but because in the «people's heads» and thus affect the innovation process, emphasizing the creation of new conditions for sustainable economic development. Thus, the success of the modern economy is described by researchers through perception, reactions and feelings–non-traditional factors for economic theory, which is a manifestation of financial and economic leadership, defining the scheme and stages of change and transformation [4], as a result of which the economy develops in a new, innovative version, providing a factor of sustainability in economic, social and environmental vectors (Fig. 2). Perception is used in analysis [5] that has a positive effect on the reality of economic development models, minimizing risks and uncertainty.

The innovative foreshortening of sustainable economic development is a new hypothesis of continuous changes by the transformation of social, economic and environmental processes through the integration of tools and phenomena necessary for the creation and dissemination of new methods provided by intellectual activity of specialists who should have the art of strategic thinking [6]. The main emphasis is on: expanding professional judgment through the activation of intellectual potential without limiting only professional knowledge, using an interdisciplinary approach to professional development; creating a team «by interests», motivating the increase of activity to self-development and self-learning, focusing on the practical result and giving responsibility for the decisions made; maximum consideration of the characteristics, thinking and implementation of the acquired knowledge of employees to determine the limitations, which can inhibit the development of the total intellectual potential in the actualization of the innovative basis of economic transformations.

Innovative transformations in the economy can be carried out on the basis of organizational and methodological principles of I-engineering, which corresponds to the essence of innovation as an objective and logical process of ensuring sustainable economic development. This is a reasonable redesign of social, economic and

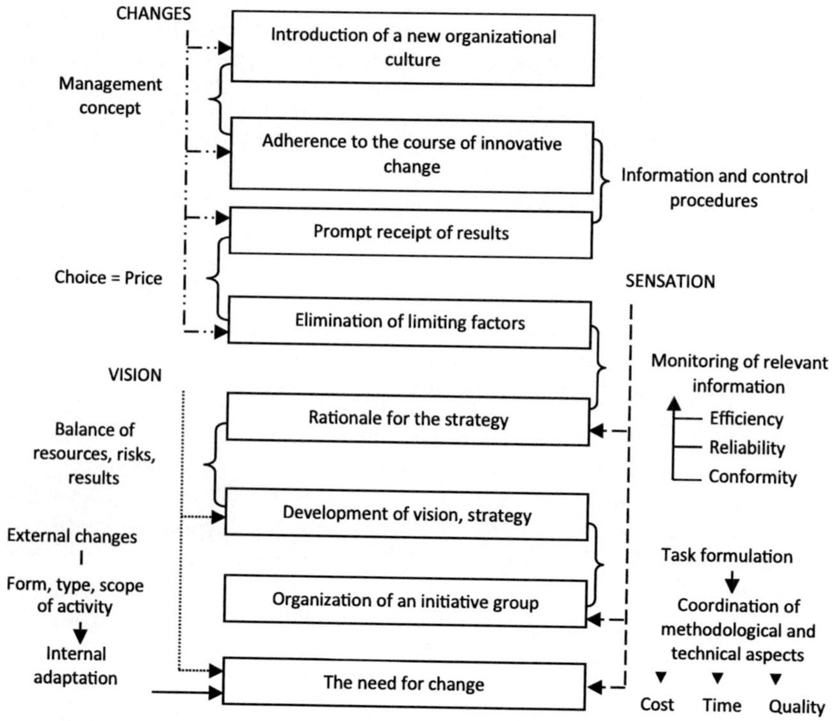

Fig. 2 Innovative foreshortening of sustainable economic development

environmental vectors using intelligent solutions with appropriate technical and technological ensuring and professional provision.

I-engineering is not a transformation caused by an attempt to save a business whose viability turned out under threat due to the influence of objective internal and external causes. This is an optimization of a strategy that works quite effectively, but needs to be changed, taking into account environmental factors and the needs of the management process. The main principle of I-engineering is efficiency, which provides instantaneous (speed) implementation of all structural transformations, which are the result of a thorough assessment of internal and external factors and business indicators, which are adjusted over time and completely replaced by new.

Under I-engineering we understand the need caused by the underestimation of human resources due to redundancy in determining the effectiveness of information and communication technologies, which corresponds to the essence of innovative economic development. I-engineering is based on the development of projects that are compiled in accordance with: indicators that characterize the evolution of the external economic environment; internal business parameters; forecast of enterprise development; analyzed initial conditions for implementation and change; selected areas of change; assessments of probable risks; personnel management policies; strategies and tactics; motivation, etc. This is a new scenario of using intellectual innovations to meet the needs of modern society and future generations, which is realized through the offer of ideas, which are characterized by a fundamentally new approach to economic development, which determines the trend of tools of responsible business.

2 Innovative Projects of the Metropolis of the Future

2.1 *The Main Drivers of Innovative Transformation of the Future*

Today, government, scientific and business structures have a common question to address–what are the likely ways to develop innovative technologies and society, trying to create a personalized model of the future with a forecast of socio-economic and environmental processes. One of the main aspects of modern development scenarios is the adaptive nature and variety of hypotheses, which focuses on a new level of development through the use of innovative tools. The concept of multivariate development is based on the scale of modern innovation, which is characterized by formal and informal nature of methods, tools and technologies, the balance in which fluctuates between normative and personalist theories, reconciling traditional and innovative views with the definition of new priorities as a basis for following different options for sustainable economic development (Fig. 3).

Multivariate development involves the expansion of models of society and economy, which today are based on innovative projects, collective knowledge, new values and views with a culture of relationships at the technological and intellectual levels. Today, according to researchers, there is a clear shift towards a new reality, where the dominant role will be in the innovative cities of the future and «wild» (by our definition) knowledge, and selfishness will be replaced by «Ecoizm», namely a project on ecology in the broadest sense words [7].

Ecoizm is one of the main drivers of innovative economic development. This is a «green» aspect of economic development, aimed at ensuring a balance between society and the environment in order to preserve natural resources for future generations. Among other drivers, the main ones are those that provide economic and social development in a new format of global innovative change.

1. Total automation of social, economic and environmental vectors of development, widespread introduction of artificial intelligence, robotics and biotechnology. There are already examples of global automated projects, in particular, the virtual economy, which allows you to form economic processes and economic relations in augmented reality, using technology platforms, computer games and social networks. It is going beyond the traditional material reality with the creation of new economic rules that are subject to informal principles, which affects the emergence of new

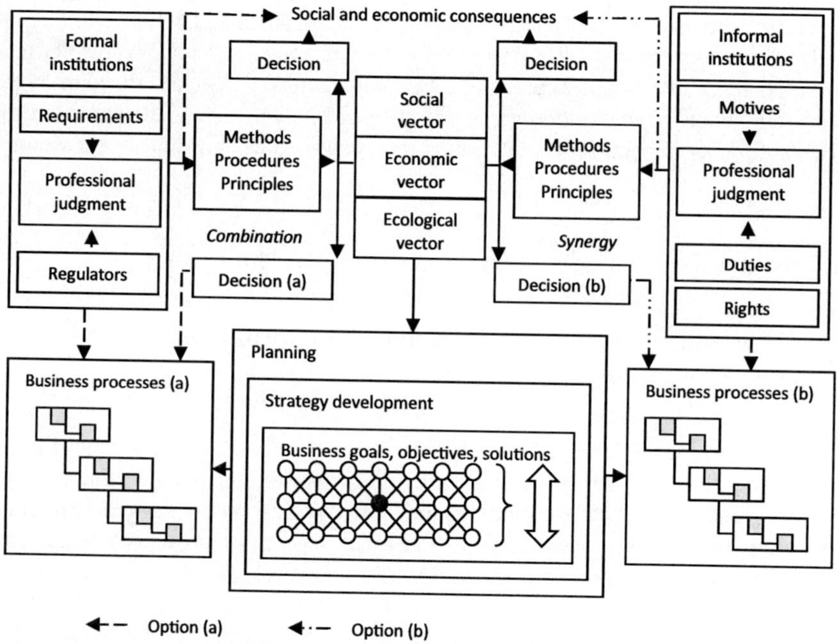

Fig. 3 Schematic interpretation of the concept of multivariate economic development

knowledge, competencies and professions. Accordingly, new educational technologies are emerging, resulting in the training of new generation professionals with advanced and binary skills and competencies, characterized by a creative approach and the ability to combine knowledge into a collective experience, forming a new hierarchy of productive teams capable of solving contextual tasks. This range of innovative knowledge is implemented in new technological sectors of the economy, which is a new generation of economic sectors with a high level of competitiveness and sustainability, given the innovative nature of their basis. This is the creation of a fundamentally new sector of the economy, which corresponds to the essence of a multimedia society with the offer of new professions, which significantly reorients the labor market.

2. Digital trend of transformation of society and economy. The digital stage of development of the modern world is characterized by the intensity of the introduction of digital technologies and digital reorientation of socio-economic processes, puts forward new requirements for the approaches used in the formation of strategies and models of development. First of all, this is due to the fact that today the external environment is changing much faster and deeper than before, and the degree of uncertainty about the impact of various factors on the future is growing. Approaches to the development of innovative strategies have evolved under the influence of new conditions of world development, the internal source of which is a person with his

psycho-physiological characteristics and the ability to transform thinking. Digitization Driving Strategy is a serious step towards radical change. Such a strategy cannot be defined as a modernization tool that integrates with the overall business strategy. Digital change is a global transformation based on digital principles with the incorporation of their basic provisions into the development strategy, business model, environmental activities and general culture of society.

3. The trend of global professional growth - modernity dictates the requirements for expanded professional knowledge and competencies, which involves the transition to the concept of learning «full life» for the formation of supra-professional competencies that meet current and future stages of economic and social development. The universal set of professional competencies should include not only professional knowledge, but also additional skills that correspond to the technologically and intellectually oriented world, in which innovation is the basic driver of change. So, necessarily determined presence of systematic and creative thinking, collaborative skills, media literacy and stress resilience is mandatory, in particular, in conditions of information overload and information noise. To perform complex tasks, the skills of predicting the future, developing parallel hypotheses and forecasting different scenarios should be acquired, which is implemented in conjunction with the skills of operating large data systems. Non-technological competencies that are unique to a person should also be acquired, which is necessary to given the importance of emotional intelligence in the formation of strategies and decision-making. This symbiosis of knowledge and skills should be realized in the integrated competence, which is plastic in terms of mastering the socio-economic processes that are constantly accelerating.

The transformation of the present and the future is carried out on the basis of innovation singularity, when changes become revolutionary and completely change the usual picture of society. Innovation should be a constant in today's world with a continuous cycle of improvement and generation of new ideas that should compete with existing strategies, models and programs of development. Also the determining factor is time, i.e. acceleration of processes of innovative development with the offer of new developments that corresponds to dynamics of the modern world. In this aspect, innovations are considered from the point of view not of modern technology, but of a perpetual motion machine that mobilizes resources and contributes to the achievement of prospects at the formal and informal levels. A distinctive feature of modern trends in innovation development is anticipation as an additional tool for in-depth understanding of the real needs of society and the economy at the future point of development. This is a higher level of knowledge management, which goes beyond the usual scope of operating structured and unstructured data. This is a combination of advanced automated and intelligent technologies in a single system that works as a set of integrated processes with the transition to a qualitatively higher level of development of society in its future. Already today, new prototypes of society's way of life are being generated, which show new models of economic and social institutions that have a digital platform and focus on new values, which are characterized by intangible character. Collectively, such changes are manifested in the development of effective mechanisms for the development of urban areas, based on the efficient use of natural resources, smart urban infrastructure, focus on quality man-made changes, ensuring the basic needs of the population and promoting sustainable development.

2.2 Innovative Solutions of New Models of Future Urbanization

Today, every state faces an important question–how to develop a new type of economy that can not only develop the material component, but also be a source for stable growth of the environment, because nature contains all the necessary wealth that is the basis of prosperity of the nation. The trends of the world economic situation are such that countries need to ensure the future, which provides for the implementation of projects that not only provide financial flows, long-term capital protection, but also create tens of thousands of jobs, economic chains, provide profits in the short and medium term [8, 9].

The generator of the future for the economy in the new business ecosystem project is the creation of developing ecological cities, which are artificially created on the basis of the latest technologies and advanced decisions. The project can be implemented gradually–from the neighborhood to the city, which provide both basic human needs (food, energy, security, not to mention respect and self-realization in work, study and leisure), and create unique, greenhouse conditions for business development.

The projects of the cities of the future combine conceptual, technological and town-planning projects, which is inherent in the research and practical nature of implementation. It is a combination not only of new ideas and technologies, but of a new culture of life, which corresponds to the concept of sustainable development, which goes beyond the project and focuses on the transformation of a normal lifestyle with the introduction of a new value system.

The idea of the concept of cities of the future lies not in the material plane–it has a humanistic character, which reflects the challenges facing humanity to not only ensure the quality of life of modern generations, and to take into account the interests and needs of future generations. Such innovative projects are hypertrophied in nature and describe global changes, which involves defining a new mission of urbanization with a reorientation of development goals and objectives, which accordingly creates a new urban reality.

The cities of the future are a new form of organization of life, which not only provides services and human needs, but also creates conditions for the development of the national economy, attracting investment, creating new jobs, protecting the environment and promising development of society. In the innovative cities of the future, economic development is ensured through the introduction of holistic models (chains) of organization of innovation processes: from the idea and research to the production and sale (introduction) of high-tech products and technologies. The living

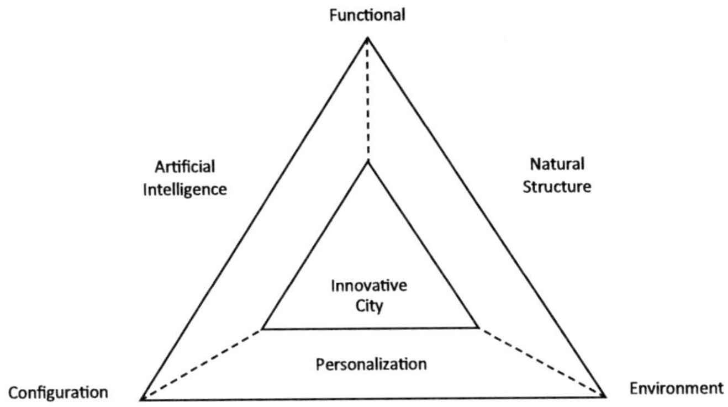

Fig. 4 Innovative dimension of designing the city of the future

environment is based on high-tech advances in science to ensure the competitiveness and profitability of business products, both in domestic and global markets.

The ecological component of innovative urbanization projects involves taking into account the basic values of the ecological vector of sustainable development, implemented through the restoration of natural areas and reorganization of inefficient urban areas, the use of adaptive technologies for secondary use of resources, implementation of alternative energy projects, modeling of innovative urban structures. Innovative city is a new strategy of sustainable development, in which the provision of improving the living standards of the population and forming of a developed system of ecological restoration, development of urban infrastructure, culture and leisure, which is modeled as an innovative administrative unit with expanded functionality (Fig. 4).

Innovative cities are characterized by accelerating the pace of life that is reflected in the multistructured urban infrastructure and the offer of new services, which are an element of new values of the population.

In these projects, a special place is occupied by artificial intelligence, which forms an extensive network of knowledge that is used for efficient and promising functioning of the city–promoting energy recovery through the use of innovative projects based on city infrastructure. Personalization of the environment means that cities will be adapted to the individual needs of the population and basic characteristics, including special tools and mechanisms to ensure environmental restoration and rational use of resources.

In general, this is an innovative model of integration of social, economic and environmental vectors, which provides for the construction of a new system of relations with the transition to a qualitatively higher level of meeting the real needs of today. Such cities of the future that formed as trends in innovative development have a stimulating effect on the environment, contribute to the innovation culture of the population, the socio-economic rise of peripheral areas with low development rates. Such innovative project is an economic generator capable of providing dozens of related industries, creating chains of technologies, products, services, thousands of jobs. This is an opportunity to provide better living conditions while reducing its cost through the introduction of new technologies on the principle of reasonable sufficiency and economic feasibility [10–20].

Modern development of urbanization is focused on dynamic transformations of cities in accordance with the basic criteria of sustainable development - environmental restoration, modernization of urban infrastructure, meeting the needs of present and future generations, efficient use of resources. An effective urban policy should be developed as a roadmap of decisions with mandatory items–environmental urban planning, optimization of administrative functions, restructuring of industrial facilities, the establishment of a comprehensive waste management system.

Ecological urban construction is to design the infrastructure of the city, taking into account the establishment of branched «green» park areas, which are the «lungs» of the city and a certain type of natural filters. Also, such a project provides for a significant transformation of transport infrastructure with the priority of alternative types of vehicles and the development of a transport system based on alternative types of energy. These are large-scale transformations that cover the entire infrastructure of the city and are considered as an innovative map of the connection with the transition to a higher level of environmental protection.

A special place in the road map of innovative urban planning belongs to the optimization of administrative functions, namely the establishment of a number of regulations that coordinate environmental issues, links with city life. In particular, these are control norms and environmental supervision, information on which should be widely available and should be integrated with large databases. Additional control procedures allow orienting city objects on observance of requirements and rules of ecological safety, and digital technologies allow operating data in real time that considerably reduces risks of non-fulfillment of ecological norms. Also environmental management consulting services should be provided that will allow establishing an effective environmental policy and rational use of resources.

Modernization of cities and ensuring their sustainable development should begin with assessing the effectiveness of the city structure with the restructuring of industrial facilities to be relocated to special areas outside the city. That is, it is a question of creation of special industrial parks which will be equipped according to needs of the industry with an emphasis on preservation of environment and dissociation with the city. Creation of separate zones of industrial development allows realizing a qualitative configuration with introduction of the newest technologies concerning rational use of resources and avoidance of pollution of environment. Accordingly, industrial facilities continue to operate, but with effective social and environmental indicators. Lands of industrial facilities can be used for effective urban redevelopment with construction for the needs of innovative development of the city, which gives additional financial investment in real estate, and therefore has a positive effect on economic indicators of urban development.

Establishment of a comprehensive waste disposal system is based on effective collaborative relations between industries from different industries based on the use of innovative technologies. So, industrial facilities can be combined into a single center, which will allow rational use of resources and carry out efficient waste disposal. For example, it can be a combined work of industrial and energy facilities, as a result of which an effective technical solution will be implemented that will satisfy the business and have an effective indicator in the ecological development of the city. Each decision must be made collectively, which will significantly reduce the potential risks and establish an effective project of business operation in terms of three vectors of sustainable development.

3 Conclusions

Development of strategies for the development of the modern economy with a focus on promising innovative trends of today is considered in article, which is aimed at the development of the economic, social and environmental sectors. This is the definition of a new direction of development with a reorientation of general conclusions and the transformation of ideas and values that are basic for the economy. The future of the economy is shown through the innovative developments of today, which are characterized by the nature of anticipating the real needs of future generations. The general conclusions from the results of scientific research are as follows:

1. Prospective changes in the strategy of sustainable development based on the use of the basic provisions of the Lean concept are investigated, which is a fundamental rethinking and general redesign of information, technological, economic, socio-cultural processes, which is a new formula for the business development. It is substantiated that Lean-development of innovative economy is the result of activation of rational and irrational human activity in the global multimedia environment, which creates better conditions for the development of sustainable development vectors, increasing the efficiency of economic, environmental and social processes by forming a new economic model.

2. The innovative perspective of sustainable economic development on the basis of intellectual innovations is defined, to which special attention is paid today, considering their high efficiency in comparison with other developments. This is the definition of a new tool for sustainable economic development, which is to propose the transformation of approaches to the development of the environmental vector, in particular, the use of ideocognitive methods that combine creative approaches with a priori knowledge. Innovative transformations in the economy are substantiated on the basis of organizational and methodological principles of I-engineering, which corresponds to the essence of innovation as an objective and logical process of ensuring sustainable economic development. This is the redesign of social, economic and environmental vectors using intelligent solutions with appropriate technical and technological provision and professional provision.

3. A schematic interpretation of the concept of multivariate economic development is proposed, based on the scale of modern innovation, which is characterized by formal and informal nature of methods, tools and technologies, the balance of which fluctuates between normative and personalistic theories, reconciling traditional and innovative views with new priorities as a basis for following different options for

sustainable economic development. Drivers of innovative development of economy are defined, which provide economic and social vectors in a new format of global innovative changes. This is a combination of advanced automated and intelligent technologies in a single system that works as a set of integrated processes with the transition to a qualitatively higher level of development of society in its future.

4. Innovative solutions of new models of future urbanization are considered, which consists not only in combining new ideas and technologies, but in forming a new culture of life, which corresponds to the concept of sustainable development, goes beyond the project and focuses on transforming the usual lifestyle with a new value system. A new strategy for the development of urbanization is identified, in which the provision of improving the living standards of the population and the formation of a developed system of ecological restoration, development of urban infrastructure, culture and leisure, which is modeled as an innovative administrative unit with expanded functionality. It is substantiated that an effective urban policy should be developed as a roadmap of decisions with mandatory items–environmental urban planning, optimization of administrative functions, restructuring of industrial facilities, the establishment of a comprehensive waste disposal system.

References

1. Imai M (1986) Kaizen: the key to Japan's competitive success. McGraw-Hill/Irwin, New York
2. Womack JP, Jones DT (1996) Lean thinking: banish waste and create wealth in your corporation. 1st edn. Productivity Press, New York
3. Tapping D (2006) The lean office pocket guide. 1st edn. MCS Media, Inc, Chelsea
4. Gothelf J, Seiden J (2017) Sense and respond: how successful organizations listen to customers and create new products continuously. Harvard Business Review Press, Boston
5. Elster J (1998) Emotions and Economic Theory. J Econ Lit 36(1):47–74
6. Dixit AK, Nalebuff BJ (2010) The art of strategy: a game theorist's guide to success in business and life. Illustrated edition. W.W. Norton & Company, New York
7. Nordström KA, Schlingmann P (2015) Urban express: 15 urban rules to help you navigate the new world that's being shaped by women and cities. Engelsk
8. Pauli G (2010) Blue economy-10 years, 100 Innovations, 100 Million Jobs. Paradigm Publications, New Mexico
9. Rifkin J (2013) Third industrial revolution. Griffin, New York
10. Bochulia T (2020) Decisions of the future in innovation-oriented development of the national economy. In: National University «Poltava Polytechnic». Conference «Building innovations–2020», pp 409–410, Poltava
11. Onishchenko S, Matkovskyi A, Puhach A (2014) Analysis of threats to economic security of Ukraine in conditions of innovative economic development. Econ Ann-XXI 1–2(2):8–11
12. Sivitska S, Vartsaba V, Filonych O (2018) Buildings energy-efficient renovation investment. Int J Eng Technol (UAE), 7(3):408–412. https://doi.org/10.14419/ijet.v7i3.2.14562
13. Drobyazko S, Bondarevska O, Klymenko D, Pletenetska S, Pylypenko O (2019) Model for forming of optimal credit portfolio of commercial bank. J Manage Inf Decis Sci 22(4):501–506
14. Iasechko S, Haliantych MK, Skomorovskyi VB, Zadorozhnyi V, Obryvkina O, Pohrebniak O (2020) Contractual relations in the information sphere. Syst Rev Pharm 11(8):301–303. https://doi.org/10.31838/srp.2020.8.46
15. Lavrentieva OO, Rybalko LM, Tsys OO, Uchitel AD (2019) Theoretical and methodical aspects of the organization of students' independent study activities together with the use of ICT and tools. Pap Presented CEUR Workshop Proc 2433:102–125

16. Onyshchenko VO, Filonych OM, Bunyakina NV, Senenko NB (2020). A new agent for removing concrete residues from the surfaces of polypropylene molds in the manufacture of paving slabs and its advantages. https://doi.org/10.1007/978-3-030-42939-3_67
17. Onyshchenko V, Sivitska S (2014) Alternative energy developing investment support in terms of energy dependence. Econ Ann-XXI 9–10:34–37
18. Onyshchenko V, Sivitska S, Cherviak A (2018) Construction industry in Ukraine credit analysis. Int J Eng Technol (UAE), 7(3):280–284. https://doi.org/10.14419/ijet.v7i3.2.14420
19. Lakhno V et al (2018) Model of managing of the procedure of mutual financial investing in information technologies and smart city systems. Int J Civil Eng Technol 9(8):1802–1812
20. Onyshchenko V, Sivitska S, Cherviak (2020) Buildings reconstruction within the new educational space project in Ukraine. https://doi.org/10.1007/978-3-030-42939-3_66

Construction Management Based on European Experience of Using Smart-City Technology

Liudmyla Boldyrieva, Shahla Alizada, Viktor Gryshko, and Mariya Hunchenko

Abstract The definitions of the basic concepts "intellectual city", "digital city", "city of knowledge" and "sustainable city", etc. have been analyzed. It has been proved that city government is crucial for solving global problems. It has been stated that the most progressive integrated "network" approach to the implementation of Smart-City technologies known from European practice is used by the United Kingdom is. It has been found out that nowadays Smart London is a kind of platform that accumulates innovations and their creators. It has been established that Smart-City is an innovation and e-service that helps residents solve issues in a variety of areas on a daily basis, from parking and charging to medical care or services payment, and the implemented Transport For London strategy has become one of the best role models. The smartest European cities rating has been studied taking into account the following criteria: human capital (development, attraction and education of talents), social cohesion (consensus between different social groups), economy, environment, governance, management, urban planning, international relations, technologies, mobility and transport (lightness). Positive and negative results obtained after the smartest European cities evaluation have been described. Examples of transition to the smart cities formation effective strategy in Ukraine have been given.

Keywords Construction · Development · Startups · Smart-City · Concept · Technologies · Management

1 Introduction

It is well known that the vast majority of Europeans live in cities where there is a wide range of urgent challenges regarding health, comfort and high quality of life.

L. Boldyrieva (✉) · V. Gryshko · M. Hunchenko
National University «Yuri Kondratyuk Poltava Polytechnic», Poltava, Ukraine

S. Alizada
Azerbaijan University of Architecture and Construction, Baku, Azerbaijan

V. Onyshchenko et al. (eds.), *Proceedings of the 3rd International Conference on Building Innovations*, Lecture Notes in Civil Engineering 181,
https://doi.org/10.1007/978-3-030-85043-2_57

Therefore, with each following year cities should become safer, greener, and more culturally beautiful with clean air and environmentally friendly transportation.

Numerous challenges to the construction and development of modern cities as complex systems require new approaches. Thus, a management model that can flexibly adapt to dynamic changes in the environment and at the same time combine the solutions of different industries is needed. One of such up to date modern models is the Smart-City concept. The Smart-City concept is a strategic landmark of European cities. However, it should be noted that the possibilities of implementing such an option are mostly made for the medium or even long-term perspective.

2 An Overview of the Latest Sources of Research and Publications

Boykova M., Ilina I. and Salazkin M. claim that the Smart-City concept is much in demand. In the publication, "The Smart City Approach as a Response to Emerging Challenges for Urban Development", the researchers concluded that "smart city" strategies in many cases continue to be based on a narrow, "technological" approach. Such approach involves the availability of intellectual infrastructure and can solve many urban problems, as well as improve the quality of urban life. However, unlike the extended, integrated approach, it does not take into account many social and economic factors and real needs of the residents. Implementation of an integrated approach involves a number of conditions capable of: firstly, management decisions integrating at different levels; secondly, predicting how changes in one system affect other systems; thirdly, focusing on interdisciplinary cooperation [1, p. 65].

Albert Meijer and Manuel Pedro Rodríguez Bolívar prove a comprehensive perspective: smart city management as a new form of human cooperation through ICT for getting the best results and more open management processes [2, p. 392]. Researchers also highlight the fact that the publications analyzed by them have a different focus on: 1) smart technologies, smart people or smart cooperation as the defining features of smart cities; 2) transformational or gradual view of changes in the city management; 3) the best results or a more open process, as a requirement of smart city management legitimacy [2, p. 392].

Tcholtchev Nikolay and Schieferdecker Ina prove that the level of cities and communities development around the world is possible due to an intelligent city based on information and communication technologies (ICT). It is ICT that ease management and processing of vast amounts of data and information for reasonable management of urban infrastructure and processes, residents involvement and promotion of new services and applications in various aspects of urban life (supply chains, mobility, transportation, energy, residents participation, public safety, interaction between residents and public administration, water resources management, parking and many other cases and areas) [3, p. 156].

The scientists team claim that social and economic situation in the cities during the COVID-19 pandemic caused an increase in inequality and a record number of unemployed. All these events led to the necessity of a radical revision of the city [4]. They offer to support a "15-min city" concept suggested by Carlos Moreno in 2016, which has a new "chrono-urbanism" perspective, namely the concept of constructing safer, more sustainable and inclusive cities [4].

In their book, Dameri Renata Paola and Rosenthal-Sabroux Camille summarized several materials from various scientific studies across the whole Europe and offered a broad vision of the smart city phenomenon and comparative research as well as reasoning on how to define a smart city, how to develop a smart strategy and how to measure whether reasonable actions really can create social value for residents and improve the quality of life [5].

The aim of the research is to study the European experience regarding construction management based on the use of Smart-City technology and analysis of the strategy of forming smart cities in Ukraine.

3 Main Results of the Study

The very concept of Smart-City encompasses several definitions, namely: intelligent city, digital city, city of knowledge, sustainable city, and so on.

Annalisa Cocchia's researches prove that Smart City and Digital City are the most frequently used terminology in the literary sources to denote the reasonableness of the city [6].

City government is crucial to solving global problems. Thus, for example, the current administrative focus on cities as centers of governance should be based on technology and innovation. E-government and innovations are crucial in urban management.

According to European practice, the most advanced integrated "network" approach to the implementation of Smart-City technologies was used by the United Kingdom. Smart-City is innovations and e-services that help residents solve tasks every day in different areas: from parking and phone charging to receiving medical care or paying for services.

London has created a public program Smart London Innovation Network to promote new technological startups and social initiatives in the field of Smart-City. In this case, the government has become as open as possible to communicate with residents. Thus, Londoners began to take an active part in all pilot projects, in particular, their thoughts and views started to be taken into account during establishing a city development plan.

At the time, Smart London is a kind of platform that accumulates innovations and their creators. Global companies are teaming up with research institutes and scientists in order to freely develop the best technological approaches, startups. Thus, Siemens has invested 30 million pounds in the development of projects on Smart technologies systematization. At its own discretion, Intel funds the charter of Connected Cities and Imperial College, which are developing new solutions concerning Smart-City in the security sector.

To improve transport infrastructure in London, a digital map of key actions was firstly created, followed by the gradual implementation of the Transport For London strategy, which is now one of the best examples to follow. The project raised investments equivalent to tens of millions of pounds and allowed: firstly, to "unload" roads; secondly, to improve traffic; thirdly, to enhance the safety and comfort of parking.

Besides this, the operational transit system Heathrow pod was launched at the London airport, and takes into account the wishes of both employees and passengers. As a result of the implementation of such a unique solution, Heathrow pod transit completely eliminated the need for regular travels on buses in the airport area, which previously made more than 70 thousand trips a year [7]. As is known, London is always open to experimental practices and leading scientific thoughts. Representatives of leading research institutes and young scientists in technical fields of universities join all existing Smart-City projects. Thus, the most striking example was the largest project developed in partnership with graduates of Imperial College London on Innovate18 railway infrastructure (an underground railway network with a unique high-speed train Crossrail will be able to transport freight and passenger on the basis of 7 innovative approaches) [7].

Thus, the experience of Great Britain is worth of borrowing and following.

As of 2019, 50 cities in Germany are already testing various options for integrating information technology into urban infrastructure, the most active are Nordrhein-Westfalen and Baden-Württemberg. The main areas of Smart-technologies implementation in Germany are city administration, work of public services improvement, mobility, energy saving, ecology and health care. Thus, for example, in Hamburg it is planned to automate almost all processes in municipal departments by 2022, in particular, residents will be able to solve almost all issues via the Internet.

Ingolstadt, Bavaria, is developing a mobility concept for a flying taxi that could be used, in particular, by rescue services. Kaiser Slauter and Karlsruhe are working on the concept of autonomous or semi-autonomous transport. Duisburg with Microsoft creates Smart Home for the elderly.

The Netherlands pays considerable attention to the blockchain. The Ministry of Economy has launched a blockchain coalition, namely: the government, corporate organizations and startups are cooperating in the development of blockchain technologies in five specific areas, which will be implemented nationwide: 1) identity identification; 2) pensions; 3) logistics; 4) issuance of diplomas and educational certificates; 5) subsidies.

The Center for Globalization and Strategy of the IESE Business School (Spain) compiled a ranking of the world's smartest cities in 2019, and the analysis itself was carried out taking into account the following criteria (see Fig. 1).

A truly smart city is one that aims to improve the quality of life of its residents while ensuring economic, social and environmental sustainability. A total of 174 cities from 80 countries were included into the IESE Cities in Motion Index 2019.

According to the evaluation results, London topped the list of the smartest cities in the world due to excellent results in almost all categories for smart cities from the

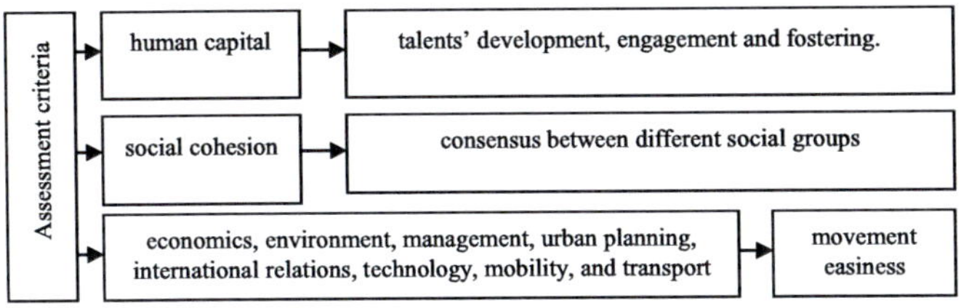

Fig. 1 Evaluation criteria regarding the smartest cities ranking identifying

criteria list, including human capital (due to the large number of business schools and universities) and conditions for international cooperation. The cities of London and Paris are also the most important financial centers in Europe.

The results revealed after the smartest cities evaluation in Europe are given in Table 1. The city of Zurich is the largest financial and scientific center of Switzerland. Apart from a large number of international banks, it is characterized by smart houses and garbage recycling system as well as the organization of public transport.

The city of Copenhagen is actively developing in the field of IT-technologies, ecology, medicine, economy, business and transport infrastructure. Launch of the City Data Exchange—an online platform with public and private information open to residents. The city's data base in Copenhagen, besides the main task which is access to the information, has a side goal—to make the capital even more environmentally friendly. Special applications encourage to think about the consequences of harmful substances emissions.

Thus, such cities as London, Amsterdam, Paris, Reykjavik, Barcelona, Copenhagen, Helsinki, Vienna and Grenoble are considered to be exemplary regarding the implementation of the Smart City concept in Europe.

It is worth noting that the first steps in the transition to an effective strategy for the formation of smart cities have already been made in Ukraine. Cities such as Kyiv, Lviv, Kharkiv, Vinnytsia, Chernivtsi, Dnipro are active initiators of the introduction of effective Smart-technologies.

Smart-technologies processes were launched in Kyiv, namely: Smart-streets, E-tickets, Kyiv Smart City mobile application, Safe City project, etc. Yes, on the street. Salyutny, 2 has 12 CCTV cameras, garbage sorting tanks, equipped bike path and free city Wi-Fi, two charging stations for electric cars, benches with USB-chargers powered by solar panels, an emergency call button, and There is also a stationary air quality monitoring system (the first in Ukraine was created within the framework of the Managing Air Quality project) [8]. The mode of operation of such a system is that the sensor measures the concentration of fine dust (PM2.5 and PM10), the level of nitric oxide (NO), sulfur oxide (SO2), ozone (O3), benzene (C6H6), nitrogen dioxide (NO2), carbon monoxide (CO), as well as temperature and humidity. In all districts of the capital daily air quality control is carried out by a mobile laboratory [8]. Thus, smart-street is the embodiment of the international practice of creating a Living Lab for startups [9–15].

Table 1 Positive and negative results of the evaluation were detected

City	Country	Results
London	Great Britain	The largest number of startups and programmers than in any other city in the world. There is an open data platform (London Datastore), which is used by more than 50,000 people every month. The weakest aspect in the British capital is currently social cohesion
Amsterdam	Netherlands	Active residents participation in the public life of the city and the possibility of wide online access to governmental services. A large number of tourists annually. It is famous for good business opportunities. An open data web site called Amsterdam Smart City was created for the city's residents. Furthermore, the platform supports an application designed to help visual impaired people and other perceptional peculiarities
Paris	France	The strengths are the economy, human capital, international relations, technologies and mobility. Actively uses open innovations, the Internet of Things (to optimize transport and pedestrians flow) and an automated subway system. In order to improve the air quality in the city, the authorities are promoting the use of bicycles and electric cars
Reykjavik	Iceland	Wins due to environmental criteria (more than 99% of electricity in the city is produced by geothermal sources). Complete rejection from combustible minerals. The program according to which it is planned to become a city with zero carbon emissions by 2040 has been approved
Geneva	Switzerland	Efficient waste recycling. Active construction of smart houses. By 2020, the government plans to reduce carbon dioxide emissions by a quarter of the current volume
Zurich	Switzerland	A large number of smart houses, an efficient garbage recycling system, public transport organization. By 2024, it is planned to lay an underground network for cargo transportation through Zurich, which will run on renewable energy sources
Stockholm	Sweden	The main city of Sweden received the highest rating for online access to governmental services. There are also no industrial enterprises in the city, which has a positive impact on the environment
Copenhagen	Denmark	The city is actively developing in the IT-technologies area, ecology, medicine, economy, business, transport infrastructure. Launching of the City Data Exchange, an online platform with public and private information open to residents. Special applications encourage to think about the consequences of harmful substances emissions

In the city of Lviv, at the municipal level, an automated transport management system with the uMuni cloud energy monitoring service is used, which analyzes traffic flow, controls traffic lights and monitors the work of dispatchers and carriers, as well as a mobile application for tourists Lviv Travel Places.

Kharkiv has its own geographic information system for managing the city economy, which operates in conjunction with an online service for processing citizens' requests.

The first service center that processes administrative and economic appeals of citizens was established in the city of Vinnytsia. Also available online service control of utilities with their evaluation, selection of contractors, calculations through the site and the application WinDim24. And a separate 24-h online center works with applications for housing and communal services, education, medicine and transport issues.

In the city of Chernivtsi, a call center has been created by analogy with the one in Kyiv, which accepts applications from citizens around the clock in order to speed up the solution of city problems. And here they use geo-applications for tourists, available by QR-codes.

The city of Dnipro is a leader in the implementation of electronic and administrative services. There are services that allow to control the budgets and state of administrative institutions, and even road inventory service, social mobile applications ("E-contact", "My police"), which bring residents closer to the "body" of services and officials in case of any problems.

4 Conclusions

Thus, the strategy of Smart-specialization, which provides for the identification of strong suits and the development of competitive advantages of the regions on the basis of the existing structure of regional economy and innovations, has become up to date in Ukraine over time. Studying the experience of EU countries and its implementation into Ukraine regions: firstly, will provide additional financial opportunities for regions support; secondly, will create new jobs; thirdly, will improve the living conditions of households. Therefore, the development and implementation of own smart specialization strategies by the regions of Ukraine still remain problematic issues. In the XXI century, informational education, informational and communicative (Internet, multimedia, hyperreality) as well as professional and communicative technologies are developing. With the help of Smart informational systems and intelligent complexes, social and economic systems can be managed.

References

1. Boykova M, Ilina I, Salazkin M (2016) The smart city approach as a response to emerging challenges for urban development. Foresight STI Govern 10(3):65–75. https://doi.org/10.17323/1995-459X.2016.3.65.75
2. Meijer A, Bolívar MPR (2016) Governing the smart city: a review of the literature on smart urban governance. Int Rev Adm Sci 82(2):392–408. https://doi.org/10.1177/0020852314564308

3. Tcholtchev N, Schieferdecker I (2021) Sustainable and reliable information and communication technology for resilient smart cities. Smart Cities 4(1):156–176. https://doi.org/10.3390/smartcities4010009
4. Moreno C, Allam Z, Chabaud D, Gall C, Pratlong F (2021) Introducing the "15-Minute City": sustainability, resilience and place identity in future post-pandemic cities. Smart Cities 4(1):93–111. https://doi.org/10.3390/smartcities4010006
5. Dameri RP, Rosenthal-Sabroux C (2014) Smart city and value creation smart city. How to create public and economic value with high technology in urban spaceю, pp 1–12. Springer, New York. https://doi.org/10.1007/978-3-319-06160-3_1
6. Annalisa C (2014) Smart and digital city: a systematic literature review. In: Dameri RP, Rosenthal-Sabroux C (eds) Smart city. How to create public and economic value with high technology in urban space, pp 13–43. Springer, Cham. https://doi.org/10.1007/978-3-319-06160-3_2
7. Smart City: How to apply world experience in Ukraine correctly (2018). Homepage https://nv.ua/ukr/biz/experts/smart-city-yak-pravilno-zastosuvati-svitoviy-dosvid-v-ukrajini-2506784.html
8. Fund KM (2017) Copenhagen has been named the smartest city on the planet. Homepage https://kfund-media.com/ru/kopengagen-pryznan-samym-umnym-gorodom-na-planete/
9. Smart Kyiv: which smart technologies the capital contain. Digitalization of the city/Our Kyiv (2019). Homepage https://nashkiev.ua/zhournal/rastouschiy-gorod/oumnyy-kiev-kakie-smart-tehnologii-est-v-stolitse.html
10. Sankov P, Dikarev K, Kushnir Y, Tkach N (2020) Modern smart city concept considering population safety issues. In: Onyshchenko V, Mammadova G, Sivitska S, Gasimov A (eds) Proceedings of the 2nd international conference on building innovations. ICBI 2019. Lecture notes in civil engineering, vol 73. Springer, Cham. https://doi.org/10.1007/978-3-030-42939-3_24
11. Chevhanova V, Chichulina K, Skryl V (2018). Effective cluster model of thermal power company management. Int J Eng Technol (UAE), 7(3):65–70. https://doi.org/10.14419/ijet.v7i3.2.14377
12. Chichulina K, Skryl V (2020) Energy service companies are an effective tool for financing energy-efficient projects. Int J Manag 11(5):252–266. https://doi.org/10.34218/IJM.11.5.2020.025
13. Iasechko S, Haliantych MK, Skomorovskyi VB, Zadorozhnyi V, Obryvkina O, Pohrebniak O (2020) Contractual relations in the information sphere. Syst Rev Pharm 11(8):301–303. https://doi.org/10.31838/srp.2020.8.46
14. Drobyazko S, Bondarevska O, Klymenko D, Pletenetska S, Pylypenko O (2019) Model for forming of optimal credit portfolio of commercial bank. J Manage Inf Decis Sci 22(4):501–506
15. Sankov PM, Zaporozhets VI, Yu OT (2020) Prospects for the implementation of Smart City technologies in light practice. In: The 14th international scientific and practical conference «Actual problems of science and practice», Stockholm

Management in the Construction Sector Using Smart Technologies: European Experience

Liudmila Boldyrieva, Alina Chaikina, and Khanlar Ganiyev

Abstract The practical experience of European countries on use and dissemination of Smart technologies in construction were studied, which shows that to change difficult situation in the construction sector of the economy it is necessary to: firstly, actively use Smart technologies in construction; secondly, gradually reduce energy consumption in the construction and housing sectors through mass construction of energy-efficient buildings and structures; thirdly, to reconstruct the already built ones. It was established that the introduction of Smart Home system is becoming more and more relevant, which allows to make everyday life more convenient: saves energy (automatic switching off of the light); saves money (automatic switching off of heating as soon as windows open); provides the best protection for the house (automatically gives an audible alarm when the intruder enters, and warns residents by sending push messages to their smartphones). Certain advantages from the use of Smart technologies in construction were described. The characteristic of the software for architectural and construction design and preparation of construction documentation was given in the article. Systems for monitoring of architectural and construction projects and planning of construction works, design of building constructions of building's engineering systems, automated design of general purpose were analyzed by authors. Architectural and construction applications for AutoCAD and applications for design and calculation of plumbing systems were described. The characteristics of programs for calculation of building structures, design and calculations of pipelines, heat exchangers, geotechnical calculations, design of infrastructure objects were given in the article.

Keywords Construction · Construction sector · Efficiency · Energy efficiency · Development · Smart technologies · Management

L. Boldyrieva · A. Chaikina (✉)
National University «Yuri Kondratyuk Poltava Polytechnic», Poltava, Ukraine

K. Ganiyev
Azerbaijan University of Architecture and Construction, Baku, Azerbaijan

V. Onyshchenko et al. (eds.), *Proceedings of the 3rd International Conference on Building Innovations*, Lecture Notes in Civil Engineering 181,
https://doi.org/10.1007/978-3-030-85043-2_58

1 Introduction

EU countries actively use Smart technologies because they increase efficiency, reduce costs, generate new sources of income, improve the daily lives of population, and most importantly, accelerate construction and development of the construction sector. In time, the most progressive cities in Europe are following the Smart path of development.

Effective Smart technologies change the preparation and maintenance of construction projects, as well as the subsequent operation of infrastructure.

Development is gaining momentum thanks to artificial intelligence (AI), IoT gadgets for the home and 5G (fifth generation of telecommunications standards). In turn, 5G provides high data rates in unmanned vehicles, virtual reality and innovative urban infrastructures. 5G also allows the implementation of services such as "Smart" parking, smart meters, CCTV cameras, traffic management, waste control and coordination of emergency services. The next stage in the development of the construction of "Smart" homes will be more accessible through the development of this technology all over the world.

According to McKinsey & Company, the construction sector is one of the least digitized sectors of the economy. In the world in the construction sector, the cost of Smart technologies does not exceed 1%.

2 An Overview of the Latest Sources of Research and Publications

Jinying Xu and Weisheng Lu from the University of Hong Kong [1, p. 157] are convinced that the current management of the life cycle of construction projects suffered from the dilemma of "lack of information" and "explosion of information". They developed and proposed the use of a closed life cycle management system that provides a constant flow of information for all stakeholders. Matej Štefanič and Vlado Stankovski make a review of technologies and applications for Smart construction. In his publication [2, p. 83] they concluded that new opportunities in construction are possible through the Internet of Things (IoT), artificial intelligence (AI) and cloud computing technologies (CT), which monitor construction, site management, occupational safety, early warning about natural disasters, resource and asset management. After all, IoT, AI and CT offer new opportunities for the development of new intelligent applications in the construction sector. The authors are convinced that they stimulate a new wave of affordable, reliable, secure and efficient Smart applications that provide various benefits, including efficient logistics and safe working conditions. Although Matej Štefanič and Vlado Stankovski [2, pp. 83–84] focus on construction, their findings may also relate to the wider context, namely Industrie 4.0, cities, communities, civil engineering, and smart homes, transport, etc.

Authors: Yuhan Niu, Weisheng Lu, Ke Chen, George G. Huang [3] investigated intelligent construction objects (ICO), the main building blocks of future construction, which are in the stage of rapid development. ICO includes construction resources: machines, devices and materials that are "Smart" with the help of appropriate technologies that ensure effective decision-making in construction, autonomy.

The conclusions and proposals of Phil Purnell and Ming K. Lim are interesting [4, p. 223], as they tried to answer the question of the use of Smart technologies in enabling construction components reuse.

The team of authors: Guerriero Annie, Kubicki Sylvain, Berroir Fabrice and Lemaire Clément [5] at the scientific conference Funchal, Portugal offered their own research on BIM-enhanced collaborative Smart technologies for LEAN construction processes.

Boton C., Kubicki S. and. Halin G. [6] carried out research in the direction of 4D/BIM Simulation for Preconstruction and Construction Scheduling.

There is reason to believe that Smart technologies have improved life in Ukraine's cities, and helped authorities to work more efficiently. It should be noted that the first steps in the transition to an effective strategy for smart cities formation have already been made in Ukraine. Cities such as Kyiv, Lviv, Kharkiv, Vinnytsia, Chernivtsi, Dnipro are active initiators of the introduction of effective Smart technologies and digital services.

The purpose of the research is to study the European experience in managing efficient Smart technologies in the construction sector and to develop practical recommendations for its improvement in Ukraine.

The theoretical and methodological basis of scientific research is a dialectical method of scientific knowledge, a systematic approach, provisions of modern economic theory, scientific works of economists who deal with problems related to the management of effective Smart technologies in the construction sector.

Scientific research is based on the use of general scientific methods, namely: theoretical generalization, system analysis and synthesis, abstraction and formalization.

3 Main Results of the Research

Hypothetically, it can be argued that the construction and operation of buildings is one of the most energy-intensive sectors of any country economy. Thus, in some EU countries, buildings and structures consume about 30% of all energy in the country.

The practical experience of the EU countries shows that to change the difficult situation in the construction sector of the economy it is necessary: firstly, actively use Smart technologies in construction; secondly, gradually reduce energy consumption in the construction and housing sectors through mass construction of energy-efficient buildings and structures; thirdly, thirdly, to reconstruct the already built ones.

Ideally, an energy-efficient house is a virtually closed system: water is supplied by groundwater and rainwater, electricity and hot water are provided by solar panels, gas is generated from sewage waste; exhaust ventilation supplies fresh air of the required temperature (heat recovery unit, and excess air heat is used to heat water).

Germany actively supports innovative heating technology, i.e. a special coating is applied to the facades of buildings, which accumulates solar energy during the day and gives it to the house at night. Systems with efficient heat exchangers, which provide warm water even in the cold season, have also become widespread.

Western European countries such as Sweden, Denmark and Norway also use technological techniques for homes to minimize or eliminate the use of external electricity.

In real time for visual monitoring and organization of work, environmental monitoring, feedback and signaling, cloud backup and demonstration of data analysis, you can use Smart technology "from head to toe" [1].

To collect data at different stages, a closed-loop life cycle management system is used for intelligent design based on the Internet of digital technologies, including such as: 3D laser scanner, drone, building information modeling (BIM), augmented reality (AR), Auto-ID, Global Positioning System (GPS), wireless sensor network (WSN), robotics, web applications and mobile digital devices.

This structure integrates the latest technologies in intelligent construction, building automation and the Internet of Things (IoT). Data is stored, shared, processed and used on a single unified platform for all stakeholders to support better decision-making and interaction throughout the project life cycle. Practitioners can use the information through a standardized interface between different programs throughout the life cycle [1].

Properly connected Internet of Things with the construction of information modeling can provide a safer, greener and more efficient building system [3].

Different degrees of operational autonomy are provided by intelligent applications, which include the use of IoT devices, such as smartphones, cameras, actuators, sensors, cars and robots. The construction of context-specific intelligent environments is supported by the advent of modular open source IoT platforms and interoperability standards, such as Open Fog Consortium [7] and Cloud Native Computing Foundation [8–13]. Using methods and tools that implement such standards, you can quickly integrate physical devices and software.

British businessman and founder of Virgin Startup Richard Branson, write in Linkedin that the construction industry is now lagging behind in the innovation race, that is why he began working with British construction company Colmore Tang, creating a powerful initiative for Constructech.

Quickly draws attention to the construction sector LEAN Management. In parallel, building information modeling (BIM) is a key approach to making project processes smoother, more transparent, and more integrated. There is a two-stage technological proposal for LEAN methods aimed at creating a BIM information system for complete management of construction processes. In the short term, on construction sites, a smart construction planner is the answer to the implantation of LEAN digital methods. A reflection of 4D/BIM methods and principles of LEAN determines the path to future LEAN/BIM IT-development [5, 11].

Smart Home systems as well becoming more and more relevant, which allows to make everyday life more comfortable and helps owner to:

- save energy (automatic switching off of light);
- save money (automatic switching off of heating as soon as windows open);
- provide the best protection of the house (automatically gives an audible alarm when the intruder enters, and warns residents by sending push messages to their smartphones).

In general, the French trust Smart Home more than, for example, the British, Americans or Austrians.

Smart-City projects are implemented in more than 20 cities in the UK, including large cities and small towns. All this took place within the framework of a single plan for digitization and automation of Great Britain, but with remarkable individual decisions for each settlement.

The "smart architecture" system is aimed, on the one hand, at improving architectural appearance of the city, and on the other, at improving physical and emotional state of citizens. Such system was created on the basis of BIM developments, technologies, management of the construction process and information supply to both customer and contractors.

Smart architectures are a universal reality and open new opportunities for the progressive development of the construction sector. It helps to create multifunctional architectural objects that unite in one complex: a residential building, shopping and entertainment halls, a fitness center, parking in a modern urban style. Such solutions will save time and effort for residents of such a complex and nearby houses.

Modern engineering solutions are used on facades, in comfort systems, as well as in automated fire safety and evacuation systems.

Also, for comfort of residents and visitors of such complexes, the "green" roof technology based on technology of the Swiss company ZinCo is widely used.

However, we should focus on certain advantages of using Smart technologies in construction, which are listed in Table 1.

Implementation of management strategy in the construction sector using Smart-technologies is interpreted (see Fig. 1).

The main sector of IT technology in the field of construction is software for the design and construction of the building, and thus—tracking the state of construction at specific stages, the so-called design systems, which aim to ensure multivariate projects through change and variation initial data, increase the quality of design, reduce number of required consumables, engineers engaged in design and construction, and funds, reduce design time.

The main sector of IT technology in the field of construction is software for the design and construction of the building, and thus—tracking the state of construction at specific stages, the so-called design systems, which aim to ensure multivariate projects through change and variation initial data, increase the quality of design, reduce number of required consumables, engineers engaged in design and construction, and funds, reduce design time.

Table 1 Benefits of using Smart technologies in construction

Use of smart technologies	Advantages
Reduce energy consumption	Passive solar heating
	Correct use of quality materials
Monolithic construction	Provides a comfortable indoor climate (cool in the warm season, and warm in the cold)
Hinged ventilated facades	Increase heat and sound insulation
	Exterior of the building at the same time becomes beautiful and interesting
"Cold roof"	Reflect sunlight
	Reduce the temperature on the roof itself and directly in the house
	The ability to reduce energy consumption by 15%
	No air conditioner needed
	Care for the environment
"Green roof"	Plantations create shade in summer and keep warm in winter
Modern window systems (Smart windows)	Provides UV protection
	Regulate, depending on the time of day, amount of light that enters the room
	Change the level of transparency and adapt to external conditions

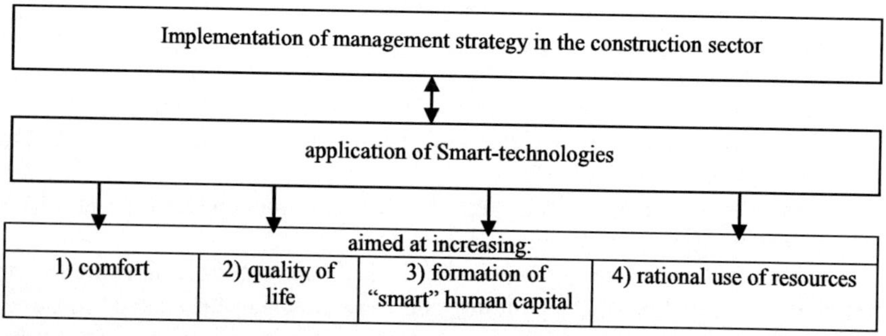

Fig. 1 Implementation of management strategy in the construction sector with the use of Smart-technologies

Software, systems, applications and programs are used for architectural design, engineering and construction calculations as shown in Table 2.

Permanent and immeasurable progress of the designed objects, modern technologies and equipment requires appearance of new and improved programs.

Table 2 Software, systems, applications and programs for architectural design, engineering and construction calculations

Software	
Software for architectural and construction design and preparation of construction documentation	Autodesk Building Design Suite; Autodesk Revit Architecture; Autodesk Factory Design Suite; Autodesk Factory Design Suite; AutoCAD Architecture
Systems for:	
Monitoring of architectural and construction projects and planning of construction works	Autodesk Navisworks
Design of building structures	Autodesk Revit Structure; Tekla Structures
Design of engineering systems of buildings	AutoCAD MEP, Revit MEP
Automated general purpose design	AutoCAD; AutoCAD LT, Autodesk 3ds MaxDesign; Std Manager CS
Applications:	
Architectural and construction applications for AutoCAD	PARCS (Parallel Asynchronous Recursively Controlled Systems), SDDC (Systems of design documentation for construction) GraphiCS
Design and calculations of plumbing systems	Plumbing designer's automated workspace, WENTSIS
Programs for:	
Calculation of building structures	SCAD Office, SCAD (Structure CAD), Comet, Kristall, Arbat, Kamin, Monolith, Section Constructor, CoKon
Design and calculations of pipelines, heat exchangers	AutoCAD Plant 3D, Autodesk Plant Design Suite, Plant 4D, START, HYDROSYSTEM, Resource, Ecolog-Shum
Geotechnical calculations	Plaxis, Plaxis Dynamics Module, Plax Flow, Plaxis 3D Tunnel, Plaxis 3D Foundation
Design of infrastructure facilities	Autodesk Civil 3D, Geoni CS

Computer-controlled software for modern construction equipment also requires appropriate software. For example, models for 3D printing, including in the field of construction, are usually distributed in STL files.

To convert a STL file to G-code (a language understood by a 3D printer), you need a "slice program" (the name is due to the fact that this program "slice" 3D model on a set of flat two-dimensional layers from which 3D printer will make physical object. Among the well-known "slicer programs" are the following: TinkerCAD, 3DTin, Sculptris, ViewSTL, Netfabb Basic, Repetier, FreeCAD, SketchUp, Simplify3D, Blender. OctoPrint and others.

Typically, required software comes bundled with a 3D printer. For example, Ultimaker and Reprap printers are equipped with Cura software, and MakerBot—MakerWare and ReplicatorG. In fact, when you buy a 3D printer, you also get licensed software "in the package". In this case, you pay in complex for equipment itself and for the use of the relevant programs, which are the objects of intellectual property rights of specific individuals (companies), i.e. for obtaining a license.

4 Conclusion

Thus, the construction sector of Ukraine should increase the level of use Smart technologies. After all, they occupy a decisive place in the process of modern construction of EU countries.

In time, from the very beginning of design work to the implementation of Smart technology development, computer-aided design systems are widely used, which provide a variety of projects and possible verification of their condition in the future, construction and computer equipment requires appropriate information and technical support, construction of buildings within the concept of "Smart Home", as well as also need software control.

European experience proves possibility and importance of the application and dissemination of Smart technologies in construction. After all, they ensure regional development, set strategic priorities and implement intellectual policies to maximize the potential of knowledge-based development.

References

1. Xu J, Lu W (2018) Smart construction from head to toe: a closed-loop lifecycle management system based on IoT. In: Conference: construction research congress, pp 157−168. https://doi.org/10.1061/9780784481264.016
2. Matej Š, Vlado S (2019). A review of technologies and applications for smart construction. Civ Eng 172(2):83−87. https://doi.org/10.1680/jcien.17.00050
3. Yuhan N, Weisheng L, Chen K, Huang GG (2016) Smart construction objects. J Comput Civ Eng 30(4). https://ascelibrary.org/doi/10.1061/%28ASCE%29CP.1943-5487.0000550
4. Purnell P, Ming KL (2018) The use of smart technologies in enabling construction components reuse: a viable method or a problem creating solution? J Environ Manage 216:214–223
5. Guerriero A, Kubicki S, Berroir F, Lemaire C (2017) BIM-enhanced collaborative smart technologies for LEAN construction processes. In: International Conference on Engineering, Technology and Innovation (ICE/ITMC), Funchal, Portugal. https://ieeexplore.ieee.org/document/8279994/authors#authors. https://doi.org/10.1109/ICE.2017.8279994
6. Boton C, Kubicki S, Halin G (2015) 4D/BIM simulation for preconstruction and construction scheduling. Multiple levels of development within a single case study. In: Creative construction conference, pp 500–505
7. Open fog consortium (2018). https://opcfoundation.org/markets-collaboration/openfog/
8. Cloud native computing foundation (2018). https://www.cncf.io/
9. Iasechko S, Haliantych MK, Skomorovskyi VB, Zadorozhnyi V, Obryvkina O, Pohrebniak O (2020) Contractual relations in the information sphere. Syst Rev Pharm 11(8):301–303. https://doi.org/10.31838/srp.2020.8.46
10. Onyshchenko V, Sivitska S, Cherviak (2020) Buildings reconstruction within the new educational space project in Ukraine. https://doi.org/10.1007/978-3-030-42939-3_66

11. Bielova AI, Zhuravska NY, Kochedikova AY (2020) Promising directions for the development of BIM technologies in Ukraine on its way to European integration. https://doi.org/10.1007/978-3-030-42939-3_53
12. Onyshchenko VO, Filonych OM, Bunyakina NV, Senenko NB (2020) A new agent for removing concrete residues from the surfaces of polypropylene molds in the manufacture of paving slabs and its advantages.https://doi.org/10.1007/978-3-030-42939-3_67
13. Onyshchenko V, Sivitska S (2014) Alternative energy developing investment support in terms of energy dependence. Econ Ann XXI 9–10:34–37

Implementation of Energy Efficient Technologies and Systems in Housing Construction

Olha Dzhyhora and Akif Gasimov

Abstract The article is devoted to the features of energy efficiency and energy saving, topical issues of the development of energy efficient construction in Ukraine. Particular attention is paid to the types of structural systems, building materials, products and equipment that meet the requirements of energy saving, environmental safety, manufacturability, efficiency, low labor intensity of construction, adaptability to the conditions of reconstruction and modernization of residential and industrial buildings. Specific heat losses in buildings and structures of Ukraine are analyzed. The structure of heat energy losses through separate enclosing structures depending on the number of storeys of the building is considered. The conditions on which the strategy of energy saving in the sphere of construction and operation of buildings and structures should be based are proposed. Energy-saving urban planning, architectural and planning solutions, constructive, engineering systems are proposed. The types of energy efficient building materials are considered. The physicomechanical and thermophysical properties of window frames are estimated. The efficiency of energy saving in the residential sector is calculated due to various technologies and systems.

Keywords Energy efficiency · Construction · Energy efficient technologies · Materials · Building structures · Heat loss · Energy supply of buildings · Housing sector

1 Introduction

In recent years, a particularly difficult situation has developed in the energy sector of Ukraine, therefore, the problem of energy saving in buildings is one of the most important. The accumulated experience in the design and construction of energy-efficient buildings indicates that energy efficiency is not a static characteristic laid

O. Dzhyhora (✉)
National University «Yuri Kondratyuk Poltava Polytechnic», Poltava 36040, Ukraine

A. Gasimov
Azerbaijan University of Architecture and Construction, Ayna Sultanova 11, Baku, Azerbaijan

V. Onyshchenko et al. (eds.), *Proceedings of the 3rd International Conference on Building Innovations*, Lecture Notes in Civil Engineering 181,
https://doi.org/10.1007/978-3-030-85043-2_59

down at the design stage, but a dynamic one that forms throughout the entire life cycle of a building.

It is worth noting that the concept of "energy efficient house" has existed in the theory and practice of construction for over 50 years, and throughout this time interest in these houses has not disappeared, and the problem of ensuring energy efficiency is even more modern and relevant.

2 Analyzing the Latest Research and Publications

Various solutions in the field of energy saving and energy efficiency are offered by such prominent scientists as E.I. Bakunin, S.N. Bulgakov, L.A. Golovanova, O. G. Khomenko, I.I. Podgorny, G.G. Farenyuk, A.S. Gorshkov, M.K. Ageev, O.D. Samarin, I.V. Petrova, I.N. Butovsky, G.S. Ratushnyak, M.V. Barabash, V. Tretton, N.V. Savitsky, L.D. Boguslavsky.

3 Formulation of the Purpose of the Article

The purpose of this work is to study the implementation of innovative energy efficient technologies and systems in modern construction.

4 Main Material

The operational energy consumption of existing residential and public buildings in Ukraine significantly exceeds similar indicators in technically developed countries with similar natural and climatic characteristics. Currently, energy-efficient structures of buildings and structures are also successfully used in Canada, the United States, most European countries, as well as in countries with a tropical climate.

It should be noted that the theoretical developments, energy-saving programs, equipment samples and experimental facilities, which have been carried out in recent years, have created realistic prerequisites for reducing the energy consumption of buildings and structures. Due to the fact that the annual increase in residential and industrial space due to new construction is about 1% of the existing space, the main energy saving potential is contained in the operational sphere and can be realized through the reconstruction and rehabilitation of existing fixed assets.

The specific heat losses in buildings and structures are analyzed (Fig. 1).

So, as can be seen from the Fig. 1, up to 40% of heat loss in buildings and structures occurs due to organized and unorganized infiltration of heated air, up to 30% is obtained due to insufficient resistance to heat transfer of enclosing structures,

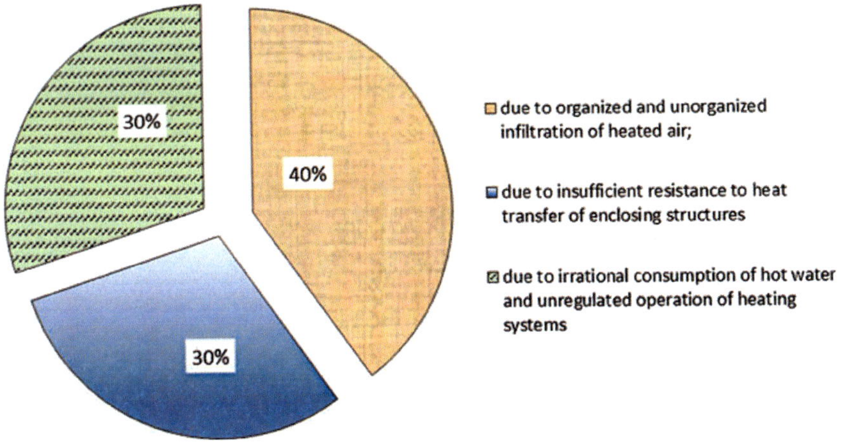

Fig. 1 The reasons for heat loss in buildings and structures of Ukraine

up to 30%—due to irrational consumption of hot water and unregulated operation heating systems.

The main reasons for the irrational consumption of heat energy are the following:

- imperfection of unregulated natural ventilation systems;
- poor quality and looseness of mating of wooden window frames and balcony doors;
- disadvantages of architectural, planning and engineering solutions for heated staircases and staircase-elevator blocks;
- insufficient thermal insulation quality of external walls, coatings, basement ceilings and translucent fences;
- lack of metering, control and regulation devices on heating and hot water supply systems;
- a highly developed network of heating external to insufficient or impaired heat insulation;
- outdated, unproductive types of boiler equipment;
- lack of an effective mechanism of material interest of energy consumers in its saving;
- extremely insufficient use of non-traditional and secondary energy sources.

The main consequence of the increased requirements for thermal protection of building envelopes was the transition to multilayer structural solutions. They allow you to achieve high resistance to heat transfer without increasing the thickness of the enclosing structures due to the action of effective insulation. In accordance with modern building codes, the required resistance to heat transfer, for example, of walls has increased 3–3.5 times compared to the old standards [1].

The structure of heat energy losses through separate enclosing structures also differs for the same type of buildings, depending on the number of storeys, material

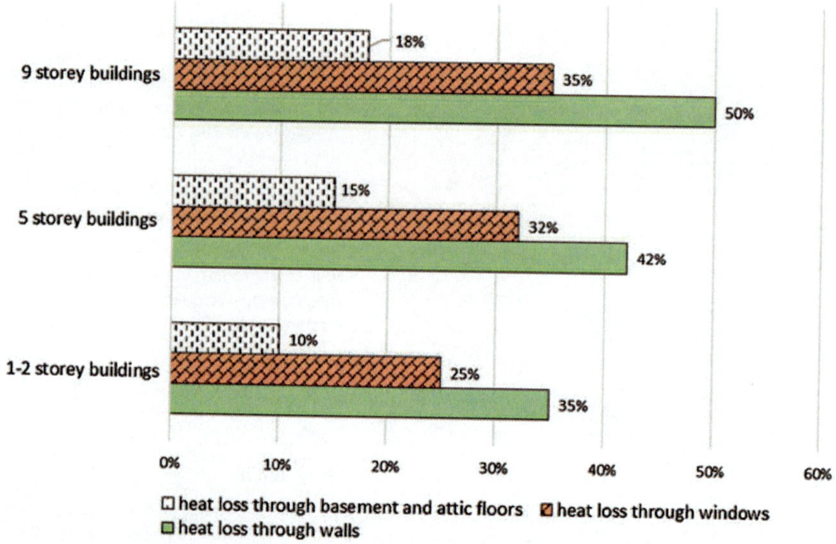

Fig. 2 The structure of heat energy losses through separate enclosing structures depending on the number of storeys of the building

of enclosing structures, year of construction, service life, quality of construction work. For example, for houses, depending on the number of storeys, it is as follows (Fig. 2):

So, walls account for 30–35% of heat loss in one- and two-story buildings, up to 42% in five-story buildings, up to 49% in nine-story buildings; heat loss through windows is 25, 32 and 35%, respectively; through basement and attic floors, an average of 10 to 18% of heat is lost. As a result, various thermal insulation materials and structures, energy-efficient facade systems, technologies for the construction of monolithic houses with permanent formwork, energy-efficient translucent structures are used in construction today.

An energy saving strategy in the construction and operation of buildings and structures should be based on:

- a systematic approach and economically sound performance of a complex sequence of interrelated and interdependent energy saving measures of urban, architectural planning, design, engineering and operational matters;
- the program-targeted method for the development and implementation of a system of energy-saving measures, focused on the maximum saving of non-renewable fuel resources with a minimum cost of funds and time to achieve this goal;
- the priority orientation of scientific, design and practical activities on energy saving to the most energy-intensive sphere of operating fixed assets through the modernization and reconstruction of operating buildings, structures, engineering systems, communications and energy facilities;

- the transition to energy-efficient design standards and construction of new buildings and structures [2].

Energy Saving Urban Planning Solutions. It is necessary to establish a moratorium on expanding the boundaries of cities for 20–30 years, their development during this period should be carried out through more rational use of territories, compaction of buildings to the standard level without the development of new suburban areas and without increasing the length of main heating pipelines, other energy networks and transport routes.

It is necessary to develop feasibility of complex use of traditional and non-centralized heating systems, including local using boiler container type, placed on the roof or near the heated buildings. It is also possible to provide a complete program for the completion of the development of residential quarters and micro-districts with the elimination of wind-through spaces and the organization of closed courtyard and intra-quarter territories.

An effective solution to increasing the level of energy saving in residential buildings can be the development of master plans, programs and business plans for the secondary development of reconstructed low-rise residential areas with insulation of the enclosing structures of existing buildings in accordance with new heat engineering standards, the transition to automated individual heating points, reconstruction of heating networks, use roof-top boilers for heating and hot water supply to increase housing space and the implementation of a set of measures for energy saving with the organization on the basis of these quarters of energy efficient zones of the municipal economy [3].

Also no less appropriate is to develop of a program for the use of underground space (underground urbanization) for the placement of parking lots, storage and auxiliary premises using the natural heat of the earth or artificial sources of air heating to a positive temperature.

Energy-Saving Architectural and Planning Solutions. A significant influence on the specific heat loss in residential and public buildings is exerted by their space-planning solutions and, in particular, the ratio of the area of enclosing structures to the total area of buildings, the ratio of the area of window openings to the area of external walls, the configuration of buildings in plan, their placement on the relief and relatively countries of the world [4–13].

Recommended solutions:

- The transition to the design and construction of residential buildings shirokokorpusnyh with a reduction of 20–30% of the specific area enclosing structures per square meter of of housing area.
- The use of wide-frame houses in the secondary development of reconstructed quarters, including the construction of secondary buildings in the place of existing two-five-storey buildings without demolition, but with the simultaneous reconstruction and extension of the life cycle to the level of new buildings.

- Construction of attic floors in buildings with existing walling increased thermal protection according to the second step standards "Thermal Engineering", thereby avoiding excessive losses of heat through the coating renovated buildings.

Energy Efficient Building Materials. There are different options for insulating the building envelope, depending on climatic conditions and the constructive decision taken at the stage of construction design. There are two main options:

1) when there is a structural layer and a layer of insulation in multi-layer walls, this is a heat engineering non-uniform enclosing structure;
2) when the insulation layer and the structural layer coincide, it is a heat engineering homogeneous enclosing structure [14].

Currently, reducing the consumption of fuel and energy resources is especially important when operating brick buildings. The successful solution of this problem is associated with an increase in the heat-shielding qualities of the outer walls. Achieving this in traditional ways leads to an increase in the material consumption of brick walls.

However, there are also less material-intensive methods, in particular, adding phenol-formaldehyde, mineral and other additives to the wall structure. The disadvantage of this method is the reduction in capital, durability and fire resistance of brick buildings. Most importantly, this violates the environmental indicators in the premises, due to the properties of raw materials and the technology of production of ceramic wall materials.

It is possible to solve the problem of increasing the level of thermal insulation by creating and using a new generation of effective porous ceramic wall materials. The experiments carried out helped to develop the main directions for achieving a higher level of heat-shielding qualities of external brick walls without increasing their thickness. This policy is based on the production of large-sized ceramic stones, reducing the density of stones and creating cavities of rational sizes. This makes it possible to significantly reduce the consumption of mortar in the masonry of the wall, to increase the porosity of the shard by introducing a complex of burnout additives into the mixture.

In areas with relatively warm climatic conditions, to provide the required thermal protection of buildings, it is sufficient to use a material such as lightweight concrete, which simultaneously performs two functions: constructive and insulation. Such structural and thermal insulation materials include polystyrene concrete (PSB), which has a high resistance to heat transfer. Among lightweight concrete, it has the lowest bulk density—300 kg/m^3 or less. From polystyrene concrete, blocks of large overall dimensions can be made, which speeds up the laying of blocks when erecting walls and partitions. The advantages of using this material in construction include low transport costs and costs associated with the use of lifting equipment and the preparation of mortar directly at the construction site.

Another energy-efficient material for enclosing structures is sand-cement wall blocks, which are made with different voids and different strength indicators. The basis of the block is made up of lightweight aggregates, which occupy up to 50%

of its volume. This is the most common heat-and-energy-saving material that allows you to perform the entire list of construction work in the construction of enclosing structures for low-rise buildings.

Energy Saving Construction Systems. The most rational types of energy-efficient external enclosing structures are multilayer composite structures of walls and coverings using effective mineral materials. When designing new and renovating existing buildings, one of three methods of insulation of external walls is used—from the outside, inside and inside.

The main reserves of heat saving can be realized by insulating existing residential buildings. Thermal insulation of external walls—the most expensive and laborious process—reduces heat loss by about 12–15%.

The most famous and widespread methods of external wall insulation include: ventilated structures for external wall insulation or, as they are called, ventilated facades (Fig. 3); non-ventilated structures for insulation of external walls using mineral wool and polystyrene plates with their fastening directly to the walls or to the frame, as well as all kinds of combinations of these options using local insulation.

The most famous are ventilated facades of the Kraspan type, which are multilayer systems that are effective in terms of their physical and construction parameters. The constructive solution of this system is shown in Fig. 4.

In facade design solutions, the following are most used as a heat insulator:

- thermal insulation boards made of mineral wool;
- extruded polyethylene foam;
- heat-insulating plates made of basalt rocks;
- plates (blocks) made of foam glass, etc.

The methods of insulation considered above refer to the blank part of the wall. However, the weakest link in a building in terms of energy efficiency is the windows.

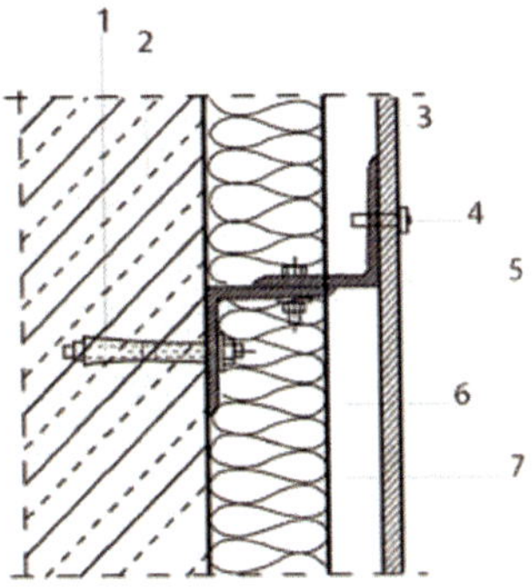

1. Anchoring; 2. Base for anchoring; 3. External wall cladding; 4. Fastening material; 5. Substructure; 6. Ventilation fence; 7. Thermal insulation

Fig. 3 Ventilated facade structure [15]

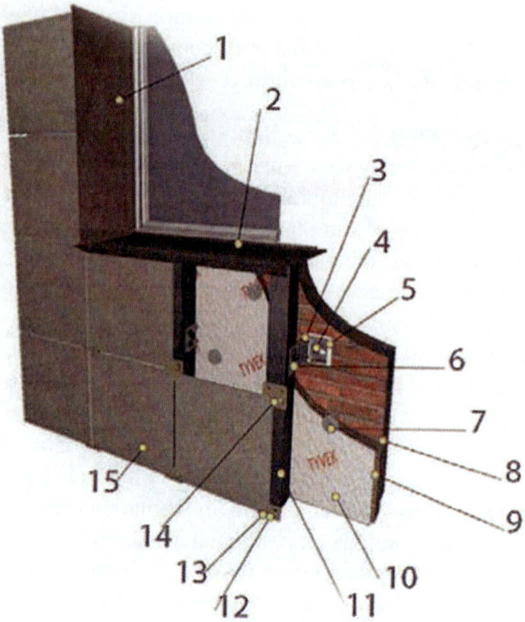

Fig. 4 Ventilated facade type Kraspan [16]

1. Shaped elements; 2. Self-tapping screw;
3. Mounting bracket; 4. Anchor for mounting the bracket;
5. Paronite gasket; 6. Self-tapping screw;
7. Dowel to strengthen the insulation
8. Wall; 9. Insulation; 10. Hydro-windproof membrane
11. Vertical fastening profile; 12. Steel rivet
13. Fastening clamp, final; 14. Fastening clip private;
15. Porcelain stoneware slab

Heat losses through windows reach 50% of the total heat losses through the enclosing structures, therefore, first of all, it is necessary to improve the heat-shielding qualities of windows. Window fillings made of wood and fiberglass with triple glazing, in the form of double-glazed windows, with double glazing and a layer of film, provide regulatory heat protection requirements. During reconstruction, a decrease in heat loss through windows can be ensured by insulating the slopes with the installation of platbands and by installing a translucent screen in the inter-glass space of a window block with separate or paired bindings.

The introduction of the screen makes it possible to restrict natural convection in the interlayers and to achieve the design mode of thermal conductivity in the windows.

With the simultaneous consideration of the lighting and heat engineering properties of structures, windows with screens are more energy efficient.

One of the directions for the development of energy saving in construction is windows with heat-reflecting glass. The use of such windows in residential construction can reduce heat loss through them up to 40% of the energy. In this case, the recoupment of additional costs does not exceed 1.5 years.

The traditional materials for making window sashes are wood, steel and aluminum. Among the polymeric materials for use in the construction of window and door blocks, glass-filled thermosetting materials based on polyester resins—polyester plastics—are most acceptable. These materials have all the positive qualities of polymers, without the disadvantages of thermoplastics. For example, polyester fiberglass plastics have the thermal conductivity of wood, the strength and durability of metal, biological resistance, moisture and weather resistance of the polymer (Table 1).

In the reconstruction of houses, a large part of the cases attic floors are built of lightweight structures and materials with high heat-shielding properties.

A promising solution for lightweight structures of attic floor frames are frames using metal-wood structures that combine the advantages of wood and metal as materials. The joint work of a metal sheet and wooden planks squeezing it can significantly reduce the weight of the structure and reduce metal consumption by 4 times while ensuring the required bearing capacity.

Options for the construction of attics with enlarged spatial blocks have been developed. Constructive solutions of volumetric block-rooms for the attic arrangement provide maximum weight reduction and the necessary rigidity of the elements for their transportation and installation.

Table 1 Comparative physical, mechanical and thermophysical properties of window frames

Physical and mechanical characteristics	Fiberglass	Glass	PVC	Steel	Aluminum	Wood (pine)
Density, t/m^3	1,6–2,0	2,2	1,4	7,8	2,7	0,46–0,53
Breaking stress in compression (extension), mN/m$_2$ (mPa)	410–1180	35	41–48	410–480	80–430	40–80
Breaking stress in bending, mN/m^2 (mPa)	690–1240	25–50	80	400	275	80
Tensile modulus, hPa	21–41	50–85	2,8	210	70	11
Flexural modulus, hPa	27–41	50–85	2,8	210	70	10
Linear expansion coefficient, ×10 °C	5–14	3,2–11	57–75	11–14	22–23	5,4–34
Thermal conductivity coefficient, W/m × °C	0,3–0,35	0,45	0,15–0,35	46	140–190	0,04–0,1

Metal frame, which is monolithic with polystyrene concrete in the plane of the floor, covering, external and internal walls. In this case, the thickness of the outer walls and floor is taken according to the thermal engineering calculation for each region, and the thickness of the interior and interroom walls (partitions)—taking into account the strength and sound insulation requirements.

The energy efficiency of mansard superstructures is ensured, in addition to efficient enclosing structures, also by choosing rational heating systems. The analysis shows that in the absence of reserve capacities, the most effective solution for heat supply to attic superstructures is the use of individual apartment boilers. This option minimizes both capital costs and annual operating costs.

Energy-Saving Engineering Systems. A significant share of the energy saving effect can be obtained by modernizing existing and introducing new engineering systems, energy sources, equipment and instrumentation for energy saving during the operation of facilities.

The priority is the following components:

- increasing the efficiency of boiler equipment;
- elimination of heat losses in the main and intra-quarter heating networks;
- modernization of heating systems and hot water supply of buildings, apartment accounting and regulation of energy consumption.

According to expert estimates, the systematic implementation of energy-saving technologies can reduce operating energy costs in the residential sector by 2.0–2.5 times. In this case, the specific share of energy saving will be as follows (Fig. 5):

So, due to the improvement of urban planning solutions—8–10%, architectural and planning solutions—up to 15%, structural systems—up to 25%, engineering

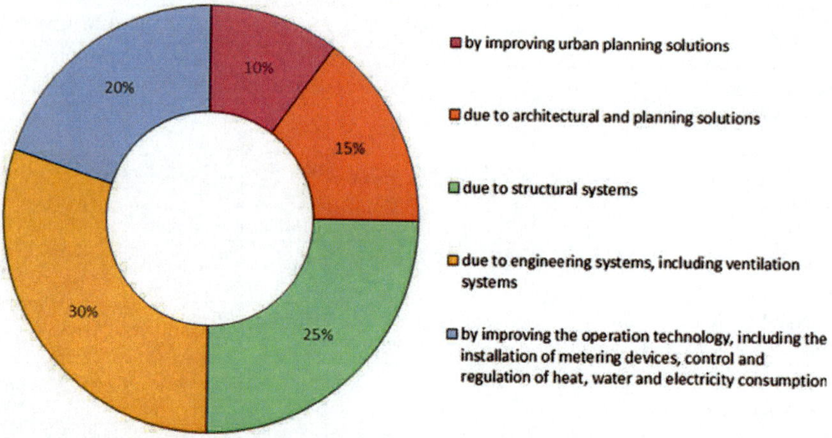

Fig. 5 Specific share of energy savings in the residential sector due to various technologies and systems

systems, including ventilation systems—up to 30%, due to the improvement of operation technology, including the installation of metering devices, control and regulation of heat, water and power consumption—up to 20%.

In order to increase the level and degree of energy saving in the residential sector, the following energy-saving technologies should be carried out and introduced:

- use of high-performance boiler equipment, including local container-type boiler houses, when placed on the roof of buildings, the need for heating networks is eliminated;
- transition to automated individual heating points with the exclusion of the use of jet mixers—pumps (elevators) with free quantitative and qualitative control of the heat carrier for front and sectional supply. Setting heating modes for daytime, nighttime, winter and autumn-spring periods, weekends, emergency heating, etc.

Transition to autonomous hot water supply systems independent of centralized heat supply using apartment-type gas or electric water heaters and a two-part tariff for payment for electricity. Up to 25% of the total possible effect of saving thermal energy can be obtained by installing apartment-based meters for the consumption of hot water (8–10%) and metering and regulation devices for heating systems, which contribute to the elimination of overheating of premises with an off-season and temporary increase in the outdoor temperature and in the room regulation of temperature during the heating season (10–12%).

When reconstructing existing houses and designing new ones, it is advisable to use fundamentally new heating systems.

The most widespread in mass housing construction are vertical one-pipe heating systems. In these systems, it is impossible to fully realize the potential energy savings. The organization of door-to-door accounting of the consumption of the coolant in these systems is technically complex and requires large material costs.

It should be noted that a significant saving of thermal energy and an increase in the level of thermal comfort in heated rooms is achieved when using horizontal heating systems with apartment-to-apartment distribution of the coolant. Thermal energy savings during the operation of the systems under consideration is 20–25% per heating season in comparison with existing vertical single-pipe heating systems.

When the total implementation of the technology upgrading of engineering systems, heating costs in public buildings for heating and heat supply or infiltrating air may be reduced by 30–40%. At the same time, one-time capital costs will be significantly (from 2 to 10 times) lower than the cost of increasing the thermal resistance of the walls. In general, it is possible to realistically bring the calculated heat losses in residential buildings to the level achieved in advanced countries—30–35 W/m^2.

5 Conclusions

Thus, it should be noted that in the next decade, at the junction of the traditional periods of exhaustion and lack of development of new energy sources, there will be energy scarcity and a sharp rise in the cost of, and the problem of energy saving will be a priority. In this regard, in the field of creation, modernization and operation of construction products, the dominant factor will be to ensure minimal heat loss in buildings through the development and use of energy-efficient space-planning and design solutions, new with a high coefficient of resistance to heat transfer of building materials and products, energy-efficient equipment and regulated, including non-traditional, power supply systems. The priority direction for the development of building materials, products and equipment will belong to energy-saving types.

Despite the fact that the implementation of technologies to increase energy efficiency during construction increases the cost of construction in comparison with traditional technologies for the construction of buildings by 5–10% of the cost of the construction object, however, the introduction of energy-saving technologies at the construction stage will not only increase the level of comfort in the premises, but it will also help in the future to save energy and reduce the cost of their use. A detailed calculation of the economic efficiency of the introduction of energy-saving technologies shows that the costs incurred at the stage are recouped within 5–8 years. This happens due to savings during the operation of houses and payment of utilities. In addition, due to energy saving during construction, more comfortable living conditions for people are simultaneously created, which increases the level of social satisfaction and security.

Based on the above, with a sufficient degree of reliability, it can be assumed that the development of structural systems, building materials, products and equipment will take place in traditional and new directions that meet the requirements of energy saving, environmental safety, manufacturability, efficiency, low labor intensity of construction, adaptability to the conditions of reconstruction and modernization of residential and industrial buildings.

References

1. Bulgakov S (2012) Energy efficient building systems and technologies. AVOK (2). https://www.abok.ru/for_spec/articles.php?nid=135. Accessed 30 Jan 2021
2. Golovanova L (2005) Energy saving in housing construction. Khabar Publishing House, Khabarovsk. State Tech. University, p 146
3. Khomenko O (2019) Energy saving technologies in business: the main electronic book. Deaf., p. 118. http://tpgnpu.ho.ua/images/my_images/doc_pdf/energhozberezhennjaj.pdf. Accessed 26 Jan 2021
4. Tsitsin, K (2013) Energy efficient technologies - the future of housing construction. Strateg. Decis Risk Manage 2:50–51 https://doi.org/10.17747/2078-8886-2013-2-50-51
5. Drobyazko S, Bondarevska O, Klymenko D, Pletenetska S, Pylypenko O (2019) Model for forming of optimal credit portfolio of commercial bank. J Manage Inf Decis Sci 22(4):501–506

6. Yurin O, Galinska T (2017) Study of heat shielding qualities of brick wall angle with additional insulation located on the outside fences. Paper presented at the MATEC web of conferences, vol 116. https://doi.org/10.1051/matecconf/201711602039
7. Yurin O, Azizova A, Galinska T (2018) Study of heat shielding qualities of a brick wall corner with additional insulation on the brick. Paper presented at the MATEC Web of Conferences, vol 230. https://doi.org/10.1051/matecconf/201823002039
8. Cherniha R, Pliukhin O (2013) New conditional symmetries and exact solutions of reaction-diffusion-convection equations with exponential nonlinearities. J Math Anal Appl 403(1):23–37. https://doi.org/10.1016/j.jmaa.2013.02.010
9. Leshchenko MV, Semko V (2015) Thermal characteristics of the external walling made of cold-formed steel studs and polystyrene concrete. Mag Civ Eng 60(8):44–55. https://doi.org/10.5862/MCE.60
10. Aliyev F, Mehdaliyev A, Mammadov N, Aliyev F (2008) Development of mathematical model of intellectual building in Baku city. In: The 2nd international conference «problems of cybernetics and informatics». Dedicated to the 50th Anniversary of the ICI in Azerbaijan, Baku, vol 1, pp 275–278
11. Mammadova GH, Mammadov NY, Akbarova SM, Feyzieva GH (2021) Expert recommendations for energy improvements in educational facilities: case study-school buildings in Azerbaijan. Paper presented at the IOP conference series: materials science and engineering, vol 1030, no 1. https://doi.org/10.1088/1757-899X/1030/1/012063
12. Mammadova G, Sharifov A, Akbarova S (2021) Estudio experimental del rendimiento térmico de la cavidad de aire de fachadas ventiladas opacas en condiciones de viento extremo: Estudio de caso bakú. Informes De La Construccion 73(561):1–8. https://doi.org/10.3989/IC.74247
13. Akbarova S (2018) Trends of energy performance certification of buildings in Azerbaijan. Int J Eng Technol (UAE) 7(3):563–566. https://doi.org/10.14419/ijet.v7i3.2.14590
14. Golovanova L, Blum E (2014) Energy-efficient building structures and technologies. Sci Notes PNU 5(4):71–77
15. Ventilated Kraspan facades. http://mastersbuilder.ru. Accessed 28 Jan 2021
16. Polystyrene concrete is a qualitatively new building material. http://elport.ru. Accessed 27 Jan 2021

Methodological Approach to Creating a Cluster in the System of Innovation-Oriented Management of the Construction Complex Development

Oksana Furmanchuk, Olena Filonych, Olena Koba, and Vitaliya Skryl

Abstract The methodological approaches to the creation of a cluster in the system of innovation-oriented management of the construction complex development are determined. An advanced model that determines the principles, conditions of activity and the expected economic effect from the introduction of the cluster is proposed. The model is aimed at ensuring the innovative development of the construction complex and reducing the cost of construction. The directions of innovation-oriented management of the construction complex development are proposed; their implementation makes it possible to obtain combinatorial advantages from the joint activities of its participants.

Keywords Construction complex · Innovation-oriented management · Cluster model

1 Construction Cluster as a Tool for Innovation-Oriented Management in Construction

Constant changes in the economy, low level of macroeconomic indicators and the life of the population necessitate strengthening the competitiveness of industries and the economic complexes, important for the state. In the context of the above, it is important to determine the sectors of the economy with highest priority that have significant impact on the socio-economic development of Ukraine and primarily need constructive changes. These undoubtedly include construction, which is involved in the creation of fixed assets in other sectors of the economy, providing the population with one of the primary public goods—housing. Today, the intensive path of development of the country and its regions is possible only if the achievements of science and technology are utilized, in particular, the strategy for implementing the principles of

O. Furmanchuk · O. Filonych · O. Koba (✉) · V. Skryl
National University "Yuri Kondratyuk Poltava Polytechnic", Poltava, Ukraine
e-mail: ev_pntu@ukr.net

V. Onyshchenko et al. (eds.), *Proceedings of the 3rd International Conference on Building Innovations*, Lecture Notes in Civil Engineering 181,
https://doi.org/10.1007/978-3-030-85043-2_60

innovative development. That is why the introduction of innovations in construction should become a priority task for the further development of the complex.

The innovative development of construction complex makes it possible to implement the investment policy of state, to determine national economic proportions, scale and rate of development of particular industries, scientific and technological progress and efficiency of investments in all sectors of economy [1].

The choice of goals and directions of economic policy, formation of the concept of development of Ukrainian society provides for the renewal of organizational structures in accordance with modern global integration trends, especially with those that combine innovation and investment processes. The analysis of the problems of formation and development of innovative organizational and economic structures indicates the intensification of the implementation of clusters, an organizational model that is widespread in the world [2–4].

Foreign experience shows that it is clustering that is the form of internal integration, capable of ensuring socio-economic development. Thus, in the USA there are about 380 clusters; their sphere of activity covers the processing industries and the service sector, producing about 61% of industrial products. The West Pannonia cluster in Hungary accounts for 9% of GDP. Finland, having only 0.5% of the world's forest resources, due to the high productivity of clusters ensured by innovative structures, provides 10% of the world's wood products exports. Austria has cross-border clusters with Hungary, Italy, Switzerland and Germany.

M. Porter used the term "cluster" in economics at the end of the last century [5, 6]. According to him, a cluster is a combination of enterprises and institutions for which the territorial commonality of its components is an additional factor in economic efficiency increasing due to the constancy of interrelations, reduction of transport costs, rational use of all types of local resources, creation of conditions for combination with territorial planning and management.

In Ukraine, the first theoretical and practical scientific developments are found in 2000 in the works of M. Voynarenko [7]. While studying the peculiarities of the implementation of cluster technologies in the Ukrainian economy, he conducted a study of the conditions for the creation and functioning of clusters, the role of state authorities in stimulating small and medium-sized businesses to incorporation. In his opinion, the cluster is a territorial and sectoral voluntary integration of enterprises that work closely with scientific institutions and local authorities in order to increase the competitiveness of their own products and economic growth. The other scientists [8–10] continued the research.

The construction cluster is a special form of the cluster model, as construction is an industry that is focused primarily on the domestic market. Its main products are services: reconstruction, modernization and new construction. The construction cluster brings together not only constructors or manufacturers of constructional materials. It includes research institutions, planners, designers, lawyers, trade associations and private entrepreneurs, auditors and consulting firms.

Effective innovation-oriented management in construction is hindered by the underdevelopment of the systemic functioning of the scientific and innovation spheres. In modern conditions, with the growth of people mobility, finance, production capacity, especially within the EU, not only advanced technologies and production are needed, but, first of all, regions that are ready to accept and ensure their activities. It is the regions that are the "engine" of economic development based on research, technology and innovation. The regions with strong innovation and investment potential do not have a particular specialization and are prone to innovations and industry flexibility, they are called the basis of sustainable development and competitiveness of the country.

Foreign experience shows that clusters are an effective mechanism for regional development. They are a kind of platform for a constructive dialogue between representatives of business, science and government bodies; they enable the increase in the involvement of the private sector, the state, research and educational institutions in the innovation process [11–13]. Slovenia is an example of efficient formation of cluster-type networks in the construction sector. Construction Cluster of Slovenia (CCS) was created in February 2004 by 14 companies. Since then, it has developed as an innovation cluster, with a significant focus on internationalization and international cooperation. Austria has significant experience in clustering the construction industry. In 2003, the Green Construction Cluster of Lower Austria (Lebensmittel Cluster Niederösterreich) was formed there. At present, its turnover is more than € 3,534,000,000, and its main activities are the reconstruction of old houses, the construction of new multi-storey passive houses, improvement of the comfort of living quarters. In Ukraine, the first construction cluster was created in Khmelnytsky on the initiative of the "Podillia Pershyi" Association; it unites more than 50 small and medium-sized enterprises specializing in the production of construction materials, design, construction, and provides a full cycle of construction. In 2007, a construction cluster was created within "Slobozhanshchina" Euroregion that covers Kharkov and Belgorod regions.

2 Models of Construction Clusters Formation in the Regions of Ukraine

The study of domestic and foreign experience of construction clusters [14–17] made it possible to develop a conceptual model of their formation in the regions of Ukraine, based on the four-aspect vision of regional development (Fig. 1).

The presented model gives only a systematic idea of the interrelation of regional development, it does not separate the aspect of education and science as the main aspect, a subsystem for the formation of other aspects of regional development. But, in the authors' vision, the interaction of science, education and industrial production is a key factor in the civilized process, further economic and social development. In Ukraine, the relevance of such interaction is enhanced by the fact that the joint participation of scientific institutions and higher educational institutions, innovative

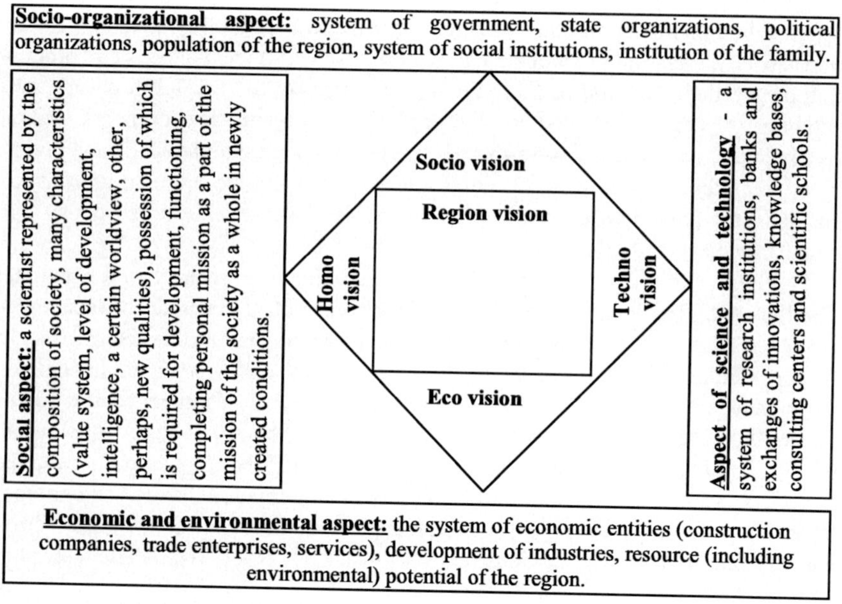

Fig. 1 The model of four-aspect vision of regional development

structures and enterprises in the training of highly qualified personnel, the implementation of priority research, the development and implementation of the latest technologies is an important condition for the implementation of an effective system of innovation-oriented management.

With this in mind, the authors proposed a cluster model that has a focal system of creation, that is, any higher educational institution in the region of Ukraine with a centuries-old history in construction is chosen as the "core" of such a cluster. The model of the four—aspect vision of regional development of Ukrainian regions, based on the determinative role of the higher educational institution, which acts as the "core" and the condition for creation and functioning of the research and production construction cluster, is shown in Fig. 2.

As a rule, an innovation cluster has a core structure, forming around a large enterprise, research center or university. Accordingly, if there are several such enterprises (centers), we are talking about a multi-core cluster. A cluster that develops effectively must include medium-sized specialized enterprises, one large industrial enterprise, a research center and an educational institution. To coordinate actions and increase the efficiency of interaction, a kind of platform is needed, that represents, on the one hand, the individual interests of each of the participants, and on the other hand, the general direction of movement in the interests of the entire cluster.

The experience of other countries shows that the most effective is the creation of innovative infrastructures based on higher educational institutions. Thus, four main

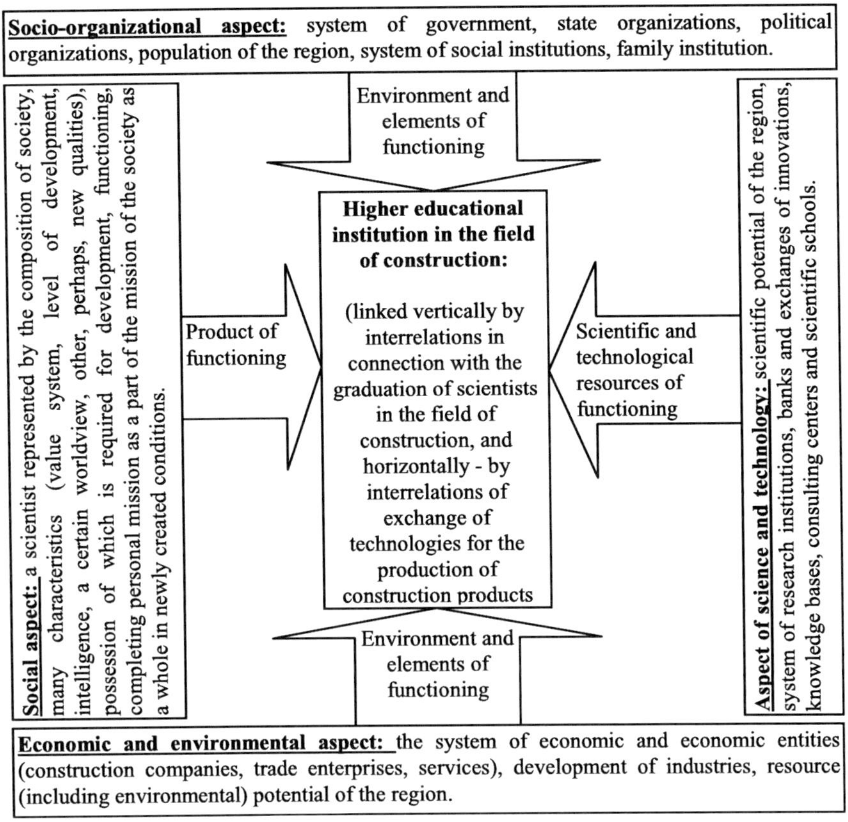

Fig. 2 The "core" of the cluster in the structure of the region's economy (compiled by the authors)

groups of university innovation structures can be identified: business incubators, technology parks, technopolises and science parks.

Business incubators are multifunctional complexes that provide various services to new innovative firms that are at the stage of emergence and formation.

Technology parks, in comparison with incubators, provide for the creation of a more diverse innovative environment. The services of technology parks are used by small and medium-sized innovative enterprises that are at various stages of commercial development of scientific knowledge, know-how and science-intensive technologies.

A technopolis, unlike previous forms, is a large modern scientific and industrial complex, that includes a university or other higher educational institution, research institutes, as well as residential areas with cultural and recreational infrastructure.

The purpose of science parks establishing, in contrast to technology parks, is the development of scientific, technical and innovative activities in a higher educational institution or a scientific institution, the effective and rational use of the available scientific potential, material and technical resources for the commercialization of scientific research results and their implementation in the domestic and overseas markets.

Of all the presented university innovation structures, a technology park is most consistent with the goal of creating an innovation cluster; that is why it is included in the core of the proposed construction cluster.

The main purpose of the technology park is to form the competitive advantages of the cluster in the region as a whole and its subjects (participants), in particular, through integration in a single information space of education, science and business, interaction between them and creating synergies to develop priority areas profits due to the coordination of activities and organization of joint structures of different legal forms (specialized educational institutions, research organizations and commercial enterprises). Thus, close cooperation between innovation-active enterprises, science, education and the local community (including public authorities, the media) in the presence of a certain coordinating element forms the so-called triple spiral—business, university and society.

In the author's vision, the final "core" of the cluster is represented, on the one hand, by the technology park, and on the other hand - by the key partner of the technology park—the profile university in the construction industry. In the context of the proposed model, the system-forming element is the technology park, which is created on the basis of the university buildings, which provides a number of key functions for the cluster as a whole.

High efficiency, sustainability and viability of the cluster are determined by the presence of participants-partners of the "core" of the cluster (technology park and university), including research centers, large construction companies in the region, service firms needed to provide quality services to innovative firms and companies specializing in profile activities of the technology park, regional authorities responsible for the implementation of innovation policy in the region.

Under this approach, within the proposed cluster there is a possibility of forming a "closed" innovation and educational chain, in which each of the elements plays a key role within its competences, and the technology park provides coordination and coherent policies aimed at achieving efficiency and competitiveness of innovative construction cluster. in general.

The technology park core is designed to serve as a link between the needs of business, the capabilities of the research center, the university, government officials on a wide range of issues, namely:

- support of development processes on the basis of the university and its separate research laboratories and approbation in the regional market of science-intensive technologies and high-tech products in the field of construction, ensuring interaction between science, education and business;
- systematic monitoring of construction projects carried out at the university and specialized research organizations, in order to identify intellectual property and technologies that have sufficient potential for commercialization;

- creation and maintenance of a database on research, personnel and other capabilities of the Faculty of Civil Engineering and adjacent departments, research units, laboratories to work with industrial construction models, primarily—with construction companies in the region—members of the technology park;
- formation of a "community" within the construction cluster; balancing individual activities in the interests of its participants, coordination and development of resources, evaluation and protection of results.

In this way, the technology park creates and develops the necessary connections between all members of the innovative construction cluster: innovative companies, research organizations, the university, large construction companies, financial institutions, authorities and the media.

In addition to the formation of a common synergetic effect for the cluster and increase its competitiveness, we can talk about a number of additional benefits that participants will receive from the proposed innovative construction cluster from the implementation of the proposed strategy:

1. For construction companies:
   - increasing competitiveness in domestic and foreign markets through the use of advanced technologies and results of scientific and technical achievements of research centers and universities that are the part of the cluster;
   - solving the personnel problem by using the educational potential of the university and other educational organizations—members of the cluster;
   - significant cost savings due to preferential terms of placement and provision of a wide range of construction services on the basis of the technology park;
   - facilitating access to the capital market, forming an attractive investment image; access to unique high-tech equipment and the ability to interact with public authorities and the media.
2. For research organizations: commercialization of the results of scientific and technical activities, increasing demand for research and scientific and technical products from enterprises that are part of the cluster.
3. For a higher education institution in the field of construction:
   - intensification of scientific and technical activities in the field of construction, commercialization of the results of scientific and technical activities and intensification of innovative activity of students, graduate students, faculty;
   - improving the quality of educational processes of the university due to the use of the innovative potential of the technology park and cluster.
4. For the construction industry as a whole: creation of an innovative construction high-tech cluster and a powerful technology park, new jobs for highly qualified specialists in the construction industry; increase of tax revenues to the budgets of different levels; development of international cooperation in the field of construction, realization of intellectual and innovative potential of the construction complex of the region.

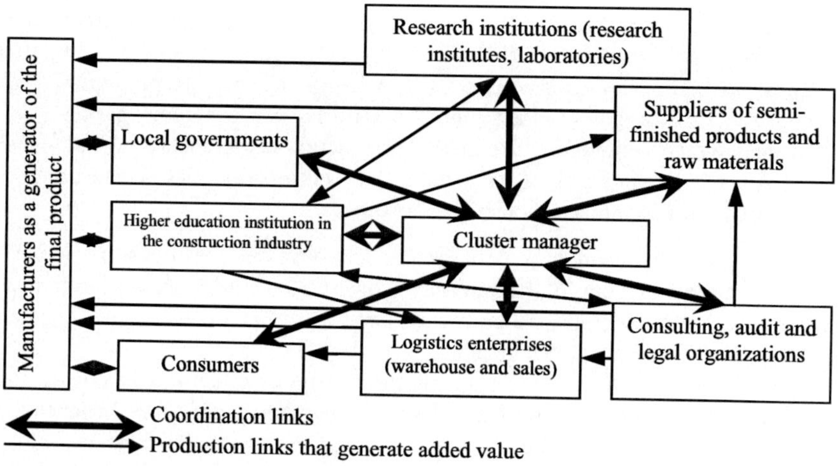

Fig. 3 The structure of relations within the research and production construction cluster (author's development) while long ones are justified

The links of the innovation cluster are provided by peculiar mechanisms that can operate within it (Fig. 3).

The levels of development of each element in a cluster cannot differ greatly. The existence of competition between the horizontal elements of the cluster has a positive effect on the efficiency of the cluster. In addition to its main production function, it provides coordination and production links within the cluster, each cluster element is able to produce "additional utility" for other elements.

For example, a higher education institution may be interested in the success of a cluster not only because its research is in demand, but also because the most successful graduates are guaranteed employment in the cluster's enterprises. Small suppliers can count on the support of larger participants in the form of short-term cheap loans (or other types of financial support). Research organizations can include in their research plan not only the large developments required for the production of the main product of the cluster, but also small case studies commissioned by other partners in the cluster.

Additional utility can not always be measured in monetary terms, but any social, organizational, even moral support increases the efficiency of the entity, and this over time leads to direct economic benefits that can be measured in monetary terms.

Due to the creation of clusters, as the most important method of innovation-oriented management of the development of the construction industry, construction companies will become potentially attractive to domestic and foreign investors. Clustering will help develop specialization, improve production quality, attract reserves, reduce production costs and maximize the overall economic effect.

In addition, the creation and effective development of innovative construction cluster in any region of Ukraine will create additional jobs, increase employment, increase wages and budget contributions, relieve social tension in society, improve the quality of life, promote innovative development of the construction industry.

3 Assessment of the Economic Efficiency of the Functioning of an Innovative Construction Cluster

The effective functioning of a research and production innovative construction cluster based on a technology park depends on many factors, the main of which are: the level of construction complex development; availability of demand for construction goods and services; cooperation of construction companies heads with government agencies; the presence of leading enterprises that could become the "core" of the cluster; human resources of operating enterprises of the construction complex; availability of educational institutions of the appropriate profile; investment attractiveness of the region; the level of development of industry, social sphere, etc.

The question of determining the economic efficiency of the technology park and the cluster association as a whole remains open. Since it is extremely difficult to assess in advance all the benefits of such cooperation, especially in the context of the lack of a regulatory aspect of this problem.

The efficiency of the construction cluster is a relative indicator that shows the ratio of results and costs associated with the use of means of production, labor, entrepreneurial skills in the provision of construction services of a certain quality in a limited resource and unlimited needs.

At the same time, the efficiency of the cluster itself is assessed, which reflects the efficiency of the process of forming new organizational entities and, accordingly, the structural elements of the cluster and the efficiency of using resources for their work according to the formula:

$$\mathrm{Ew.cl.} = \sum_{\mathrm{i=1}}^{\mathrm{n}} \mathrm{Ecl.i} \Big/ \sum_{\mathrm{i=1}}^{\mathrm{n}} \mathrm{Bcl.i} \quad (1)$$

where *Ecl.i*–the effect of the *i*-th participant from working in a cluster;

Bcl.i–costs incurred during the cluster period;

$i = 1, \ldots, n;$

n–number of enterprises.

Assessment of economic efficiency of the proposed innovative construction cluster based on the technology park, which is based on the use of fixed and working capital, labor and financial resources, leads to an increase in the level of soundness of management decisions, cost optimization [18], which allows full use of resource potential and opportunities in the construction services market. The overall effect of the implementation of the cluster on the basis of the technology park, the model of which was proposed, is to optimize the costs of the construction industry in the region by effectively managing them and increasing the profitability of innovation in the regions of Ukraine.

4 Conclusion

The successful implementation of the cost optimization mechanism in the system of innovation-oriented management of the development of the construction complex is possible only if the complex interaction of all its elements at all levels of the complex management. Managers must clearly understand the structure of the cost mechanism at all levels of management, since the price of construction projects depends on its functioning and the interaction of all its constituent elements. In the construction industry of Ukraine, the formation of product prices is influenced by many different factors, but the regional authorities due to the introduction of an innovative construction cluster have the opportunity to find reserves to reduce the total cost of construction, which is the result of cost formation. After all, the main reason for the high cost of construction projects is the imperfection of cost management, which is manifested in the inefficient organization of resource and cost management in the construction industry of the regions, which leads to high and uncompetitive prices for construction sites.

References

1. Onyshchenko S, Yehorycheva S, Furmanchuk O, Maslii O (2019) Ukraine construction complex innovation-oriented development management. In: Proceedings of the 2nd international conference on building innovations, Lecture notes in civil engineering, vol 73, pp 687–700 (2019)
2. Rocha H, Kunc M, Audretsch DB (2020) Clusters, economic performance, and social cohesion: a system dynamics approach. Reg Stud 54:1098–1111
3. Litzel N (2017) Does embeddedness in clusters enhance firm survival and growth? An establishment-level analysis using CORIS data. Reg Stud 51:563–574
4. Abdesslem BA, Chiappini R (2019) Cluster policy and firm performance: a case study of the French optic/photonic industry. Reg Stud 53:692–705
5. Porter M, Michael E (2000) Location, competition, and economic development: local clusters in a global economy. Econ Dev Q 14:15–34
6. Delgado M, Porter M, Stern S (2015) Defining clusters of related industries. J Econ Geogr 16(1):1–38
7. Voynarenko MP (2000) The concept of clusters - the way to revival of production at the regional level. Yekonomist. 1:29–33
8. Heyets VM (2008) Clusters and network structures in the economy - a very interesting topic, but not yet fully explored… Ekonomist 10:10–11
9. Kuzmin O, Zhezhukha V (2010) Clusters as a factor of innovative development of enterprises. Econ Ukraine 2:14–23
10. Merkulov NN (2006) Using a cluster approach in the formation of industrial policy. Econ Ind 1:81–86

11. Braune E, Mahieux X, Boncori AL (2016) The performance of independent active SMEs in French competitiveness clusters. Ind Innov 23(4):313–330
12. Liberati D, Marinucci M, Tanzi M (2016) Science and technology parks in Italy: Main features and analysis of their effects on the firms hosted. J Technol Transf 41:694–729
13. European observatory for clusters and industrial change, https://ec.europa.eu/growth/industry/policy/cluster/observatory_en. Accessed 10 Dec 2021
14. Mykytas M, Terenchuk S, Zhuravska N (2018) Models, methods and tools of optimizing costs for development of clusterized organizational structures in construction industry. Int J Eng Technol 7(3.2):250–254
15. Karapetjan ET (2017) Mizhnarodnyj dosvid vykorystannja klasternogho pidkhodu u rozvytku integhracijnykh procesiv. Visnyk Ternopiljsjkogho nacionaljnogho ekonomichnogho universytetu 2:69–82
16. Anića I-D, Corrocherb N, Morrison A, Aralicaa Z (2019) The development of competitiveness clusters in Croatia: a survey-based analysis. Eur Plan Stud 27(11):2227–2247
17. Laur I (2019) Cluster initiatives within the European context: stimulating policies for regional development dreams. New Technol Based Firms New Millennium 11:147–170
18. Koba O, Yu M, Biba V (2018) Construction work cost determination. Int J Eng Technol 7(3.2):169–175

Features of Strategic Analysis of Externals Factors of the Building Organization: Methodological Aspect

Volodymyr Grishko, Inna Kulchii, Alla Glebova, and Turkan Hasanova

Abstract The article considers the types of modified PEST analysis (PESTLE, SLEPT, DESTEP, PESTPMED, STEP, STEEPLE, PESTLIED), which are important in the development of projects and development programs. It is determined which tasks can be solved with the help of strategic analysis. It is the strategic analysis that answers the question of how to achieve the desired in 3–5 years. The features of strategic analysis of building organizations are certain. It is proposed to use the LoNGPESTLE method—analysis for the implementation of international projects, which will minimize risks and increase profitability. Its advantages are: simplicity, the ability to carry out individually or collectively, allows you to capture many thoughts at once; identify factors that may effect, for a construction organization, this is important in the implementation of projects using funds from international organizations (and/or) the involvement of innovative technologies. It can be used regularly with a fixed time to track the dynamics of changes in strategy indicators.

Keywords PEST-analysis · Strategic analysis · LoNGPESTLE · STEP · Building organization

1 Introduction

In the conditions of the digital economy, the role and value of the building industry are considerable. Because the environment we create affects the physical and psychological well-being of people who live and interact from birth to death. Therefore, the introduction of innovations and environmental standards cannot only meet the needs of modern society in housing but also ensures the formation of an effective social infrastructure that affects the quality of life. The last 10 years have been difficult for

V. Grishko · I. Kulchii · A. Glebova (✉)
National University «Yuri Kondratyuk Poltava Polytechnic», Pershotravnevyj Ave. 24, Poltava 36011, Ukraine

T. Hasanova
Azerbaijan University of Architecture and Construction, Ayna Sultanova 11, Baku, Azerbaijan

V. Onyshchenko et al. (eds.), *Proceedings of the 3rd International Conference on Building Innovations*, Lecture Notes in Civil Engineering 181,
https://doi.org/10.1007/978-3-030-85043-2_61

the construction industry of Ukraine. The difficult political, socio-economic situation in the domestic market, the unfolding of the global financial crisis, and the armed conflict in Ukraine have led to an increase in the number of threats.

The presence of internal problems in the industry: constant changes in the State Architectural and Construction Inspectorate, which lead to delays in permits and commissioning of buildings, high level of corruption, lack of qualified personnel, low level of investment attraction, slow introduction of innovations in the construction process, high level of moral and physical depreciation of fixed assets of construction companies, slow introduction of energy and resource-saving technologies. All this leads to a decrease in the competitiveness of the industry and the need not only to address these issues but also to develop an anti-crisis strategy taking into account current trends and pandemics. Because it was the COVID-19 pandemic that created new threats, but also opportunities for construction development, especially in regional markets. At the same time, it demonstrated the relevance of strategic analysis methods for building companies.

In the scientific literature, the study of strategic analysis, its methodology was engaged in such scientists: Ansoff I [1], A. Thompson, A. Strickland [8], Gerasymchuk V., Azoev G. [2], Dovgan l., Shershneva Z., Solovyov V. [11], Nemtsov V., Onyshchenko V. [12–16] etc. The methodology of strategic analysis is being improved taking into account globalization and the growth dynamics of environmental factors, so it needs further research.

2 Main Body

In world practice, strategic analysis demonstrates the possibilities of forming strategies for the development of corporations, the conduct of competitive activity is on international markets, development of effective anti-crisis strategies.

In Ukraine of practical workers of strategic management and analysis did not yet occupy a high level. Building organizations in Ukraine are difficult to apply strategic management for lack of experience, dynamic terms of menage, absence of only methodological basis, lack of funds for innovation, lack of education. At the same time, these factors determine the need for strategic analysis. Strategic analysis is a key factor in the success of management. It allows you to study the organization and its work environment to develop an operational strategy. It improves the efficiency of operational management and forms the key success factors. Thus, strategic analysis allows you to set the goals to be pursued to get the desired result and the tasks to be performed.

At the same time, improvement plays an important role. Improvement is one of the constant tasks in the construction organization, which requires constant, systematic implementation. Therefore, to develop it is necessary to periodically conduct strategic analysis. Then it will help to build organizations to plan in good time and define, what directions need perfection. Therefore, the following features (Fig. 1) characterize the strategic analysis of the building organization.

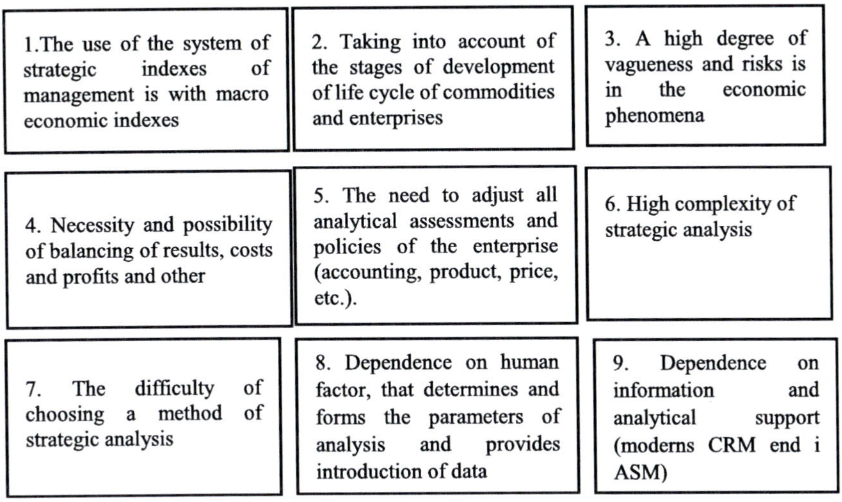

Fig. 1 Features of strategic analysis of building organization. Source: Compiled by the authors

Thus, in the process of organization of strategic analysis of building organization it may be to decide next basic tasks (Fig. 2).

This is ensured by adhering to the following principles of strategic analysis (Fig. 3).

During the strategic analysis of the construction organization, the main elements (indicators) are determined, which are then used in external and internal reports (Fig. 4).

It allows providing the possibility to elect direction of further development of building organization. Without the result of strategic analysis of strategy, id est reasonable plan it will be difficult to realize the pre-arranged actions and attain a desirable effect. The strategic analysis gives answers to three basic questions:

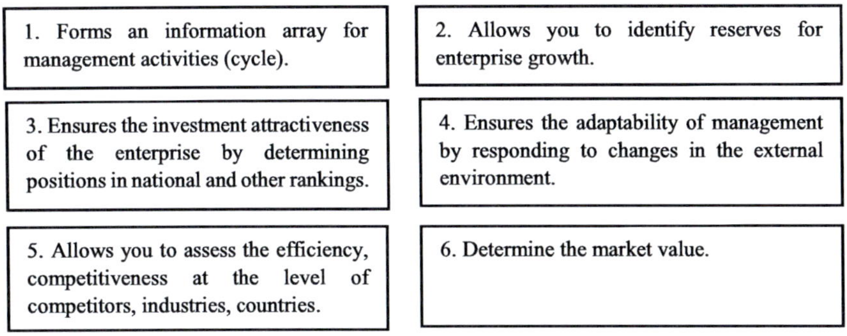

Fig. 2 Basic tasks that are decided by the strategic analysis of building organization. Source: Generalized by authors based on source elaboration [2, 4, 9]

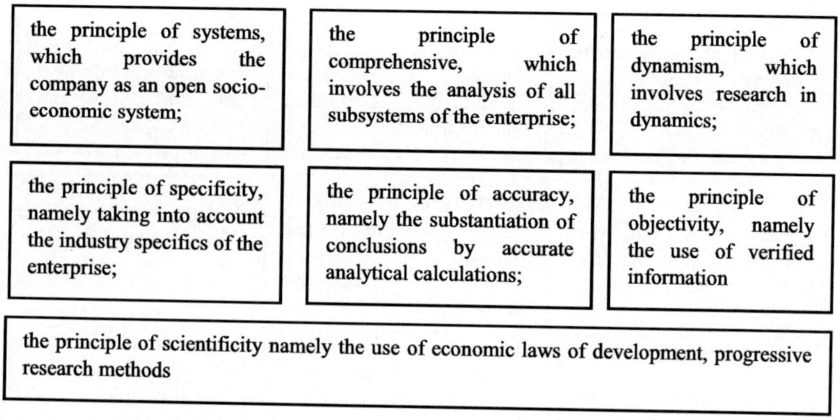

Fig. 3 Principles of realization of strategic analysis of building organization. Source: Generalized by authors based on source elaboration [1, 2, 4, 8, 9]

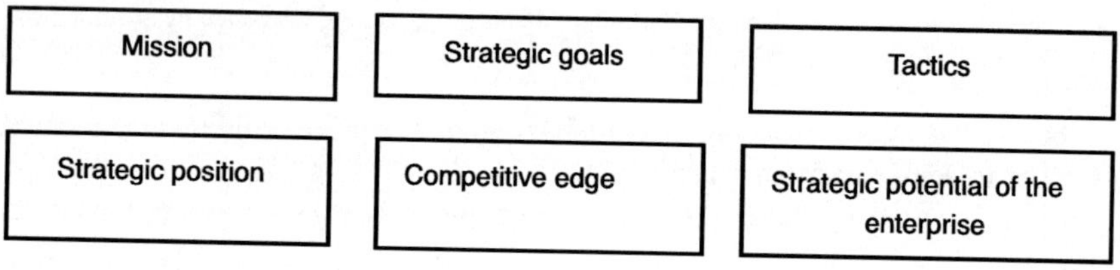

Fig. 4 Basic elements (results) of strategic analysis of building organization. Source: Generalized by authors based on source elaboration [3, 8]

1) In what position is a company now?
2) What position must a company be in through a certain interval of time (through three, five, ten years)?
3) What do exist ways of achievement of desirable position?

Examining the organization of strategic analysis at a construction company, we found that an important role is played by methodological support (Table 1). It involves the use of general scientific methods. Also, the use of various applied techniques, which reflects their uniqueness and difference from traditional methods of analysis.

The analysis of scientific works revealed that different approaches to classification are used for strategic analysis. In this case, we offer them to group in three basic groups of methods of strategic analysis: macro environments, meso environments, and microenvironments. Such an approach allows distinguishing the strategic alternatives of development of building organization, especially when the strategy of development is developed and project requests are formed on the realization of strategic projects with bringing in of investors and under back money. This will minimize risks and increase business transparency.

Table 1 Methodical support of the process of strategic analysis of the enterprise

Types of analysis	Methods of analysis
Strategic analysis of macro environment	Segmentation methods; PEST analysis; Econometric modeling; Extrapolation forecasts
Strategic analysis of the meso environment	Enterprise life cycle analysis; Analysis of input and output barriers of the industry; Analysis of strategic groups of competitors; Benchmarking; Cluster analysis; Methods of expert assessments; SWOT analysis; SPACE analysis; Compiling an environment profile; Monitoring of weak signals; Matrix methods of strategic analysis;
Strategic analysis microenvironment	Enterprise life cycle analysis; Product life cycle analysis; SWOT analysis; Time series analysis; Methods of predicting bankruptcy; Network analysis methods; Analysis of key success factors

Source: Generalized by authors based on source elaboration [3–9].

Among the most common methods of strategic analysis of environmental factors is the method of a PEST analysis. PEST analysis is a simple and widely used tool that helps analyze political, economic, socio-cultural, and technological changes in your business environment. The founder of PEST analysis is Francis J. Aguilar, a professor of management at Harvard University. He developed this macro-environment analysis tool for scanning the business environment (1967). The factors of PEST analysis will depend on the importance to the company depending on its industry and the products and services it provides (Table 2).

It is important to mark that during realization of evaluation of factors, then specific gravity (V) is determined on the degree of importance from 0,001 to 0,9. A total sum must present 1. Points are set by experts: 0 - a factor does not have an influence; 1 - a factor exists, but does not have an influence; 2 - the factor exists and may affect the results; 3 - exists and has no direct impact on the result of the enterprise; 4 - exists and has a direct impact on the result of the enterprise; 5 - has a significant impact on the activities of the enterprise and its results.

Table 2 PEST analysis

political	specific weight, V	point, P	product, P	social	specific weight, V	point, P	product, P
total				total			
Middle estimation of influence of political factors				Middle estimation of influence of social factors			
The strength of the influence of political factors				The strength of the influence of social factors			
Econo-mic	specific weight, V	point, P	product, P	Tech-nological	specific weight, V	point, P	product, P
total				total			
Middle estimation of influence of economical factors				Middle estimation of influence of technological factors			
The strength of the influence of economical factors				The strength of the influence of technological factors			

This helps you understand the «big picture» of the forces of change you are undergoing, and take advantage of the opportunities they present. It is important to consider PEST analysis as part of external analysis for strategic planning or strategic analysis. Another option is a PESTLE analysis, to which legal and environmental factors are added. If people want to add another layer locally, nationally, or globally, they can use another option, namely the so-called LoNGPESTLE. There are the following variations of PEST analysis (Fig. 5).

PESTLE / PESTEL: political, economic, social, technological, legal, environmental

DESTEP: Demographic, Economic, Sociocultural, Technological, Ecological and Political/Legal.

SLEPT: political, economic, socio-cultural, technological, legal

PESTPMED: demographic, economic, sociocultural, technological, ecological and political, legal, international

STEEPLE: Social/Demographic, Technological, Economic, Environmental, Political, Legal, Ethical.

STEP: political, economic, social, technological

PESTLIED: Political, Economic, Socio-Cultural, Technological, Legal, International, Environmental, Demographic

Fig. 5 Types of PEST analysis. Source: Compiled by the authors

The realization of PEST analysis for building an organization will be useful for four principal reasons (Fig. 6).

Based on of PEST analysis for building organizations that aim to go out to the international market it is recommended to use LoNGPESTLE is an analysis that was formed based on PESTLE—to the analysis. In contrast, it provides an analysis at three levels (Table 3).

LoNGPESTLE structure—the analysis allows to classifying economic, political, social, technical, legal, and ecological factors taking into account geography.

Its advantages are: simplicity, the ability to carry out individually or collectively, allows you to capture many thoughts at once; identify factors that may influence and engage ordinary professionals. The peculiarity of its holding is the geographical factor, which must be clearly defined and discussed.

And the last method is STEP analysis (Table 4), which is a kind of PEST analysis. This method envisages the analysis of key factors that influence the activity of building organization. Him it may be to apply on regular basis with the fixing of time, that allows watching the dynamics of changes of indexes, and also their influence. It allows estimating the influence of factors of environment of building organization.

Thus, several modified methods are used in world practice to study environmental factors, in particular, PEST analysis: PESTLE, SLEPT, DESTEP, STEP, LoNG-PESTLE, PESTPMED. They allow a more detailed analysis of environmental factors to identify strategic alternatives and minimize risks. As the external environment

It helps you notice business or personal opportunities, and also gives you the opportunity to warn of significant threats.	It reveals the direction of change in your business environment. It helps you shape what you do so that you work with the changes, not against them.
This helps you avoid starting projects that may not be implemented for reasons beyond your control.	This helps you to free yourself from unconscious assumptions when you enter a new country, region or market. This helps you develop an objective view of the new environment.

Fig. 6 Advantages of using PEST analysis. Source: Compiled by the authors

Table 3 LoNGPESTLE analysis

Levels	Factors					
	Political	Economic	Social	Technological	Legal	Ecological
Local						
National						
Global						

Source: Generalized by authors based on source elaboration [5–7].

Table 4 STEP analysis matrix

Factor	Events/factors	Threats/opportunities	Probability of occurrence event or manifestation factor	Importance factor or event	Impact on organization	Action/program
Economic						
Political						
Social						
Technological						

Source: Generalized by authors based on source elaboration [7].

in Ukraine is very dynamic and difficult to predict, their role and significance are important and necessary. Therefore, it is advisable to expand the practice of using PEST analysis, especially in the implementation of foreign economic activity of construction companies. Therefore, it is expedient to suggest using LoNGPESTLE, PESTLE, SLEPT, DESTEP, STEP, PESTPMED. In Ukraine, the analysis of the macro-environment will allow us to develop more accurate cash flow forecasts and improve the position of construction companies. Will reduce the number of unsuccessful construction projects. It will allow the introduction of innovations in the process of housing construction, commercial real estate, infrastructure.

3 Conclusions

Thus, further research is needed by the questions of development of strategy of changes of building organization based on factors that will be set in the total PEST analysis or PESTLE, SLEPT, DESTEP, STEP, PESTPMED and their intercommunication with motive forces of the market. The importance of the decision of these questions consists in the necessity of the permanent monitoring of influence of factors of environment of enterprise in an order to envisage character and degree of their influence. It will identify trends and tendencies, respond to them and implement strategic changes. It will reduce the number of errors in the implementation of international construction projects.

References

1. Ansoff I (2009) Strategic management. Classic edition, 344 p
2. Azoev G (2000) Competitive advantages of the company. Printing house "News"
3. Davydov GM (2011) Strategic analysis: training manual. Knowledge
4. David F (2006) Strategic management: concepts and cases, 11th edn. Prentice Hall
5. PESTLE analysis of Adidas. http://pestleanalysis.com/pestle-analysis-adidas. Accessed 29 Jan 2021
6. PESTLE analysis of Starbucks. http://pestleanalysis.com/pestle-analysis-ofstarbucks. Accessed 29 Jan 2021
7. PESTLE analysis of Uber. http://pestleanalysis.com/pestle-analysis-uber/. Accessed 29 Jan 2021
8. Thompson A, Strickland J (1998) Strategic management: art of development and implementation strategy. Banks and exchanges
9. Thompson AA, Strickland AJ (2006) Strategic management: concepts and situations for analysis. Williams Publishing House
10. Whellen T, Hunger D (1992) Strategic management and business policy. Addison Wesley. Publishing Company Inc., N.Y.
11. Bondar-Pidhurska OV, Solovyov VP (2017) The strategy of sustainable innovative society-oriented development of Ukrainian economy (by the example of mineral resource industry). Sci Bull Nat Min Univ 4(160):122–132

12. Onyshchenko VO, Filonych OM, Bunyakina NV, Senenko NB (2020) A new agent for removing concrete residues from the surfaces of polypropylene molds in the manufacture of paving slabs and its advantages. https://doi.org/10.1007/978-3-030-42939-3_67
13. Onyshchenko V, Sivitska S (2014) Alternative energy developing investment support in terms of energy dependence. Econ Ann XXI 9–10:34–37
14. Onyshchenko V, Sivitska S, Cherviak A (2018) Construction industry in Ukraine credit analysis. Int J Eng Technol (UAE) 7(3):280–284. https://doi.org/10.14419/ijet.v7i3.2.14420
15. Iasechko S, Haliantych MK, Skomorovskyi VB, Zadorozhnyi V, Obryvkina O, Pohrebniak O (2020) Contractual relations in the information sphere. Syst Rev Pharm 11(8):301–303. https://doi.org/10.31838/srp.2020.8.46
16. Onyshchenko V, Sivitska S, Cherviak A (2020) Buildings reconstruction within the new educational space project in Ukraine. https://doi.org/10.1007/978-3-030-42939-3_66

Assessment of the Regional Energy Efficiency Potential of the Housing Sector of Ukraine

Olha Komelina, Svitlana Shcherbinina, and Mahabbat Mammadov

Abstract This article develops a methodological approach to the integrated assessment of potential opportunities for energy efficiency providing for the residential sector in the regional context. Regional comparison of energy efficiency potential is based on values that characterize technical, economic, and social components of energy efficiency in the residential sector. The regions are ranked according to the integrated values and clusters are formed according to the level of energy efficiency assessment, which characterizes the differentiation of regional effectiveness of organizational and economic supplying of energy efficiency in the residential sector. The results, attained during the estimation of potential opportunities of energy efficiency management give the possibility to assess not only the action efficacy of all government branches to implement energy-efficient technologies but also to identify opportunities to apply political instruments to support the practice of stable building, based on sustainable development principles.

Keywords Energy efficiency potential · Integral assessment · Ranking of regions by integrated rating score · Ensuring energy efficiency of the residential sector

1 Introduction

The issue of energy efficiency and the economic use of energy resources has been important for a long time for many countries around the world. More and more of them are trying to solve the energy efficiency problem by introducing the latest technologies. Ukraine is doing its best not to stand aside of those tendencies. The detected reserves of energy saving and energy efficiency of the residential sector of Ukraine's economy according to a recent evaluation by scientists and practitioners, vary from 50 to 70% of the current level of energy consumption.

O. Komelina (✉) · S. Shcherbinina
Poltava National Technical Yuri Kondratyuk University, Poltava, Ukraine

M. Mammadov
Azerbaijan University of Architecture and Construction, Baku, Azerbaijan

V. Onyshchenko et al. (eds.), *Proceedings of the 3rd International Conference on Building Innovations*, Lecture Notes in Civil Engineering 181,
https://doi.org/10.1007/978-3-030-85043-2_62

Therefore, there is a need to develop a methodological approach that would allow assessing the potential opportunities for energy efficiency of the residential sector in the regional context.

2 An Overview of Recent Research Sources and Publications

Scientific and methodological approaches to the energy efficiency of the residential sector of the economy are actively studied by scientists and practitioners all over the world, and their systematization of generalization allows identification of modern mechanisms and devices to improve organizational and economic guarantees of energy efficiency of the residential sector in Ukraine.

Among all the variety of extremely important problems of energy efficiency maintenance of residential sector in foreign countries the following should be included: scientific and methodological approaches to energy efficiency maintenance of the residential sector: assessment of energy poverty in low-income households, which includes climatic, building, socio-economic characteristics of the country (Spain) [1]; studying the difference in energy efficiency depending on the group of consumers and its type [2]; application of a nonlinear methodology and interdisciplinary approach to the study of energy consumption in households, in particular analysis of the covariance structure, to isolate direct and indirect consequences of household and housing characteristics for total annual domestic energy consumption (USA) [3]; study of the utility market potential, the role and tasks of its players, demand modeling and energy resources prices forecasting [4]; development of energy consumption management for efficient use of resources in terms of technical progress [5]; conceptual approaches to housing fund renovation and partial reconstruction in context of European energy policy, thermal modernization of housing fund, as well as combining energy efficiency issues with ecological, financial, social and cultural goals of housing management (Sweden) [6]; increasing energy efficiency in the non-profit housing sector (Netherlands) [7]; problems of effective energy applying and energy efficiency measures in the construction sector as a tool to reduce energy consumption and to improve local ecological sustainability (Malaysia) [8]; assessment of the impact of municipal construction structures and their actions in energy efficiency stable building and renovation field [9]; mechanisms to stimulate energy efficiency during the projecting and construction of residential and public buildings, including the availability of energy efficient housing for the population; impact of energy efficient measures in residential buildings on the cost of habitation (USA) [10]; problems of management innovation-oriented development the construction industry of Ukraine [11]; methodical approach to determining the minimum mediate costs for 1 m^2 of affordable housing at the level of Ukrainian regions [12].

3 Purpose of the Article

The main purpose of the paper is to develop and implement a methodological approach to estimating the regional potential of energy efficiency, which allows to apply comparative assessment of the performance of the management and economic allowance of energy efficiency and use of energy efficiency potential in the housing sector in Ukraine (including Donetsk and Luhansk regions), to rank the regions according to the integrated rating score.

4 Methodical Approach to Assessing the Regional Potential of the Residential Sector

Theoretical Aspects

The concept of potential in the modern economy is interpreted ambiguously, there are a lot of theoretical approaches to determining its essence and constitution. It is most frequently used in the study of market opportunities of economic entities, the formation and evaluation of current and future opportunities that ensure the effectiveness of activity and create circumstances for the development of a competitive environment. At the same time potential performs: as a set of available instruments and resources; includes sources, opportunities, tools, reserves, which could be involved in solving certain issues [13]; characteristics of available productive forces in any field [14]; present opportunities, resources, reserves, instruments, etc. [15, 16].

The potential of energy efficiency is identified by authors as a set of resources, possibilities, and tools, which can be used to reduce the cost of energy resources in residential buildings while ensuring the optimal level of comfort for the population.

The regional potential of energy efficiency characterizes local features of energy consumption and energy efficiency in regions, it should be considered as a set of resources, possibilities, and tools, which can be executed to decrease energy expenditures in residential buildings while ensuring the optimal level of comfort for the population.

The introduction of such a concept allows considering a methodological approach to the integrated assessment of potential opportunities for energy efficiency of the residential sector in a regional context as an important tool for improving the organizational and economic support of energy efficiency of the residential sector in the national economy.

Regional differentiation of organizational and economic support of energy efficiency of the residential sector in the national economy will indicate the effectiveness of energy efficiency management of buildings in Ukraine in the regional context. In this study, based on the available statistics of the State Statistics Service of Ukraine [17], the integrated rating of the regions according to their energy efficiency potential was calculated.

Regional comparison of energy efficiency potential is carried out on eight indicators: total area of housing area, in thousand m^2; the current number of population in regions, in thousand people; disposable income of the population in regions of Ukraine, in million UAH; the number of buildings, equipped with units of commercial heat metering (residential) by regions, %; use of natural gas by regions, million m^3; use of heat by regions, Gkal; electricity use, million kWh; the total amount of subsidies allocated to reimburse the cost of housing and communal services, thousand UAH.

The algorithm of rating assessment of energy efficiency potential is accomplished in three stages. The first stage is the calculation of the sum of the relative deviations of the indicators, which characterize certain activities of economic entities in the region, from the best values of these indicators by the formula:

$$S_j = \sum_{i=1}^{n} W_{ij}, \tag{1}$$

where $W_{ij} = \left(\frac{B_{ij} - B_{\min_i}}{B_{\max_i} - B_{\min_i}}\right)$ for stimulator indicators;

$W_{ij} = \left(\frac{B_{\max_i} - B_{ij}}{B_{\max_i} - B_{\min_i}}\right)$ for destimulator indicators;

S_j—rating evaluation of energy efficiency potential of the j-th region for each of the indicators;

B_{ij}—value *i*-th indicator of *j*-th region, $1 \leq i \leq n$;

$B_{\max_i}$, $B_{\min_i}$—a minimal and maximal value of indicators.

The second stage. Determining the arithmetic mean of the sum of the rating of the energy efficiency potential of the assessment region for each of the indicators according to the formula

$$S_{j_{avg}} = \frac{S_j}{n}, \tag{2}$$

$S_{j_{avg}}$—the arithmetic mean of the sum of energy efficiency potential ratings by n-th indicators;

n—the number of indicators on which the calculation is performed.

The third stage. Determination of the integrated rating score of the energy efficiency potential of the region using formula

$$S_{rj} = \sum \left(S_{j_{avg}} \cdot q_i\right), \tag{3}$$

where S_{rj}—integrated rating score of the energy efficiency potential of the j-th region;

q_i—the significance of *i*-th indicators group.

Region with the highest S_{rj} value has the biggest energy efficiency potential.

The Results

The regions are ranked according to the integrated indicator and clusters are formed according to assessment level of energy efficiency potential, which characterizes differentiation of regional effectiveness of organizational and economic support of energy efficiency in the housing sector of the national economy, the bigger the energy efficiency potential is, the lower energy efficiency considering the usage of energy resources is in the residential sector and the larger volume of energy resources can be saved, by increasing the effectiveness of organizational and economic support of energy efficiency in the residential sector of the national economy at the regional level (Fig. 1).

Dnipropetrovsk, Donetsk, Kharkiv, Odesa regions, and the city of Kyiv perform the highest energy efficiency potential.

The result of energy efficiency potential assessment of buildings allows evaluating not only the effectiveness of all branches of the government in the implementation of energy-efficient technologies but also to identify opportunities to introduce political instruments to support sustainable construction practice based on stable development principles.

This approach requires balancing all factors of sustainability—economic, energy, ecological and social.

Despite some positive changes in Ukraine's energy efficiency in general, the level of effectivity in regions of the country has changed in different ways, which requires taking in account certain regional specifics and, accordingly, setting adequate targets for regions to realize the potential, which would take into account national goals and regional specifics at the same time.

At the same time, it should be noted that, despite this, in Ukraine, a criteria base for energy efficiency assessment has not been defined yet, as well as methodical developments for its measurement and comparison (it is possible to adequately compare different countries and regions of the world in terms of energy efficiency, considering climatic, territorial, infrastructural and other features). In this case, the only, relatively universal and commensurate for international and regional comparisons of energy efficiency is the energy intensity of GDP taking into account purchasing power parity (PPS).

Scientific Novelty

A methodological approach to integrated assessment of regional energy efficiency potential has been developed, which characterizes the local features of energy consumption and energy efficiency of the regions of Ukraine, and can also be used to identify potential opportunities and priorities for energy efficiency management of buildings.

Practical Relevance

Firstly, the obtained results of the research allow making adjustments of energy-saving policy and strategies of social and economic development of the regions of

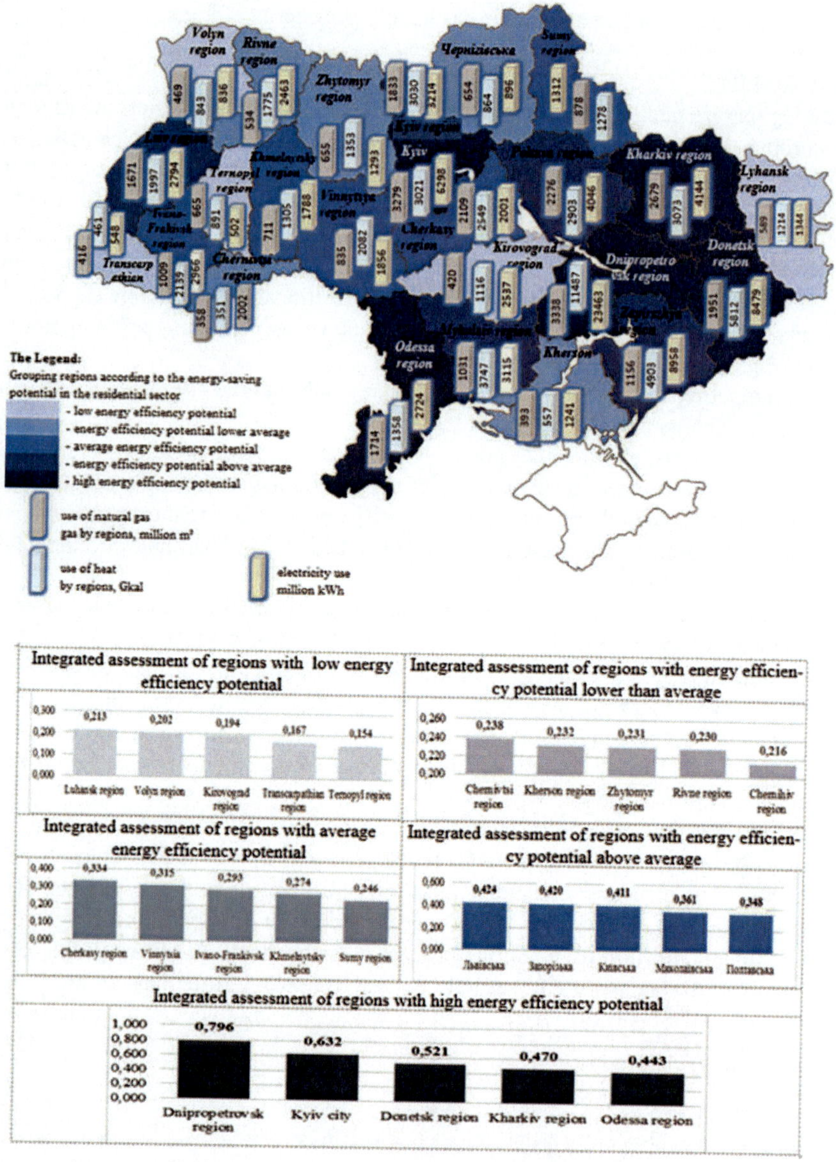

Fig. 1 Assessment of energy efficiency potential in the residential sector of Ukraine in the regional context, 2018 (developed by the author, data on the Autonomous Republic of Crimea are missing)

Ukraine on a systematic scientifically substantiated basis, as well as be taken into account in the development of relevant program documents. The practical use of the obtained results of the implementation of this approach makes it possible to determine the differentiation of regional effectiveness of organizational and economic support of energy efficiency of the residential sector in Ukraine.

5 Conclusion

The obtained results make it possible to optimize the content of regional and local strategies of sustainable energy development from a standpoint of energy efficiency, implementation of effective mechanisms to stimulate the population to introduce energy-saving technologies on a basis of the integrated approach, in terms of which along with creating the legislation, it is necessary to take into account the economic interests of housing owners and investors. This fundamental moment was understood in all developed countries of the world.

References

1. Sanchez CG, Gonzalez JN, Aja AH (2018) Energy poverty methodology based on minimal thermal habitability conditions for low income housing in Spain. Energy Build 169:127–140. https://doi.org/10.1016/j.enbuild.2018.03.038
2. Gillingham K, Palmer R (2013) Bridging the energy efficiency gap: policy insights from economic theory and empirical evidence. Rev Environ Econ Pol 8:18–38. https://doi.org/10.2139/ssrn.2206995
3. Estiri H (2015) The indirect role of households in shaping US residential energy demand patterns. Energy Pol 86:585–594. https://doi.org/10.1016/j.enpol.2015.08.008
4. Sioshansi FP (2016) Future of utilities – utilities of the future: how technological innovations in distributed energy resources will reshape the electric power sector, 1st edn. Academic Press, 492 p
5. Kayakutu G, Mercier-Laurent E (2016) Intelligence in energy, 1st edn. ISTE Press – Elsevier, 252 p
6. Femenias P, Mjornell K, Thuvander L (2018) Rethinking deep renovation: the perspective of rental housing in Sweden. Cleaner Prod 195:1457–1467
7. Filippidou F, Nieboer N, Visscher H (2017) Are we moving fast enough? The energy renovation rate of the Dutch non-profit housing using the national energy labelling database. Energy Policy 109:488–498. https://doi.org/10.1016/j.enpol.2017.07.025
8. Shaikh PH, Mohd NB, Nor SAA, Nallagownden P, Elamvazuthi I, Shaikh MS (2017) Building energy for sustainable development in Malaysia: a review. Renew Sustain Energy Rev 75:1392–1403. https://doi.org/10.1016/j.rser.2016.11.128
9. Hakkinen T, Rekola M, Ala-Juusela M, Ruuska A (2016) Role of municipal steering in sustainable building and refurbishment. Energy Procedia 96:650–661. https://doi.org/10.1016/j.egypro.2016.09.123
10. Im J, Seo Y, Cetin KS, Singh J (2017) Energy efficiency in U.S. residential rental housing: adoption rates and impact on rent. Appl Energy 205:1021–1033. https://doi.org/10.1016/j.apenergy.2017.08.047

11. Onyshchenko S, Yehorycheva S, Furmanchuk O, Maslii O (2019) Ukraine construction complex innovation-oriented development management. In: Proceedings of the 2nd international conference on building innovations, pp 687–700
12. Komelina OV, Shcherbinin LN, Shcherbinina SA, Ivanyuk BM (2019) Methodical approach to optimization of housing cost in the housing market of Ukraine. In: Proceedings of the 2nd international conference on building innovations, pp 609–616
13. Khomiakov VI, Bakum IV (2007) Upravlinnia potentsialom pidpryiemstva. Kondor, Kyiv
14. Busel V (2002) Velykyi tlumachnyi slovnyk suchasnoi ukrainskoi movy. Kyiv
15. Borisov AB (2003) Bolshoi ekonomichieskii slovar
16. Krasnokutska NS (2005) Potentsial pidpryiemstva: formuvannia ta otsinka: Navch. posib. Kyiv
17. Derzhavna sluzhba statystyky Ukrainy. http://www.ukrstat.gov.ua/. Accessed 11 Oct 2020

Economic Reforms and Development Strategies as Providing Sustainable Development

Fidan Mammadova, Mahabbat Mammadov, Alla Kariuk, and Emil Mammadov

Abstract The article examines the economic reforms carried out in Azerbaijan, the goals outlined by industries according to Strategic Plan of the "Strategic Road Map for the National Economy and Main Sectors of the Economy", aimed at implementing these reforms, development strategies, and reforms carried out in accordance with State Programs. It has been noted the importance of these reforms in the national economy sustainable and constant development. The article describes the development of the non-oil sector in the framework of general, sectoral and regional programs adopted in Azerbaijan to achieve sustainable development, diversification of the economy, development of infrastructure aimed at ensuring socio-economic development. Thus, the level of gas supply infrastructure in the regions, the level of gasification, the level of industrialization, the creation and development of infrastructure in existing industrial zones, as well as the implementation of clustering measures were reviewed and determined that in 2019, 46 enterprises were registered as residents and more than 8,000 permanent jobs have been created. At the same time, during the 4th stage of the industrial revolution, the implementation of innovations in the field of ICT in Azerbaijan, the creation of new opportunities for sustainable development of the country as a result of digitalization and the formation of a digital economy were introduced.

Keywords Reforms · Strategic Roadmap · Politics · Sustainable development · Regional development · Development strategies

F. Mammadova (✉) · M. Mammadov
Azerbaijan University of Architecture and Construction, Ayna Sultanova 11, Baku, Azerbaijan

A. Kariuk
National University «Yuri Kondratyuk Poltava Polytechnic», Pershotravnevyj Ave. 24, Poltava 36011, Ukraine

E. Mammadov
Azerbaijan Mortgage and Credit Guarantee Fund, Bulbul Avenue 40, Baku, Azerbaijan

V. Onyshchenko et al. (eds.), *Proceedings of the 3rd International Conference on Building Innovations*, Lecture Notes in Civil Engineering 181,
https://doi.org/10.1007/978-3-030-85043-2_63

1 Introduction

Effective integration into the world economic system depends on the implementation of effective economic reforms that ensure the sustainable development of each country and the Development Strategies, State Programs, Action Plans that are part of it. The study of the experience of developed countries in this area shows that countries that develop Development Strategies in line with modern requirements, ensure its implementation and carry out appropriate reforms in stages, achieve sustainable development. Azerbaijan is also developing and implementing new reforms in priority areas of the economy and the strategies that condition these reforms.

2 Main Part

2.1 Purpose of the Article

Thus, as a result of the economic policy pursued in Azerbaijan over the past 15 years, necessary measures to diversify the economy, ensure dynamic and sustainable development of the non-oil sector, accelerate the development of competitive industries in all areas during the implementation of Azerbaijan's Strategic Road Map 2016–2020 The goal is to ensure the sustainable development of the national economy until 2025 and to strengthen its competitiveness during this period. At the same time, the main goal for the period up to 2025 is to improve social welfare and build a strong competitive and inclusive economy based on development of high technology and the economy most optimal and suitable structure, which will ensure the highest human index. To this end, the country continues to implement measures covering all sectors of the economy in accordance with the above-mentioned development concepts in order to achieve sustainable development of the economy and, consequently, to ensure the production of competitive products. As a result, the implementation of targeted measures to diversify the economy in our country has accelerated the development of the non-oil sector in recent years. The measures taken to liberalize the business environment in the country have been of particular importance in ensuring the sustainable development of these sectors. With the aim to achieve the objectives and goals set out in the Strategic Roadmap for the national economy and its 11 economic sectors, the priority of measures to ensure economic reforms in the country should be identified in stages, evaluated and their implementation in a logical sequence [1, 5, 12].

2.2 Research Methodology

Systematic conceptual approach and analytical generalization methods were used in the research process. The research database consists of the materials of the

State Statistics Committee of the Republic of Azerbaijan, the Ministry of Economy of the Republic of Azerbaijan, the scientific-practical journal of Construction Economics and Management and the International Scientific Conference—Infrastructure Support of Diversified Economy.

2.3 Results, Discussions

Reforms and development strategies in Azerbaijan are the basis for sustainable development. The economic policy developed and implemented under the leadership of President Ilham Aliyev is a guarantee for achieving the above-mentioned goals and sustainable development of the country. It should be pointed out that these goals achievement is possible due to implementation of measures aimed at accelerating the development of the non-oil sector, diversification of the economy, sustainable and balanced socio-economic development of the regions within the general, sectoral and regional development programs adopted in the country, which has special significance.

For this purpose, one of the priority directions envisaged in the four state programs designed to promote socio-economic development of the Republic of Azerbaijan regions for 2004–2023 is the implementation of measures for the reconstruction and development of energy infrastructure in the country's regions is made. As a result of these measures, 56.2 thousand km of gas pipelines have been laid and repaired in the regions over the past 15 years, more than 1,800 settlements have been supplied with natural gas and the level of gasification in the regions has increased from 41 to 93.2% [3].

Analysis of the final report for 2019 of the fourth program on socio-economic development of the regions covering 2019–2023, including the state programs that play an important role in the implementation of economic reforms, development strategies and their implementation, shows that during this period the country 61 settlements and 37 new housing estates were gasified, as a result, 15,338 subscribers were provided with natural gas and the total level of gasification in the country reached 96%. The analysis of the mentioned state program shows that the newly laid or overhauled gas pipelines were 1376.7 and 487.7 km, respectively, and these measures were effectively implemented in the economic regions. Thus, 128.7 and 84.4 km of newly built and overhauled gas pipelines in Baku, respectively; Absheron 90, 28.58 km; Ganja-Qatar—342.57, 104.32 km; Sheki-Zagatala—84.24, 15.15, 9.99 km; Lankaran—60.32, 59.97 km; Guba-Khachmaz—281.37, 33.19 km; Aran—395.8, 109.3 km; Upper Karabakh—0.2.29 km; Mountainous Shirvan—143.61, 55.81 km and in the last 15 years more than 1800 settlements have been supplied with natural gas. Coincidently, special attention is given to the industrialization and non-oil industry development, which is an integral part of the policy of economic diversification in Azerbaijan. For this purpose, the "State Program for the Development of Industry in the Republic of Azerbaijan for 2015–2020" was developed and currently the implementation of this program has ensured the creation of new value-added industries, raised the level of self-sufficiency in many industrial products. in turn, has a significant impact on the emergence of sustainable development in the country [2, 8, 9]. To this end, to ensure the diversified development of the country's industrial sector in 2019, support the creation of new processing industries, production of export-oriented and import-substituting competitive science-intensive products, application of innovative technologies, creation of modern infrastructure in existing industrial zones, appropriate measures are also being taken to support clustering and the efficient operation of industrial enterprises. Implementation of these measures is considered as one of the main objectives of the Public Investment Program, but at the same time sustainable and balanced development of the country's economy, especially the non-oil sector, regional development, ensuring economic security, systematic and effective implementation of investment projects aimed at environmental protection, is one of the main goals of the program. Investment programs are prepared in accordance with the goals and directions of socio-economic development of the country. Thus, in preparing the investment program for 2019, in accordance with the country's socio-economic development of, the development of infrastructure and ensuring their sustainable use, stimulating investment in the non-oil sector and regional development, optimal distribution of investment between regions, implementation of investment priorities in social spheres directed.

At the same time, the study of the above report shows that in 2019, the implementation of measures in various sectors of industry continued. One of the mechanisms aimed at industrial development is the creation of industrial parks and industrial districts, as this year 46 enterprises were registered as residents. started production and as a result, more than 8,000 permanent jobs were created.

One of the main goals of economic reforms in the country is to create a modern infrastructure in the economy. The study of implementation of the above-mentioned development strategy, state programs in the field shows that the infrastructure of the country's economy is constantly modernized, advanced modern infrastructure is created. During the fourth stage of the industrial revolution, our country's favorable geographical location, rich use of rich natural and human resources creates new opportunities for sustainable development of the country as a result of international ICT innovations, digitalization and the formation of the digital economy [4, 6, 7, 10]. In addition, the "Strategic Roadmap for the National Economy and Key Sectors of the Economy" prepared on the basis of an in-depth analysis of the current economic situation and approved by the Presidential Decree on December 6, 2016 are the main objectives of the Strategic Plan to ensure the sustainability of economic reforms in the country. As a result of reforms in these areas, Azerbaijan has undergone rapid and dynamic development in recent years in terms of economic reforms, and its achievements are highly valued by international organizations and international financial institutions.

Meanwhile, it should be considered that the study of reports on the work carried out and currently carried out under various state programs aimed at socio-economic development shows that the measures envisaged in the development and management of the country's economy. The activities of the sectors are also of particular interest. Therefore, regular reports should be reviewed to assess the level of activity of the private sector within the programs and its coordination with other institutions, and relevant proposals and recommendations should be prepared to improve public-private partnership. Coordinated activities of the public and private sectors in the implementation of measures envisaged in the state programs to ensure the achievement of the goals set for the development of sectors in the Strategic Roadmap to ensure the diversified development of the country's economy, which ultimately forms the basis for a strong economy. These areas of activity play an important role in attracting local and foreign investment, modern technologies and the effective use of modern management practices to create enterprises that produce competitive products and ensure their production of competitive products in accordance with international standards. It should be noted that the sectoral programs and four state programs for 2004–2023, aimed at developing the infrastructure of the economy of the country's regions, provide for a wide range of measures, which determine the areas of activity of the sectors. In order to achieve the above-mentioned goals, the state programs prioritize the implementation of measures to improve the infrastructure of the country's non-oil sector by ensuring the interaction of public and private sectors, ensuring rapid development of entrepreneurship, stimulating export-oriented production, increasing employment. They are aimed, among other factors, at the development of the country's economy and the acceleration of its effective integration into the world economic system. In all four programs aimed at socio-economic development of the regions, measures are being implemented in stages in the field of infrastructure, roads and transport, electricity, gas and heat supply, which are important in improving the infrastructure of the economy of economic regions. In our opinion, the role of coordinating the activities of both public and private sectors within the limits of the programs in achieving the goals set out in the 11 roadmaps developed in the Strategic Roadmap for 12 sectors should be regularly assessed, and improvements should be made at each subsequent stage.

It should be noted that the ongoing reforms and state programs, development concepts, strategies and plans implemented in Azerbaijan and the sustainable economic development achieved and strengthened as a result of comprehensive measures to ensure macroeconomic stability in the country is highly valued by international economic and financial organizations. Necessary decisions have been made to diversify the economy in Azerbaijan, ensure the development of the non-oil sector, stimulate the production of competitive production, improve the investment and business environment, constantly improve the regulatory environment of the business environment to modern standards, support the private sector. The annual improvement in the results of sustainable reforms aimed at the formation, implementation and improvement of the business environment is constantly reflected in the reports of international financial institutions [10–12].

In the "Global Competitiveness 2017–2018" report published by the World Economic Forum, the Azerbaijani economy has moved up two places in terms of competitiveness, ranking 35th among 137 countries, and first in the CIS since 2009. In recent years, Azerbaijan results in reports prepared by the World Economic Forum are constantly improving. Thus, according to the "Global Competitiveness" reports, our country has moved up 34 places—from 69th place among 117 economies in 2005 to 35th place among 137 economies in 2017. As a result of these indicators, Azerbaijan has left behind a number of G-20 member states. Thus, in the Global Competitiveness Report 2017–2018, Indonesia, which is a member of the G-20, ranks 36th, the Russian Federation 38th, India 40th, Italy 43rd, Mexico 51st and Turkey 53rd, South Africa is 61st, Brazil is 80th and Argentina is 92nd. According to the Global Competitiveness Index, Azerbaijan ranks higher than the neighboring countries as the most competitive economy in the region, but in the report, the countries of the region, Georgia, rank 67th, and Armenia 73rd (Fig. 1).

The "Global Competitiveness Report 2018/2019" lists Azerbaijan as the country with the highest level of social equality in the world. According to the level of electricity supply to the population, Azerbaijan has risen to the first place in the world with a maximum score of 100%. According to the report, Azerbaijan ranks 31st among 140 countries in the ranking of business dynamics (Fig. 2).

Reforms implemented in the country under the leadership of President Ilham Aliyev within the framework of state programs, development concepts, strategic development plans have ensured that Azerbaijan has achieved new records in the overall ranking in the "Doing Business 2019" report, and its position on most indicators has risen. In the "Doing Business 2019" report which was published by the World Bank and the International Finance Corporation, our country was included

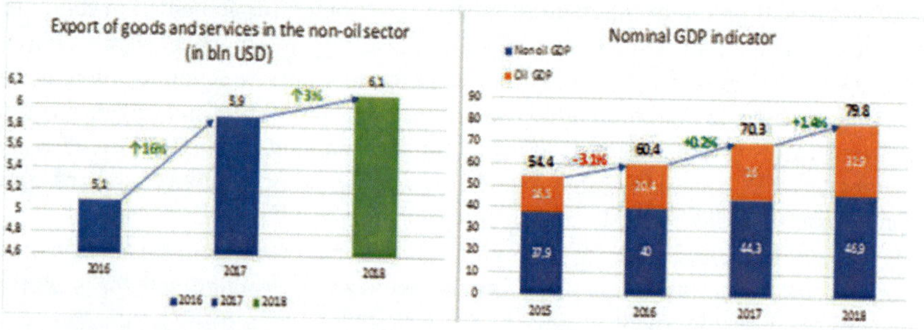

Fig. 1 Azerbaijan international rankings

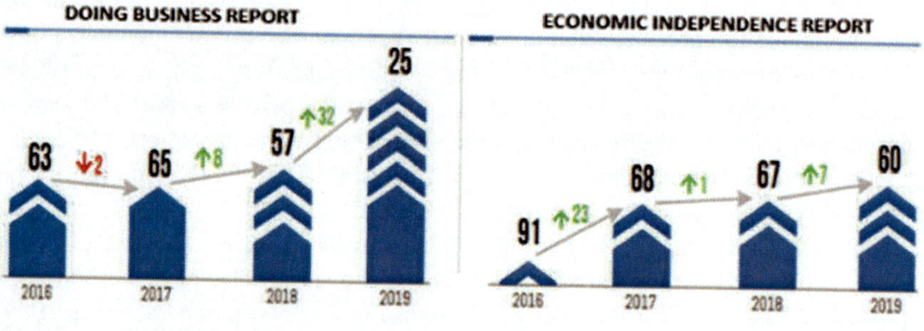

Fig. 2 The results of reforms in Azerbaijan

in the list of the 10 most reformist countries in the world and declared the most reforming country in the world.

In the "Doing Business 2020" report, the continuous implementation of reforms in the above-mentioned areas in our country was assessed with 78.5 points, and Azerbaijan rose from 34 to 28th place among 191 countries. As a result of the progress made in this position, our country has again been assessed by the World Bank as "one of the 10 most reformist countries in the world."

Thus, it should be noted that these assessments of the International Financial Institutions are a real result of sustainable reforms implemented in Azerbaijan in recent years in the framework of the above Development Strategies, State Programs, Strategic Plans.

2.4 Results

We believe that the measures taken as a result of the carried-out reforms in the country and implemented development strategies will allow to achieve the following organizational, managerial and socio-economic advantages:

- Ensures the diversified development of the country's industrial sector;
- Ensures sustainable development and competitiveness in the industrial sector
- Ensures the implementation of targeted projects in the non-oil sector
- Allows wide application of innovative technologies
- Creates conditions for the creation of modern infrastructure in industrial zones, as well as to support the effective operation of clustering
- Ensures sustainable and consistent implementation of investment projects in the field of environmental protection;
- Development of business environment, creation of business incubators will stimulate the development of economic regions of the country;
- Sustainable economic reforms implemented within the framework of state programs, development strategies and concepts have resulted in ensuring macroeconomic stability;
- The state regulation of entrepreneurial activity, the application of advanced methods in reforms is highly valued by the World Financial Institutions, which allows us to assess it as a real result of the regular and sustainable implementation of reforms in Azerbaijan.

2.5 Practical Importance

The results obtained from this research can be used while preparing of Strategic Development Plans of relevant organizations and institutions.

3 Conclusions

The study shows that the reforms implemented in various sectors of the economy of Azerbaijan form the solid basis for sustainable development.

References

1. Strategic roadmap for the national economy and key sectors of the economy. Strategic plan, presidential decree of 6 December 2016. https://www.turan.az/ext/news/2020/1/free/analytics/en/121631.htm
2. Infrastructure provision of the diversified economy: main trends and directions of development. In: Materials of the international scientific conference 2019. Publishing and Printing Center, AzUAC, Baku, 593 p. https://unctad.org/system/files/official-document/der2019_en.pdf
3. State program of socio-economic development of the regions of the Republic of Azerbaijan in 2019–2023, 9-month report for 2019. https://www.azernews.az/nation/161283.html
4. Economics and management of construction. Sci Pract Mag 2019(3). Baku-2019
5. Kozachenko H, Onychshenko S, Masliy O (2018) Region building complex social and economic security threats. Int J Eng Technol (UAE) 7(3):215–218. https://doi.org/10.14419/ijet.v7i3.2.14405
6. Sivitska S, Vartsaba V, Filonych O (2018) Buildings energy-efficient renovation investment. Int J Eng Technol (UAE) 7(3):408–412. https://doi.org/10.14419/ijet.v7i3.2.14562
7. Onyshchenko V, Sivitska S (2014) Alternative energy developing investment support in terms of energy dependence. Econ Ann XXI 2014(9/10):34–37
8. Onyshchenko V, Sivitska S, Cherviak A (2020) Buildings reconstruction within the new educational space project in Ukraine. https://doi.org/10.1007/978-3-030-42939-3_66
9. Onyshchenko V, Sivitska S, Cherviak A (2018) Construction industry in Ukraine credit analysis. Int J Eng Technol (UAE) 7(3):280–284. https://doi.org/10.14419/ijet.v7i3.2.14420
10. Drobyazko S, Bondarevska O, Klymenko D, Pletenetska S, Pylypenko O (2019) Model for forming of optimal credit portfolio of commercial bank. J Manage Inf Decis Sci 22(4):501–506
11. Iasechko S, Haliantych MK, Skomorovskyi VB, Zadorozhnyi V, Obryvkina O, Pohrebniak O (2020) Contractual relations in the information sphere. Syst Rev Pharm 11(8):301–303. https://doi.org/10.31838/srp.2020.8.46
12. Svistun L, Glushko A, Shtepenko K (2018) Organizational aspects of development projects implement at the real estate market in Ukraine. Int J Eng Technol (UAE) 7(3):447–452. https://doi.org/10.14419/ijet.v7i3.2.14569

The Role of Financial and Investment Potential in Achieving Economical Equilibrium in Construction

Inna Miniailenko, Volodymyr Byba, Elshad Yusifov, and Anna Pavelieva

Abstract The article considers the categories of "equilibrium" and its main types, "economical equilibrium", equilibrium macroeconomic models. It is proved that equilibrium can be achieved through economy crises, which serve as specific tool for rebalancing and overcoming disequilibrium.

There have been formed the main directions of the state economic policy for overcoming system contradictions of economic destabilization in Ukraine. The article stresses the key role of construction in creating economical equilibrium and identifies the ways to build Ukraine's development potential. The authors of the article have analyzed indicators of construction development of Ukraine, determined the main tendencies of changes and carried out modeling of the equilibrium position of the Ukrainian construction industry. The author's approach to modeling the coefficient of economic equilibrium in construction has been worked out in this scientific study.

Keywords Equilibrium · Economical equilibrium · Principles · Construction · Investments · Potential

1 Introduction

The problem of using the existing potential is closely related to the most important aspect of the economic performance—achieving economical equilibrium, which is a strategic goal of national development.

It is common knowledge that the economy of any country develops cyclically, there are periodic fluctuations in economic and business activities. Economic growth is replaced by a sharp drop in gross domestic product (the GDP), economic upturn

I. Miniailenko · V. Byba (✉) · A. Pavelieva
National University "Yuri Kondratyuk Poltava Polytechnic", Pershotravnevyj Avenue 24, Poltava, Ukraine
e-mail: bibavv1@ukr.net

E. Yusifov
Azerbaijan University of Architecture and Construction, 5 Ayna Sultanova, Baku, Azerbaijan

V. Onyshchenko et al. (eds.), *Proceedings of the 3rd International Conference on Building Innovations*, Lecture Notes in Civil Engineering 181,
https://doi.org/10.1007/978-3-030-85043-2_64

is occasionally punctuated with the crisis and recession. Such changes are common to each state, with the same consecution and regularity. But it should be noted that in different countries, the growth and crisis cycles (or economic cycles or boom-bust cycles) differ in regularity, duration and causes of occurrence.

For a long time, macroeconomic has paid attention to the study of business cycles of the economy upon certain statistical factors, but it did not take into account the problem of economical equilibrium.

The general equilibrium theory is the basis of scientific research of many Western European scientists. The founder of the importance of economical equilibrium and its relationship to crises is Karl Marx, who formulated the ultimate principle according to which economic slowdowns are considered as departures from equilibrium position. The fundamental premise of this study is the fact that the constant tendency of certain industries or economic activities to equilibrium is the reaction to its constant violation.

But before proceeding to the methodological study of this category, it is necessary to analyze the very concept of equilibrium and to determine its types and classification features.

Thus, according to the economic encyclopedia, equilibrium is physical system state in which its characteristics do not change over time [1]. Despite the fact that equilibrium is a constant, there are statistical and dynamical equilibrium.

Statistical equilibrium is peculiar to the systems in which the particle motion ceases. Dynamic equilibrium is characterized by the motion of the system at certain constant characteristics.

The state of equilibrium is characteristic of any science. Let us consider its certain types and determine their defining characteristics (Table 1) [1].

According to the considered types of equilibrium, it becomes possible to determine their main characteristic, namely the constancy of values and a certain balance that ensures their proper functioning. But the question arises as to whether the constancy of certain indicators is good for the economy or it implies the state of stagnation, depression that is setback in production.

2 Main Body

Economical equilibrium is the state of the economic system in which the proportions in the national economy provide the best possible reconcilement of objectives of economic development and available resources, supply and demand, commodity and cash flows, accumulation and consumption, savings and accumulation and other elements and indicators of the system, and as a result—no economic crises. The two sides of economical equilibrium are the action of the forces of equilibrium and disequilibrium, in the process of which equilibrium is ensured through disequilibrium. This means that equilibrium is optimal dynamic imbalance which plays both positive and negative roles in the economy.

Table 1 Types of equilibrium

Types	The content
1. Chemical equilibrium	It is the state in which the chemical reaction proceeds at the same rate as the counter reaction, resulting in no change in the amount of each component
2. Thermodynamic equilibrium	It is the system condition in which its internal processes do not lead to changes in macroscopic parameters, in particular, such as pressure
3. Thermal balance	It is the condition of material objects that are in thermal contact, characterized by complete absence of heat exchange between them, which clearly means the equality of their temperatures
4. Mechanical equilibrium	It is the state of a body at rest or moving uniformly, in which the sum of forces and moments acting on the body equals to zero
5. Biological equilibrium	It is the balance of people and animals
6. Ecological equilibrium	This is the relative balance of stability of the species composition of living organisms, their number and productivity, spatial distribution, seasonal changes, biotic circuit of substance and other biological processes in natural or man-made ecological systems
7. Economical equilibrium	It is the situation in which all factors influencing the economy fully balance each other in such a way that the variable does not change as a result
8. Nash equilibrium in the game theory	It is the optimal policy for all game players aimed to ensure that no player will make a profit by changing his/her strategy until everyone else changes his/her strategies
9. Market equilibrium	It is the market situation in which the quantity of supply is equal to the quantity of demand

The disequilibrium arises objectively due to the continuous development of STP (scientific-technological progress), due to differences in time, in quantitative, qualitative and structural aspects of certain macroeconomic indicators. The accumulation of disequilibrium forces leads the economy to stress and crisis. At the same time, disequilibrium is necessary because it means that the economy is not overloaded or stretched to the limit, but there are still reserves for manoeuvres and restructuring of the economy.

The disequilibrium increases during the uncertainty period in the economy, due to foolish economic policy or the wrong socio-economic course.

Forms of economical equilibrium are the presence of optimal correspondence between the interests of economic subjects (economic entities), on the one hand, and unbalanced development (milk-bar economy), disagreement, strife—on the other hand.

There are statistical and dynamical, partial and complete, microeconomic and macroeconomic equilibrium, and so on.

Partial equilibrium is quantitative and qualitative correspondence of two related parties of the economy (for example, correspondence of production and consumption), general equilibrium is concerted, planned and proportional development of all areas, economic realms, markets and branches of economy and so on.

The condition of partial equilibrium is equilibrium in individual markets, while the condition of general equilibrium is equilibrium in all markets. When there exists a market self-equilibrating apparatus, state and supranational regulation, its equilibrium is achieved mainly through the mechanism of price fluctuations, and the equilibrium of the whole economic system is achieved primarily due to state and supranational regulation, which involves the use of price mechanism.

The conditions of partial equilibrium were studied by A. Marshall, market equilibrium in conditions of imperfect competition was studied by J. Robinson and E. Chamberlin, general economic equilibrium was examined by L. Walras.

In general, the dialectical method of research contradicts the general equilibrium theory, as it lays emphasis on the action of the law of the unity and struggle of opposites in the process of development, within which the unity of the two sides of the contradiction in numerical terms is relative (as it is carried out through modifications of quantity within single quality, without violating the measure, that is without a leap), while in qualitative terms the unity of the two sides of the contradiction and the struggle of opposites occur through mutual transition from quantity to quality, qualitative transformations within the single entity, through the formation of a new quantitative force of development through new quality etc., that is through the operation of the law of quantitative and qualitative changes. The concept of "economical equilibrium" largely coincides with the concept of "economic optimum".

There is a sound approach to the definition of equilibrium as a balanced and proportional economy: aggregate demand and aggregate supply, aggregate production and aggregate consumption. It is possible to achieve macroeconomic equilibrium, but only temporarily, as it is of dynamic nature and operates more on the principle of "equilibrium-disequilibrium". However, it is of interest for society that departures from equilibriums be minimal and short-dated. Therefore, the society arrived at the conclusion that it is necessary to regulate macroeconomic equilibrium in the process of implementing appropriate economic policy. To assess such equilibrium, it is recommended to use the final figures of macroeconomic development: the gross domestic product (the GDP).

The general macroeconomic equilibrium is sustainable development of all economic realms, concurrent compliance with the market as a whole.

Macroeconomic equilibrium presupposes the achievement of proportionality as expediency, orderliness and conformity of reproducible relations, parties and phases. In this case, local, individual proportions are interconnected, they complement each other and motivate equilibrium or balanced development of the entire economic system.

There is a close relationship between equilibrium, proportionality and robustness. If equilibrium is the desired goal of macroeconomic development, then proportionality and robustness are the elements of the mechanism used for achieving it.

Ignoring any of these concepts always leads to one-sided results, to contradictions and imbalances of economy and social life.

The following prerequisites are required for the general economic equilibrium:

- compliance of aggregate demand with aggregate supply in all types of market (merchandise market, labor market, capital market);
- efficient use of all economic resources, which means the absence of large scale unemployment, high inflation rate, non-distributed stocks and their deficit;
- compliance of the national agenda with the existing economic opportunities;
- reconciliation of production and consumption with the orientation of their development in the interest of consumers;
- compliance of income with expenses.

The research of the problems of economical equilibrium, creation of concepts, theories, models has lasted for already two centuries.

There are the following models of economical equilibrium (Table 2) [2].

The theory of general equilibrium also promotes the research of economic welfare. That is, any departure from the equilibrium state, which occurred under the influence of both external and internal factors, involves the forces that tend to equilibrium.

As noted, the achievement of equilibrium position is carried out through economic crises, which are a specific tool for overcoming imbalances of economy and restoration of equilibrium.

If one looks at the crisis impartially, it will become clear that the crisis is as necessary for a developing system as the stable state. The crisis should be seen as a defining moment in the development of the system, which gives more space for a new round of economic changes.

A crisis is the critical stage in the functioning of any system, when it falls under the influence from the outside or inside, which requires a qualitatively new response from the system. The main feature of the crisis is that it threatens to destroy the system (partially or completely) [3].

The crisis is one of the forms of economic development, during which there appears space for the growth and establishment of something new, and out-of-fashion machinery and technologies, setup for production and job arrangement are eliminated [4]. Simultaneously with overcoming the obsolete, there is inevitably a setback in production, rise in unemployment, decline in personal income, which negatively affects people's living conditions.

Ukraine, as an independent state for more than thirteen years of its history, has been in various crises. The country has already weathered one debt crisis in 1998–1999 and, starting in the second half of 2008, another financial and economic crisis, accompanied by a fall of living standards as a result of severe economic free fall.

Considering current crisis developments in Ukraine, we can identify such key factors as the bank crisis, incoherent fiscal and monetary policy of the state, as well as the political situation in the country. The basis for the growth of macro-financial unsteadiness in Ukraine was primarily the implemented ill-considered economic policy, which did not serve the strategic objectives and opportunities for the development of the national economy [5].

Table 2 The models of macroeconomical equilibrium

Models	Representatives	Principle
1. The classical model	François Quesnay (1694–1774), Jean-Baptiste Say (1767–1832)	François Quesnay in his work "Tableau économique" (1758) made the first attempt to create a model of equilibrium in the form of schematic representation of the process of reproduction on a simple scale and circulation of gross national product, and its implementation between representatives of three classes: manufacturing one, land holders and industrialists. He substantiated the concept proceeding from the development of the economy on the basis of laissez-faire, spontaneous market pricing, equivalence of trading-off and governmental non-interference with business In his "A Treatise on Political Economy" (1803), J. B. Say created the model of economical equilibrium "demand is supply". According to this model, the production of goods and their supply (merchandise offering) create their own demand. Under these conditions, supply and demand in the economy are always balanced, and imbalances in the form of commodity deficiency or production overage are impossible

(continued)

Table 2 (continued)

Models	Representatives	Principle
2. Marxian economic model	Karl Marx (1818–1883)	The creation of a Marxian model of macroeconomical equilibrium is associated with the research of Karl Marx (1818–1883), who in the second volume of "Capital: The Process of Circulation of Capital" substantiated the general conditions for the proportional development of reproduction on a simple scale and extended reproduction, explained the essence of reproduction of productive forces and production relations, and all its phases: production, distribution, trading-off and consumption. When the conditions of proportionality are violated, disproportion and imbalance arise
3. Neoclassical model	Marie-Esprit-Léon Walras (1834–1910)	Leon Walras developed a model of general economical equilibrium, which shows how market-driven economy ensures equality between aggregate demand and aggregate supply. According to the "Walras law", economic agents have no receipt of funds from abroad, as well as no "put-up demand". Therefore, when at constant prices supply and demand coincide in all markets, then supply and demand automatically coincide even in the single market due to budgetary restrictions of economic agents. L. Walras showed that the balance of households and consumer goods markets is congruent with the balance of firms and production factors

(continued)

Table 2 (continued)

Models	Representatives	Principle
4. Keynesian model	John Maynard Keynes (1883–1946)	The Keynesian model of general equilibrium is based on the priority of aggregate demand, on the regulatory role of the quantity traded and on the ability to ensure equilibrium in the presence of unemployment, inflation or setback in production. The state must pursue an active policy of expansion, influencing the level of aggregate demand by cutting taxes, increasing government spendings and interest rate reduction. J. M. Keynes discovered the income multiplier effect—when the overall rise in national income exceeds government expenditures for the encouragement of aggregate demand, which leads to expansion of production volume and higher level of employment
5. The monetarist model	Milton Friedman (mid-twentieth century)	The founder of modern monetary economics proposed to take into account in the model of equilibrium instead of the state policy of expansion the policy of restricting the personal income, the level of aggregate demand, monetary benefits to the population, government spendings, money supply growth. At the same time, there may be rise in unemployment and decline in inflation. In such state of the economy, the state must pursue a policy of deregulation, that is giving greater freedom to develop exchange relations (market relations)

Thus, based on the above, we can say that the basis of economic destabilization in Ukraine were the following long-term system contradictions:

a) excessive customer share and insufficient profit accumulation under the conditions of underdeveloped internal market, which led to the need to cover surplus of domestic demand due to proactive import growth, and thus to a steady decrease in balance of foreign trade and risks of currency destabilization;
b) institutional underdevelopment of the financial system and tightness of refinancing mechanisms of the banking system, which led to the active entry of commercial banks into outer loan facilities markets, where the use of consumer credits focused primarily on the purchase of foreign ware and on the mortgage fraction of market;
c) unfavorable conditions for long-term investing, which led to the predominant investment attractiveness of the sectors with high volume of transactions and the quickest return of capital, priority rates of investment in the sectors that redistribute the national output (the output aggregate), over manufacturing investment, which increased the risks of instability, inflation and failing financial health;
d) the concentration of export specialization of Ukraine on a small number of family groups of goods with the growth of economic publicity, which formed high dependence of economic dynamics and financial position on overseas markets [5].

The considered contradictions are shown in Fig. 1.

In recent years, not only have there been no systematic actions to overcome these contradictions, but also a number of miscalculations in economic policy have been made, due to the orientation of economic policy to gain the loyalty with the voters under the conditions of accelerated repetition of the political cycle and negligence of the development and structural modernization goals:

- the high level of fiscal exaction of resources for consumption needs was maintained;
- there was a delay in the introduction of effective measures for the regulation of external borrowings of the national economic agent market participants;
- decisions on the ways of effective spending of surplus fiscal revenues over the planned ones were not made in time, so a significant part of funds in the single treasury account was frozen, which led to current demonetization of the economy during the formation of risky inflationary "money overhang".

The growing political crisis was another active negative factor that contributed to destabilization of the economic situation. While the manifestations of the crisis required prompt coordinated professional decisions of the branches of government, the political crisis reduced the efficiency of the latter, increased the pessimism of market participants in assessing the prospects of crisis phenomena development, disorganized their actions in the framework of antirecession policy (turnaround programme), increased inflation expectations, led to further decline in investment attractiveness of the economy.

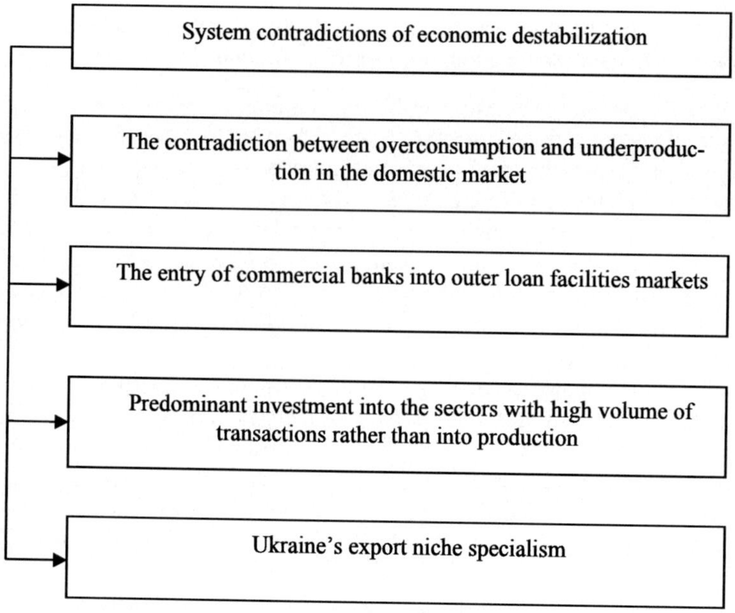

Fig. 1 System contradictions of economic destabilization in Ukraine

Thus, it can be said that intensification of the internal political crisis in Ukraine caused marked aggravation of structural maladjustments in the national economy, exposed to the light of the day the incapacity of the economic growth model, which had been established in the previous years [5].

That is, the antirecession policy of any state must be implemented at three levels:

a) maintaining the competitiveness of the key economic realms;
b) prevention and avoidance of crises in economically and socially significant areas;
c) reduction of the negative outcomes of business failures, the fullest satisfaction of the interests of their employees and creditors [6].

Sustainable economic development, economic growth should be the main goal of anti-cyclical measures. In view of this, anti-cyclical measures should be assumed taking into account the country's economic development strategies. Based on this, according to our reckoning, approaches to anti-cyclical measures for the economy should be focused not only on overcoming the effects of the crisis, but, above all, on ensuring transformational changes in the economy and on overcoming structural maladjustments.

In international practice, the development of construction infrastructure projects is considered as an important stimulator of economic activity. In particular, new construction infrastructure projects provide for the modernization of transport infrastructure, power industry, including alternative sources of energy.

Thus, in the United States, The American Recovery and Reinvestment Act of 2009 (Recovery Act) was signed in order to overcome the financial crisis of 2008 and to provide two-year funding of construction, maintenance and modernization of schools and hospitals, environmental projects, health information technologies, construction and reconstruction of roads and bridges, projects aimed at landscape gardening of America, investment in rail transit, energy efficiency of public buildings at the federal, state and local levels.

Two bailout and stimulus packages were adopted in Germany. They provide additional financing for community infrastructure projects, performance improvement programs, innovative and investment programs in the transport sector.

The government of the Republic of China has provided priority support for such industries as construction, transport (control augmentation, service safety, staff training and preparation of information systems), energetics (performance improvement of power stations and nuclear power plants, technologies for remote electricity supply of objects), fuel energy industry, and infrastructure development of the areas affected by earthquakes.

The material and technical basis for overcoming all types of crises and passing cycles is overcoming structural maladjustments in the economy; as well as academic and technological improvement of production, its transition to a new stage of progress, so these are the determinative factors for both overcoming the crisis and reaching the forefront of economic and social development.

The nature of the crisis determines not so much the support of industries and enterprises with obsolete technologies, as the advanced search, justification of upcoming sectors and industries that meet the requirements and needs of innovative development. The actual practice of many countries confirms that structural changes are achieved through the priority development of industries that use high technologies [7].

In the current crisis resolution, along with scientific and technological factors, economic and financial resources are of great importance. Among them, undoubtedly, one of the leading roles is played by the market (Fig. 2).

All this proves that crisis resolution is associated primarily with systematic interaction of public and private regulation and management of economical equilibrium.

The article uses the concept of "equilibrium" and it is interpreted as a momentary characteristic of individual systems and is considered to be the main concept in determining the directions of the state's economic advance as a whole.

The correctly chosen ratios of the components of the potential will allow to determine, plan and forecast the main macroeconomic indicators.

Equilibrium of the economy as a system is such a state when the output of products is formed in accordance with the supply-and-demand equilibrium, which operates at the macrolevel.

The basic principles of the equilibrium of economics include:

- demand is balanced by supply;
- interrelation of commodity prices and factor prices (or input prices);

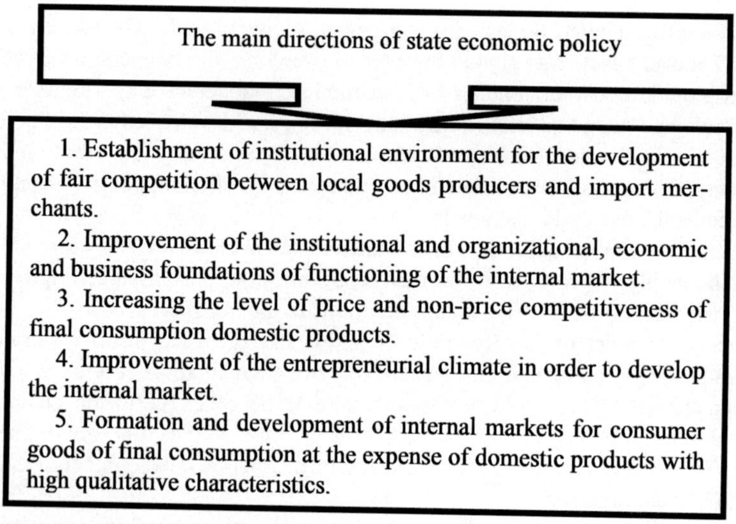

Fig. 2 The main directions of state economic policy

- the existence of equilibrium prices, when demand is balanced by supply;
- consideration of equilibrium as a basis for studying economic growth and building the system of intersectoral linkages.

Let us consider the author's methodological approach to determining the economical equilibrium in construction. We have chosen this very area, because economic recovery in the construction sector of the economy implies surmounting the crisis.

Construction is the economic activity cost line that plays a significant role in creating conditions for local development and for the dynamic development of the nation's economy. The enterprises of Ukraine's construction sector provide other branches of the national economy with new capital funds and carry out construction and installation works, render capital improvements, carry out repairs, technical reequipment and maintenance of facilities in operation.

The dramatic economic environment in today's Ukraine has had a negative impact not only on the development of the country's construction industry, but also on certain industries.

The study [8] analyzed the state of the Hungarian construction industry, noted its leading role in ensuring the economic development of the state. There have been identified the causes of occurrence of crisis phenomena in this area, in particular, insufficient provision with experienced labor forces and fluctuations in prices for construction materials. The analysis of trends in the construction industry, which suffered the most during the crisis of 2008, showed that in order to improve the situation it is advisable to implement large-scale innovation and investment infrastructural construction projects in terms of public private partnership (public-private collaboration).

The research work [9], which indicates the impact factors influencing the level of construction activity and the spheres of interaction between the state and the market, is representative of the abovementioned theses on the importance of the development of the construction industry for surmounting the crisis.

The potential for development of the construction industry and influence of construction development potential on resource efficiency were also analyzed in the research paper [10, 11], which showed that all the regions of China have certain specific nature of capacity building in this area and identified their peculiarities and differences.

The growth of the construction industry will cause the country's economic advance as a whole. Therefore, it is very important to assess the state of the construction industry and development trends in Ukraine (Table 3).

The operating analysis of the indicators of construction enterprises in Ukraine points to the growth of volume of construction products sold over the researched period. The growth of key performance indicators in the construction sphere implies the beginning of overcoming the economy crisis. Thus, cost effectiveness of operative businesses in the construction field in 2019 amounted to 4.7%, which is 1.7% less than in 2018. In 2010, this criterion was 1.5% and it indicates unprofitableness of the industry, which had been growing until 2015 (Fig. 3).

Table 3 The analysis of construction development indicators in Ukraine for 2010–2019

Indicators	Years				
	2010	2015	2017	2018	2019
Construction output of operative businesses in construction fields, %	−1,5	−7,6	1,6	3	4,7
Financial result of construction, UAH million	4418,2	−25,074,1	−3535,8	6433,9	16,106,5
The ratio of construction companies that suffer damages, %	37,5	23,9	24,8	22,8	22,8
The construction output sold, UAH million	101,991,1	150,540,5	236,497,2	308,805,9	362,772,2
The amount of man-power employed in construction, K people	524	282,5	293,7	310,4	331,3
The ratio of employees in construction companies, %	90,1	87,8	87,8	87,1	87,0
The ratio of employees in private entities, %	9,9	12,2	12,2	12,9	13
Structure of construction output sold according to types of enterprises, %: - enterprises; - private entities	100 94,8 5,8	100 94,9 5,1	100 93,6 6,4	100 93,7 6,3	100 93,3 6,7
Indices of construction output, % against the previous year	94,6	87,5	126,4	108,6	123,6

The source compiled by the authors according to the State Statistics Committee of Ukraine.

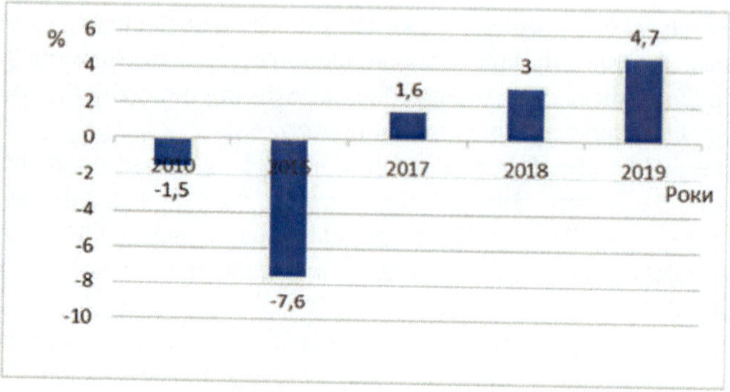

Fig. 3 Dynamics of cost effectiveness of construction operative businesses in Ukraine for 2010–2019

It should be noted that in 2018–2019, only 22.8% of construction companies made losses, and in 2010 37.5% of construction companies were unprofitable. It is also important to note that the least profitable year for construction companies was 2010, although, in general, the cost effectiveness (profitability) of the industry was 4.7%.

It is noteworthy that the share of the volume of products sold, belonging to enterprises and private entities (respective individuals), which work in construction, has been increasing slightly.

The largest number of employed workers is at the enterprises, their share is about 87%. The ratio of employees in private entities (respective individuals) for the study period increased from 9.9% in 2010 to 13% in 2019.

Construction companies are characterized by fluctuations in the index of construction output from 94.6% in 2010 to 123.6% in 2019 (Fig. 4).

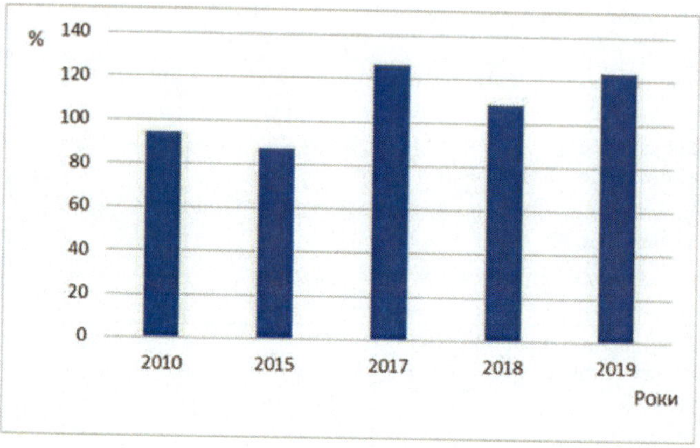

Fig. 4 Indices of construction output in Ukraine for 2010–2019

Table 4 The models of macroeconomical equilibrium

Construction output according to types	2010	2015	2016	2017	2018	2019
1. Buildings	45,6	50,3	51,7	50	47,3	46
1.1 residential buildings	16,1	24,2	243,4	22,5	20,8	18,3
1.2 non-residential (nondomestic) buildings	29,5	26,1	27,3	27,5	26,5	27,7
2. Engineering constructions	54,4	49,7	48,3	50	52,7	54
Total	100	100	100	100	100	100

It should be noted that for the studied period of 2010–2019, construction of engineering structures has the largest share in the structure of construction output—about 54%. The decrease in the share of engineering structures is observed in 2015–2016, which is due to the slowdown in economic development of Ukraine.

In order to predict the country's economic crisis and to determine the optimal structure of the country's economy, it is necessary to improve the method of determining the equilibrium in any sector of the national economy.

Let us specificate the structure of construction output according to types (Table 4).

According to the above-mentioned principal features and parameters, we will model the equilibrium position of Ukraine's construction, taking into account the equilibrium supply and demand for the sector of the national economy.

Let us illustrate the function by an example. Based on the data of the State Statistics Committee of Ukraine, we define a strictly determined descriptive model of the equilibrium state of the economy, using the following Keynesian demand function:

$$C = a + b * Y, \tag{1}$$

where C is the innovative demand, which in our case is equated to the nominal capital (basic stock) in construction;

a—autonomous consumption, we take this value equal to the investment potential (investment in construction);

Y—supply equal to Gross Domestic Product (GDP), which provides a type of economic activity, including construction;

b—is the equilibrium coefficient that characterizes the relationship between equilibrium production and demand.

Application of the descriptive approach in modeling is explained by the need to empirically identify various dependencies in the economy, to find out statistical regularities of economic behavior of indices, to study the probable evolution of economic processes. Examples of descriptive models are production functions (resource-production ratios), in particular the Cobb-Douglas function and the buying interest function, that was used in this study.

When plotting the curve of resource-production ratio of the country's economy, Gross Domestic Product (GDP) according to types of economic activity is often taken for the value of annual output of products. Certain types of potential in value terms are considered as resources. The above mentioned function is as follows:

$$P = a_0 P_1^{a1} P_2^{a2}, \tag{2}$$

P_1, P_2—certain types of potential (ultimate potential resources, capital resources, innovation potential, investment potential);
a_0—coefficient of production function (resource-production ratio);
a_1, a_2—power indicators, $a_1 + a_2 = 1$.

The demand function in its dialectical nature is quite close to the production function (resource-production ratio). The properties of production functions (resource-production ratios) include the following characteristics:

- there is no production without input requirements;
- the output of products leaps with the growth of cost outcome of at least one resource.

The demand functions of Fischer, Modigliani, Friedman, Keynes can be used in scientific and practical work. Common to these functions is the existence of such values as the size of public production, which is expressed through the production function (resource-production ratio).

The demand function has the same properties as the production function (resource-production ratio). Therefore, they are used simultaneously, in combination.

The authors have identified the methodology and have developed the methodology for unlocking the existing potential, which provides economical equilibrium, with the aim of ensuring the country's sustainable development. The algorithm for achieving the determined goal is shown in Fig. 5.

According to the developed method, the equilibrium coefficient (b) is determined by the following formula:

$$b = (c - a)/Y, \tag{3}$$

c—demand (nominal capital);
a—finance and investment potential;
Y—the volume of gross domestic product according to the types of economic activity.

But this function requires certain limitations, without which it is impossible to say that it will help to characterize the equilibrium of the economic system.

Firstly, all changes introduced into the function must be measured in actual prices, bearing in mind that the equilibrium coefficient is a relative number. That is, the impact of inflation factors will be neutralized. Thus, we can compare the obtained indicators, analyze and predict their value for the future.

Secondly, we assume that the national economy is resilient and self-regulatory, including the elements of state planning.

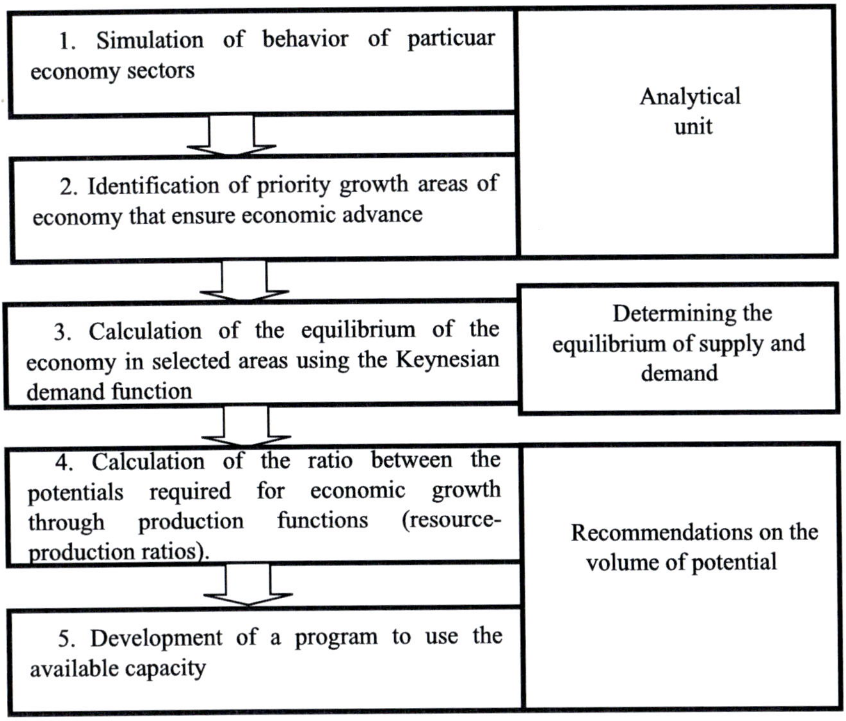

Fig. 5 Algorithm for determining economical equilibrium for increasing the use of existing potential for development

Thirdly, when determining the economical equilibrium of the construction industry, it is necessary to take into account that construction itself is a crucial sector in the structural and investment restructuring of the national and regional economy. The authors of this work note that the result of modeling the economy of the state as a whole was identification of the highest priority areas for its further development, which will improve macroeconomic indicators, ease and overcome crises.

Our task is to find the equilibrium coefficient "b". Therefore, using the methodological approaches mentioned above, we will calculate the equilibrium coefficients of construction in Ukraine according to the data of State Statistics Committee of Ukraine (Table 5).

During the period under study, there is an increase in financial and investment potential and the result of functioning—gross domestic product. According to the calculation of the equilibrium coefficient, it is necessary to state the fluctuations of this indicator.

The dynamic study of the equilibrium coefficient value has been undertaken using a graph that characterizes the quantitative and qualitative equilibrium position of the permanent buildings and facilities construction industry (Fig. 6).

Table 5 Modeling the equilibrium position of the construction industry in Ukraine for 2012–2019

Indices	2012	2013	2014	2015	2016	2017	2018	2019
GDP, UAH million	1,459,096	1,522,657	1,586,915	1,988,544	2,385,367	2,983,882	3,560,596	3,974,564
GDP for foreign business operations (construction), UAH mln	184,751	174,158	162,551	188,595	240,327	326,496	428,010	437,202
Investments, UAH million	293,692	267,728	219,420	273,116	359,216	448,462	578,726	623,979
Investments in construction, UAH mln	44,895	46,318	36,057	43,464	44,444	52,176	55,994	62,347
Fixed assets, UAH million	11,410,035,6	10,401,324	13,752,117	7,641,357	8,177,408	7,733,905	9,610,000	9,574,186
Fixed assets in construction, UAH million	80,107	82,646	64,352	62,090	72,810	78,704	91,715	117,057
Equilibrium coefficient for construction (b)	0,1906	0,2086	0,1741	0,0988	0,1180	0,0813	0,0835	0,1251
Average value (b_{aver})								0,1350
Deviation (b_i-b_{aver})	0,0556	0,0736	0,0391	−0,0362	−0,0170	−0,0537	−0,0515	−0,0098

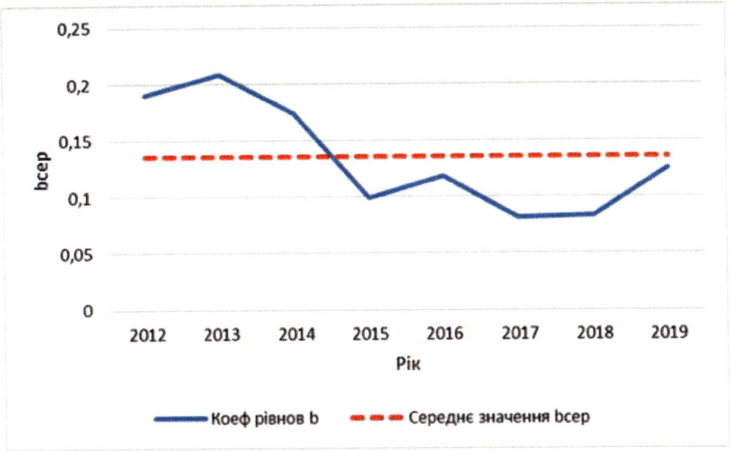

Fig. 6 Dynamics of the equilibrium coefficient of construction in Ukraine for 2010–2019

Dynamics is characterized by business cycle fluctuations, which can be explained by the general economic situation in the country. Thus, in particular, in 2012–2013 we observed an increase in the equilibrium coefficient, which can be explained by some stability in the capital construction market. The 2013–2014 cycle is characterized by a decrease in the equilibrium coefficient, which is the consequence of troubled political circumstances and volatile economic environment in Ukraine, although this value is still above average. However, a rather negative trend was observed in 2015, then there was a gradual increase in 2016 and another decrease in 2017–2018.

2019 is a year of gradual capacity building of the country's development potential. But this ratio, judging by the cyclical nature observed in retrospective, may reduce again, even without reaching the average level. Therefore, it is necessary to develop a plan of anti-recessionary measures for the development of this industry. In particular, the draft of the provision of the law "On Prevention of the Impact of the Crisis on the Development of the Construction Industry and Residential Construction", aimed at supporting the construction industry, infrastructure projects, stimulation of residential construction under recessional conditions and at enforcement of the rights of the citizens, who need the state support.

3 Conclusion

The developed method of determining the economical equilibrium should be used to predict the optimal structure of the region's economy and the national economy as a whole and the formation of priorities for sustainable development.

Further analysis should be made, as mentioned above, using production functions, the calculation methodology of which is expanded in the study [11–15]. This would enable identifying the optimal amount of financial and investment potential needed to ensure immediate-action equilibrium and gradual economic growth.

References

1. Mocherny SV et al (eds) (2000) Economic encyclopedia in three volumes, vol. 1. Publishing House "ACADEMIA", 864 p
2. Bazilevich VD, Bazilevich KS, Balastrik LO (2007) Macroeconomics. Bazilevich VD (eds). Znannia, 703 p
3. Mochernyi SV (1995) Economic dictionary-handbook. Femina, 368 p
4. Chukhno A (2010) Modern financial and economic crisis: nature, ways and methods of overcoming it, part 1. Econ Ukraine 1:4–18
5. Zhalilo YA Economic crisis in Ukraine: measurements, risks, prospects. http://old.niss.gov.ua/book/vidannya/kpyza2009.htm
6. Kolinets L (2010) Anti-cyclical measures for the economy under the conditions of world economic crises: orientation on realization of the economic development strategy. In: Economic analysis. Collection of Scientific Works of the Department of Economic Analysis of Ternopil National Economic University, vol 5, pp 276–279
7. Chukhno A (2010) Modern financial and economic crisis: nature, ways and methods of overcoming it, part 2. Econ Ukraine 2:4–13
8. Kozma G, Marincsák M, Kulcsár B (2015) The transformation of the Hungarian construction industry and spatial tendencies between 2000 and 2012. Int Rev Appl Sci Eng 6(1):77–83. https://doi.org/10.1556/1848.2015.6.1.11
9. Alkay E, Watkins C, Keskin B (2018) Explaining spatial variation in housing construction activity in Turkey. Int J Strateg Prop Manag 22(2):119–130. https://doi.org/10.3846/ijspm.2018.443
10. Liu B, Chen Y, Wang R, Shen Y, Shen Q (2016) Different interaction mechanisms of market structure in the construction industry TFP from the spatial perspective: a case study in China. KSCE J Civ Eng 20(1):23–33. https://doi.org/10.1007/s12205-015-0444-7
11. Komelina OV, Vasiuta VV, Miniailenko IV (2018) Spatial development of construction: modeling and its financial and investment support. Int J Eng Technol 7(3.2):183 – 190. http://www.springer.com/lncs. Accessed 21 Nov 2016
12. Drobyazko S, Bondarevska O, Klymenko D, Pletenetska S, Pylypenko O (2019) Model for forming of optimal credit portfolio of commercial bank. J Manage Inf Decis Sci 22(4):501–506
13. Onyshchenko V, Sivitska S (2014) Alternative energy developing investment support in terms of energy dependence. Econ Ann XXI 9–10:34–37
14. Onyshchenko, V., Sivitska, S., Cherviak A (2020) Buildings reconstruction within the new educational space project in Ukraine. https://doi.org/10.1007/978-3-030-42939-3_66
15. Onyshchenko V, Sivitska S, Cherviak A (2018) Construction industry in Ukraine credit analysis. Int J Eng Technol (UAE) 7(3):280–284. https://doi.org/10.14419/ijet.v7i3.2.14420

Economic and Mathematical Modeling of Innovative Development of Enterprises in the Construction Industry

Volodymyr Onyshchenko, Natalia Ichanska, Vitaliia Skryl, and Oksana Furmanchuk

Abstract The article is devoted to the issue of creating an economic and mathematical model of innovative development of enterprises in the construction industry. The analysis of economic and mathematical models carried out in the article allowed to reveal the general patterns of application of these techniques in the construction industry, to allocate their advantages and disadvantages. Analysis of the state of the construction industry showed its inertia. However, the conservative construction industry is undergoing radical changes. It was found that the innovative development of the construction industry is not only a specific modern technology or new building material, but also a fairly broad engineering organizational and legal implementation. Analysis of modern innovative developments in the construction industry has identified the construction of energy-efficient houses. The paper considers in detail two economic and mathematical models of innovative development of enterprises in the construction industry. It has been established that functional models reflect informational, physical and temporal construction processes. The use of a functional model will help to invest wisely and take into account the interests of the customer, investor, as well as the creditor. The balance model makes it possible to achieve balanced conditions for innovative development of enterprises in the construction industry.

Keywords Innovations · Innovative development · Investments · Economic and mathematical models · Construction industry

1 Basic Concepts of Redistribution of Forces in Axial Elements

Modeling, being a powerful tool for analyzing problematic situations in the economics and substantiation of management decisions, traditionally attracts the

V. Onyshchenko · N. Ichanska · V. Skryl (✉) · O. Furmanchuk
National University «Yuri Kondratyuk Poltava Polytechnic», Pershotravnevyj Avenue 24, Poltava 36011, Ukraine

V. Onyshchenko et al. (eds.), *Proceedings of the 3rd International Conference on Building Innovations*, Lecture Notes in Civil Engineering 181,
https://doi.org/10.1007/978-3-030-85043-2_65

attention of the leading expert scientists and practitioners. The analysis of the application of economic and mathematical models showed that their application is increasingly penetrating the multilevel decision-making processes and they cover almost all areas of economics, management and social life. Nowadays there is considerable experience of creating economic and mathematical models of innovative development of the enterprises in the construction industry in the international practice. Thus, the works [1–3] note that Research and Development (R&D) is a substantial activity in the modern economy of knowledge and therefore funding R&D activities is a challenging task for private and public institutions. The difficulty in evaluation of R&D projects is mainly the inherent uncertainty that has to be dealt with. In addition, when we have funding programs, various policy constraints for the allocation of funds must be also taken into account. The authors [4–6] analyze existing mathematical models, and in their turn propose new economic and mathematical models. The proposed mathematical model for calculation of socio-economic benefits of the built heritage is spatial technique enabling to capture the synergetic effects of co-existing heritage objects using sum of scaled kernel functions that can be used in cost-benefit analysis in different contexts. Also, experts [7–11] note that it is innovative approaches in any economic sector that can contribute to its development. Precisely because of implementation of inovative start-ups it is possible to create and shape new industries and generate considerable economic and societal impacts. Accordingly, a variety of policy initiatives are aimed at promoting the establishment, growth and impact of innovative start-ups. Designing such policies is a challenging task. The use of innovative technologies in the construction industry is considered in works [12–21] and it is believed that in the process of designing and solving problems related to engineering, the ability to find innovative solutions is becoming more and more relevant. Following this principle, engineers should acquire transversal competences through such courses as Project Evaluation and Risk Management, in which they learn to solve current problems through innovation.

The project evaluation supposes an approach to the real world, since an idea is evaluated in the different phases of the logical cycle of a project, culminating with its techno-economic analysis and investment. Following the analogy, engineers develop and evaluate their ideas and present them to a group of professors and professionals posing as investors.

The results present a holistic approach to the evaluation of ideas and the capacity of a team to develop engineering projects. The study performs an assessment methodology for an open-innovation concept, and the investor's exercise may contribute to practical strategies to foster creative engineers.

This actualized the need to develop an economic and mathematical model of innovative development of enterprises in the construction industry.

2 Effective Statically Indeterminate Reinforced-Concrete Beams and Their Design Baselines

Nowadays models of economic processes and phenomena are very diverse. They are closely correlated with the functions of cognition of phenomenological, explanatory, prognostic and normative science—and do not always require the use of a strict formalized apparatus. At the initial stages of research, descriptive, conceptual, logical-structural economic-mathematical models are usually formed, which can be described at a qualitative level, in terms of the analyzed subject area. And this is not their disadvantage. Qualitative economic and mathematical models allow to conduct research using a specific language of the subject area and its meaningful interpretations. Built on the basis of factual data, meaningful economic and mathematical models determine the logic of a process or phenomenon and provide an opportunity to conduct further analytical study based on certain hypotheses about their structural, static and dynamic characteristics.

An undoubted priority of recent times should be the orientation of the economics to innovative development. The application of economic and mathematical modeling in various areas of innovative economy is schematically presented in Fig. 1.

Usually economic and mathematical modeling of the construction industry can be used on three levels.

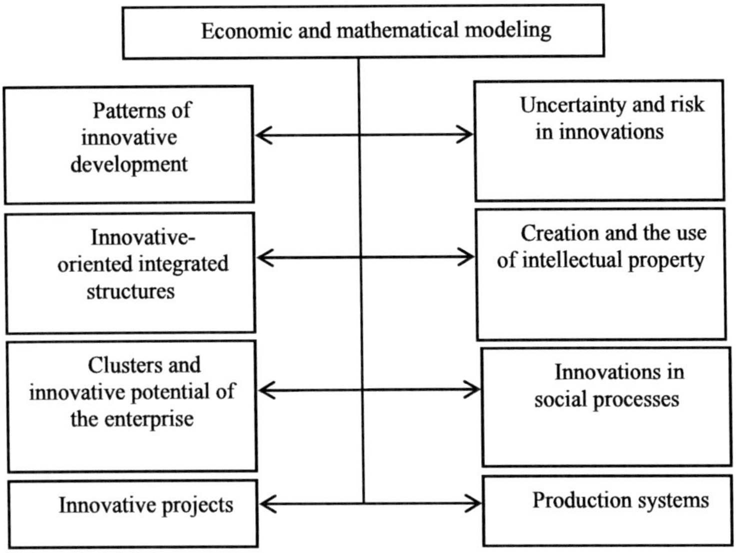

Fig. 1 Application areas of of economic and mathematical modeling in the innovative development of the construction industry

The first level is reduced to the formalization of certain economic concepts, ie quantitative assessment of phenomena and processes that have not previously been subjected to economic descriptions. In particular, for the construction industry, we can present the following dependence of all components of the economic-mathematical model:

$$CI = \sum_{1}^{n} bl. + \sum_{1}^{n} en.st., \tag{1}$$

where CI- construction industry;
$bl.$—volume of buildings;
$en.st.$—volume of engineering structures.

Due to the economic and mathematical structures of the first level it becomes possible to answer the question—how much?

The second level of economic-mathematical modeling makes it possible to reveal the existing connections and dependencies in the processes that take place, and helps to answer the question—what are the relationships? Usually economic and mathematical models of the second level are functional (i.e. have a linear relationship $y = C + kX$) and correlation (characterizes the types of communication that do not have a functional relationship $y = f(x_1, x_2, \ldots, x_n)$.

The third level of economic and mathematical modeling is associated with deductive (imaginary) construction of the model, which reflects the essence of the phenomena and processes that occur in a particular system under study, and allows you to answer the question—what is the reason?

It is worth noting that the three levels of economic and mathematical modeling correspond to a significant number of mathematical methods that can be used in the practice of innovative development in the construction industry. In general, the models that can be used to study the innovative development in the construction industry can be divided into two groups. The first model includes econometric models (functional, algorithmic, simulation, theoretical, empirical, probabilistic), the second model—the balance ones.

To apply the economic-mathematical model of the construction industry, it is necessary to determine its condition. Thus, a critical problem of the construction industry is its conservatism and sluggishness in the introduction and distribution of new technologies. This is intrinsic not only of Ukrainian enterprises, but also to the world construction companies. The construction industry is one of the last in the ranking of innovation and investment active industries of the leading economic states. In the special literature construction has been labeled as "laggard industry" for a long time. The main argument that confirms the fairness of this characteristic is that many studies refer to the extremely low share of the R&D component (research and development) in the overall cost structure of construction companies. For example, according to the US Department of Energy in the US, construction companies invest 0.3–0.4% of total sales in innovative development, while on average in other industries only 3–4% are deducted for this purpose. Similar ratings exist in most construction companies in Western Europe.

This characteristic doesn't fully reflect the real innovative picture in the construction industry. Thus, it is necessary to make a significant correction to the fact that a significant part of new technological developments implemented in the construction industry, comes there from other industries—metallurgy, forestry and woodworking, chemical industries and so on.

However, even with this necessary adjustment, the general assessment of the construction industry, as one that is not particularly innovative, seems quite objective. The inertia of the construction industry is determined by several factors. First of all, it is a long-term operation of buildings and constructions, during which the deficiencies of the applied technology may appear. Therefore, the construction industry is extremely careful in choosing new materials, technologies or construction methods. The second reason of conservatism is a great builders' responsibility for the result, because the use of inappropriate technology or design errors can pose an immediate danger to the lives of a considerable number of people.

However, due to a number of factors, the situation has changed significantly over the last decade. The conservative construction industry is likely to be forced to abandon its well-established traditions and make a series of radical changes. For example, the rapid introduction of building information modeling (BIM) techniques in all key stages of the construction cycle and other advanced IT technologies has already significantly changed the face of the industry.

The lack of innovation in the construction industry is due to a lower level of globalization than in industry. For this reason, innovation in the construction industry in all developed countries is also carried out mainly by large enterprises, construction holdings, and networks.

Also at the centre of attention of the innovative development of the construction industry is the issue of the possibility of reducing costs. During developing an investment project in the construction industry from the conception to the operational stage, three main stages of implementation must be taken into account: pre-investment, investment and operational. Each of these steps, in turn, are divided into stages, the implementation of which must take into account activities such as counselling, design and production. In the pre-investment stage, several activities are carried out that will partly influence the next, investment phase. Once the investment opportunity studies have been completed and the viability of the project has been determined, the investment and planning and execution phases are followed.

Under these conditions, more attention is paid to the final evaluation of the business project and the investment period as the most expensive. In order to reduce the cost of limited resources, it is necessary to understand the sequence of actions in developing a business plan—from the concept to operation, it is necessary to understand the meaning and functional roles of key investment project participants: an investor, a customer, a general contractor, a project designer, trading banks, finance institutions, equipment suppliers, insurance and engineering companies, consulting firms. The consultancy and project work carried out by the main participants in the investment project should take place at all stages of the investment cycle. The success or failure of an investment project depends on financial and economic conditions, marketing research, technical characteristics and their interpretation, especially in the development of a feasibility study. Expenses should not be a barrier to appropriate pre-investment appraisal and evaluation of a project, as this considers a significant contingencies.

All investment-related activities begin with the identification of investment opportunities and their outcome is the mobilization of investment funds. Direct investors, both public and private, are particularly interested in learning about emerging investment opportunities in the market. In order to have such information, it is necessary to study the external environment, to carry out marketing research of this sector of the economy. This approach to identifying investment potential sometimes includes the collection of information, as well as the analysis of resources and comparisons of the studied parameters. They analyse such aspects as environmental, consumer demand, import substitution, interaction with other industries, general investment climate, export opportunities.

Taking into account the pace of construction in Ukraine, it is necessary to create an economic and mathematical model of investment activity in the construction industry, which will help to make investments correctly with minimum risks for all parties. In considering the problem of modelling investment in the construction industry, three main actors can be identified: the investor, the lender and the developer. Each of these participants has a common goal: construction of a new residential facility, taking into account certain requirements, such as construction time, costs, energy efficiency, comfort and ergonomics, reliability and durability, safety and economy, etc. The main goal of the investor is to acquire real estate through the provided investments, and the sooner the investment is made, the more profitable the value of 1 sq. m., which price may more than double over the life of the project. Rising inflation makes it more profitable to invest surplus funds through a lending facility than to accumulate them. The lender invests in the construction of the dwelling, receiving the maximum return on each invested UAH. It is in the interest of the developer to invest consistently in the construction of the facility, and the investor's main objective is to recover the invested funds through the acquisition of new flats, while the lender wishes to maximize the economic benefits.

Timely investment in the construction of the accommodation facility is required to meet the required deadlines and reduce overhead costs. Delays in the construction schedule have resulted in additional financial costs as well as a significant increase in overhead costs. Thus, it is necessary to attract additional financing, i.e. to increase the cost of 1 sq. m. of constructing accommodation. Since for the investor the price remains unchanged from the moment of the conclusion of the contract, all additional costs must be reimbursed by the developer. It should be considered that the construction of the facility is carried out with the help of credit, so that the size of the fund can be increased. That's why its size should be set at the beginning of construction. The developer is also encouraged to increase interest rates on the use of credit owing to the extended construction period. This situation, which may entail financial costs for the three participants in the investment cycle, up to the time when the construction works are stopped for an indefinite period, for the developer—bankruptcy, for the investor—loss of investment with existing obligations to the creditor [22].

Let's analyse the main existing models of investment in the construction industry. The construction industry, namely its residential sector, can be divided into two functional parts:

- housing reproduction (improvement of the quality of the housing stock through overhaul and upgrading the conditions of the dwellings to normal operational conditions);
- introduction of innovative housing.

Innovative construction is the introduction of innovations, ensures qualitative increase of efficiency of construction and maintenance processes. Such construction is in a great demand by the market, i.e., it has practical application. Innovation is not all novations or novelty, but the one that significantly increases the efficiency of the system. Innovative building technology or material must meet one or more of the criteria:

- simplify and accelerate the construction process;
- reduce construction costs or operating costs;
- increase energy efficiency of the facility;
- increase the life cycle of a building/structure.

Based on this definition, it can be concluded that the efficiency of innovation depends directly on its profitability and relevance to developers. Building power-efficient houses is a good example.

In the era of cheap gas heating, few thought about building a well-insulated house. Simply put, the developer could afford to heat not only the house, but the street as well. Now, due to the constant increase of price on energy products and the high cost of connecting to a pipe with «blue fuel», the situation has changed radically. More and more developers are thinking about building well-insulated houses where heat losses are minimized. Due to the fact that the heat that flows out through the blow through walls, the cracks in the windows or the doors pour money down the drain. Hence the growing interest of private developers in recuperation, insulation, energy efficient foundations of type ISW (insulated Swedish wall), additional insulation of walls and roofs. That is, to the end product or to innovative technology.

But the housing sector as a whole requires investment resources from the above sources, with financing from private investors playing an important role. The main goal is to apply mechanisms of interest to private investors, in which individual and societal needs for improved housing are jointly agreed upon. On the basis of the economic interests of owner investors and tenant investors who operate in the new housing market, investors use the right to rent at favourable rates or to take ownership of a part of the living space after completion of the full renovation-reconstruction works. A favourable condition for the economic interest of investors is the ratio of production costs to the economic benefits obtained, taking into account alternative options. From the standpoint of the tenant investor, the alternative is to rent the house at full cost, but without repair. In the case of an investor-owner who has certain objectives (a portfolio investor whose main objective is the efficient placement of his or her own funds and assets), the alternative is to invest in banks at interest rates and to earn average income. In the case of an investor-owner whose purpose is to obtain housing (a strategic investor), the alternative is to purchase similar space on the housing market. In any case, the returns of these investors should be lower than the alternative returns in these areas.

Another important aspect to be taken into account is the relationship between the investor and the city authorities, which is based on the compensation principle and represents the public interest of the citizens in the construction industry. In such a situation, the analysis of these mechanisms makes it possible to describe the behaviour of investors in the home reproduction market by means of optimal models represented in different productions: linear, non-linear, parametric and probabilistic. Thus, using these mechanisms, depending on the required initial version for setting the problem, the economic and mathematical model should take into account such factors as borrowed funds and bank loans for the implementation of an innovative project, costs arising from the temporary resettlement of residents of the reconstructed building, the probability of price risk associated with the sale of the renovated space in the housing market, etc.

Functional model of a private investor who invests its assets into construction industry, takes into account the compensation costs in the relationship between the investor and the city authorities and will be based on the following aspects:

- the market value of housing sales (due to the presence of realtors in the market), which is usually inflated;
- market value of 1 sq. m. of housing which is higher than the standard cost of construction;
- the actual cost of 1 sq. m of living space, which decreases with the increasing number of the building floors [23].

It is important for the investor to change the main economic values: the size of their own investment, the cost of land for new construction, the number of floors of buildings, the area of housing built, which is transferred under the contract to the municipal housing stock.

We use the following notation:x—the number of floors in the house; $m_2 y$—the size of the area of housing transferred to the municipal fund; S_1—the size of the area occupied by the land for new construction; S_2—the size of the area of land allocated for new development, $S_2 > S_1$; Z—the price of 1 sq. m. of land; M—the cost of construction of 1 sq. m. of housing of this type; m_1—purchase price of 1 sq. m. of housing of this type; m_2—the sale price of 1 sq. m. of housing of this type; k k—cash investments. We take for unknown quantities the number of floors of the object and the area of the housing stock. Thus, the main inequalities of the economic-mathematical model will take the following form:

$$(\mathrm{m}_1 - M s_1 x)x - m_1 y \geq z s_2, \tag{2}$$

$$Ms_1x \leq k, \tag{3}$$

$$Ms_1x + zs_2 \leq (s_1x - y)m_1, \tag{4}$$

$$\max\{(m_2 - M)(s_1x - y)\}, \quad x = 1, 2 \ldots - whole \tag{5}$$

$$y \geq \overline{y} > 0, \tag{6}$$

$$\max\{(m_2 - M)(s_1x - y)\}, \tag{7}$$

The restriction on cash financial resources comes out of the formula (3); inequality (4) reflects the cost (the expenditures of construction and purchase of land should not exceed the cost of purchasing the same area that remains with the investor after the transfer of housing to the municipal housing stock); condition (5) is the integer of the variable; inequality (6) shows the lower limit of the size of the transmitted area; (7)—a criterion for maximizing the profit that the investor receives from the sale of built housing [21].

In canonical form, the optimal integer problem of determining x and y has the form:

$$Ms_1x \leq k, \tag{8}$$

$$(\mathrm{m}_1 - Ms_1x)x - m_1y \geq zs_2, \tag{9}$$

$$x = 1, 2 \ldots - whole \tag{10}$$

$$y \geq \overline{y} > 0, \tag{11}$$

$$\mathrm{F} = (m_2 - M)(s_1x - y), \tag{12}$$

Problems (8)–(12) can be solved as a Due to its relative simplicity, it is easy to find that the optimal solution is achieved at the point (where $(k/(Ms_1); \overline{y})$, where $\mathrm{F}_{\text{опт}} = (m_2 - M)(k/(M - \overline{y}))\rightarrow$ max [22]. The existence of a non-empty set of admissible solutions is determined by the condition:

$$\overline{y} \leq \left(\frac{k}{M}\right)\left(1 - \frac{M}{m_1}\right) - \frac{zs_2}{m_1} = A, \tag{13}$$

In addition, the optimal solution exists when:

$$\overline{y} \le \left(\frac{k}{M}\right) - \varepsilon/(m_2 - M) = B, \tag{14}$$

for any small $\varepsilon > 0$.

From comparison (13), (14) we obtain a sufficient condition for the existence of an optimal solution:

$$\overline{y} \le min\{A, B\}, \tag{15}$$

Functional model (8)–(12) can take an extended form, taking into account the demand and supply of housing, the risks associated with the discrepancy of the studied factors [24–26].

The balance model of innovative development of enterprises in the construction industry should weigh the balance of factors taken into account by investors, due to the scenarios of enterprise development. Among them are the following:

1. General economic situation, prospects of construction and financial market for the investor.
2. Exposition of the project. Features of the project location and the investor's decision on how to implement it.
3. Consumer characteristics of housing construction facilities under construction. The conditional type of housing that is being built is determined by dividing it into categories: economy class housing of mass demand, comfort class housing of improved quality, and housing of the highest class of maximum value.
4. Planned technical and economic indicators.
5. Participation of private investors (households and legal entities) in financing housing built on the basis of equity construction and other forms of collective investment.

Thus, the strategy of choice by the investor-developer will be understood as a set of project indicators, options for its implementation and financing.

$Xqj \in \{0;1\}$, where $q = \overline{1, G}$; $j \in j_g$ (Boolean variable which takes the following values: «1»—if the project Πqj, included in the previous list, is accepted by the investor for implementation, «0»—in case of refusal);

y_{ij}—the number (pcs.) i—constructed residential buildings of type j ($j \in j_g$), intended for sale at prices c^T_{gj} on the open market;

$\beta^c_{gj} > 0$, the share of the project financed from the private partner (%);

Ω—the amount of the attracted loan (UAH);

p^c—value of return on equity (%);

p^3—the value of return on borrowed capital (%).

Thus, the following set of parameters will describe the behavior of the investor-builder $\{x_{qj}, y_{ij}, \beta^c_{qj}, \left(q = \overline{1, G}, j \in J_q, i = \overline{1, n}\right), \Omega, p^c, p^3\}$.

The condition for achieving the balance of the above factors will be achieved through the strategy $e = \{x_{qj}, x_{ij}, \beta^c_{qj}, \mu, p^c, p^3\}$, described by the following equations:

$$b_{ij}^{m} + y_{ij} + z_{ij} = b_{ij}^{0}, \left(i = \overline{1,n};\ j \in J_q;\ j : x_{qj} = 1\right), \tag{16}$$

$$z_{ij} \le b_{ij}^{u}; \left(i = \overline{1,n};\ j \in j_q\right), \tag{17}$$

where b_{ij}^{m}—the number of i—constructed residential buildings of type j ($j \in J_q$);

b_{ij}^{0}—the total number (pcs.) of residential objects *i*—type, in the construction object type *j*;

z_{ij}—the number of constructed housing *i* in the object of type j, intended for sale to private investors in the housing market;

b_{ij}^{u}—number (pcs.) *i*—of constructed residential objects of type j ($j \in J_q$), for which it is planned to conclude agreements on participation in share construction or other financing agreements.

Balance Eq. (16) indicates the coincidence of the total number of constructed housing facilities (b_{ij}^{0}), which were planned before the implementation of the project Πqj (in this case $x_{qj} = 1$) with the fund of residential real estate transferred to the municipality (b_{ij}^{m}), which is implemented in the housing market under construction (y_{ij}) and transferred to private investors participating in share construction (z_{ij}).

Inequality (17) indicates the maximum amount of housing that can be transferred to the ownership of private investors, taking into account the burdens of the investor-developer, arising at the stage of commissioning.

The study of economic and mathematical modeling of innovative development of enterprises in the construction industry allowed us to draw the following conclusion.

3 Conclusion

Innovative development of enterprises is the construction of new, state-of-the-art housing, using the latest technologies, which is what the investor wants to finance. Demand for housing, its price and quality, which is usually measured by the level of energy efficiency, is a decisive factor for investors today. The construction of innovative, energy-efficient housing is currently the most present in the structure of the population's needs. At the same time, the use of economic and mathematical models for the innovative development of construction companies will form the most profitable model of activity for the investor. Two models were considered in the work. Functional mathematical model reflects the information, physical and temporal processes in construction, which are performed during technological processes. Balance model is designed to form a balance of the main factors influencing the development of enterprises in the construction industry. The obtained models provide an opportunity to analyze and select such parameters that will allow construction companies to adjust their activities for effective operation.

References

1. Kankhva V, Silka D (2016) Formation of the mechanism of facilitating and enhancing of economic security of investment projects in construction of underground facilities. Procedia Eng 165:1337–1342. https://doi.org/10.1016/j.proeng.2016.11.861
2. Mayadunne S, Park S (2016) An economic model to evaluate information security investment of risk-taking small and medium enterprises. Int J Prod Econ 182:519–530. https://doi.org/10.1016/j.ijpe.2016.09.018
3. Mavrotas G, Makryvelios E (2020) Combining multiple criteria analysis, mathematical programming and Monte Carlo simulation to tackle uncertainty in research and development project portfolio selection: a case study from Greece. Eur J Oper Res 291:794–806. https://doi.org/10.1016/j.ejor.2020.09.051
4. Rudokas K, Landauskas M, Gražulevičiūtė-Vilneiškė I, Viliūnienė O (2019) Valuing the socio-economic benefits of built heritage: local context and mathematical modeling. J Cult Herit 39:229–237. https://doi.org/10.1016/j.culher.2019.02.016
5. Cioccolanti L, Tascioni R, Arteconi A (2018) Mathematical modelling of operation modes and performance evaluation of an innovative small-scale concentrated solar organic Rankine cycle plant. Appl Energy 221:464–476. https://doi.org/10.1016/j.apenergy.2018.03.189
6. Pignotti M, Suárez-Taboada M, Vázquez C (2019) A new mathematical model for pricing a mine extraction project. Nonlinear Anal Real World Appl 50:8–24. https://doi.org/10.1016/j.nonrwa.2019.04.007
7. Audretsch D, Colombelli A, Grilli L, Minola T, Rasmussen E (2020) Innovative start-ups and policy initiatives. Res Policy 49(10):104027. https://doi.org/10.1016/j.respol.2020.104027
8. Hagspiel V, Kort PM, Nunes C, Pimentel R, Støre K (2021) Capacity optimization of an innovating firm. Int J Prod Econ 233:108021. https://doi.org/10.1016/j.ijpe.2020.108021
9. Kostrzewski M, Kosacka-Olejnik M, Werner-Lewandowska K (2019) Assessment of innovativeness level for chosen solutions related to Logistics 4.0. Procedia Manuf 38:621–628. https://doi.org/10.1016/j.promfg.2020.01.080
10. Onyshchenko S, Yehorycheva S, Furmanchuk O, Maslii O (2020) Ukraine construction complex innovation-oriented development management. https://doi.org/10.1007/978-3-030-42939-3_68
11. Liana P, Maya C (2018) Crowd funding as a tool for financing the construction of social facilities. Int J Eng Technol 7(3.2):329–333
12. Zizka L, McGunagle DM, Clark PJ (2021) Sustainability in science, technology, engineering and mathematics (STEM) programs: authentic engagement through a community-based approach. J Cleaner Prod 279:123715. https://doi.org/10.1016/j.jclepro.2020.123715
13. Kozachenko H, Onyshchenko S, Masliy O (2018) Region building complex social and economic security threats. Int J Eng Technol 7(3.2):79–85. https://doi.org/10.14419/ijet.v7i3.2.14405
14. Žižlavský O (2014) Net present value approach: method for economic assessment of innovation projects. Procedia Soc Behav Sci 156:506–512. https://doi.org/10.1016/j.sbspro.2014.11.230
15. Buonomano A, De Luca G, Montanaro U, Palombo A (2016) Innovative technologies for NZEBs: an energy and economic analysis tool and a case study of a non-residential building for the Mediterranean climate. Energy Build 121:318–343. https://doi.org/10.1016/j.enbuild.2015.08.037
16. Bernardo Llamas M, Storch D, de Gracia LF, Mazadiego JP, Alonso J (2019) Assessing transversal competences as decisive for project management. Thinking Skills Creativity 31:125–137. https://doi.org/10.1016/j.tsc.2018.11.009
17. Arranz B, Ruiz-Valero L, González MP, Sánchez SV (2020) Comprehensive experimental assessment of an industrialized modular innovative active glazing and heat recovery system. Energy 212:118748. https://doi.org/10.1016/j.energy.2020.118748
18. Bertsch M, Ceseri M, Felici G, Natalini R, Santoro M, Sgalambro A, Visconti F (2014) Mathematical desk for Italian industry: an applied and industrial mathematics project. Procedia Soc Behav Sci 108:79–95. https://doi.org/10.1016/j.sbspro.2013.12.822

19. Souza MT, Ferreira IM, Guzi E, de Moraes L, Senff AP, de Oliveira N (2020) 3D printed concrete for large-scale buildings: an overview of rheology, printing parameters, chemical admixtures, reinforcements, and economic and environmental prospects. J Build Eng 32:101833. https://doi.org/10.1016/j.jobe.2020.101833
20. Nima SS, Heidarpour A, Xiao-Ling Z, Al-Mahaidi R (2018) A component-based model for innovative prefabricated beam-to-hybrid tubular column connections. Thin-Walled Struct 132:265–275. https://doi.org/10.1016/j.tws.2018.08.021
21. Cornaro C, Zanella V, Robazza P, Belloni E, Buratti C (2020) An innovative straw bale wall package for sustainable buildings: experimental characterization, energy and environmental performance assessment. Energy Build 208:109636. https://doi.org/10.1016/j.enbuild.2019.109636
22. Golikov AP (2009) Economic and mathematical modeling of world economic processes, 3rd edn. (re-edn.) And add. - K .: Knowledge, 222 p
23. Vlasyuk OS (2011) Economic and mathematical modeling of processes of socio-economic development of Ukraine. K .: DNNU "Acad. Finn. management", 520 p
24. Michels VO, Shilyuk PS, Goiko AF, Bondar VP (2006) Economic and mathematical methods and models in construction. Millennium, 380 p
25. Chichulina K, Skryl V (2020) Energy service companies are an effective tool for financing energy-efficient projects. Int J Manag 11(5):252–266. https://doi.org/10.34218/IJM.11.5.2020.025
26. Chevhanova V, Chichulina K, Skryl V (2018) Effective cluster model of thermal power company management. Int J Eng Technol (UAE) 7(3):65–70. https://doi.org/10.14419/ijet.v7i3.2.14377

Risks and Threats to Economic Security of Enterprises in the Construction Industry Under Pandemic Conditions

Svitlana Onyshchenko, Alina Hlushko, Oleksandra Maslii, and Vitaliia Skryl

Abstract The paper actualises the problem of identification of risks and threats to economic security of enterprises in the construction industry under pandemic conditions. It proves that ensuring timely identification of potential or real threats to economic security of enterprises in the construction industry is essential to develop a set of preventing measures for preventing or minimising the causes of their manifestation. A statistical analysis indicates the impact of COVID-19 pandemic and the quarantine measures upon the development of national economics in general and construction industry in particular. The paper describes the negative tendencies in the Ukrainian construction industry market. It proves that an efficient tool for preventing and effective overcoming the negative consequences of pandemic is timely identification of risks and threats to enterprise security in the construction industry. It suggests an improvement to multi-criteria classification of risks and threats to enterprises in the construction industry, which enables detailing the preconditions of threats and their sources, and rather more systematic approach to their predicting, forecasting their negative impact, identifying the quality structure of risks as well as taking into consideration the possibility of system risks development. The authors developed a structure model that enables identifying the predominant factors and preconditions of threats to appear, which allows removing the threats in an early stage of their development, predicting the probability of threats to appear, and taking preventive measures for their timely removing.

Keywords Pandemic · Risks · Threats · Economic security · Construction industry

S. Onyshchenko · A. Hlushko (✉) · O. Maslii · V. Skryl
National University «Yuri Kondratyuk Poltava Polytechnic», Pershotravnevyj Ave 24, Poltava 36011, Ukraine

V. Onyshchenko et al. (eds.), *Proceedings of the 3rd International Conference on Building Innovations*, Lecture Notes in Civil Engineering 181,
https://doi.org/10.1007/978-3-030-85043-2_66

1 Pandemic as a Real Threat to Economic Security of Enterprises in the Construction Industry

Scale and impact of the COVID-19 pandemic have radically changed the world economy as well as the paradigms of social and economical development of each country. The viral danger became a challenge not only to the healthcare systems but to the very ability of state authorities and local governments, businesses to face and withstand such serious challenges and threats.

Construction companies are an integral part of the national economy, an effective indicator of its growth or decline, and the object of external and internal risks and threats. The COVID-19 pandemic has demonstrated the imperfection of crisis management. The business entities proved to be unprepared for an effective fight against the large-scale pandemic [1], which created a number of diverse threats to the economic security of enterprises in the construction industry.

Timely assessment of threats, provoked by the pandemic, aims at changing the model of responding to challenges to the economic security of enterprises. At the same time, ensuring timely identification of potential or real threats to economic security of enterprises in the construction industry is essential to develop a set of preventing measures for preventing or minimising the causes of their manifestation [2]. In this context, the problem of identification of risks and threats to the economic security of construction industry under the pandemic conditions is undeniably relevant.

A considerably large number of studies, carried out by both Ukrainian and overseas scholars, focus on the problems of ensuring the resilience of economic entities, including the construction industry, against the negative impact of a broad range of threats, within the context of ensuring the socio-economic security of the state [3–9, 10, 11]. However the crisis phenomena in the economic system of Ukraine demonstrate that Ukrainian enterprises have appeared unprepared against external threats as well as there is a necessity of identifying the threats which have been caused by the pandemic in order to prevent or efficiently overcome its negative consequences.

During 2016–2019 the level of economic security of Ukraine, even under the conditions of new challenges the country had faced, was characterised by restrained dynamics. It was confirmed by a slight increase in the key macroeconomic indicators. In 2016 the real GDP increased by 2.4%, in 2017 its grows rate consisted 2.5%, in 2018 it increased by 3.4%. In 2019 the real GDP decreased by 0.2 percentage points compared to the previous year and amounted to 3.2% [12]. The growth of the level of economic independence and stability of the country's economic system is also confirmed by the improvement of Ukraine's position in international rankings. For instance, in the international ranking Doing Business-2020 Ukraine took 64th place among 190 countries (as in the ranking of the previous year it held 71st place) [13]. It was the best result in recent years (see Fig. 1).

In recent years, the economic system of Ukraine has been exposed to many risks and threats. Their negative impact reflected in the unstable foreign economic situation, unfavourable investment climate, low growth of aggregate demand in the

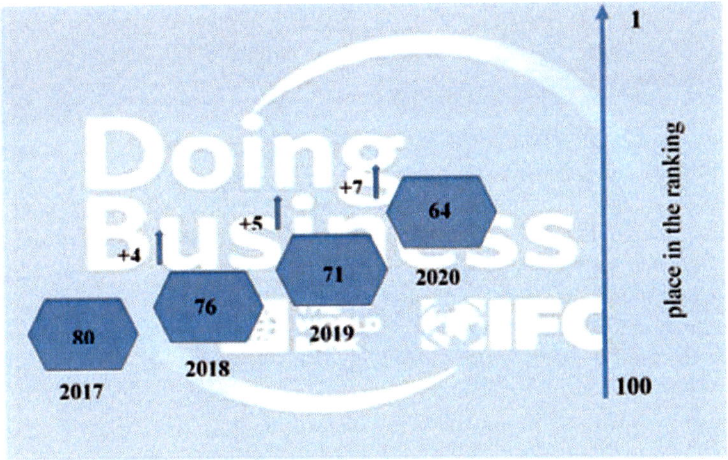

Fig. 1 The place dynamics of Ukraine in the international ranking Doing Business

domestic market, negative trade balance [14]. The COVID-19 pandemic has become the biggest real threat of recent decades to the economic security of Ukraine as well as of the entire world community. The sharp decrease in economic activity as a result of quarantine measures led to a slowdown in economic development with a predictable decrease in GDP, a reduction in consumer activity and production in general, including negative tendencies in the construction industry.

According to the data of European Construction Industry Federation [15], the impact COVID-19 caused the decrease of activity in the construction industry of most FIEC member-countries. The largest decrease was in Ireland (35%), Spain (23%), Italy (20%), France (15%), Slovakia (13.3%), Turkey (10.6%) and Belgium (10.5%). Meanwhile, some countries have managed to keep the pace in the construction industry or even increase its production volumes. For instance, Greece increased by 5%, Denmark by 2.7%, Germany by 1%. According to *European Economic Forecast: Autumn 2020 (Institutional Paper 136 | November 2020)*, issued by European Commission, investments in the construction industry in the Euro area decreased by 6.7% in 2020 compared to 2019. The temporary shutdown of construction sites and a significant reduction in granting building permits severely affected the sector in the first half of 2020. The uncertainty forced the companies and enterprises to cut their costs and revise their plans, delaying or even cancelling the construction projects.

In 2021 and 2022 the European Commission forecasts an increase in investment by 5.2 and 2.9%, respectively.

The FIEC forecast varies across member countries. In some countries they forecast deepening the crisis: from 1% fall in Austria in 2020 compared to 2019 to −4.5% in 2021/2020, from −5.7 to −10% in the Czech Republic, respectively. In some countries the fall is expected to get slowing down: from −35 to −20% in Ireland, from −8 to −0.7% in Luxembourg, from −13.3 to −10% in Slovakia. There are countries that hope to reverse the negative trend and increase the pace of construction.

Belgium, having experienced the fall of −10.5% in 2020/2019 aims to reach +9.2% in 2021/2020, Italy—having −20% is going to achieve +3%, Turkey strives to rise from −10.6 to +6.5%.

In Ukraine, due to the COVID-19 pandemic and quarantine restrictions from March to May 2020, 10% of construction companies had stopped their work during the quarantine and still did not resumed it, 23% of companies paused their activities for some time, 56% of companies worked without interruption. 75% of construction companies announced the postponement of new projects, more than 30% of companies announced staff reductions. The number of transactions fell by more than 50 in 43% of respondents; by a quarter in 15; 10% reported that they had only a few transactions; and 3% said they had no transaction at all. Only 5% of construction companies could boast that their sales remained the same as in the previous year; yet the same number of companies got the state investments. 36% of construction companies plan to reach pre-quarantine profits within 9–12 months, 28% plan doing that in the period up to six months. 20% companies are planning to do this within two to three months as soon as the quarantine measures will be over [16].

State Statistics Service of Ukraine official data confirm the negative tendencies in the construction industry [12]. Thus, in the period from January to August 2020, comparing it with the same period in 2019, the index of construction products did not reach the pre-crisis level, having amounted only up to 97.2% (see Fig. 2).

Under threat are the construction companies that operate at the expense of real estate buyers and finance the next stages of construction at their expense. Decline in sales can stop construction, which may cause a situation that will painfully affect the investors who bought real estate at the initial stage of construction. According to experts, 15–20% of new buildings across the country are at risk. In the nearest future one should expect a chain of construction shutdowns as well as a number of bankruptcies of construction companies. Developers are massively launching promotions and offering discounts up to 15%, but it does not help, as in comparison with the previous year demand has fallen by 70–80% [16].

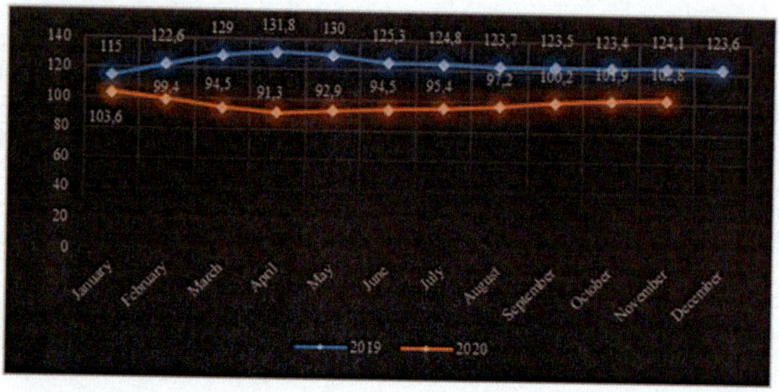

Fig. 2 Indices of construction products (in % to the corresponding period of the previous year, cumulative total)

Considering everything said above, the necessity of identifying risks and threats caused by the COVID-19 pandemic to enterprise security in the construction industry in order to prevent or efficiently overcome its negative consequences is undeniable.

2 Typology of Risks and Threats to Enterprise Economic Security in the Construction Industry

Security of economic entities plays a key role in strengthening the state security, which is an essential precondition for the stable economical growth and development of any state. There is a high probability of economic risks in the business sector. In the construction industry, for instance, there are risks due to the conflict of interests between business entities and both the external and internal environmental factors. Conflicts of interest, accordingly, are often a threat to the activities and development of the primary economic actors.

The specifics and number of threats to the functioning of both individual entities in the construction industry and the industry itself as a whole is determined by the variety of entity types, forms of their organization and legislation, areas and kinds of their activities. Providing a clear typology, therefore, can contribute to understanding of the essential characteristics of risks as well as facilitate the taking of management decisions that aim at their localization.

First of all, threats to the economic security in the construction industry should be considered, dividing them according to the areas in which they may occur into: (counter) productive, industrial, political, economical, financial, innovative, and legislative.

Productive (technical and technological) threats are the risks associated with the difficulties for the construction companies in accessing the economic resources, means of production, and technologies; insufficiency or critical deterioration of material and technical resources; breaking or lost of production ties; similar equipment or technology introduced by rivalling companies; errors in technical or technological policy [17, 19].

Industrial risks are rather external to the system of enterprise economic security. They are characterized by the following threats: a sharp increase in competition; "aging" and "decline" of the industry (associated with the relevant stage of its life cycle), which are accompanied by declining demand and market capacity; reducing in the attractiveness of investment, etc.

Political risks are to some extent the subject to public administration, but they often cause significant damage to individual enterprises in the construction industry or the entire industry as a whole [18]. They are mostly related to the excessive politicization of the actions of government officials, which diverts their attention from the performance of their functions and duties (or may provoke inaction during the political crisis). All this may cause delays in problem solving (including the problems related to the management of business economic security), postponing in taking the necessary decisions that aim at developing a full-fledged institutional framework for the economy and industry development, slowing down the procedures for granting permits, etc. All this may considerably reduce the business activity of the people.

Economic and financial risks are, as a rule, provoked by unfavourable market conditions, which make the construction companies to digress from planned financial and economic results; inefficient financial planning and poor asset management; ill-considered marketing policy; underestimation of financial risks; force majeure or (even) criminal acts.

It should be noted that another important aspect is the study of financial risks in the construction industry under conditions of economic and financial crises [20]. This conclusion can be made taking into account two following regularities: 1) the interdependence of risk and financial performance of the business entity; 2) different effect and strength with which the risks impact different enterprises.

It is clear that under conditions of economic crises, the business risk—in the construction industry in particular—increases significantly. As a rule it happens due to rising costs, getting rather more complicated access of economic entities to financial and credit resources, deteriorating financial and economic conditions of counterparties and consumers of goods (services), reducing economic activity, introducing additional costs and losses of business entities due to changes in fiscal or credit policies, particular components of administrative methods of state regulation.

Legal risks to the enterprise economic security in the construction industry include defects in the legislation or the system of its implementation, errors that come from inadequate qualification of lawyers or attorneys in the enterprise legal service or poor planning; ineffective defence of business interests in case of conflict situations.

There are also certain particular risks for business entities by type of their economic activities [21]. Thus, among the specific risks for enterprises in the construction industry there can be mentioned the decrease in effective demand for residential and non-residential real estate, worsening of the economic efficiency and growth of receivables/payables, critical increase in project payback period, loss of markets, inflation and fiscal pressure, cuts in investment, raiding and crime in the construction sector, bureaucratization and corruption in obtaining building permits, etc.

Considering all we have said above, the classification of threats to enterprise economic security in the construction industry can be represented as follows (see Fig. 3).

It should also be noted that there is a fundamental difference between traditional risks and systemic threats to the enterprise economic security in the construction industry. Traditional risks can also be called natural. They are constant companions of any business activities of any business entity. Traditional risks do not come from any subjective actions or measures, and, therefore, do not affect the critical mass of enterprises in a negative way [7]. The general (traditional) classification of enterprise economic security risks in the construction industry may focus on following features: the way the object is impacted or affected; the scale of (probable) losses from the threat; threat target; the area of activity of the object, etc.

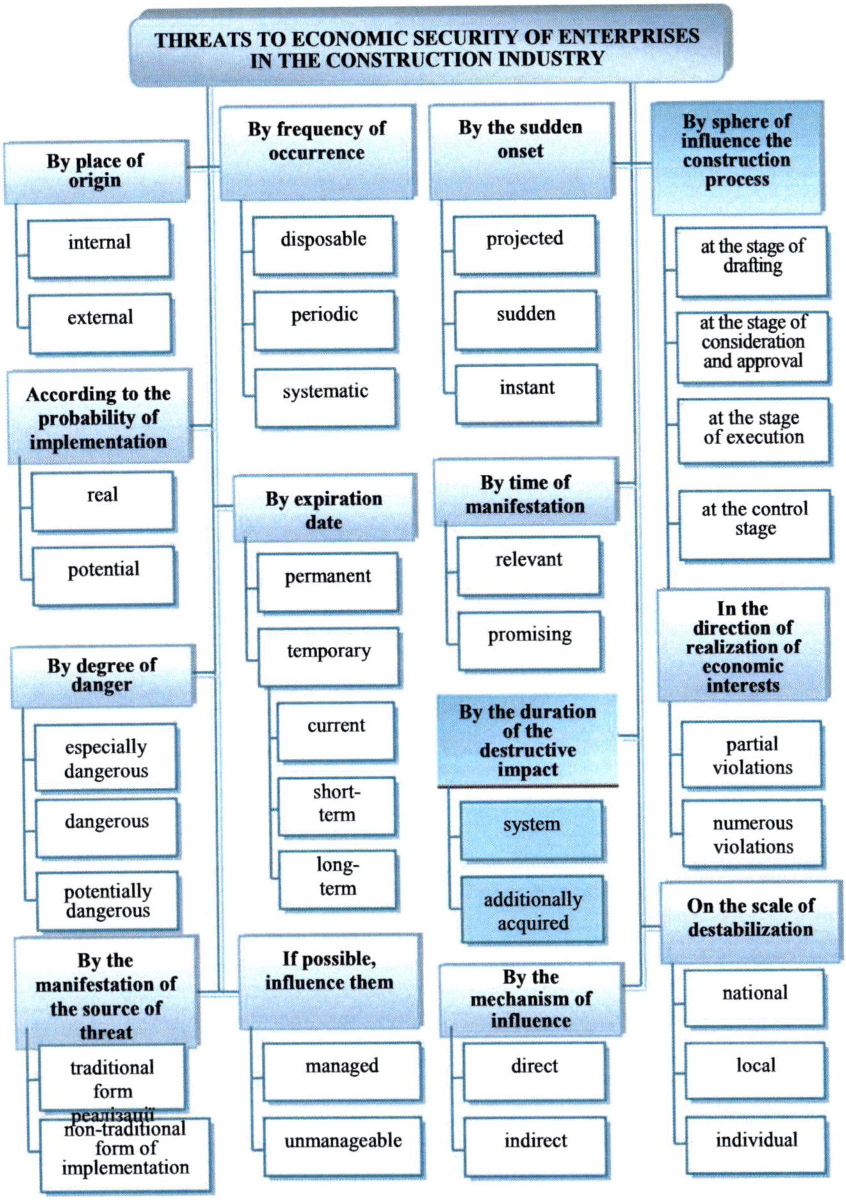

Fig. 3 Multi-criteria classification of the threats to enterprise economic security in the construction industry Source: created by the authors

Systemic threats in the construction industry are caused by illegal actions of certain subjects or even government officials that create an unfavorable economic or legal environment. Their consequences are rather more global [22]. They hinder the activities of almost all business entities and, therefore, disable not only the development of business sector but the very functioning of it.

In addition to the threats to economic security mentioned in Fig. 3, the systemic threats to enterprise business security may be identified by groups as follows:

1) market (it includes the threats related to the shadowing of economic relations and markets, their monopolization, criminalization of business and public relations, problems with the purchase and sale of business, property and resources, the unfavorable investment and innovation environment);
2) state and political (due to the presence of cooperation between the state and business representatives, affiliated structures, raiding, non-transparent access to state property, resources and markets, instability of the "game rules");
3) resource (related to the insecurity of private property rights, high transaction costs, excessive tax pressure) [23–26].

It should be added that the risks to the enterprise economic security in the construction industry may differ not only in their content and origin, but also in frequency. In this regard, it seems advisable to divide them into: disposable, which may occur suddenly, but pose a threat over a long period of time (or constantly) and have a negative impact on vital business characteristics (significant changes in legislation, increased competition between developers, etc.); multiple, which affect the activities of the enterprise or their combination either stochastically or with a certain period of occurrence (temporary severance of relations with the supplier or contractor, etc.) [7].

Risk classification according to their predictability (manageability) reflects the possibility of localizing their consequences. Manageable unfavourable events may be dealt with and improved by public administration bodies or the enterprise managers. Moreover, the means of improvement can be developed in course of the enterprise economic security monitoring and to be expressed in a set of strategic, tactical and operational management actions and institutional transformations.

Unpredictable (uncontrolled) threats are not the subjects to the directly management, correction or improvement. They reflect the threats caused by the development of market relations in the country, and therefore public administration must necessarily take them into account in their activities. Examples of unmanageable threats are: the seizure of market share by foreign companies or multinational corporations, the dynamism of changes in legislation, high levels of corruption, and so on.

It should be noted that in addition to the possibility of forecasting, it may seem advisable to divide the risks of enterprises economic security in the construction industry on such a classification basis as the possibility of their prevention by public administration [27]. Therefore, some of them can be predicted, and the implementation of adequate preventive measures can eliminate the critical negative consequences. Among such risks, most of them are internal and concern the inadequate level of management of primary economic agents professional training, heads of public authorities, negative market and economic changes, which are just emerging or have not reached a level of critical destructive impact on personal security yet, society, enterprise, entrepreneurship, industry and economy in general.

Risks that cannot be prevented include those that are more force majeure. This is, in particular, the effect of natural factors on the construction complex. To minimize their negative effects, business entities should firstly use insurance practices.

The risks of construction companies economic security, which can be partially prevented, include significant systemic transformations that occur (develop) on the basis of objective irrational development (recession) of the economy, sectoral changes (deformations) and other general macroeconomic parameters. It is difficult to change radically the direction of such phenomena development, however, we believe that the list of public administration tasks should include the negative consequences of such risks mitigation [7].

The given above risks typology corresponds to a wide range of the category "construction companies economic security". It reflects a significant number of threats sources to sustainability of the studied system and, thus, requires a comprehensive approach to activities that must be carried out to strengthen it. However, an integral classification of construction companies' safety risks should also include the following: the headline, the origin and nature and the consequent effect.

Thus, in accordance with the nature of systematic consideration of economic security of entrepreneurship category, the areas of risks negative impact are the vital interests of this system subjects.

By origin, of economic security of construction business entities risks should be divided into internal (due to defects in the existing business security system) and external (due to the influence of factors outside this system and not due to the actions of its subjects). However, it should be noted that internal negative manifestations and threats as a result, may arise at micro (level of person, enterprise), mezzo and macro levels of economic management hierarchy and may be caused by imperfect public administration. At the same time, external threats are mostly a manifestation of the shadow market sector, foreign markets and integration actions, where public administration bodies participate [24, 28]. We add that, the difference between internal and external risks is whether they are caused by actions (inaction) of the economic entities economic security or institutions of its external environment.

In accordance with such classification feature as the level and nature of the impact of enterprises economic security factors in construction industry, it is necessary to distinguish: minor risks (that significantly affect financial losses of construction industry enterprise); possible (that affect interests of a significant number of enterprises); critical (the interests of the industry as a whole).

These types of risks that undermine the principles of proper functioning of the construction industry economic security can be considered traditional (classical). At the same time, the problem of strengthening economic entities economic security is exacerbated during periods of economic downturn, slowing the pace of business development, reducing the quality of life of the population, that characterize the construction business in a pandemic functioning. Therefore, the typology of the so-called "systemic" risks to the construction companies economic security is especially relevant.

In particular, such type of risk is classified: according to the consequences of institutional conflict of interests; by forms of illegitimate property formation; by forms of the system of public administration corruption; on obstacles to the shadow economy legalization; by the preconditions of unfair competition; by the negative consequences of economic transformation [7].

In contrast to the traditional security risks of economic entities during socio-economic crises, the effect of objective risks increases, which destroy both economic and psychological preconditions for doing business and competitive business environment characteristics in general. In particular, such risks may be caused by institutional interests conflict (public administration and business representatives; civil servants and public and political associations; business and society, individuals, etc.).

We add that public administration bodies material and legal preferences granting to the subjects of construction business often leads to distortion of a healthy competitive environment formation mechanism. This process often results in the concentration of economic and financial resources in a small number of market participants, which distorts the level playing field.

We note that in transformations of economic systems, there are often prerequisites for business entities risks, including the construction industry. They are associated with the possibility of illegitimate, "opaque" ownership, which in combination with instability, lack of planned and justified (resource provision), step-by-step (with appropriate adjustment) economic policy of the state create preconditions for the management system formation, where market instruments are not effective enough, do not create transparent and equal rules of economic behavior for all business entities [29, 30]. It should be noted that the main directions (forms) of property rights illegitimate formation and, thus, systemic risks to the entrepreneurship functioning and development are as follows: fictitious bankruptcy, lobbying for state resources access, fictive business creation, etc.

Moreover, ownership redistribution through bankruptcy procedure or through the creation of fictitious enterprises results in such negative consequences as dispersal of property, reducing of the level of corporate control over state and public property through resale to "related" structures. Such procedures are often used in the construction business to eliminate competitors, evade taxes or debt obligations.

We studied the economic essence of threats, the processes of their emergence, the nature of manifestation and impact on the economic security of Ukraine. As a result, modelling of the process of their emergence, spread and impact on construction enterprises economic security system was carried out. To quantify the parameters that characterize the process of emergence, spread and impact of threats on the enterprises functioning economic and mathematical modelling tools were used (Formula 1).

$$M_{I3E} = \mathrm{S}_{\mathrm{B3}} \cap \mathrm{f} : \mathrm{N}_{\mathrm{B3}} \Big(\bigcup_{i}\bigcup_{j} a_{ij} \cup \bigcup_{k}\bigcup_{j} a_{0hj}\Big) = \mathrm{S}_{\mathrm{B3}} \Big((\text{ПЕ} \cap \text{ДР} \cap \text{ЗФ}) \leftrightarrow \bigcup_{\mathrm{i}=1}^{9} \mathrm{A}_{\mathrm{i}} \leftrightarrow E\Big) \cap \mathrm{f} : \mathrm{N}_{\mathrm{B3}} \quad (1)$$

where M_{I3E} is a model of the emergence, spread and impact of threats on the economic security of the enterprise;

S_{B3}—areas of threat;
N_{B3}—consequences of the threats impact;
ПЕ—the current financial and economic condition of the enterprise;
ДР—shortcomings in the regulation of economic activity;
ЗФ—external factors affecting the activities of the construction industry;
E—the functioning of the real estate market under pandemic conditions.

Despite the significant number of variables and conditions that must be taken into account in the process of analyzing the emergence, spread and impact of threats, only a small part of them can be considered essential for describing economic security of enterprise. Therefore, in this case, it is advisable to simplify the description of the economic security system on the basis of which the model is built. The key task here is to identify the dominant factors (variables, their parameters and constraints) on the basis of correlation and regression analysis and to study the predictive dynamics of the identified threats.

Thus, the source of threats are the factors and conditions that provoke the emergence of phenomena and processes in the external or internal environment of the economic security system. Their analysis and timely detection should be the basis of economic security mechanism. This approach, in contrast to the situational response, makes it possible to prevent the emergence of threats and to neutralize them at the initial stage.

The proposed structural model of emergence, spread and impact of threats on national economy economic security is based on the following principles:

Segmentation—the identification of threats is carried out by areas of their emergence (S_{B3}) within each component of national economy economic security;
objectivity—the existence of a threat to economic security must be proven by economic and mathematical methods of analysis of statistical indicators;
dynamics—detection and prediction of threats is based on trends that shape the current state of the national economy;
completeness—all factors, phenomena, events and processes that are sources of threats are subject to consideration; based on the possible negative consequences of their impact on the economic security system of the national economy;
significance—only those threats are subject to modelling that objectively hinder the realisation of national economic interests and are catalysts for destructive changes in the system of national economy economic security.

In the absence of a situational response to existing threats at this stage, their accumulation occurs, which leads to structural changes both in the economic security system of the state and in the state economic system as a whole.

Other significant systemic risks of construction industry development under the conditions of transformation processes are: insufficient level of competitive environment formation that leads to creation of unequal conditions of functioning in the market [7]. With a high level of corruption, including corruption in the system of public administration as well, the following systemic risks of construction enterprises economic security are formed: assistance to affiliated structures, criminal lobbying, investment from the state budget, creation of pseudo-enterprises related to government and management officials, etc. Corruption often leads to the criminalization of society and the economy, which in future creates the preconditions for institutionalization of structures that use corruption technologies to replace public administration bodies in the provision of certain services, or in some way affect the qualitative content of relevant state externalities.

Summarizing the study, we can conclude that the conducted classification of risks allows to detail the preconditions and sources of threats, more systematically approach their prediction and forecasting the consequences of adverse effects, especially under condition of COVID-19 pandemic, determining the qualitative structure of risks, and consider the possibility of systemic risks development in construction industry enterprises.

The innovative areas for further development, to which the construction industry should be directed, are: renovation, energy efficiency and environmental standards improvement, as well as digitalization of all stages of project implementation.

3 Conclusion

It is justified that in the context of the COVID-19 pandemic destructive impact on the national security system and its individual components, including the construction industry as an integral part of the national economy of the country, an effective indicator of its growth or decline, the identification of risks and threats to the economic security of enterprises in the construction industry acquires strategic importance. A detailed statistical analysis indicates the impact of the COVID-19 pandemic and quarantine measures on the development of national economy in general and the construction industry in particular.

The suggested improvement to the multi-criteria classification of risks and threats to enterprises in the construction industry, which enables detailing the preconditions of threats and their sources, and rather more systematic approach to their predicting, forecasting their negative impact, identifying the quality structure of risks as well as taking into consideration the possibility of system risks development. Considering the negative trends in the Ukrainian construction market, it is proved that timely identification of risks and threats to the safety of enterprises in the construction industry is an effective tool for preventing and effective overcoming the negative consequences of pandemic. The authors developed a structure model that enables identifying the predominant factors and preconditions of threats to appear, which allows removing the threats in an early stage of their development, predicting the probability of threats in the future, and taking preventive measures for their timely removing. Modeling the process of the threats emergence, spread and impact on the enterprise economic security allowed creating a structural model. The model enables identifying the dominant factors and conditions causing the threats to appear in the areas of economic activity. All this allows either to manage and remove the threats at a very initial stage or to minimize their negative impact on the enterprise economic security system, taking into consideration the defects and shortcomings of economic activity regulation and the influence of external factors.

References

1. Onyshchenko S, Yanko A, Hlushko A, Sivitska S (2020) Conceptual principles of providing the information security of the national economy of Ukraine in the conditions of digitalization. Int J Manage (IJM) 11(12):1709–1726 (2020). https://iaeme.com/MasterAdmin/Journal_uploads/IJM/VOLUME_11_ISSUE_12/IJM_11_12_157.pdf.https://doi.org/10.34218/IJM.11.12.2020.157
2. Onyshchenko S, Hlushko A (2020) A pandemic is a real threat to Ukraine's socio-economic security. Actual trends of modern scientific research. In: Abstracts of the 3rd international scientific and practical conference. MDPC Publishing. Munich, Germany, pp 190–195
3. Rangel A (2019) Why enterprises need to adopt 'need-to-know' security. Comput Fraud Secur 2019(12):9–12. https://doi.org/10.1016/S1361-3723(19)30127-7
4. Allen M (2019) Enterprise security risk management, chap 2. In: The chief security officer's handbook, leading your team into the future, pp 19–33. https://doi.org/10.1016/B978-0-12-818384-7.00002-1
5. Moral-García S, Moral-Rubio S, Fernández EB, Fernández-Medina E (2014) Enterprise security pattern: a model-driven architecture instance. Comput Stand Interfaces 36(4):748–758. https://doi.org/10.1016/j.csi.2013.12.009
6. Varnaliy Z, Onishchenko S, Masliy A (2016) Threat prevention mechanisms of Ukraine's economic security. Econ Ann XXI 159(5–6):20–24. https://doi.org/10.21003/ea.V159-04
7. Onishchenko S, Pugach A (2015) Threats to economic security of Ukraine: essence, evaluation and prevention mechanism. PoltNTU, Poltava, Ukraine
8. Weippl E, Leithner M (2012) Enterprise security for the executive. Comput Secur 31(4), 636. https://doi.org/10.1016/j.cose.2012.03.001
9. Bayuk JL (2009) Enterprise security for the executive: setting the tone from the top
10. Wu X, Zheng W, Chen X, Wang F, Mu D (2020) CVE-assisted large-scale security bug report dataset construction method. J Syst Softw 160:110456. https://doi.org/10.1016/j.jss.2019.110456
11. Svystun L, Glushko A, Shtepenko K (2018) Organizational aspects of investment and construction projects implementation at the real estate market in Ukraine. Int J Eng Technol 7(3.2):447–452 (2018). https://doi.org/10.14419/ijet.v7i3.2.14569
12. Official site of the State Statistics Service of Ukraine. Homepage: http://www.ukrstat.gov.ua/
13. Doing Business 2020. International Bank for Reconstruction and Development. The World Bank Homepage: https://openknowledge.worldbank.org/bitstream/handle/10986/32436/9781464814402.pdf
14. Glushko AD (2013) Directions of efficiency of state regulatory policy in Ukraine. World Appl Sci J Pak Int Digit Organ Sci Inf 27(4):448–453. https://doi.org/10.5829/idosi.wasj.2013.27.04.13656
15. Official site the European Construction Industry Federation. Homepage https://www.fiec.eu/
16. Public organization "Center for Applied Research". Representation of the Konrad Adenauer Foundation in Ukraine. Impact of COVID-19 and quarantine restrictions on the economy of Ukraine. Office study. July, 2020. Homepage https://www.kas.de/documents+2020.pdf/b7398098-a602-524d-7f88-6189058f69d3?version=1.0&t=1597301028775

17. Onyshchenko S, Yehorycheva S, Furmanchuk O, Maslii O (2020) Ukraine construction complex innovation-oriented development management. Proceedings of the 2nd international conference on building innovations, pp 687–700. https://doi.org/10.1007/978-3-030-42939-3_68
18. Ramalingam D, Arun S, Anbazhagan N (2018) A novel approach for optimizing governance, risk management and compliance for enterprise information security using DEMATEL and FoM. Procedia Comput Sci 134:365–370. https://doi.org/10.1016/j.procs.2018.07.197
19. Kim S, Kim S, Lee G (2009) Structure design and test of enterprise security management system with advanced internal security. Futur Gener Comput Syst 25(3):358–363. https://doi.org/10.1016/j.future.2006.04.019
20. Kozachenko H, Onyshchenko S, Masliy O (2018) Region building complex social and economic security threats. Int J Eng Technol. 7(3.2):79–85. https://doi.org/10.14419/ijet.v7i3.2.14405
21. Saleh MS, Alfantookh A (2011) A new comprehensive framework for enterprise information security risk management. Appl Comput Inf 9(2):107–118. https://doi.org/10.1016/j.aci.2011.05.002
22. Chase RE (2015) Is enterprise risk management (ERM) leaving security behind?, chap 10. In: Security leader insights for risk management, lessons and strategies from leading security professionals, pp 41–46. https://doi.org/10.1016/B978-0-12-800840-9.00010-1
23. Pooley J (2017) Information security in the modern enterprise. Computer and information security handbook. Elsevier, pp 3–11. https://doi.org/10.1016/B978-0-12-803843-7.00001-6
24. Booker R (2006) Re-engineering enterprise security. Comput Secur 25(1):13–17. https://doi.org/10.1016/j.cose.2005.12.005
25. Contos BT (2007) Enterprise security management (ESM). Enemy at the water cooler. Elsevier, pp 69–98. https://doi.org/10.1016/B978-159749129-7/50008-4
26. Chen H, Chau M, Li S-H (2011) Enterprise risk and security management: data, text and Web mining. Decis Support Syst 50(4):649–650. https://doi.org/10.1016/j.dss.2010.08.026
27. Akinosho TD, Oyedele LO, Bilal M, Ajayi AO, Delgado MD, Akinade OO, Ahmed AA (2020) Deep learning in the construction industry: a review of present status and future innovations. J Build Eng 32:101827. https://doi.org/10.1016/j.jobe.2020.101827
28. Perera S, Samudaya Nanayakkara MNN, Rodrigo SS, Weinand R (2020) Blockchain technology: is it hype or real in the construction industry? J Ind Inf Integr 17:100125. https://doi.org/10.1016/j.jii.2020.100125
29. Onyshchenko SV, Matkovskyi AV, Puhach AA (2014) Analysis of threats to economic security of Ukraine in conditions of innovative economic development [Analiz zahroz ekonomichnii bezpetsi Ukrainy v umovakh innovatsiinoho rozvytku ekonomiky]. Econ Ann XXI 1–2(2):8–11
30. Iasechko S, Haliantych MK, Skomorovskyi VB, Zadorozhnyi V, Obryvkina O, Pohrebniak O (2020) Contractual relations in the information sphere. Syst Rev Pharm 11(8):301–303. https://doi.org/10.31838/srp.2020.8.46

Increasing Information Protection in the Information Security Management System of the Enterprise

Svitlana Onyshchenko, Alina Yanko, Alina Hlushko, and Svitlana Sivitska

Abstract The article actualizes the problem of management and strengthening of enterprises information security. It is proved that in the conditions of information and technological development it is the information that has become a factor that forms the competitiveness of many business entities. A statistical analysis of the level of cybercrime and financial losses from crimes in cyberspace. It is substantiated that construction of an effective information security management system based on international standards of the ISO series is the key to the successful operation of enterprises. The method of error correction in the system of residual classes is determined by the basis of improving the information security of economic entities.

The method of enterprise data errors correction in the system of residual classes is developed in the article, which is based on the application of corrective properties of L-codes, which are formed when using mutually pairwise difficult bases. This method enables expanding the class of corrected errors that are corrected, which expands the corrective capabilities of L-codes. Examples of performing the correction operation are given, as well as the features of the implementation of the device for detecting data errors are described.

Keywords Computer system for processing enterprise data · Conditional alternative sets · Data protection · Error correction · Information · Information security · Linear codes · System of residual classes

1 Protection of Information in the Information Security Management System

In terms of information technology development, characterized by on the one hand the intensification of modern information technology in all sectors of the national economy, and on the other hand, increasing the scale and frequency of cyber attacks,

S. Onyshchenko · A. Yanko · A. Hlushko (✉) · S. Sivitska
National University «Yuri Kondratyuk Poltava Polytechnic»,
Pershotravnevyj Avenue 24, Poltava 36011, Ukraine

V. Onyshchenko et al. (eds.), *Proceedings of the 3rd International Conference on Building Innovations*, Lecture Notes in Civil Engineering 181,
https://doi.org/10.1007/978-3-030-85043-2_67

the emergence of new risks and security threats to businesses and the state as a whole, the issue of increasing the level of information security becomes particularly relevant.

Noting the role and importance of existing research on information security [1–4], it is obvious that in the context of intensification of informatization and progressive development of the IT sector, one of the most pressing issues is the problem of cyber attacks, which necessitates increased protection of information.

Information security of enterprises characterizes the state of their access to information, its security and storage, efficiency of use, business intelligence, information and analytical work with external and internal entities, the ability of information and analytical system of economic entities to develop.

Analysis of the internal and external environment of enterprises is the first and integral function of management [5]. Therefore, the quality of information and analytical support directly affects the effectiveness of management and further development of the enterprise. But for this, the system of information and analytical support of the process of economic entities economic security managing should include the following elements: information, a set of indicators, indicators, methods of assessment and analysis of information security [6].

Of course, the system of analytical information for management decisions is characterized by complexity. And there is a tendency to complicate the relationships in the information flow. At the same time there is a systematic increase in the amount of information, its redundancy in terms of lack of information to make optimal management decisions. Information on the business entities security is quite heterogeneous [7]. All this complicates its use in the management of security and viability of the entity. In most businesses, information that managers use to ensure their safety comes predominantly from internal sources. A specialized analytical group or security service is created, the functional responsibilities of which include all or part of the information support.

The key issue is the concept of essential information. Information is considered significant, the non-provision or distortion of which may influence the decisions of its users. The main objectives of information security management of economic entities are to ensure timely detection of information loss channels, threats and their level of importance, types of information theft, methods of their actions; ensure prompt response to threats; create conditions for the maximum possible compensation for damage; avoidance of economic and industrial espionage.

Thus, information security of enterprises is designed to protect their interests and their staff from the misuse of inside information and protection of trade secrets, which often constitute a significant part of the intellectual property of the business entity.

However, despite attempts to protect trade secrets and other sources of information, it is often the case that information is disseminated to persons who have no right to do so. Under modern business conditions, this phenomenon is quite common, because information about know-how, technology, methods of conducting business and management of individual business processes forms the competitiveness of many businesses [8]. And in conditions of constant competition, even a drop of the necessary information can dramatically change market conditions for an individual business entity, and for another—to turn into losses and damages. That is why, in Western Europe and the United States, 20% loss of confidential information can lead to bankruptcy [9].

Current dependence of enterprises on information systems and their services means that businesses are becoming increasingly vulnerable to information security threats. Interaction of public and private networks, as well as sharing of information resources increases the difficulties of access control and ensuring guarantees of services and security of information and communication systems and networks.

Intensification of the processes of economic activity digitalization creates preconditions for increase in cases of unauthorized use of computers, telecommunications systems, computer networks and telecommunications networks [10], i.e. cybercrime.

Back in the 1990s, computer fraud in the United States caused more than $10 billion in annual damage [11]. In the UK, where cybercrime had quadrupled at the time, the Confederation of British Industrialists estimated that the annual damage was £5 billion.

Today, financial losses from cybercrime are growing every year. According to a study conducted by experts from the Center for Strategic and International Studies (CSIS) and McAfee, a company that develops anti-virus software, the level of losses from cybercrime in the global economy is growing rapidly: if in 2014 they amounted to 345–445 billion dollars (0,6% of world GDP), then in 2016 it was already 445–600 billion dollars (0,8% of world GDP), in 2017—about 1.5 trillion dollars [12]. Global losses from the hacker attack using the NotPetya virus program alone amounted to $850 million, of which $300 million was the financial loss of the national economy of Ukraine (0,4% of GDP) [13].

Hacker attacks in 2020 cost the world economy more than a trillion dollars or 820 billion euros, which is 50% higher than in 2018 (see Fig. 1).

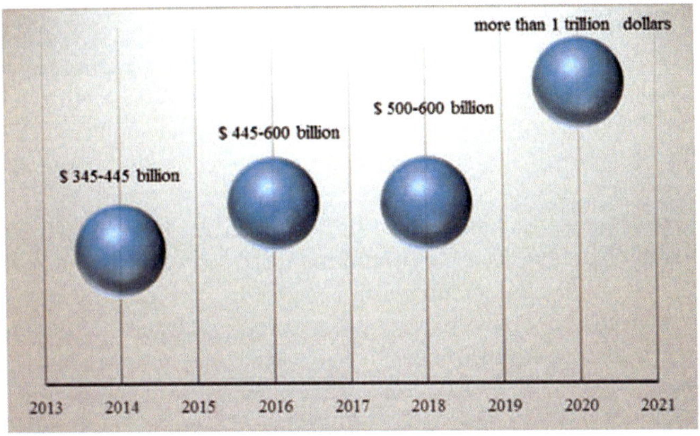

Fig. 1 Estimated average cost of cybercrime

Establishing the real scale of financial losses from the actions of cybercriminals is extremely difficult and almost impossible. However, the lack of official statistics does not diminish the relevance and importance of information security. After all, information is one of the universal types of resources that are necessary both for the decision-making process and for the formulation of strategic, tactical and operational tasks of economic development at the macro-, meso- and micro-levels.

According to domestic and foreign experts, solving the problems of investigating this type of crime is a difficult task for law enforcement agencies both in our country and abroad. Domestic and foreign criminologists classify "computer" crime as hyper latent. According to various estimates, law enforcement officers become aware of only 10–20% of such crimes [14].

A key aspect of protection against cybercrime is preparation and identification of vulnerabilities, as well as resilience in terms of interoperability with common management systems.

Fundamental in the field of information security management are the International Standards ISO/IEC 27,001 and ISO/IEC 27,002 (it was called ISO/IEC 17,799 until 2007). They are a model of the management system, which determines the overall organization of processes, data classification, access systems, planning areas, employee responsibility, the use of risk assessment, etc. in the context of information security. Thus, due to the implementation of the ISO/IEC 27,001 standard, business entities have the opportunity to assess their risks, implement control measures to mitigate them, control risks, improving, if necessary, information protection. The ISO/IEC 27,002 standard is used to establish a system of effective information protection and to improve information protection methods.

To date, there are more than 40 international standards of the ISO/IEC 27,000 series, covering everything from creation of a common dictionary (ISO/IEC 27,000), risk management (ISO/IEC 27,005), security in cloud technologies (ISO/IEC 27,017 and ISO/IEC 27,018) to forensic methods used to analyze digital evidence and investigate incidents ISO/IEC 27,042 and ISO/IEC 27,043 respectively) [15]. They enable businesses to constantly update themselves in the fight against cybercrime.

Building an effective information security management system based on international ISO series standards in modern information and communication systems and networks is an important task for every business entity. To counter external attacks, it is necessary not only to have effective means of protection, but also to know their system of operation, settings and weaknesses of operating systems [16].

Thus, strengthening the information security of enterprises is based on ensuring reliability, confidentiality, integrity of information resources. Modern methods and means of information and communication and digital technologies cannot fully ensure productive and reliable processing of ever-growing arrays of information. The existing positional binary number system, which operates in modern information technology, has shortcomings, and existing methods of unauthorized access, hacker attacks, viruses and other types of hacking and violation of the integrity of information are built using binary position code. In this aspect, correction of errors in the system of residual classes can rightly be determined as the basis for improving the information security of economic entities.

2 Features of Error Correction in the SRC

It is possible to significantly increase the reliability and credibility of processing economic data only based on the use of new machine arithmetic. Because existing positional numeral system (PNS) has a number of disadvantages. Execution of an arithmetic operation involves a sequence of elementary operations on operands (digits) according to the rules defined by its content. These operations cannot be completed until the values of all the results of the operations have been obtained, taking into account all the interconnections between the operands (digits). As a result, modern computer systems that use binary code to represent and process all types of data, i.e. based on PNS, have a significant disadvantage—the presence of inter-bit connections between the operands being processed, which affects the architecture of computer systems and methods of arithmetic operations, complicates the equipment and limits the speed of arithmetic operations.. And most importantly, due to inter-bit connections, one error in the bit leads to a block of errors, and the existing methods of error correction in the PNS are very difficult to implement and strongly affect the redundancy of the machine code [17].

In the system of residual classes (SRC), each number is represented in the form of several small-digit positional numbers, which are remainders from dividing the initial number into mutually simple bases. In the usual positional binary system, operations (for example, the addition of two numbers) were performed sequentially by bit, starting from the youngest. This creates a transfer to the next most significant digit, which determines the bitwise processing sequence. The possibility of paralleling this process appeared in the SRC: all operations on the remainders on each basis are performed separately and independently (in parallel), therefore, due to their low bitness, easily and quickly [18].

Also the SRC has a valuable property of the independence of the residuals from each other on the basis of the adopted system [19]. The independence of the residuals makes it possible to build computer system for processing enterprise data (CSPED) in the form of a set of information—independent, parallel computing paths (separate "small" computing paths of information processing, operating on a specific module m_i in the SRC, independently). Thus, CSPED have a modular design that enables maintenance and troubleshooting without interrupting the solution of the computational problem. Errors that occur due to failures of binary circuits in an arbitrary computational path on the basis m_i do not spread to neighboring paths (remain within one residual), which makes it possible to increase the reliability of calculations in the CSPED. It does not matter whether there was a single or multiple error, or even a package of errors with a length of not more than $m_i - 1$ binary digits. Thus, the error that occurred in an arbitrary path m_i of the CSPED in the SRC, or will remain in this path until the end of the calculations, or in the process of further calculations will be eliminated (for example, if after a failure in the residual a_i the intermediate result is multiplied by a number zero having a zero digit based on m_i). In this case, with the help of SRC you can build a system of error correction with the introduction of minimal redundancy, which uses the dynamics of the computational process, introducing the concept of an alternative set of numbers. The set of bases of SRC $m_{i1}, m_{i2}...m_{ik}$ by which numbers A_1, $A_2...A_k$ differ from the incorrect operand Ã, is called an alternate set of numbers, and is denoted $\overline{\mathrm{W}}\left(\tilde{\mathrm{A}}\right)$. The basic idea of determining the erroneous residual (basis) $a_i = a_i + \Delta\, a_i$ is that for the resulting sequence of incorrect operands $\tilde{\mathrm{A}}_i = (i = \overline{1, p})$ in the dynamics of the computational process, without interrupting the solution of the problem, sequentially in time are determined by conditional alternative sets (CAS):

$$\overline{W}(\tilde{A}) = \overline{W}_{i-1}(\tilde{A}) \wedge \overline{W}_i(\tilde{A}). \tag{1}$$

For some time, CAS is charged to the erroneous basis (or to two bases m_i and m_n). After that, the known methods are used to correct the distorted residual a_i. A feature of this method of error correction is the ability to correct errors without stopping calculations, which is important for CSPED that operate in real time. A detailed study of the features of the SRC allows us to conclude that the devices that operate in the SRC, are easily controlled and easily diagnosable objects. The noted feature of CSPED functioning in SRC promotes development of effective methods of control and diagnostics [20].

This independence offers wide opportunities for constructing not only new machine arithmetic, but also a fundamentally new scheme for the implementation of CSPED, which in turn significantly expands the use of machine arithmetic.

Thus, this property of the SRC makes it possible to implement a unique system of control and correction of errors in the dynamics of the computational process with the introduction of minimal code redundancy without stopping calculations, which is very important for economic structures operating in real time [21].

3 Implementation of L-codes in the SRC

Currently, possibilities of R-codes for error correction in the SRC are being intensively investigated. This is due to the clear and simple structure of R-codes without deep interconnections, and really effective corrective properties, with a relatively simple procedure for their construction for any required minimum code distance.

Also for comprehensiveness it is necessary to consider other types of codes in non-positional systems, namely in the SRC, which are found in the literature under the name—linear codes (L-codes). When considering these L-codes in various scientific sources, it's described not by quantitative characteristics, but by qualitative ones.

Currently, no one is deeply researching the corrective capabilities of linear codes in combination with the properties of SRC. Developments in this direction would reveal broad and effective corrective properties, which necessitates the improvement of SRC-based data processing systems and the use of these systems to increase the reliability of CSPED.

The sum, difference, and product of any vectors of a linear code are code words. In this case, non-code words cannot be associated with any natural numbers. We show that error correction in the SRC with the help of L-codes leads to hardware redundancy equivalent to reservation. To this end, we consider two known theorems [22].

Theorem 1. To correct an error in the residual at an arbitrary base m_i of the number $A = (a_1, \ a_2, \ ..., a_n)$, specified in the system of residual classes with bases $m_1, m_2, \ ..., \ m_n$, it is necessary that:

$$(d_{ik} - 1)(d_{ij} - 1) \geq m_i - 1 - (K_{d_{i\,k}} + K_{d_{ij}} - K_{[d_{i\,k}, d_{ij}]}), \tag{2}$$

where $d_{ik} = (m_i, \ m_k)$, $d_{ij} = (m_i, \ m_j)$, $K_{d_{i\,k}}$ id the number of divisors, multiples of d_{ik};

$K_{d_{i\,j}}$ is the number of divisors, multiples of d_{ij};

$K_{[d_{i\,k}, d_{ij}]}$ is the number of divisors, multiples of the lowest common multiple (LCM) $[d_{i\,k}, d_{ij}]$ of the divisors d_{ik} and d_{ij}, $i \neq j$.

Proof. Calculate the values a_{ij}, a_{ik}, a_{jk}. If the error occurred at the base m_i, then $a_{ik} = 0$, $a_{ij} \neq 0$ and $a_{ik} \neq 0$. The number of different combinations of a_{ij}, a_{ik} is $(d_{ij} - 1) \cdot (d_{ik} - 1)$, where $(d_{ij} - 1)$ is the number of possible values of a_{ij} $(a_{ij} \neq 0)$, $(d_{ik} - 1)$ is the number of possible values of a_{ik} $(a_{ik} = 0)$, and the number of possible values of base errors a_{ik} is $m_i - 1$ $(\Delta a_i \neq 0)$ minus the number of undetected errors [22]. The number of undetected errors consists of the number of errors, multiples of the divisor $d_{ik} - K_{d_{ik}}$ and multiples of the divisor $d_{ik} - K_{d_{ik}}$. Thus, the number of possible values of detectable errors is:

$$m_i - 1 - (K_{d_{i\,k}} + K_{d_{ij}} - K_{[d_{i\,k}, d_{ij}]}). \tag{3}$$

To ensure compliance with the possible values of the errors on the basis m_i it is necessary to fulfill inequality (2).

Q.E.D. The necessary condition of Theorem 1 is sufficient if different values of the error values Δa_i correspond to different product values $a_{ik} \cdot a_{ij}$, and vice versa. Indeed, in this case there is a one-to-one correspondence between the possible values Δa_i and the values of the product $a_{ik} \cdot a_{ij}$, which determines the possibility of uniquely determining the magnitude of the error [23].

Based on Theorem 1, we compose an error correction algorithm for an arbitrary base m_i:

1. Determine the number of distorted residual. To do this, we calculate the values:

$$\begin{aligned} a_1 - a_2 &= a_{12} (\text{mod} d_{12}) \\ a_2 - a_3 &= a_{23} (\text{mod} d_{23}) \\ &\dots \\ a_{n-1} - a_n &= a_{n-1n} (\text{mod} d_{n-1n}) \\ a_n - a_1 &= a_{n2} (\text{mod} d_{n1}) \end{aligned} \tag{4}$$

If all residuals are $a_{ij} = 0(\mathrm{mod} d_{ij})$, then the number A is correct. If the error occurred at the base m_i, then $a_{ij} \neq 0$ and $a_{ik} \neq 0$, thus, the number being tested $\widetilde{A} = (a_1, a_2, ..., \widetilde{a}_i, ..., a_n)$ is incorrect.

2. By the values of a_{ij} and a_{ik} appeal to the block of error constants, where we select the appropriate value of Δa_i.

3. We perform the correction of the number $\widetilde{A}$ in and we get the correct number $A = \widetilde{A} - \Delta A$, i.e.

$$A = (a_1, a_2, ..., a_i, ..., a_n). \tag{5}$$

If in the abbreviated SRC due to the exclusion of the base on which the error occurred, it is possible to unambiguously represent a number A, then instead of determining by the values of a_{ij} and a_{ik} the value of the error Δa_i, we will directly calculate the values of the correct remainder a_i.

Consider this error correction algorithm:

1. Calculate the value of residuals $a_{12}, a_{23}, ..., a_{n1}$.
2. Determine the number of distorted balance. Let the error occurred at the base m_i. In this case, this basis is excluded, and the number A is presented on the bases $m_1, m_2, ..., m_n$, i.e.

$$A = (a_1, a_2, ..., a_{i-1}, a_{i+1}, ..., a_n). \tag{6}$$

3. Perform a convolution of the number A into positional code.
4. Determine the true value of the distorted residual:

$$a_i = A - [A/m_i]m_i, \tag{7}$$

where $[x]$ is the whole part x, not exceeding x. Corrected number $A_{cor} = (a_1, a_2, ..., a_i, ..., a_n)$.

Let us determine the conditions under which it is possible to exclude some bases from the SRC. To do this, we present the bases of the original SRC in the canonical form:

$$\begin{aligned} m_1 &= \beta_{11}^{a_{11}} \beta_{12}^{a_{12}} \ldots \beta_{1l_1}^{a_{1l_1}} \\ m_2 &= \beta_{21}^{a_{21}} \beta_{22}^{a_{22}} \ldots \beta_{2l_2}^{a_{2l_2}} \\ &\ldots \\ m_n &= \beta_{n1}^{a_{n1}} \beta_{n2}^{a_{n2}} \ldots \beta_{nl_n}^{a_{nl_n}} \\ M &= \beta_1^{a_1} \beta_2^{a_2} \ldots \beta_k^{a_k} \end{aligned} \tag{8}$$

To uniquely determine the number A, specified in the SRC with bases $m_1, m_2, ..., m_n$, and lying in the range $[0, M)$ it is possible to exclude only those bases for which $\beta_m = \beta_{il_i}, (m = \overline{1, k}, i = \overline{1, n})$. It is necessary that $a_m \geq a_{il_i}$.

Thus, the necessary and sufficient conditions for error correction are determined by eliminating the distorted base. These conditions are simultaneous fulfillment of equality and inequality:

$$\beta_m = \beta_{i\,l_i},\ a_m \geq a_{i\,l_i}. \quad (9)$$

Above, the algorithm for detecting and correcting errors in the SRC by means of L-codes was described. Let be $(a_k - a_{k+1}) \bmod d_{kk+1}$ when calculating values, it is determined that $a_{i-1\,i} \neq 0$, $a_{i\ i+1} \neq 0$, and all other values are:

$$a_{k\,k+1} = (a_k - a_{k+1}) \bmod d_{k\,k+1} = 0. \quad (10)$$

Then it is stated that the number A is incorrect, and the error is present in the remainder of the base m_i, i.e.

$$\widetilde{A} = (a_1,\ a_2,\ \ldots, \widetilde{a_i},\ \ldots,\ a_n). \quad (11)$$

Referring by the values $a_{i-1\ i}$ and $a_{i\ i+1}$ to the block of error constants, we determine the error value Δa_i and then we determine the true value of the residual:

$$a_{i\ cor} = \widetilde{a_i} - \Delta a_i. \quad (12)$$

The corrected number will appear as:

$$A_{cor} = (a_1,\ a_2,\ \ldots, a_{i\ cor},\ \ldots,\ a_n). \quad (13)$$

To correct an error with the help of the developed correction method, it is necessary that the error Δa_i is not at the same time divisible by two dividers $d_{i-1\ i}$ and $d_{i\ i+1}$, which limits the class of corrected errors [24].

Thus, there is a need to develop effective methods and algorithms to expand the class of possible correctable errors.

The method of correction of one-time errors, allowing to correct errors that are multiples of one of the dividers $d_{i-1\ i}$ or $d_{i\ i+1}$, is as follows [22].

Let a SRC be set with mutually not simple bases, i.e. greatest common divider (GCD) $(m_1,\ m_2,\ \ldots, m_n) \geq 2$.

And let a number be given in the SRC $A_{cor} = (a_1,\ a_2,\ \ldots,\ a_n)$.

We define all values $a_{k\ k+1}$, i.e. $a_{12},\ a_{23},\ a_{34},\ \ldots,\ a_{n-1\ n},\ a_{n\ 1}$. Without breaking the generality of reasoning, we assume that $a_{i\ i+1} \neq 0$, and all other values are $a_{k\ k+1} \neq 0$. Because:

$$a_{i\ i+1} = (a_i - a_{i+1}) \bmod d_{i\ i+1} \neq 0. \quad (14)$$

error may be present only in residues on the bases m_i or m_{i+1}. In this regard, two hypotheses are possible:

- an error is present in the residual a_i;
- an error is present in the residual a_{i+1}.

Before we consider the error correction process by the proposed method, we formulate and prove a theorem, the result of which we use in determining the convergence process for the totality of numbers of the form:

$$A^{(k_i)} = (a_1,\ ...,\ a_{i-1},\ a_{i\,k_i},\ a_{i+1}, ...,\ a_n), \tag{15}$$

to the correct number:

$$A^{(\rho)} = (a_1,\ ...,\ a_{i-1},\ a_{i\,\rho},\ a_{i+1}, ...,\ a_n). \tag{16}$$

First consider the lemma.

Lemma. The sum, difference, and product of any L-code code words are also code words.

Theorem 2. Let in the ordered $(m_{i-1} < m_i;\ i = \overline{1,\ n})$ system of residual classes with bases $m_1,\ m_2,\ ...,\ m_n$ an incorrect (distorted in one residue) number be given $\widetilde{A} = (a_1,\ a_2,\ ...,\ a_{i-1},\ \widetilde{a_i},\ a_{i+1}, ...,\ a_n)$ and let $\Delta a_i = \widetilde{a_i} - a_i = k_i d_{i-1\,i}$.

Then in the set of values $a_{ik_i} = (\widetilde{a_i} - k_i d_{i-1\,i}) \bmod m_i$ there is a single value of $a_{i\rho}$, at which the number:

$$A^{(\rho)} = (a_1,\ a_2,\ a_{i\,\rho},\ ...,\ a_n), \tag{17}$$

is the correct number, where $d_{i-1\,i}(m_{i-1},\ \ m_i)$, and k_i may take values $k_i = 1,\ 2, ...,\ m_i/d_{i-1\,i} - 1$.

Proof. We show that there is such a value of $a_{i\rho_1}$, at which the number $A = (a_1,\ a_2,\ ...,\ a_{i\,\rho},\ ...,\ a_n)$ is the correct number. By the condition of the theorem, the error Δa_i is a multiple of the divisor $d_{i-1\,i}$. The expression $k_i d_{i-1\,i}$ contains all possible multiples of $d_{i-1\,i}$.

Thus, there will be at least one value of $k_i = \rho_1$, at which:

$$\Delta a_{i\rho_1} = \rho_1 d_{i-1\,i}, \tag{18}$$

and

$$a_{1\rho_1} = \widetilde{a_i} - \Delta a_{i\rho_1}. \tag{19}$$

We show that $A^{(\rho_1)}$ is the only correct number from the set of numbers of the form $A^{(k_i)}$.

Suppose there is such a value:

$$a_{1\rho_2} = \widetilde{a_i} - \rho_2 d_{i-1\,i}, \tag{20}$$

at which the number $A^{(\rho_2)}$ is also correct. Then, in accordance with lemma, the number:

$$A^{(\rho_1)} - A^{(\rho_2)} = (0, ..., a_{i\,\rho_1} - a_{i\,\rho_2}, \; ..., \; 0), \tag{21}$$

is correct. If the number $A^{(\rho_1)} - A^{(\rho_2)}$ is correct, then in accordance with lemma we have:

$$\begin{aligned}(\rho_2 - \rho_1)d_{i-1\,i} &\equiv 0(\mathrm{mod} d_{1-i})\\ (\rho_2 - \rho_1)d_{i-1\,i} &\equiv 0(\mathrm{mod} d_{2-i})\\ &\ldots\\ (\rho_2 - \rho_1)d_{i-1\,i} &\equiv 0(\mathrm{mod} d_{n-i})\end{aligned} \tag{22}$$

If $i \neq n$, then the only correct number $A^{(\rho_1)} - A^{(\rho_2)}$ is the zero code word. This is due to the fact that $d_{i-1\,i} \neq 0$ and $d_{i-1\,i}$ is not equal to the GCD of the dividers $d_{1i}, \; d_{2i}, \; ..., \; d_{ni}$.

Moreover, inequality $d_{i-1\,i} \neq [d_{1i}, \; d_{2i}, \; ..., \; d_{ni}]$ contradicts the condition of arbitrary choice of bases $m_1, \; m_2, \; ..., m_n$. Therefore, the following equality holds $A^{(\rho_1)} - A^{(\rho_2)} = (0, \; 0, ..., 0, \; ..., \; 0)$.

Thus, $\rho_1 = \rho_2$, that confirms the uniqueness of existence ρ_1, at which:

$$A^{(\rho_1)} = (a_1, \; a_2, \; ..., \; a_{i\,\rho_1}, \; ..., \; a_n), \tag{23}$$

is correct. Q.E.D.

Thus, the developed method of error correction in the SRC allows to extend the class of corrected errors. This greatly expands the corrective possibilities of the *L*-codes in the class of deductions.

Consider the operation of the device for detecting errors using L-codes, in accordance with the above algorithm. This device contains the input register, modulo adders m_i and d_{1i} $(i = \overline{2, \; n})$ and $(n - 1)$—the input element OR (see Fig. 2).

The algorithm of operation of this device corresponds to the error detection algorithm developed above [25].

As can be seen from the considered materials of the performance of the error correction operation, using the L-codes, the error detection process is implemented extremely simply.

The time of error detection for the SRC given by any system of bases is always equal to three conditional time ticks and does not depend (as is observed for *R*-codes) on the number n of information bases [26].

The above-discussed variants of devices for detecting errors in the SRC make it possible to guarantee the detection of a number A, distortion, however, this does not determine the number of the base on which the residue was distorted [22].

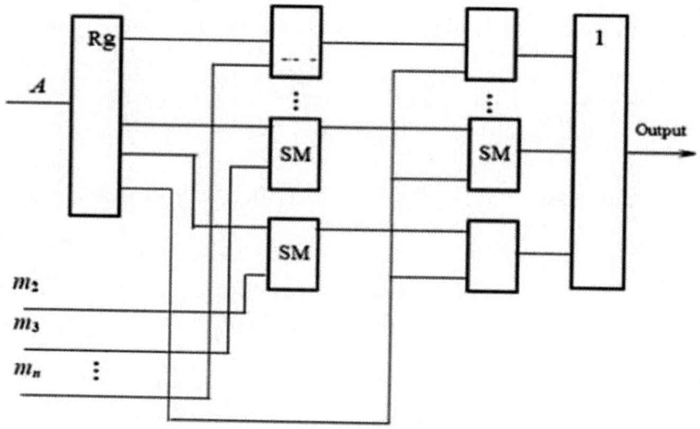

Fig. 2 Device for detecting errors

4 Conclusion

The need to eliminate threats and destructive factors of progressive information and technological development of the national economy became the basis for substantiation of conceptual and practical provisions for the formation of a secure information environment. It is proved that strengthening the information security of the business entity should be based on ensuring the reliability, confidentiality, integrity of information. It is established that modern methods and means of information and communication and digital technologies cannot entirely provide full protection and reliable processing of ever-growing arrays of information. Therefore, the basis for ensuring the information security of enterprises is determined by improvement of methods of errors diagnosis and correction in the system of residual classes.

Thus, the error correction algorithms in the SRC with mutual in pairs non-simple bases make it relatively easy to implement a procedure for detecting and correcting one-time errors [27]. The considered scheme of detecting and correcting one-time errors makes it possible to localize the erroneous base and correct the error in one residual in just five conventional time ticks for any number of the SRC bases. The main advantages of the *L*-codes in the SRC is the simplicity of the procedure for detecting the location of an error and its localization. By the simplicity of decoding schemes, the *L*-codes have no analogues, both in the PNS and in the SRC.

References

1. Antonyuk V Mechanisms of state response to modern challenges and threats to information security. http://www.dy.nayka.com.ua/?op=1&z=747
2. Baranov O (2014) On the interpretation and definition of "cybersecurity." Inf Law 2(42):54–62

3. Lipkan V (2011) Modern content of information operations against Ukraine. Curr issues Int Relat 102(1):34–43 (2011). http://nbuv.gov.ua/UJRN/apmv_2011_102%281%29__7
4. Lytvynenko O (2017) Information component in the modern hybrid war against Ukraine: challenges and threats. Ukrainian Stud Almanac 19:171–174
5. Glushko A, Marchyshynets O (2018) Institutional provision of the state regulatory policy in Ukraine. J Adv Res Law Econ 9(3):941–948 (2018). ASERS Publishing House. https://doi.org/10.14505/jarle.v93(33).18.
6. Onyshchenko SV, Matkovskyi AV, Puhach AA (2014) Analysis of threats to economic security of Ukraine in conditions of innovative economic development. Econ Ann XXI 1–2(2):8–11
7. Svystun L, Glushko A, Shtepenko K (2018) Organizational aspects of investment and construction projects implementation at the real estate market in Ukraine. Int J Eng Technol 7(3.2):447–452. https://doi.org/10.14419/ijet.v7i3.2.14569
8. Kozachenko H, Onyshchenko S, Masliy O (2018) Region building complex social and economic security threats. Int J Eng Technol 7(3.2):79–85. https://doi.org/10.14419/ijet.v7i3.2.14405
9. Kavun SV, Pylypenko AA, Ripka DO (2013) Economic and information security of enterprises in the system of consolidated information: a textbook. Kharkiv: View. KhNEU, 364 p
10. Onyshchenko S, Hlushko A (2020) Conceptual principles of information security of the national economy in terms of digitalization. Soc Econ KhNU 59:14–24
11. Zhivko Z, Zhivko M, Ortynsky V, Kernytsky I (2009) Economic security of enterprises, organizations and institutions. Alerta, 544 p
12. Economic impact of cybercrime – no slowing down. Report of the Center for Strategic and International Studies (CSIS), 2018. Homepage https://www.csis.org/analysis/economic-impact-cybercrime
13. The analysis of regulatory impact of draft regulations of the Cabinet of Ministers of Ukraine "On amendments to the Rules of protection of information in information, telecommunication and information-telecommunication systems". State service of special communication and information protection of Ukraine. Homepage http://195.78.68.84/dsszzi/control/uk/publish/article?showHidden=1&art_id=288142&cat_id=38837&ctime=1522850564776
14. Bilenchuk P, Malyi M (2019) Cybersworld in the new millennium. Who are they: cybercriminals, cybercriminals, cyberterrorists? Legal Bull Ukraine 39:14–15
15. Official site the International Organization for Standardization. Homepage https://www.iso.org/home.html
16. Onyshchenko S, Yanko A, Hlushko A, Sivitska S (2020) Conceptual principles of providing the information security of the national economy of Ukraine in the conditions of digitalization. Int J Manage (IJM). 11(12), 1709–1726 (2020). Homepage https://iaeme.com/MasterAdmin/Journal_uploads/IJM/VOLUME_11_ISSUE_12/IJM_11_12_157.pdf. https://doi.org/10.34218/IJM.11.12.2020.157
17. Akushsky IY, Yuditsky DI (1968) Machine arithmetic in residual classes. Moscow: Sov. Radio, 440 p (in Russian)
18. Torgashov VA (1973) System of residual classes and the reliability of a computer. M.: Sov. Radio, 118 p (in Russian)
19. Krasnobayev V, Yanko A, Koshman S (2017) Algorithms of data processing in the residual classes system. In: 4th international scientific-practical conference problems of infocommunications, science and technology (PIC S&T), Kharkov, pp 117–121
20. Krasnobayev VA, Yanko AS, Koshman SA (2016) A Method for arithmetic comparison of data represented in a residue number system. Cybern Syst Anal 52(1):145–150
21. Hariri A, Navi K, Rastegar R (2005) A simplified modulo (2n−1) squaring scheme for residue number system. In: EUROCON 2005: the international conference on "computer as a tool", Belgrade, 2005, pp 615–618. https://doi.org/10.1109/EURCON.2005.1630004
22. Krasnobayev A, Kuznetsov A, Yanko A, Koshman S, Zamula A, Kuznetsova T (2019). Data processing in the system of residual classes. Monograph. ASC Academic Publishing, 208 p. ISBN 978-0-9989826-6-3 (Hardback). ISBN 978-0-9989826-7-0 (Ebook)

23. Krasnobayev V, Kuznetsov A, Kononchenko A, Kuznetsova T (2019) Method of data control in the residue classes. In Proceedings of the 2nd international workshop on computer modeling and intelligent systems, CMIS-2019, Zaporizhzhia, Ukraine, 15–19 April 2019, pp 241–252
24. Yatskiv V, Tsavolyk T, Yatskiv N (2017)The correcting codes formation method based on the residue number system. In: 2017 14th international conference the experience of designing and application of CAD systems in microelectronics (CADSM), Lviv, 2017, pp 237–240. https://doi.org/10.1109/CADSM.2017.7916124
25. Krasnobayev A, Kuznetsov A, Lokotkova I, Dyachenko A (2019) The method of single errors correction in the residue class. In: 2019 3rd international conference on advanced information and communications technologies (AICT), Lviv, Ukraine, 2019, pp 125–128. https://doi.org/10.1109/AIACT.2019.8847845
26. Roshanzadeh M, Saqaeeyan S (2012) Error detection & correction in wireless sensor networks by using residue number systems. Int J Comput Netw Inf Secur 4(2):29–35. https://doi.org/10.5815/ijcnis.2012.02.05
27. Krasnobayev A, Kuznetsov A, Yanko A, Kuznetsova T (2020) The analysis of the methods of data diagnostic in a residue number system. In: 2st international workshop on cyber hygiene & conflict management in global information networks, Kyiv

Contribution Ukraine's Sustainable Energy Development (Modelling and Forecasting)

Volodymyr Onyshchenko, Olha Komelina, and Igor Shchurov

Abstract Formation and development peculiarities of Ukraine's sustainable energy development and its energy efficiency potential use have been determined. The characteristics of sustainable energy development model are complexity, multilevel, multifactorality, the presence of specific peculiarities and interdependencies of its components, their functions and parameters. The methodology for assessing energy efficiency potential requires non-standard approaches use in the study of external and internal factors impact on the national economy effectiveness to the criterion of energy efficiency. It is important to take into account the dynamics of changes in the vector and the strength of the factors influence depending on the world energy markets, domestic goods and services energy intensity, markets' structure changes and competition conditions. The strategic objectives of Ukraine's sustainable energy development are to increase the level of energy efficiency and ensure the country's energy security.

Keywords Sustainable development · Sustainable energy development · Potential · Energy efficiency · Model

1 Introduction

Problems of energy saving and energy efficiency of the national economy have been widely covered in the works of foreign and domestic scientists. Analytical assessment of recent research in this area shows the diversity of methodological and methodical approaches to the formation of a sustainable energy development model. As a rule, in these researches, the study and structuring of the national economy energy efficiency potential is carried out on the basis of a resource approach to assess the national economy energy efficiency potential [1–10]. In general, scientific works of domestic and foreign scientists highlight different approaches, methods, mechanisms for implementing energy and resource-saving measures at micro, meso and

V. Onyshchenko · O. Komelina (✉) · I. Shchurov
National University «Yuri Kondratyuk Poltava Polytechnic», Poltava, Ukraine

V. Onyshchenko et al. (eds.), *Proceedings of the 3rd International Conference on Building Innovations*, Lecture Notes in Civil Engineering 181,
https://doi.org/10.1007/978-3-030-85043-2_68

macro levels, as well as various approaches to energy efficiency management [5, 6]. However, these publications do not provide a systematic understanding of the formation peculiarities of a national economy sustainable energy development model, taking into account its energy efficiency potential [1–3].

Modern domestic and foreign scientists works referring creating a methodology of sustainable energy development require some adaptation to the national economy realities taking into account the actual information on energy resources availability and efficiency, forecasting changes concerning their potential increase, application of strategic management and planning tools.

The purpose of the article is to investigate the peculiarities of forming a Ukraine's sustainable energy development model taking into account the its energy efficiency potential.

Bibliography Review

The concept of sustainable development in modern interpretation began to be actively used in scientific publications, world governmental policy, international documents in the 1980s. In 1992, at the UN World Summit (Rio de Janeiro) with the participation of more than 180 countries, many international organizations and leading scientists, the "Agenda for the XXI Century" was adopted. It created the basis for the formation of sustainable development doctrine as a set of ideas, concepts, positions and postulates of various sciences, in particular philosophy, sociology, economics, ecology, which have already formed the basis of UN documents and individual countries. The main task of the transition to the sustainable development model is to expand the opportunities for economic growth of countries through changes in the relationship system between human being and nature as well as the introduction of a coordinated global strategy for human survival, which aims to ensure long-term continuous development that meets the needs of the people living today, without harming the needs of future generations.

Important aspects of sustainable development are covered in the works of foreign researchers: Jovane F., Yoshikawa H. [11], Onishi A. [12], Blinc R., Zidan'sek A., Slaus I. [13], Clark G. [14], Jegatheesan V., Liow JL, Shu L., Kim SH, Visvanathan C. [15], Goncz E., Skirke U., Kleizen H., Barber M. [16], Hughes B., Johnston P. [17], Kotabe M., Murray J. [18]. The works of domestic scientists Doguntsov S, Dolishny M, Tregobchuk V [19, 20], are connected to the development of basic conceptual provisions of Ukraine's sustainable development.

In the context of Ukraine's European integration strategy implementation and its transition to a sustainable development model, a number of reforms has also been carried out. The National Report "Sustainable Development Goals: Ukraine" provides a vision of the guidelines for Ukraine's Sustainable Development Goals achievement, which were approved at the UN Summit on Sustainable Development in 2015. The increase of energy efficiency level and green energy share in the country's energy balance are strategic objectives of sustainable development of Ukraine.

Currently, the EU countries have developed a pack of documents in the field of energy efficiency ("Energy 2020. A strategy for competitive, sustainable and secure energy: Communication from the Commission to the European Parliament, the Council, the European Economic and Social Committee and the Committee of the Regions"; "EU Energy Efficiency Policy - Achievements and Outlook"; "EU Energy trends to 2030" [22–24]. The European Commission has adopted a strategy "European Green Deal" in order to make EU's transition to climate-neutral development by 2050. This was taken into account during the formation of the concept and legal framework for sustainable energy development in Ukraine, the Energy Strategy development by 2035 "Security, Efficiency, Competitiveness" and its modernization by 2050, the implementation of active energy reform, regulatory renewal provision for Ukraine and the EU's energy systems integration in 2023.

Main material. Sustainable development national economic model formation is carried out under the influence of economic laws and objective laws at a certain stage of economic development, and the globalization processes which take place in the world economy determine its transformation. An effective national model of sustainable development is able to ensure an appropriate level of population's welfare at a certain stage of social development. In modern conditions, such model formation factors are objective economic, environmental, innovation and investment, technical and technological, social, political, cultural and other processes. In Ukraine, energy security and energy efficiency are key factors in the transition to a sustainable functioning model of the national economy, and its features determine the content of sustainable energy development national model.

Perspective tasks of Ukraine's sustainable energy development on the basis of green energy transition are:

- wide implementation of "clean" technologies and on this basis ensuring the transformation of industrial production and export structure with the predominance of goods with high added value;
- increase of the economy energy efficiency to the European level and decrease of the energy consumption scale;
- maximum deployment of renewable energy sources (hereinafter—RES) and electrification;
- transition to environmentally friendly transport;
- introduction of a "circular" economy (closed cycle economy);
- development of "intelligent" networks and communications;
- expansion of bioenergy and natural carbon sequestration technologies;
- absorption of CO_2 emissions due to carbon capture, storage and utilization technologies.

The main features of sustainable energy development model are complexity, multilevel, multifactorality, the presence of specific peculiarities and interdependencies of its components, their functions and parameters. The methodology for assessing energy efficiency potential requires non-standard approaches use in the study of external and internal factors impact on the national economy effectiveness to the criterion of energy efficiency and its energy security.

In order to assess the Ukraine's sustainable energy development model, the macroeconomic indicators of the national economy were studied, the dynamics of vector change and the strengths of key factors influence of its formation were determined (Table 1). In order to assess the peculiarities of such a model formation, energy efficiency indicators and macroeconomic indicators were used.

Analytical data show that in 2007–2019 the economy of Ukraine functioned unstable under internal and external factors influence. This significantly limited the country's transition to sustainable energy development. At the same time, the available statistical base does not allow to take into account the influence of such factors

Table 1 Dynamics of the main macroeconomic indicators of Ukraine (2007–2019, in actual prices, UAH million)

Indicator	2007	2010	2013	2016	2019
Changes in commitments and net wealth					
Net savings	104,146	82,311	−48,293	143,958	51,888
Capital transfers received from other countries	136	95	192	230	72
Capital transfers paid to other countries	−121	−31		−26	−24
Total (changes in net wealth at the cost of savings and capital transfers)	**104,161**	**82,375**	**−48,101**	**144,162**	**51,936**
Changes in assets					
Gross fixed capital formation	198,348	195,927	263,661	368,691	252,071
Fixed capital consumption	−73,071	−115,338	−200,903	−313,522	−122,346
Working capital inventories change	4685	3616	−13,761	148,581	−123,678
Acquisition excluding disposal of valuables	285	375	208	929	156
Acquisition excluding disposal of non-produced non-financial assets		−1414	671	−2144	−24
Net lending (+), net borrowing (−)	−26,086	−791	−97,977	−58,373	45,757
Total	**104,161**	**82,375**	**-48,101**	**144,162**	**51,936**
Energy consumption and energy efficiency					
Final consumption of FER, thousand tonnes of oil equivalent)	85,956	74,004	69,558	51,649	49,359
Energy efficiency indicator (NW increase/FER consumption)	1,212	1,113	−0,692	2,791	1,052
GDP, UAH billion	720,7	1120,6	1522,7	2385,4	3974,6
Energy efficiency indicator (GDP/FER costs)	0,008	0,015	0,022	0,046	0,081
FER final consumption, thousand tonnes of oil equivalent	85,956	74,004	69,558	51,649	49,359
Energy efficiency indicator (NW increase/FER costs)	1,212	1,113	−0,692	2,791	1,052

Source Compiled according to data [25, 26].

as changes in world energy markets, energy intensity of domestic goods and services, changes in the structure of these markets and their competition conditions.

Energy efficiency is the most important criterion for the necessary potential formation for sustainable development of the state. Energy efficiency means the optimal use of fuel and energy resources (FER), taking into account the existing levels of technological development and current environmental requirements. The main indicator of energy efficiency is the specific energy consumption per unit of useful product in all spheres of human activity (economy, technology, life). In the national economy, this indicator is the energy intensity of gross domestic product (GDP). The energy efficiency indicator of GDP is a value that is the inverse of energy intensity, which means the ratio of GDP to the total value of used FER

$$Eei(1) = \frac{GDP}{FER\text{costs}} \tag{1}$$

The GDP indicator cannot be considered as a well-founded indicator of the final result. It does not take into account the social effect of economic activity and the energy factors impact. Some scientists [25] suggest considering the magnitude of the increase in national wealth (NW) to the cost of all energy resources as an indicator of energy efficiency. NW is a more comprehensive indicator, which includes an assessment of natural resources, human (labor resources) as well as social and productive capital of the country; is a comprehensive indicator of life quality, considering income, environmental situation, living potential level, person's moral satisfaction state in life and society.

The energy efficiency indicator according to the second approach is determined by the following way:

$$Eei(2) = \frac{NWincrease}{FER\text{costs}} \tag{2}$$

This indicator more fully reflects the final effectiveness of energy activity—the growth of the NW as an indicator of life quality allows to conduct more reasonable cross-sectoral comparison, and most importantly—to assess the multiplier effect not only of energy costs, pricing, technological modernization, export and import policy, but also the choice of priority areas in the structural as well as social and environmental policy of the country.

In order to identify the formation features of Ukraine's sustainable energy development model, the dynamics of the main Ukraine's macroeconomic indicators for the period of 2007–2019 (Table 1) were considered taking into account its energy efficiency.

The dynamics of change in the studied indicators reflects the lack of a stable and effective state policy to ensure sustainable energy development (Fig. 1).

In order to identify the peculiarities of Ukraine's sustainable energy development model formation, considering the potential of its energy efficiency, appropriate calculations have been conducted and factors that affect the final consumption of FER in the

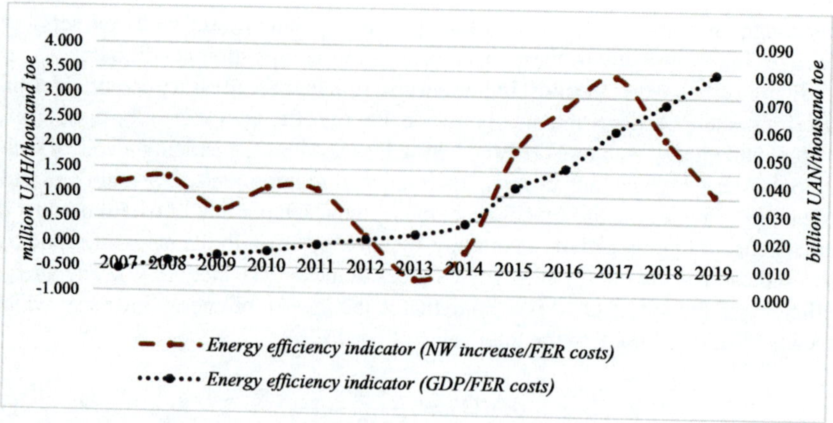

Fig. 1 Dynamics of energy efficiency indicators of Ukraine for 2007–2019

national economy have been selected. The most significant factor is "Capital investments"—an important catalyst for production processes, stable economic development of entrepreneurship and the country in general. The nature of the dynamics of capital investments volume attracted to the national economy, as well as "environmental investments" volume (capital investments and current expenditures for environmental protection) provide an opportunity to assess the threats and opportunities for sustainable energy development. Indicator "household income" shows features of households investment potential formation referring the introduction of energy efficient technologies and energy consumption.

The number of selected factors for constructing Ukraine's sustainable energy development model, considering the potential of its energy efficiency necessitates the use of multiple correlational and regressional analysis to quantify the bonds between statistics that characterize individual social and economic processes (Table 2).

The mathematical form of the relationship between factors and output indicators (regression analysis) has been established and the closeness of this relationship (correlation analysis) has been determined.

The effective indicator Y—final consumption of fuel and energy resources (thousand tons o. e.)—is presented in the form of a power multifactor model

$$Y = a_0 \cdot X_1^{a_1} \cdot X_2^{a_2} \cdot X_3^{a_3} \tag{3}$$

where X1—capital investments and current expenditures for environmental protection, UAH billion;

X2—household income, UAH billion;
X3—capital investments, UAH billion.

The model is built using the MS Excel spreadsheet editor and the "Solution Search" add-on. The solution of dependence (3) is equation:

Table 2 Dynamics of factors change that have an impact on the final consumption of FER in the economy of Ukraine

Period	Capital investments and current expenditures for environmental protection, UAH billion	People's income, UAH billion	Capital investments, total, UAH billion
2007	9,69	623,29	222,68
2008	12,18	845,64	272,07
2009	11,07	894,29	192,88
2010	13,13	1101,18	189,06
2011	18,49	1266,75	259,93
2012	20,51	1457,86	293,69
2013	20,38	1548,73	267,73
2014*	21,93	1516,77	219,42
2015*	24,59	1772,02	273,12
2016*	32,49	2051,33	359,22
2017*	31,49	2652,08	448,46
2018*	34,39	3248,73	578,73
2019*	43,74	3699,35	623,98

Note * Excluding the temporarily occupied territory of the Autonomous Republic of Crimea, Sevastopol and part of the temporarily occupied territories of Donetsk and Luhansk regions.
Source Compiled according to data [25].

$$Y = 659799,2 \cdot X_1^{-0,293} \cdot X_2^{0,074} \cdot X_3^{-0,06} \tag{4}$$

The results of the calculations are presented in Figs. 2 and 3.

The obtained results allow to conclude about Ukraine's sustainable energy development model formation. For multifactor power regression (2) Multiple correlation coefficient is 0.89, which indicates a close relationship of Y with factors X1, X2,

	Y_f	X_1	X_2	X_3	Y_r	$(Y_f-Y_r)^2$	$(Y_f-Y_a)^2$	a_0	a_1	a_2	a_3	Forecast quality
2007	85956	623,29	9,69	222,68	85698,21	66456,72	442455044,38	659799,20	-0,293	0,074	-0,060	0,0595
2008	83282	845,64	12,18	272,07	78760,87	20440648,25	337112197,30					0,0633
2009	67555	894,29	11,07	192,88	77919,63	107425579,72	6935929,99					0,0962
2010	74003	1101,18	13,13	189,06	72566,32	2064038,64	82475737,99					0,0802
2011	75852	1266,75	18,49	259,93	68237,29	57983823,07	119478352,69					0,0763
2012	73107	1457,86	20,51	293,69	65078,41	64458326,97	67004299,22					0,0822
2013	69558	1548,73	20,38	267,73	63960,14	31336067,19	21498202,22					0,0908
2014*	61459	1516,77	21,93	219,42	64072,03	6827942,93	11988107,22					0,1163
2015*	50830	1772,02	24,59	273,12	60797,91	99359165,45	198567120,38					0,1700
2016*	51648	2051,33	32,49	359,22	57283,90	31763314,81	176182739,15					0,1647
2017*	49911	2652,08	31,49	448,46	53226,70	10993887,67	225311646,30					0,1763
2018*	51458	3248,73	34,39	578,73	49889,08	2461507,34	181262725,30					0,1659
2019*	49359	3699,35	43,74	623,98	47340,95	4072515,09	242187814,92					0,1803
2020*		3700	44	624	**47321,50**	**439253273,84**	**2112459917,08**					**0,1171**
												88,29%

$$Y = a_0 \cdot X_1^{a_1} \cdot X_2^{a_2} \cdot X_3^{a_3}$$

R=	0,89
$F_{розр}$	20,95
$F_{табл}$	4,26

Fig. 2 The results of the final fuel and energy resources consumption modeling on the basis of multifactor power dependence

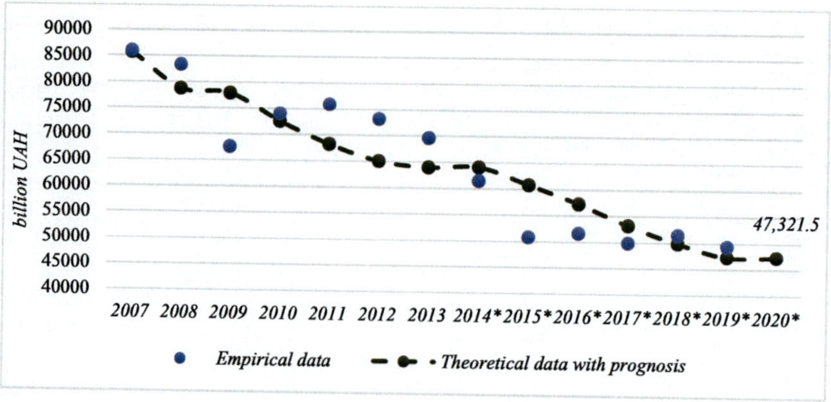

Fig. 3 Graphic representation of the final fuel and energy resources consumption forecasting results on the basis of multifactor power dependence

X3, as well as the proximity of the selected mathematical model to the selected data. According to Fisher's criterion F (4,26) = 20,395, which significantly exceeds the critical tabular value and indicates the importance of the relationship. Thus, the constructed model has a fairly high quality forecast (88.3%).

Positive coefficients values of independent variables of the constructed sustainable energy development model indicate an increase in final consumption of fuel and energy resources with increasing influence factors, negative on the contrary, a decrease. The conducted calculations made it possible to forecast the final consumption of fuel and energy resources on the basis of multifactor dependence and to determine the main dependencies in the Ukraine's sustainable energy development model formation.

2 Conclusions

The effect of natural and economic laws of sustainable energy development model formation is closely related to many different key (permanent) and random factors, which indicates its objective propensity to the equilibrium loss risk. The slow "green" energy transition may lead to a deepening of crisis processes in the economic development of Ukraine and increase the negative social and economic consequences for citizens. Ukraine's transition to a climatically neutral economy requires the state to attract investments of up to 5% of GDP annually. Households play a huge role in these processes. Households demand for energy equipment, heating, vehicles and their investments into energy efficiency also affect the formation of the country's economy technological structure, the development of renewable energy, the use of "smart" energy management systems etc.

Thus, Ukraine's transition to the sustainable energy development model requires the improvement of regulatory policies and economic incentives mechanisms, environmental supervision and monitoring, improvement of overall methodological approaches to regulating the polluting enterprises activities. Further study of sustainable energy development model formation requires a systematic study of the relationship of economic, environmental and social indicators, identification of existing laws, conducting the necessary assessments and conclusions. This will create an objective basis for improving the management system at the macro and micro levels to ensure energy security and the transition to energy efficient and energy saving use and consumption of energy resources with innovative technologies implementation.

References

1. Kurbatova T, Khlyap H (2015) State and economic prospects of developing potential of non-renewable and renewable energy resources in Ukraine. Renew Sustain Energy Rev 52:217–226. https://doi.org/10.1016/j.rser.2015.07.093
2. Denysiuk S, Kotsar O, Chernetska Y (2016) Ukraine's energy efficiency. Best projects' ideas. K., Igor Sikorsky Polytechnik Institute, 79 p
3. Komelina OV, Shcherbinina SA (2018) Methodology of estimation of energy reserves and energy efficiency of the housing fund of Ukraine. Market Manage Innov 1:382–390
4. Onyshchenko S, Yehorycheva S, Furmanchuk O, Maslii O (2019) Ukraine construction complex innovation-oriented development management. In: Proceedings of the 2nd international conference on building innovations, pp 687–700. https://doi.org/10.1007/978-3-030-42939-3_68
5. Bohoslovska A, Komelina A, Shcherbinina S (2018) Priorities of sustainable building concept and evaluation of possibilities for energy efficiency management of buildings (as in the case of Ukraine). Int J Eng Technol 7(3.2):46–53 (2018)
6. Zaverbnyi A (2018) Ukraine's energy and economy complementarity in the European integration conditions. Lviv Polytechnik, Lviv, 308 p
7. Energy efficiency system in Ukraine. Project for discussion (2018). http://www.minregion.gov.ua/wp-content/uploads/2018/09/GIZ-brochure_17-09.pdf.7
8. Peyramale V, Wetzel C (2017) Analyzing the energy-saving potential of buildings for sustainable refurbishment. Procedia Environ Sci 38:162–168. https://doi.org/10.1016/j.proenv.2017.03.098
9. Gassar A, Young Yun G (2017) Energy saving potential of PCMs in buildings under future climate conditions. Appl Sci 7(12):1219. https://doi.org/10.3390/app7121219
10. Boeglel A, Singh D, Sant G (2015) Energy saving potential in Indian households from improved appliance efficiency. Prayas Energy Group, Pune. http://www.prayaspune.org/peg/media/k2/attachments/energy_saving_potential_from_indian_households_from_appliance_efficiency_108A01.pdf
11. Jovane F, Yoshikawa H, Alting L, Boër CR, Westkamper E, Williams D, Tseng M, Seliger G, Paci AM (2008) The incoming global technological and industrial revolution towards competitive sustainable manufacturing. CIRP Ann 57(2):641–659
12. Onishi A (2005) Futures of global interdependence (FUGI) global modeling system. J Policy Model 27(1):101–135
13. Blinc R, Zidan'sek A, Slaus I (2007) Sustainable development and global security. Energy 32(6):883–890
14. Clark G (2007) Evolution of the global sustainable consumption and production policy and the United Nations Environment Programme's (UNEP) supporting activities. J Clean Prod 15(6):492–498

15. Jegatheesan V, Liow JL, Shu L, Kim SH, Visvanathan C (2009) The need for global coordination in sustainable development. J Clean Prod 17(7):637–643
16. Goncz E, Skirke U, Kleizen H, Barber M (2007) Increasing the rate of sustainable change: a call for a redefinition of the concept and the model for its implementation. J Clean Prod 15(6):525–537
17. Hughes B, Johnston P (2005) Sustainable futures: policies for global development. Futures 37(8):813–831
18. Kotabe M, Murray J (2004) Global sourcing strategy and sustainable competitive advantage. Ind Mark Manage 33(1):7–14
19. Sneddon C, Howarth R (2006) Sustainable development in a post-Brundtland world. Ecol Econ 57(2):253–268
20. Doroguntsov S, Ralchuk O (2001) Sustainable development - a civilizational dialogue of nature and culture. Visn NAS Ukraine 10:17
21. Dolishniy M, Tregobchuk V, Gutsulyak G et al (1999) The state of the environment and its impact on the Lviv region labor resources, Chernivtsi, pp 149–150
22. Energy 2020. A strategy for competitive, sustainable and secure energy: Communication from the Commission to the European Parliament, the Council, the European Economic and Social Committee and the Committee of the Regions: Doc.com (2010) 639 final. Brussels, 10.11.2010 [Electronic resource]
23. EU Energy Efficiency Policy. Achievements and Outlook (2010) Brussels: European Parliament, Directorate general for internal policies, 213 p
24. EU Energy trends to 2030 – Update 2009 (2010) Brussels: European Commission, Directorate-General for Energy and Traansport, 180 p
25. Denisyuk SP, Borichenko OV (2015) Theoretical foundations of energy management systems construction in Ukraine. Energy Econ Technol Ecol 1:7–17
26. State Statistics Service of Ukraine. http://www.ukrstat.gov.ua/. Accessed 10 Jan 2021

Impact of Innovation and Digital Technologies on the Financial Security of the State

Volodymyr Onyshchenko, Svitlana Yehorycheva, Oleksandra Maslii, and Nadiia Yurkiv

Abstract The article reveals the content of the economy digitalization as a challenge to the financial security of the state. The main advantages and additional opportunities for the national economic system from the introduction of information and telecommunication technologies are investigated. A comparative analysis of the level of implementation of innovations and digital technologies in Ukraine and other countries worldwide is carried out. It is substantiated that efficiency of the financial system in the face of the challenges associated with digital transformation is crucial in the development and maintenance of socio-economic stability of the state. A meaningful description of the financial security of the state is given, taking into account the concept of the information society. Risks, preconditions and factors of emergence of threats to financial security of the state in the conditions of digitalization are revealed. A multi-criteria description of threats to the financial security of the state in the process of digital transformation is given. The low level of protection against digital threats in Ukraine is revealed due to technological unpreparedness for introduction of innovations and digital technologies. It is established that ignoring the negative impact of fundamentally new threats, such as the cryptocurrency market, electronic payment services, cybercrime in the financial system and spread of misinformation, makes it impossible to ensure financial security. It is substantiated that indicators of digitalization processes should be taken into account in order to identify modern threats, objectively assess their impact and form priority areas for ensuring the financial security of the state.

Keywords Financial security of the state · Innovations · Digital technologies · Digitalization · Risks · Threats

V. Onyshchenko · S. Yehorycheva · O. Maslii (✉)
National University "Yuri Kondratyuk Poltava Polytechnic", Pershotravnevyj Avenue 24, Poltava 36011, Ukraine

N. Yurkiv
National Institute for Strategic Studies, Pirogova Street 7-A, Kyiv 01054, Ukraine

V. Onyshchenko et al. (eds.), *Proceedings of the 3rd International Conference on Building Innovations*, Lecture Notes in Civil Engineering 181,
https://doi.org/10.1007/978-3-030-85043-2_69

1 Introduction

Under modern conditions, introduction of innovations and digital technologies in all spheres of the national economy is transformed from a simple method of improving the quality of life into a driver of global economic development, which provides increased business efficiency. Digitalization of the economy based on real-time data exchange, in the progressive development of the knowledge economy, informatization and computerization of public relations provides opportunities for significant competitive advantages, increased productivity and reduced production costs, creating new business models and new forms of business, significant economic growth.

The concept of information society [2–5, 7] from the standpoint of economic and financial security [8–11] at different levels of the social hierarchy, along with undeniable advantages and additional opportunities is a catalyst for new challenges, risks and threats, which without the appropriate regulatory framework to be adopted at the stage of digital economy and ensure a high level of competition, efficiency of the institutional environment, the formation of skills in modern information and communication technologies, improving the quality of training and stimulating lifelong learning, can cause significant damage and create obstacles to the development of the digital economy and ensuring the financial security of the state.

Dynamism of the environment, associated with the processes of globalization and digitalization, necessitates constant broad monitoring of risks, challenges and threats to the financial security of the state. Comprehensive analysis of trends in the financial system of the state, timely response to negative changes, adaptation to new conditions, systematization of existing threats, counteraction to existing and elimination of possible risks on the basis of preventive approach is especially relevant in modern conditions. There is a need to introduce modern tools and technologies for analysis, detection and control of risks and threats to financial security of the state, especially in the context of digitalization.

2 Research of Theoretical Foundations of Innovation and Digital Technologies Impact on the State Financial Security

Introduction of information and communication technologies is the fourth industrial revolution, which forms a new technological way, based on production of equipment that uses information and communication technologies and related software. In general, the transformation processes in the context of digitalization can be divided into social and economic. However, this division is conditional, because formation of a new social (digital) environment through the development of new ways of communication is closely interconnected and leads to the emergence of new activities, as well as digitalization of certain sectors of the economy. On the other hand, it was rapid development of computer technology, information and telecommunication technologies that served as a catalyst for the formation of society, strives for the improvement and interaction of business, science and technology.

In today's world, the term "digitization" is used in both narrow and broad sense. In a narrow sense, the process approach under digitalization means conversion of information into digital form, which usually reduces various costs (tangible, intangible, labor), to create additional opportunities, in particular, for economic development etc. In a broader sense, digitalization is interpreted as a modern trend of global socio-economic development, based on transformation of information into a digital form.

Adhering to the broader definition of digitalization, the World Bank experts note that the digitalization of the economy not only stimulates economic growth, but also significantly accelerates its pace. This is the basis of the most common and fairly broad definition of the digital economy as a result of digitalization of economic relations, proposed by the World Bank, according to which the digital economy is considered a new paradigm of accelerated economic development based on real-time data exchange [12].

In 1995, an American computer scientist Nicolas Negroponte in his book "Being Digital" [13] formed the concept of electronic economics and introduced the term "digitalization". The term "digital economy" was first proposed in 1995 by Canadian economist and business consultant Don Tapscott in his book "Digital Economy" [14], who highlighted the following major implications of the digitalization of the global economy: reduction of transaction costs; emergence of new business models; exclusion of intermediaries due to direct relationship between the manufacturer and the buyer of products (goods, work, services).

Digitalization is one of key characteristics of a new digital economy, a new type of socio-economic structure, which is gradually formed within the post-industrial period of economic development through implementation of achievements of scientific and technological progress and innovative methods of management, intellectualization of human capital, use of advanced new technologies, accelerated development of knowledge-intensive sectors of the economy, giving priority to the production of knowledge and services and formation of mentality of a creative, effective, rational business [18].

The level of innovations and digital technologies implementation in the country is reflected by the system of indicators. These indicators, according to the World Bank experts, include indicators of access to technologies, the type and content of digital technologies introduced in the country and the willingness to introduce the latest technologies [19]. Analysis of the key indicators that reflect the level of innovations and digital technologies implementation (see Fig. 1) shows that Ukraine, in comparison to other countries, is introducing digital technologies at a much slower pace, and in terms of use of robots in industry is at one of the lowest positions.

However, readiness to introduce innovations and digital technologies actualizes the problem of the financial security of the state. Effectiveness of the financial system functioning in the face of challenges associated with digital transformation is decisive in matters of development and ensuring the socio-economic stability of the state.

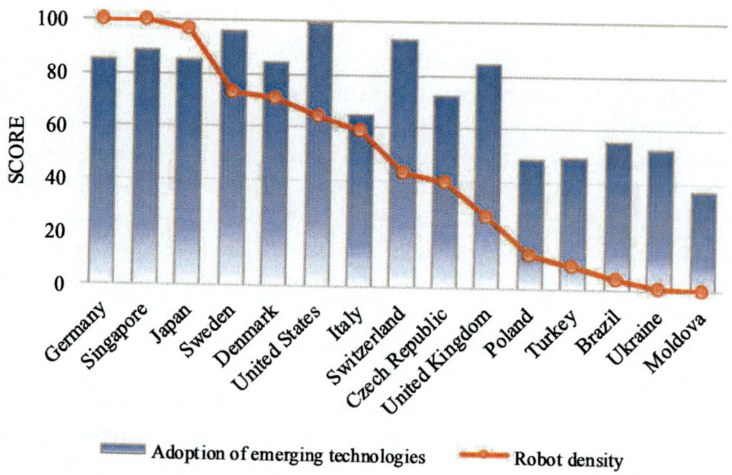

Fig. 1 The level of innovations and digital technologies spread in different countries

Continuous monitoring and control of risks to financial stability through current and strategic analysis of indicators of the financial system development, in particular the availability, placement and use of internal and external financial resources at the state level, the formation of state and local budgets, balance of payments etc., is the key element of ensuring financial security of the state based on the proactive approach and prevention of the implementation of threats.

An untimely response to the existing challenges of society informatization or its lack leads to emergence and spread of a threat; it is a stage of extreme aggravation of contradictions, the most specific and immediate form of danger or a set of conditions and factors that endanger the interests of citizens, society and the state in the financial sphere [16].

Summarizing the results of research of Ukrainian and foreign scholars on the problems of ensuring financial security at the macro level (A. Baranovskyi, S. Varnaliy, V. Muntiyan, S. Onyshchenko and others) we are inclined to think that financial security of the state is the degree of financial interests of the state protection, the status of the financial, monetary, budgetary, tax, currency, banking, investment, payment and stock systems; it is characterized by balance, resistance to internal and external negative influences, ability of the state to effectively form and rationally use financial resources sufficient to meet its needs by fulfilling obligations and ensuring socio-economic development [8].

Emergence of imbalances under the influence of digitalization by components of financial security of the state, including such components as budgetary, debt, monetary, foreign exchange, banking and non-banking financial sector, is a source of threats, the degree of influence of which can only be determined on the basis of constant monitoring of the system of indicators and their comparison with standard values. It should be noted that not all scientists share this approach, defining other components of the financial security of the state, that include other areas of emergence of threats, such as security of insurance market, corporate sector and the like.

No threat emerges by itself; it is preceded by risks and challenges. Thus, risk is understood as the possibility and the likely scale of consequences of an adverse event. Risk in the information society is an integral part of any system functioning, including the financial security of the state, and it can have such economic consequences as negative, zero or positive impact. In turn, objectivity of the risk itself lies in the fact that it is a form of qualitative and quantitative expression of existing uncertainty of the environment. At the same time, risk is always associated with the choice of certain alternatives and calculation of their result probability; this manifests its subjectivity [17].

It should be noted that a threat occurs only when a risk event has occurred or has significant chances of occurring, but its negative consequences are delayed in time, and there is an opportunity to counteract the negative influences. This process is answer to challenge. Thus, threats to financial security of the state in the context of introduction of information technologies are existing or potentially possible processes, phenomena, conditions and factors that create a danger and hinder the realization of national interests in the financial sphere.

Digitalization carries a number of risks that lead to challenges associated with the adaptation of macroeconomic processes in the transformation of their own form, namely the transition to a digitalized computer network. That is why a timely response to challenges of digital transformation and use of preventive principle in ensuring the financial security of Ukraine is of particular relevance. However, analysis of the indicators of financial security of Ukraine, determined by current guidelines for calculating the level of economic security, shows that impact of digital transformation in Ukraine is not reflected by any of the indicators, in particular, the indicators of the shadow economy in the financial sector, which has also undergone the process of digitalization. This creates significant difficulties in taking into account the challenges of digital transformation when identifying threats to the financial security of Ukraine.

Summarizing the results of the previous studies, we can highlight the main opportunities and advantages of digitalization of the economy, presented in Fig. 2.

Thus, for financial security of the state, which aims to protect and harmonize the national economic interests in financial sphere at the state level, there is a complex process of monitoring risks, responding to challenges in implementation of innovations and information technologies and counteracting the spread of threats and reducing or neutralizing their impact. Complexity of this process in the current conditions of financial system of Ukraine is expressed in the slow adaptation to global digitalization processes and the emergence of new threats that require, in addition to timely detection, monitoring and evaluation of their possible impact.

Threats in the context of digitalization, in contrast to key factors of the market environment, such as supply or demand, can be latent or artificial. That is why constant systematization of real threats has function of preventing future crises by anticipating and finding potential threats, testing hypotheses in new areas of financial activity, monitoring risks, since constant development of the economy causes risks and threats in conservative industries. Therefore, broad classification and constant

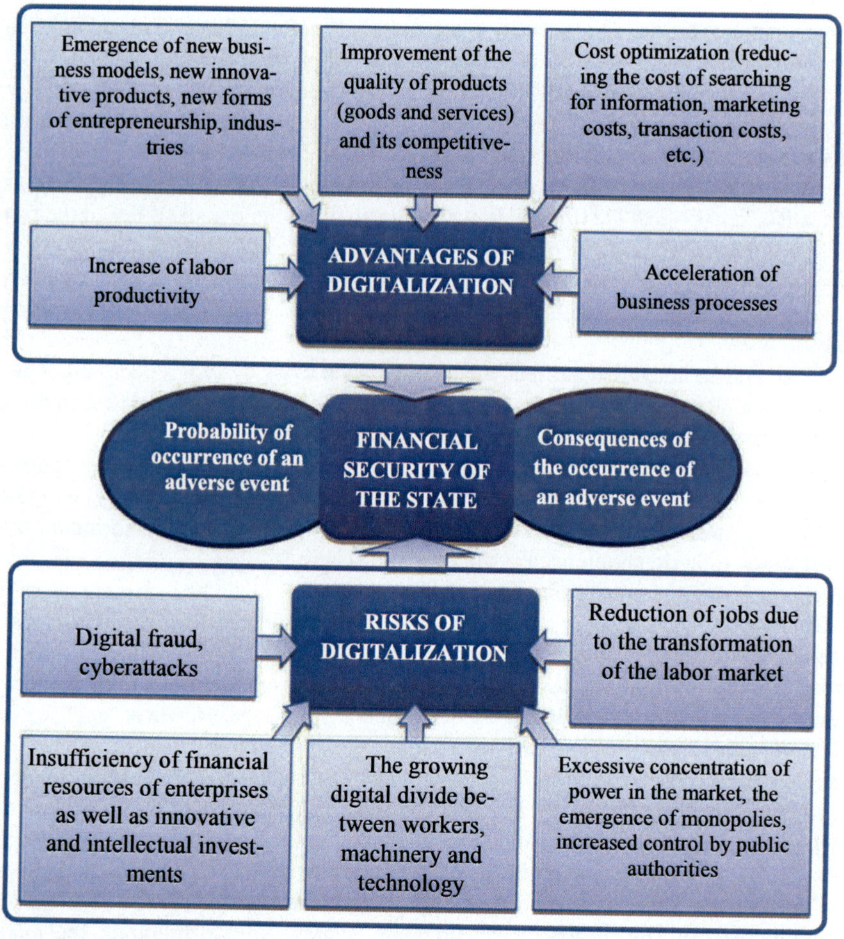

Fig. 2 Impact of digitalization on financial security of the state

systematization of threats should become the basis for such activities to identify and prevent threats under digital transformation, which the state and the financial security of the state should be aware of and ready for targeted action to combat or overcome problematic phenomena.

In terms of today's conditions, it is important to take into account peculiarities of financial system of the state functioning while studying threats to its financial security. Thus, there are almost "classic" threats to financial security of the state, in particular: imperfection of fiscal policy and misuse of budget funds, inefficiency of the tax system, massive tax evasion, significant amounts of public and state-guaranteed debt, problems with its service, sharp changes in prices and the exchange rate of the national currency, a significant difference in the ratio of income of the most and least well-off population and insufficient social protection of certain groups, low capitalization of the banking system, small long-term bank lending and high interest rates on loans, dependence of economic reform on foreign loans, low level of investment activity, growth of the "shadow" economy, increase in its criminalization, illegal outflow of foreign currency abroad [11]. At the same time, there are threats related to digital transformation, in particular: cyber-terrorism and cybercrime, insufficient efficiency of state regulation of the financial and credit sphere in terms of digital transformation, and others.

The degree of internal and external threats influence on the financial system is constantly changing due to the dynamic conditions of society activity, national and world economic systems, which are primarily influenced by the new for historical discourse process of digitization of social and financial relations, as well as incredible size of production globalization, politics and economics. It is difficult even to determine unambiguously whether these are internal threats to financial security condition of the state that have really greater influence, since more and more often destructive influence of external threats can significantly outweigh the danger from internal ones and complex influence, most often, leads to aggregate danger strengthening.

The process of digital transformation, in addition to e undeniable benefits, is a challenge to the financial security of the state and can be a catalyst for additional threats, the impact of which is maximized due to systemic imbalances that have accumulated and deepened over time and have not been overcome in recent years. Thus, shortcomings of the banking and legal system enable legalizing funds obtained by criminal means, using cryptocurrencies that facilitate the process of "money laundering". Another problem in this sector is online fraud targeting bank accounts and personal data, which are accessed due to operational deficiencies. Thus, ignoring the challenges of digitalization creates a situation where negative phenomena in the financial security of the state come into synergy, and therefore, in addition to identifying and counteracting existing threats, measures should be taken to prevent new threats through active research and institutional support of cryptocurrencies and cryptocurrency, digitization of business processes, protection of the media environment from "fake news", digital terrorism, fraud, piracy, etc.

Protection level of Ukraine from digital threats according to the international rating The Network Readiness Index is worth considering, with Ukraine in 2020 ranked 64th from 139 countries getting 49.43 points of 100 possible, 26 positions lower than Latvia, 31 positions than Poland and much lower than European countries with high people incomes level. Such indicators pose a threat to the beneficial development of digitalization and mean the creation of cross-border threats to Europe, which could worsen Ukraine's relations with the IMF. This indicator is affected worst by political and legal support and, thus, state and economic adaptation to digitalization processes in whole.

The sub-indexes value The Network Readiness Index of Ukraine in 2020 is presented in Fig. 3.

As we can see from Fig. 3 the highest negative impact on the formation of a low value of technological readiness of Ukraine integrated index has a low economic impact of participation in digital economy (26.17 points of 100 possible); unwillingness to introduce new technologies in the future (30.5 points of 100 possible); total

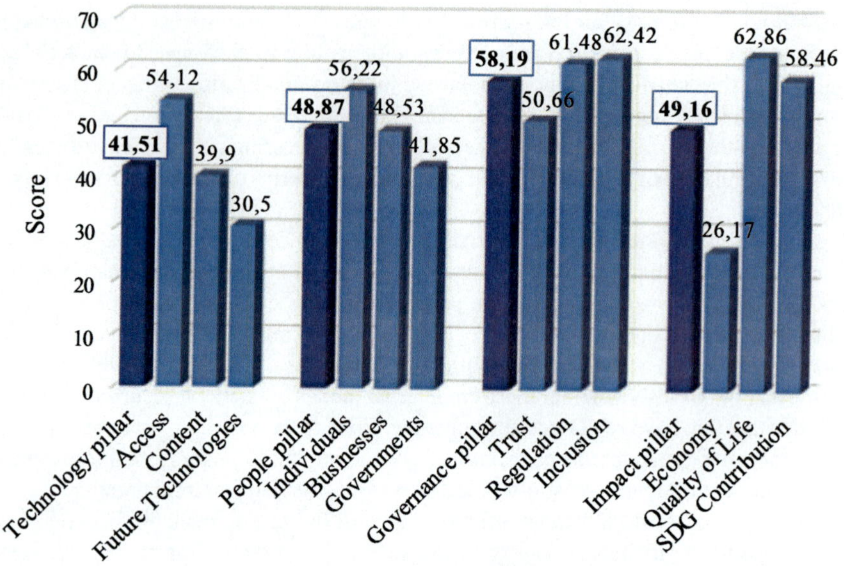

Fig. 3 The level of protection for digital threats in Ukraine

amount of innovations and digital technologies produced in countries (39.9 points of 100 possible). Compared to other sub-indexes, the state influence and assistance in the development of information technology is higher, but its value is generally not high—58.19 points of 100. Thus, the level of protection for digital threats in Ukraine is quite low, that confirms the necessity of the economy security-oriented management and the formation of new approaches to ensuring the financial security of the state in modern conditions.

A non-classical factor for influencing on financial security related to the introduction of information technology can be identified as misinformation, which is assessed by a number of indicators, such as the amount of suspicious financial transactions being reported to the national financial intelligence unit. Misinformation negatively affects the efficiency of the economy, GDP in particular, as a significant part is generated in illegal economic relations. The public danger of misinformation is reduced by turning undetected misinformation into detected misinformation, and the high probability of detecting misinformation deters people from spreading it. An important factor in misinformation counteracting is to protect the quality of information provided by the authorities and security institutions, and to increase public confidence in such data. It should ensure the trust and discipline of citizens in the implementation of security arrangements, but government officials must also fully adhere to the principles.

Therefore, under post-industrial globalization, new threats to financial system are associated with the introduction of modern technics and technological progress that requires information transparency and disclosure, market dynamics, moral and systemic risk. Ignoring these threats leads to them taking hold as well as shadow institutionalization, which completely eliminates the influence of economic regulators and requires more systematic countermeasure for them.

One of the first real threats facing financial security institutions is inefficiency and obsolescence of the very approach to identifying and countering threats, since new sectors of economic cooperation such as the cryptocurrency market, electronic payment services, cybercrime in the financial system and distribution are not considered. Therefore, threats in these areas have an unlimited impact on debt, budget and monetary security, which was especially reflected during the coronavirus pandemic. Creation of a new reformed methodological approach to the definition of financial threats only, taking into account new world trends and mass phenomena should be such counteraction. At the same time, the structure of financial security needs to be changed as a response to new challenges.

The European coalition countries and the United States are paying special attention to cybersecurity, which in recent decades has become an integral part of national and economic security. About two-thirds of cyberattacks occur in the financial sector [20], due to particular attractiveness of potential benefit, such as user data theft, solicitation of payments for information recovery, and due to low security of individuals, because of their differentiation and non-secure technical connection. Along with the impetus for the active digitalization of financial relations caused by the pandemic, the urgency of digital security issues has reached a new level and requires immediate actions. Thus, an important disincentive is the anonymity of the Internet, which makes it difficult to detect and combat crime. Ukraine is defined as a country with a high historical level of cyber threats and cybercrime [20]. Therefore, combating digital risks and threats is an important strategic goal for Ukraine to improve investment attractiveness, to solve domestic problems and to consolidate with international investors. A 30%-increase in digitalization in 2020 especially requires a large-scale state strategy deployment.

In conclusion, it should be noted that countering risks and threats is impossible within one country and should be expanded at the international level. Most countries with transition economy are affected by the development of stronger economies, but if they want to catch up with the leaders, they have to solve their problems faster. However, stronger partners may block reforms that are moving too fast in order to maintain their status quo. Thus, governments and financial institutions need to act under conditions of pandemic, globalization and trade expansion, taking into account both the actions of domestic shadow counterparties and the external aggressive environment. Most developing countries are trying to overcome external problems by integrating with stronger economies, as well as by adapting their fiscal policies to specific conditions.

Countering digital risks and threats is an important strategic goal for the state aimed at improving investment attractiveness, solving domestic problems and consolidating with global financial security. Effective countering the spread of threats is based on monitoring financial security of the state, development and implementation of appropriate strategic, tactical and operational management measures. This monitoring should be extended to hitherto ignored, in terms of financial security, areas of digital infrastructure and software. Since unforeseen threats associated with the introduction of digital technologies are not directly affected, they usually cannot be prevented, we can only develop the measures to reduce their impact indirectly affect related financial security factors.

3 Conclusion

The basis of information society concept, which is intensively implemented in modern conditions due to quarantine restrictions and additional opportunities for digital transformation, which allows to minimize negative impact of the pandemic and to adapt to new challenges, is the need for harmonious integration of people and technologies that continue to evolve and improve with the emergence of technological innovations. To ensure the effectiveness of this integration, it is necessary to implement modern security-oriented management mechanisms at the level of business structures and the state as a whole to address the problems of their inclusion and to ensure financial security.

In the process of digitalization, threats to financial security of the state become stronger, the price of information increases significantly and a new infrastructure for financial activities is created. Therefore, measures to prevent threats to state financial security are especially important in the context of digital infrastructure and software protection.

The primary problem of state financial security at the present stage is the failure of existing institutions to take into account the challenges of digital transformation, inefficiency and outdated approach to identification of threats and negative security factors. The outdated list of state financial security indicators does not reflect the current threats in the context of digitalization and new sectors of economic cooperation, such as cryptocurrency market, electronic payment services, cybercrime in the financial sector and dissemination of false data. That is why, threats in these areas have unlimited impact on debt, budget and monetary security as parts of financial one. Therefore, to identify the current threats to financial security in the economic security of the state an indicator approach must be used, but with respect to the processes of digitalization.

References

1. Europe and the Global Information Society (1994). Recommendations to European Council. Brussels
2. Machlup F (1962) The production and distribution of knowledge in the United States. Princeton University Press, Princeton
3. Porat M (1977) The information economy. nine volumes. Office of Telecommunication, US Department of Commerce, Washington

4. Castells M, Himanen P (2002) The information society and welfare state—the Finnish Model. Oxford University Press
5. Webster F (2002) Theories of the information society, 2nd edn. Routledge, London and New York
6. The 2012 PREDICT report (2012) An analysis of ICT R&D in EU and beyond. JRC, European Commission
7. Dubov D, Ozhevan O, Gnatyuk S (2010) Information society in Ukraine: global challenges and national opportunities. Analytical report. NISS, Kyiv, Ukraine
8. Baranovskyi O (2000) Financial security in Ukraine (assessment, methodology and mechanism for maintenance). Institute for Economic Forecasting of Ukrainian National Academy of Sciences, Kyiv, Ukraine
9. Blank I (2002) Management of financial security of the enterprise. Nika-Center, Kyiv, Ukraine
10. Varnaliy Z, Burkaltseva D, Sayenko O (2011) Economic security of Ukraine: problems and priorities of strengthening. Znannia Ukrainy, Kyiv, Ukraine
11. Onishchenko S, Matkovskyi A, Puhach A (2014) Analysis of threats to economic security of Ukraine in conditions of innovative economic development. Econ Ann XXI 1–2(2):8–11
12. World development report 2016: Digital dividends. A World Bank Group flagship report, Washington. https://openknowledge.worldbank.org/bitstream/handle/10986/23347/210671RuSum.pdf?sequence=16
13. Negroponte N (1995) Being digital. Alfred A. Knopf, New York, USA
14. Tapscott D (1995) The digital economy: promise and peril in the age of networked intelligence. McGraw-Hill
15. Tapscott D (2014) The digital economy anniversary edition: rethinking promise and peril in the age of networked intelligence. McGraw-Hill
16. Sergunin A (2005) International security: new approaches and concepts. Political studies 6:126–137
17. Bogoyavlensky S (2010) Risk management in socio-economic systems. Publishing house SPbGUEF, SPb
18. Ligonenko L, Khripko A, Domansky A (2018) Content and mechanism of digitalization strategy formation in business organizations. Int Sci J "Internauka" 22. https://www.internauka.com/uploads/public/15464416163127.pdf
19. The Network Readiness Index 2020 (2020) Portulans Institute. https://networkreadinessindex.org/wp-content/uploads/2020/11/NRI-2020-V8_28-11-2020.pdf
20. Onyshchenko S, Yehorycheva S, Furmanchuk O, Maslii O (2020) Ukraine construction complex innovation-oriented development management. In: Lecture notes in civil engineering, vol 73, pp 687–700
21. Kopp E, Kaffenberger L, Wilson C (2017) Cyber risk, market failures, and financial stability. IMF working paper 185. Western Hemisphere and Monetary and Capital Markets Departments

Alternative Energy Construction in Ukraine: Analysis and Economic Feasibility

Volodymyr Onyshchenko, Svitlana Sivitska, Anna Cherviak, and Volodymyr Datsenko

Abstract In modern terms, solving the problem of improving Ukraine's energy security is considered through the possibility of using the alternative energy market potential, the prerequisite for effective regulation of which is to determine strategic priorities for its development, creating an appropriate institutional environment and regulatory framework.

The state and prospects of alternative energy development and direct construction of renewable energy generation facilities in Ukraine are analyzed. The structure of energy consumption according to the categories of energy consumers is studied. The logical sequence of direct and indirect impact of construction of renewable energy sources on the economic stability and security of the country is determined. It was analyzed the growth tendency of using non-traditional energy in context of the state, households and comparison of the energy security level in European countries.

One of the main ways to restore and further develop the country's economy is its transition to using and generation of «green energy», as RES-related projects will be cost-effective and help attract investment.

Keywords Alternative energy · Construction of power plants · Renewable sources · Non-traditional energy sources · Energy security · Economic security

1 First Section

1.1 A Subsection Sample

At the beginning of the XXI century for mankind, the constant provision of energy needs without harming environment has become extremely important. Given this, using of alternative energy sources has global perspective for the further successful civilization development. Currently, there are world phenomena that violate the stability of civilized development of society:

V. Onyshchenko · S. Sivitska · A. Cherviak (✉) · V. Datsenko
National University «Yuri Kondratyuk Poltava Polytechnic», Poltava, Ukraine

V. Onyshchenko et al. (eds.), *Proceedings of the 3rd International Conference on Building Innovations*, Lecture Notes in Civil Engineering 181,
https://doi.org/10.1007/978-3-030-85043-2_70

- traditional energy sources are exhausted;
- cost increase of extraction;
- intensive environmental pollution;
- destruction of the biosphere;
- excessive amounts formation of organic waste from industrial, agricultural and domestic origin.

In this context, the development of alternative energy is relevant and requires intensification of investment processes in this area.

A significant contributions in problems of energy security assessment and ways to ensure it research were made by follow scientists V. Barannik, S. Bevz, Z. Varnaliy, D. Voloshin, A. Galchinsky, V. Geets, V. Gorbulin, G. Darnopikh, O. Zakrevsky, I. Zaremba, M. Zemlyany, D. Zerkalov, Y. Kolesnyk, V. Ksyonzenko, A. Sukhorukov, O. Serdyuchenko, A. Shevtsova, V. Shlemko, A. Shchokin, O. Yuspin and others, these problems are covered in hearings of foreign scholars, such as: L. Abalkin, L. Goncharenko, P. Kanigin, V. Kryukov, K. Samsonov, V. Senchagov, A. Tatarkin, D. Sher and others. Problems of development oil and gas complex were dealt by P.P. Borshchevsky, Y.V. Vytvytsky, I.V. Diak, E.I. Kryzhanivsky, N.A. Mamontova, A.V. Nikonov, V.O. Onyshchenko, V.L. Saprykin, I.I. Temnenko and other.

Ukraine's national security directly depends on purposeful state policy to protect its national interests in the political, economic, social and environmental spheres.

The national economy's development and effective functioning directly depends on ensuring economic security. In turn, energy security is its component, which level is characterized by the availability of natural resources.

One of the main factors limiting Ukraine's economic security is its high energy dependence on from energy imports. In today's conditions, it is undeniable that without solving the problem of energy supply, it is impossible to successfully implement economic, scientific, technical and social programs aimed at ensuring stability of the national economy. Therefore, ensuring the energy security of the state is determined by the priority of state policy at the present stage of development of the Ukrainian economy.

Today, energy sources are divided into traditional and alternative. The traditional minerals include—oil, gas, coal which are non-renewable. It is their exhaustion and impact on the planet's ecology (gas emissions from the combustion of minerals) that motivate to find alternative solutions in energy supply.

According to the Law of Ukraine «On Alternative Energy Sources» [1], alternative energy sources are renewable energy sources, which include solar, wind, geothermal, hydrothermal, aerothermal, wave and tidal energy, hydropower, biomass energy, organic gas waste, sewage treatment plants, biogas and secondary energy resources, which include blast furnace and coke oven gases, methane gas, degassing of coal deposits, conversion of waste energy potential of technological processes.

Alternative energy [1]—the field of energy, which provides electrical production, thermal and mechanical energy from alternative energy sources.

Ukraine has sufficient potential for the development of renewable energy sources and the replacement of traditional fuel and energy resources (Fig. 1).

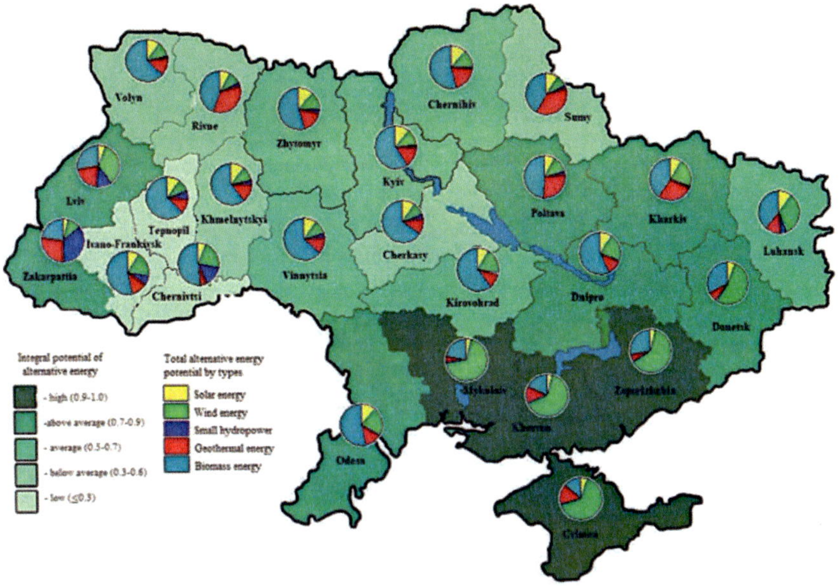

Fig. 1 Map of integrated and aggregate potential of alternative Ukraine energy by regions and types

The development of alternative energy and power plants construction will have a direct impact on the country's economy, and RES-related projects will help address energy issues to communities and stimulate economic processes, creating jobs, boosting business, replenishing budgets at various levels (Fig. 2).

The National Commission State Regulation of Energy and Utilities [2] has electricity consumption statistics data. Electricity consumption for all consumer groups decreased, except agricultural consumers and population category, as can be seen from Table 1. This trend is related to the pandemic and quarantine restrictions that have been introduced in Ukraine. Thus, work in almost all industries was switched to remote or «online» mode, so population electricity consumption increased. Agricultural work did not require significant contact with population and need in food (cereals, grain, wheat, etc.) only increased.

Thus, even during the global crisis, energy consumption did not decrease significantly, and RES market participants provided electricity and received passive income.

Thus, as for the installed capacity of RES, according to Ukrenergo [3] since 2017 there is a tendency to increase. The increase in energy MW in the solar power sector is particularly clear, especially the sharp jump in 2019 when it increased from 1224.8 MW in 2018 to 3537.3 MW in 2019 (Fig. 3).

According to the National Commission State Regulation of Energy and Utilities (NCRECP) in 2019, «green» tariffs were set for 511 electricity facilities that produce electricity from alternative energy sources.

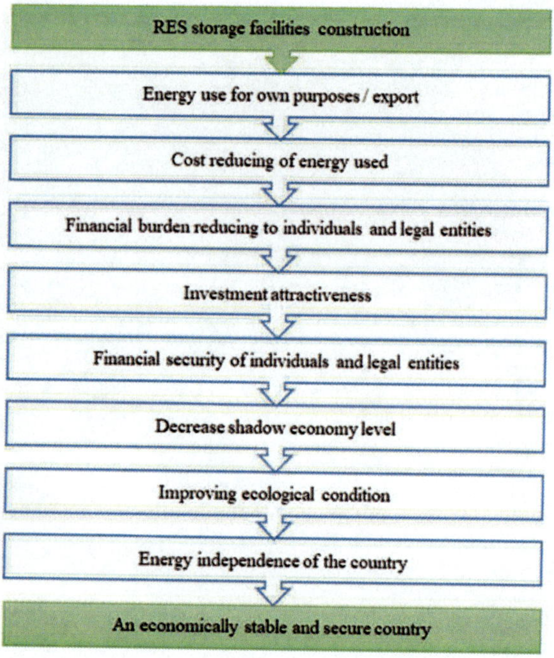

Fig. 2 The sequence of connection between RES facilities construction and country's economy

Table 1 Electricity consumers groups in 2017–2020

Consumer groups/year	2017	2018	2019	2020
Net consumption	118927,05	122143,54	120219,40	117906,90
Industry	50952,02	52023,10	51154,60	49307,40
Agricultural consumers	3642,09	3867,79	3710,00	3796,90
Transport	7044,00	6955,04	6603,30	5712,50
Construction	891,78	964,39	967,40	956,80
Com.-life. I lived	15016,29	15506,41	15066,20	14195,80
Other industry. Consumed	6361,02	6880,06	7482,00	7383,30
Population	35019,86	35946,77	35236,00	36554,10

The total installed capacity of electricity facilities that produce electricity from alternative energy sources and which have a «green» tariff in 2019 is 4,249.6 MW (of which wind farms—637.1 MW, SES—3,537.3 MW, biomass/biogas—71.3 MW, micro-, mini- and small hydropower plants—3.9 MW). At the same time, in 2019, due to the entry into force of the new electricity market and need to obtain licenses in accordance with new licensing conditions for electricity production, there was an additional adjustment of total electricity facilities installed capacity power projects based on the power of current generators located on the site [2].

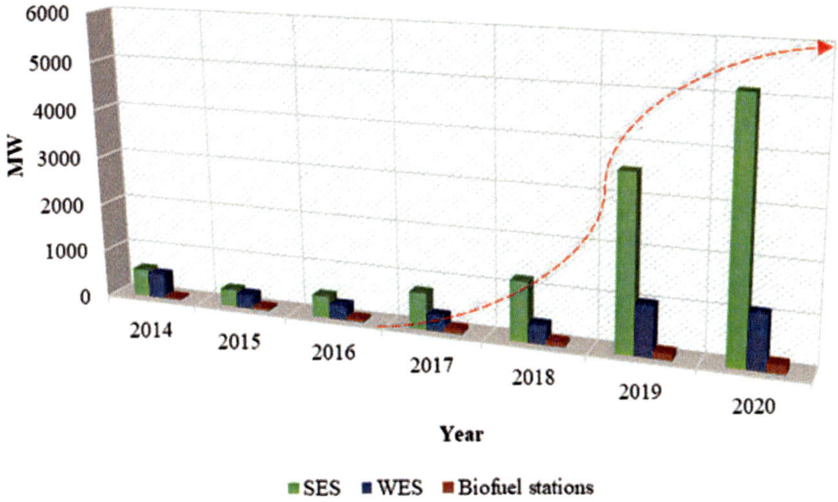

Fig. 3 Installed capacity of renewable energy sources for 2014–2020

The construction and commissioning of new electricity facilities producing electricity from alternative energy sources will increase employment in energy sector and increase the number of new jobs.

As 01.01.2020, the energy sector related to the production of electricity from alternative energy sources employs 7,840 workers [2]. The dynamics of the number of economic entities and electricity facilities that produce electricity from RES is shown in Fig. 4.

According to the Ukraine State Tax Service [4] in 2019, participants in the alternative energy market paid 19.8 billion UAH to the budgets of all levels, which is on

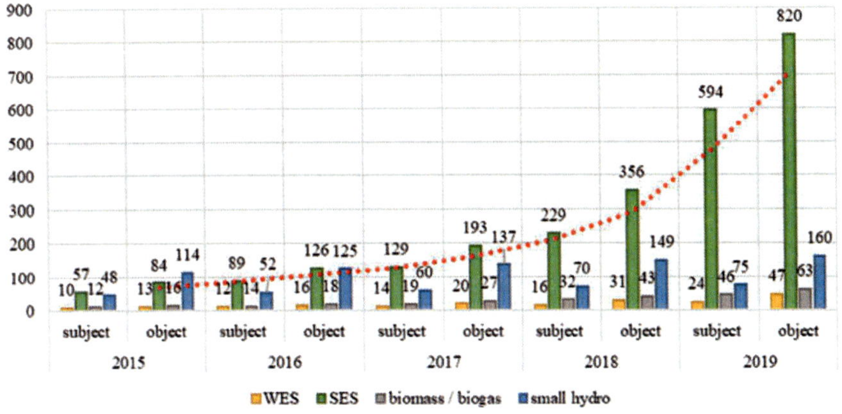

Fig. 4 Subjects and objects figures dynamics for the RES electricity generation

three times more than paying taxes to the coal sector. In 2009–2019 period operators of RES installations paid 93.6 billion UAH to the state budget, which is on 45.4 billion UAH more than the level of tax payments by coal mining and heat generating enterprises.

According to the Law of Ukraine «On the electricity market» [5], the consumer has the right to install the following generating units designed for electricity production:

- household consumers in their private households—generating units designed for generating units with an installed capacity not exceeding 50 kW, intended for the production of electricity from solar energy and/or wind energy;
- other consumers, including energy cooperativesenerating units designed for generating units with an installed capacity not exceeding 150 kW, designed to produce electricity from solar and/or wind energy, biomass, biogas, hydropower, geothermal energy.

In this case, the production of electricity is carried out without a license.

Private households sell generated electricity in excess of the monthly consumption of such a private household at a «green» tariff to the universal service provider based on contract concluded between them for the purchase and sale electricity at a «green» tariff, which is executed by the parties in the form of an annex to the contract on supply of electricity by the universal service provider.

In total, in 2019, USP purchased electricity generated by generating units from private households in the amount of 284,007.8 thousand kWh at a cost of 1,496.7 million UAH (excluding VAT). As of the end of 2019, there are 22,462 private households with generating units designed to produce electricity from solar energy and/or wind energy, which is 14,983 or 3 times more than in 2018 [2].

The largest number of generating units, according to the operative data of the National Commission for Electricity Generation, intended to electricity production from solar radiation and/or wind energy (according to the number of concluded contracts for purchase and sale of electricity at a "green" tariff by private households), is established in Dnipropetrovsk—2,704 or 12.0% of the total number of generating plants, Ternopil—2,070 or 9.2%, Kyiv—1,797 or 8.0%, Ivano-Frankivsk—1,774 or 7.9%.

In complicated times, the «green transition» is one of the ways to rebuild countries, taking into account economic, social and environmental priorities. Thus, «green» energy has attracted more than 1.24 billion Euros to Ukraine investment in 2020, even in difficult pandemic time. In particular, according to the Ukraine State Agency Energy Efficiency and Energy Saving [6], last year 46 million Euros was invested in energy efficiency projects through the «warm loans» program (35 million Euros loans) and energy service (122 EU-CO agreements worth 11 million Euros), 1.2 billion Euros are invested in 1.6 GW of new capacity that generates electricity from renewable sources (Fig. 5).

In Ukraine, it is quite difficult to obtain funding for RES projects. It is explained, first of all, by the imperfection of the regulatory framework, which is designed to create favorable conditions for work in the Ukrainian market of renewable energy. Representatives of companies operating in the Ukrainian alternative energy market

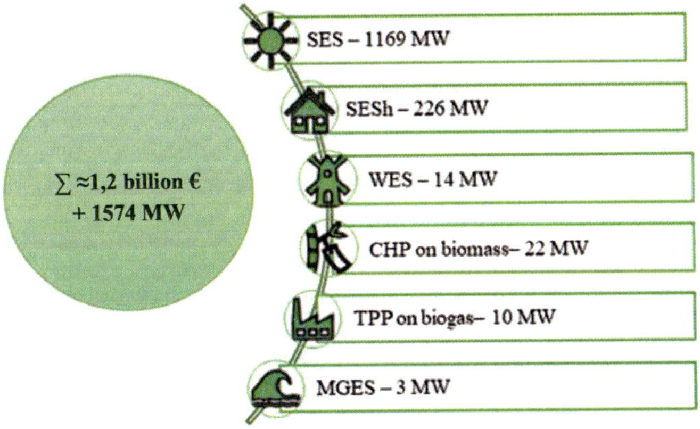

Fig. 5 Ukraine economy investments in 2020 due to the development of renewable energy

note that the current legislation is imperfect and needs significant additions and clarifications. In particular, in the field of alternative energy it is necessary to more clearly define the mechanisms and instructions for granting preferences to companies operating in the alternative energy market of Ukraine [7].

The development of the construction industry has a significant impact on the social and economic security of the country and each region [8]. Therefore, attracting investment is extremely important in the context of community development. The main stages of the investment process include [9]:

1. the emergence of objective needs and opportunities for investment;
2. justification of the investing feasibility;
3. forecasting and planning of investment activities taking into account the current rules and legislation;
4. preparation of investment projects and substantiation of their efficiency;
5. checking the real conditions of project financing;
6. project investment insurance;
7. implementation of investment projects, formation and control of results.

The share of renewable energy sources in electricity generation in Europe in 2020 for the first time in history exceeded the share of fossil fuels, according to the German Institute Agora Energiewende [10]. According to analytical data, in 2020 38% of all electricity was generated from renewable sources, and 37% from fossil fuels. This is due to the expansion of using energy from wind and solar, the volume of which has almost doubled since 2015. By 2020, one-fifth of electricity in the EU was generated by wind and solar power plants.

Alternative energy sources are actively used in construction, financial institutions are developing special lending programs, «warm loans» [11, 12]. Also, a project of the New Education Space for buildings was introduced in Ukraine [13–15]. All these measures are related to energy efficiency, environmental and economic security of the country.

The development of the alternative energy sector is a long-term energy and environmental priority for Ukraine, as provided by domestic legislation and participation in international agreements. In the long run, using RES is a guarantee of Ukraine's energy independence and energy network stability.

Ukraine has good economic reasons to move quickly and consistently towards the proper use of its financial capabilities and investment resources. At the services market, everything must be done transparently and reliably, and the government, in turn, must do everything possible to provide a legal framework for investors and market participants. Thanks to free and fair competition, Ukraine will quickly overcome the pandemic and move to sustainable economic growth.

References

1. Law of Ukraine «On Alternative Energy Sources», electronic resource. https://zakon.rada.gov.ua/laws/show/555-15#Text
2. National Commission State Regulation in the Spheres of Energy and Utilities, electronic resource. https://www.nerc.gov.ua/
3. National Energy Company Ukrenergo, electronic resource. https://ua.energy/vstanovlena-potuzhnist-energosystemy-ukrayiny/#01-2020
4. Ukraine State Tax Service, electronic resource. https://tax.gov.ua/
5. Law of Ukraine «On the electricity market», electronic resource. https://zakon.rada.gov.ua/laws/show/2019-19#n1177
6. Ukraine State Agency Energy Efficiency and Energy Saving, electronic resource. https://saee.gov.ua/
7. Sivitska SP (2014) Strategic directions of investing in alternative energy in the context of national economy development. Youth Econ Digest 1(1):76–80. http://nbuv.gov.ua/UJRN/med_2014_1%281%29__17
8. Varnaliy Z, Onyshchenko S, Zavora T (2018) Construction complex development influence on region social and economic security. Int J Eng Technol (UAE) 7(3):469–472. https://doi.org/10.14419/ijet.v7i3.2.14574
9. Onyshchenko V, Sivitska S (2014) Alternative energy developing investment support in terms of energy dependence. Econ Ann XXI 9–10:34–37
10. German Institute Agora Energiewende, electronic resource. https://www.agora-energiewende.de/ueber-uns/
11. Onyshchenko V, Sivitska S, Cherviak A (2018) Construction industry in Ukraine credit analysis. Int J Eng Technol (UAE) 7(3):280–284. https://doi.org/10.14419/ijet.v7i3.2.14420
12. Drobyazko S, Bondarevska O, Klymenko D, Pletenetska S, Pylypenko O (2019) Model for forming of optimal credit portfolio of commercial bank. J Manage Inf Decis Sci 22(4):501–506
13. Onyshchenko V, Sivitska S, Cherviak A (2020) Buildings reconstruction within the new educational space project in Ukraine. https://doi.org/10.1007/978-3-030-42939-3_66
14. Glushko A, Marchyshynets O (2018) Institutional provision of the state regulatory policy in Ukraine. J Adv Res Law Econ 9(3):941–948. https://doi.org/10.14505/jarle.v93(33).18
15. Sivitska S, Vartsaba V, Filonych O (2018) Buildings energy-efficient renovation investment. Int J Eng Technol (UAE) 7(3):408–412. https://doi.org/10.14419/ijet.v7i3.2.14562

Problems and Factors of Construction Business Innovation and Investment Development

Liana Ptashchenko, Lyudmyla Svystun, Maya Chechelashvili, and Kateryna Shtepenko

Abstract The article is dedicated to consideration of theoretical and applied aspects of the main balance factors of construction business successful innovation and investment development. The analysis of scholars' works enabled presenting the components of a construction enterprise innovation and investment development, in particular process, resource and potential. Considering the relationship between innovation potential and development, the author's definition is suggested, where construction business innovative development is an activity based on constant search and use of new methods, technologies and areas of enterprise potential realization in changing external and internal market of construction products and the adopted competitive strategy of activity and business development. Analysis of the problems associated with conservatism and slowness in introduction and dissemination of new technologies in the construction industry is performed. Despite the fact that energy-efficient technologies are being actively introduced in the industry, "smart homes" are being built etc., analytical report of the Global Innovation Index states that the share of R&D expenditures in construction is only 2.9%. One of the reasons for such a low rate is that a significant part of new technological developments implemented in construction, originate from other industries, namely metallurgy, forestry and woodworking, chemical industries and others. It is established that innovation inertia of Ukrainian construction companies is determined by buildings and structures prolonged operation, during which the faults of the used technology may be revealed as well as builders' high responsibility for the result, since construction products must be safe for people's lives. In order to implement the strategy of construction company innovative development, it is suggested to create the success factors of innovation and investment development, which are balanced and to some extent interdependent. Adequate assessment these factors impact will enable choosing the right direction

L. Ptashchenko (✉) · L. Svystun · K. Shtepenko
National University "Yuri Kondratyuk Poltava Polytechnic", Pershotravnevyj Avenue 24, Poltava 36011, Ukraine
e-mail: lianaptaschenko63@meta.ua

M. Chechelashvili
Georgian Technical University, Kostava street, 77, Tbilisi, Georgia

V. Onyshchenko et al. (eds.), *Proceedings of the 3rd International Conference on Building Innovations*, Lecture Notes in Civil Engineering 181,
https://doi.org/10.1007/978-3-030-85043-2_71

of construction business innovation and investment development and construction industry in general under pandemic and growing economic selfishness.

Keywords Innovation and investment development · Construction business · Potential · Construction · Technologies · Innovation activity

1 Introduction

The world economy and national economies of most countries are in crisis due to quarantine measures taken to combat COVID-19. Thus, the reduction in gross domestic product of the United States is estimated at an average of 3.3% per year. In countries with developed tourism, GDP is projected to fall even more. For example, in Greece and Cyprus - up to 6%, Montenegro - by 8%, in Lithuania and Latvia - by 7%. In Ukraine, the forecast for GDP decline is 7–8%, and according to the latest data - by 4.5% [1]. Such trends lead to increase in economic selfishness in society, inhibit restrain of investment processes and, consequently lead to economic crises of enterprises, individual industries and national economies. Lack of investment financing of Ukrainian enterprises negatively affects their innovation activity and makes it impossible to bring Ukraine closer to European Union standards of economic development. Problems of investment and innovation nature that arise as a result of the pandemic, actualize the study, which is dedicated to theoretical and applied aspects of balance of innovation and investment development of construction industry under pandemic and growing economic selfishness.

2 An overview of the Latest Sources of Research and Publications

The works of such scientists are devoted to the problems of investment-innovation and safety-oriented development of construction industry: Torkatyuk V.I., Drill N.V., Khoroshko I.O., Zheleznyakova I.L., Onyshchenko V., Sivitska S., Trach R. V., Onyshchenko S., Yehorycheva S., Matkovskyi A., Puhach A., Svistun L., Glushko A., Lukianov A. M., Kysh L. M. [2–10]. The problems of innovation-oriented management in construction enterprises focusing by Onyshchenko S., Yehorycheva S., Furmanchuk O., Maslii O., Svistun L., Glushko A. [5, 8]. Onyshchenko V. and Sivitska S. are investigating financing of construction industry in Ukraine [7]. Torkatyuk V.I., Drill N.V., Khoroshko I.O., Lukianov A. M. considered the principles of innovative development of construction companies [2–4].

3 Main Material

Governments of economically developed countries are pursuing a course of innovative development. Yet in Ukraine, this development is hampered by a number of objective and subjective reasons. And those innovations that are implemented, are unbalanced in different sectors of the economy. For example, construction is given much less attention than industry or IT technologies in the context of monitoring and stimulating innovation and investment activities [2].

Much less work is dedicated to the issues of construction industry innovative development than other industries. However, importance of the construction industry in the system of national production and the dynamics of its development in recent years indicates the problem importance. Lag of the Ukrainian economy from the average European level of economic development is so significant that even large-scale deployment of construction activity will not be effective, since outdated logistical and technological base requires significant investment in innovation.

The main objective of construction industry innovative development is to create competitive advantages in the strategic perspective, which create safe and comfortable environment for human life, which meets high world quality standards, to ensure sustainable socio-economic development of the country. Achieving competitive advantages should be based on innovative re-equipment of the construction industry, formation of innovative competencies, engineering schemes for managing the construction site life cycle, use of information modeling to increase productivity, reduce energy, material consumption and cost of construction products. Construction industry makes a significant contribution to national economy modern infrastructure creation and development, on which functioning of industries, enterprises of the real sector depends. Competitiveness of construction industry is an important factor for its sustainability ensuring, which guarantees the population employment, comfort of living environment, quality of housing, and has a crucial role in reducing greenhouse gas emissions. Further development of construction industry is associated with innovative efforts and innovation activity, which should result in new building materials, construction design technologies, construction, operation of buildings and structures, access to new markets for construction services (exports), as well as transition to new management technologies (engineering companies creation) etc. Innovative development of construction industry is based on implementation of national, regional and corporate innovation programs, projects for development of its innovative potential and innovative culture of construction production [3].

It should be noted that in most cases in the research by domestic and foreign scholars the term "innovation and investment development" is used when referring to the appropriate type of development at macro level and research on mechanism of scientific and technological progress in the country's economic development, formation of so-called knowledge economy, search for new sources of economic growth, construction of a state innovative model of development etc. In this case, innovation and investment type of development is understood as a way of economic growth based on constant and systematic innovations aimed at significant improving of all aspects of economic system, periodic regrouping of the forces due to the logic of scientific and technological progress, goals and objectives of the system, possibility of using certain resource factors in the creation of innovative goods and formation of competitive advantages; an innovative one is such a model of development that is directly based on obtaining new scientific results and their technological implementation in production, providing GDP growth mainly through production and sale of science-intensive products and services [4].

As a rule, the term "innovation and investment development" has long been used not only at the macro but at the micro level as well. The existing definitions of "innovation and investment development of the enterprise" concept presented in the scientific works of scientists are heterogeneous (Table 1) [9].

The analysis of scholars' works enabled presenting the components of construction business innovation and investment development (Fig. 1).

Considering the relationship between innovation potential and development, let us define that innovative development of a construction company is its activity based on constant search and use of new methods, technologies and areas of realization of the company's potential in terms of changing conditions of external and internal construction products market within the vision and accepted competitive strategy of the enterprise activity and development.

It was considered previously that construction companies characteristic is their conservatism and slowness in introduction and dissemination of new technologies. In the ranking of leading economic states innovation and investment active industries, construction complex took one of the last places. In the special literature, the label "laggard industry" was used for construction. The main argument was that many studies refer to extremely low share of research and development in the overall cost structure of construction companies [3].

As stated in the analytical report of the Global Innovation Index, the share of R&D expenditures in construction is only 2.9%, while ICT equipment and electronic equipment is 23.5%, pharmaceuticals and biotechnology - 18.8%, cars - 15, 6%, software and ICT services - 14.4% (Fig. 2).

Statistics confirms low innovation activity in the construction industry. Thus, during 2017–2019, the highest share of innovative enterprises was at the enterprises of information and telecommunications (22.1%), processing industry (21.9%), financial and insurance activities (21.7%), activities in the field of architecture and engineering (20.1%). Whereby the share of enterprises with technological innovations was higher than the national average among processing enterprises (15.6%), electricity, gas, steam and air conditioning supply (12.6%), as well as enterprises engaged in activities in the spheres architecture and engineering, research and development, advertising - 13.2%; with non-technological innovations - among enterprises of financial and insurance activities (18.0%), information and telecommunications (17.3%), processing industry (15.3%) [18].

However, in the recent years the situation with innovations in the construction industry has been somewhat improving. Energy-efficient technologies are being introduced, "smart homes" are being built etc. In addition, it is necessary to make a significant correction to the fact that a considerable part of new technological

Table 1 Scholars approaches to interpretation of "innovation and investment development of the enterprise" definition

Author	"Innovation and investment development of the enterprise" definition	Definitions
O. Adamenko [11]	Enterprise activity that is based on constant search of new methods and means of consumer needs satisfaction and increase in managing efficiency	New methods and means of meeting consumers' needs, efficiency increase; innovations introduction in various spheres of activity
I. Borysova [12]	Creating attractiveness in the stock market in terms of return on investment risk, i.e. increase in business value through innovation management	Investment risk profitability, business value
H.Gumba [10]	Development of a system of factors and conditions necessary for innovation and investment process implementation, i.e. innovation and investment potential	Innovation and investment process; innovation and investment potential
S. Illiashenko [13]	Economic management process that is based on continuous search and use of new ways and areas of enterprises potential realization in terms of changing environmental conditions	Economic management process; potential realization; activity modification; market outlets
O.Moros [14]	Process of finding and creating new products and processes based on the use of all available tools and capabilities of the enterprise, which leads to qualitative changes	Qualitative changes, creation of new products and processes
T. Piliavoz [15]	The process of purposeful, consistent enterprise movement to a balanced innovation and investment state under the influence of synergistic action of external and internal factors that determine the enterprise organizational and functional system stability under the market economy	Purposeful, consistent movement towards a balanced state of innovation and investment; the result of quality, which depends on intensity and speed of innovation and investment processes
I.Fedylova [16]	This enterprise, where the source of development is innovation, that develops	Development through innovation

developments that are implemented in construction, originate from other industries - metallurgy, forestry and woodworking, chemical industries and others.

Due to the influence of a number of factors, the situation has changed profoundly over the last decade. Conservative construction industry is likely to be forced to abandon its established traditions and go through a series of radical changes. Thus,

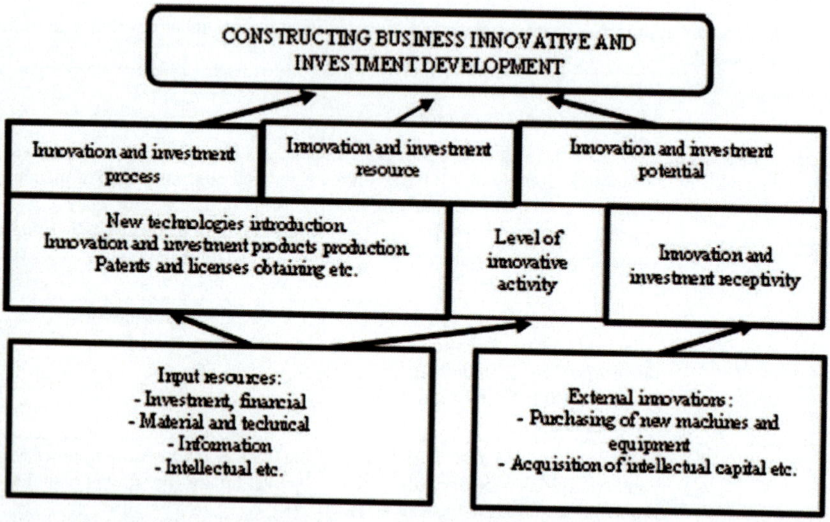

Fig. 1 Construction business innovation and investment development components

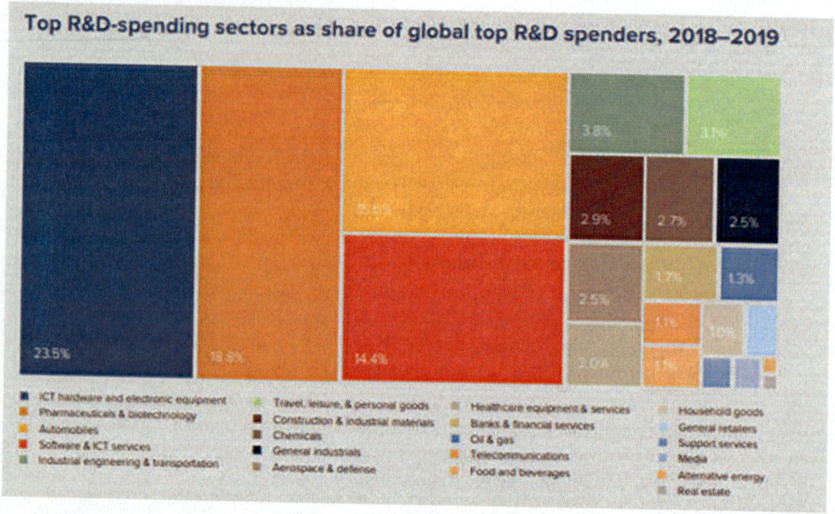

Fig. 2 Top R&D-spending sectors as share of global top R&D spenders, 2018–2019 [17]

rapid implementation of computer methods of information modeling (BIM) of all the key stages of the construction cycle and other advanced IT-technologies already has a significant impact on the innovative component of the industry.

In just a few decades, the very set of materials and technologies used in the field may change almost beyond recognition [19].

Despite the positive changes in the innovative development of construction industry in economically developed countries, most Ukrainian construction companies are not particularly prone to innovation. The inertia of the construction complex enterprises is determined by several factors. First of all, it is a buildings and structures prolonged operation, during which the faults of the applied technology may be revealed. Therefore, construction companies are extremely careful in choosing new materials or methods of construction. The second reason for conservatism is high responsibility of builders for the result, because the use of inappropriate technology or design errors can pose an immediate danger to the lives of a large number of people.

Besides, the lack of innovation activity of construction companies is explained with lower level of globalization than in industries, mainly due to the length of the construction cycle, existence of a large number of small and medium size enterprises that are reluctantly conservative since they cannot invest in research, and do not have the necessary competence to evaluate and use high-tech innovations. For this reason, innovations in the construction industry in all the developed countries are also carried out mainly by large enterprises, construction holdings and network associations. Network organizational structure enables providing services in several areas of construction activities simultaneously, to involve different suppliers of construction materials and equipment in construction works, to provide interaction of subjects of construction production by means of information and communication technologies and information modeling.

The pace and scale of technological progress in the industry will depend on degree and speed of transition to automated methods of construction and mass implementation of robotics and technology with minimal human intervention. Thus, many experts today agree that one of the key trends in the coming decades in the construction industry should become accelerated transition from traditional technologies of buildings construction directly on construction sites (on-site manufacturing) to prefabricated (offsite) housing construction and then - to practically conveyor production of houses from unified panel or modular components designed by means of computers.

In order to choose a certain strategic direction of innovation and investment development, enterprise managers need to determine and assess the impact of success factors of innovation and investment development (Fig. 3).

Based on the assessment of the influence factors, the company can choose adequate areas of innovation and investment development.

According to the strategic direction of innovation and investment development, a construction company can choose several areas:

- balanced innovation and investment development, which is applied in terms of gradual technical changes;
- offensive innovation and investment development, which is used under rapid technical changes, when it is necessary to achieve advance or maintain market leadership through the use of new achievements of scientific and technological progress;

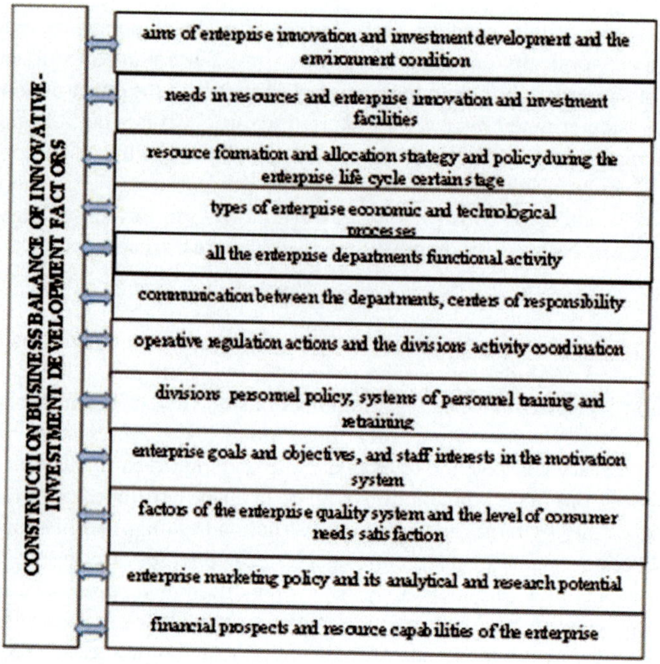

Fig. 3 Balance of the main factors of construction business innovation and investment successful development

– absorbing innovation and investment development, through nominal, rather than real innovation transformations [20].

Therefore, challenges facing the construction industry on the way to improving its efficiency can be formed via the agreed upon areas, the model of which is shown in Fig. 4.

Among the effective investment programs, the greatest multiplier effect have those related to residential construction, since this sector of economy has the closest relationship with other industries. In addition to positive impact on economic growth, residential construction performs an important social function: providing housing for the country population, and thus able to indirectly affect the country's demographic situation.

Despite the fact that energy-efficient technologies are being actively introduced in the industry, "smart homes" are being built etc., the analytical report of the Global Innovation Index states that the share of R&D expenditures in construction is only 2.9%. One of the reasons for such a low rate is that a significant part of new technological developments that are implemented in construction originate from other industries - metallurgy, forestry and woodworking, chemical industries and others.

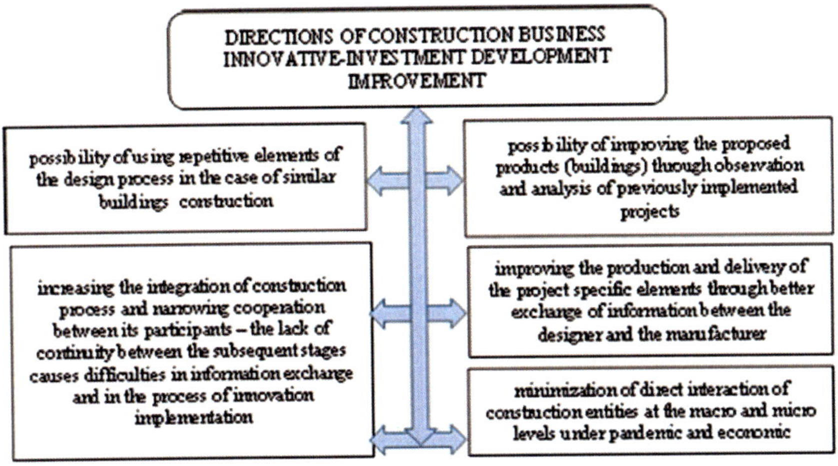

Fig. 4 Consistency in improving the construction business innovation and investment development

4 Conclusions

Global crises and pandemics make adjustments to the economic policy of development and support of various sectors of the national economy, including construction. Technological renewal of the construction industry on the basis of innovations is necessary for the of competitive advantage development in the strategic perspective due to increasing in global competition in the construction services market, acceleration of innovation and technological development and reindustrialization of the world economy, as well as requirements dictated by new technologies in the field of construction materials production, energy saving, energy efficiency, environmental issues and pandemics.

It is determined that innovative development of the construction business is an activity based on constant search and use of new methods, technologies and areas of the company's potential realization in terms of changing conditions of external and internal construction products market within the vision and agreed upon competitive strategy of the enterprise activity and development.

It is established that Ukrainian construction companies inertia to innovation is determined by the prolong operation of buildings and structures, during which there may be faults of the technology used, and high responsibility of builders for the result, since construction products must be safe for people's lives. In order to implement the construction business strategy of innovative development, it is suggested to create the success factors of innovation and investment development, which are balanced and to some extent interdependent. Adequate assessment of these factors impact will enable choosing the construction business and the construction industry in general right direction of innovation and investment development in terms of pandemic and growing economic selfishness.

In order to choose a certain direction of innovation and investment development, managers of a construction company must determine and assess the influence of the factors on success of innovation and investment development. Further research by the authors involves adapting existing models for estimating such factors to the construction business.

References

1. Povernennia do normalnosti: yak YeBRR bachyt vidnovlennia yevropeiskoi ekonomiky, http://www.eurointegration.com.ua. Accessed 14 May 2020
2. Torkatyuk VI, Drill NV, Khoroshko IO, Zheleznyakova IL (2020) Chornomordenko T.V., Kukhtin K.V. Directions of innovation process in construction of Ukraine, https://core.ac.uk/download/pdf/11335231.pdf. Accessed 14 Dec 2020
3. Trach RV (2017) Information modeling as one of the key factors of innovative development of a construction enterprise. Sci Bull Uzhhorod Natl Univ 13, Part 2, 1:129–132
4. Lukianov AM, Pleshkanovska AM (2017) Intelektualne misto – standart maibutnoho chy innovatsiinyi rozvytok miskykh terytorii. Mistobuduvannia ta terytorialne planuvannia 65:336–341
5. Onyshchenko S, Yehorycheva S, Furmanchuk O, Maslii O (2020) Ukraine construction complex innovation-oriented development management. In: Onyshchenko V, Mammadova G, Sivitska S, Gasimov A (eds) Proceedings of the 2nd International Conference on Building Innovations: ICBI 2019. Springer, Cham, pp 687–700
6. Onishchenko S, Matkovskyi A, Puhach A (2014) Analysis of threats to economic security of Ukraine in conditions of innovative economic development. Econ Ann-XXI 1–2(2):8–11
7. Onyshchenko V, Sivitska S, Cherviak A (2018) Construction industry in Ukraine credit analysis. Int J Eng Technol (UAE) 7(3):280–284
8. Svistun L, Glushko A, Shtepenko K (2018) Organizational aspects of development projects implement at the real estate market in Ukraine. Int J Eng Technol (UAE) 7(3):447–452
9. Soroka AM (2019) Innovatsiinyi rozvytok yak mekhanizm upravlinnia konkuentospromozhnosti telekomunikatsiinykh pidpryiemstv. Ekonomika. Menedzhment. Biznes 1:97–102
10. Kysh LM (2017) Stratehichnyi innovatsiinyi rozvytok pidpryiemstv v Ukraini. Prychrnomorski ekonomichni studii. 23:90–95
11. Adamenko OA (2010) Kontseptualni zasady innovatsiinoho rozvytku pidpryiemstv. Naukovi pratsi Natsionalnoho u-tu kharchovykh tekhnolohii 35:5–10
12. Borysova YS (2011) Osnovy innovatsyonnoho razvytyia predpryiatyi. Izvestyia PHPU im. V.H. Belynskoho 24:225–229
13. Illiashenko SM (2003) Upravlinnia innovatsiinym rozvytkom: problemy, kontseptsii, metody. VTD Universytetska knyha, Sumy
14. Moroz OS (2012) Formuvannia systemy pokaznykiv dlia otsiniuvannia innovatsiinoho rozvytku pidpryiemstva. Ekonomika Kryma 3(40):263–266
15. Piliavoz TM (2012) Innovatsiinyi rozvytok pidpryiemstva yak vazhlyvyi aspekt rozvytku ekonomiky. Innovatsiina ekonomika 4:185–190
16. Fedulova IV (2010) Synerhetychna evoliutsiina model innovatsiinoho rozvytku pidpryiemstva. Nauk. pr. Nats. un-tu kharch. tekhnolohii 36:114–118
17. Global Innovation Index (2020). https://www.wipo.int/edocs/pubdocs/en/wipo_pub_gii_2020.pdf. Accessed 01 Nov 2021
18. Derzhavna sluzhba statystyky (2021). http://www.ukrstat.gov.ua/ Accessed 11 Jan 2021
19. Holovatiuk VM (2017) Innovatsiinyi rozvytok Ukrainy v konteksti yevropeiskoi intehratsii. Nauka ta naukoznavstvo 3:3–22
20. Amosha OI, Salomatina LM (2017) Innovatsiinyi rozvytok promyslovykh pidpryiemstv u rehionakh: problemy ta perspektyvy. Ekonomika Ukrainy 3:20–34

Social, Financial and Ecological-Energy Criteria for Making Management Decisions in Construction

Lyudmyla Svystun, Iuliia Samoilyk, and Konul Aghayeva

Abstract The construction industry functioning on an up-to-date stage of economic development requires the implementation of energy efficiency and environmental friendliness principles, which is in line with global trends in sustainable development. Transformation processes have been encouraged construction companies to adapt their management decisions to meet the needs of construction products consumers. It has been really important under condition of the urbanization level growing.

Main trends and features of the construction industry of Ukraine and European countries have been analyzed; problems of the industry functioning during the pandemic have been identified. On the empirical research basis, financial and ecological-energy criteria of the housing choice by buyers in Ukraine have been substantiated, and also priorities changes in dynamics have been investigated. The mathematical function for making managerial decisions in construction has been formed.

Also the ways for financial problems decision in building sphere taking into account possibilities of the financial market have been developed. The energy-efficient decisions in the field of real estate exploitation have been proved.

Keywords Construction · Energy efficiency · Ecology · Social-financial indicators · Management

L. Svystun (✉)
National University "Yuri Kondratyuk Poltava Polytechnic", Pershotravnevyj Avenue 24, Poltava 36011, Ukraine

I. Samoilyk
Poltava State Agrarian University, 1/3 Skovorody Street, Poltava 36003, Ukraine

K. Aghayeva
Azerbaijan University of Architecture and Construction, Ayna Sultanova 11, Baku, Azerbaijan

V. Onyshchenko et al. (eds.), *Proceedings of the 3rd International Conference on Building Innovations*, Lecture Notes in Civil Engineering 181,
https://doi.org/10.1007/978-3-030-85043-2_72

1 Introduction

The construction industry is an integral part of the economic complex, an indicator of the growth or the decline of the economy. It is closely connected with other industries, their development through the production facilities creations, storage and offices, logistics and engineering networks. Construction companies' further development dynamics prospects are an important research area under transformation processes conditionals occurring around the world during the pandemic.

These processes significantly affect the economy of each country. Thus, not only external factors of influence, such as political, macroeconomic, globalization, but also internal guidelines for the functioning of enterprises are changing. An enterprise's activities vectors have been directed to the energy-saving, which saves the internal financial resources of the business entity and energy efficiency i.e. the most efficient use of all available resources provides.

A reduction of economic activity and incomes of economic entities influences a reduction of employment and incomes of citizens. At the same time, the trend of rising energy prices continues. Under such conditions, the dwelling's energy efficiency increase and the level of financial costs for its maintenance reduction are the key advantages of the proposals in the residential real estate market. This poses to make new challenges to the construction industry, which has suffered the most during the pandemic. Competitive advantages will be given only to those construction companies that offer to customers housing that meets modern environmental, energy efficiency criteria, will be economical and at the same time meets the social needs of people. That is, management decisions of construction companies must be adapted to modern pandemic conditions.

2 An overview of the Latest Sources of Research and Publications

The problems of energy efficiency and energy saving in the housing sector have been engaged by such economists: H. Elsharkawy and P. Rutherford, M. Villca-Pozo, B. Serrano-Lanzarote, I.G. Hamilton, B.Coyne, R.Lawrence and C.Keime, B.Lin, V.Brigilevich, V.Volkov, S. Sivitska, Komelina O. [1–11]. In particular, the studies of Spanish scientists regarding the quantitative determination of the energy saving potential in the Spanish housing fund, the formation of a strategy for its energy renewal and the introduction of tax benefits for the modernization of energy efficiency in housing in Spain are interest [6, 7]. Elsharkawy H. and Rutherford P. presented the results from an extensive pre- and post-retrofit home energy use and performance survey of 150 properties located in Nottingham's Aspley. Their research seeks to inform and improve the uptake and delivery of future housing retrofit initiatives [5]. The reasons for the high level of energy consumption in Ukrainian housing have been covered in the work by Sivitska S., Vartsaba V. and Filonych O. [8].

The issue of construction efficiency in rural areas has been partially explored in our previous work [9]. Many Ukrainian scientists study foreign experience of improving the energy efficiency in housing and technologies of rehabilitation and construction of energy efficient buildings. Mykytenko V. [10] study tools and spheres of energy efficiency state regulation. Komelina O.V., Samoilyk I.V., Boldyrieva L.M., Krapkina V.V. [11] comprehensively investigated energy efficiency in the use process of machines and equipment in construction. Problems of financing the construction business were studied by Onyshchenko V., Krekoten I., Svistun L. [12, 13]. Varnaliy Z. and Onyshchenko S. [14, 18, 19] studied construction complex development influence on region social and economic indicators.

Consequently questions of a constructions efficiency increasing are very important. But this problems needs further investigation taking into account current trends in the energy and construction spheres. Especially, the criteria for making decisions about construction, as well as the acquisition of real estate need to be studied in detail.

3 The Purpose of Scientific Research

The purpose of scientific research is to summarize scientific approaches to basic criteria for making managerial decisions in construction, identifying the main trends in the construction industry in the world with a focus on European countries, analyze empirical research results to rank socio-financial and ecological-energy criteria for making managerial decisions in construction, creating an appropriate model and the recommendations for improving criterions compliance to customers' expectations.

4 Main Body

The construction is one of the most dynamic industries for many countries. The construction development level shows the country's investment attractiveness, the technology development, innovation activity. The construction industry on the one hand is an economic development indicator, and on the other hand is an of social security indicator. Housing class, the new buildings number growth dynamics, the difference between construction infrastructure in rural areas, medium-sized cities and metropolises, the housing availability for different segments of the population affects other socio-economic spheres, namely industry, labor market, other areas of construction, financial and investment sphere, technologies, etc. In general, the world is growing rapidly. Before the coronavirus pandemic construction industry costs grew approximately to 12 trillion U.S. dollars, during next period this indicator will grow by three percent per annum. These constructions costs include building projects in residential and commercial real estate and also it is costs in infrastructure

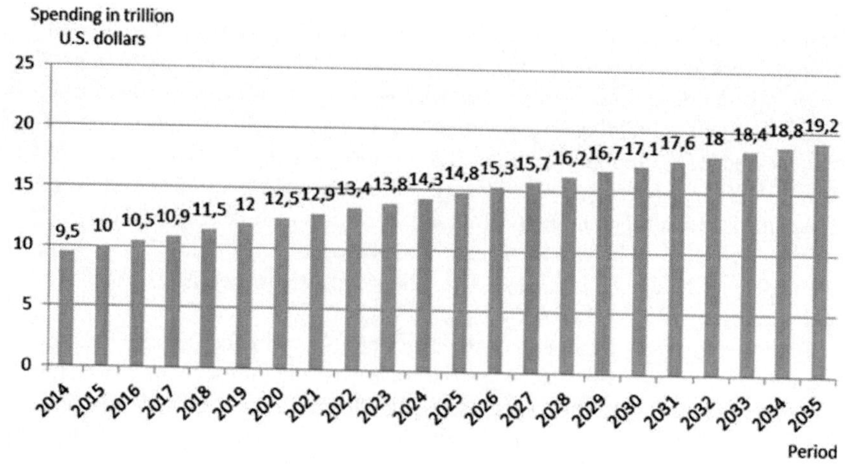

Fig. 1 A Construction industry spending worldwide from 2014 to 2019, with forecasts from 2020 to 2035 [15]

and industrial structures. Construction costs include labor and materials, architectural and engineering work, and taxes data spending (see Fig. 1).

Since 2014, there has been an increase in total construction costs from $ 9.5 trillion U.S. dollars to 12 trillion U.S. dollars. By analytics forecasts to 2035, construction will grow to 19.2 trillion U.S. dollars. Every year the differences between the level of construction development in terms of rural and urban areas become more noticeable. The level of urbanization in all regions of the world is growing significantly. At the same time, the American continents are the most urbanized (see Fig. 2).

The construction industries in the United States are one of the largest around the world. During 2019 and 2020, Buenos Aires shows one of the largest growths in the city's constructions industry. In 2021, Asia Pacific will have the highest construction industry production in the world.

Rises in the European construction industry carry out in recent years also. The EU construction market has been showing positive dynamics since 2014 after a prolonged decline in 2008–2013. Albania is showing significant indicators of the development of the construction industry. Also, the leaders by the EU construction market growth in 2019 were Malta (3.2 times), Ireland (3.1 times), Cyprus (3 times), and Hungary (2.9 times) (Table 1).

The construction pace has hardly changed in the countries of old Europe, in particular in the United Kingdom, the Netherlands, Norway, and Sweden. Slightly lower construction growths are in Eastern Europe: Lithuania, Slovakia, Poland, Bulgaria, Slovenia, Latvia and Hungary. Despite the increase in the tax press and the mass withdrawal of production abroad, the Greek market has been shown a positive pace in recent years. Three major European economies, namely Germany, France and Italy, have been shown a general trend of improving the situation in the construction industry. The German construction market has remained either more or less stable.

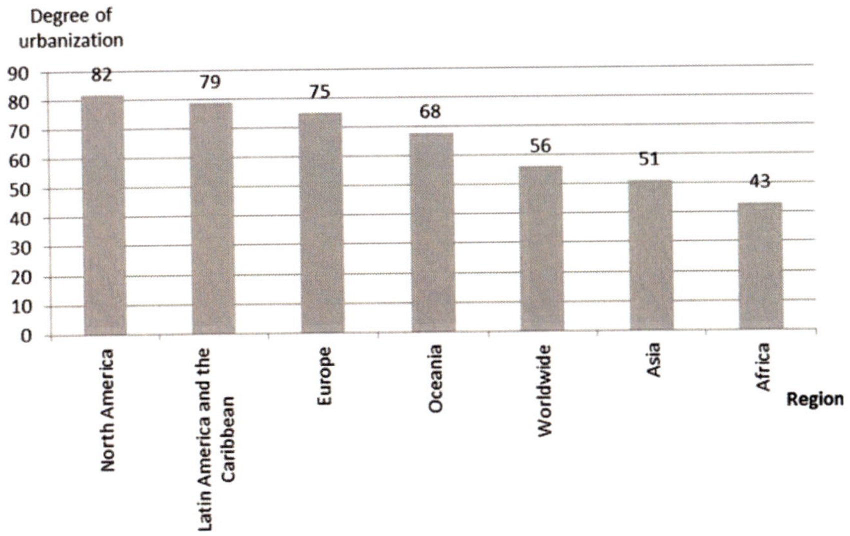

Fig. 2 Degree of urbanization (percentage of urban population in total population) by continent in 2020 [15]

France, after a long decline for a decade, showed growth, which was the beginning of the recovery process.

Favorable conditions are created for Ukrainian companies focused on the export of construction materials to enter foreign markets. In addition to such large and stable markets as Germany, Austria, Sweden, for Ukraine it is interesting and not so large, but fast-growing markets of Eastern Europe with a lower level of competition during periods of growth.

The construction industry affects the efficiency of the country's governance system. It belongs to the material production sphere and ensures the creation and restoration of fixed assets of economic entities and the population. The development of the construction industry contributes to the growth of building materials production, equipment, mechanical engineering, metallurgy, petrochemistry, woodworking, transport, energy, which leads to economic growth and solving many social problems.

The construction industry in Ukraine has been received a very big boost in recent years. According to the results of 2019, Ukrainian enterprises performed construction works in the amount of UAH 181,697.9 million UAH (Table 2).

The index of construction products in relation to 2018 was 123.6% (see Fig. 3). New construction, re-construction and technical re-equipment accounted for 74.2% of the total volume of construction work performed capital and current repairs (25.8%). According to the sub-bags of the year, the construction of engineering structures increased by 31.8%, buildings - by 25.1%.

The construction industry leaders are Kharkiv, Dnipropetrovsk, Poltava, Kyiv, Lviv, Odessa regions. However, despite the positive changes, there are still problems

Table 1 Building permits in European countries – annual data [16]

Countries	2013	2014	2016	2017	2018	2019
Albania	…	…	614,6	1083,3	1583,3	1500
Malta	68,5	74,4	190,2	248,8	326,5	316,3
Ireland	55,2	56,8	122,3	159,3	223,1	308,6
Cyprus	129,5	89,3	114,1	154,5	194	301,1
Hungary	59,2	77,4	257,1	312,3	302,4	291,1
Portugal	90,4	84,9	139,2	172,5	245,8	287,7
Serbia	76,2	77,4	126	168,9	184,8	238,6
Greece	128,8	104,6	103,5	123,9	175,9	234,4
Croatia	110,2	112,1	133,1	179,6	168,9	221
Spain	86,6	92,4	123,7	154,4	191,8	216,3
Bulgaria	71,1	91,8	105,2	144,6	205,8	189,9
Latvia	197,5	112,1	122	143,1	167,6	167,7
Czechia	84	90,9	104,2	123	127,2	152,4
Poland	72,4	82,5	112,2	133,8	138,2	144,5
Estonia	54,6	70,5	107,7	141	125,1	143,6
Luxembourg	90,8	145,7	120,6	123,6	137,3	134,8
Denmark	51,1	71,6	129,6	129,4	158,7	134,3
Italy	124,4	109	103,9	120,8	127,4	128,4
Slovenia	117,6	99,3	109,2	116,8	136,2	122,2
Austria	90,6	95,9	115,6	130	108,2	121,8
Belgium	107,1	119,1	111,4	110	136,1	121,5
Finland	84,2	94,5	126,2	150,3	135,9	121,3
Germany	90,1	93,2	115,6	113,1	115,4	119,4
Slovakia	74,7	81,1	114,6	104,7	116,6	115,5
Lithuania	88,3	83	122	119,4	122,5	113,8
France	106,3	93,1	114,7	122,1	114,1	110

Table 2 The volume of construction work performed by type of construction products in Ukraine, 2014–2019 [17]

Years	Construction total		Building		including				engineering structures	
					residential		non-residential			
	million UAH	%	million UAH	%	million UAH	%	million UAH	%	million UAH	%
2014	51108.7	100	24856.5	48.63	11292.4	22.09	1356.1	26.54	26252.2	51.37
2015	57515	100	28907.5	50.26	13908.8	24.18	14998.7	26.08	28607.5	49.74
2016	73726.9	100	38106.4	51.69	18012.8	24.43	20093,6	27.25	35620.5	48.31
2017	105682.8	100	52809.6	49.97	23730	22.45	29079.6	27.52	52873.2	50.03
2018	141213.1	100	66791.6	47.30	29344.8	20.78	37446.8	26.52	74421.5	52.70
2019	181697.9	100	83589.3	46.00	33208.8	18.28	50380.5	27.73	98108.6	54.00

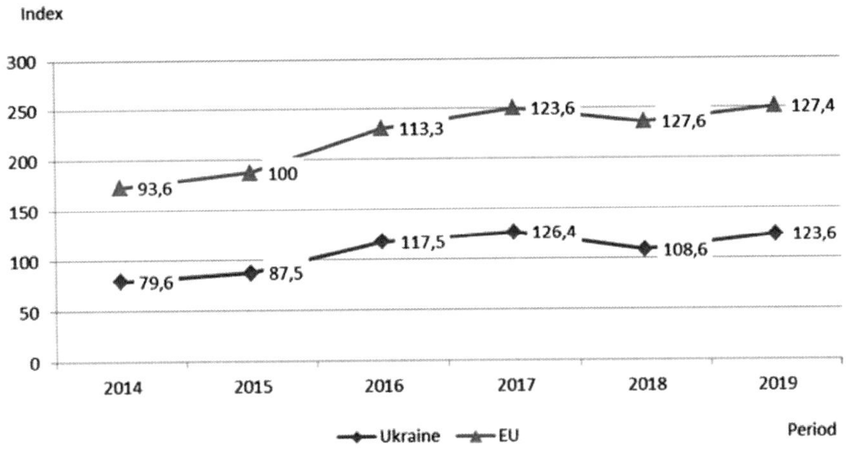

Fig. 3 The construction growth index in the European Union and Ukraine, 2014–2019, % [16, 17]

with registration, lending, investment, quality control in the construction industry. The construction industry in Ukraine includes housing construction, construction of industrial facilities, repair and construction of roads. In today's conditions, the construction industry of Ukraine operates mainly through the construction of residential buildings, as this component of the construction industry is in demand. Industrial, social and communal construction is practically not carried out due to lack of investment.

In order to develop proposals for effective management decisions in the construction sector and level the threatening trends of reduced construction during the pandemic by developing an appropriate anti-crisis management strategy, a survey has been conducted to identify current priorities in housing selection.

The conducted empirical research included a survey of different age groups respondents (of working age) and social groups from the metropolis (Kyiv), medium-sized city (Poltava) and rural areas of Poltava region. The majority of respondents belong to the age groups of 36–46 years (21.4%) and 46–60 years (26.5%) in an almost equal ratio of men and women. Among the respondents, 37.7% belong to families of 3 people and 76% are working with a stable income. As for the level of income, in Kyiv the majority of respondents have an average monthly income of 20–25 thousand UAH per person, in Poltava, it is 10–15 thousand UAH, and in rural areas, it is up to 10 thousand UAH.

Respondents over the age of 45 mostly have their own housing (over 82%), and young people, especially in Kyiv, mostly live in rented housing. Respondents from cities live in apartments (94%), but only 77% would prefer an apartment if there was a choice. Others would move to their own homes. In rural areas, on the contrary, the population lives and wants to continue living in private homes.

Respondents from Poltava and the region in 96% of cases would prefer their own housing as opposed to rent, and in Kyiv there were only 83%. In general, the mentality

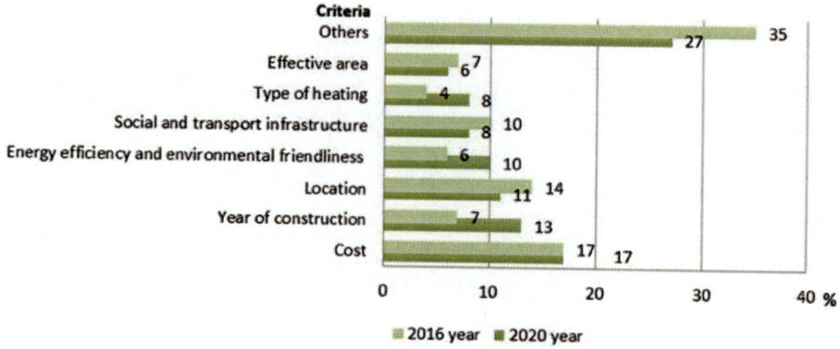

Fig. 4 Criteria for choosing a home when buying it in Poltava according to the 2016 and 2020 survey

and the idea of a successful life dictate to Ukrainians the need for their own housing. Given the need to choose housing, the vast majority would prefer a new building rather than a secondary market (this percentage is slightly lower in rural areas).

The study was conducted during 2016–2020. During this time, respondents' preferences regarding the key priorities of housing choice for the last 5 years have been changed. The following criteria were proposed for respondents: cost, area, planning, year of construction, location, environmental friendliness of the neighborhood and area, social and transport infrastructure, distance to work, energy efficiency and environmental friendliness of materials and structures, type of heating, maintenance costs (utility bills), the availability and size of the adjacent territory, the security of the neighborhood, etc. Thus, for a typical Ukrainian medium-sized city in 2016, the main criteria for purchasing housing were cost and location (see Fig. 4).

Also the availability of developed infrastructure and the area and layout of the apartment were important for respondents. Instead, in 2020, along with cost and location, the key criteria for a potential buyer were the energy efficiency and environmental friendliness of the materials and structures of a residential building and the type of heating. Cost, location, planning, infrastructure and distance to work are priority for a big city in 2016 (see Fig. 5).

And in 2020, along with cost and location, the emphasis has shifted to the energy efficiency and environmental friendliness of the materials and structures of a residential building, the type of heating and the reliability of the developer. In both cases, the year of construction is also important; everyone would prefer to buy housing in a new building. In 2016 also the leaders of the rating were the area of apartments, year of the construction, type of heating (see Fig. 6).

In 2020 from the sixth to the second place came the criterion of energy efficiency and environmental friendliness of the houses materials and structures, and the third place from the fifth moved the type of heating. Also, villagers are concerned about the cost of maintaining the house and the level of heating costs.

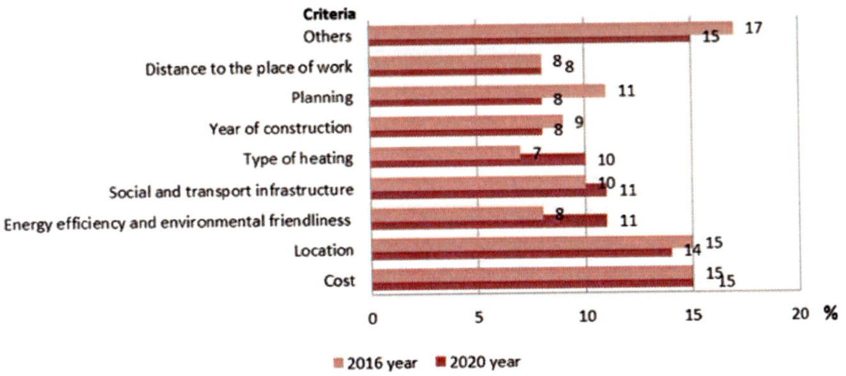

Fig. 5 Criteria for choosing a home when buying it in Kyiv according to the 2016 and 2020 survey

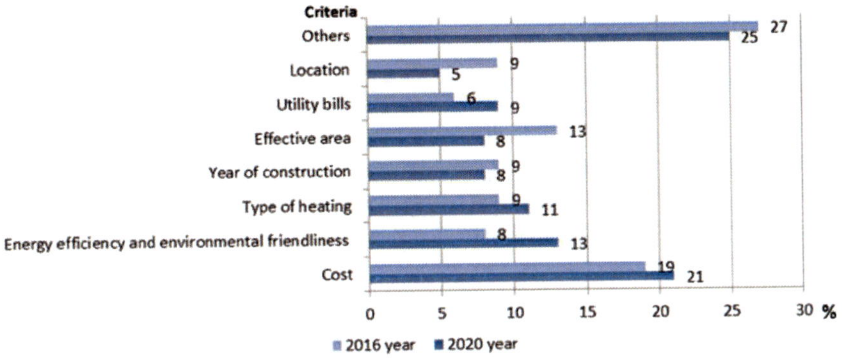

Fig. 6 Criteria for choosing housing when buying it in rural areas according to the 2016 and 2020 survey

In general, it is advisable to conclude that the energy efficiency and environmental friendliness of real estate along with its location for residents of large cities compared to small ones are very important.

In all regions, due to the trend of rising energy prices in recent years, the area and type of heating, which depends on the level of operating costs and utility bills, is becoming more important. In the background are planning of the house, security, the adjacent territory.

At the same time, there are differences in preferences for the type of heating in urban and rural areas. In cities, district heating still predominates with a decreasing trend (about 64% is it district heating, 22% is it individual and 14% is it autonomous), and in rural areas individual and autonomous dominate (63 and 30%, respectively). The preference for electricity and alternative sources is growing as a desirable type of energy source for heating homes, especially in individual homes. Electricity and alternative sources are the most desirable types of energy for heating homes, especially in individual homes.

Thus, on the survey basis, a mathematical function (Eq. 1) for making management decisions in construction (for the medium-sized city) can offer:

$$F(X_1, ...X_2) = 0.17\,X_1 + 0.13\,X_2 + 0.11\,X_3 + 0.1\,X_4 + 0.08\,X_5 + 0.08\,X_6 + 0.07\,X_7 \quad (1)$$

X_1 – Cost,
X_2 – Year of construction,
X_3 – Location,
X_4 – Energy efficiency and environmental friendliness,
X_5 – Social and transport infrastructure,
X_6 – Type of heating,
X_7 – Effective area

Each of the factor features is an integrated indicator that takes into account the characteristics of the region. For example, the criterion of "location" is a formed rating of districts. The criterion of "energy efficiency" is formed on the basis of building materials rating and so on. The criterion of the construction year is represented by the number of the actual operation years of the building after commissioning.

Thus, financial indicators occupy the first place when making management decisions to purchase real estate.

An important criterion when choosing real estate is the ability to purchase it in the mortgage. The vast majority of respondents (68%) would prefer to buy a home on credit under favorable mortgage terms. However, modern conditions of mortgage lending in Ukraine do not allow making this tool available to Ukrainian consumers in contrast to European countries [13].

At the same time, the state construction financing reduction (reduction of investments, deterioration of the investment climate) is one of the main problems hindering the development of the construction industry. The following measures are needed to solve the problems of construction financing in Ukraine:

- expansion of the project financing and the self-financing opportunities for construction companies and introduction of an integrated scheme of the housing financing, which consists in using project financing of developers at the first stage and attracting funds from individuals in the second stage of construction;
- attracting funds from citizens for the construction of industrial and social projects through the issuance of securities;
- de-shadowing of the capital of construction companies;
- creating favorable conditions for lending to construction companies;
- development of a long-term programs to expand the possibilities of financing construction by state and regional authorities.

The implementation of these measures will contribute to the development of the construction industry, which, in turn, will increase the investment attractiveness of construction companies in Ukraine.

Issues of energy efficiency and environmental friendliness, when choosing housing, are very important and is becoming more significant every year. For Ukraine, the benefits of energy saving are especially important due to its high energy dependence and energy consumption. The domestic economy is energy-deficient, satisfying its energy needs through its own production only in part. For the most part, the available housing stock of Ukraine belongs to the most energy-intensive class.

Modern new buildings largely meet the requirements of energy efficiency and environmental friendliness, but the share of new buildings in Ukraine remains insignificant. In Ukraine, the issue of energy saving is especially acute in the global financial crisis that arose during the pandemic. Regarding energy saving in the housing and communal sector, the strategic task is to use energy efficient materials, modern engineering networks and systems, equipment, energy metering and control devices at the stage of construction of new facilities, and modernization and insulation of existing housing. An important role in improving the energy efficiency of construction is played by the comprehensive use of alternative energy sources, which also meets the needs of society in the environmental friendliness of building structures and the environment. All this should be taken into account when developing a strategy for managing a construction company and making decisions at the level of state regulation in the construction sphere of the country.

5 Conclusion

Thus, under conditions of transformation processes taking place all over the world during the pandemic period, the priorities of the construction industry enterprises change significantly. There has been a direction of real estate owners' needs to increase energy efficiency and reduce maintenance costs, which saves financial resources. Under such conditions, the key advantages in the construction market will have developers who provide appropriate proposals, that is, they offer to customers housing that is economical, environmentally friendly, energy efficient and meets the social and aesthetic needs of people. That is, management decisions in construction companies must be adapted to the needs of consumers. And such needs differ significantly in large-scale settlements.

Improving the energy efficiency of the construction complex will solve a number of pressing problems: reducing the cost of housing through more rational and efficient use of energy resources; reducing the country's energy dependence by reducing imports of scarce fuel and energy resources; improving the environment.

References

1. Hamilton IG, Summerfield AJ, Shipworth D et al (2016) Energy efficiency uptake and energy savings in English houses: a cohort study. Energ Build 118:259–276. https://doi.org/10.1016/j.enbuild.2016.02.024
2. Coyne B, Lyons S, McCoy D (2018) The effects of home energy efficiency upgrades on social housing tenants: evidence from Ireland. Energ Effi 11(8):2077–2100. https://doi.org/10.1007/s12053-018-9688-7
3. Lawrence R, Keime C (2016) Bridging the gap between energy and comfort: post-occupancy evaluation of two higher-education buildings in Sheffield. Energ Build 130:651–666. https://doi.org/10.1016/j.enbuild.2016.09.001
4. Lin B, Liu Y, Wang Z, Pei Z, Davies M (2016) Measured energy use and indoor environment quality in green office buildings in China. Energy Build 129:9–18. https://doi.org/10.1016/j.enbuild.2016.07.057
5. Elsharkawy H, Rutherford P (2018) Energy-efficient retrofit of social housing in the UK: Lessons learned from a community energysaving programme (CESP) in Nottingham. Energ Build 172:295–306. https://doi.org/10.1016/j.enbuild.2018.04.067
6. Serrano-Lanzarote B, Ortega-Madrigal L, García-Prieto-Ruiz A, Soto-Francés L, Soto-Francés V-M (2016) Strategy for the energy renovation of the housing stock in Comunitat Valenciana (Spain). Energ Build 132:117–129. https://doi.org/10.1016/j.enbuild.2016.06.087
7. Villca-Pozo M, Gonzales-Bustos J-P (2019) Tax incentives to modernize the energy eficiency of the housing in Spain. Energ Policy 128:530–538. https://doi.org/10.1016/j.enpol.2019.01.031
8. Sivitska S, Vartsaba V, Filonych O (2018) Buildings energy-efficient renovation investment. Int J Eng Technol (UAE) 7(3):408–412. https://doi.org/10.14419/ijet.v7i3.2.14562
9. Samoilyk I, Svystun L (2019) The prospects of the level increase of rural housing energy efficiency. Visnyk Cherkaskoho natsionalnoho universytetu imeni Bohdana Khmelnytskoho. Seriia ekonomichni nauky (Bulletin of the Cherkasy Bohdan Khmelnytsky National University. Economic Sciences) 2:113–128 https://doi.org/10.31651/2076-5843-2019-2-113-128
10. Mykytenko V (2006) Energy efficiency of the national economy: socio-economic aspects. Visnyk NAN Ukrainy (Bulletin of the National Academy of Sciences of Ukraine) (10):17–26
11. Komelina OV, Samoilyk IV, Boldyrieva LM, Krapkina VV (2020) The Management of organizational processes of the transport use in construction. In: Onyshchenko V, Mammadova G, Sivitska S, Gasimov A (eds) Proceedings of the 2nd International Conference on Building Innovations. ICBI 2019. Lecture Notes in Civil Engineering, vol 73. Springer, Cham pp 601–608. https://doi.org/10.1007/978-3-030-42939-3_59
12. Onyshchenko V, Sivitska S, Cherviak A (2018) Construction industry in Ukraine credit analysis. Int J Eng Technol (UAE) 7(3):280–284. https://doi.org/10.14419/ijet.v7i3.2.14420
13. Krekoten I, Svistun L, Khudolii Y (2018) Prospects of using security mechanism in the process of rehabilitation the building sphere in Ukraine. Int J Eng Technol (UAE) 7(3):219–224. https://doi.org/10.14419/ijet.v7i3.2.14406
14. Varnaliy Z, Onyshchenko S, Zavora T (2018) Construction complex development influence on region social and economic security. Int J Eng Technol (UAE) 7(3):469–472. https://doi.org/10.14419/ijet.v7i3.2.14574
15. Construction industry spending worldwide from 2014 to 2019, with forecasts from 2020 to 2035. https://www.statista.com/statistics/788128/construction-spending-worldwide/. Accessed 21 Jan 2021
16. Eurostat. Data Browser. Building permits - annual data. https://ec.europa.eu/eurostat/databrowser/view/sts_cobp_a/default/table?lang=en. Accessed 21 Jan 2021
17. State Statistics Service of Ukraine. Statistical information. Construction. http://www.ukrstat.gov.ua. Accessed 21 Jan 2021

18. Svistun L, Glushko A, Shtepenko K (2018) Organizational aspects of development projects implement at the real estate market in Ukraine. Int J Eng Technol (UAE) 7(3):447–452 https://doi.org/10.14419/ijet.v7i3.2.14569
19. Shcherbak V, Bryzhan I, Chevhanova V, Svistun L, Hryhoryeva O (2020) Impact of forced migration on the sustainable development of rural territories. Global J Environ Sci Manage 6(4):481–496. https://doi.org/10.22034/gjesm.2020.04.05

Prerequisites for Shadowing and Corruption in the Construction Business of Ukraine

Zakhariy Varnalii, Svitlana Onyshchenko, Oleksandra Maslii, and Olena Filonych

Abstract The article reveals the content of the shadow economy and corruption as the main obstacles to construction business effective running in Ukraine. The main economic activity shadowing forms of manifestation in the construction business are studied. The prerequisites and factors of shadow economy emergence phenomenon and corruption as one of the most actual social problems of modern Ukraine are revealed. It was found that corruption covers most spheres of public life, reduces entrepreneurial activity and leads to the spread of organized crime, creates social tension and generates feeling of despair among the population. Based on a study of modern construction, it was found out that this is one of the key sectors of the national economy, which provides expanded reproduction and renewal of the main assets of the country and allows to obtain a multiplier effect from investing in construction. As a result of Ukraine's construction functioning, the investment policy of the state is implemented, the scale and pace of individual industries development, scientific and technological progress and the efficiency of investment in all sectors of the economy are determined. It is substantiated that to restore the potential of the construction industry, to intensify small and medium-sized businesses in construction, and for capital investment, a system of measures to combat shadow processes in construction and implementation of state policy to de-shadow the economy and a targeted program to prevent and combat corruption in Ukraine are needed.

Keywords Construction business · Shadow economy · Shadowing · Corruption · Raiding · De-shadowing of the economy · Public policy

---

Z. Varnalii
Taras Shevchenko National University of Kyiv, Vasylkivska Street, 90-A, Kyiv 03022, Ukraine

S. Onyshchenko · O. Maslii (✉) · O. Filonych
National University "Yuri Kondratyuk Poltava Polytechnic",
Pershotravnevyj Avenue 24, Poltava 36011, Ukraine

V. Onyshchenko et al. (eds.), *Proceedings of the 3rd International Conference on Building Innovations*, Lecture Notes in Civil Engineering 181,
https://doi.org/10.1007/978-3-030-85043-2_73

1 Introduction

At the present stage of development, small and medium-sized businesses occupy an important place in national economies. Construction is an important and strategic priority for the development of each country's national economy. The conditions for running construction business in Ukraine are determined by the growth of numerous threats and dangers, the leading places among which are occupied by the shadow economy, corruption and raiding. The processes of shadowing exist as an integral part of the economic system, which is common to all countries of the world. According to international experts, the global volume of the shadow sector is 10–12 trillion US dollars of value added annually, which is not reported by businesses, and the size of the global shadow economy is equivalent to the US economy, which has one of the largest GDPs in the world [14].

The existing diversity, complexity, and sometimes inconsistency of laws and regulations allow business entities to limit the provision of reliable information about their activities and create conditions for its shadowing, which inevitably leads to criminalization of the economy, corruption, reduced investment activity and innovation development, which significantly affects the level of economic development of the state.

Shadowing and corruption in business under modern conditions have become a complex multi-component phenomenon that acquires characteristics of systemacy and requires definition of priority areas of overcoming it, formation of effective public policy in this sphere and a civilized institutional environment [7, 8]. From these positions of determining the preconditions, factors and forms of manifestation of these phenomena in the construction industry the problem of research was identified.

The study of theoretical and practical aspects of existence of shadow sector of economy and corruption in the world occupies a significant place in the works of such well-known experts [3–6, 9, 10]. Since shadow economy, by its definition, includes all unrecorded economic activities, it is extremely difficult to establish correct size of shadow economy and preconditions in each country.

Construction business has recently lost many opportunities and increased potential due to the progressive growth of corruption and raiding. It is due to the development of the construction industry that the state's investment policy is implemented, economic proportions, scales and rates of individual industries development, scientific and technological progress and efficiency of investments in all sectors of the economy are established. The business in the construction industry includes various areas of activity. Large construction business is more active in the field of capital construction, i.e. it builds shopping centers; neighborhoods with all the infrastructure; factory buildings; departmental buildings of various institutions. Small business mainly builds camps; cottages. In addition to capital construction (construction of industrial and other facilities, preparation of construction sites, work on building equipment, work on construction completion, applied and experimental research and developmnt etc.), the construction business provides services for repair and construction work, reconstruction, and also the production and sale of building materials.

It is corruption, according to numerous analytical reports and surveys that is the biggest problem for both foreign and domestic construction businesses. The obvious problems caused by high levels of bribery and abuse are the destruction of society's values, total distrust to public authorities, which ultimately leads to negative economic consequences, increasing inequality and slowing down personal income growth [1]. This is evidenced by the result of the Corruption Perceptions Index 2020, according to which Ukraine, as in 2019, is recognized as the most corrupt country in Europe, ranks 117th out of 180 countries, along with Egypt, Nepal, Sierra Leone, Swaziland and Zambia. At the bottom of the list are Syria (14 points out of 100), Somalia (12) and South Sudan (12), the countries where the political situation has been unstable for a long time and where governments only partially control the country's territory [2].

Raiding is an unfriendly takeover of companies and redistribution of property and corporate rights. At the same time, unfriendly takeovers are often reduced to the seizure of enterprises under the guise of legal or pseudo-legal grounds [18]. It is the improvement of Ukraine's legislation in terms of protection of private property rights that will ensure sustainable development of economic ties and compliance with the required level of security of business structures.

2 Research of Theoretical and Methodological Foundations of Shadow Economy and Corruption in Construction

Currently, the study of shadow economy in the sectors of the economy is at the stage when researchers have not formulated a single conceptual apparatus to define the shadow economy and its constituent segments. In studies of this phenomenon a wide variety of English terms are used, such as "hidden economy", "underground economy", "informal economy", "second economy", "unofficial economy", "shadow economy", and so on.

In French-language publications, the term "underground economy" (l'economie souterraine) predominates, while in German-language publications, the "shadow economy" (Schattenwirtschaft) is used more often. It is also possible to find "color" terminology, while the "good", "white" economy is opposed to the "black" or "gray" one [15]. Three or four terms mainly dominate ("black", "underground", "hidden", "shadow"), although the total number of terms to name this sector of the economy is more than a dozen.

All these comparative terms are not accidental. They reflect rather not the lack of unity in scientists' understanding of the conceptual essence of this issue, but they reflect various aspects and structural components of the shadow sector of the economy. These terms emphasize the internal qualitative complexity of this phenomenon of public life [10]. From a methodological point of view, it is important to define as clearly as possible the structure of the shadow sector of economy and to find out which are the most characteristic types of economic activity and income, that can be part of shadow sector of the economy, and which are official.

In each country, the informal sector of the economy has its own specifics. In developed countries, these are mostly small businesses and criminal parasitic forms of activity. In countries where excessive tax pressure is practiced, the first place belongs to the concealment of the actual volume of business transactions and profits [11]. As a result of tax evasion, the budgets of developed countries do not add from 15 to 30% of the total amount of planned revenues annually, and in developing countries, this index reaches 50% [12].

Along with illegal economic activity, there are completely legal ones, but hidden from taxation. Any economic activity that under certain conditions is not taken into account by official statistics, evades taxes and is not included in GDP is considered criminal. To summarize, we can identify the following three components of the shadow economy, which are mentioned by foreign researchers: Underground economic activity; Hidden economic activity; Criminal economic activity.

This approach is consistent with the concept of the shadow economy proposed in SNA-1993 [13], which considers the shadow economy as three components: hidden production, unofficial, informal activities or invisible economy, illegal economic activity.

Hidden production includes legal economic activity, the indicators of which are hidden in order to evade taxes.

According to the SNA methodology, informal sector of economy is a part of the household sector created by unincorporated enterprises, i.e. enterprises owned by individuals or households that produce goods and services both for their own consumption and for sale on the market. The largest amounts of shadowing that are difficult to calculate are in informal employment in trade, construction and agriculture.

Illegal institutions include economic entities engaged in activities prohibited by law (production and sale of weapons, drugs, etc.) and persons who are not entitled to a certain type of activity without professional education or licenses.

The difficulty of describing the structure of the shadow economy of Ukraine, in our opinion, consists in the imperfection of socio-economic relations, high entropy of the economic environment, due to the diversity of multilevel economic relations, unformed market relations, weak order and inconsistency of legal norms governing relations in economic sphere.

From our research of the shadow economy structure we can conclude that it is part of the real life of modern society, exists in any economic system with state organization, but in different systems it has its own specifics and is not qualitatively homogeneous in its structural components, and in the final consequences of their impact on socio-economic processes and relations.

Construction business in terms of financial and economic activities does not differ significantly from the activities of enterprises in other sectors of the economy, although there are some differences. Shading in the construction industry occurs at the stage of business processes, current financial and economic activities, accounting and reporting, and especially informal employment. The activities of construction companies are regulated and controlled by the state. The activities of construction companies are licensed, and the beginning of the construction process is also controlled by the state through permitting procedures for design and construction. Numerous approvals of different services, which are required, provoke and promote corruption in this area. In addition, the authorities in some cases establish or recommend the order of pricing in the construction industry. Thus, the activity of a construction company depends much on relations with state and local authorities. This entails a number of threats: the use of administrative opportunities for profit and administrative opportunities for pressure from competitors.

Shadow employment and shadow wages are the phenomena that have become firmly entrenched in public life and have become significant. This is not only a serious social but also an economic problem of Ukraine. It is in the construction business that most employees, usually construction workers, are involved without proper registration. As a rule, their salary is paid "in envelopes", this payment to unregistered employees is not guaranteed and remains a matter of employer's integrity.

Based on the generalization and systematization of existing approaches to the interpretation of the concept of "shadow economy", the author's concept of "shadowing of the construction business" is proposed. It is a complex multicomponent phenomenon that acquires characteristics of systemicity and requires definition of priority areas for its overcoming, formation of effective state regulatory policy in the field of construction and a civilized institutional environment.

The study of the peculiarities of formation and cyclical nature of shadow processes according to spheres of formation made it possible to establish interdependent factors of the "shadow economy" phenomenon emergence. Economic factors are the following: excessive tax pressure, reduced state control over the budget sphere, lack of effective institutional support, lack of financial resources and other economic problems that have led to high unemployment, spread of illegal employment, increasing population stratification on the basis of income level, as well as outflow of skilled labor from the legal sector of the economy. Social factors: low wages, deformation of the labor force and its non-compliance with the needs of the labor market, low social and territorial mobility of the population, low legal culture of citizens, deformation of legal awareness and moral consciousness, impunity for criminal structures. Political factors are: insecurity of business entities from abuse by the authorities, lack of political will to fight corruption, insufficient scientific and organizational and legal support for the fight against criminalization. This makes it possible to determine the effectiveness of existing methods and mechanisms for economic activity shadowing.

To the problem of shadow economy existence the lack of a single and accurate method of quantifying the level of shadow economy is added, which leads to significant differences in calculating the size of shadow economy.

Shadow economy exists in all countries of the world. Its main differences in any country are the size, factors, forms, level of socio-legal control over it and level of its implementation.

In this regard, at the international level, the amount of shadowing economy quantitative assessment is carried out by the method of the World Bank. It is based on a unified empirical method - Multiple Indicators Multiple Causes (MIMIC), proposed

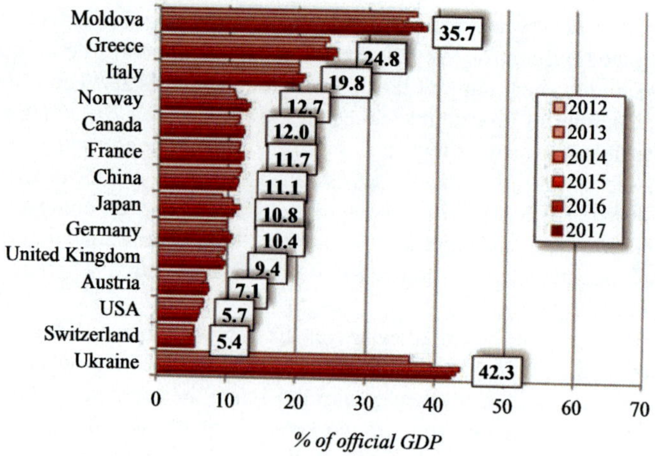

Fig. 1 The level of the shadow economy in some OECD countries

by F. Schneider [9], where the level of shadow economy is calculated by determining the conditions of economic activity, the level of disadvantage of which determines the level of the shadow economy.

According to periodic assessment of the shadow economy at worldwide level, the size of the shadow economy in the most developed economies ranges from 14.0 to 16.0% of GDP, while in developing countries - from 32.0 to 35.0% of GDP. Shadow activity is much larger in Latin and Central America, as well as in Africa (more than 40.0% of GDP), while in the Middle East and Asia it is 25.0–35.0% of GDP [9]. For example, this figure is: in Belgium - 16.5%, Great Britain - 9.4%, Germany - 10.4%, Greece - 24.8%, Denmark - 11.7%, Spain - 20.3%, Italy - 19.8%, Canada - 12.0%, the Netherlands - 8.8%, Portugal - 16.1%, the United States - 5.7%, France - 11.7%, Switzerland - 5.4%, Austria - 7.1%, Japan - 10.8%, China - 11.1% (see Fig. 1). The average performance is about 17% of GDP [9].

Preliminary calculations for 2020, which have not been officially announced yet, indicate that such trends continues and the shadowing of the world economy level gradually increases due to the crisis. According to the World Bank's calculations, at the initial stage of Ukraine's formation as an independent state, its shadow economy accounted for 10.0–11.0% of GDP. However, calculations in the process of political and economic transformations in Ukraine have created optimal conditions for the development of the economy shadow sector. Thus, in 1990–1994, the level of the shadow economy relatively to GDP doubled every two years and reached its maximum in 1997, rising to 65.6% of GDP, and exceeded it during the global financial crisis of 2008–2009, when the level of the shadow economy according to the World Bank was 68.7% (see Fig. 2). According to preliminary calculations of the Ministry of Economic Development and Trade of Ukraine at the beginning of 2021, the level of the shadow economy in Ukraine was 31.0% of official GDP.

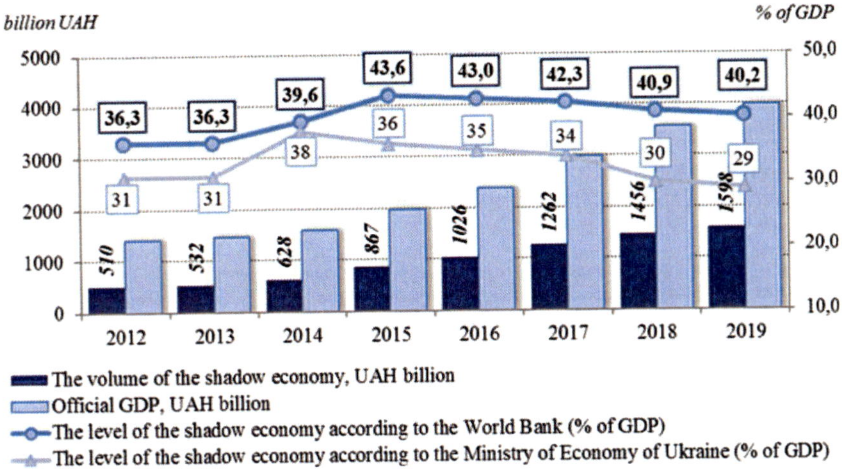

Fig. 2 The level of the shadow economy in Ukraine in relation to the level of official GDP by the world bank methodology

However, according to foreign experts, the level of shadow activity in Ukraine ranges from about a third of GDP, which is recorded by statistics, to a level comparable to the size of official sector [14]. Thus, according to the latest calculations of Austrian economist F. Schneider, which are widely used in international comparisons, the level of the shadow economy in Ukraine is the highest in Europe and is 42.3% of official GDP [9]. This indicator is widely used in international comparisons. However, its dynamics do not always accurately analyze the impact of all factors on the level of the shadow economy, so it is constantly improving.

Estimation of the level and analysis of the national economy shadowing trends as a whole and by individual types of economic activity is carried out by using indirect methods of estimating its level including extrapolation based on excess cash demand, unsupportable electricity consumption and labor market trends. However, it should be noted that the MIMIC method used by the World Bank is convenient for comparing the level of the economy shadowing between different countries, but ineffective for determining its nominal value [14]. Therefore, in Ukraine at the official level since 2009 the shadow economy is assessed by the Ministry of Economic Development, Trade and Agriculture of Ukraine on a regular basis once a quarter of the year in accordance with the Methodology for calculating the integrated level of the shadow economy, approved by the Ministry of Economics of 18.02.2011 № 123 [16]. The method provides for the calculation of the shadow economy level by four methods covering a certain sphere of national economy (with a correspondingly different part of illegal sector, whereby the results differ significantly), with further determination of the level of the shadow economy integrated indicator, which serves as a comprehensive indicator of such phenomenon.

We note, that the World Bank shadow economy of Ukraine estimates are close to the calculations of the Ministry of Economy, with subtraction of the value indicator "economy not directly observed", which is added by the State Statistics Service of Ukraine to the official GDP. The part of the economy, which is not directly observed, according to the calculations of the State Statistics Service of Ukraine, in recent years is from 15.0 to 18.0% of GDP. Although the methodology of this indicator is consistent with international standards, its value is perceived by experts to be somewhat underestimated, given the results obtained by the World Bank [9]. In addition, this indicator is not used in international comparisons. It is explained by the fact that not all countries evaluate the economy, which is not directly observed, for all its components, and therefore their comparisons may have unfounded conclusions.

The shadow economy, that is relatively low at between 10.0–15.0%, acts as a "cushion" for the corporate sector and households, as it mitigates economic fluctuations at the beginning and during the economic crisis. However, in countries where the shadow sector level exceeds 30.0%, the further growth of shadow activity in such a period may provoke significant macroeconomic imbalances and shake the mechanism of economic security of the state. Thus, during the global financial crisis of 2008–2009 and in 2020, a certain part of business in order to maintain production and reduce losses pursues a policy of minimizing costs by avoiding paying taxes, and as a result, the part of the shadow sector of Ukraine for this period increased for more than 10% and nearly reached 40.0% of GDP only by official estimates [17]. Such a high value of this indicator had a significant destabilizing effect on the economy of Ukraine, aggravating negative trends in economy. At the same time, we note that during 2010–2011, in the conditions of acute phase of the crisis completion and transition to the recovery phase, less than half of this part of business returned to legal sector, which according to the optimistic scenario should take place in 2021–2022. In the acute phase of 2013–2014 crisis in Ukraine, such a scenario was not implemented due to the continuation of constant local fighting and military aggression in eastern Ukraine.

We note that the range of informal activities in different sectors of economy differs significantly depending on the production and economic activities. For example, the service sector, in particular retail business, home services or agriculture, where there is no need for significant capital expenditures, is characterized by the highest level of shadowing. Whereas activities that require a high level of skill and significant capital expenditures are usually within the formal sector of the economy. However, according to calculations using official data from the State Statistics Service (see Table 1) in Ukraine, the opposite situation is observed in the construction sector, where shadow employment reaches almost 50% in 2019, and in 2015 did not differ significantly from shadow employment in agriculture and reached for 62% of total employment in this sector of economic activity [19].

According to the results of calculations by using official statistics, the highest level of shadow employment was found in agriculture - more than 60.0% of total number of people employed in industry work in shadow. High level of shadowing is in the field of construction and real property business - nearly 50%, in the field of transport and communications, as well as in the field of services, mainly due to temporary accommodation, catering service and retail trade, where the capacity of illegal transfer by various estimates range from 20.0 to 30.0%. Despite the fact,

Table 1 The level of shadow employment in Ukraine and by individual sectors of the national economy, % of total employment

Indicator	2014	2015	2016	2017	2018	2019
Shadow employment in Ukraine	25.1	26.2	24.3	22.9	21.6	20.9
Shadow employment by economic sectors:						
Industry	8.6	9.9	8.4	7.2	6.7	6.9
Agriculture	68.2	66.0	63.6	62.5	62.3	64.4
Construction	61.0	62.0	57.2	55.8	52.9	49.3
Transport and communication	15.8	16.5	13.1	11.8	14.0	11.8
Trade	31.8	32.3	28.4	24.2	21.1	19.8
Temporary accommodation and catering service	33.1	34.9	30.1	26.1	22.7	22.6
Other types of economic activity	7.0	7.4	7.7	7.5	7.2	6.8

that the level of shadow wages in construction sphere is 2–3 times higher than, for example, in agriculture, the negative fiscal consequences for the economic security of Ukraine from the existence of such a great size of the shadow economy in the construction industry is higher than in other sectors of national economy.

In addition to official statistics for 2018 in the Public Report of the Head of the State Labor Service of Ukraine [20] by the results of inspections and surveys in different sectors of Ukrainian economy that have the highest risk of undeclared work, the second place is obtained by construction companies with an average potential 4035 people employed in the shadows. It follows the service sector, where there are 4789 people thereby (see Fig. 3).

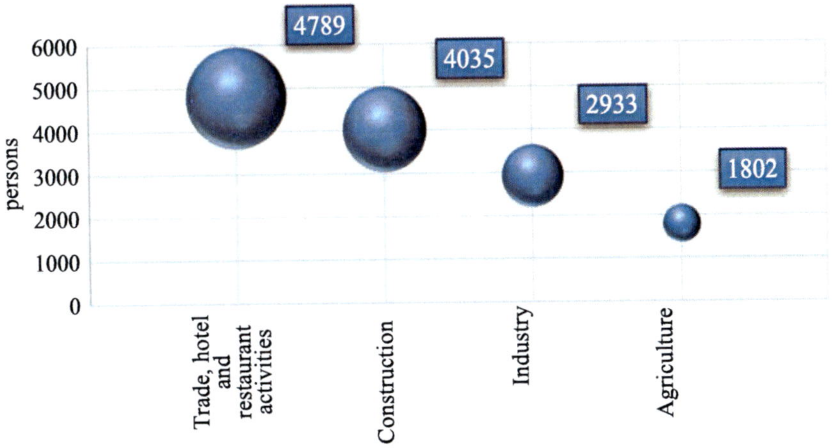

Fig. 3 The risk of using undeclared work by sectors of the economy of Ukraine based on the results of the state labor service of Ukraine inspections

In the field of construction for the first quarter of 2020, the share of the shadow sector, by the Ministry of Economic Development, Trade and Agriculture of Ukraine calculations, increased by 9 percentage points up to 36% compared to the first quarter of 2019. This fact, according to the calculation methodology, is due to an increase in losses of enterprises in the industry by 1.8 times (up to 1.33 billion UAH in January-March 2020), got by a significant slowdown in housing construction (due to reduced economic activity in the world and supply chain disruption due to the implementation of preventive quarantine restrictions).

The indicators of the shadow sector in construction business still remain quite high. The influence of shadow factors on economic life becomes so notable that the contradictions between informal and formal modes turn from secondary to the main socio-economic contradiction [21]. Further shadow sector expansion of the economy threatens its general subordination, so this problem is becoming of national importance.

To increase the reliability of the results, it is advisable to improve the method of level of the shadow economy calculating in Ukraine, as by international estimates of the shadow sector real level in Ukraine significantly exceed the level, found by the Ministry of Economy, Trade and Agriculture of Ukraine. Thus, the main methodological approaches to the level of the shadow economy calculating used by the Ministry of Economy of Ukraine do not take into account such factors of the economy shadowing as imperfect institutional environment, legal framework inconsistency and shortcomings in economic activity regulation. Based on the fact that all the above factors force companies to move from the legal sector to the shadow one, their impact should be taken into account by improving approaches to the shadow economy level estimating by introducing additional adjustment factors or using a separate method in calculating the integrated indicator of the shadow economy. It should also be noted that the method of level of the shadow economy calculating in Ukraine also does not take into account the factor of budget allocations misuse, which according to experts and based on research by Austrian economist F. Schneider has a significant impact on the formation of shadow capital (it is given in details in the source [14]). Therefore, taking into account foreign and domestic experience, the identified shortcomings in methodological recommendations for calculating the shadow economy level in Ukraine should be taken into account when improving the relevant methodology in order to obtain more reliable results of calculations for it.

The method of software modeling tries to cut through this shortcoming. The shadow economy is defined as a non-observable and inaccessible for direct measurement variables. Modeling uses observable indicators to obtain estimates of unobserved variables. To estimate the long-term level of the shadow economy, we use classical equation of demand for money, modified with the demand for them from the shadow economy. Most theories of demand for money lead to a model as in Eq. (1).

$$\frac{M^d}{P} = g(Y, R), \tag{1}$$

In the Eq. (1): M^d, P, Y and R – mean nominal money, price level, variable and the vector of return on various assets.

In empirical studies, Eq. (1) is often approximated in (semi-)logarithmic form as in Eq. (2).

$$m_t^d - p_t = \beta_0 + \beta_1 \gamma_t + \beta_2 R^{own} + \beta_3 R^{out} + \beta_4 \Delta p_t \quad (2)$$

In the Eq. (2): β_1 shows the long-term elasticity of money demand to the variable level, and β_2, β_3, β_4 are long-term semi-elastic to the own R^{own} and alternative R^{out} of nominal money yield and inflation rates. Demand for funds, accompanied by an additional variable in Eq. (2).

And since this variable is not observational, approximations are used. The ratio of the official shadow economy is a function of such variables as the growth rate of the measured product, inflation rates and their change. Inflation leads to an increase in tax payments when taxpayers move to another tax category as a result of an increase in nominal rather than real amounts of income, and, therefore, inflation characterizes the desire of taxpayers to participate in shadow economic activities. In addition, since inflation affects different sectors of the economy to varying degrees, it leads to changes in the distribution of income, which, in turn, can lead to neglecting the tax laws.

It is proved that the choice of the shadow economy assessment method is a very situational phenomenon. The need to assess certain parameters of the shadow economy depends on the tasks and research areas. According to calculations, the budget is formed at various levels of management, considering possible losses from illegalization of economic activities. To increase the degree of public policy flexibility at various levels of the management hierarchy to the impact of external and internal factors, it is necessary at the stage of budget formation to develop measures that enable to compensate negative changes as quickly as possible and with the least losses.

Of no less importance in Ukraine remains the problem of corruption, which is not limited to individual countries and is gaining international scale. And the fight against corruption is one of the most pressing social problems of our time, solution of which is extremely difficult and important for many countries. For Ukraine, corruption, as a component of the shadow economy, has become a factor that really threatens the constitutional order of the state and its national security.

The extent of corruption is difficult to assess. This is primarily due to the fact that it (like other types of shadow economic activity) is generally hidden from official statistical accounting. Thus, measuring corruption is an important element of any national and regional anti-corruption strategy. Regular surveys of corruption extent are a source of data on the factors that give rise to corruption in its various forms and help to develop adequate anti-corruption measures. Measuring corruption in itself is a powerful policy tool: research attracts and strengthens public awareness of the corruptiondangers and helps it put pressure on the government [1].

Analyzing the world rating "Corruption Perceptions Index 2020" [2], which is annually formed by the international anti-corruption organization Transparency

Fig. 4 Dynamics of the Corruption Perceptions Index (CPI) of Ukraine for 2014–2020

International, it is worth noting that Ukraine is rated as the most corrupt country in Europe, having only 33 points out of 100 possible and taking 117th place among 180 studied countries of the world on par with some African countries such as Nepal, Egypt, Sierra Leone, Swaziland and Zambia. However, during 2020, Ukraine climbed six steps in the global rankings, adding three points compared to the fall in positions in 2019, when the country took 126th place against 120th place out of 180 countries in 2018 (see Fig. 4).

Analyzing the structure of the corruption perception index in Ukraine, the largest growth during 2020 is observed in the data of Global Insight Country Risk Ratings - by 13 points, reflecting the risks that people/companies in the country l face bribery or other corrupt business practices from receiving large contracts to daily work. The Ukraine Index in 2020 improved thanks to the launch of the Supreme Anti-Corruption Court with appropriate jurisdiction, the restart of the National Agency for the Prevention of Corruption, the granting of the right to independently carry out secret investigative actions for NABU, the return of responsibility for illicit enrichment, strengthening the protection of whistleblowers and the necessary changes in the field of public procurement [23]. It was these events that finally "completed the chain" of creating an anti-corruption infrastructure in Ukraine, which could only partially work towards this [2]. However, the delay in the implementation of a full-fledged judicial reform, constant pressure on anti-corruption institutions, attempts to disrupt and neutralize achievements in the field of public procurement do not significantly improve position of Ukraine in the rating and eliminate the corruption threat to the construction business.

Imperfect legislation, lack of reliable insurance and government guarantees also hinder the open attraction of domestic and foreign investments to the economy, at the same time, prerequisites are created for the penetration of organized crime into the sphere of legal entrepreneurship (investments in production, construction, real estate, etc.) [22]. For the same reason, imperfect legislation and corruption in state bodies, profitable projects (for construction, import of raw materials or goods, etc.)

through various corruption schemes in state tenders are often fought by criminals. It is important for state institutions to return investor confidence, by improving the current legislation, developing and adopting appropriate regulatory legal acts, it is necessary to provide domestic and foreign capital owners with a profitable, transparent and reliable investment policy, to convince them to reduce the risk of investing in the Ukrainian economy.

The study of the level of corruption in the construction industry was carried out by the "Ukrainian Real Estate Club" with the support and advice of the Business Ombudsman Council in Ukraine from April to June 2019. More than 80% of developers believe that the source of corruption in the construction sector is associated with the activities of the State Architectural and Construction Inspection. According to the results of the survey, the most corrupt one remains state regulatory bodies, and most often corruption is recorded at the stages of project implementation: from obtaining permits to putting objects into operation. Thus, most of the surveyed construction and development companies - more than 58% in Kyiv and more than 43% in the regions - noted that manifestations of corruption are mainly concentrated at the stages of obtaining approvals and permits from regulatory authorities. The most common form of corruption, according to construction companies, is monetary compensation: 91.7% of respondents [24].

Corruption and shadowing in the field of investment in construction on land plots cause conditions for deformation and destruction of competition mechanisms in the land market and distort the mechanisms of cost formation in the sectors of the economy. For example, high bribes for the acquisition or allotment of a land plot for construction unreasonably overstate prices in the residential real estate markets [27–30]. Consequently, the actual problem of land appraisal in today's conditions is set aside for construction, because due to the imperfection of the existing methodological foundations for the appraisal of soils and the monetary appraisal of land, along with the absence of normative documents for the economic appraisal of land plots, conditions are created for underestimating their value when selling and lease and, as a result, a shortfall in income of the State and local budgets.

To reduce the level of shadowing and corruption in the construction business of Ukraine, it is necessary to implement the anti-corruption recommendations that are provided annually by Transparency International, to introduce measures to legalize the construction business, to implement effective public policy instruments of foreign countries, considering certain components of shadowing: corruption, economic cybercrime, "money laundering" through offshore zones and others.

As the studies have shown, the shadow capital amnesty is possible and effective only under the condition of a stable political and economic system. In addition, the level of distrust in the state is quite high not only among the population, but also among business, financial and political groups.

Consequently, the emphasis must be made on improving the institutions of legal management in construction, the most important are the following components:

1. Regulatory and methodological base and organizational forms of small and medium-sized businesses shadowing volumes operational monitoring at the

sectoral, regional and subregional levels of management based on the proposed methods of selective micro-slowdown and aggregation of resulting estimates.
2. The system of counteracting monopolization and strict control over the observance of fair competition based on the gradual harmonization of competition, trade and regulatory policies with those applied in the EU countries.
3. Rational deregulation and regulation of small and medium-sized businesses as a way to reduce the transaction costs of business in the legal and increase in the shadow sectors of the economy.
4. Formation of a system for harmonizing the interests of business entities and society through the development of associative formations of entrepreneurs, consumers, associations of citizens and strengthening their influence on the legislative and executive authorities, on economic agents, including on improving and unconditional compliance with tax, regulatory and antimonopoly legislation.

Thus, construction is a specific activity. The construction business is the most profitable capital investment, which brings a high return on invested capital and a stable income. Phenomena such as shadowing, raiding, corruption in construction are widespread enough, they are more of a norm of activity than an exception to the rule. Moreover, they are highly latent.

It is necessary to transfer the provision of administrative services in construction to an electronic format and to introduce automatic registration in the register of permits. Also it should be introduced an exclusively judicial procedure for appealing the right to carry out preparatory, construction work under the specified permits and cancellation of declaration of the object readiness for operation [25, 26].

3 Conclusion

In the conceptual, interdisciplinary and scientific-applied components of theoretical and applied research of the prerequisites for shadowing and corruption in the construction business of Ukraine, epistemological and logical aspects should be distinguished. The former are means of carrying out research, while the latter reflect the essence of the knowledge obtained as a result of research. With a productive epistemological toolkit and its correct application, the gained knowledge becomes an appropriate scientifically verified information fund for solving the problems of countering shadow processes in the construction and implementation of state policy of shading the economy and preventing corruption. The de-shadowing of the economy is a combination of macro- and micro-level economic, organizational, managerial, technical, technological and legal state measures to create economic, legal and social prerequisites for the self-interested and proactive return of relationships among participants in the financial and economic turnover of things, rights, actions from shadow to legal (official) economic turnover, as well as building the organizational and legal infrastructure of a preventive influence on the elimination of the causes and conditions conducive to the reproduction of sources of shadow activity of business entities. The priority in the strategy of de-shadowing the Ukrainian economy should be to accelerate the development and implementation of the entire range of measures necessary to achieve this goal. Considering the real state of the domestic economy, first of all, it is referred to all concerns anti-crisis measures, structural and innovative transformations in the country in the short and long term.

The most effective methods of shading and decriminalizing the economy are associated with the active use of economic and legal levers in the regulation of economic processes. Solving the problem of shadowing and corruption requires concerted efforts of the state in many areas of state policy. A real reduction in the shadow sector can occur only if costs decrease and the efficiency of the legal economy grows, and parity conditions are created for the functioning of various forms of business, including construction. It enhances the competitiveness of legal entrepreneurial activity and makes it unprofitable to conceal its results. An indisputable condition for such a transformation should be the weakening of fiscal pressure, the elimination or significant reduction of administrative barriers to business development, creates motivation for the transfer of shadow operations to a legal basis. Effective counteraction to the process of corruption and shadowing in the construction business of Ukraine, as well as the transformation of the shadow component into the legal economy, promote the development of domestic business, guarantee the economic security of Ukraine, its independence, sovereignty, progressive socio-economic development.

References

1. Cabinet of Ministers of Ukraine: Fight against corruption. https://www.kmu.gov.ua/diyalnist/reformi/verhovenstvo-prava-ta-borotba-z-korupciyeyu/borotba-z-korupciyeyu
2. Transparency International: Corruption Perceptions Index 2020. https://images.transparencycdn.org/images/2020_Report_CPI_EN.pdf
3. Dreher A, Schneider F (2006) Coruption and the Shadow Economy: an Empirical Analysis, CESifo Working Papers Series 1613, CESifo Group Munich
4. Elgin C, Oztunali O (2013) Institutions, Informal Economy and Economic Development, Working Papers 2013/03, Bogazici University, Department of Economics
5. Frey R, Van Hannemann P (2003) The Analysis of the second economy. The Economics of the shadow economy, London
6. Gatti R, Honorati M (2008) Informality Among Formal Firms: Firm-Level, Cross-Country Evidence on Tax Compliance and Access to Credit. Policy Research Working Papers Series 4476, The World Bank (2008)
7. Johnson S, Kaufmann D, Zoido-Lobaton P (1998) Regulatory Discretion and the Unofficial Economy. Am Econ Rev 88(2):387–392
8. La Porta R, Shleifer A (2008) The Unofficial Economy and Economic Development. NBER Working Papers 14520, National Bureau of Economic Research
9. Medina L, Schneider F (2019) Shedding Light on the Shadow Economy: A Global Database and the Interaction with the Official One. CESifo Working Paper, 7981. Munich, Germany. https://www.cesifo.org/DocDL/cesifo1_wp7981.pdf
10. Mandibura B (1999) The shadow component of the population living standards of Ukraine. Ukraine: aspects of labor 5:32–38

11. Cherednichenko V (1999) Shadow economy: problems of development and implementation of projects to combat economic crime. Bull Natl Acad Public Adm President Ukraine 2:296–301
12. Petrenko S (2004) Shadow economy: Causes and impact on the legalized sector of the economy. Bulletin of the Accountant and Auditor of Ukraine 3–4:40–42
13. Mocherny S (2001) Methodology of economic research. World, Lviv, Ukraine
14. Onishchenko S, Pugach A (2015) Threats to economic security of Ukraine: essence, evaluation and prevention mechanism. PoltNTU, Poltava, Ukraine
15. Varnaliy Z (2006) Shadow economy: essence, features and ways of legalization: Monograph. NISS, Kyiv, Ukraine
16. Methodical recommendations of shadow economy level calculation. Order of the Ministry of Economy of Ukraine 123, February 18, 2011. https://zakon.rada.gov.ua/rada/show/v0123665-09#Text
17. Ministry for Development of Economy, Trade and Agriculture of Ukraine: Trends in the shadow economy in Ukraine in January-March 2020. https://www.me.gov.ua/Documents/Download?id=699fa73c-084f-431a-9491-47ad5ffbaf09
18. Varnaliy Z, Mazur I (2007) Raiding in Ukraine: preconditions and ways to overcome it. Strateg Priorities 2(3):129–136
19. State Statistics Service of Ukraine: Statistic information. http://www.ukrstat.gov.ua/
20. Public report of the Head of the State Labor Service of Ukraine issues for 2018. http://dsp.gov.ua/wp-content/uploads/2019/03/zvit.pdf
21. Onyshchenko S, Yehorycheva S, Furmanchuk O, Maslii O (2020) Ukraine construction complex innovation-oriented development management. In: Proceedings of the 2nd International Conference on Building Innovations pp 687–700. https://doi.org/10.1007/978-3-030-42939-3_68
22. Index of Economic Freedom. http://www.heritage.org/index/ranking
23. National Agency on Corruption Prevention: Methodology of the standard survey on the level of corruption in Ukraine. https://nazk.gov.ua/wp-content/uploads/2020/10/12.01-do-p.8.pdf
24. Onyshchenko SV, Matkovskyi AV, Puhach AA (2014) Analysis of threats to economic security of Ukraine in conditions of innovative economic development [Analiz zahroz ekonomichnii bezpetsi Ukrainy v umovakh innovatsiinoho rozvytku ekonomiky]. Econ Ann-XXI 1–2(2):8–11
25. 80% of developers called the Architectural and Construction Inspectorate a source of corruption. Economic truth.https://www.epravda.com.ua/news/2019/07/10/649552/
26. Ptashchenko L, Chechelashvili M (2018) Crowd funding as a tool for financing the construction of social facilities. Int J Eng Technol (UAE) 7(3):329–333
27. Kunitsky, O.: War on construction corruption: siege and assault. UNIAN https://www.unian.ua/economics/other/10755357-viyna-z-budivelnoyu-korupciyeyu-osada-ta-shturm.html
28. Shading of land relations. UA Economic. http://www.uaeconomic.com/ulens-1707-1.html
29. Varnaliy, Z., Onyshchenko, S., & Zavora, T. (2018). Construction complex development influence on region social and economic security. International Journal of Engineering and Technology(UAE), 7(3), 469–472. doi:https://doi.org/10.14419/ijet.v7i3.2.14574
30. Kozachenko H, Onychshenko S, Masliy O (2018) Region building complex social and economic security threats. Int J Eng Technol (UAE) 7(3): 215–218 https://doi.org/10.14419/ijet.v7i3.2.14405

Analysis of the Role and Place of the Building Materials Industry in the Development of Azerbaijan's Economy

Elshad Yusifov, Ahmad Sarkarli, Volodymyr Byba, and Tetiana Kushnirova

Abstract The article is devoted to the analysis of the features of the sectoral structure of the construction materials industry in Azerbaijan, its place in the manufacturing industry, the economic results of state and private enterprises with the corresponding industry.

It includes a graphical analysis of various results obtained in the production of construction materials in the country, including the number of enterprises, industrial production, number of employees, an average monthly salary in the industry, the amount of domestic and foreign investment, and other indicators.

Therefore, in order to improve the efficiency of organizational and technological solutions in the companies of producing construction materials, it is necessary to determine the factors affecting its production and prepare scientifically justified recommendations. Relevant recommendations and suggestions were made in order to ensure the sustainable development of the industry.

Keywords Competitive environment · Private sector · Profitability · Non-oil sector · Processing industry · Construction materials

1 Introduction

In modern economic conditions competition has become global. Therefore, the policy of diversification of the economy is being implemented in our country to ensure a sustainable development through a competitive economy and economic entities, to maintain its longevity. As part of the diversification policy, it is important to improve

E. Yusifov (✉)
Azerbaijan University of Architecture and Construction, 5 Ayna Sultanova, Baku, Azerbaijan

A. Sarkarli
Azerbaijan State Oil and Industry University, 34 Azadliq Avenue, Baku, Azerbaijan

V. Byba · T. Kushnirova
National University Yuri Kondratyuk Poltava Polytechnic, Pershotravnevyj Avenue 24, Poltava, Ukraine

V. Onyshchenko et al. (eds.), *Proceedings of the 3rd International Conference on Building Innovations*, Lecture Notes in Civil Engineering 181,
https://doi.org/10.1007/978-3-030-85043-2_74

the structure of the processing industry, to give priority to industry, especially the processing industry, to become a key factor in ensuring economic growth in the country as a whole, as well as improving the quality of production and services.

As one of the important sectors of the processing industry, the production of construction materials is considered one of the key links in the national industrial system. A rich raw material base in the relevant field differs in terms of development, as well as the potential to organize and implement export-oriented production, employment in the country and other important aspects. The production of construction materials plays the role of the material supply base of the construction industry, which is the main leading sector of the economy after the oil and gas industry. The leading position of the construction industry in the country's economy depends mainly on the production of qualitative construction materials, taking into account important factors such as quality, price, innovation, energy and science in the production process. Thus, the directions we have listed are also reflected in the State Program for the Development of Industry in the Republic of Azerbaijan for 2015–2020 and the Strategic Roadmap for the Development of the National Economy. From this point of view, it is necessary to study an issue of competitive factor's increase in the relevant industry.

The production of construction materials serves to ensure the implementation of social programs in Azerbaijan. This is one of the important points that makes the relevant sector competitive. Therefore, in modern conditions, an importance and efficiency of the construction materials industry, as well as competitiveness must increase significantly.

2 Main Part

2.1 Purpose of the Article

The main purpose of the article is to determine the place of the construction materials industry in the processing industry of the Azerbaijani economy. Also analysing the features of the development of the construction materials industry in Azerbaijan and to make recommendations for improving companies of this industry.

2.2 Research Methodology

The methodological basis of the research is a systematic approach aimed at studying and solving the works of domestic and foreign scientists dealing with the problems of improving the efficiency of construction of high-rise monolithic reinforced concrete buildings.

During the research, methodological guidelines of leading scientific institutions dealing with the problem of improving the efficiency of the construction of monolithic reinforced concrete buildings were used, as well as actual materials of a number of construction contractors performing large volumes of monolithic reinforced concrete works.

The method of expert assessments, probability theory and mathematical statistics, correlation and regression analysis, graph-analytical method, methods of economic and mathematical modeling ere used as a research metodology.

2.3 Results

In a competitive environment, both in the domestic market and internationally, delivering a product that the consumer wants to buy as soon as possible is a very important factor in terms of gaining an advantage over competitors. To make it possible, the proper use of an equipment, production time and the process of transporting products must be carefully considered [7].

In developed countries, the appearance and quality of products as a result of production are made possible by improving the characteristics of competition [2].

Increasing a macroeconomic efficiency in a foreign economic activity depends on eliminating the factors that negatively affect competitiveness of local businesses and Azerbaijani products. The processing industry is the basis of a success in the developed and newly industrialized countries [1].

The current adoption of the processing industry as a priority area allows to focus on the activities of existing economic entities in this area, as well as making decisions on the establishment of new production and processing enterprises in promising areas. The processing industry has a lot of work to do to meet domestic demand at the expense of a local production, and the implementation of this process depends on making optimal decisions based on a comprehensive analysis of all areas.

In general, the global rating organization S&P pays special attention to the following factors in determining the competitive positions of enterprises of construction materials [8]:

- market position and development prospects;
- diversity of geographical location, customers and access markets;
- market consolidation;
- pricing capacity;
- range of products offered and brand strength;
- efficiency of operations related to production and sales activities;
- condition of capital expenditures and price base.

The production of construction materials occupies a special place in the processing industry of the Azerbaijani economy. It is also considered one of the most promising areas of the processing industry. The number of enterprises operating in the relevant industry is 236, most of them are private enterprises. Relevant enterprises are represented by 12% in the processing industry. Also, there are more than 1,800 registered individual entrepreneurs engaged in the production of construction materials. The average number of employees is 12,000, and the average monthly salary is about 600 manats. This is higher than the average monthly salary in the country. At a time when world oil prices were falling, Azerbaijan also saw a decline in revenues, and in those years the production of construction materials declined, but in 2018 there was an increase. The volume of production is more than 730 million manats. Of course, this figure is below the potential, but the growth trend over the years indicates the prospect.

In the building materials industry, there are not only competitive and diverse production facilities, but also various forms of ownership. In modern economic conditions, one of the positive indicators of the processing industry in Azerbaijan is that the number of privately owned manufacturing enterprises often exceeds the number of state-invested enterprises. For a market economy to be fully formed and to achieve its specific goals, the state must act as a controller rather than a participant. The policy of stimulating the private sector should be preferred and efforts should be made to improve the structural structure of production.

Most of the enterprises operating in the field of construction materials production are private enterprises. This, in turn, means meeting one of the requirements of a market economy (Fig. 1) (Table 1).

Construction materials enterprises in Azerbaijan are located almost in all regions of the country, and this factor is very important at a time when regional development is defined as a goal. Unlike other industries, the fact that the raw material base is rich throughout the country is a factor. Also, industrial enterprises producing construction materials in exports are achieving development indicators. Thus, Norm LLC, Holcim Azerbaijan OJSC, Matanat A Company, Metak, Gazakh Cement Plant, Gilan Construction Materials Plant, Fakhraddin K LLC and other enterprises have access to foreign markets [5].

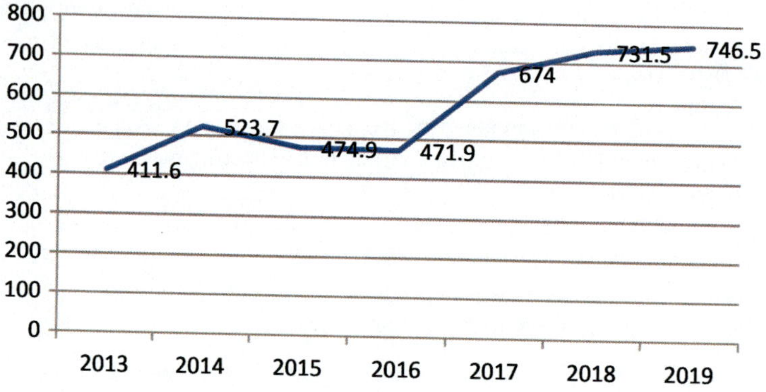

Fig. 1 Volume of production of construction materials for 2013–2019 (million manats)

Table 1 Amount of construction materials industry enterprises registered in Azerbaijan by type of ownership, (units)

Years	For all types of property	Processing industry	Building materials production companies	Belonging to the state	Private sector
2010	2650	1909	280	19	261
2011	2508	1791	244	15	229
2012	2514	1795	220	11	209
2013	2527	1764	209	11	198
2014	2534	1762	230	12	218
2015	2583	1778	239	10	230
2016	2561	1775	224	10	214
2017	2582	1826	229	10	220
2018	2837	2034	236	9	227
2019	3169	2330	254	10	244

We can present the field structure of the industry of production of construction materials in Azerbaijan as an industry with the following features:

- current indicators of industry development in the country;
- technical and technological development in the country;
- freedom of the national economy;
- labor productivity situation.

Figure 1 shows the growth dynamics of the construction materials industry in Azerbaijan. Here are the indicators for the recent years. The figure shows that between 2013 and 2018, production volumes were accompanied by changes. In 2015 and 2016, due to an objective reason, i.e. the fall in world energy prices, the decline in revenues in Azerbaijan, and thus the reduction of direct investment in the construction sector, reduced the demand for construction materials. This, in turn, has led to the entry of industrial enterprises into the market with demand. In 2017 and 2019, on the contrary, there was an increase of 203 and 15 million manats compared to the previous year.

Some of the technical and commercial issues faced by local companies when marketing their products can be presented in the following form:

- high interest rates when raising funds required for anorganization of activities in the form of loans;
- an import of an equipment necessary for efficient development of material and technical resources;
- Insufficient expenditures on staff training, market research and scientific research, and in some cases constitute the bulk of total expenditures.

It should be noted that the fall in crude oil prices on the world market has highlighted the importance of the refining industry in the economy with the disruption of export–import relations in a resource-dependent economy. The experience of modern developed and newly industrialized countries shows that a competitive economy is impossible without the leadership of the processing industry.

Sustainable development, increasing employment, developing the export capacity of the national economy, competitive prices for science and education, investment in human capital and products for local consumers in relation to existing wages. It is very important to set the development of the processing industry as the main goal. A protection of national interests is provided by the ability and readiness of the government to develop and implement economic projects in the development of the national economy, to create mechanisms to support the socio-political stability of a society.

The development of industry in modern times is measured by the current state of the processing industry. From this point of view, enterprises in the field of production of construction materials (industry) in Azerbaijan have not fully used their production potential. In our opinion, the solution of certain issues will create new opportunities for achieving new economic goals.

The composition of competition policy in the industrial sphere should be structured in accordance with market demand and structure. Industrial enterprises aiming at economic growth, first of all, pay attention to the competitiveness of projects when making correct assessments and efficient use of resources [9].

Based on the facts found in the study of international experience, we can say that according to the accepted rule, in countries with diversified economies, the processing industry should be 20% or more of GDP. The processing industry in Azerbaijan is quite developed. The production of construction materials is also considered as an integral part of the processing industry.

Figure 2 shows the representation of enterprises in the field of construction materials production in the industrial sphere of the country with the following indicators, as well as the percentage of investments in a relevant industry.

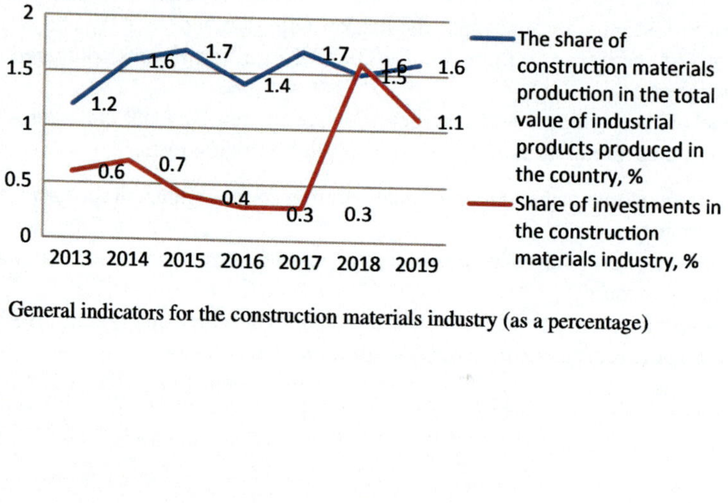

Fig. 2 General indicators for the construction materials industry (as a percentage)

Increasing production will, of course, lead to increased profitability. A constant change in the level of an income once again confirms the importance of the development of the processing sector – one of the most important industries in Azerbaijan. Based on the experience of countries with advanced economic achievements, we can say that a sustainable and competitive economy is inconceivable without industry. Satisfactory prices for daily consumer goods in the national consumer market and the development of industry, especially the processing industry, are very important for other purposes. From this point of view, the development of industry means the development of the processing industry.

In order to increase the business environment in the Republic of Azerbaijan and furthermore improve the position of our country in international rankings, the industry must be developed. For this purpose, the country's leadership takes various measures and issues orders [6].

It should be noted that this industry consists of two main areas: mining and processing. The extractive industry has developed to a high level in Azerbaijan and has become the main driving force of the economy.

Processing industry is one of the economic sectors in the Republic of Azerbaijan to bring the products obtained through the mining industry to a state that can be consumed in various fields by new forms and methods. This area of an industry, as a complex area of a complex nature, achieves efficient production results by combining hundreds of areas.

The industry of production of construction materials is one of the 4 largest sectors in the grouping of processing industries and occupies an important place in the Azerbaijani economy in terms of potential opportunities. The share of construction materials in the country's GDP is small. From this point of view, the following conclusions can be drawn when analyzing economic processes and regularities:

- significant share of the extractive industry in the country;
- weak activity in the processing industry as a result of ensuring revenues from raw material-based exports;
- specialization, mainly in the tourism and other services, is one of the factors contributing to the weak development of the processing industry.

Industrial enterprises producing construction materials in Azerbaijan play an important role in the policy of economic diversification. The non-oil sector is the most important factor in achieving sustainability and effectiveness in an economic development. The organization and maintenance of economic activity at an increasing pace can be possible only as a result of sustainable economic activity.

The need to diversify the economy underscores the importance of the events in the world economy and, consequently, in the national economy. Comprehensive efforts should be made to increase the share of various industries, including construction materials, in GDP. As mentioned above, our country has almost all the opportunities to increase the production of construction materials.

The non-oil sector proves the importance of the construction materials industry with a comparable indicator. It is possible to stimulate the creation of new enterprises in the field in order to achieve both economic growth and employment, thereby

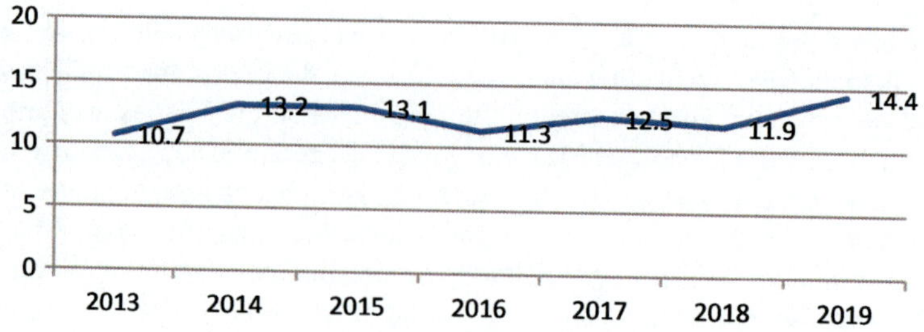

Fig. 3 Amount of workers in the construction materials industry (thousand people)

increasing the competitiveness of the country's economy by ensuring perfect competition. A perfectly competitive enterprise can raise profits from production when the current market price is higher than the final cost of production. It is also possible to increase the relevant indicators by improving the legal framework for the struggle of individual enterprises for qualified personnel, affordable supply of raw materials, the mass of buyers and sales markets (Fig. 3).

As we have already mentioned, the production of construction materials is the main industry which supplies materials to the construction sector. Thus, the fact that the construction sector is one of the leading sectors makes it necessary to pay attention to the results of economic activity of production enterprises engaged in the production of construction materials, which perform the function of material supply of that sector. From this point of view, in the table below it would be expedient to consider the indicators of production of the main types of products by enterprises in kind and to analyze their growth rate.

The establishment of enterprises producing construction materials, as well as industrial complexes will ensure the sustainability and sustainability of the economy by reducing the dependence on the oil industry, ensuring the future economic development of the country.

Based on the results of the analysis using the information received in the sector, to formulate a certain direction of activity for the future period and participate in decision-making in the relevant sector of the industry.

In particular, an income of the population operating in this field shows the importance of this field for the country. We can say that the higher the profitability in the industry, the higher the added value created in the country's economy. In this regard, if we look at the average monthly wages of workers in Fig. 4, it is clear that the average monthly wages of employees in the construction materials industry in the analyzed years were lower than in the general industry. It should be noted that until 2015, 1 USA dollar was worth 0.78 manats in Azerbaijan, now it is 1.70 manats.

A high level of wages suggests that the construction materials industry plays an important role in economic development, including improving the living standards of the population and especially workers in this field. Based on the years described here, the average monthly salary in the country in general and the salary in the construction materials industry in general are shown. It is clear from Fig. 5 that the salaries paid by

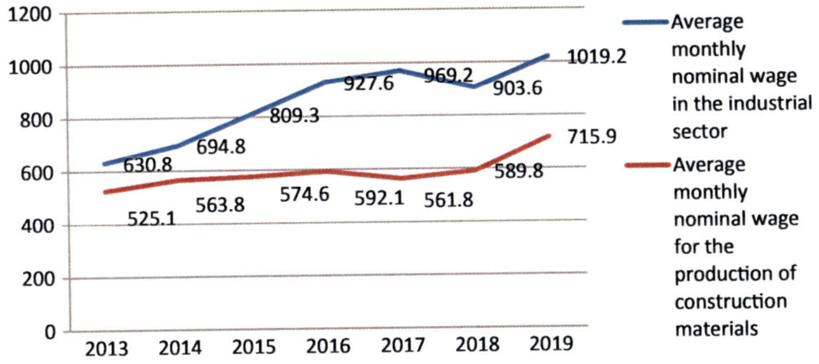

Fig. 4 Average monthly salary of industrial workers (in manats)

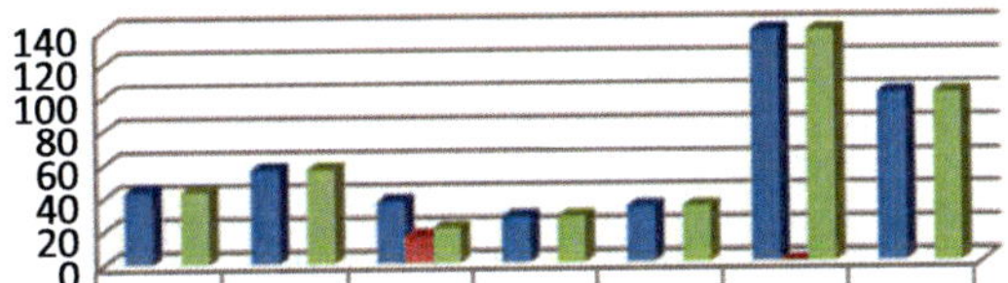

	2013	2014	2015	2016	2017	2018	2019
In general, in the construction materials industry	43.3	56.5	37.3	27.6	33.3	138.4	100.4
Foreign investment			16.1			0.1	
Domestic investments	43.3	56.5	21.2	27.6	33.3	138.3	100.4

Fig. 5 Investments in fixed assets in the construction materials industry (million manats)

non-state enterprises in the production of construction materials from state-owned enterprises in 2018 alone were more than 250 manats. This once again underscores the need for the development of private industrial enterprises, especially the private sector.

In industry, it is virtually impossible to ensure the competitiveness of the national economy without significant investments in fixed assets from internal and external sources [4].

Attracting foreign investment is even more important. Due to the spread of a new technology and progressive management and marketing techniques, the country's technological development accelerates as the volume of foreign investment increases [3]. Along with a foreign investment, leads to the import of more advanced technological equipment, as well as the expansion of export opportunities. Figure 6 shows the volume of investments in the construction materials industry across the country.

Attracting and protecting investments serves to create new enterprises in the industrial sector, as well as to increase the production capacity of existing enterprises.

2.4 Scientific Novelty

As a result of the analysis of the economic activity of the enterprises, it was revealed that they have various shortcomings, and in order to increase the competitiveness of the enterprise by eliminating them, the following is proposed:

- support ways to increase product sales with modern opportunities;
- continuous improvement of product quality;
- cost reduction;
- regular professional development of staff;
- application of innovative development policy.

2.5 Practical Importance

In addition to the above mentioned, the acquisition of a cheap raw material base, as well as the reduction of transport and logistics costs is one of the key factors ensuring a competitive advantage.

3 Conclusions

Compared to industrialized countries, production in the construction materials industry in Azerbaijan is more material-intensive and energy-intensive. The material cost of the products produced in the relevant industry complicates the process of using price competition. In this case, the reduction of energy costs is seen as an important step in increasing the competitiveness of local building materials industry products. It should be noted that energy saving in the building materials industry can be considered as one of the most important factors to ensure a competitive advantage.

References

1. Aliyev A, Shakaraliyev A, Dadashov I, (2003) World economy: problems of modern period, monography. "Baku University" Publishing House, Baku, 455 p
2. Berge B (2001) The Ecology of Building Materials, Translated from Norwegian by Filip Henley With Howard Liddell, Architectural Press, 453 p
3. Yusifov E, Sarkarli A (2019) The role of joint venture in the development of Azerbaijan industry. Building Innovations - 2019, Poltava, Ukraine, 23–24.05. pp 619–623
4. Gradov AP (2008) National Economy, 2nd edition/Translated from Russian. University of Economics Publishing House, Baku 275 p
5. Industry of Azerbaijan (2020) Official publication, Statistical Yearbook, Baku, 252 p
6. Doghan Oİ, Marangoz M (2003) Facts and an application affecting the competitive power of enterprises in the domestic and foreign markets. J Inst Soc Sci 5(2):139

7. Statistical Yearbook of Azerbaijan (2020), Baku, 788 p
8. S&P Global Ratings, Criteria, Corporates, Industrials: Key Credit Factors for The Building Materials Industry, 15 p (2013)
9. Publications of Ministry of Employment and the Economy, İndustrial competetiveness approach means to guarantee üconomic growth in Finland, 59 p (2013) https://www.stat.gov.az/source/industry/ - official site of TheState Statistical Comittee of Republic of Azerbaijan

Construction Companies Vat Taxation Features in Ukraine and European Union

Iryna Pasichna, Inna Kulchii, Volodymyr Zadorozhnyy, and Viktoriya Smahliuk

Abstract Features of value added tax (VAT) taxation of construction companies are considered in the research. The authors examine the current tax legislation of Ukraine and some European Union countries. It is determined that development of the national economy in general and the construction industry in particular depends on many factors, including tax structures used by the state in the legal regulation of tax relations. Considerable attention is paid to tax benefits analysis of construction companies, including tax exemptions, application of preferential tax rates, reduction of the object of taxation, establishment of simplified taxes, etc. Direct and indirect taxation benefits of construction companies may be provided by the tax legislation. It is determined that any tax benefits stimulate the development of the construction industry. The purchasing power of citizens is declining, profits of construction companies are falling, and economic development is slowing down in conditions of coronavirus spread. Therefore, the state needs to support business, including construction. It is necessary to reduce the tax burden on construction companies by establishing tax benefits. Reducing the amount of VAT will have an effect of stabilization of the buyers' purchasing power, improving the performance of construction companies. Ukraine should use EU countries experience; find alternative sources of revenue to the state budget—non-tax revenues, including investments in business and production. Stimulation of the construction industry is an important tool for the development of the state economy, which ensures the development of other sectors of the national economy.

Keywords VAT · Tax legislation · Tax privileges · Construction companies · EU experience

---

I. Pasichna (✉) · I. Kulchii · V. Zadorozhnyy · V. Smahliuk
National University Yuri Kondratyuk Poltava Polytechnic,
Pershotravnevyj Avenue 24, Poltava 36011, Ukraine

V. Onyshchenko et al. (eds.), *Proceedings of the 3rd International Conference on Building Innovations*, Lecture Notes in Civil Engineering 181,
https://doi.org/10.1007/978-3-030-85043-2_75

1 Introduction

Company in any industry seeks to minimize taxable income by increasing costs. Construction companies are no exception. For the development of the construction industry in the current conditions in Ukraine, it is important to provide support and create a favorable climate on the part of the state for the operation of construction companies. The state can influence through tax legislation in order to stimulate industry development. Tax legislation provides for a variety of methods for influencing the company's activities, in particular, exemption from taxation, reduction of the object of taxation, the use of preferential tax rates, etc.

This research is dedicated to analysis of taxation of construction industry companies in Ukraine and EU countries. The challenge is to ensure the dynamic innovative development of the construction industry, which will give impetus to other sectors of the economy, to protect the housing rights of its citizens, which will promote positive economic and social situation in society.

Using the experience of EU countries on the one hand will support the development of construction companies in Ukraine, on the other hand will ensure the harmonization of tax legislation of Ukraine to EU requirements. Today, the tax system of Ukraine still does not contribute to the formation of a socially oriented competitive market economy. The state of the tax system cannot be compared with the EU countries in economic development, level of corruption, level of shadow economy, tax culture, state support of business, etc.

2 Aims

The purpose of the paper is to investigate the tax legislation of Ukraine and the EU to analyze features of vat taxation of construction companies, use positive experience to stimulate the development of the construction industry, as well as to develop a strategy for effective tax management in conditions of coronavirus spread. This investigation should serve as further scientific research that is necessary to develop a strategy for effective public management of the construction industry in particular and the national economy in general.

3 Methods of Study

The general dialectical method and special theoretical research methods are used in the paper. In particular, the dialectical method is used to understand the legal nature of VAT, tax benefits. The research used comparative legal method, according to which the legislation of the European Union is analyzed and the experience of legal regulation of tax relations in some countries is studied. The method of scientific abstraction is used to determine the features of taxation of construction companies; system analysis method is used to study the tax system of Ukraine and certain EU countries; statistical methods—used to analyze VAT rates and tax benefits.

4 Main body

The process of tax harmonization faces the solution of two opposites in essence problems. The essence of the first is that the formation of a single internal market involves the unification of tax legislation. This is especially true for the profits of enterprises and the income of investors, because only the solution of this problem allows the capital to move freely between EU member states and to avoid barriers that hinder the development of scientific and technological integration and the dissemination of the latest advances in science, technology and technology between the states. On the other hand, the harmonization of taxes, its rates, the provision of various investment privileges affects the domestic policy of the state, objectively reduces the amount of taxes received in the revenue part of the state budget. In addition, the government's influence on domestic producers is decreasing [7, 17, 18].

The process of tax harmonization is necessary not only to bring tax legislation into line with European standards. It is necessary for Ukraine to form an effective tax policy and in the future to be able to compete with the countries of the European Union [19, 20].

According to the Visa Liberalization Action Plan, one of the issues is the direction of development and implementation of tax systems and their institutions in accordance with international and EU standards, covering tax reform, including the adoption and implementation of VAT legislation [12].

Value added tax is widely used in the European Union. This tax is added to the value added of goods and services. It is paid by consumers who buy goods and services in the European Union. Therefore, exported goods or services provided abroad are generally not VAT subject. On the contrary, imported goods are taxed for equal and fair competition in the EU market [13].

VAT is one of the main revenues of the budget. The tax is paid by citizens as the main consumers of goods and services. High VAT rates and rapid inflation can be a factor holding back economic development. Rising prices reduce the purchasing power of the population and lead to negative economic consequences in the current conditions of the spread of coronavirus in Ukraine and abroad.

VAT applies to most countries of the European Community. This tax is used to create a single revenue part of the Community budget and amounts to 1.4% of the taxes collected by each member state of the Community. Indirect taxation mechanisms within the EU are constrained by significant differences in value added tax rates (from 14%—in Germany to 22%—in Denmark). Therefore, one of the main achievements in the formation of the single European market is the agreement on the unification of the indirect taxation system and the introduction of an average rate of value added tax at 15% [5].

In accordance with the Strategy for the sustainable competitiveness of the construction sector and its enterprises (2012) that improves training, tendering and financing in the construction sector the construction sector plays an important role in the European economy. It creates 20 million jobs in large and small businesses. Intermediate products are used in construction, including technical equipment, raw materials and others. The construction industry will significantly affect the development of the economy as a whole [6].

Legal regulation of tax relations in the Member States of the European Union (EU) has features that depend on the type of taxation. EU member states retain some degree of autonomy in the area of collection of direct taxes in line with the national tax systems of the EU common economic policy. Joint actions of the Member States of the European Union on indirect taxes are laid down in Regulation (EC) No. 1286/2013 of the Council and of the European Parliament of 11 December 2013 on the establishment of a program of actions to improve the functioning of the taxation system for the period 2014–2020 in the European Union (Fiscalis 2020) and repealing Decision No. 1482/2007/EC [8].

Council Directive 2006/112/EC of 28 November 2006 on the common system of value-added tax provides for a list of supplies of goods and services to which reduced rates may apply. In particular, the list includes the construction, repair and renovation of housing within the limits of social policy, which may be subject to reduced rates of value-added tax. According to Annex III of the Council Directive 2006/112/EC of 28 November 2006 list of supplies of goods and services to which the reduced rates referred to in article 98 may be applied is defined. According to Article 98 of the Council Directive, Member States may apply either one or two reduced rates. The reduced rates shall be defined as a percentage of the taxable amount, which may not be less than 5%. Each reduced rate shall be so fixed that the amount of VAT resulting from its application is such that the VAT deductible under Articles 167 to 171 and Articles 173 to 177 can normally be deducted in full (Article 99) [2].

As a rule, value added tax legislation in the EU countries defines a basic, reduced and preferential tax rate. Zero value added tax is applied to individual tax transactions. Countries stimulate the development of the construction industry in different ways by applying a zero or reduced tax rate. In the Czech Republic Parliament has resolved Act of 1 April 2004 on Value-Added Tax No. 235/2004 Coll. In § 47 of this law, the following tax rates for taxable transactions are defined: base rate 21%, first reduced rate 15%, second reduced rate 10%. According to § 47 of the Act on Value-Added Tax the first reduced tax rate applies to the provision of construction or assembly work carried out on a completed construction if it is a residential or social housing construction, which includes the construction of a multi-apartment building, the construction of a family home in accordance with the legislation governing the land cadastre, including that established on a land plot forming a functional unit with this building of an apartment building or family home; living space; a room used in conjunction with a living space located in the same building, which is firmly attached to the ground [14]. In addition to § 15 of the Law of the Czech National Council on income tax of November 20, 1992, provides that the taxable base deducts an amount equal to the interest payable in the tax period for a savings loan for construction [15].

Provision of § 92E of the VAT Act came into effect on January 1, 2012, under which the reverse charge mechanism has been expanded to the construction and assembly works. Thanks to the implementation of the reverse charge in the building industry, there was a significant extension of the group of taxpayers, who must apply the reverse charge. VAT Act provides that the reverse charge mechanism contained in § 92a can be applied only to the realized taxable fulfillments with the place of the fulfillment in the home country. Voluntary registrations to VAT are typical in relation to the provision of construction works related to housing, where there is the application of a reduced VAT rate. However, the purchases of construction materials are performed in the standard VAT rate. Therefore, it is advantageous to become a VAT payer also for persons liable to tax which do not reach the legal limit for the compulsory VAT registration. The reverse charge mechanism is not applicable if the taxpayer provides assembly and construction works to recipient who is not VAT payer, and also in the case, when the building contractor and provider of assembly works are not liable to VAT. Tax liability in the construction sector was also significantly influenced by the change to the reduced tax rate of VAT from 10 to 14%. As consequence of imposing tax liability on the purchaser (a person who receives taxable supplies) tax liability shifted from the construction sector to other branches [4].

In Slovenia Value-Added Tax is paid from construction work, including repairs, cleaning, maintenance, reconstruction and demolition of real estate. Value-Added Tax is charged and paid at a general rate of 22% of the taxable amount and it is the same for the supply of goods and services. According to Article 41 of the Value-Added Tax Act adopted on October 27, 2006, Tax shall be paid at a lower rate of 9,5% of the taxable amount for the supply of goods and services referred to in Annex I, which is annexed to this Act and its constituent part. Appendix I determine the List of supplies of goods and services, entered into the reduced price of Value-Added Tax, which includes flats, housing and other facilities intended for permanent residence and parts of these buildings that are not part of a social policy, including construction, reconstruction and repair [16].

In the United Kingdom, which withdrew from the European Union, the value added tax shall be charged in accordance with the An Act to consolidate the enactments relating to value added tax, including certain enactments relating to VAT tribunals — Value Added Tax Act of 5th July 1994. VAT shall be charged at the rate of 20% and reduced rate is 5%. If the taxable person supplies construction services then the supply is zero-rated. According to Article 30 of Value Added Tax Act, a supply of goods or services is zero-rated by virtue of this subsection if the goods or services are of a description for the time being specified in Schedule 8 or the supply is of a description for the time being so specified. Group 5 "Construction of buildings, etc." of Schedule 8 includes constructing a building, the supply in the course of the construction, the supply to a in the course of conversion of a non-residential building or a non-residential part of a building, the supply of building materials to a person to whom the supplier is supplying services [10].

In Italy, the procedure for paying value-added tax is regulated by Presidential Decree of the Republic of October 26, 1972 633—Creation and regulation of value-added tax. According to Article 79 of this Decree "Application of the tax in the building sector" for the sale of buildings or production units by construction companies, as well as for the services rendered by mortgage contracts relating to construction, the buildings themselves, the value-added tax rate is reduced to six percent. For certain services dependent on contracts related to the construction of buildings, the value-added tax rate is applied at the level of 4% PARTE II (part I of the Decree). Part II of the Decree refers to goods and services taxed at a rate of 10%. These services include services rendered depending on contracts related to construction work on urbanization, subway construction and other transport lines, fixed line transport lines; industrial enterprises and distribution networks of heat and electricity from solar photovoltaic and wind power sources; sewage treatment plants, also intended for connection to sewage networks, intercommunication and the variety of supplies associated with them [3].

Taxpayers apply reduced VAT rates to construction-related transactions not just in social construction. Reduced VAT rates are applied in EU countries for construction work on new buildings (%): Belgium – 6, 12; Ireland – 13.5; Spain – 4, 10; Italy – 4, 10; Luxembourg –3; Poland – 8; Portugal – 6; Slovenia – 9.5. There are also reduced VAT rates (%) for supplies of new buildings (Ireland – 13.5; Spain – 10; Italy – 4, 10; Luxembourg – 3; Hungary – 5; Poland – 8; Slovenia – 9.5) [11].

Tax legislation of Ukraine provides for tax incentives for construction customers, as well as individuals who pay income tax. According to Article 166.3.8 of the Tax Code of Ukraine, the taxpayer has the right to include in the tax deduction the taxpayer's expenses for the construction (purchase) of affordable housing determined by law, including the repayment of a residential mortgage loan granted for such purposes and the interest on it.

Taxation of construction companies has a number of features that depend on many factors. First of all, the company's taxation depends on the type of construction financing: through joint investment institutions, construction financing funds, mortgage lending, etc. In particular, according to Article 197.15 of the Tax Code of Ukraine provides for the exemption from value-added tax of operations for the supply of construction works on affordable housing and housing that is being built on the state funds. In addition, it is determined that exempt from taxation the supply of housing (housing stock), in addition to their first supply (clause 197.1.14, item 197.1 of Article 197 of the Tax Code of Ukraine). The tax credit includes the amount of tax paid/accrued in the case of operations for the construction of non-current assets (paragraph 198.1 of Article 198 of the Tax Code of Ukraine). VAT payers that are developers and implementers target program of construction and directly engaged in construction work have the right to apply the tax exemption privileges on VAT treatment of supplies of construction works on construction of housing.

Besides, tax legislation has established features of taxation of collective investment institutions that are used not only in the activities of construction companies. According to Article 141.6.1 of the Tax Code of Ukraine, common investment funds are exempt from taxation [9]. Taking into account the relevant rules governing the procedure for paying corporate income tax, it should be noted that there is no separate preferential income tax procedure for construction companies. Given the relevant rules governing the procedure for payment of corporate income tax, it should be noted that a separate preferential procedure for corporate income tax for construction companies is absent in Ukraine.

Construction companies need state support to stimulate development. In our opinion, one of the types of such support is the use of the experience of European countries in applying a reduced VAT rate and tax exemptions. Clearly, under conditions of coronavirus spread to public authorities, it will be difficult to do. It's necessary to find alternative sources of revenue to the state budget – non-tax revenues, including investments in business and production. Besides scientists recommend that self-reporting tax system is highly activated so that the costs of value added tax collection reduced [1].

European experience should be implemented in Ukrainian legislation. These changes will be in line with EU Council Directive 2006/112/EU of 28 November 2006 [2] on the common system of value added tax, which provides for the application of a reduced VAT rate to the construction, repair and reconstruction of housing. But it's necessary to develop a strategy for the gradual amendment of tax legislation. VAT needs to be reduced gradually by setting temporary standards and assessing the economic and social consequences.

5 Conclusions

Prerequisite for effective reform in our country is experience of foreign countries investigation in organizing the tax legislation reforms. A comparative analysis of the tax legislation of Ukraine and individual member states of the European Union showed that states apply different ways of stimulating development of the construction industry by establishing privileges for both direct and indirect taxation of building enterprises. In Ukraine, there is a direct indication of exemption from value added tax for the supply of construction and installation work for the construction of affordable housing and housing built at public expense.

In the EU, direct taxes are regulated independently by the Member States, which certainly needs further research. With regard to indirect taxation, in particular value added tax, as a general rule, EU Member States have the right to apply reduced rates of value added tax to construction companies. Laws on zero or reduced VAT rates for companies that build housing under state social programs do not apply in all EU countries. Nevertheless, the level of economic development, including the construction industry in the EU, is much higher than in other countries, which in turn provides adequate funding for the construction industry by the state. It is necessary to expand the list of benefits for the taxation of construction companies to stimulate the development of the construction industry. Ukraine needs to use the European experience in applying a reduced VAT rate to construction companies that perform work to ensure construction under the State Budget program and for companies that engage in construction work on new buildings. But reduction of the tax rate should be gradual and cautious, as revenues from this tax to the budget now account for more than a third of all revenues. This, in turn, will reduce the tax burden on construction companies that build social housing; reducing the cost of housing, which is of great importance in the current context of the spread of coronavirus and reducing the citizens' purchasing power.

Stimulating such a priority sector as construction through preferential taxation will give impetus to the development of other important sectors of the economy: energy, engineering, metallurgy, transport and more.

References

1. Alizadeh M, Motallabi M (2016) Studying the effect of value added tax on the size of current government and construction government. Procedia Econ Financ 36:336–344. https://doi.org/10.1016/S2212-5671(16)30045-4
2. Council Directive 2006/112/EC of 28 November 2006 on the common system of value added tax. https://eur-lex.europa.eu/legal-content/EN/TXT/?uri=uriserv:OJ.L_.2006.347.01.0001.01.ENG&toc=OJ:L:2006:347:TOC
3. Decreto del Presidente della Repubblica del 26 ottobre 1972 n. 633 – Istituzione e disciplina dell'imposta sul valore aggiunto. http://www.garanziaviaggi.it/files/DPr_633_1972_74ter.pdf
4. Grásgruber M, Otavová M, Semerád P (2013) Impacts of the application of the reverse charge mechanism of the value added tax. Acta Univ Agric Silvic Mendelianae Brun 61(7):2133–2141. https://doi.org/10.11118/actaun201361072133
5. Kucheriavenko M (2013) Tax law of Ukraine. Law, Kharkiv, Ukraine
6. Strategy for the sustainable competitiveness of the construction sector and its enterprises. https://eur-lex.europa.eu/legal-content/EN/TXT/?uri=CELEX%3A52012DC0433
7. Ostapenko V (2019) Management of tax system adopting to European standards. Dev Manage 17(1):51–62. https://doi.org/10.21511/dm.5(1).2019.05
8. Regulation (EU) No 1286/2013 of the European Parliament and of the Council of 11 December 2013 establishing an action program to improve the operation of taxation systems in the European Union for the period 2014–2020 (Fiscalis 2020) and repealing Decision No 1482/2007/EC. https://eur-lex.europa.eu/legal-content/EN/TXT/?qid=1556974885566&uri=CELEX:32013R1286
9. Tax code of Ukraine. December 02, 2010 N 2756-VI. https://zakon.rada.gov.ua/laws/show/2755-17
10. Value Added Tax Act of 5th July 1994. http://www.legislation.gov.uk/ukpga/1994/23/data.xht?view=snippet&wrap=true
11. VAT rates applied in the Member States of the European Union. https://ec.europa.eu/taxation_customs/sites/taxation/files/resources/documents/taxation/vat/how_vat_works/rates/vat_rates_en.pdf
12. Visa liberalization with Moldova, Ukraine and Georgia. European Commission (2017). https://ec.europa.eu/home-affairs/what-we-do/policies/international-affairs/eastern-partnership/visa-liberalisation-moldova-ukraine-and-georgia_en
13. What is VAT? https://ec.europa.eu/taxation_customs/business/vat/what-is-vat_en
14. Zákon České národní rady o dani z přidané hodnoty ze dne 1. dubna 2004 č. 235/2004 Sb. https://www.zakonyprolidi.cz/cs/2004-235
15. Zákon České národní rady o daních z příjmůze dne 20 listopadu 1992. č. 586/1992 Sb. https://www.zakonyprolidi.cz/cs/1992-586
16. Zakon o davku na dodano vrednost z dne 27.10.2006. http://www.pisrs.si/Pis.web/pregledPredpisa?id=ZAKO4701
17. Iasechko S, Haliantych MK, Skomorovskyi VB, Zadorozhnyi V, Obryvkina O, Pohrebniak O (2020) Contractual relations in the information sphere. Syst Rev Pharm 11(8):301–303. https://doi.org/10.31838/srp.2020.8.46

18. Onyshchenko V, Sivitska S, Cherviak A (2018) Construction industry in Ukraine credit analysis. Int J Eng Technol (UAE) 7(3):280–284. https://doi.org/10.14419/ijet.v7i3.2.14420
19. Dmitrievna GA (2013) Directions of efficiency increase of state regulatory policy in Ukraine. World Appl Sci J 27(4):448–453. https://doi.org/10.5829/idosi.wasj.2013.27.04.13656
20. Glushko A, Marchyshynets O (2018) Institutional provision of the state regulatory policy in Ukraine. J Adv Res Law Econ 9(3):941–948. https://doi.org/10.14505/jarle.v93(33).18

Author Index

V. Onyshchenko et al. (eds.), *Proceedings of the 3rd International Conference on Building Innovations*, Lecture Notes in Civil Engineering 181,
https://doi.org/10.1007/978-3-030-85043-2